NUCLEAR REACTOR ENGINEERING

by

SAMUEL GLASSTONE

Consultant, United States Atomic Energy Commission

and

ALEXANDER SESONSKE

Professor of Nuclear Engineering
Purdue University

PREPARED UNDER THE AUSPICES OF
THE DIVISION OF TECHNICAL INFORMATION,
UNITED STATES ATOMIC ENERGY COMMISSION

VNR VAN NOSTRAND REINHOLD COMPANY
New York   Cincinnati   Toronto   London   Melbourne

VAN NOSTRAND REINHOLD COMPANY REGIONAL OFFICES:
New York Cincinnati Chicago Millbrae Dallas

VAN NOSTRAND REINHOLD COMPANY INTERNATIONAL OFFICES:
London Toronto Melbourne

Copyright © 1967 by Litton Educational Publishing, Inc.

ISBN: 0-442-02725-7

Manufactured in the United States of America

Published by VAN NOSTRAND REINHOLD COMPANY
450 West 33rd Street, New York, N. Y. 10001

Published simultaneously in Canada by
VAN NOSTRAND REINHOLD Ltd.

15 14

PREFACE

The purpose of this book is to describe the fundamental scientific and engineering principles of nuclear reactor systems. The main emphasis is on reactor design and behavior related to the fission process and its associated radiations rather than to those engineering areas that are not unique to nuclear reactors. The book is intended primarily as a text for a two-semester introductory course at about the senior or first-year graduate level. The treatment is therefore such as to meet the requirements of students who have essentially completed an undergraduate engineering curriculum.

The design of a nuclear reactor system is a very complex procedure involving the cooperation of several groups of specialists, with each group responsible for a particular aspect of the design, such as neutronics, stability and control, heat removal, radiation shielding, components, etc. Because these individual aspects are dependent to a great extent upon the others, it is necessary that the engineer acquire a broad, general perspective of the whole field of reactor engineering before specializing in a particular area. One of the objectives of the present book is to provide such a perspective.

There has been very considerable progress in nuclear reactor technology since the predecessor book in this field was written in 1955. For example, at that time it was possible to describe within the space of fifty pages the characteristics of all the reactors about which information was available, including several in the design stage. At present, however, a mere listing of the known nuclear reactors would take many pages and a large volume would be needed to describe them. Furthermore, in several instances a number of different satisfactory approaches have been developed for the solution of the technological problems encountered in reactor design. Consequently, this book cannot provide a complete and detailed description of what is now a complicated and mature technology but it should be useful as a guide to the basic concepts in related areas, as indicated above. The book should also prove of interest to those in other branches of engineering who may wish to know something about the impact of nuclear energy on their own activities.

In preparing this book we have received assistance and suggestions from

iii

PREFACE

many individuals to whom we wish to acknowledge our indebtedness. Their names are listed separately below. In addition, our thanks are due to several members of the A. E. C., Division of Technical Information, in particular J. G. Gratton, Chief, and J. D. Cape and G. K. Ellis, Scientific Publications Branch, for their help in securing unpublished materials, in obtaining reviews of the draft manuscript, and in many other ways, and C. R. Bruce and W. J. Pearson, Publishing Branch, Oak Ridge, for their work on the illustrations. We are also grateful to Dr. N. E. Bradbury, Director and P. F. Belcher, Assistant Director, Los Alamos Scientific Laboratory, for their courtesy in making available to us the excellent facilities of the Laboratory, and to Dr. D. B. Hall, Division Leader, and Drs. R. P. Hammond and R. M. Kiehn of K (Reactor) Division for their helpful cooperation. Finally, A. Sesonske wishes to thank his wife for her continued help and encouragement.

Los Alamos, New Mexico
August, 1962

SAMUEL GLASSTONE
ALEXANDER SESONSKE

ACKNOWLEDGMENT

The authors wish to thank the following who contributed in various ways to the review of the draft manuscript:

Adler, K. L., Atomics International
Alexander, L. G., Oak Ridge Natl. Lab.
Arnold, E. D., Oak Ridge Natl. Lab.
Ashley, R. L., Atomics International
Babb, A. L., University of Washington
Bernsen, S. A., Advanced Technology Labs.
Billington, D. S., Oak Ridge Natl. Lab.
Blanco, R. E., Oak Ridge Natl. Lab.
Blizard, E. P., Oak Ridge Natl. Lab.
Brand, G. E., Atomics International
Brinkman, J. A., Atomics International
Budney, G. S., Atomics International
Bump, T. R., Argonne Natl. Lab.
Burnett, T. J., Oak Ridge Natl. Lab.
Bush, S. H., Hanford Atomic Products Operation, G. E. Co.
Cohen, E. R., Atomics International
Chapman, R. H., Oak Ridge Natl. Lab.
Cheverton, R. D., Oak Ridge Natl. Lab.
Davis, F. W., Southwest Research Inst.
Davis, M. V., Atomics International
DeBear, W. S., Atomics International
Denham, R. S., Atomics International
Dickinson, R. W., Atomics International
Dietrich, J. R., General Nuclear Engineering Corporation
DiNunno, J. J., AEC, Div. of Licensing and Regulation
Dunenfeld, M., Atomics International
Dunning, G. M., AEC, Div. of Operational Safety
Eggen, D. T., Atomics International
Etherington, H., Allis-Chalmers Manufacturing Company
Falk, E. D., Atomics International
Fillmore, F. L., Atomics International

Fisher, W. L., Atomics International
Foster, K. W., Atomics International
Fromm, L. W., Argonne Natl. Lab.
Gimera, R. J., Atomics International
Glasgow, L. E., Atomics International
Golan, S., Atomics International
Gresky, A. T., Oak Ridge Natl. Lab.
Griffin, C. W., Atomics International
Gross, E. E., Oak Ridge Natl. Lab.
Grossman, N., AEC, Div. of Compliance
Grotenhuis, M., Argonne Natl. Lab.
Guthrie, C. E., Oak Ridge Natl. Lab.
Gylfe, J. D., Atomics International
Hallett, W. J., Atomics International
Iskenderian, H., Argonne Natl. Lab.
Jarrett, A. A., Atomics International
Kann, W. J., Argonne Natl. Lab.
Keen, R. T., Atomics International
Kelber, C., Argonne Natl. Lab.
Kerze, F., AEC, Div. of Reactor Development
Kolba, V. M., Argonne Natl. Lab.
Kuhn, D. W., AEC, Div. of Operations Analysis and Forecasting
Lafyatis, P. G., Oak Ridge Natl. Lab.
Lane, J. A., Oak Ridge Natl. Lab.
Lang, J. C., Atomics International
Lemcoe, M. M., Southwest Research Inst.
Lennox, D. H., Argonne Natl. Lab.
Lesch, F. R., Atomic Power Development Associates Inc.
Link, L. E., Argonne Natl. Lab.
Lowell, E. G., Atomics International
Lyon, R. N., Oak Ridge Natl. Lab.
MacFarlane, D. R., Argonne Natl. Lab.
Mahlmeister, J. E., Atomics International

Marable, J. H., Oak Ridge Natl. Lab.
Marron, J. F., Atomics International
Mattern, K. L., Atomics International
McArthy, A. E., Argonne Natl. Lab.
Meem, J. L., University of Virginia
Mims, L. S., Atomics International
Moon, D. P., Argonne Natl. Lab.
Murbach, E. W., Atomics International
Naymark, S., Atomic Products Division, G. E. Co.
Nelson, L., Oak Ridge Natl. Lab.
Nestor, C. W., Oak Ridge Natl. Lab.
Osborn, R. K., University of Michigan
Pennington, E., Argonne National Lab.
Persiani, P., Argonne Natl. Lab.
Prohammer, F. G., Argonne Natl. Lab.
Richards, R. B., Atomic Products Division, G. E. Co.
Roberts, H., Alco Products, Inc.
Rosen, F. D., Atomics International
Ross, D. M., AEC, Div. of Operational Safety

Rossin, A. D., Argonne Natl. Lab.
Royden, H. N., Atomics International
Shimazaki, T., Atomics International
Siegel, S., Atomics International
Sinizer, D. I., Atomics International
Spencer, S. C., Atomics International
Stelle, A. M., Atomics International
Stockdale, W. G., Oak Ridge Natl. Lab.
Sturm, R. G., Strum-Krouse, Inc.
Sutton, C. R., International Nickel Co.
Tobias, M., Oak Ridge Natl. Lab.
Uhrig, R. E., University of Florida
Ullmann, J. W., Oak Ridge Natl. Lab.
Weber, C. E., Atomics International
Weills, J. T., Argonne Natl. Lab.
Wett, J. S., AEC, Div. of Reactor Development
Whan, G. A., University of New Mexico
Young, H. D., Argonne Natl. Lab.
Zack, J. F., Atomics International

Since it was not possible to incorporate all the suggestions made by reviewers, the authors accept sole responsibility for the material appearing in this book.

S. G.
A. S.

TABLE OF CONTENTS

Chapter 1

INTRODUCTION TO NUCLEAR REACTOR ENGINEERING

INTRODUCTION

THE WORLD'S ENERGY RESOURCES

1.1. The discovery of nuclear fission in 1939 was an event of epochal significance, because it opened up the prospect of an entirely new source of power, utilizing the internal energy of the atomic nucleus. The basic materials that can be used for the release of nuclear energy by fission are the elements uranium and thorium. Minerals containing these elements are widely distributed in the earth's crust, so that, as will be apparent shortly, they represent a very large potential source of power.

1.2. For the past half century fossil fuels, namely, coal, oil, and natural gas, have supplied the major portion of the world's energy requirements. It has long been realized, however, that in the not too distant future these sources of energy will be largely exhausted. At the present time the total energy consumption, for all countries, is about 1×10^{17} Btu per year. Since the world's population is steadily growing and the power use per capita is increasing as well, the rate of energy utilization by the year 2000 could well be five to ten times the current value. According to one estimate, the known coal, oil, gas, and oil shale which can be extracted at no more than twice the present cost would be equivalent to roughly 4×10^{19} Btu [1].* This means that within about 100 years the world's economically useful reserves of fossil fuels may approach exhaustion.

IMPORTANCE OF FISSION ENERGY

1.3. Even when allowance is made for errors in the foregoing estimates, the conclusion is inevitable that new sources of power must be found during the next 50 years or so if the earth is to support the growing population with some

---

* The numbers in brackets apply to the references given at the end of the chapter.

increase in living standards. Two such sources have been considered: solar energy and nuclear energy. Although the idea of making more direct use of the sun's energy is very attractive, the development of large-scale processes appears some years away. Nuclear energy may be made available either by fission of the heaviest elements or by fusion of very light nuclei. The problems associated with the controlled release of fusion energy are formidable and are currently being investigated. Nuclear fission, on the other hand, has already been established as a practical means for the production of energy which will be economically competitive with energy from fossil fuels in the near future. The objective of this book is to explain the basic principles involved in the design and operation of the devices known as *nuclear reactors* in which fission energy is released.

1.4. The total amount of basic raw materials, uranium and thorium, in the earth's crust, to a depth of three miles, is very large, possibly something like 10^{12} tons. However, much of this is present in minerals containing such a small proportion of the desired element that extraction would be prohibitively expensive. In fact, the known high-grade ore reserves are believed to be no more than 2×10^6 tons. Assuming that technological advances will reduce the cost of recovery from moderately low-grade ores to $100 or less per pound of metal, it has been estimated that the world reserve, based on admittedly incomplete mappings, is 20 million tons of uranium and 1 million tons of thorium [2]. As will be seen in due course, it is unlikely that the whole of this material can be utilized economically in fission, but in favorable circumstances perhaps one-third might be so used.

1.5. It will be shown below (§ 1.46) that 1 pound of material capable of undergoing fission can produce roughly 3×10^{10} Btu of energy. Consequently, the energy available from the economically recoverable 21×10^6 tons of uranium and thorium is approximately 4×10^{20} Btu. This energy reserve is therefore many times greater than that from the fossil fuels and so it would, if properly developed, represent a very significant contribution to the world's energy resources.

ATOMIC STRUCTURE

ATOMIC NUMBER AND MASS NUMBER

1.6. The operation of a nuclear reactor depends upon various interactions of neutrons with atomic nuclei. In order to understand the nature and characteristics of these reactions it is desirable to consider briefly some of the fundamentals of atomic and nuclear physics [3, 4].

1.7. An atom consists of a positively charged *nucleus* surrounded by a number of negatively charged particles, called *electrons*, so that the atom as a whole is electrically neutral. Atomic nuclei are built up of two kinds of primary particles, namely, *protons* and *neutrons,* often referred to by the general name of *nucleon.* The proton carries a single unit positive charge, equal in magnitude to the electronic charge. It is, in fact, identical with the nucleus of a hydrogen atom, i.e., a hydrogen atom minus its single electron. The neutron is very slightly heavier than the proton (§ 1.13) and, as its name implies, it is an electrically neutral particle carrying no charge. All atomic nuclei, with the exception of those of ordinary hydrogen, contain one or more neutrons in addition to protons.

1.8. For a given element, the number of protons present in the atomic nucleus, which is the same as the number of positive charges it carries, is called the *atomic number* of the element and is usually represented by the symbol Z. It is identical with the ordinal number of the element in the familiar periodic table of the elements. Thus, the atomic number of hydrogen is 1, of helium 2, of lithium 3, and so on, up to 92 for uranium, the element of highest atomic weight existing in nature to any appreciable extent. A number of heavier elements have been made artificially; of these, plutonium, atomic number 94, is important in connection with the release of nuclear energy.

1.9. The total number of nucleons, i.e., of protons and neutrons, in an atomic nucleus is called the *mass number* of the element and is denoted by A. The number of protons is Z, as stated above, and so the number of neutrons in the atomic nucleus is $A - Z$. Since the masses of both neutron and proton are close to unity on the atomic mass scale, it is evident that the mass number is the integer nearest to the atomic weight of the species under consideration.

ISOTOPES

1.10. It is the atomic number, i.e., the number of protons in the nucleus, which determines the chemical nature of an element. This is so because the chemical properties depend on the (orbital) electrons surrounding the nucleus, and their number must be equal to the number of protons, since the atom as a whole is electrically neutral. Consequently, atoms with nuclei containing the same numbers of protons, i.e., with the same atomic number, but with different mass numbers, are essentially identical chemically, although they frequently exhibit marked differences in their nuclear characteristics. Such species, having the same atomic number but different mass numbers, are called *isotopes.* They are, in general, indistinguishable chemically, but have different atomic weights.

1.11. A particular isotope of a given element is identified by writing the mass number after the name or symbol of the element. For example, the common isotope of oxygen, which has a mass number of 16, i.e., 8 neutrons and 8 protons in the nucleus, is variously represented as oxygen-16, O-16 or O^{16}. In some instances it is convenient, although not necessary, to include the atomic number; this is then added to the symbol as a subscript, thus $_8O^{16}$.

1.12. The element oxygen, which prior to 1962 served as the basis of the atomic weight scale, actually occurs in nature as three isotopes, namely, O^{16}, O^{17}, and O^{18}, the two latter being present in relatively small proportions. This situation resulted in the development of two different systems for the representation of atomic weights. On the conventional atomic weight scale, a value of exactly 16.00000 was assigned to the average (weighted) mass of the atoms of the three isotopes of oxygen in the proportions present in the atmosphere. However, another scale, sometimes called the physical atomic weight scale, was based on the assignment of a mass of exactly 16.00000 to the atom of the oxygen-16 isotope. Atomic masses on this scale are thus about 0.028 per cent larger than the corresponding conventional atomic weights. As of January 1, 1962, all atomic weights are to be expressed on a single scale which assigns a value of exactly 12.00000 to the common isotope of carbon, C^{12}. This means a decrease of 37 parts per million in the chemical atomic weights and 318 parts per million in the physical atomic weights formerly in use. However, since it will take several years for the new values to get into the literature, the few physical atomic weights given in this text are based on the older (oxygen-16) scale.

1.13. In nuclear physics and related fields, the masses of atoms, of nuclei, and of nuclear particles are invariably expressed on the so-called physical scale. The atomic mass unit (or amu) is then defined as exactly one-sixteenth of the mass of the O^{16} atom.* In terms of a more practical unit, 1 amu may be represented by $1/N_a$ gram, where N_a, the Avogadro number, is the number of individual atoms in exactly 16.00000 grams (1 gram atom) of the O^{16} isotope, i.e., 0.6025×10^{24}. Hence 1 amu is equivalent to 1.660×10^{-24} gram. The mass of a single proton is 1.007596 amu, i.e., 1.6725×10^{-24} gram, whereas that of a neutron is 1.008986 amu, i.e., 1.674×10^{-24} gram. The electron mass is only 0.000549 amu or 9.11×10^{-28} gram, so that nearly the whole of the mass of an atom is due to the protons and neutrons present in the nucleus.

1.14. At the present time, uranium is the most important element for the release of nuclear energy by fission. It exists in nature in at least three isotopic forms, with mass numbers 234, 235, and 238, respectively. The proportions in which these isotopes occur in natural uranium, together with their atomic masses in amu, are given in Table 1.1. It is seen that uranium-238 is by far the most abundant isotope, but there is always present a little over 0.7 per cent of uranium-235. Both of these isotopes are significant for the production of nuclear energy, although, as will be seen in due course, it is chiefly the latter isotope that can be utilized directly for the release of fission energy. The lightest of the three isotopes given in the table, namely, uranium-234, occurs in such small proportions in uranium minerals that it can be ignored for all practical purposes.

---

* With the adoption of the carbon-12 scale, the atomic mass unit will be defined as exactly one-twelfth of the mass of the C^{12} atom. The resulting changes in the values given here will be negligible for the present purpose.

TABLE 1.1. ISOTOPIC COMPOSITION OF NATURAL URANIUM

Mass Number	Weight Per Cent	Isotopic Mass (amu)
234..............	0.0058	234.1140
235..............	0.711	235.1175
238..............	99.283	238.1252

1.15. Another element which is important from the nuclear energy standpoint is thorium, atomic number 90. This occurs in nature almost entirely as a single nuclear species, with mass number 232. There are traces of other isotopic forms, but their proportions are negligible.

NUCLEAR ENERGY AND NUCLEAR FORCES

BINDING ENERGY

1.16. The direct determination of nuclear (or isotopic) masses, by means of the mass spectrograph and in other ways, has shown that the actual mass is always less than the sum of the masses of the constituent nucleons. The difference, called the *mass defect*, which is related to the energy binding the particles in the nucleus, can best be determined in the following manner. For electrical neutrality, an atom must contain Z electrons outside the nucleus, in addition to the Z protons and $A - Z$ neutrons in the nucleus. If m_p, m_n, and m_e represent the masses of the proton, neutron, and electron, respectively, the sum of the masses of the constituents of an atom is $Zm_p + Zm_e + (A - Z)m_n$. Suppose the observed (isotopic) mass of the atom is M; then

$$\text{Mass defect} = [Z(m_p + m_e) + (A - Z)m_n] - M$$
$$= Zm_H + (A - Z)m_n - M, \tag{1.1}$$

where $m_p + m_e$ has been replaced by m_H, the mass of the hydrogen atom. Since m_H is 1.008145 and m_n is 1.008986 amu, the mass defect can be evaluated for any nuclear species (or *nuclide*) for which the isotopic mass is known from experiment.

1.17. According to the concept of *the equivalence of mass and energy*, based on the special theory of relativity, the mass defect is a measure of the energy which would be released if the individual Z protons and $A - Z$ neutrons combined to form a nucleus.* Conversely, it is numerically equal to the energy which would have to be supplied to break apart the nucleus into its constituent nucleons. Thus, the energy equivalent of the mass defect is called the *binding energy* of the nucleus.

* The Z electrons contribute a small amount of binding energy, but this is largely allowed for in the replacement of $m_p + m_e$ by m_H in equation (1.1).

1.18. If m is the decrease in mass accompanying any particular process, then the equivalent amount of energy E released is given by the Einstein equation

$$E = mc^2, \tag{1.2}$$

where c is the velocity of light. Expressing m in grams and c as 2.998×10^{10} cm/sec, the energy E in ergs will be

$$E \text{ (ergs)} = m \text{ (gram)} \times 8.990 \times 10^{20}. \tag{1.3}$$

In the nuclear energy and related fields, energies are usually stated in terms of the *electron volt* unit, represented by ev; this is the energy acquired by a unit (electronic) charge which has been accelerated through a potential of 1 volt. Since the electronic charge is 1.602×10^{-19} coulomb, it follows that 1 ev is equivalent to 1.602×10^{-19} joule or 1.602×10^{-12} erg, and 1 million electron volts (or 1 Mev) is equivalent to 1.602×10^{-6} erg. Hence, energy in ergs may be converted into energy in Mev upon dividing by 1.602×10^{-6}, so that from equation (1.3),

$$E \text{ (Mev)} = m \text{ (gram)} \times 5.614 \times 10^{26}.$$

Finally, the mass difference m is conveniently expressed in amu if the right side of this equation is multiplied by 1.660×10^{-24}; the result is

$$E \text{ (Mev)} = 931m \text{ (amu)}. \tag{1.4}$$

1.19. The numerical value of the nuclear binding energy in Mev can thus be obtained upon multiplying by 931 the mass defect in amu, as given by equation (1.1). A more useful quantity is the average binding energy per nucleon, which is equal to the total binding energy (B.E.) divided by the number of nucleons, i.e., by the mass number A; hence, from equations (1.1) and (1.4),

$$\frac{\text{B.E.}}{A} = \frac{931}{A} [1.008145Z + 1.00898(A - Z) - M].$$

Consequently, the mean binding energy per nucleon in any nucleus can be determined if the isotopic (atomic) mass M is known.

Example 1.1. Determine the binding energy per nucleon in (a) tin-120 for which M is 119.9401 amu and (b) uranium-235.

(a) The atomic number of tin is 50; hence, for the nuclide of mass number 120,

$$\frac{\text{B.E.}}{A} = \frac{931}{120} [(1.00814 \times 50) + (1.00898 \times 70) - 119.9401]$$

$$= 8.50 \text{ Mev per nucleon.}$$

(b) The atomic number of uranium is 92; the isotopic mass of uranium-235 is given in Table 1.1 as 235.1175; consequently,

$$\frac{\text{B.E.}}{A} = \frac{931}{235} [(1.00814 \times 92) + (1.00898 \times 143) - 235.1175)$$

$$= 7.59 \text{ Mev per nucleon.}$$

1.20. Values of the binding energy per nucleon in various nuclides, obtained in the manner just described, are plotted in Fig. 1.1 as a function of the mass number. Most of the points are seen to fall on or close to a single curve. This curve shows that the binding energy per nucleon is relatively low for nuclei

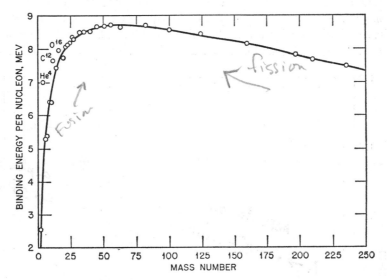

FIG. 1.1. Variation of binding energy per nucleon with mass number

of small mass number but increases, with increasing mass number, to a broad maximum at about 8 Mev in the mass region of roughly 50 to 75; subsequently the binding energy per nucleon decreases steadily. It can be readily shown (see § 1.43) that a process in which a nucleus (or nuclei) of lower binding energy per nucleon, i.e., less stable, is converted into others of higher binding energy, i.e., more stable, must be accompanied by the release of energy. Hence, it is apparent from Fig. 1.1 that energy should be obtainable by the combination (or fusion) of the lightest nuclei or by the splitting (or fission) of those of high mass number.

NUCLEAR STABILITY

1.21. If the number of neutrons in each nucleus of the 270 or so known stable nuclides is plotted as ordinate against the corresponding number of protons as abscissa, a series of points is obtained, as shown in Fig. 1.2. Points on the diagonal line in the figure represent equal numbers of neutrons and protons. It is seen that, in many stable nuclei of low mass number, up to 40, the numbers of neutrons and protons are equal or approximately so. In other words, the neutron/proton ratio is exactly or slightly larger than unity. With increasing mass (or atomic) number, however, a nucleus is stable only if it contains more neutrons than protons. Thus, for the heaviest stable nuclei, with atomic numbers of 80 or more, the neutron/proton ratio has increased to about 1.5.

1.22. Before attempting an interpretation of the foregoing results, it should be noted that certain nuclei exhibit exceptional stability. These contain the so-called *magic numbers* of 2, 8, 20, 50, or 82 protons or 2, 8, 20, 50, 82, or 126 neutrons, as indicated by the broken lines in Fig. 1.2. Nuclei having magic numbers of both protons and neutrons are said to be "doubly magic"; examples

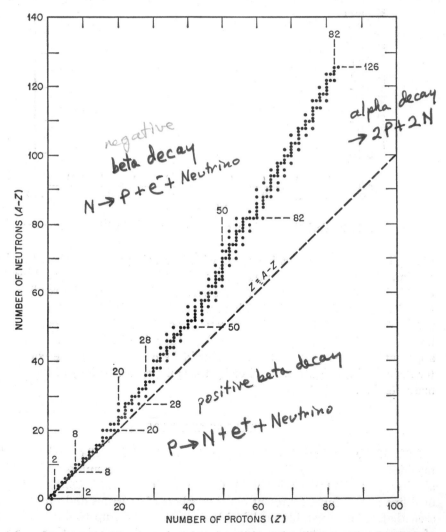

Fig. 1.2. Numbers of neutrons and protons in stable nuclei. (The short dashed lines indicate magic numbers of neutrons and protons)

are $_8O^{16}$, $_{20}Ca^{40}$, and $_{82}Pb^{208}$. In illustration of the stability associated with the magic numbers, it may be noted that tin, which has 50 protons in its nucleus, exists in ten stable isotopic forms. In general, it is apparent from Fig. 1.2 that magic nuclides are common in nature.

1.23. The special stability of the magic nuclei is attributed to the presence of completed (or closed) shells of neutrons or protons (or both). Just as the inert elements helium, neon, argon, etc., which contain closed shells of electrons, have great chemical stability, so the species with magic numbers of nucleons exhibit unusual nuclear stability. In addition to the magic numbers, which represented closed main shells, there are semi-magic numbers, e.g., 6, 14, and 28, which correspond to one or more closed shells plus a completed sub-shell. The isotopes of the abundant elements carbon and silicon fall in this category.

1.24. Apart from the details associated with magic numbers, the general character of Fig. 1.2 can be accounted for by postulating that two types of forces exist between nucleons. First, there are attractive forces of approximately equal magnitude among the nucleons, i.e., protons attract other protons, neutrons attract neutrons, and protons attract neutrons to about the same extent. These are characteristic intranuclear forces operative over very short distances, on the order of 10^{-13} cm only. One consequence of the short range is that the total attractive force in a nucleus is roughly proportional to the number of nucleons present. This is the basic reason for the approximate constancy of the binding energy per nucleon over a large range of mass numbers (Fig. 1.1). In addition to the short-range (attractive) forces, there are the conventional electrostatic (or coulomb) repulsive forces between the positively charged protons that are capable of acting over relatively large distances. The total electrostatic repulsive force between all the protons in the nucleus is proportional to the square of their number, i.e., to Z^2. There are other repulsive forces among the nucleons, but these are not significant for the present purpose.

RADIOACTIVITY

1.25. Provided the atomic (or mass) number is low, the repulsive force among the protons is small. Hence, since the proton-proton, neutron-neutron, and proton-neutron forces are roughly equal, a neutron/proton ratio close to unity is to be expected for stability. But with increasing atomic number the electrostatic repulsion between the protons, which varies as Z^2, becomes more and more important. In order to maintain stability, the nuclei must now contain an increased proportion of neutrons, so that the attractive neutron-neutron and neutron-proton forces can compensate for the rapidly growing repulsive forces between the protons. There is a limit, however, to the excess of neutrons over protons which a nucleus can contain and still remain stable. Consequently, the elements of atomic number 84 or larger have no stable isotopes; although

alpha particles = 2P + 2N (helium nucleus)
beta particles = e⁻

elements 84 (polonium) through 92 (uranium) exist in nature, they are unstable and exhibit the phenomenon of *radioactivity*.

1.26. Radioactive nuclides undergo spontaneous change at a definite rate which varies with the nature of the nuclide. The unstable nucleus emits a characteristic particle (or radiation) and is thereby transformed into a different nucleus, which may (or may not) also be radioactive. Nuclides which owe their instability to their high mass numbers emit either positively charged *alpha particles*, which are identical with helium nuclei and consist of two protons and two neutrons, or negatively charged *beta particles*, which are the same as ordinary electrons. The nucleus itself does not contain electrons, and in radioactive beta decay the electron arises from the spontaneous conversion of a neutron into a proton and an electron; thus,

$$\text{Neutron} \rightarrow \text{Proton} + \text{Electron (Beta particle)} + \text{Neutrino.}$$

The additional neutral particle, with essentially zero mass, called a *neutrino*,* carries off some of the energy liberated in the radioactive transformation.

1.27. It follows from the foregoing considerations that the product (or daughter) nucleus of alpha decay has two protons and two neutrons less than the parent nucleus, so that its mass number is four units less. On the other hand, in beta decay the daughter nucleus has one neutron less and one proton more than its parent, but the mass number is unchanged.

1.28. Radioactivity can arise from a cause other than high atomic number. It is seen that the points representing stable nuclei in Fig. 1.2 fall within a narrow range of neutron/proton ratios. Any nuclide having a composition outside this range is radioactive. If the nucleus lies above the stability range, i.e., there are too many neutrons for stability, for the given atomic number, the nuclide exhibits beta activity, similar to that described above. A neutron is replaced in the nucleus by a proton, so that the neutron/proton ratio decreases; the daughter nucleus will be more stable than its parent, although not necessarily completely stable.

1.29. In the event that a nuclide contains too few neutrons to yield a stable nucleus with a given number of protons, it could become more stable by emitting an alpha particle, thereby increasing the neutron/proton ratio. However, this rarely occurs with unstable nuclides of low and intermediate mass number; with but few exceptions, alpha decay is observed only with the heaviest nuclides. One alternative is for the nucleus to capture an orbital electron which then combines with a proton to form a neutron (plus a neutrino); the net result is an increase in the neutron/proton ratio. More commonly, however, positive beta decay occurs, i.e., the emission of a positive electron (or positron) as a result of the transformation

$$\text{Proton} \rightarrow \text{Neutron} + \text{Positive electron} + \text{Neutrino.}$$

* Strictly speaking, this particle is an *antineutrino*, but the distinction is of no consequence here.

Although many radioactive species are known which exhibit positive beta decay, they are not encountered in the operation of nuclear reactors. Consequently, such radionuclides will not be discussed further in this book.

1.30. In many cases, although not always, radioactive decay is associated with the emission of *gamma rays*, in addition to an alpha or beta particle. Gamma rays are penetrating electromagnetic radiations of high energy, essentially identical with x-rays. In fact, the only difference between gamma rays and x-rays is that the former originate from an atomic nucleus whereas the latter are produced by processes outside the nucleus. Gamma rays occur in a radioactive change when the daughter nucleus is formed in what is called an *excited state*, i.e., a state in which it has a higher internal energy than the normal (or *ground*) state of that nucleus. The excess energy is then released almost instantaneously as gamma radiation. It may be mentioned that gamma rays accompany other nuclear processes in which nuclei are produced in excited states.

NEUTRON REACTIONS

1.31. Although neutrons generally occur bound in nuclei, it is possible to obtain them in the free state (§ 2.71). Such free neutrons can interact in various ways with nuclei. The neutron-nuclei reactions of present interest fall mainly into three general categories, namely, scattering, capture, and fission. As a general rule, the first step in such interactions is that the nucleus absorbs the neutron to form a *compound nucleus* in an excited state of high internal energy. In *scattering reactions*, the compound nucleus rapidly expels a neutron with a lower kinetic energy than the absorbed neutron, the excess energy remaining on the residual nucleus. If this additional energy is in the form of internal energy, so that the nucleus is in an excited state, the phenomenon is referred to as *inelastic scattering*. On the other hand, if the extra energy which the nucleus has acquired is solely kinetic in nature, a form of *elastic scattering* has occurred. The term "scattering" is used to describe these reactions because the direction of motion of the neutron remaining after the interaction with a nucleus is generally different from that prior to the interaction.

1.32. When first liberated, free neutrons usually possess high kinetic energies, in the million electron volt range, and so they are called *fast neutrons*. However, as a result of scattering collisions with various nuclei in the medium through which the neutrons move, they can lose much of their kinetic energy and become *slow neutrons*, with energies of an electron volt or less. Ultimately, the kinetic energy can be reduced to such an extent that the average is much the same as that of the atoms (or molecules) of the medium. Since the value of the kinetic energy then depends on the temperature, it is called the *thermal energy*. Neutrons whose energies have been reduced to this extent are called *thermal neutrons*. At ordinary temperatures, the most probable energy of such neutrons is only 0.025 ev.

1.33. If, instead of expelling a neutron, the excited compound nucleus formed by the absorption of a neutron emits its excess energy in the form of gamma radiation, the process is referred to as *radiative capture* or, in brief, as *capture*. The residual nucleus, having an additional neutron, is thus an isotope of the original nucleus but one unit higher in mass number. Radiative capture processes, represented by the symbol (n, γ), are very common and almost invariably occur more readily with slow than with fast neutrons.

1.34. Several radiative capture reactions are of importance in connection with the operation of nuclear reactors, and two of immediate interest will be mentioned here. The first is the capture of neutrons by uranium-238, the most abundant naturally occurring isotope of this element (see Table 1.1); the (n, γ) process in this case may be represented as

$$_{92}U^{238} + {_0}n^1 \rightarrow {_{92}}U^{239} + \gamma.$$

The resulting nucleus, uranium-239, is radioactive and decays with the emission of a negative beta particle, indicated by $_{-1}\beta^0$ (charge -1, mass number zero); thus,

$$_{92}U^{239} \rightarrow {_{-1}}\beta^0 + {_{93}}Np^{239},$$

the product, Np^{239}, being an isotope of an element of atomic number 93, called *neptunium*, which does not normally exist on earth to any detectable extent. Neptunium-239 is also beta active and decays fairly rapidly, according to the process

$$_{93}Np^{239} \rightarrow {_{-1}}\beta^0 + {_{94}}Pu^{239},$$

to form the isotope Pu^{239} of the element of atomic number 94, called *plutonium*, which occurs in nature in the merest traces only.

1.35. A series of processes similar to the one just described is initiated by the (n, γ) reaction with the naturally occurring thorium-232; thus,

$$_{90}Th^{232} + {_0}n^1 \rightarrow {_{90}}Th^{233} + \gamma,$$

so that the product is the isotope thorium-233. This undergoes two successive stages of beta decay, the first being

$$_{90}Th^{233} \rightarrow {_{-1}}\beta^0 + {_{91}}Pa^{233},$$

where Pa^{233} is the symbol for protactinium-233, and the second

$$_{91}Pa^{233} \rightarrow {_{-1}}\beta^0 + {_{92}}U^{233},$$

so that the product is uranium-233, an isotope of uranium which is not found in any appreciable amount in nature.

1.36. The third important class of interaction between neutrons and nuclei mentioned in § 1.31 is *fission* or, more precisely, *nuclear fission*. Since the fission process is basic to the operation of nuclear reactors, it will be treated below in greater detail than were the other two types of neutron-nucleus interactions.

NUCLEAR FISSION

THE FISSION PROCESS

1.37. Fission occurs only with certain nuclei of high atomic (and mass) number, and the large value of Z^2, and hence the repulsive force within the nucleus (§ 1.24), is an important contributory factor. When fission occurs, the excited compound nucleus formed after absorption of a neutron breaks up into two lighter nuclei, called *fission fragments*. If the neutron is one of low kinetic energy, i.e., a slow neutron, the two fragment nuclei generally have unequal masses. That is to say, symmetrical fission by slow neutrons is rare; in the majority of slow-neutron fissions the mass ratio of the fragments is approximately 2 to 3.

1.38. Only three nuclides, having sufficient stability to permit storage for a long time, namely, uranium-233, uranium-235, and plutonium-239, are fissionable by neutrons of all energies, from thermal values (or less) to millions of electron volts. Of these nuclides, uranium-235 is the only one which occurs in nature; the other two are produced artificially from uranium-238 and thorium-232, respectively, in the manner described in § 1.34 *et seq.*, i.e., by neutron capture followed by two stages of radioactive decay. Several other species are known to be capable of undergoing fission by neutrons of all energies, but they are highly radioactive and decay so rapidly that they have no practical value for the release of nuclear energy.

1.39. In addition to the nuclides which are fissionable by neutrons of all energies, there are some that require fast neutrons to cause fission; among these, mention may be made of thorium-232 and uranium-238. For neutrons below about 1-Mev energy, the only reaction is radiative capture, but above this threshold value, fission also occurs to some extent. Since fission of thorium-232 and uranium-238 is possible with sufficiently fast neutrons, they are known as *fissionable* nuclides. In distinction, uranium-233, uranium-235, and plutonium-239, which will undergo fission with neutrons of any energy, are referred to as *fissile* nuclides. Moreover, since thorium-232 and uranium-238 can be converted into the fissile species, uranium-233 and plutonium-239, respectively, they are also called *fertile* nuclides.

1.40. The importance of fission, from the standpoint of the utilization of nuclear energy, lies in two facts. First, the process is associated with the release of a large amount of energy per unit mass of nuclear fuel and, second, the fission reaction, which is initiated by neutrons, is accompanied by the liberation of neutrons. It is the combination of these two circumstances that makes possible the design of a nuclear reactor in which a self-sustaining fission chain reaction occurs with the continuous release of energy. Once the fission reaction has been started in a few nuclei by means of an external source of neutrons, it can be maintained in other nuclei by the neutrons produced in the reaction. It should be noted that it is only with the fissile nuclides mentioned above that a self-sustaining chain is possible. Thorium-232 and uranium-238 cannot support a fission chain because the fission probability is small even for neutrons with energies in excess of the threshold of 1 Mev, and inelastic scattering soon reduces the energies of many neutrons below the threshold value.

1.41. The liberation of neutrons in the fission reaction can be explained as follows. In the compound nucleus U^{236} formed when a uranium-235 nucleus captures a neutron, the ratio of neutrons to protons is nearly 1.57; consequently, when this nucleus splits into two parts, with mass numbers in the range of roughly 95 to 140, the average neutron to proton ratio in the instantaneous products must have the same value. It is seen from Fig. 1.2, however, that this ratio is too large for stability in nuclei of intermediate mass (cf. § 2.178). Consequently, if these nuclei, produced in fission, have sufficient excitation energy, they can expel neutrons, thereby tending to become more stable.

1.42. The actual number of neutrons released in this manner is too small, however, to confer stability on the resulting fission fragments. The latter still have too high a ratio of neutrons to protons and so, in accordance with the arguments in § 1.25, they are radioactive, exhibiting negative beta decay. The fission fragments undergo, on the average, three stages of radioactive decay before stable nuclei are formed. The general term *fission products* is applied to the complex, highly radioactive, mixture of nuclides consisting of the fission fragments and their various decay products.

FISSION ENERGY

1.43. The amount of energy released when a nucleus undergoes fission can be calculated by determining the net decrease in mass, from the known isotopic masses, and utilizing the Einstein mass-energy relationship. A simple, but instructive although less accurate, alternative procedure is the following. Disregarding the neutrons involved, since they have a negligible effect on the present calculation, the fission reaction may be represented (approximately) by

Uranium-235 → Fission product A + Fission product B + Energy.

In uranium-235, the mean binding energy per nucleon is about 7.6 Mev, as seen in Example 1.1, so that it is possible to write

$$92\,p + 143\,n \rightarrow \text{Uranium-235} + (235 \times 7.6) \text{ Mev,}$$

where p and n represent protons and neutrons, respectively. The mass numbers of the two fission product nuclei are mostly in the range of roughly 95 to 140, where the binding energy per nucleon is, as in tin-120, for example, about 8.5 Mev; hence,

$$92\,p + 143\,n \rightarrow \text{Fission products A and B} + (235 \times 8.5) \text{ Mev.}$$

Upon subtracting the two binding energy expressions, the result is

$$\text{Uranium-235} \rightarrow \text{Fission products} + 210 \text{ Mev.}$$

1.44. The fission of a single uranium-235 (or similar) nucleus is thus accompanied by the release of over 200 Mev of energy. This may be compared with about 4 ev which are released by the combustion of an atom of carbon-12. Hence, the fission of uranium yields something like 2.5 million times as much energy as the combustion of the same weight of carbon. Alternatively, it may be stated that 1 pound of fissile material should be capable of producing the same amount of energy as 1400 tons of 13,000 Btu/lb of coal.

1.45. In order to convert the fission energy values into practical units, it should be recalled that 1 Mev is equal to 1.60×10^{-6} erg, and this is equivalent to 1.60×10^{-13} watt-sec. Hence the total energy (200 Mev) available per fission is about 3.2×10^{-11} watt-sec, so that 3.1×10^{10} fissions are required to release 1 watt-sec of energy. In other words, fissions at the rate of 3.1×10^{10} per sec produce 1 watt of power.

1.46. One gram atom, i.e., the atomic weight expressed in grams, of any element contains the Avogadro number, i.e., 0.602×10^{24}, of individual nuclei, and if all of these undergo fission, the energy liberated would be (0.602×10^{24}) $(3.2 \times 10^{-11}) = 1.9 \times 10^{13}$ watt-sec, i.e., 5.3×10^6 kw-hr. This is the total amount of heat that would be released by the complete fission of 233 grams of uranium-233, or 235 grams of uranium-235, or 239 grams of plutonium-239. Neglecting the relatively small differences between these weights, the results in Table 1.2 may be regarded as applying to the heat produced by the fission of 1 pound of any of these materials. A useful fact to remember is that the power production corresponding to the fission of 1 gram of material per day would be roughly 10^6 watts, i.e., 1 megawatt. The mass consumed is, however, greater than 1 gram because some of the fissile nuclei are lost as a result of non-fission capture reactions (cf. § 2.159).

TABLE 1.2.   HEAT LIBERATED BY ONE POUND
OF FISSILE MATERIAL

0.9×10^{13} cal
1.0×10^7 kw-hr
2.8×10^{13} ft-lb
3.6×10^{10} Btu

Example 1.2. The cumulative exposure of a nuclear fuel in a reactor (or "burnup") is commonly expressed in terms of the megawatt-days (Mw-days) of energy per metric ton (or tonne), i.e., 1000 kg or 10^6 g, of total uranium in the fuel. An alternative, used by metallurgists, is to state the exposure in terms of fissions per cm^3 of fuel. Derive a conversion factor between these two units. A uranium dioxide fuel, with a density of 10.2 g/cm^3, has a burnup of 10,000 Mw-days/tonne; what are the fissions per cm^3?

Since 1 megawatt = 10^6 watts, 1 day = 8.64×10^4 sec, and 1 tonne = 10^6 g,

$$1 \text{ Mw-day/tonne U} = 10^6 \times 8.64 \times 10^4/10^6 = 8.64 \times 10^4 \text{ watt-sec/g U.}$$

Furthermore, 3.1×10^{10} fissions produce 1 watt-sec; hence,

$$1 \text{ Mw-day/tonne U} = 8.64 \times 10^4 \times 3.1 \times 10^{10} = 2.68 \times 10^{15} \text{ fissions/g U.}$$

$$= (2.68 \times 10^{15})(\text{g U/cm}^3) \text{ fissions/cm}^3.$$

Consequently, to convert Mw-days/tonne into fissions/cm^3, the burnup in Mw-days/tonne is multiplied by the factor 2.68×10^{15} and also by the number of grams of uranium per cm^3 of the fuel.

Assuming the uranium in the dioxide is mainly uranium-238; then $238 + 32 = 270$ g of UO$_2$ contain 238 g U. The density of the UO$_2$ is 10.2 g/cm^3, which represents $(10.2)(238)/270 = 8.99$ g U/cm^3. Hence,

$$10,000 \text{ Mw-days/tonne} = (10^4)(2.68 \times 10^{15})(8.99)$$

$$= 2.41 \times 10^{20} \text{ fissions/cm}^3.$$

1.47. The major proportion—over 80 per cent—of the energy of fission appears as kinetic energy of the fission fragments, and this immediately manifests itself as heat. Part of the remaining 20 per cent or so is liberated in the form of instantaneous gamma rays from excited fission fragments and as kinetic energy of the fission neutrons. The rest is released gradually as energy carried by the beta particles and gamma rays emitted by the radioactive fission products as they decay over a period of time. This decay energy ultimately appears in the form of heat as the radiations interact with and are absorbed by matter. The distribution of the fission energy for uranium-235, which may be regarded as applying approximately to all three of the important fissile species, is given in Table 1.3.

TABLE 1.3.   APPROXIMATE DISTRIBUTION OF FISSION ENERGY

	Mev
Kinetic energy of fission fragments	165
Instantaneous gamma-ray energy	7
Kinetic energy of fission neutrons	5
Beta particles from fission products	7
Gamma rays from fission products	6
Neutrinos	10
Total fission energy	~200

1.48. The 10-Mev energy of the neutrinos accompanying the beta radioactivity is not available for power production because the interaction between these particles and matter is extremely weak. Thus, essentially all the neutrinos escape from a fission reactor carrying their energy with them.

CRITICAL MASS

1.49. Since two or three neutrons are liberated in each act of fission whereas only one is required to maintain a fission chain, it would seem that once the fission reaction were initiated in a given mass of fissile material, it would readily sustain itself.   However, such is not the case because not all the neutrons produced in fission are available to carry on the fission chain.   Some neutrons are lost in nonfission reactions, mainly radiative capture, with the various extraneous materials present and even with the fissile species itself, whereas other neutrons escape entirely from the system undergoing fission.   The fraction of neutrons lost by escape through the geometric boundaries can be reduced by increasing the size (or mass) of the fissile material.   The minimum quantity of such material that is capable of sustaining a fission chain, once it has been initiated by an external source of neutrons, is called the *critical mass*.

1.50. The critical mass of material required for a reactor depends upon a wide variety of conditions, although for any specific reactor system it always has a definite value.   Thus, the critical mass of uranium-235 may range from less than 1 kg (kilogram), for a system consisting of a solution in water of a salt of uranium containing about 90 per cent of the fissile isotope, to some 200 kg (or more) present in over 30 tons of natural uranium embedded in a matrix of graphite.   Natural uranium alone, containing about 0.7 per cent of uranium-235, can never become critical, no matter how large its mass, because too high a proportion of the fission neutrons are lost in nonfission reactions.

NUCLEAR FISSION REACTORS

GENERAL FEATURES OF NUCLEAR REACTORS

1.51. In spite of numerous variations in the design and components of nuclear reactor systems, there are, nevertheless, a number of general features which all such systems possess in common, to a greater or lesser extent.   In outline (Fig. 1.3), a reactor consists of an active *core* in which the fission chain is sus-

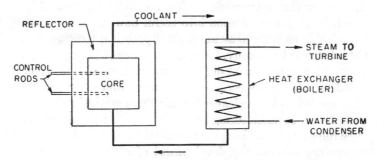

FIG. 1.3.   Schematic representation of nuclear reactor system

tained and in which most of the energy of fission is released as heat.   The core contains the nuclear fuel, consisting of a fissile nuclide and often a fertile material in addition.   If it is desired, as is often the case, that most of the

varying the neutron density in the core.   This is done either by moving rods of a material that captures neutrons readily, i.e., a *neutron poison*, or by displacing part of the reactor core or reflector.   Insertion of a poison, e.g., cadmium or boron, results in a decrease in neutron density and hence in the reactor power; withdrawal of the poison is accompanied by an increase in both neutron density and power level.   Similarly, displacement of a portion of the reflector or of the core may permit neutrons to escape, thus causing a decrease in the neutron density, whereas its replacement will produce an increase in the density.

REACTOR TYPES

1.56. Nuclear reactors can be classified in various ways, but the most fundamental distinction is that based on the kinetic energy (or speed) of the neutrons causing most of the fissions in the given reactor.   Nearly all the neutrons liberated in fission have high energies and so, if no moderator is present in the reactor core or reflector, the majority of fissions are produced by fast neutrons. A nuclear reactor in which this is the case is called a *fast reactor*.   The fuel material for such reactors must contain a significant proportion—about 10 per cent or more—of a fissile nuclide.   The remainder must be a substance of high or intermediate mass number, since elements of low mass number would slow down the neutrons.   As far as possible, too, elements must be avoided which can cause inelastic scattering, and hence slowing down, of neutrons of moderately high energy.

1.57. If a fertile species is present in the fast reactor core or in what is called a *blanket* surrounding the core, it will be converted into fissile material by neutron capture (§ 1.34 *et seq.*).   Wasteful (or parasitic) capture of fast neutrons is relatively small, and if the loss of neutrons by escape can be kept to a minimum, it is possible for more than one neutron to be available, per fission, for the conversion of fertile into fissile nuclei.   In these circumstances it is possible for more fissile material to be produced by neutron capture than is consumed by fission.

1.58. When the fissile nuclide produced is identical with that used to maintain the fission chain, the reactor is called a *breeder*.   A fast reactor utilizing plutonium-239 as the fuel and uranium-238 as the fertile species can act as a *power-breeder*, generating power and, at the same time, producing more plutonium-239 than is consumed.   An analogous fast power-breeder reactor is possible with uranium-233 and thorium-232 as fissile and fertile nuclides, respectively.   However, it appears that breeding in this system can be achieved in reactors of another type, to be described below, which have some advantages over fast reactors.

1.59. It is possible for a fast reactor to employ uranium-235 as the fissile species, to maintain the nuclear chain reaction, and uranium-238 as the fertile material which is converted into plutonium-239. Such a *converter* reactor is not strictly a breeder, even though the quantity of plutonium-239 produced may exceed that of the uranium-235 consumed. If the latter situation applies, the reactor is sometimes called a *pseudo-breeder*.

1.60. The reserves of uranium-235, the only fissile material existing in nature, are small. Hence, there may ultimately come a time when essentially all the available uranium-235 is exhausted. The further utilization of the remaining uranium-238 and of thorium-232 will then depend on the use of plutonium-239 or uranium-233 to maintain the fission chain. It is for this reason that true breeders, which both produce and consume these fissile nuclides, are important. But until sufficient of these materials has been accumulated to permit of their use as nuclear fuels, it is probable that pseudo-breeders, based on uranium-235, will play an important role.

1.61. If the reactor core contains a considerable proportion of a moderator, the high energy of the fission neutrons will be rapidly decreased to the thermal region. Most of the fissions in such a reactor, called a *thermal reactor*, will then be caused by thermal (or slow) neutrons. Thermal reactors have the advantage over fast reactors in greater flexibility of design. There is a reasonable choice of both moderators and coolants, as well as of fuel materials. Depending on the nature of the fuel and moderator, a thermal reactor may be quite small or relatively large. Fast reactors, on the other hand, are generally small, so that the heat removal becomes a problem. According to circumstances, the fuel in a thermal reactor may range from natural uranium, with 0.7 per cent of uranium-235, to material enriched to the extent of 90 per cent (or more) in the fissile isotope.

1.62. Many thermal reactors include a fertile species, so that they are converters. However, in certain circumstances, where reactors of compact design are required, the fuel consists of essentially pure uranium-235. Reactors of this latter type are called *burners*, since they consume fissile material without replacing it. Such reactors can be justified only where special situations make their use mandatory, e.g., in a submarine.

1.63. Because of the high probability that slow neutrons will be captured in nonfission reactions in uranium-235 or plutonium-239, less than one neutron is available per fission, on the average, for the conversion of fertile into fissile nuclei, after allowing for inevitable losses of neutrons by escape. Hence, the quantity of plutonium-239 produced in a thermal converter reactor does not usually exceed the amount of this isotope or of uranium-235 consumed in maintaining the fission chain. Consequently, it is difficult to design a thermal breeder or pseudo-breeder based on either uranium-235 or plutonium-239 as the fuel.* This situation does not apply, however, when uranium-233 is the fissile

---

* A thermal breeder (or pseudo-breeder) of this type is theoretically possible by decreasing the loss of escaping neutrons in a large reactor and taking advantage of the fission of uranium-238 by fast neutrons.

material, and so thermal breeders involving this nuclide, with thorium-232 as the fertile species, are of interest.

THE REACTOR FUEL CYCLE

1.64. The production of the reactor fuel, its utilization in the reactor, and the recovery of unused fissile and fertile materials from the spent fuel constitute the *fuel cycle.* This cycle represents an important aspect of reactor design, largely because of its influence on the economics of nuclear power. In the first place, considerable processing effort is required to convert the raw material, e.g., uranium ore, into the pure state suitable for use as fuel; the latter may be in the form of uranium metal, as oxide or carbide, or as a salt, e.g., uranyl sulfate, capable of yielding a fuel solution in water. Furthermore, many reactor designs require uranium which has been, at least partially, enriched in the fissile uranium-235, i.e., in a proportion greater than that present in the element as found in nature. An isotope separation process must, therefore, be included in the fuel materials preparation to meet this requirement.

1.65. The useful lifetime of the fuel in a reactor may be limited by dimensional changes in solid fuel elements, by the accumulation of fission product poisons, especially in thermal reactors, and by depletion of the fissile material. As a general rule, the fuel will require replacement when only a few per cent of the total fissile and fertile species have been consumed. The unused materials must, therefore, be recovered and recycled, and this involves a complex chemical procedure. In spite of the low utilization of fissile material, the intense radioactivity of the fission products (§ 1.42) introduces a special problem in the treatment of the spent fuel. A "cooling" period, which may be as long as 100 days, must generally be allowed for the radioactivity to decay to a sufficient extent to permit further processing. Since this delay adds to the inventory (or rental) cost of the fuel, consideration must be given to procedures which require only a short cooling period.

1.66. The chemical and metallurgical processes whereby the uranium, plutonium, and the fission products are separated may be quite complicated, depending on the form of the reactor fuel. For thermal reactors, in particular, a high degree of decontamination, i.e., removal of fission products, is desirable in order to minimize the proportion of neutron poisons present. If the fuel is to be used in a fast reactor, this matter is less serious because of the small probability, in general, of nonfission capture of fast neutrons. However, in any event, the removal of radioactive fission products is always necessary to the extent that will permit fabrication of the recovered material into the required form without constituting a health hazard.

1.67. Finally, mention may be made of the formidable problem involved in the disposal of large amounts of highly radioactive fission product wastes from the fuel processing plants. In many cases, the waste solutions must first be concentrated and then shipped in heavily shielded containers to suitable disposal sites.

HISTORY OF REACTOR DEVELOPMENT

1.68. In order to provide a general background for some of the succeeding chapters, the history of nuclear reactor development will be outlined. Soon after the discovery of fission was reported in February 1939, it was widely realized that this reaction opened up the possibility for the practical release of nuclear energy. In fact, as early as May 1939, a patent for a system designed to achieve this end, using heavy water as the moderator, was filed with the

FIG. 1.4.   The first "pile" during construction; shows 18th layer containing fuel lumps and 19th layer of graphite only

Swiss Patent Office.   However, it was not until December 2, 1942, that the world's first self-sustaining nuclear fission chain was realized in the United States at the University of Chicago.   The fuel consisted of 40 tons of uranium, in the form of lumps of metal and oxide, distributed throughout a matrix of 385 tons of graphite, which served as moderator and reflector.   The nominal operating power was 2 kw, so that the ambient air provided sufficient cooling.   The structure was made by gradually piling up graphite blocks, one layer upon another, and inserting the uranium in holes in the blocks (Fig. 1.4).   It was consequently called a "pile," a name which was retained for several years until the preferable term "nuclear reactor" came into general use.

1.69. When it was originally planned, the purpose of studying the uranium-graphite system was to determine whether a chain reaction could be realized, with the object of employing such a reaction to produce an atomic explosion. However, while the work was in progress, it appeared that the neutron chain could also be used to make fissile plutonium-239, in the manner described in § 1.34 *et seq.*, for use in atomic weapons.   For this purpose it would be necessary to construct large reactors operating at high power, and early in 1942 plans were initiated for the Hanford reactors to produce plutonium-239 in appreciable quantities.   In the meantime, in order to secure sufficient of this isotope to permit a study of its properties and to obtain experience with the operation of reactors, it was decided to build an experimental reactor at Oak Ridge, with an initial design power of 1000 kw.   Like the Chicago pile, this reactor, which commenced operation in November 1943, used graphite as moderator and reflector; however, the fuel was in the form of cylinders (or slugs) of metallic uranium jacketed in aluminum and forced air cooling was used to remove the heat generated by fission.   It is of interest to mention that this reactor, the oldest in existence, is still operating at Oak Ridge National Laboratory at a power of 4000 kw.   It serves as a valuable experimental tool for research requiring a strong neutron source.

1.70. The Hanford,Wash., production reactors were originally designed to use helium gas under pressure as the coolant.   But the problems of supplying the large quantities of this relatively rare gas and of preventing leakage were so great that the use of helium was abandoned in favor of water as coolant.   The difficulties of constructing a large graphite-moderated and water-cooled reactor were considerable, but they were overcome and the first Hanford reactor attained criticality in September 1944.   Subsequently, a number of additional reactors of the same general type were built at Hanford for plutonium production.   No use is made of the fission heat carried off by the cooling water.   The more recent production reactors at Savannah River, S. C., employ heavy water as the moderator, since this permits a higher efficiency of conversion of uranium-238 into plutonium-239.   In both the United Kingdom and France, the original plutonium reactors were graphite-moderated and air cooled, and the heat produced, as in the American reactors, was wasted.   The later designs, however, are dual-purpose reactors; they are cooled by carbon dioxide under pressure and yield both plutonium-239 and useful power.

1.71. Until about the middle of the 1950's, nearly all the reactors constructed, apart from those intended for the production of plutonium, were experimental devices to be used either for physical research or to provide information for the design of power reactors. Among those intended for research, two of the earliest novel types were completed in May 1944. One was the so-called "water boiler" reactor at Los Alamos, N. M., in which the fuel was partially enriched uranium sulfate dissolved in water, the solvent being the moderator. This was both the first homogeneous reactor, i.e., one in which the fuel and moderator are uniformly mixed, as well as the first to employ uranium enriched in the fissile isotope, uranium-235, as the fuel. The other research reactor, started about the same time, was the CP-3 (Chicago Pile No. 3) at the Argonne Laboratory. This was the first reactor to be moderated by heavy water. The moderator was circulated through an external heat exchanger where it was cooled by ordinary water. Both the water boiler and the CP-3 have served as prototypes for research reactors constructed in the United States and elsewhere.

1.72. Largely because of the facilities for the production of heavy water and the availability of ample supplies of natural uranium, an early interest was demonstrated in Canada for reactors employing these materials. The low-power Zero Energy Experimental Pile (ZEEP) was completed in April 1945, and this was soon followed by the NRX reactor at Chalk River, Ontario. For some time after its completion in August 1947, it had the highest thermal neutron flux (product of neutron density and velocity) of any known reactor. The NRX reactor was novel in the respect that, although moderated by heavy water, the coolant was ordinary water circulating in annular channels about the cylindrical fuel rods of natural uranium metal.

1.73. Another type of design, which was utilized in several subsequent reactors, was introduced with the Materials Testing Reactor (MTR), constructed at the National Reactor Testing Station, Arco, Idaho. This reactor, which commenced operation in March 1952, was designed to provide a high neutron flux—higher than that of the NRX reactor—for experimental purposes, in particular for the study of the effects of nuclear radiations on reactor materials. The MTR has two special design features: first, it uses ordinary water, circulating through a cylindrical tank containing the reactor core, as moderator and coolant, and, to some extent, as reflector; and second, the fuel elements are slightly curved "sandwich" plates consisting of an alloy of aluminum and enriched uranium clad on both sides with aluminum. A number of parallel plates are brazed into aluminum side members to form a long box-like assembly, roughly 3 in. by 3 in. across. The plate-type fuel elements provide a large heat-removal area, making operation possible at fairly high power (40 Mw).

1.74. Before the MTR commenced operation, preliminary tests with a mockup had established the convenience and efficiency of the fuel elements described above. The use of these MTR-type elements facilitated the construction of the Bulk Shielding Reactor (BSR) at Oak Ridge National Laboratory toward the end of 1950. The core consisted of 16 to 25 of the box-like assemblies, arranged in the form of a parallelepiped with spaces in between for the control rods. The whole was suspended to a depth of some 20 ft in water which served as moderator, coolant, and reflector, and also as a radiation shield. Operation at powers up to 100 kw is possible using free-convection cooling, so that forced circulation of the water is not necessary. The BSR was nicknamed the "swimming pool" reactor because of the large tank of water in which it was contained. The great simplicity and convenience of the pool design has led to its adoption in a number of reactors for use in university and industrial-type research laboratories.

1.75. In all of the foregoing reactors a moderator is employed to slow down the neutrons, and so they are essentially thermal reactors. The first fast-neutron reactor commenced operation at Los Alamos, N. M. in November 1946; the fuel was plutonium-239, the first time this material was utilized in a reactor. Mercury was the coolant and the operating power was 25 kw. There was, of course, no moderator, since the objective was to show that a controlled fission chain could be realized in which most of the fissions resulted from the absorption of fast neutrons. The second fast reactor was the Experimental Breeder Reactor (EBR-I) completed at Arco in 1951. It was designed by the Argonne National Laboratory to provide information concerning the possibility of breeding with fast neutrons, with the simultaneous production of electric power, and to utilize a liquid metal (sodium-potassium alloy) as the high-temperature (660°F) coolant. The EBR-I is, strictly, a pseudo-breeder, because the earlier cores used uranium-235 as the fuel and uranium-238 as the fertile material. Nevertheless, it served to prove the feasibility of breeding. The EBR is of some interest in having achieved the first known generation of electricity from nuclear energy in December 1951.

1.76. Although research reactors are undoubtedly valuable experimental tools, the prime significance of nuclear reactors lies in the possibility of producing power either for special or normal purposes. In the United States, it was early realized that a compact nuclear power plant would have great advantages for submarine propulsion, since it would make possible long journeys without the necessity for surfacing or refueling. The reactor designed for this purpose employed water under pressure as the moderator-coolant and highly enriched uranium metal, clad in zirconium, as the fuel. The use of these materials permitted the design of a small reactor of high power. The pressurized water is circulated through an external heat exchanger which produces steam for operating the turbines. A prototype of the submarine reactor started operation at Arco in March 1953 and produced substantial amounts of electric power after May 31, 1953. The U. S. S. Nautilus, the first nuclear-powered submarine, commenced its sea trials in January 1955.   It has proved so effective that many other sub-marines have been—and are being—built with propulsion plants of the same general design.

1.77. As a result of experience gained in the successful operation of the sub-marine reactors, the first central-station nuclear power plant in the United States, at Shippingport, Pa., was designed to use a Pressurized Water Reactor (PWR).   The water pressure in the reactor vessel is 2000 psia and the steam produced in the heat exchanger has a temperature of about 490°F and a pressure close to 600 psia.   To decrease the cost of the power produced, only a small number of the fuel elements are highly enriched in uranium-235 (as an alloy with zirconium), the remainder being of normal uranium (as the dioxide).   The change in the character of the fuel necessitates a larger core, but the increased size of the reactor is not a drawback for a land-based facility.   The Shippingport PWR went into operation on December 2, 1957, and three weeks later it attained its design power output of 60 Mw electrical (230 Mw thermal).   Other pres-surized-water reactors, using a single type of slightly enriched uranium dioxide fuel element, have been constructed.   Some are large, for central-station power plants, while others are compact designs of relatively low power for use in re-mote locations, e.g., in Arctic and Antarctic regions, where conventional fuel costs are extremely high.

1.78. In a modification of the pressurized-water concept, steam is produced directly by utilizing fission heat to boil water within the reactor core, rather than in an external heat exchanger.   Following upon a number of Boiling Water Reactor Experiments (BORAX) at Arco, which were initiated in 1953, the Experimental Boiling Water Reactor (EBWR) was built at the Argonne Na-tional Laboratory.   It first went critical on December 1, 1956 and generated electricity on December 23, 1956; its full design power of 5 Mw electrical was reached six days later.   Steam is produced in the reactor vessel at a pressure of 600 psig and, after separating the moisture, is fed to a turbo-generator.   The advantage of the BWR over the PWR design is that the former can produce steam of equal quality at a lower system pressure and without the use of an external steam generator.   Several boiling-water reactors have been developed for large-scale power production.

1.79. It has been mentioned (§ 1.51 *et seq.*,) that a number of different fuel materials, moderators, and coolants are available for use in reactors.   It is evi-dent, therefore, that numerous combinations, leading to a variety of reactor designs, are possible.   Some of these may have basic incompatibilities, but among the many that appear to be practical, there is no obviously "best" design for a power reactor.   Every concept which has been proposed so far has drawbacks as well as advantages.   Consequently, in the early stages of power reactor development, each country has followed along lines determined either by previous technological experience or by the availability of materials.

1.80. In the United States, for example, the first central-station nuclear power plants utilized the pressurized-water design, because of satisfactory experience with the submarine reactor coupled with the availability of enriched uranium. In the United Kingdom, on the other hand, the experience with the air-cooled, graphite-moderated production reactors at Windscale, which were deemed to be less hazardous than the water-cooled Hanford reactors, led to the development of the Calder Hall dual-purpose design. These reactors use natural uranium fuel, graphite moderator, and carbon dioxide gas as coolant. Apart from their use in submarines, pressurized-water reactors have attracted no interest, one reason given being the high cost of enriched uranium. A number of central-station power plants in the United Kingdom, for power only, have followed the same basic design principle as the Calder Hall reactors. The first power reactor (5 Mw electrical) in the U.S.S.R., operated in 1954, was a water-cooled, graphite-moderated system using slightly enriched fuel, possibly as a result of experience with plutonium production reactors of the Hanford type. Another power reactor (100 Mw electrical) of similar design was started in Siberia in 1958. Finally, as was indicated earlier, in Canada, reactors employing heavy water as moderator and natural uranium as fuel have attracted attention, chiefly because of accessibility of the necessary materials.

1.81. In spite of the progress that has been made, it is generally appreciated that nuclear reactor technology is still in its infancy. A great deal more research and development will be necessary before the long-range potentialities of different reactor types can be properly assessed, especially in regard to their prospects for producing economical power. Because of the possible significance of breeding in making the maximum use of nuclear fuel resources, extensive work is being carried out in various countries on power-breeders utilizing fission by fast neutrons. The Enrico Fermi Atomic Power Plant, near Monroe, Mich., is to be the world's first large-scale power installation to employ a fast breeder reactor. In addition, various other concepts are being studied. In the United Kingdom, for example, advanced gas-cooled reactors, using enriched fuel and designed for operation at high temperatures, are receiving prime consideration. However, in the United States a number of promising types, in addition to pressurized-water, boiling-water, and gas-cooled, high-temperature systems, are being investigated. Among these mention may be made of boiling-water reactors with nuclear superheating for improved efficiency, reactors using organic liquids as moderator-coolants to avoid the pressurization necessary with water, and designs with graphite as moderator and liquid sodium as coolant to permit high-temperature operation at normal pressures. Central-station plants of all these types have been constructed or are under construction in the United States to learn more about the technology of different reactor designs.

1.82. The foregoing review has covered some of the main aspects of reactor development since 1942. Although the highlights have been indicated, it has not been possible to treat all the reactor types which have been considered, investigated, or actually constructed. Several promising ideas, such as the

Homogeneous Reactor Experiments, which used aqueous (or heavy water) solutions of a uranium salt as fuel, have been abandoned, temporarily at least, and have not been mentioned. It is felt, however, that sufficient has been given here to provide some general orientation concerning the past, present, and immediate future of the nuclear reactor program. More details concerning various reactor types will be found in Chapter 13, with a broad assessment of their economic potentials in Chapter 14.

SYMBOLS USED IN CHAPTER 1

A mass number

amu   atomic mass unit (1.66×10^{-24} gram)

c velocity of light (3.00×10^{10} cm/sec)

E energy

ev   electron volt (1.60×10^{-12} erg)

Mev   million electron volts (1.60×10^{-6} erg)

M isotopic mass

m mass

m_e mass of electron

m_H mass of hydrogen atom

m_n mass of neutron

m_p mass of proton

N_a Avogadro number (0.602×10^{24})

n neutron

Z atomic number

β beta particle

γ gamma radiation

REFERENCES FOR CHAPTER 1

1. P. C. Putnam, "Energy in the Future," Ch. 6, D. Van Nostrand Co., Inc., Princeton, N. J., 1953; see also R. D. Nininger, *et al.*, "Energy from Uranium and Coal Reserves," U. S. AEC Report TID-8207 (1960).
2. P. C. Putnam, Ref. 1; see also R. D. Nininger, *et al.*, Ref. 1; *Power Reactor Technology*, **3**, No. 1, 1 (1959); **4**, No. 2, 1 (1961).
3. S. Glasstone, "Sourcebook on Atomic Energy," 2nd Ed., D. Van Nostrand Co., Inc., Princeton, N. J., 1958.
4. I. Kaplan, "Nuclear Physics," 2nd Ed., Addison-Wesley Publishing Co., Inc., Reading, Mass., 1962.

PROBLEMS

1. The heat of combustion of carbon is 7860 calories per gram. Calculate the energy in electron volts produced by the combustion of 1 atom of carbon. (Compare the result with the energy liberated in the fission of 1 nucleus of uranium-235.)

2. Determine the number of electron volts per second consumed by a 100-watt lamp.

3. Calculate the decrease of mass in grams accompanying the complete fission of 100 lb of uranium-235.

4. The heat of combustion of anthracite coal is 11,500 Btu/lb. Estimate the decrease of mass in grams that would be associated with the combustion of 100 lb of coal. (Compare the result with that of the preceding problem.)

5. A reactor produces heat at the rate of 100 Mw. How many atoms of uranium-235 suffer fission per second? If the fuel in the core contains 30 kg of uranium-235, what fraction of the fissile material has been used up after 30 days of operation? How much fuel would be required in the same time by an equivalent coal-fired station? If coolant, having a specific heat of 0.52 Btu/(lb)(°F), enters the reactor core at 500°F and leaves at 518°F, what would be the coolant circulation rate in lb/hr?

6. If about 25 atoms of stable fission product gases are produced per 100 atoms of uranium undergoing fission, calculate the volume of gas, at standard temperature and pressure, released during a 1-month operation of a reactor with a thermal power of 300 Mw.

7. Calculate the energy in Mev released in each of the three following fusion reactions·

$$_1H^2 + {}_1H^2 \rightarrow {}_2He^3 + {}_0n^1$$
$$_1H^2 + {}_1H^2 \rightarrow {}_1H^3 + {}_1H^1$$
$$_1H^3 + {}_1H^2 \rightarrow {}_2He^4 + {}_0n^1,$$

using the masses $_1H^1 = 1.008145$; $_1H^2 = 2.014740$; $_1H^3 = 3.017005$; $_2He^3 = 3.016986$; $_2He^4 = 4.003874$; $_0n^1 = 1.008986$ amu. The first two reactions occur simultaneously at the same rate and the H^3 (tritium) formed in the second reaction then interacts rapidly with a deuterium (H^2) nucleus. Estimate the total amount of energy that would be obtainable theoretically from 1 lb of deuterium as a result of the three fusion reactions given above. Compare the result with the fission energy available from 1 lb of uranium-235.

Chapter 2

NUCLEAR REACTIONS AND RADIATIONS

INTRODUCTION

2.1. In the preceding pages a general review was presented of the processes taking place in a nuclear fission reactor. The purpose of the present chapter is to discuss in more detail some of these processes, especially radioactive decay, radiative capture, and fission. These reactions are accompanied by the emission of nuclear radiations of various kinds, e.g., alpha and beta particles, neutrons, and gamma rays. Consequently, a study of the interactions of such radiations with matter is of interest. It should be noted that it is the practice in this connection to use the general term "radiation" to include both material particles and true electromagnetic radiation. The following sections will deal with the general characteristics of nuclear reactions and radiations insofar as they are significant for reactor design and operation.

RADIOACTIVITY

RADIOACTIVE ISOTOPES

2.2. The naturally occurring forms of the great majority of elements are stable, but a few of high atomic weight, from polonium (atomic number 84) onward, e.g., radium (88), thorium (90), and uranium (92), consist entirely of unstable, radioactive isotopes. In addition, the elements thallium (81), lead (82), and bismuth (83) exist in nature largely as stable isotopes, but also to some extent as radioactive species. The unstable substances undergo spontaneous change, i.e., radioactive disintegration or radioactive decay, at definite rates. As seen in Chapter 1, the decay is associated with the emission from the atomic nucleus of an electrically charged particle, either an *alpha particle*, i.e., a helium nucleus, or a *beta particle*, i.e., an electron. In many instances, gamma radiation accompanies the particle emission. Frequently the products of decay are themselves radioactive, expelling either an alpha or a beta particle. After a number of stages of disintegration, an atomic species with a stable nucleus is formed.

2.3. Besides the radioactive substances referred to above, there have been produced in recent years a total of about 800 so-called "artificial" radioisotopes of all the known elements. These have been obtained either by bombardment of stable elements with charged particles in cyclotrons, etc., by the capture of neutrons (§ 1.33), or as a result of nuclear fission. A few of them expel alpha particles, but a large proportion, including most of the fission products, are negative-beta emitters. A number of artificial radioactive species emit positive beta particles, i.e., positrons or positively charged electrons, but such substances play essentially no part in nuclear reactors.

RATE OF RADIOACTIVE DECAY

2.4. For a given radioactive species, every nucleus has a definite probability of decaying in unit time; this decay probability has a constant value characteristic of the particular radioisotope. It remains the same irrespective of the chemical or physical state of the element at all readily accessible temperatures and pressures. In a given specimen, the rate of decay at any instant is always directly proportional to the number of radioactive atoms of the isotope under consideration present at that instant. Thus, if N is the number of the particular radioactive atoms (or nuclei) present at any time t, the decay rate is given by

$$\frac{dN}{dt} = -\lambda N, \tag{2.1}$$

where λ, called the *decay constant* of the radioactive species, is a measure of its decay probability. Upon integration between any arbitrary zero time, when the number of radioactive nuclei of the specified kind present is N_0, and a time t later, when N of these nuclei remain, it is found that

$$N = N_0 e^{-\lambda t}. \tag{2.2}$$

Radioactive decay is seen to be an exponential process, the actual decay rate being determined by the decay constant λ and by the number of the particular nuclei present.

2.5. The reciprocal of the decay constant, represented by t_m, is called the *mean life* (or *average life*) of the radioactive species; thus,

$$t_m = \frac{1}{\lambda}. \tag{2.3}$$

It can be shown that the mean life is equal to the average life expectancy of the nuclei present at any time.

2.6. The most widely used method for representing the rate of radioactive decay is by means of the *half-life*. It is defined as the time required for the number of radioactive nuclei of a given kind (or for their activity) to decay to half its initial value. Because of the exponential nature of the decay, this time is independent of the amount of the radioisotope present. Thus, if N is set equal to $\frac{1}{2}N_0$ in equation (2.2), the corresponding time $t_{1/2}$, which is the half-life, is given by

$$e^{-\lambda t_{1/2}} = \tfrac{1}{2}$$

or

$$t_{1/2} = \frac{\ln 2}{\lambda} = \frac{0.6931}{\lambda}. \qquad (2.4)$$

The half-life is thus inversely proportional to the decay constant or, by equation (2.3), directly proportional to the mean life, i.e.,

$$t_{1/2} = 0.6931 t_m. \qquad (2.5)$$

The half-lives of known radioactive species range from a small fraction, e.g., about a millionth, of a second to billions of years.

2.7. Since the number of nuclei (or their activity) decays to half its initial value in a half-life period, the number (or activity) will fall to one-fourth by the end of two periods, and so on (Fig. 2.1). In general, the fraction of the initial

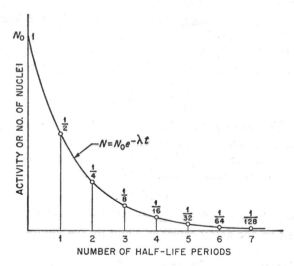

Fig. 2.1. Representation of exponential radioactive decay in terms of half-life periods

number of nuclei (or activity) remaining after n half-life periods is $(\frac{1}{2})^n$. After seven half-life periods, the amount of radioactive material has decreased to less than 1 per cent of its initial value.

2.8. The half-lives of a number of substances of interest in the nuclear energy field are quoted in Table 2.1. On the left side are given the half-lives of species which exist in nature, while on the right are those of artificial substances. Thus, thorium-233 is formed from thorium-232 by neutron capture (§ 1.35); it then decays to protactinium-233, which in turn yields uranium-233. Similarly, ura-

TABLE 2.1. CHARACTERISTICS OF RADIOACTIVE SPECIES

Naturally Occurring			Artificial		
Species	Activity	Half-Life	Species	Activity	Half-Life
Thorium-232	Alpha	1.39×10^{10} yr	Thorium-233	Beta	23.5 min
Uranium-238	Alpha	4.51×10^{9} yr	Protactinium-233	Beta	27.4 days
Uranium-235	Alpha	7.13×10^{8} yr	Uranium-233	Alpha	1.62×10^{5} yr
			Uranium-239	Beta	23.5 min
			Neptunium-239	Beta	2.33 days
			Plutonium-239	Alpha	2.44×10^{4} yr

nium-239 results when uranium-238 captures a neutron (§ 1.34); the successive decay products are then neptunium-239 and plutonium-239.

2.9. It is of interest to note that the three fissile species, namely, uranium-233, uranium-235, and plutonium-239, and the two fertile substances thorium-232 and uranium-238, all of which are alpha emitters, have long half-lives and so are comparatively stable. On the other hand, the substances produced when the fertile nuclei capture neutrons, i.e., thorium-233 and uranium-239, and their immediate decay products, i.e., protactinium-233 and neptunium-239, which are all beta emitters, have relatively short half-lives. It means that these particular fertile and fissile isotopes can be kept for many years without appreciable loss. However, since the intermediate stages following upon the capture of neutrons by fertile material have short half-lives, they will decay almost completely within a few days to form fissile species.

RADIOACTIVE EQUILIBRIUM

2.10. In a series of decay stages, such as those which follow nonfission neutron capture by thorium-232 or uranium-238, or as represented by the members of a natural radioactive series, or as exhibited by many fission products, each radioactive member of the series decays in accordance with equation (2.1) with its own specific value for the decay constant. Such a series may be represented by

$$A \xrightarrow{\lambda_A} B \xrightarrow{\lambda_B} C \xrightarrow{\lambda_C} D \xrightarrow{\lambda_D} \cdots \rightarrow X,$$

where A may be the parent of a natural radioactive series, or the species formed when a neutron is captured, e.g., thorium-233 or uranium-239, or it may be a radioactive fission product; the stable end-product of the series is represented by X. Consider any member of the series, other than the first or last, such as B, for example. While the rate of decay is given by equation (2.1), it must be remembered that, at the same time, B is being formed by decay of A; hence, at any specified time, the net rate of change of B with time is given by

$$\frac{dN_{\mathrm{B}}}{dt} = \lambda_{\mathrm{A}} N_{\mathrm{A}} - \lambda_{\mathrm{B}} N_{\mathrm{B}}, \tag{2.6}$$

the first term on the right representing the rate of formation of B by decay of A, and the second term the rate of decay of B.

2.11. Upon rearrangement, this yields

$$\frac{dN_{\mathrm{B}}}{dt} + \lambda_{\mathrm{B}} N_{\mathrm{B}} = \lambda_{\mathrm{A}} N_{\mathrm{A}},$$

which is a linear differential equation of the first order. If both sides are multiplied by the integrating factor $e^{\lambda_{\mathrm{B}} t}$, the left side becomes a complete differential; thus,

$$\left(\frac{dN_{\mathrm{B}}}{dt} + \lambda_{\mathrm{B}} N_{\mathrm{B}} \right) e^{\lambda_{\mathrm{B}} t} = \lambda_{\mathrm{A}} N_{\mathrm{A}} e^{\lambda_{\mathrm{B}} t}$$

so that

$$d(N_{\mathrm{B}} e^{\lambda_{\mathrm{B}} t}) = \lambda_{\mathrm{A}} N_{\mathrm{A}} e^{\lambda_{\mathrm{B}} t} \, dt. \tag{2.7}$$

By equation (2.2),

$$N_{\mathrm{A}} = N_{\mathrm{A}_0} e^{-\lambda_{\mathrm{A}} t}, \tag{2.8}$$

where N_{A_0} is the initial amount of A. If this is substituted into equation (2.7), integration yields

$$N_{\mathrm{B}} e^{\lambda_{\mathrm{B}} t} = \frac{\lambda_{\mathrm{A}} N_{\mathrm{A}_0}}{\lambda_{\mathrm{B}} - \lambda_{\mathrm{A}}} e^{(\lambda_{\mathrm{B}} - \lambda_{\mathrm{A}}) t} + C, \tag{2.9}$$

where C is the integration constant. If N_{B_0} is the initial amount of B, then setting t equal to zero in equation (2.9) gives the value of C as

$$C = N_{\mathrm{B}_0} - \frac{\lambda_{\mathrm{A}} N_{\mathrm{A}_0}}{\lambda_{\mathrm{B}} - \lambda_{\mathrm{A}}},$$

and hence equation (2.9) leads to the expression

$$N_{\mathrm{B}} = \frac{\lambda_{\mathrm{A}} N_{\mathrm{A}_0}}{\lambda_{\mathrm{B}} - \lambda_{\mathrm{A}}} (e^{-\lambda_{\mathrm{A}} t} - e^{-\lambda_{\mathrm{B}} t}) + N_{\mathrm{B}_0} e^{-\lambda_{\mathrm{B}} t}. \tag{2.10}$$

Thus, equations (2.8) and (2.10) give the amounts of the parent A and daughter B present at any time t, in terms of their initial amounts and the respective decay constants.

2.12. If the parent has an appreciably longer half-life than the daughter nuclide, i.e., λ_{A} is less than λ_{B}, and N_{B_0} is zero, equation (2.10) reduces to the approximate form

$$N_{\mathrm{B}} \approx \frac{\lambda_{\mathrm{A}}}{\lambda_{\mathrm{B}} - \lambda_{\mathrm{A}}} N_{\mathrm{A}_0} e^{-\lambda_{\mathrm{A}} t},$$

since, after a sufficient time, $e^{-\lambda_{\mathrm{B}} t}$ can be neglected in comparison with $e^{-\lambda_{\mathrm{A}} t}$. Upon introducing equation (2.8), it follows that

$$\frac{N_{\mathrm{B}}}{N_{\mathrm{A}}} \approx \frac{\lambda_{\mathrm{A}}}{\lambda_{\mathrm{B}} - \lambda_{\mathrm{A}}},$$

i.e., the ratio of the amount of daughter, B, to that of the parent, A, becomes constant. This expression describes the condition of *transient equilibrium* in which the absolute amounts of A and B are changing but their ratio remains the same.

2.13. In the case that the parent has a very long half-life, so that λ_A is very small in comparison with λ_B, the $e^{-\lambda_B t}$ terms in equation (2.10) may be disregarded and, further, $\lambda_B - \lambda_A \approx \lambda_B$. Consequently, it can be shown from equation (2.10) that, after a period of time,

$$\frac{N_B}{N_A} \approx \frac{\lambda_A}{\lambda_B} \quad \text{or} \quad \lambda_A N_A \approx \lambda_B N_B,$$

which represents the state of *secular equilibrium*. Not only is the ratio of B to A fixed, but the absolute amount of B also remains constant. This follows since $\lambda_A N_A$ is approximately equal to $\lambda_B N_B$; in other words, A is decaying to form B at the same rate as B is decaying into C, so that the net amount of B is unchanged. If C is radioactive, it will also attain a condition of secular (or radioactive) equilibrium in the course of time and its amount will become constant. Thus, all the members of a radioactive series, except the last, will ultimately be in radioactive equilibrium with one another, provided the parent of the series has a very long half-life.

Example 2.1. One of the natural radioactive series begins with uranium-235 (half-life 7.13×10^8 yr) which emits an alpha particle to form thorium-231 (half-life 25.6 hr), the latter being a beta emitter. Starting with uranium-235 from which the decay products have been removed by chemical separation, determine (*a*) the atomic ratio of thorium-231 to the initial uranium-235, (*b*) the ratio of the total activity (rate of particle emission) to the initial value, after 50 hr. (*c*) What is the atomic ratio of the two nuclides mentioned when radioactive equilibrium is attained?

(*a*) Since it is not obvious what approximations may be used at 50 hr, it is best to start with equation (2.10). Uranium-235 is A and thorium-231 is B, so that N_{B_0} is zero; hence,

$$N_B = \frac{\lambda_A N_{A_0}}{\lambda_B - \lambda_A} (e^{-\lambda_A t} - e^{-\lambda_B t}).$$

From equation (2.4), $\lambda = 0.693/t_{1/2}$; expressing time in hours,

$$\lambda_A = \frac{0.693}{(7.13 \times 10^8)(365)(24)} = 1.11 \times 10^{-13} \text{ hr}^{-1}$$

$$\lambda_B = \frac{0.693}{25.6} = 2.71 \times 10^{-2} \text{ hr}^{-1}.$$

Since λ_A is so small, because of the long half-life of uranium-235, $\lambda_B - \lambda_A$ is essentially equal to λ_B, and the first exponential term is unity. Consequently, the required atomic ratio is

$$\frac{N_B}{N_{A_0}} = \frac{1.11 \times 10^{-13}}{2.71 \times 10^{-2}} [1 - e^{-(2.71 \times 10^{-2})(50)}]$$

$$= 4.1 \times 10^{-12}(1 - 0.258) = 3.04 \times 10^{-12}.$$

(The values of exponentials can be obtained from Fig. A.1 in the Appendix.)

(b) The activity (or rate of particle emission) of any radionuclide is equal to λN, and so the required activity ratio is

$$\frac{\lambda_A N_{A_0} + \lambda_B N_B}{\lambda_A N_{A_0}} = 1 + \frac{\lambda_B}{\lambda_A}\left(\frac{N_B}{N_{A_0}}\right)$$

$$= 1 + \frac{2.71 \times 10^{-2}}{1.11 \times 10^{-13}}(3.04 \times 10^{-12})$$

$$= 1.74.$$

(c) At radioactive equilibrium, A is decaying to form B at the same rate as B is decaying, i.e., $\lambda_A N_A = \lambda_B N_B$; hence, the equilibrium ratio is

$$\frac{N_B}{N_A} = \frac{\lambda_A}{\lambda_B} = \frac{1.11 \times 10^{-13}}{2.71 \times 10^{-2}}$$

$$= 4.1 \times 10^{-12}.$$

2.14. Although in a reactor the situation is somewhat different, because the parent is formed by fission or by nonfission neutron capture, it can be shown that a similar condition of equilibrium is attained, provided the power (or neutron density) remains constant. Use is made of this fact to determine the neutron density by exposure of a material which becomes radioactive as a result of radiative capture (§ 2.80). When the reactor is shut down or the radioactive material is removed, the equilibrium condition is disturbed. The rate of change in the amounts of the various members of the series can then be determined by solving a set of differential equations of the form of equation (2.6).

RADIOACTIVITY UNITS: THE CURIE

2.15. The *curie*, which was based on the estimated activity of a gram of radium, is now defined as the quantity of any radioactive species decaying at a rate of 3.70×10^{10} disintegrations per second (dis/sec). A sample of radioactive material is said to have an activity of 1 curie when it disintegrates at this rate. Subsidiary units, for smaller quantities of active material, are the *millicurie* (1 mc), which is a one-thousandth part of a curie, i.e., 3.70×10^{7} dis/sec, and the *microcurie* (1 μc), a one-millionth of a curie, i.e., 3.70×10^{4} dis/sec. For large amounts of radioactive substances two other units are often employed. These are the *kilocurie*, which is a thousand curies, and the *megacurie*, i.e., a million curies.

2.16. It was shown above that a specimen containing N atoms (or nuclei) of a given radioisotope decays at the rate λN, and, since λ is equal to $0.693/t_{1/2}$, where $t_{1/2}$ is the half-life, it follows that

$$\text{Rate of decay} = \frac{0.693N}{t_{1/2}} \text{ dis/sec,} \qquad (2.11)$$

if $t_{1/2}$ is in seconds. The number of individual atoms (or nuclei) present in 1 gram-atom, i.e., the atomic weight in grams, is the Avogadro number, 0.602×10^{24} (§ 1.46). Hence, if g grams is the weight of the radioisotope, of atomic weight A, present in the specimen, equation (2.11) leads to

$$\text{Rate of decay} = \frac{(0.693)(0.602 \times 10^{24})g}{At_{1/2}} \text{ dis/sec}, \tag{2.12}$$

and the number of curies present is determined upon dividing by 3.70×10^{10}. It can be readily seen that the mass of a particular radioisotope having an activity of 1 curie depends upon both its atomic weight and its radioactive half-life. For an isotope that decays slowly, i.e., one with a long half-life, the mass per curie will, apart from the effect of atomic weight, be greater than for a rapidly decaying species.

Example 2.2. The average adult human body contains 250 g of normal potassium, of which 0.012 per cent is the radioactive beta emitter potassium-40 (half-life 1.3×10^9 yr). Calculate the rate of production of beta particles in the body from the decay of potassium-40. What is the activity in microcuries?

$$\text{Amount of K}^{40} \text{ in the body} = (250)(0.012 \times 10^{-2}) = 0.030 \text{ g}$$
$$= 0.030/40 = 7.5 \times 10^{-4} \text{ g-atom}$$
$$= (7.5 \times 10^{-4})(0.602 \times 10^{24}) = 4.5 \times 10^{20} \text{ nuclei}.$$

$$\text{Decay constant } (\lambda) \text{ for K}^{40} = \frac{0.693}{1.3 \times 10^9} = 5.3 \times 10^{-10} \text{ yr}^{-1}.$$

$$\text{Decay rate (rate of beta-particle emission)} = (4.5 \times 10^{20})(5.3 \times 10^{-10})$$
$$= 2.4 \times 10^{11} \text{ per yr}$$

$$= \frac{2.4 \times 10^{11}}{(365)(24)(3600)} = 7.6 \times 10^3 \text{ per sec.}$$

Since 1 curie represents a decay rate of 3.7×10^{10} disintegrations per second,

$$\text{Number of curies} = \frac{7.6 \times 10^3}{3.7 \times 10^{10}} = 2.1 \times 10^{-7}$$

$$= 0.21 \text{ microcurie } (\mu c).$$

2.17. Some confusion and error have arisen in the determination of the activity of a material in curies because of a failure to distinguish between the rate of disintegration and the rate of emission of particles. If a certain radioisotope emits a single type of radiation, e.g., beta particles only, then the rates of decay and of particle emission are the same. However, there are cases in which two particles are formed, e.g., a beta particle and a gamma-ray photon (§ 2.20), and not necessarily in equal numbers. In such instances the total rate of particle emission is not a measure of the activity in curies. The latter is defined in terms of the actual rate of disintegration, and data based on particle emission may require correction for various reasons.

2.18. Before the discovery of nuclear fission, a 1-curie amount of radioactive material was considered to be very highly active. The operation of nuclear reactors has, however, led to the formation of relatively large quantities of intensely radioactive fission products. Consequently, activities of the order of thousands or even millions of curies, i.e., kilocuries or megacuries, are now not unusual for the fuel elements immediately after removal from a reactor. Expressions for the activity of fission products at various times are given later in this chapter (§ 2.178 *et seq.*).

2.19. It may be mentioned here that a unit of a different type, which is related to the effect produced by (or energy absorbed from) the radiation emitted by a radioactive material, is used in health physics work. This unit is called a "roentgen"; it will be considered more fully in Chapter 9.

GAMMA RAYS

2.20. In a variety of nuclear reactions, including many radioactive decay processes, a product nucleus is formed in an excited state having internal energy in excess of the ground state. Within a short time, the excess energy is emitted as gamma rays (§ 1.30). According to the quantum theory of radiation, the energy of the gamma rays (and of other electromagnetic radiations) is emitted as *photons;* these may be regarded as unit "particles" of the radiation concerned. The energy of the photon is equal to the difference in energy between the two states involved in the transition. In Fig. 2.2, for example, are shown two tran-

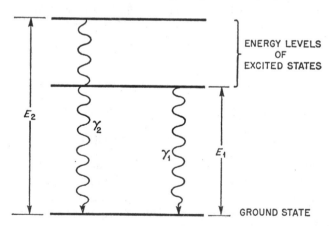

FIG. 2.2. Transitions from two different excited states of a given nucleus

sitions from different excited states to the ground state of a particular nucleus; the photon energies are E_1 and E_2, respectively.

2.21. The relationship between the energy, E, carried by a photon and the wave length, λ, of the radiation is derived from quantum theory. According to Planck's equation,

$$E = h\nu = hc/\lambda,$$

(handwritten at top: $\dfrac{6.62 \times 10^{-27}}{1.60 \times 10^{-6}} \times 3.00 \times 10^{10}$ *)*

where h is a constant (6.62×10^{-27} erg-sec) and ν is the radiation frequency in cm^{-1}; λ is the corresponding wave length in centimeters and c is the velocity of light (3.00×10^{10} cm/sec). With these values of h and c, the energy is given in ergs, but by making use of the relationship between ergs and Mev (§ 1.45), it is readily found that

(handwritten: $\lambda = \dfrac{hc}{E}$ *)*

$$\lambda(\text{cm}) = \frac{1.25 \times 10^{-10}}{E},$$

(handwritten: $1\,Mev = 1.60 \times 10^{-6}\,erg.$ *)*

where E is expressed in Mev.

2.22. It has become the common practice to describe gamma radiation in terms of its photon energy. For example, the expression "1-Mev gamma rays" refers to radiation with photons carrying 1 Mev of energy. Nuclear excitation energies are generally in the range from 0.1 to 10 Mev (§ 2.90); it follows, therefore, that the energies of gamma rays are usually of similar magnitudes. The corresponding wave lengths are thus in the region of 10^{-9} to 10^{-11} cm, i.e., 0.1 to 0.001 A.

2.23. If the transition from an excited state to the ground state takes place directly, as in Fig. 2.2, one photon will carry the whole of the excitation energy. In many cases, however, the transition from a higher excited state to the ground state has a very low probability; such a transition is said to be "forbidden." One or more intermediate stages are then involved in the transition, leading to the emission of a succession (or cascade) of gamma rays of lower energy, as indicated at the right of Fig. 2.3. However, the total energy of the gamma rays

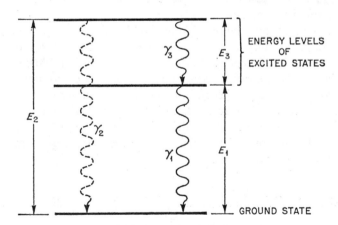

FIG. 2.3. Allowed and forbidden transitions

must be the same, and equal to the energy difference between initial and final energy levels, irrespective of whether the transition occurs in one or several stages, i.e., $E_2 = E_1 + E_3$. As a general rule, the transition from a higher to a lower energy state, accompanied by gamma-ray emission, occurs within a small fraction of a second, although the so-called "forbidden" transitions often take much longer.

2.24. In nuclei of low mass number, the first excited state is at least 1 Mev above the ground state; the separation between the lower excited levels is of the same order of magnitude, but it decreases with increasing energy, i.e., the levels become closer together. For nuclei of larger mass number the minimum excitation energy is less than 1 Mev and in the heaviest nuclei the first excited state may be only about 0.1 Mev above the ground state. The magic nuclei, however, constitute an exception to this rule. Bismuth-209, for example, which has 126 neutrons in its nucleus, behaves like a light element, with an energy separation of the order of 1 Mev between the lower levels.

BREMSSTRAHLUNG

2.25. Apart from the fact that x-rays frequently have lower energies, i.e., somewhat longer wave lengths, the essential difference between gamma rays and x-rays is that the latter are produced outside the atomic nucleus. The characteristic x-rays, which, as their name implies, have definite energies (and wave lengths) characteristic of the particular element concerned, result from transitions between electronic energy levels of the atom. These radiations are, however, of little significance for the present purpose. Of greater interest are the continuous x-rays, called *bremsstrahlung*, literally "braking radiation," which are produced when electrons (or beta particles) of high speed lose their energy in passing through matter.

2.26. As a general rule, the fraction of the kinetic energy of the electrons converted into radiation in this manner increases with the energy of the electron and with the atomic number of the material in which it is slowed down. The energy of the resulting x-rays covers a large range; the maximum is close to that of the electrons (or beta particles), but the average energy is much less. When electrons of energies of 1 Mev or more interact with elements of high atomic number, e.g., lead, some of the resulting bremsstrahlung, although originating outside the nucleus, are indistinguishable in their behavior from gamma rays arising from nuclear transitions.

INTERACTION OF ALPHA AND BETA PARTICLES WITH MATTER

IONIZING RADIATIONS

2.27. In its passage through matter, a fast-moving charged particle, such as an alpha or a beta particle, will occasionally approach close enough to an atom (or a molecule) for the electrical interaction to be sufficient to remove an external (or orbital) electron from the atom. The residue of the atom, after removal of the electron, is a positively charged ion, and the system of the separate electron and ion so formed is called an *ion-pair*.* Electrically charged particles are thus able to cause ionization, and, as they traverse matter, they leave behind a number of ion-pairs in their path. For this reason, alpha and beta particles are frequently referred to as *ionizing radiations*.

2.28. The intensity of the ionization produced by a moving charged particle is expressed by the *specific ionization;* this is the number of ion-pairs formed per centimeter of path in a given material. For charged particles of the same mass, the specific ionization increases with the charge. For particles of the same energy, those of higher mass produce a higher specific ionization. Since they move more slowly, the heavier particles (of given energy) spend more time in the vicinity of a given atom or molecule of the material through which they pass, thus increasing the probability of ionization.

2.29. The specific ionization due to an alpha particle, mass about 4 amu (§ 1.13), is thus considerably greater than that caused by a beta particle, mass 0.00055 amu, of the same energy. In their passage through air at atmospheric pressure, alpha particles produce approximately 50,000 to 100,000 ion-pairs per cm, whereas beta particles of similar energy form about 30 to 300 ion-pairs per cm of path.

2.30. The energy required to produce an ion-pair varies from one medium to another. Consequently the specific ionization produced by a charged particle depends on the nature of the medium through which it passes, as well as on the charge, energy, and mass of the particle. In its passage through air, a charged particle loses, on the average, about 34 ev of energy for every ion-pair produced. This energy varies somewhat with the nature of the particle, but the figure quoted may be accepted for most purposes. It is seen, therefore, that an alpha particle having an initial energy of 5 Mev will be capable of forming something like 1.5×10^5 ion-pairs in air. Since the specific ionization is roughly 5×10^4 ion-pairs per cm, the ionization will occur over a path length (or *range*) of about 3 cm. The effective range for a beta particle of the same energy will be nearly a thousand times greater.

2.31. Inasmuch as gamma rays and x-rays do not carry an electric charge, they cause relatively little direct ionization. Nevertheless, as will be seen subsequently, in their passage through matter these rays eject electrons and the latter, being electrically charged, are able to produce considerable ionization. Gamma rays and x-rays thus mainly cause ionization indirectly; in addition, they can cause electronic excitation directly or indirectly. Neutrons, which are also electrically neutral, behave somewhat similarly, although a different mechanism is involved. The direct ionization (and excitation) caused by alpha and beta particles and the indirect ionization (and excitation) due to gamma rays,

---

* In some interactions the energy transferred from the radiation to the atom or molecule may be insufficient to cause ionization, i.e., complete removal of an electron, but may be sufficient to raise the electron to a higher energy level, so that the atom or molecule is in an excited electronic state.

x-rays, and neutrons are important in connection with methods of measuring or counting these nuclear radiations and also with their effects on living tissue.

2.32. The majority of the separated positive ions and electrons formed, directly and indirectly, when nuclear radiations traverse matter, will sooner or later reunite to produce neutral atoms or molecules. This process is accompanied by the liberation of energy in the form of heat. Hence essentially all the energy of nuclear radiations is ultimately degraded into heat. Some of the charged particles may become involved in chemical action; in this event, a portion of the energy of the radiation is converted into chemical energy.

ABSORPTION OF ALPHA PARTICLES

2.33. If the specific ionization caused in air by alpha particles from a given source is measured at various distances from the source, a curve of characteristic shape, called a *Bragg curve*, is obtained. A typical Bragg curve is shown in Fig. 2.4. It is seen that the specific ionization increases with increasing distance from the source, at first slowly and then rapidly; after passing through a maximum, it drops sharply to zero. The increasing specific ionization is accounted for by the fact that, as the alpha particle traverses its path, producing ion-pairs as it proceeds, its energy and, consequently, its speed decrease steadily. The specific ionization, which increases with decreasing speed, as stated above, thus increases with distance from the source. Near the end of its path, when the alpha particle is moving relatively slowly, the specific ionization reaches a maximum.

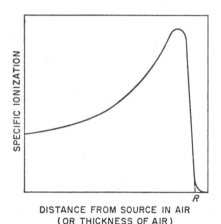

DISTANCE FROM SOURCE IN AIR
(OR THICKNESS OF AIR)

FIG. 2.4. Specific ionization and range of alpha particles (Bragg curve)

2.34. Beyond a certain point, the energy of the alpha particle has become so low that it captures one and then two electrons, being thus converted into a neutral helium atom no longer capable of producing ionization. This explains the decrease in the specific ionization shown in Fig. 2.4. The fact that this decrease is relatively sudden, and not gradual, is of interest; it implies that all the alpha particles from a given source have practically the same energy, and all cease to produce ionization after having traveled the same distance. Strictly speaking, if all the alpha particles behaved exactly alike, the end portion of the curve would be vertical, but minor differences account for the slight slope. The extrapolated distance, represented by R, at which alpha particles from the given source effectively cease to produce ionization, is called the *range* of the particles.

2.35. The range of alpha particles in air (or in other material) depends on the source of the particles, since the energy varies accordingly; for alpha particles of radioactive origin, the ranges are from 2.5 cm (1.0 in.) for thorium-232 to 8.6 cm (3.4 in.) for thorium C'. There is an inverse relationship between the half-life of the radioisotope and the energy, and hence the range, of the alpha particles. Thus, the shorter the half-life, the greater the alpha-particle range. Some of the data for alpha-particle emitters of interest in nuclear reactors are given in Table 2.2. Uranium-235 produces two groups of alpha particles with slightly different energies and ranges.

TABLE 2.2. HALF-LIVES, ENERGIES, AND RANGES IN AIR

Radioelement	Half-Life (yr)	Energy (Mev)	Range in Air	
			cm	in.
Thorium-232	1.39×10^{10}	4.0	2.5	1.0
Uranium-238	4.51×10^{9}	4.2	2.7	1.06
Uranium-235	7.13×10^{8}	4.4 and 4.6	2.9 and 3.1	1.15 to 1.2
Uranium-233	1.62×10^{5}	4.8	3.3	1.3
Plutonium-239	2.41×10^{4}	5.1	3.6	1.4

2.36. In absorbing materials more dense than air, the specific ionization is, in general, greater because of the greater concentration of atoms; the range of alpha particles of given energy is, consequently, less than in air. The relative *stopping power* of an absorber is defined by

$$\text{Relative stopping power} = \frac{\text{Range of alpha particle in air}}{\text{Range of alpha particle in material}}. \quad (2.13)$$

It is essentially independent of the alpha-particle energy, provided, of course, that the ranges refer to particles from the same source.

2.37. The relative stopping power of water is about 1000, for aluminum 1600, and for lead it is over 5000. Paper and animal tissue have roughly the same stopping power as water, i.e., around 1000. If an alpha particle has a range of, say, 2 in. in air, which is relatively high, the range in paper or living tissue would be about 0.002 in., i.e., 2 mils. Thus, an ordinary sheet of paper, having a thickness of roughly 4 mils, would stop essentially all alpha particles of radioactive origin.

2.38. Another method for representing the absorbing effect of a medium is by means of the *thickness density*,* frequently referred to, in brief, as the "thickness." It is the product of the linear range of the given particles in the medium and the density of the medium; thus, if the range is R cm and the density is ρ g/cm³, then

---

* It is also sometimes called the *areal density* because it is the mass per unit area of material.

$$\text{Thickness density} = R \times \rho \text{ g/cm}^2 \qquad (2.14)$$
$$= 1000R \times \rho \text{ mg/cm}^2,$$

according as the value is expressed in g/cm² or mg/cm². The latter is generally used because the numbers are then of a reasonable magnitude. Physically, the thickness density is the mass per unit area of absorbing medium required to stop (or absorb) the given alpha particles, i.e., having a linear thickness equal to the range of the particles.

ABSORPTION OF BETA PARTICLES

2.39. The passage of beta particles through matter has some features in common with the behavior of alpha particles, e.g., production of ion-pairs at the rate of about 34 ev per ion-pair in air, but there are some important differences. As stated earlier, the smaller mass of the beta particle means that the specific ionization is less than that due to an alpha particle of the same energy. Furthermore, all the alpha particles from a given source have essentially the same energy, or they fall into two or three groups of definite energy. Beta particles, on the other hand, have a continuous distribution of energies, i.e., a continuous energy spectrum, up to a definite maximum for each particular source (Fig. 2.5).

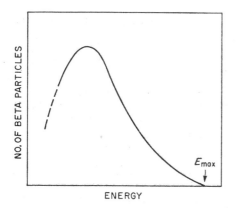

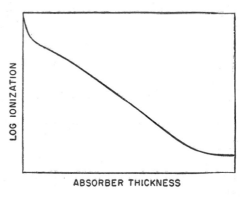

FIG. 2.5.   Energy spectrum of beta   FIG. 2.6.   Absorption of beta particles
                particles

This maximum is known, from the isotopic masses of the parent and daughter nuclides, to be equal to the total energy of the radioactive change. The reason why few (if any) beta particles have this amount of energy is that, in any given transition, the available energy is divided between the beta particle and the accompanying neutrino (§ 1.26). On the average, the beta particles carry about one-third, and the neutrinos the remaining two-thirds, of the total energy.

2.40. Because of their relatively large mass, alpha particles do not, on the whole, undergo any marked change of direction in their passage through matter. In other words, the majority travel in straight lines, thus leading to a fairly definite range for a given energy. Beta particles, however, are subject to considerable scattering, with frequent changes in direction as a result of electrostatic interactions with atomic nuclei and electrons. Consequently, beta particles which have passed through the same thickness of a given absorber may come out in widely different directions, so that they will actually have traversed paths of different lengths in the material.

2.41. The combined effect of the continuous energy spectrum of beta particles and their scattering means that these particles do not have a definite range, as do alpha particles from a given source. However, due to a fortuitous combination of circumstances, which are too complex for complete theoretical analysis, it is found experimentally that the ionization caused by beta radiation from a given source falls off in a roughly exponential manner with distance. The general form of the plot of the logarithm of the ionization produced against the thickness of the absorbing material is shown in Fig. 2.6; except for small or large thicknesses, the curve is approximately linear. At large absorber thicknesses the curve becomes almost horizontal, indicating a more or less constant ionization. This "tail" of the curve is due to the presence of the highly penetrating bremsstrahlung (§ 2.25) resulting from the loss of energy by the fast-moving beta particles in their passage through the absorbing medium.

2.42. The energy loss as bremsstrahlung per unit path length by beta particles (or electrons) is approximately proportional to the square of the atomic number of the absorber and to the energy of the particles. The association of bremsstrahlung with the passage of beta particles of given energy through matter is thus most marked with elements of high atomic number. The radiations cover a range of energies up to a maximum equal to the initial (maximum) energy of the beta particles. If gamma rays are emitted from the beta-particle source they will, of course, also contribute to the "tail" in Fig. 2.6. After passage through considerable thicknesses of material, both gamma radiation and bremsstrahlung will be absorbed.

2.43. Beta particles do not have a definite range, in the same sense as do alpha particles. Nevertheless, it is possible to specify a more or less definite thickness of absorber which will reduce almost to zero the ionization, other than

TABLE 2.3. APPROXIMATE RANGES OF BETA PARTICLES IN AIR

Energy (Mev)	Range	
	meters	ft
0.1........	0.11	0.36
0.5........	1.5	4.9
1.0........	3.7	12
2.0........	8.5	28
3.0........	13	43

that due to bremsstrahlung, produced by beta particles of given energy. The approximate ranges in air of beta particles of various maximum energies are given in Table 2.3. It may be noted that the average maximum energy of the beta particles from fission products is about 1.2 Mev; the absolute maximum energy probably does not exceed 3 Mev and is appreciably less in most cases.

2.44. As for alpha particles, the approximate range of beta particles in an absorbing material is frequently expressed in terms of its thickness density in grams per cm², defined by equation (2.14). The values for aluminum as absorber have been determined experimentally for beta particles from various sources, and the results fall on or very close to the curve shown in Fig. 2.7; the ordinates are

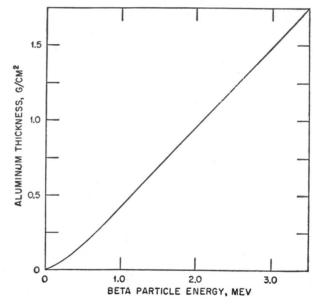

Fɪɢ. 2.7.   Aluminum thickness density (g/cm²) as function of beta-particle energy

the thickness densities of aluminum in grams per cm² required to absorb beta particles with maximum energies indicated by the abscissae. Within the range of about 0.8 to 3 Mev, the relationship between the absorption thickness density $(R \times \rho)$ of aluminum and the maximum energy (E_m) of the beta particles from the given source is linear and may be represented by the expression

$$R \times \rho \ (g/cm^2) = 0.54 E_m \ (Mev) - 0.15. \tag{2.15}$$

2.45. In the absence of other data, it may be supposed, as a first approximation, that the absorption thickness density is independent of the nature of the absorbing material; this is based on the assumption that the linear range of the beta particles in any medium is inversely proportional to the density. In this event, the values of $R \times \rho$ given by equation (2.15) would be the same for all absorbers.   The (approximate) linear range in any material can then be obtained if the density is known.

Example 2.3.   Estimate the maximum range in concrete of density 2.8 g/cm³ of beta particles from fission products.

As stated in § 2.43, the energy of the beta particles from fission products does not exceed 3 Mev; hence, from equation (2.15), the corresponding maximum value of $R \times \rho$ is 1.47 g/cm². Since ρ is 2.8 g/cm³, it follows that

$$R = 1.47/2.8 = 0.53 \text{ cm.}$$

The maximum range is thus about 0.53 cm or 0.21 in. in concrete.

2.46. Positive beta particles, i.e., positrons, behave like negative beta particles in their interaction with matter.   There is, however, another factor to be considered in the former case.   Within a very short time of its liberation, a positron is likely to combine with an electron, of which large numbers are always present in matter as the outer electrons of atoms.   The positively charged positron and negatively charged electron neutralize one another; the particles are thereby annihilated and energy is liberated in the form of radiation, called *annihilation radiation.*   The total mass of a positron and an electron is 0.00110 amu and, by equation (1.4), this is equivalent to 1.02 Mev.   To be consistent with the principle of the conservation of momentum, this energy is usually divided equally between two photons moving in opposite directions.   The energy of the annihilation radiation, which has properties similar to gamma radiation, is thus mainly 0.51 Mev, although there may be a small amount of 1.02-Mev energy photons.

ČERENKOV RADIATION

2.47. Charged particles of high energy emit visible (electromagnetic) radiation in their passage through a transparent medium, provided their velocity is greater than the velocity of light in that medium.   This radiation is called *Čerenkov radiation.*   The bluish glow which is often observed surrounding the cores of reactors cooled and moderated by water is the Čerenkov radiation generated by Compton electrons (§2.53) produced by fission-product gamma rays.

INTERACTION OF GAMMA RAYS WITH MATTER

INTRODUCTION

2.48. Although x-rays, bremsstrahlung, and annihilation radiation are not strictly gamma rays, since they do not arise from nuclear transitions, they are essentially identical with gamma rays in their fundamental nature.   As far as their interaction with matter is concerned, the only differences that may arise are the result of the higher energies, in general, of the gamma radiations.   The discussion in the following paragraphs may, therefore, be taken as applicable to all electromagnetic radiations of high energy, e.g., 0.01 to 100 Mev.

2.49. There are several ways in which gamma rays interact with an absorbing material; three, namely, the photoelectric effect, the Compton effect, and pair production, are important and will be considered here in some detail. Certain other reactions due to gamma rays will be mentioned elsewhere (§ 2.73).

PHOTOELECTRIC EFFECT

2.50. In the photoelectric effect, a gamma-ray photon, with energy greater than the binding energy of an orbital electron in an atom, interacts with the latter in such a way that the whole of the gamma-ray energy is transferred to an electron which is consequently ejected from the atom. If E is the energy of this gamma-ray photon and B is the binding energy of the electron in the atom, the difference, i.e., $E - B$, is carried off as kinetic energy by the ejected electron. The photoelectron, as it is called, behaves like a beta particle of the same energy in its passage through matter. For gamma rays of high energy, the photoelectrons are mainly expelled in the forward direction, i.e., in the same direction as the incident gamma rays, but for low-energy rays, the emission is largely in the direction at right angles.

2.51. The extent of the photoelectric interaction depends on both the energy, E, of the gamma radiation and the atomic number, Z, of the absorbing material. As a rough approximation,

$$\text{Probability of photoelectric interaction} \approx \text{constant} \times \frac{Z^n}{E^3},$$

where n varies from 3, for gamma rays of low energy, to 5, for high-energy rays. It is apparent, therefore, that the photoelectric effect increases with increasing atomic number of the absorber and with decreasing energy of the gamma rays. In actual practice, it is found that photoelectric absorption of gamma radiation is important only for energies less than about 1 Mev, and then only for absorbers of high atomic number (cf. Fig. 2.10).

2.52. Following the expulsion of the photoelectron, another electron, from an outer orbit, takes its place in the atom; this transition is accompanied by the emission of characteristic x-rays. These are of low energy compared with the original gamma radiation. The expulsion of the x-ray photon frequently causes the ejection of an outer electron, called an *Auger electron*, in a type of photoelectric (Auger) effect; the photon thereby loses all of its energy. The Auger electron will escape from the atom and will dissipate its energy as a result of interactions similar to those experienced by a beta particle. It is evident, from the foregoing discussion, that the photoelectric effect leads to the virtually complete absorption of the gamma-ray photon; it may be replaced, to some extent, by x-ray photons of low energy, but few of these escape as such from the absorbing material.

COMPTON EFFECT

2.53. In a Compton interaction a gamma-ray photon makes an elastic (or "billiard-ball") collision with an electron of the absorbing material. Such an electron behaves as if it were free, because its binding energy is much less than the photon energy. In the collision both momentum and energy are conserved, and part of the energy of the incident photon is transferred to the electron. Another (scattered) photon of lower energy then moves off in a new direction, so that it is deflected, i.e., scattered, from its initial path (Fig. 2.8). The relation between the energy E of the incident photon, E' of the scattered photon, both in Mev, and the scattering angle θ, is given by*

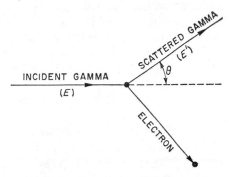

Fig. 2.8. Compton scattering of gamma-ray photons

$$E' = \frac{0.51}{1 - \cos \theta + 0.51/E}. \tag{2.16}$$

2.54. If the scattering angle is small, $\cos \theta \approx 1$, and then E' is approximately equal to E. This means that the scattered photons of higher energy, close to the energy of the incident photons, proceed in a nearly forward direction. On the other hand, for $\theta = 90°$, $\cos \theta = 0$, and then

$$E' = \frac{0.51E}{E + 0.51} < 0.51 \text{ Mev.}$$

Consequently a photon scattered at right angles cannot have energy greater than 0.51 Mev.

2.55. The fraction of the initial energy carried by the scattered photon for different scattering angles is derived from equation (2.16) as

$$\frac{E'}{E} = \frac{0.51}{E(1 - \cos \theta) + 0.51}.$$

For a specified value of θ, this fraction decreases with increasing energy of the incident photon. In other words, for a given scattering angle, the greater the energy of the incident photon, the smaller the fraction carried by the scattered photon, and the larger the fraction (and amount) of energy lost by the gamma ray in the Compton interaction.

---

* The quantity 0.51 Mev, which appears in this equation, is m_0c^2, where m_0 is the mass of the electron at rest, and c is the velocity of light. It is the energy equivalent, according to the Einstein equation, of the rest mass of the electron (cf. § 2.46).

2.56. Since the Compton effect involves interaction between a photon and an electron, its magnitude is dependent on the number of orbital electrons in the atom of the absorber; this is the same as the atomic number. The Compton interaction is thus directly proportional to the atomic number of the absorber, so that, like the photoelectric effect, it is more significant for materials of high atomic number. The energy dependence of the Compton effect is given by the Klein-Nishina formula [1]. This is somewhat too complicated to reproduce here, but it can be stated that the Compton interaction decreases monotonically with increasing energy of the gamma radiation (cf. Fig. 2.10). As a very rough approximation, it is possible to write

$$\text{Probability of Compton interaction} \approx \text{constant} \times \frac{Z}{E}.$$

2.57. There is a significant difference between the photoelectric and Compton effects to which attention must be called. The photoelectric effect is a true absorption process; that is to say, the photon is absorbed, as stated above. In the Compton process, however, there is merely a decrease in the photon energy, the extent of this decrease being greater the larger both the initial energy and the scattering angle. A photon which is involved in a Compton interaction is replaced by another one, although the latter will have a somewhat lower energy and may be moving in a different direction. Hence, starting with a single photon of high energy, there may be several Compton collisions in a sufficiently thick absorbing medium. The photon is then said to undergo multiple scattering. If it does not escape, the scattered photon will ultimately be absorbed as a result of photoelectric interaction which becomes increasingly probable as the energy decreases (§ 2.51).

PAIR PRODUCTION

2.58. When a gamma-ray photon with energy in excess of 1.02 Mev passes near the nucleus (and to some extent the outer electrons) of an atom, the photon can be annihilated in the strong electrical field with the formation of an electron-positron pair. Since the energy equivalent of the total mass of an electron and a positron is 1.02 Mev (§ 2.46), this is the minimum energy necessary for the production of the pair of particles. Any energy of the gamma-ray photon in excess of 1.02 Mev appears mainly as kinetic energy of the electron and positron, with a small fraction transferred to the atomic nucleus. The particles produced tend to travel in the forward direction, the effect becoming more evident with increasing gamma-ray energy.

2.59. The extent of pair production by gamma radiation of energy E (in Mev) is related to the atomic number Z of the absorber by

$$\text{Probability of pair production} \approx \text{constant} \times Z^2(E - 1.02),$$

so that it increases with the atomic number of the absorbing material and with increasing photon energy in excess of 1.02 Mev. Actually this expression allows only for pair formation in the electrical field of the nucleus, but in addition the process occurs, to some extent, in the fields of the orbital electrons.   This contribution also increases with the atomic number. Since both the photoelectric and Compton effects decrease with increasing gamma-ray energy, whereas pair production increases, it is evident that the latter process will become of major importance at high energies.   For absorbers of high atomic number it becomes the dominant type of interaction for gamma rays with energies in excess of about 5 Mev.

2.60. As in the photoelectric effect, pair production results in the absorption of the gamma-ray photon.   It is true that some of the electrons and positrons formed will neutralize one another and produce annihilation radiation, consisting, in general, of two 0.51 Mev photons (§ 2.46).   However, this energy is relatively low and, furthermore, the radiation has an isotropic distribution, i.e., it is spread uniformly in all directions.   The amount of radiation energy continuing in the forward direction is thus small and for many practical purposes, e.g., in shielding calculations, it may be assumed that the photon is completely absorbed in pair production.

ATTENUATION OF GAMMA RAYS

2.61. In their passage through matter the gamma-ray photons are removed so that the intensity falls off in an exponential manner. The reason is that the extent of loss in a small thickness dx of matter, at any point in the medium, is proportional to the radiation intensity at that point and to the thickness traversed, i.e.,

$$dI = -\mu I\, dx \qquad \text{or} \qquad \frac{dI}{I} = -\mu\, dx, \qquad (2.17)$$

where I is the intensity, expressed as photons (or Mev) per cm^2 per sec, and the proportionality constant, μ, usually given in cm^{-1} units, is called the *linear attenuation coefficient* of the absorber for the given radiation.*   If a narrow collimated beam of monoenergetic gamma rays of intensity I_0 passes through a thickness x cm of absorber, the intensity I_x of the emergent beam is obtained by integration of equation (2.17); the result is

$$I_x = I_0 e^{-\mu x} \qquad (2.18)$$

or

$$\log \frac{I_x}{I_0} = -0.4343\, \mu x. \qquad (2.19)$$

---

* It has become the general practice to refer to the loss of photons from a gamma-ray beam as "attenuation."   The term "energy absorption" is used when considering the energy removed from the radiation in its passage through matter (§ 7.27).

It should be noted that if μ is the total attenuation coefficient, which includes the scattering (Compton effect) and absorption (photoelectric and pair-production effects), equation (2.19) is applicable only to a narrow, collimated (parallel) beam of gamma radiation. If there were no scattering of photons, the equation could also be used for a broad collimated beam, with μ representing the coefficient for absorption only. For noncollimated beams, e.g., those arising from a point source, which spread outward, allowance must be made for the spreading, as will be seen in Chapters 9 and 10.

2.62. One of the consequences of the exponential attenuation of gamma rays, as represented by equations (2.18) and (2.19), is that, although the *amount* of radiation absorbed by a specified thickness of material is proportional to the initial intensity, the *fraction* absorbed (or emerging) is independent of this intensity. Thus, it requires the same thickness of absorber to decrease the intensity of the gamma rays of a given energy from 1 to 0.1 per cent of its initial value as is required to reduce it from 100 to 10 per cent. Another consequence is that, theoretically, an infinite thickness of material would be necessary to absorb gamma radiation completely, i.e., to make $I_x = 0$. Nevertheless, in practice, e.g., in shielding, a finite thickness can reduce the intensity to an amount that is relatively insignificant.

ATTENUATION COEFFICIENT AND GAMMA-RAY ENERGY

2.63. The linear attenuation coefficient varies with the energy of the gamma rays, because the modes of interaction of this radiation with matter, as described above, are all energy dependent in different ways. By measuring the intensity of a narrow collimated beam of monoenergetic gamma rays before (I_0) and after (I_x) passage through a known thickness (x cm) of absorber, and using equation (2.19), the linear attenuation coefficient (μ cm^{-1}) can be determined experimentally (Fig. 2.9). By combining experimental data with theoretical considerations, it is possible to divide the observed attenuation coefficient for a given gamma-ray energy into three parts, representing the contributions made by the photoelectric effect (μ_{pe}), by the Compton effect (μ_c), and by pair production (μ_{pp}), respectively. The results for an absorbing material of fairly low atomic number (aluminum) and

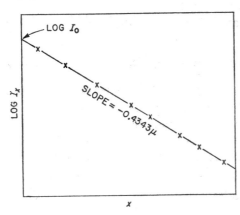

FIG. 2.9. Decrease of gamma-ray intensity with thickness of absorber

for one of high atomic number (lead) are given in Fig. 2.10, for gamma-ray energies up to 10 Mev.

2.64. It will be observed that the total attenuation coefficient of lead is at least six times that of aluminum; the reason is that all three types of absorption interaction between photons and matter increase with increasing atomic number of the absorber.   For aluminum the photoelectric effect is very small and makes no significant contribution for photon energies in excess of about 0.2 Mev.   For this element, the Compton effect is dominant up to energies of approximately 4 Mev or more.   Because both μ_{pe} and μ_c decrease with increasing energy, whereas μ_{pp} increases for energies exceeding 1.02 Mev, the total attenuation coefficient may be expected to exhibit a minimum at a particular energy.   Such a flat minimum is evident in Fig. 2.10 for lead, at a gamma-ray energy of about

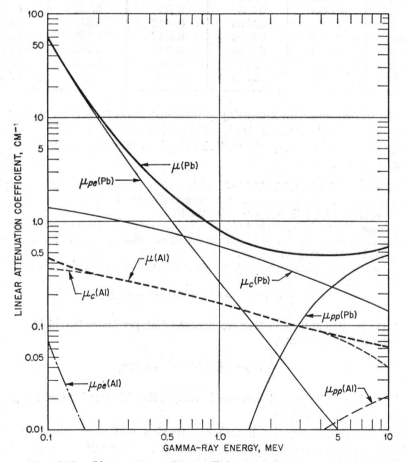

FIG. 2.10.   Linear attenuation coefficients of aluminum and lead

3 to 4 Mev. It undoubtedly also exists for aluminum as absorber, at a much higher energy.

2.65. The linear attenuation coefficients of water, concrete (density 2.35 g/cm^3), aluminum, iron, and lead, for gamma rays with energies from 0.5 to 10.0 Mev, are given in Table 2.4 [2]. In each case the attenuation coefficients

TABLE 2.4. LINEAR ATTENUATION COEFFICIENTS IN CM^{-1} UNITS

Energy (Mev)	Water	Concrete	Aluminum	Iron	Lead
0.5	0.0966	0.204	0.227	0.651	1.64
1.0	0.0706	0.149	0.166	0.468	0.776
1.5	0.0575	0.121	0.135	0.381	0.581
2.0	0.0493	0.105	0.117	0.333	0.518
3.0	0.0396	0.0853	0.0953	0.284	0.477
4.0	0.0339	0.0745	0.0837	0.259	0.476
5.0	0.0301	0.0674	0.0761	0.246	0.483
8.0	0.0240	0.0571	0.0651	0.232	0.520
10.0	0.0219	0.0538	0.0618	0.231	0.554

are seen to decrease with increasing gamma-ray energy, at least up to energies of about 5 Mev. This means that, within this range, the higher the energy the greater the thickness of a given material required to remove a specified fraction of the radiation.

Example 2.4. A 2-in. thick shield of lead is used to attenuate the gamma-ray photons (1.17 Mev and 1.33 Mev energy) in a narrow collimated beam from cobalt-60. What fraction of the initial radiation penetrates the lead?

The linear attenuation coefficient of lead is appreciably different for the two gamma-ray energies, namely 0.70 cm^{-1} for 1.17 Mev and 0.62 cm^{-1} for 1.33 Mev. Hence, the radiations must be considered separately.

For the 1.17-Mev photons, using equation (2.18) and converting in. into cm,

$$\frac{I}{I_0} = e^{-\mu x} = e^{-(0.70)(2.54 \times 2.0)} = 0.029.$$

This is the fraction of the 1.17-Mev component which penetrates the shield.

For the 1.33-Mev photons,

$$\frac{I}{I_0} = e^{-(0.62)(2.54 \times 2.0)} = 0.045.$$

In spite of the relatively small difference in energy, the difference in penetrating power is quite marked.

2.66. If the ordinary (or linear) attenuation coefficient, given in cm^{-1}, is divided by ρ, the density of the absorber in grams per cm^3, the result, i.e., μ/ρ, is called the *mass attenuation coefficient*, expressed in cm^2 per gram. This coefficient is of special interest in the energy ranges, e.g., 0.1 to 10 Mev for the lightest elements and 0.2 to 3 Mev for those of medium mass number, where the Compton effect is dominant.   Since the number of electrons present in a given quantity of matter is roughly proportional to the mass, the Compton interaction will vary approximately as the mass.   Hence, in the specified energy ranges, the mass

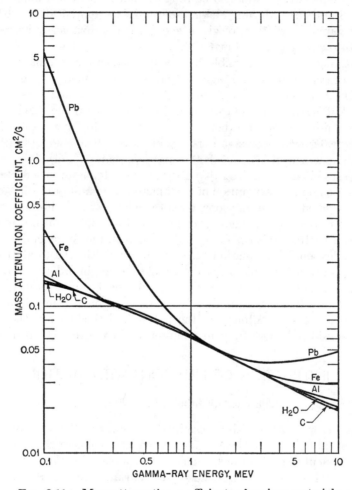

Fig. 2.11.   Mass attenuation coefficients of various materials

attenuation coefficient is essentially independent of the nature of the absorbing material, as can be seen from Fig. 2.11.

2.67. Upon dividing and multiplying the exponent in equation (2.18) by ρ, the result is

$$I_x = I_0 e^{-(\mu/\rho)(x\rho)}, \qquad (2.20)$$

where the first factor in the exponent is now the mass attenuation coefficient defined above and the second is the thickness (or areal) density of the absorber, expressed in grams per cm² (or mg per cm²), as in § 2.38. Hence, when the circumstances are such that the mass attenuation coefficient is (approximately) constant for different materials, the thickness density which attenuates the radiation to a specified extent will also be roughly independent of the nature of the absorber. In other words, the thickness of material required to absorb the same fraction of gamma radiation, of given energy, is approximately inversely proportional to the density of the material.

2.68. Although it is not possible to specify a range for gamma rays, as it is with alpha or beta particles, a rough comparison can be made of the penetrating power of the various radiations. Since μ for 1-Mev gamma rays in water is 0.0706 cm⁻¹, it is readily found from equation (2.19) that 65 cm (26 in.) of water would be required to decrease the radiation intensity to 1 per cent of its initial value. The range of 1-Mev beta particles in water is 0.39 cm (0.16 in.), and for 1-Mev alpha particles it is about 0.0004 cm (less than 0.2 mil).

2.69. It should be emphasized that the various attenuation coefficients given above are based on the assumption of a collimated gamma-ray beam, so that the Compton scattered photons are considered to be lost. Since this is by no means true in most engineering, including nuclear reactor, applications, a method for correcting for scattering is desirable. The procedure most generally employed is to use the exponential equation (2.18) with μ equal to the total attenuation coefficient, as given in Table 2.4, for example. Then, in order to allow for the photons which have been scattered back into the beam, a *buildup factor* is included. The value of the buildup factor depends on various circumstances, e.g., the energy of the photons and the nature and thickness of the absorber. It will be considered more fully in Chapter 10 in the discussion of shielding problems.

INTERACTION OF NEUTRONS WITH MATTER

THE PRODUCTION OF NEUTRONS

2.70. Before proceeding with a description of some of the reactions between neutrons and atomic nuclei, brief reference will be made to the methods used for producing neutrons. Not only are neutron sources of various kinds required for experimental purposes, but they also play an important role in the startup of nuclear reactors.

2.71. Neutrons are readily obtained by the action of alpha particles on some light elements, e.g., beryllium, boron, or lithium. The reaction may be represented by

$$_4Be^9 + {}_2He^4 \rightarrow {}_6C^{12} + {}_0n^1,$$

where, as stated in Chapter 1, the subscripts give the atomic number, or positive charge, and the superscripts are the respective mass numbers. Alternatively, it may be written in the abbreviated form Be^9 $(\alpha, n)C^{12}$, indicating that a Be^9 nucleus, called the *target nucleus*, interacts with an incident alpha particle (α); a neutron (n) is ejected and a C^{12} nucleus, referred to as the *recoil nucleus*, remains.

2.72. The chief alpha-particle emitters which have been used in (α, n) sources, together with beryllium, are radium-226, polonium-210, and plutonium-239. The last-mentioned nuclide has some advantages over radium, which has the drawback of being a strong gamma emitter, and over polonium which has a relatively short half-life. The neutron yield per gram of alpha emitter is, however, less for a plutonium-beryllium source than for one consisting of radium and beryllium. In all cases, the neutrons have high energies covering a wide range, e.g., from 1 to 10 Mev or more. Such sources are said to be polyenergetic.

2.73. The action of gamma rays of moderate energy (about 2 Mev) on certain nuclei, notably deuterium, i.e., heavy hydrogen, and beryllium, yields essentially monoenergetic neutrons. The reactions, which are of special interest in connection with the operation of nuclear reactors, are

$$_4Be^9 + {_0}\gamma^0 \rightarrow {_4}Be^8 \text{ (or 2 } {_2}He^4) + {_0}n^1$$

and

$$_1H^2 + {_0}\gamma^0 \rightarrow {_1}H^1 + {_0}n^1.$$

These are described as (γ, n) reactions, since a gamma-ray photon is the incident particle and a neutron is expelled. Sources based on (γ, n) reactions are called *photoneutron sources*. They have the drawback of requiring shielding from the gamma radiations when being handled.

2.74. The (γ, n) reaction will occur only if the energy of the gamma rays is at least equal to the binding energy of the neutron in the target nucleus. It is because the binding energy is exceptionally low in deuterium (2.2 Mev) and beryllium (1.6 Mev) that these substances are generally used in (γ, n) neutron sources. To obtain neutrons from other elements requires gamma rays of at least 6 to 8 Mev energy. For photons of a given energy, the neutrons obtained are monoenergetic, the energy being equal to the difference between the photon energy and the neutron binding energy in the target nucleus.

NEUTRON REACTIONS: ABSORPTION

2.75. Reactions of neutrons with nuclei fall into two broad classes, namely, scattering and absorption. In scattering reactions, the final result is merely an exchange of energy between the two colliding particles, and the neutron remains free after the interaction. In absorption processes, on the other hand, the neutron is retained by the nucleus and new particles are formed. The most important absorption reactions from the nuclear reactor standpoint are radiative capture and fission, as mentioned in Chapter 1. There are, however, a few neutron absorption reactions of different types which are of interest and these will be described below.

2.76. In considering absorption reactions it is convenient to distinguish between reactions of slow neutrons and of fast neutrons. There are four main kinds of slow-neutron reactions; these involve capture of the neutron by the target followed by either (1) the emission of gamma radiation (n, γ); (2) the ejection of an alpha particle (n, α); (3) the ejection of a proton (n, p); or (4) fission (n, f). Of these, the radiative capture, i.e., (n, γ), process is the most common, for it occurs with a wide variety of elements. The (n, α) and (n, p) reactions with slow neutrons are limited to a few isotopes of low mass number, whereas fission by slow neutrons is restricted to certain nuclei of high mass number.

2.77. There are reasons for believing that all absorption reactions as well as many, but not all, scattering reactions take place in two stages; the first stage is the capture of the neutron by the target nucleus to form a compound nucleus, as mentioned in § 1.31. This may be a nucleus of a familiar isotope or it may be an unstable (radioactive) species; but, in any case, immediately after its formation, the compound nucleus is in a high-energy (excited) state. Within a very short time the excited compound nucleus undergoes the second stage of the reaction, involving either (a) expulsion of a particle, e.g., a neutron (in scattering), a proton, or an alpha particle, (b) the emission of a gamma-ray photon, or (c) breaking up into two more or less equal parts, i.e., fission.

2.78. The excitation energy of the compound nucleus arises from the binding energy of the added neutron in the given nucleus, i.e., roughly 8 Mev (§ 1.20), and also from the kinetic energy of the captured neutron which may range from several Mev for a fast neutron down to a fraction of an electron volt for a thermal neutron (§ 1.32). If the excitation energy of the compound nucleus exceeds the binding energy of the least firmly held nuclear particle (or nucleon), there is a possibility that the nucleon will be expelled. An energy level of this type is referred to as a *virtual state* of the compound nucleus, to distinguish it from a *bound state* in which the excitation energy is not sufficient to permit expulsion of a nucleon.

2.79. Although the emission of a neutron from a virtual state is energetically possible, the probability of its occurrence may be very small. This is because the nuclear excitation energy of a nucleus is rapidly distributed among several of its constituent nucleons, and the chance that a single nucleon will acquire sufficient energy, roughly 8 Mev, as a result of collisions within the nucleus, is thus not large. If the excess energy of the virtual state is not much more than about 8 Mev, the most probable behavior will then be the emission of the excitation energy as a gamma-ray photon or cascade of photons. The resulting reaction is then radiative capture. However, if the available excitation energy of the virtual state is sufficiently large or the number of nucleons which share the energy is small, the probability of nucleon expulsion increases. The escape of a neutron is more likely than that of a proton, because in the latter case emission is hindered by the presence of an electrostatic potential barrier due to the positive charge.   Nevertheless, under suitable conditions expulsion of a proton or an alpha particle can occur from a virtual state of a compound nucleus, leading to (n, p) or (n, α) reactions.

median to low energy

RADIATIVE CAPTURE REACTIONS

unlikely if magic number nucleus

2.80. In radiative capture or (n, γ) reactions the excited compound nucleus emits its excess energy as gamma radiation, referred to as *capture gamma rays*, eventually leaving the compound nucleus in its lowest energy (or ground) state. The process may be represented symbolically by the equation

$$_Z X^A + {}_0 n^1 \rightarrow [_Z X^{A+1}]^* \rightarrow {}_Z X^{A+1} + \gamma,$$

where X^A is the target nucleus having an atomic number Z and a mass number A. The product X^{A+1} of the radiative capture reaction is seen to be an isotope of X, since it has the same atomic number, but with a mass number one unit greater. According to circumstances, the nuclide X^{A+1} may or may not be radioactive. If it is radioactive, then it will most likely be a negative-beta emitter, since the capture of a neutron will have produced a nucleus in which the neutron/proton ratio is too large for stability for the given atomic number.

2.81. Essentially all the elements, from hydrogen to uranium, exhibit radiative capture to a greater or a lesser extent.   However, as may be expected, nuclei having magic numbers (§ 1.22) of neutrons, e.g., $_2 He^4$, $_{40} Zr^{92}$, and $_{83} Bi^{209}$, show little tendency to capture neutrons.   Two important examples of radiative capture, by uranium-238 and thorium-232, respectively, were described in § 1.34 *et seq.*   In addition, it has been mentioned that uranium-235 and plutonium-239 exhibit radiative capture, in competition with fission, especially for neutrons of intermediate and low energies.   Other examples of reactions of this type will be considered in due course.

2.82. As seen in § 2.78, the excitation energy of the compound nucleus formed by neutron capture is equal to the binding energy of the neutron plus its kinetic energy.   This excess energy is generally emitted as several photons covering a range of energies (see Chapter 10); in a few cases, e.g., hydrogen and carbon-12, all the excess energy appears as a single photon.   Except for hydrogen, where the binding energy is 2.2 Mev, the total capture gamma-ray energy is generally about 6 to 8 Mev.

EMISSION OF ALPHA PARTICLES

2.83. Slow-neutron reactions accompanied by the emission of a charged particle, e.g., an alpha particle (n, α) or a proton (n, p), are rare.   The reason is that a positively charged particle can be expelled from a nucleus only if it has sufficient energy to overcome an electrostatic (or coulombic) potential, in addition to the energy required to detach it from the nucleus.   It is only for a few elements of low atomic number, for which the nuclear electrostatic repulsion is small, that charged-particle emission is possible after capture of a slow neutron.

2.84. The interactions of slow neutrons with lithium-6, the less common, naturally occurring isotope of lithium, and with boron-10, the rarer, stable isotope of boron, lead to the ejection of an alpha particle. Both of these reactions have a special interest in the present connection. The (n, α) reaction with boron-10 may be written as

$$_5B^{10} + _0n^1 \rightarrow (_5B^{11})^* \rightarrow _3Li^7 + _2He^4,$$

where $_2He^4$ represents an alpha particle, i.e., a helium nucleus, mass number 4 and atomic number 2. The charged particles produced in this reaction are ejected in opposite directions with relatively high energy, so that they produce considerable ionization in their passage through a gas. This is the basis of a method for detecting and counting slow neutrons (Chapter 5). Because boron undergoes the (n, α) reaction very rapidly with slow neutrons, this element, like cadmium, is used for reactor control.

2.85. The other (n, α) process which occurs readily with slow neutrons is that with lithium-6, i.e.,

$$_3Li^6 + _0n^1 \rightarrow (_3Li^7)^* \rightarrow _1H^3 + _2He^4.$$

The residual (recoil) nucleus is here H^3, a negative beta-active, hydrogen isotope of mass number 3, called *tritium*. This isotope has attracted attention because of its possible use in the so-called "hydrogen bomb," and in fusion reactions for the release of nuclear energy. It can be produced by the action of slow neutrons on lithium in a fission reactor.

REACTIONS WITH FAST NEUTRONS

2.86. Relatively few reactions of fast neutrons with atomic nuclei, other than scattering and fission, are important for the study of nuclear reactors. Although many such fast-neutron reactions are known, their probabilities are usually so small that they have little effect on reactor operation. One process of this type, however, should be mentioned; it is the (n, p) reaction of fast neutrons with oxygen-16, namely,

$$_8O^{16} + _0n^1 \rightarrow (_8O^{17})^* \rightarrow _1H^1 + _7N^{16}.$$

The product is nitrogen-16, which is beta active, having a half-life of 7.3 sec, and emitting, in addition, a gamma ray of high energy (about 6 Mev). This process occurs in reactors using either air or water as coolant, since both of these materials contain oxygen. Even though most of the fissions would be produced by thermal neutrons, a considerable number of fast neutrons, liberated in fission, are present. These will interact with oxygen-16 nuclei to yield the highly radioactive, although short-lived, nitrogen-16. The decay product of the latter is ordinary oxygen-16.

2.87. Provided the energy is available, the expulsion of a charged particle from the excited compound nucleus, formed as a result of neutron capture, is more probable than the emission of radiation. Thus (n, α) and (n, p) reactions of nuclei with fast neutrons, having energies of 1 Mev or more, frequently occur more readily than the (n, γ) reaction. If neutrons of sufficiently high energy are used, two or more neutrons or protons may be expelled from the compound nucleus. For incident neutrons of about 10 Mev, such reactions as $(n, 2n)$ and (n, np) have been observed, and for still higher energies, $(n, 3n)$, $(n, 2np)$, etc., processes are possible.

INELASTIC SCATTERING

2.88. When a fast neutron undergoes inelastic scattering (§ 1.31), it is first captured by the target nucleus to form an excited (virtual) state of the compound nucleus; a neutron of lower kinetic energy is then emitted, leaving the target nucleus in an excited (bound) state (Fig. 2.12). In other words, in an inelastic scattering collision, some (or all) of the kinetic energy of the neutron is converted into excitation (internal) energy of the target nucleus. This excess energy is subsequently emitted as one or more photons of gamma radiation, called *inelastic-scattering gamma rays*.

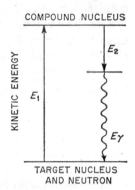

2.89. Let E_1 be the total kinetic energy of the neutron and target nucleus before collision and E_2 the kinetic energy after collision; then if E_γ is the energy emitted as gamma radiation, it follows that

$$E_1 = E_2 + E_\gamma.$$

Fig. 2.12. Energy transitions in inelastic scattering

It is evident that in inelastic scattering kinetic energy is not conserved. Nevertheless, there is conservation of momentum, so that, if E_γ were known, the mechanics of the process could be solved. This is not necessary, however, for the present purpose. What is important to note is that E_1 must be at least equal to E_γ if inelastic scattering is to occur. Since the kinetic energy of the target nucleus is, in general, negligible in comparison with that of the neutron, it follows that, in an inelastic collision, the initial energy of the neutron must exceed the minimum excitation energy of the target nucleus.

2.90. For elements of moderate and high mass number, the minimum excitation energy, i.e., the energy of the lowest excited state above the ground state, is usually from 0.1 to 1 Mev. Hence, only neutrons with energy exceeding this amount can be inelastically scattered as a result of nuclear excitation.* With

---

* For inelastic scattering which is possible with neutrons of low energy, see § 2.102.

decreasing mass number of the nucleus there is a general tendency for the excitation energy to increase, so that the neutrons must have higher energies if they are to undergo inelastic scattering. The threshold energy for such scattering in oxygen, for example, is about 6 Mev, and in hydrogen the process does not occur at all. Exceptions to the foregoing generalizations are the magic nuclei; heavy nuclei of this type, e.g., lead (82 protons) and bismuth (126 neutrons), behave like light nuclei with respect to inelastic scattering.

2.91. Another general rule relating to inelastic scattering is that the relative probability of its occurrence, as against radiative capture or other processes following neutron absorption, increases with increasing neutron energy. This is because the separation (or spacing) of the excited levels of a nucleus is smaller at high excitation energies; there are consequently more excited states, in a given energy range, which the nucleus can occupy after expulsion of a neutron. The probability of the emission of a neutron by the compound nucleus increases correspondingly.

2.92. The energy of the inelastic-scattering gamma rays depends, of course, upon the value of E_γ in the particular case, and upon whether it is emitted as one or more photons. For inelastic scattering by elements of low mass number, the total gamma-ray energy must be high, e.g., several Mev, whereas for heavy elements it will usually be lower.

ELASTIC SCATTERING

2.93. It will be apparent from the foregoing discussion that inelastic scattering of the type associated with nuclear excitation is restricted to neutrons with energies in excess of at least 0.1 Mev for collisions with the heaviest nuclei, and to those neutrons with even larger energies when lighter nuclei are involved. Consequently, neutrons having energies less than about 0.1 Mev cannot lose energy as a result of inelastic collisions. The situation in regard to elastic scattering is quite different; kinetic energy is conserved, and there are no restrictions upon the transfer of this energy between the neutron and the nucleus. Provided the kinetic energy of the neutron exceeds that of the nucleus, there is a possibility that some kinetic energy will be transferred from the former to the latter, and vice versa.

2.94. Elastic collisions of neutrons with nuclei are of two general types. One involves capture of the neutron to form a compound nucleus followed by emission of another (scattered) neutron; in the other there is apparently no compound nucleus formation (§ 2.145). In either case, the struck nucleus remains in its lowest energy (ground) state and the interaction with the nucleus can be treated as a "billiard ball" type of collision. The behavior can thus be analyzed by means of the familiar laws of mechanics, based on the principles of the conservation of both energy and momentum.

2.95. After a sufficient number of elastic scattering collisions, the velocity of a neutron is reduced to such an extent that it has approximately the same average kinetic energy as the atoms (or molecules) of the scattering medium. The energy depends on the temperature of the medium, and so it is called thermal energy (§ 1.32). Thermal neutrons are thus neutrons which are in thermal equilibrium with the atoms (or molecules) of the medium in which they are present. A particular thermal neutron undergoing collisions with the nuclei of the ambient medium may gain or lose energy in any one collision. But, if a large number of thermal neutrons diffusing in a nonabsorbing medium are considered, there is no net energy change for all the neutrons.

THE MAXWELL-BOLTZMANN DISTRIBUTION

2.96. In a weakly absorbing medium, the kinetic energies of thermal neutrons will be distributed statistically according to the Maxwell-Boltzmann distribution law, as derived from the kinetic theory of gases; this may be expressed in the form

$$\frac{dn}{n} = \frac{2\pi}{(\pi kT)^{3/2}} e^{-E/kT} E^{1/2} \, dE, \tag{2.21}$$

where dn is the number of neutrons with energies in the range from E to $E + dE$, n is the total number of neutrons in the system, k is the Boltzmann constant, and T is the absolute temperature. The equation may be written in a slightly different form by letting $n(E)$ represent the number of neutrons of energy E *per unit energy interval*. Then $n(E)dE$ is the number of neutrons having energies in the range from E to $E + dE$, which is equivalent to dn in equation (2.21). The latter may thus be written as

$$\frac{n(E)}{n} = \frac{2\pi}{(\pi kT)^{3/2}} e^{-E/kT} E^{1/2}, \tag{2.22}$$

where the left side represents the fraction of the neutrons having energies within a unit energy interval at energy E. The right side of equation (2.22) can be evaluated for various E's at a given temperature, and the Maxwell-Boltzmann curve obtained in this manner, indicating the variation of $n(E)/n$ with the kinetic energy E of the neutrons, is shown in Fig. 2.13.

2.97. It is seen from Fig. 2.13 that, although some thermal neutrons have very small, and others very large, energies, a considerable proportion have energies in a fairly narrow range. It has become the practice to give the energy of thermal neutrons as kT for the particular absolute temperature T. This is actually the kinetic energy corresponding to the most probable velocity per unit velocity interval, based on the Maxwell-Boltzmann distribution. Since the Boltzmann constant k is 1.38×10^{-16} erg per °C, i.e., 8.6×10^{-5} ev per °C or 4.8×10^{-5} ev per °F, it follows that the so-called average energy of thermal neutrons is $8.6 \times 10^{-5} T_K$ or $4.8 \times 10^{-5} T_R$ ev, where T_K and T_R are the absolute temperatures of the scattering medium on the Kelvin and Rankine scales, respectively.

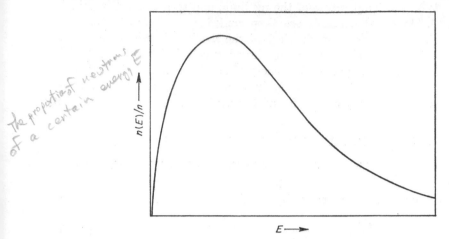

The proportion of neutrons of a certain energy E

FIG. 2.13.   Maxwell-Boltzman distribution of energy

2.98. Since the neutron energy is kinetic, provided the velocity does not approach that of light, it is equal to $\frac{1}{2}mv^2$, where m is the neutron mass and v its velocity. The value of m is 1.67×10^{-24} gram and so the neutron energy is $0.83 \times 10^{-24}v^2$ ergs or $5.2 \times 10^{-13}v^2$ ev, with v in cm per sec; consequently, it is found that

$$v = 1.4 \times 10^6 \sqrt{E} \text{ cm/sec,}$$

where E is the neutron energy expressed in electron volts. If this result is combined with the expressions for the thermal-neutron energy derived above, the speed of thermal neutrons is given as $1.3 \times 10^2 \sqrt{T_K}$ or $0.97 \times 10^2 \sqrt{T_R}$ meters per sec. The values of the energies (in electron volts) and velocities (in m/sec) for thermal neutrons at several temperatures are given in Table 2.5.

TABLE 2.5.   SPEEDS AND ENERGIES OF THERMAL NEUTRONS

Temperature		Energy (ev)	Speed (m/sec)
(°C)	(°F)		
20	68	0.025	2200
200	392	0.041	2800
400	752	0.058	3400
600	1112	0.075	3800
800	1472	0.092	4200

2.99. The Maxwell-Boltzmann distribution of thermal neutrons is based on a highly idealized model of elastic collisions in a gaseous medium between two kinds of particles, nuclei and neutrons, which do not combine with one another. In practice there are several reasons why these conditions are not applicable, so that there are deviations from the ideal Maxwell-Boltzmann distribution appropriate to the temperature of the medium. Consider, for example, the situation in a thermal reactor. Neutrons, mostly of high energy, are produced in the fission process and they are slowed down, mainly in elastic collisions with moderator nuclei. The slow neutrons are then absorbed in fission and radiative capture reactions.

2.100. As a consequence of the slowing down process, the proportion of neutrons of higher energy is greater than required for a Maxwell-Boltzmann distribution. In agreement with theoretical considerations (§ 3.76), the number of neutrons of a specified energy is inversely proportional to the energy value, that is to say, the energy distribution satisfies a $1/E$ law in the higher energy (*epithermal*) region. On the other hand, because the slowed-down neutrons are more readily absorbed, the proportion of such neutrons is less than expected for a strictly thermal distribution at the temperature of the moderator; such a distribution is said to be "hardened." Because neutrons are always absorbed to some extent while approaching thermal equilibrium with the moderator, they cannot be completely thermalized. In other words, there are actually no "thermal" neutrons in a moderator. However, the smaller the extent of absorption, the more nearly is true thermal equilibrium approached.

2.101. Both theoretical considerations and experimental evidence indicate that the neutron distribution in the lower energy range is close to that for a Maxwell-Boltzmann system corresponding to a temperature somewhat higher than the actual moderator temperature, as shown in Fig. 2.14 [3]. The actual distribution in the "thermal" region is then described by an effective neutron temperature, T_n, which exceeds the moderator temperature, T_m. The smaller the ratio of neutron absorption to scattering in the slowing down medium, the more closely does T_n approach T_m. At energies above the so-called thermal region, i.e., for epithermal neutrons, the distribution is generally proportional to $1/E$, as implied by the straight line at the right of Fig. 2.14.

2.102. Another cause of departure from an exact Maxwell-Boltzmann distribution is the occurrence of inelastic scattering collisions in which some of the kinetic energy of the neutrons is converted into internal energy of the moderator atoms (or molecules). Inelastic scattering of the type leading to an excited nuclear state, which has been discussed earlier (§ 2.88), has some influence. But it is not very significant, since it affects only neutrons of high energy and these do not have a Maxwell-Boltzmann distribution in any event. More important are the inelastic collisions in which there is a change in the vibrational energy of atoms or in the rotational and vibrational energies of molecules. Such collisions, like absorption of neutrons, generally produce a hardening of the energy distribution in the thermal region. In crystalline solids, the distribution can also be affected as a result of the diffraction of neutrons by the crystal lattice.

2.103. In the foregoing discussion of the Maxwell-Boltzmann distribution of neutrons, it has been tacitly assumed that the system is infinitely large, so that

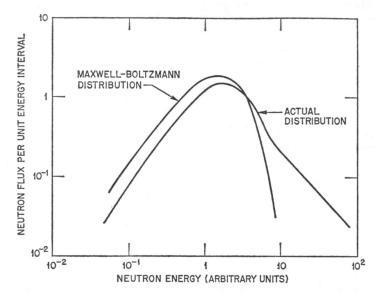

FIG. 2.14. Comparison of actual and theoretical neutron energy distribution

no neutrons are lost by escape. In a system of finite size some neutrons will inevitably be lost in this manner and the energy distribution will be thereby affected, since escape is equivalent to absorption. However, fast neutrons will escape more readily than those of lower energy, so that the thermal-neutron distribution will not be greatly affected by escape.

STRUCTURAL CHANGES CAUSED BY NEUTRON INTERACTIONS

2.104. Both fast and thermal neutrons, upon interacting with materials, can cause important changes in their mechanical properties. The effects will be described in greater detail in Chapter 7, but it is important to recognize that the interaction processes occur simultaneously with some of the neutron reactions, such as scattering and capture, treated earlier in this chapter.

2.105. If a high-energy neutron enters a crystalline lattice, e.g., of a metal or an ionic compound, there is a certain probability that the neutrons will be scattered by the lattice nuclei. A target atom (or ion) involved in such a collision will usually be displaced from its normal (stable) position in the lattice leaving behind a vacancy. The scattered neutron can then proceed to collide with other nuclei and produce more displaced atoms. If a normal site is not readily available, a displaced atom may occupy an intermediate, less stable location, called an interstitial position.   The result of the neutron collisions is thus the formation of more or less permanent defects in the solid.   Since it requires only about 25 ev of energy to move an atom from its normal position in a metal lattice, it is evident that a 1-Mev neutron, for example, might produce many defects.   If these defects are sufficiently common, there may be a marked change in the physical and mechanical properties of the material.

2.106. In addition to the effects of fast neutrons described above, the capture of thermal neutrons in (n, γ) reactions may produce significant changes in some materials.   Since momentum must be conserved in these reactions, the emission of a gamma-ray photon is accompanied by the recoil of the residual nucleus. The recoil energies may be as large as hundreds of electron volts and so are sufficient to produce a significant number of atomic displacements.   This effect is important in materials which have a large probability of capturing thermal neutrons; that is to say, they have large capture cross sections (§ 2.108).

2.107. Crystalline organic substance and noncrystalline materials, e.g., plastics, suffer radiation effects through other mechanisms, chiefly the breaking of chemical (covalent) bonds.   The threshold energy for such breakage is also of the order of 25 ev, and so both fast and thermal neutrons can produce physical changes as a result of scattering and capture processes, respectively.

CROSS SECTIONS FOR NEUTRON REACTIONS

SIGNIFICANCE OF CROSS SECTIONS

2.108. The description of the interaction of neutrons with atomic nuclei can be made quantitative by means of the concept of *cross sections*.   If a given material is exposed to the action of neutrons, the rate at which any particular nuclear reaction occurs depends upon the number of neutrons, their velocity, and the number and nature of the nuclei in the specified material.   The cross section of a target nucleus for any given reaction is a property of the nucleus and of the energy of the incident neutron.

2.109. Suppose a uniform, parallel beam of I neutrons per cm² impinges perpendicularly, for a given time, on a thin layer, δx cm in thickness, of a target material containing N atoms (or nuclei) per cm³, so that $N \, \delta x$ is the number of target nuclei per cm².   Let C be the number of individual processes, e.g., neutron captures, occurring per cm².   The nuclear cross section σ for a specified reaction is then defined as the average number of individual processes occurring per target nucleus per incident neutron in the beam; thus,

$$\sigma = \frac{C}{(N \, \delta x)I} \text{ cm}^2/\text{nucleus}. \tag{2.23}$$

Because nuclear cross sections are frequently in the range of 10^{-22} to 10^{-26} cm² per nucleus, it is the general practice to express them in terms of a unit of 10^{-24} cm² per nucleus, called a *barn*. Thus, a nuclear cross section of 2.7×10^{-25} (or 0.27×10^{-24}) cm² would be written as 0.27 barn.

2.110. The significance of the cross section may be seen by rearranging equation (2.23) into the form

$$(N \, \delta x)\sigma = \frac{C}{I}. \quad \substack{\text{— neutron captures per cm}^2 \\ \text{— impinging neutrons per cm}^2} \qquad (2.24)$$

If every neutron falling on the target reacted, then I would be equal to the number of nuclei taking part in the reaction; hence, the right side of equation (2.24) represents the fraction of the incident neutrons which succeed in reacting with the target nuclei. Thus $(N \, \delta x) \, \sigma$ may be regarded as the fraction of the surface capable of undergoing the given reaction; in other words, of 1 cm² of target surface, $(N \, \delta x) \, \sigma$ cm² is effective. Since 1 cm² of the surface contains $N \, \delta x$ nuclei, the quantity σ cm² is the *effective area per single nucleus* for the given reaction. It is this interpretation of σ that leads to the use of the term "cross section."

MACROSCOPIC CROSS SECTION

2.111. The cross section σ for a particular process, which applies to a single nucleus, is frequently called the *microscopic cross section*. Since the target material contains N nuclei per cm³, the quantity $N\sigma$ is equivalent to the total cross section of the nuclei per cm³; this is called the *macroscopic cross section* of the material for the process. Representing the latter by Σ it is therefore defined as

$$\Sigma = N\sigma \text{ cm}^{-1}, \qquad (2.25)$$

with dimensions of a reciprocal length.

2.112. If the target material is an element of atomic weight A and density ρ grams per cm³, then ρ/A is the number of gram atoms per cm³. The number of atomic nuclei per cm³ is obtained upon multiplying by N_a, the Avogadro number (0.602×10^{24}), which gives the number of individual atoms (or nuclei) per gram atom; thus,

$$N = \frac{\rho}{A} N_a, \qquad (2.26)$$

so that, from equation (2.25),

$$\Sigma = \frac{\rho N_a}{A} \sigma. \qquad (2.27)$$

Since microscopic cross sections are invariably tabulated in barns, i.e., in units of 10^{-24} cm², equations (2.25) and (2.27) will give correct values for Σ if σ is in barns, and N and N_a are in terms of 10^{24} nuclei; thus, N_a is taken to be 0.602.

2.113. For a compound of molecular weight M and density ρ, the number N_i of atoms of the ith kind per cm³ is given by a modification of equation (2.26), namely, *the number of atoms of the ith kind per cm^3*

density

atoms per molecule

$$N_i = \frac{\rho N_a}{M} \nu_i, \qquad (2.28)$$

molecular wt.

where ν_i is the number of atoms of the kind i in a molecule of the compound. The macroscopic cross section for this element in the given target material is then

$$\Sigma_i = N_i \sigma_i = \frac{\rho N_a}{M} \nu_i \sigma_i, \qquad (2.29)$$

where σ_i is the corresponding microscopic cross section. For the compound, the macroscopic cross section is expressed by

$$\Sigma = N_1 \sigma_1 + N_2 \sigma_2 + \cdots + N_i \sigma_i + \cdots$$

$$= \frac{\rho N_a}{M} (\nu_1 \sigma_1 + \nu_2 \sigma_2 + \cdots \nu_i \sigma_i + \cdots). \qquad (2.30)$$

Example 2.5. The microscopic cross section for the capture of thermal neutrons by hydrogen is 0.33 barn and for oxygen 2×10^{-4} barn. Calculate the macroscopic capture cross section of the water molecule for thermal neutrons.

The molecular weight of water (M) is 18 and the density (ρ) is 1.0 g/cm^3; the molecule contains 2 atoms of hydrogen and 1 of oxygen. Hence, by equation (2.30), recalling that the Avogadro number (N_a) is 0.602 $\times 10^{24}$ and the barn is 10^{-24} cm^2,

$$\Sigma_{H_2O} = \frac{1 \times N_a}{18} (2\sigma_H + \sigma_0)$$

barn

$$= \frac{0.602 \times 10^{24}}{18} [(2)(0.33) + 2 \times 10^{-4}](10^{-24})$$

$$= 0.022 \text{ cm}^{-1}.$$

2.114. For a mixture, either of elements or compounds or both, which contains several different nuclear species, the macroscopic cross section is given by

$$\Sigma = N_1 \sigma_1 + N_2 \sigma_2 + \cdots + N_i \sigma_i + \cdots, \qquad (2.31)$$

where the values of N_1, N_2, etc., are dependent upon the composition of the mixture as well as on the atomic (or molecular) weights and densities of the constituents. In determining the macroscopic cross section in a mixture, a distinction must be made between a so-called homogeneous system, in which there is an intimate, uniform mixture or solution of the constituents, and a heterogeneous system, in which one or more components are distributed throughout another in the form of relatively large pieces, e.g., lumps, rods, plates, etc. In the former case, the N values are the numbers of nuclei per unit volume of the whole mixture, whereas in the latter the N's are based on the volume of the particular material being considered. Hence, in a heterogeneous system the macroscopic cross section of each constituent has the same value as in the unmixed state.

Example 2.6. Disregarding the uranium-234, natural uranium may be taken to be a homogeneous mixture of 99.28 weight per cent of uranium-238 (absorption cross section 2.71 barns) and 0.72 weight per cent of uranium-235 (absorption cross section 683 barns).

The density of natural uranium metal is 19.0 g/cm³. Determine the total macroscopic and microscopic absorption cross sections of this material.

Each cubic centimeter of the metal contains $(19.0)(0.9928)$ g U-238 and $(19.0)(0.0072)$ g U-235; hence,

$$N_{238} = \frac{(19.0)(0.9928)(0.602 \times 10^{24})}{238}$$

$$= 4.77 \times 10^{22} \text{ nuclei/cm}^3$$

and

$$N_{235} = \frac{(19.0)(0.0072)(0.602 \times 10^{24})}{235}$$

$$= 3.5 \times 10^{20} \text{ nuclei/cm}^3.$$

Consequently,

$$\Sigma \text{ (natural U)} = N_{238}\sigma_{238} + N_{235}\sigma_{235}$$
$$= (4.77 \times 10^{22})(2.71 \times 10^{-24}) + (3.5 \times 10^{20})(683 \times 10^{-24})$$
$$= 0.368 \text{ cm}^{-1}.$$

This is the required macroscopic absorption cross section. The microscopic cross section is given by

$$\sigma \text{ (natural U)} = \frac{\Sigma \text{ (natural U)}}{N_{238} + N_{235}}$$

$$= \frac{0.368}{4.80 \times 10^{22}} = 7.7 \times 10^{-24} \text{ cm}^2$$

$$= 7.7 \text{ barns.}$$

CROSS-SECTION DETERMINATION BY THE TRANSMISSION METHOD

2.115. The *transmission method* for the experimental determination of cross sections is based on measurements of the attenuation of a neutron beam after passage through a slab of target material of finite thickness. Suppose a collimated beam of neutrons strikes perpendicularly a specified area of material of appreciable thickness (Fig. 2.15). Consider a thin layer, of thickness dx, parallel to the surface; then from the arguments presented above, it follows that $N\sigma\,dx$ is the fraction of the neutrons falling on this layer which react. This may be set equal to $-dI/I$, where $-dI$ is the decrease in the neutrons per cm² as the result of passing through the thickness dx of target material. Consequently,

$$-\frac{dI}{I} = N\sigma\,dx,$$

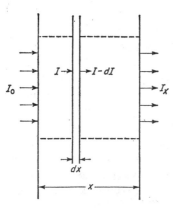

Fig. 2.15. Attenuation of collimated neutron beam in passage through a slab

and integration over the thickness x of the material gives

$$I_x = I_0 e^{-N\sigma x}$$

$$= I_0 e^{-\Sigma x}, \tag{2.32}$$

where I_0 is the number of incident neutrons falling on a particular area, and I_x is the number which succeed in passing through x cm of the material over the same area.*

2.116. The experimental arrangement for the measurement of cross sections by the transmission method consists of a neutron source and a detector, between which is placed a slab of the material being investigated. By means of a suitable collimating shield, the neutron beam passing through to the detector is restricted to a relatively small solid angle. The purpose of the shield is to prevent, as far as possible, neutrons which have been scattered in the material from reaching the detector. The only neutrons which succeed in reaching the detector are those which have neither been scattered nor absorbed in the material under study. The cross section determined in this manner is thus the total cross section. The neutron intensity as observed at the detector with the slab of material removed is proportional to I_0, whereas I_x is the corresponding value with the slab interposed between the same source and detector. For the determination of the cross section by means of equation (2.32), it is not necessary to know the absolute values of I_0 and I_x, but only their ratio which is obtained in the manner described. A method for determining cross sections for specific reactions is given in § 2.119.

RATES OF NEUTRON REACTIONS

2.117. Consider a neutron beam in which n is the *neutron density*, i.e., the number of neutrons per cm³; if v is the neutron velocity, then nv is the number of neutrons falling on 1 cm² of target material per sec. Since σ cm² is the effective area per single nucleus, for a given reaction or reactions (§ 2.110), then Σ is the effective area (in cm⁻¹) of all the nuclei per cm³ of target. Hence, the product Σnv gives the number of interactions (between neutrons and nuclei) per cm³ of target material per sec. Since each nuclear interaction involves one neutron,

$$\text{Rate of neutron interaction} = \Sigma nv \text{ neutrons/(cm}^3\text{)(sec).} \tag{2.33}$$

2.118. This is a result of considerable importance, for it gives the number of neutrons per second involved in any interaction (or interactions) with 1 cm³ of material for which Σ is the macroscopic cross section. It is sometimes written in a slightly different form by introducing the *neutron flux* in place of the neutron

---

* Comparison of equations (2.18) and (2.32) shows that the linear attenuation coefficient (μ) for gamma rays is equivalent to a macroscopic cross section. In fact, μ/N is often referred to as the (microscopic) cross section for gamma-ray attenuation.

density.*   The neutron flux is defined as the product of the neutron density and the velocity, i.e.,

$$\phi = nv,$$

so that it is expressed in units of neutrons per cm^2 per sec.   It is equal to the total distance in centimeters traveled in 1 sec by all the neutrons present in 1 cm^3, and is therefore sometimes referred to as the *track length*.   Upon substituting ϕ for nv in equation (2.33), it follows that

$$\text{Rate of neutron interaction} = \Sigma\phi \text{ neutrons}/(\text{cm}^3)(\text{sec}), \qquad (2.34)$$

a result which will find frequent application in later sections.

CROSS-SECTION DETERMINATION BY ACTIVATION METHOD

2.119.   An illustration of the use of equation (2.34) is provided by the *activation method* for the determination of specific capture cross sections.   It is particularly applicable when a stable (nonradioactive) nuclide captures neutrons and forms a radioactive product which has a relatively low absorption cross section.   If the product is represented by A, the rate at which its concentration, expressed as nuclei per cm^3, increases with time when the target material is exposed to a flux ϕ of neutrons is given by

$$\frac{dA}{dT} = \Sigma\phi - \lambda A,$$

where the first term on the right represents the rate of formation of A by neutron capture and the second term is its rate of radioactive decay, λ being the decay constant of A.   After a sufficient period of exposure, a steady state is attained in which the rates of formation and decay of A are equal; the steady-state concentration A_0 is then given by

$$A_0 = \frac{\Sigma\phi}{\lambda}. \qquad (2.35)$$

The rate of emission of particles by A in the steady state, which is equal to λA_0 per cm^3 and hence to $\Sigma\phi$, is called the *saturation activity* of the material in the given neutron flux.

2.120.   If the activated material is now removed from the neutron flux, it will decay in the normal exponential manner and the concentration of A at any time, t, after removal, i.e., $A(t)$, is expressed by

$$A(t) = A_0 e^{-\lambda t}. \qquad (2.36)$$

From measurements of $A(t)$, made by means of instruments which determine the rate of beta-particle emission, it is possible to use equation (2.36) to obtain both A_0 and λ.   Insertion of these values into equation (2.35) then gives the product $\Sigma\phi$, and if the neutron flux is known, the macroscopic cross section Σ can be

* A further discussion of the significance of neutron flux is given in § 3.13.

evaluated.   It should be noted that this (activation) cross section is that for the specific process leading to the formation of the radioactive nuclide whose activity was measured.

2.121. By making use of a material with a known activation cross section, the procedure described above is utilized to determine neutron flux.   This is not necessarily the total flux, but that of the particular neutrons, e.g., thermal neutrons, to which the cross section applies.   In utilizing this method a thin foil of the neutron absorber is exposed to the flux for an adequate length of time, so that it may be assumed that the steady state has been attained.   A thin foil is used in order that the neutron flux shall not be appreciably disturbed and also that the flux shall be uniform throughout the absorber.   The material should be one having a fairly high activation cross section for the neutrons being studied, allowing the saturation activity to be large enough to be determined with fair accuracy.   Furthermore, the disintegration rate of the radioactive product should not be too large, or decay will be too rapid after removal from the neutron flux.   Materials suitable for determining neutron flux by the foil-activation technique are indium, gold, silver, manganese, and rhodium.   Indium is particularly useful because its half-life of 54.2 min is long enough for accurate counting, but sufficiently short for a good approximation to saturation activity to be attained within a reasonable time.

2.122. If it is uncertain whether the foil has been exposed to the neutron flux long enough for the steady state to be reached, a modified form of equation (2.36) is used.   Both sides of the rate equation in § 2.119 are multiplied by $e^{\lambda T}$ and, after rearrangement, the result is

$$d(Ae^{\lambda T}) = \Sigma\phi e^{\lambda T}\, dT,$$

where T is the exposure time.   Since A is zero when T is zero, integration gives

$$\lambda A(T) = \Sigma\phi(1 - e^{-\lambda T}),$$

where the left side is the activity (or rate of particle emission) per cm³ of foil after an exposure time T.   If the foil is withdrawn from the neutron flux and measured at a time t later, the activity will be $\lambda A(T)e^{-\lambda t}$.

Example 2.7.   A 1-cm square indium foil having an areal density of 1 mg/cm² is exposed to a flux of thermal neutrons for 20 min.   At a time 3 hr after removal, the indium-116 (half-life 54.2 min) formed had a decay rate (corrected for counter efficiency) of 30,000 counts/min.   Determine the value of the neutron flux.   The activation cross section of indium-115 for thermal neutrons at ordinary temperature is 130 barns.

The mass of the foil is $1 \times 1 \times 10^{-3}$ g and since ordinary indium contains nearly 96.0 per cent of indium-115, the number of these nuclei present is $(10^{-3})(0.602 \times 10^{24})(0.960)/115$ $= 5.03 \times 10^{18}$ nuclei.   If V cm³ is the volume of the foil, then Σ is

$$(5.03 \times 10^{18})(130 \times 10^{-24})/V = 6.54 \times 10^{-4}/V \text{ cm}^{-1}.$$

The activity of a volume V cm³ of foil exposed to the neutron flux for 20 min $(= T)$ and measured 3 hr, i.e., 180 min $(= t)$ later, i.e., 30,000 counts/min, is thus given by

$$30,000 = V\Sigma\phi(1 - e^{-\lambda T})e^{-\lambda t},$$

where $V\Sigma$ is 6.54×10^{-4} cm^{-1} and λ is $0.693/54.2 = 1.28 \times 10^{-2}$ min^{-1}. Hence, the thermal neutron flux is

$$\phi = \frac{30,000}{(6.54 \times 10^{-4})[1 - e^{-(1.28 \times 10^{-2})(20)}][e^{-(1.28 \times 10^{-2})(180)}]}$$

$$= 2.04 \times 10^9 \text{ neutrons/(cm}^2)(\text{min})$$

$$= 3.40 \times 10^7 \text{ neutrons/(cm}^2)(\text{sec}).$$

MEAN FREE PATH

2.123. An alternative approach to the determination of neutron reaction rates is to consider the *mean free path*, λ,* i.e., the average total (or scalar) distance a neutron will travel before undergoing a particular interaction. Since the velocity, v, is the distance a neutron travels per second, the average number of interactions per second is v/λ. For a beam containing n neutrons per cm^3, the number of interactions per cm^3 per sec is thus nv/λ. In other words,

$$\text{Rate of interaction} = \frac{nv}{\lambda} \text{ neutrons/(cm}^3)(\text{sec}). \tag{2.37}$$

Upon comparing equations (2.34) and (2.37) it is seen that

$$\lambda = \frac{1}{\Sigma}, \tag{2.38}$$

so that the neutron mean free path for a specified reaction is the reciprocal of the macroscopic cross section for that reaction. If Σ is expressed in cm^{-1}, as it usually is, then λ will be in cm.

2.124. Equation (2.32) for the attenuation of a collimated beam of neutrons by passage through a target material can be rewritten in an alternative form upon replacing Σ by $1/\lambda$, in accordance with equation (2.38); the result is

$$I_x = I_0 e^{-x/\lambda}. \tag{2.39}$$

If the thickness x of the material is equal to λ, then $I_x/I_0 = 1/e$. Hence, after traversing a thickness λ of the target material, a fraction $1/e$ of the incident neutrons has not been involved in the particular process under consideration. If λ is the mean free path for all interactions, then the intensity of the neutron beam is reduced to a fraction $1/e$ of its initial value after passage through this thickness of absorber.†

2.125. When a neutron can take part in several different processes with a given target nucleus, e.g., (n, γ), (n, α), (n, f), inelastic scattering, and elastic scattering, there is a specific cross section and mean free path for each process.

---

* The conventional symbol λ for the mean free path is unfortunately the same as that used for the radioactive decay constant in the preceding section.

† A distance formally equivalent to λ is sometimes referred to as the "relaxation length" for the nuclear radiations (see § 10.73).

The equations derived above are quite general and apply to any reaction in which neutrons (or other particles) are involved. It is possible to define a total absorption cross section, or even a total cross section for both absorption and scattering, which is the sum of the individual cross sections. An equation of the form of (2.32) will be applicable, and I_x/I_0 will give the fraction of the initial neutrons escaping interaction.

POLYENERGETIC NEUTRON SYSTEMS

2.126. In the derivations given above, it has been assumed, for simplicity, that all neutrons have the same velocity. Since cross sections, in particular those for neutron absorption, are strongly energy dependent, there will be a different value for each neutron energy. For a polyenergetic beam of neutrons, equation (2.32) would thus become more complicated. If it were possible to divide the initial neutron beam into a number of definite energy groups, indicated by the subscripts 1, 2, 3, etc., then the initial intensity I_0 would be given by

$$I_0 = I_{01} + I_{02} + \cdots + I_{0i} + \cdots ,$$

and then, for the ith group,

$$I_{xi} = I_{0i}e^{-N\sigma_i x},$$

so that

$$I_x = \sum_i I_{xi} = \sum_i I_{0i}e^{-N\sigma_i x}, \tag{2.40}$$

where the large sigma implies summation over all values of i. In the limit, when there is a continuous distribution of neutron energies, the summation in equation (2.40) would be replaced by integration over the whole energy range.

2.127. In deriving this result it has been tacitly assumed, as it was in fact in equation (2.32), that, when passing through the thickness x of medium, the neutron velocities (or energies) do not change appreciably, so that the cross sections remain constant. This condition is particularly applicable to thermal neutrons, and, since there is then a fairly definite (Maxwell-Boltzmann) energy distribution, it is possible to derive an average cross section, as will be shown below. In this case, equation (2.40) can be considerably simplified.

2.128. If, in a polyenergetic neutron system, $n(E)$ is the density of neutrons of energy E per unit energy interval, then, as seen in § 2.96, $n(E) dE$ is the number in the energy range from E to $E + dE$. The total neutron flux ϕ, for neutrons of all energies (or velocities), is then given by

$$\phi = \int_0^\infty n(E)v \, dE \text{ neutrons/(cm}^2)(\text{sec}), \tag{2.41}$$

where the integration limits of zero and infinity are formal only and are meant to imply that integration is carried out over the whole range of neutron energies. The velocity v is that corresponding to the kinetic energy E. The rate of interaction of the polyenergetic neutron system can then be expressed, using equation (2.34), as

$$\text{Rate of neutron interaction} = \int_0^\infty \Sigma(E)n(E)v \, dE \quad \text{neutrons/(cm}^3\text{)(sec)}, \quad (2.42)$$

where $\Sigma(E)$ is the macroscopic cross section for the process for neutrons of energy E.

2.129. For the given system, it is possible to define an average macroscopic cross section $\bar{\Sigma}$ so that

$$\text{Rate of neutron interaction} = \bar{\Sigma}\phi \quad \text{neutrons/(cm}^3\text{)(sec)}, \quad (2.43)$$

where ϕ is the total neutron flux derived above. Upon combining equations (2.41), (2.42), and (2.43), it is seen that

$$\bar{\Sigma} = \frac{\int_0^\infty \Sigma(E)n(E)v \, dE}{\int_0^\infty n(E)v \, dE} \quad \text{cm}^{-1}. \quad (2.44)$$

For thermal neutrons having a Maxwell-Boltzmann distribution, $n(E)$ in the foregoing equations would be defined by equation (2.22), with n equal to the total number of neutrons per cm³. An expression similar to equation (2.44) relates the mean microscopic cross section $\bar{\sigma}$ to $\sigma(E)$.

2.130. In order to perform the indicated integrations, it is necessary to express $\Sigma(E)$ or $\sigma(E)$ as a function of the neutron energy E. It will be seen in § 2.137 that for many absorbers the cross sections in the thermal region vary as $1/v$ and, consequently, as $1/E^{1/2}$. Upon combining this result with equations (2.22) and (2.44), it is found that the average value $\bar{\sigma}_{\text{th}}$ for the absorption of thermal neutrons is equal to the cross section for neutrons of energy $4kT/\pi$, i.e., with energy 1.273 times the thermal neutron energy kT (§ 2.97). Alternatively, if σ_{kT} is the absorption cross section for neutrons having actual energy kT, it follows from the $1/E^{1/2}$ law that

$$\bar{\sigma}_{\text{th}} = \frac{\sqrt{\pi}}{2} \sigma_{kT} \quad \text{or} \quad \bar{\sigma}_{\text{th}} = \frac{\sigma_{kT}}{1.128}. \quad (2.45)$$

for the Maxwellian temperature T.

2.131. As a general rule, the values of σ_{kT} are tabulated for a temperature of 293°K, i.e., 20°C, for which kT is 0.0253 ev and the neutron velocity is 2200 meters per sec. If this temperature is represented by T_0, and $\sigma(T_0)$ are the tabulated cross sections, then the σ_{kT} at any temperature T, represented by $\sigma(T)$ is given by

$$\sigma(T) = \sigma(T_0) \left(\frac{T_0}{T}\right)^{1/2},$$

provided the $1/v$ law is obeyed. Hence, by equation (2.45), the average thermal cross section $\bar{\sigma}_{\text{th}}(T)$ at the temperature T is

$$\bar{\sigma}_{th}(T) = \frac{\sigma(T_0)}{1.128}\left(\frac{T_0}{T}\right)^{1/2}.$$

2.132. For some materials of reactor interest, e.g., uranium-235, plutonium-239, and uranium-238, the cross sections exhibit departure from strict $1/v$-behavior, and this relationship between $\bar{\sigma}_{th}(T)$ and the tabulated cross sections, $\sigma(T_0)$, is in error. An empirical correction is then applied, by means of the non-$1/v$ factor $g(T)$, which is a function of temperature [4]; thus,

$$\bar{\sigma}_{th}(T) = g(T)\frac{\sigma(T_0)}{1.128}\left(\frac{T_0}{T}\right)^{1/2}.$$

Some values of $g(T)$ for a range of temperatures are given in Fig. 2.16 [5].

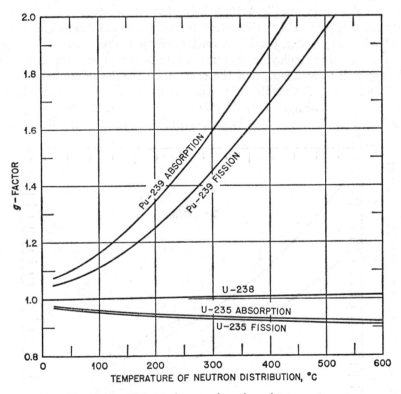

FIG. 2.16.   Values of g as a function of temperature

2.133. The foregoing results are strictly applicable to a completely thermalized system in which there is a Maxwell-Boltzmann distribution of the neutrons corresponding to the moderator temperature. As seen in § 2.99 *et seq.*, this is an ideal situation which cannot be attained, although it can be approached under suitable conditions for neutrons in the so-called thermal region. Two general methods have been used to represent the effect of departure from strict thermalization.

2.134. One procedure is to modify the $g(T)$ factor by the addition of the quantity $rs(T)$, where r is a measure of the neutron flux in the region where the distribution depends upon $1/E$ and $s(T)$ is a correction for departure from $1/v$ behavior of the cross sections in this region. The value of $g(T)$ to be used is presumably that for the effective neutron temperature, T_n, as defined in § 2.101, rather than that of the moderator, T_m. However, for well-moderated systems, in which the compositions and cross sections are such that the ratio of absorption to scattering reactions is relatively small, it is a good approximation to take T_n as equal to T_m and to neglect the additional correction term $rs(T)$. The equation for $\bar{\sigma}_{th}(T)$ given above may then be used for calculational purposes with $g(T)$ values obtained from Fig. 2.16, T being the temperature of the moderator. This approach is reasonably satisfactory for most thermal reactors in which the moderator is heavy water, beryllium (or its oxide), or graphite.

2.135. In reactors employing ordinary water as moderator, the neutrons are usually not well thermalized, and the corrections for departure from the ideal Maxwell-Boltzmann distribution of the neutrons and for non-$1/v$ behavior of the cross sections must be applied in a different manner. In these cases the treatment is based on the theoretical Wigner-Wilkins approach [6], which takes into

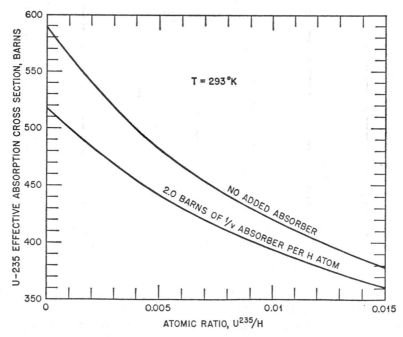

Fig. 2.17. Thermal cross sections of uranium-235 at 293°K averaged over Wigner-Wilkins spectra

account the detailed interactions of the neutrons with the nuclei in the slowing-down medium, combined with experimental cross-section measurements over a range of neutron energies. The results obtained in this manner for the thermal average absorption cross section, $\bar{\sigma}_{th}(T)$, for uranium-235, at 293°K, over a range of U^{235}/H atomic ratios, are shown by the upper curve in Fig. 2.17. The effect of an additional 2 barns per H atom of another absorber, e.g., structural material, etc., in the reactor core, is indicated by the lower curve [7]. Similar data are available at other temperatures, but for the present purpose it will be sufficient to assume that the thermal cross section varies as $1/T^{1/2}$, as in § 2.131.

VARIATION OF CROSS SECTIONS WITH NEUTRON ENERGY

EXPERIMENTAL RESULTS

2.136. The problem of the complete determination of cross sections for neutron reactions is a very complex one; not only do the values depend on the neutron energy, but they vary from one isotope to another of the same element and change with the nature of the reaction. In many instances, one reaction predominates, and this simplifies the situation; the same is true when one particular isotope has a much larger cross section than others of the same element. Most measurements have been made with naturally occurring materials, and the data are useful for calculations of the rates of processes occurring in nuclear reactors. For some elements, however, cross sections have also been derived for individual isotopes, such as uranium-235, uranium-238, etc.

2.137. For many elements, especially those of mass number exceeding 100, an examination of the variation of the absorption cross sections with neutron energy reveals the existence of three regions.* There is, first, a low-energy region, where the cross section decreases steadily with increasing neutron energy. The absorption cross section σ_a then varies inversely as the square root of the neutron energy and, since the energy is kinetic in nature, σ_a is inversely proportional to the neutron velocity. This is called the $1/v$ region, and the neutrons are said to obey the $1/v$ law.

2.138. Following the $1/v$ region for slow neutrons, the elements under consideration exhibit a *resonance region*, usually for neutrons of roughly 0.1 to 1000 ev energy. This region is characterized by the occurrence of peaks where the absorption cross section rises fairly sharply to high values for certain neutron energies and then falls again. Some elements, e.g., cadmium and rhodium, have only one high resonance peak, while others, such as indium, silver, gold, and uranium-238, have two or more peaks. The cross sections at the resonance peaks are sometimes very large, e.g., more than 2×10^4 barns for cadmium-113 at a neutron energy of 0.17 ev and 3.4×10^6 barns for xenon-135 at 0.7-ev

---

* Since scattering cross sections are usually small, the total cross section, i.e., absorption plus scattering, shows the same trend.

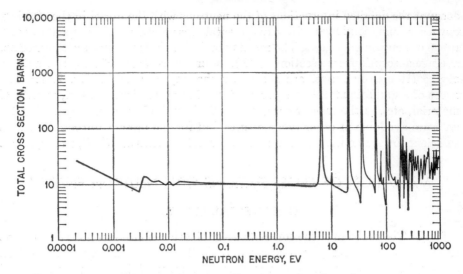

FIG. 2.18. Total cross section of uranium-238 as function of neutron energy

energy. The total cross section of uranium-238 and the total and fission cross sections of uranium-235, as functions of the neutron energy, are shown in Figs. 2.18 and 2.19, respectively. The marked resonance structure is clearly apparent.

2.139. Immediately beyond the resonance region as defined above, minor resonance peaks may occur but they are difficult to resolve with present instruments. Apart from these possible resonances, the nuclear cross sections decrease

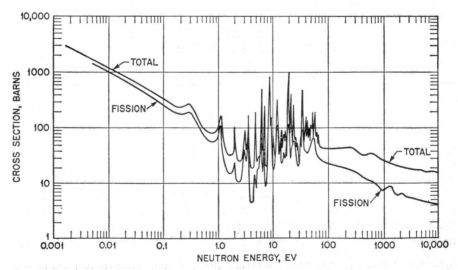

FIG. 2.19. Total and fission cross sections of uranium-235 as function of neutron energy

steadily with increasing neutron energy.   At energies in excess of about 10 kev, there is what is called the *fast-neutron region*.   The cross sections are usually low, being less than 10 barns in most cases and becoming even smaller for energies of the order of 0.1 Mev or more.   The absorption cross sections are then similar in magnitude to the geometrical cross section of the nucleus, i.e., 2 to 3 barns (§ 2.148).   Some nuclei exhibit resonance behavior in the fast-neutron region, but the peaks are not very high.

THEORETICAL INTERPRETATION: RESONANCE ABSORPTION

2.140. There are good reasons for believing that an atomic nucleus can be stable (or quasistable) only when its energy corresponds to that of a particular quantum state.   Every nucleus has several such states, the lowest being the stable or ground state, whereas the others are various excited quantum states or energy levels.*   When a nucleus captures a neutron and the energy of the resulting compound nucleus is equal (or very close) to one of the quantum states of this nucleus, the probability of capture is exceptionally high.   This is the accepted explanation of resonance absorption with its associated large capture cross sections.

2.141. The resonance effect may be illustrated by reference to Fig. 2.20, the lines at the right indicate schematically the (virtual) quantum levels of the compound nucleus (§ 2.78).   The line marked E_0, at the left, represents the energy of the compound nucleus formed when the target nucleus absorbs a neutron of zero kinetic energy.   The excitation energy of the compound nucleus at E_0 is then just (numerically) equal to the binding energy of the neutron.   An examination of Fig. 2.20 shows that the energy E_0 does not correspond to any of the quantum levels of the compound nucleus.   However, if the neutron has a certain amount of kinetic energy, sufficient to bring the energy of the compound nucleus up to E_1, the nucleus will be in one of its quantum states.   Thus, when the neutron has kinetic energy $E_1 - E_0$, resonance absorption occurs, and the nuclear cross section is exceptionally large.   There will, similarly, be resonance

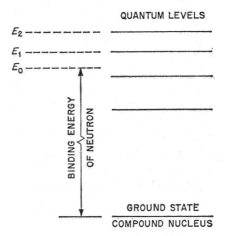

Fig. 2.20.   Formation of excited compound nucleus and quantum energy levels

* For small excitation energies, as are of interest here, the quantum levels are fairly widely spaced and distinct; at high energies the spacing becomes very much closer and the quantum states are virtually continuous.

absorption for neutrons of kinetic energy $E_2 - E_0$, so that the total energy E_2 is equal to that of another quantum level of the compound nucleus, as seen in Fig. 2.20.*

2.142. It is important to point out that not all elements show the type of behavior described above. Most elements of low mass number, as well as several of high mass number, do not exhibit resonance absorption in the region from 0.10 to 1000 ev, at least not to any appreciable extent. There may be resonances at high neutron energies, but these are usually not very significant. The total neutron cross sections, including both absorption and scattering, are small, being of the order of a few barns over the whole energy range, from thermal values to several Mev. Outstanding exceptions to this general behavior are lithium-6 and boron-10, both of which undergo the (n, α) reaction with neutrons of low and intermediate energies. The cross sections, especially for slow neutrons, are quite high and the $1/v$ law is obeyed at energies up to about 0.1 Mev.

SCATTERING CROSS SECTIONS

2.143. With the exception of hydrogen, for which the value is as high as 20 barns in the chemically unbound state,† the elastic scattering cross sections of nearly all elements lie in the range from about 2 to 10 barns for neutrons of low energy. In most cases the cross sections do not vary markedly with neutron energy, although there may be a general tendency for the values to decrease for neutrons of high energy. At very high neutron energies, the elastic scattering cross sections are of the same magnitude as the geometrical cross sections of the respective nuclei, as will be seen subsequently.

2.144. At moderate and low neutron energies, a type of elastic scattering called *resonance scattering* occurs to some extent [9]. The target nucleus captures a neutron to form a compound nucleus; the latter then expels a neutron, leaving the target nucleus in its ground state but having exchanged some kinetic energy with the neutron. It is to be expected from theoretical considerations that the resonance scattering cross section should be independent of neutron energy when the latter is less than the resonance value. At resonance a moderate maximum should occur and subsequently the cross section should decrease steadily with increasing neutron energy.

2.145. In addition to resonance elastic scattering, another type, called *potential scattering*, has been postulated from theoretical considerations [9], although for practical purposes the two types are taken together. Potential scattering does not necessarily involve compound nucleus formation, but, in the language

---

* A detailed treatment of resonance absorption is given in Ref. 8 where an expression is derived for the cross section as a function of neutron energy applicable, particularly, in the region of a resonance peak.

† In the chemically bound state, as in solid paraffin, the scattering cross section increases to 100 barns for neutrons of very low energy, e.g., 0.01 ev.

of wave mechanics, it results from interaction of the neutron wave with the potential at the nuclear surface. With increasing neutron energy (or velocity) the potential scattering cross section decreases steadily. Except, perhaps, at or near a resonance, potential scattering is generally more important than resonance scattering.

2.146. In view of the fact that inelastic scattering is observed, in general, only for neutron energies exceeding about 0.1 Mev, and then only for the heavier elements, it is evident that the cross section will be zero for low and moderate neutron energies.* At a particular energy, which depends upon the nature of the nuclear species, the inelastic scattering cross section becomes appreciable and then decreases gradually with increasing neutron energy. Hydrogen is exceptional in its behavior, for, as stated earlier, it does not, as far as is known, exhibit inelastic scattering of neutrons except perhaps at extremely high energies.

2.147. It will be apparent from the foregoing remarks that the total scattering cross section, including both elastic and inelastic scattering, will rarely exceed a few barns and will not vary much with neutron energy. The very lightest nuclear species, namely, hydrogen and deuterium, have somewhat higher scattering cross sections, and, in the bound state, the values increase with decreasing energy at very low neutron energies.

CROSS SECTIONS AT HIGH NEUTRON ENERGIES

2.148. At high neutron energies, generally in excess of 1 Mev, the cross sections both for absorption plus inelastic scattering and for elastic scattering approach the geometrical cross section of the nucleus. This means that, since resonance scattering is negligible, the total cross section for absorption and inelastic scattering, on the one hand, and for elastic scattering, on the other hand, each tend toward πR^2, where R is the nuclear radius. The total cross section (σ_t) for interaction of the nucleus with a high-energy neutron thus approaches the limit of

$$\sigma_t = 2\pi R^2. \tag{2.46}$$

It has been found in various ways that the radii of atomic nuclei, except those of very low mass number, may be represented approximately by

$$R \approx 1.4 \times 10^{-13} A^{1/3} \text{ cm,}$$

where A is the mass number of the nucleus. Hence the total microscopic cross section is given by

$$\sigma_t \approx 1.25 \times 10^{-25} A^{2/3} \text{ cm}^2/\text{nucleus}$$

$$\approx 0.125 \, A^{2/3} \text{ barns.} \tag{2.47}$$

For an element of mass number 125, for example, the limiting value of σ_t would be about 3 barns; for a mass number of 216 it would be 4.5 barns.

---

* See, however, § 2.102.

Example 2.8. Estimate the average (total) distance a fast neutron will travel in uranium-235 (density 19.0 g/cm³) before it undergoes a nuclear interaction.

The distance required is the mean free path, taking into consideration all possible interactions. According to equation (2.47), the total cross section is $(1.25 \times 10^{-25})(235)^{2/3} = 4.76 \times 10^{-24}$ cm²/nucleus. Then, from equation (2.27), the corresponding macroscopic cross section is

$$\Sigma_t = \frac{(19.0)(0.602 \times 10^{24})(4.76 \times 10^{-24})}{235} = 0.23 \text{ cm}^{-1}.$$

Hence, the mean free path, which is equal to $1/\Sigma_t$, is about 4.4 cm.

THERMAL-NEUTRON CROSS SECTIONS

2.149. The scattering (σ_s) and absorption (σ_a) cross sections for neutrons of velocity 2200 meters per sec (§ 2.131) for a few elements of interest in connection with nuclear reactors are given in Table 2.6; data for other elements will be

TABLE 2.6. 2200-METERS/SEC NEUTRON CROSS SECTIONS

Element	σ_s(barns)	σ_a(barns)	Element	σ_s(barns)	σ_a(barns)
Aluminum....	1.4	0.24	Hydrogen....	38–100	0.33
Beryllium....	7.0	0.10	Iron........	11	2.62
Bismuth.....	9	0.034	Lead........	11	0.17
Boron.......	4	755	Nitrogen.....	10	1.88
Cadmium....	7	2450	Oxygen......	4.2	0.0002
Carbon......	4.8	0.0034	Sodium......	4	0.53
Deuterium...	7	0.0005	Uranium.....	8.3	7.68
Helium......	0.8	0.007	Zirconium....	8	0.185

found in the Appendix [10]. It will be noted that the scattering cross section for hydrogen is quoted as 38 to 100 barns; as seen above, the value depends upon whether the atom is in the free state or chemically bound in a molecule and also upon the precise thermal-neutron energy. For unbound hydrogen, the appropriate value is 38 barns, but for the chemically bound state it increases to about 100 barns, as the neutron energy decreases from about 0.1 ev.

2.150. The large absorption cross sections of boron and cadmium account for the use of these elements in neutron-absorbing control rods. The small values for deuterium (heavy hydrogen), beryllium, and carbon (graphite), on the other hand, point up the advantage of these substances as moderators. Similarly, the relatively small absorption cross sections of aluminum, zirconium, and iron (or steel) make possible their use as structural materials. Lead and bismuth, and magic number nuclei (§ 1.22), in general, have low absorption cross sections.

Although of little value for structural purposes, these elements might find other applications in reactor design, e.g., as coolants.

THE FISSION PROCESS

MECHANISM OF NUCLEAR FISSION

2.151. A helpful insight into some of the characteristics of the fission process is obtained by the use of the liquid-drop model of the atomic nucleus. A nucleus resembles a drop of liquid in the respect that each constituent particle interacts equally with its nearest neighbors. As a first approximation, therefore, the internal (or binding) energy of the nucleus is proportional to the number of nucleons, i.e., to the mass number. Furthermore, the nuclear radius varies as $A^{1/3}$, as seen above, and so the effective volume of a nucleus is also proportional to its mass number. The internal energy should thus vary directly as the nuclear volume. However, in a nucleus, as in a liquid drop, particles on the surface will have a smaller number of near neighbors than do those in the interior. It is necessary, therefore, to reduce the binding energy estimated above, i.e., the volume energy, by an amount that increases with the surface area of the nucleus. In addition to these energy terms, allowance must be made, in examining the behavior of a nucleus, for the electrostatic repulsion between the protons.

2.152. Consider a drop of liquid to which a force is applied so that it is set into oscillation; the system passes through a series of stages, of which the most significant are shown in Fig. 2.21. The drop is at first spherical, as at A; it is

A   B   C   D   E

Fig. 2.21. Liquid-drop model of fission

then elongated into an ellipsoid, as at B. Although the volume remains constant, the surface area has increased, but provided the volume energy exceeds the surface energy, the drop will return to its original form. However, if the deforming force is sufficiently large, the drop will acquire a shape similar to a dumbbell, as at C. In this state, the surface energy will generally exceed the volume energy which provides the cohesive force for the liquid drop. Consequently, the drop will not return to its initial shape but will rather split into two droplets. These will, at first, be somewhat deformed, as at D, but finally they will become spherical.

2.153. The situation in nuclear fission may be thought of as being analogous to that just considered. A target nucleus absorbs a neutron and forms an excited compound nucleus. The excitation energy of the latter is then equal to the binding energy of the neutron plus any kinetic energy the neutron may have had before its capture.   As a result of this excess energy, the compound nucleus may be considered to undergo a series of oscillations, in the course of which it passes through a phase similar to Fig. 2.21, B.   If the oscillation energy is insufficient to cause further deformation beyond B, the attractive forces will compel the nucleus to return to its original form.   The excess energy is then removed by the expulsion of a gamma-ray photon from the excited compound nucleus.

2.154. If, however, the compound nucleus has enough energy to permit it to pass into the dumbbell phase (Fig. 2.21, C), the restoration of the initial state A becomes improbable, because the surface energy (plus the electrostatic repulsion energy) exceeds the volume energy.   Consequently, from C the system passes rapidly to D and then to E, representing fission into two separate nuclei—the fission fragments—which are propelled in opposite directions as a result of the electrostatic repulsion between them.   The excess energy which the compound nucleus must have in order to permit it to deform into the state C is called the *critical energy* for fission.   If this energy is available, e.g., as excitation energy following neutron capture, then fission will usually occur.   If this amount of energy is not available, fission is not possible, at least not at any appreciable rate.

2.155. The critical energy for fission may also be considered with the aid of a potential energy curve, as in Fig. 2.22.   At the extreme right, at *E*, two fission

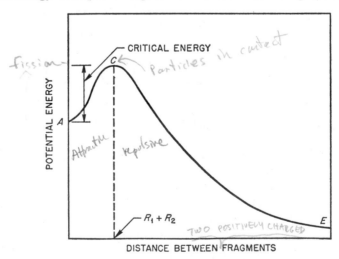

FIG. 2.22.   Critical energy for fission

fragment nuclei are supposed to be far apart, so that the potential energy of the system is virtually zero.   As the nuclei are brought closer together, there is an increase in potential energy due to electrostatic repulsion of their positive charges.   When the fragments reach the point C, where they are roughly in contact, the attractive forces become dominant and the potential energy decreases toward A.   The latter point may be regarded as corresponding to the ground state of the compound nucleus, formed when the target nucleus captures a neutron; that is to say, it represents the energy of the compound nucleus without the excitation energy resulting from neutron capture.   In order for fission to occur at a reasonably rapid rate, the system must pass from A to E, and it can only do so, in general, if the compound nucleus gains sufficient energy to raise it to the level of C.   Thus, the energy difference between A and C represents the critical energy for fission.

2.156. According to calculations based on the liquid drop model [11], the critical energy for fission should decrease as the value of Z^2/A increases, Z being the atomic number and A the mass number of the nucleus.   A qualitative argument leading to this conclusion is that repulsion between the nucleons, which favors fission, varies as Z^2, whereas the attraction is approximately proportional to A, so that fission should occur more easily as Z^2/A increases.   When Z^2/A is less than about 35, the critical energy is so large that neutrons (or other particles) of very high energy would be required to cause fission.   But for nuclei having Z^2/A values of more than 35, the critical energy is down to 6 Mev or less, which is of the order of magnitude of the binding energy of a neutron, and hence of the excitation energy that will accompany the capture of a low-energy neutron.

2.157. The values of the critical energy* and the neutron binding energy for a number of nuclides of interest are given in Table 2.7.   It is seen that for the

TABLE 2.7.   CRITICAL ENERGY FOR FISSION

Target Nucleus	Z^2/A	Critical Energy (Mev)	Neutron Binding Energy (Mev)
Thorium-232........	34.9	5.9	5.1
Uranium-238........	35.6	5.9	4.8
Uranium-235........	36.0	5.8	6.4
Uranium-233........	36.4	5.5	6.7
Plutonium-239......	37.0	5.5	6.4

fissile nuclides, uranium-235, uranium-233, and plutonium-239, the neutron binding energy exceeds the critical energy for fission.   Hence, the capture of a neutron of zero energy would provide sufficient excitation energy to permit the compound nucleus to undergo fission.   This explains why these nuclides are fissionable by neutrons of all energies.   With thorium-232 and uranium-238, however, the situation is different: the neutron binding energies are about 1

* The results quoted in Table 2.7 are the minimum energies of photons which will cause fission of the respective nuclei.

Mev less than the respective critical energies. Consequently, if the compound nucleus is to have sufficient excitation energy to permit fission to take place, the captured neutron must have at least 1 Mev of energy.

2.158. The difference in the neutron binding energies between uranium-235, uranium-233, and plutonium-239, on the one hand, and thorium-232 and uranium-238, on the other hand, arises from the fact that the former contain odd numbers of neutrons, whereas the latter have even numbers. The addition of a neutron to an odd-neutron nucleus to form one with an even number of neutrons is associated with a binding energy that is about 1 Mev greater than for the corresponding change from an even-neutron to an odd-neutron nucleus, in the same mass region.

FISSION CROSS SECTIONS

2.159. The fission cross sections for uranium-233, uranium-235, and plutonium-239, which are fissionable by slow neutrons, may be expected to vary with energy in a manner similar to that for the radiative capture processes (see Fig. 2.19). At low neutron energies, the energy dependence of the cross sections is approximately in accordance with the $1/v$ law, and at high energies the values become quite small, approaching the geometrical cross sections (§ 2.139).

2.160. The 2200 meters per sec neutron cross sections of uranium-235, plutonium-239, uranium-238, and natural uranium, which are important for nuclear reactor work, are recorded in Table 2.8. It is seen that about 84 per cent of the

TABLE 2.8. 2200-METERS/SEC NEUTRON CROSS SECTIONS OF FISSILE
AND FISSIONABLE NUCLEI

Nucleus	Fission (barns)	Radiative Capture (barns)	Total Absorption (barns)
Uranium-233........	527	54	581
Uranium-235........	577	106	683
Plutonium-239......	742	287	1029
Uranium-238........	—	2.71	2.71
Uranium (natural)...	4.2	3.5	7.7

thermal neutrons absorbed by uranium-235 cause fission, whereas only 72 per cent of those absorbed by plutonium-239 are effective in this connection.

FISSION RATE AND REACTOR POWER

2.161. It was shown in § 2.118 that the rate of any reaction involving monoenergetic neutrons can be expressed as $\Sigma\phi$, where Σ is the macroscopic cross section for the process and ϕ is the flux of the monoenergetic neutrons. As applied to fission this becomes

$$\text{Fission rate} = \Sigma_f\phi \text{ fissions/(cm}^3)(\text{sec}), \qquad (2.48)$$

where

$$\Sigma_f = N\sigma_f \quad \text{and} \quad \phi = nv,$$

N being the number of fissile nuclei per cm^3 and σ_f cm^2 per nucleus the fission cross section; n is the neutron density, i.e., neutrons per cm^3, and v is the neutron velocity in cm per sec. Although equation (2.48) is strictly applicable to fissions by neutrons with a single velocity, it can be used to determine the fission rate in a thermal reactor; σ_f is then the average fission cross section for thermal neutrons (§ 2.131), and v is their mean velocity. In general, if σ_f and v are appropriate average values, equation (2.48) can still be used to determine the fission rate in a reactor in which appreciable fissions are caused by neutrons of various energies.

2.162. In a reactor of volume V cm^3, there will occur $V\Sigma_f\phi$ fissions per sec. Since it requires a fission rate of 3.1×10^{10} fissions per sec to produce 1 watt of power (§ 1.45), provided the reactor has been operating for some time, it follows that the power P of a nuclear reactor in watts is given by

$$P = \frac{V\Sigma_f\phi}{3.1 \times 10^{10}} \text{ watts.} \qquad (2.49)$$

Thus, for a reactor of given active volume, in which fissions are caused by neutrons in a specified range, the power is proportional to the product of the macroscopic fission cross section and the neutron flux. As the reactor operates, Σ_f, which is equal to $N\sigma_f$, must decrease somewhat, since there is a gradual decrease in N, the number per cm^3 of fissile nuclei. For practical purposes, however, N and, hence, Σ_f, may be regarded as constant; consequently, the power level is proportional to the neutron flux. The power output of a nuclear reactor is thus altered by changing the neutron flux or neutron density. Similarly, the neutron flux or neutron density is widely used as a measure of the power level.

2.163. For practical use, equation (2.49) may be put into an alternative form as follows: the product $V \times N$ gives the total number of fissile nuclei, and upon division by the Avogadro number (0.602×10^{24}) and multiplication by the atomic weight (235) of the fissile substance the result is the mass, g, of the latter in grams. Thus,

$$\text{Mass of fissile material} = g = \frac{235\,VN}{0.602 \times 10^{24}} \text{ grams.}$$

From equation (2.49), it follows that

$$P = \frac{NV\sigma_f\phi}{3.1 \times 10^{10}}$$

$$= 8.3 \times 10^{10}g\sigma_f\phi \text{ watts.} \qquad (2.50)$$

If the mass of uranium-235 is expressed as w lb, then equation (2.50) becomes

$$P = 3.7 \times 10^{13} w \sigma_f \phi \text{ watts.*} \tag{2.51}$$

Since the results are given to two significant figures only, they will be equally applicable when uranium-233 or plutonium-239 is the fissile material.

Example 2.9. A water-moderated reactor contains 52,000 lb of uranium dioxide (UO_2) enriched to the extent of 3.4 per cent in uranium-235; the atomic ratio U^{235}/H is 0.005. What is the average thermal neutron flux for a thermal power of 390 Mw with the average moderator temperature of 510°F (265°C)? Structural and other materials contribute 1 barn per H atom of absorption to the system.

From Fig. 2.17, the thermal average absorption cross section of uranium-235 at 293°K, with $U^{235}/H = 0.005$ and 1 barn of additional absorber, is estimated to be 460 barns. The ratio of fission to absorption cross sections may be taken to be the same as in Table 2.8 for pure uranium-235; hence, at 293°K,

$$\sigma_f(293°) = (460)(577/683) = 389 \text{ barns.}$$

At the reactor moderator temperature of 265°C, i.e., 538°K, the value is

$$\sigma_f(538°) = (389)(293/538)^{1/2} = 287 \text{ barns.}$$

The weight of uranium-235 in the reactor is given by

$$w = (5.2 \times 10^4)(238/270)(0.034) = 1.56 \times 10^3 \text{ lb.}$$

Hence, from equation (2.51), with P equal to 390×10^6, i.e., 3.9×10^8 watts, it follows that

$$\phi_{\text{th}} = \frac{3.9 \times 10^8}{(3.7 \times 10^{-11})(1.56 \times 10^3)(287)}$$

$$= 2.4 \times 10^{13} \text{ neutrons/(cm}^2)(\text{sec}).$$

FISSION NEUTRONS AND GAMMA RAYS

2.164. The neutrons released in fission can be divided into two categories, namely, *prompt neutrons* and *delayed neutrons.* The former, which constitute more than 99 per cent of the total fission neutrons, are released within 10^{-14} sec (or less) of the instant of fission. The emission of the prompt neutrons ceases immediately after fission has occurred, but the delayed neutrons continue to be

TABLE 2.9. AVERAGE NUMBER OF NEUTRONS LIBERATED PER
NEUTRON ABSORBED IN FISSION

Fissile Nucleus	ν (thermal)	$d\nu/dE_n$ (Mev^{-1})
Uranium-233.........	2.50	0.115
Uranium-235	2.43	0.135
Plutonium-239.......	2.90	0.111

* It should be noted that in this equation σ_f is expressed in cm^2 per nucleus, if the value is in barns, then the corresponding equation would be $P = 3.7 \times 10^{-11} w \sigma_f \phi$ watts.

expelled from the fission fragments over a period of a few hours, their intensity falling off rapidly with time.   The average (total) numbers of neutrons, ν, liberated for each thermal neutron absorbed in a fission reaction are given in Table 2.9; the coefficients in the last column show the rate of increase of ν with the energy, E_n Mev, of the neutron causing fission [12].   It will be noted that the values of ν are not integers; this is because the excited compound nucleus splits in many different ways.   Although the number of neutrons expelled in any particular act of fission must be an integer, the average is not necessarily a whole number.

2.165. The release of the prompt neutrons probably takes place somewhat in the following manner.   The excited compound nucleus formed by the capture of a neutron first breaks up into two nuclear fragments, each of which has too many neutrons for stability, as well as the excess (excitation) energy of about 6 to 8 Mev required for the expulsion of a neutron.   The excited, unstable nuclear fragment frequently expels one or more neutrons—the prompt neutrons—within a very short time of its formation.   Some of the instantaneous gamma rays accompanying fission are apparently emitted at the same time.

2.166. The energy spectrum of the prompt fission neutrons has been studied

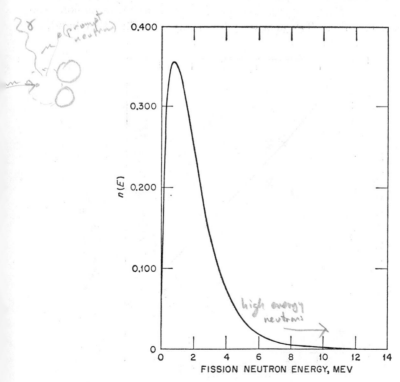

FIG. 2.23.   Fission neutron energy spectrum

from about 0.075 to 17 Mev, and it has been found that the results can be fitted quite adequately by the formula   *for U^{235}*

number of neutrons of energy E

$$n(E) = \sqrt{\frac{2}{\pi e}} \sinh \sqrt{2E}e^{-E} = 0.484 \sinh \sqrt{2E}e^{-E}, \qquad (2.52)$$

where $n(E)$ is the number of neutrons with energy E Mev per unit energy (Mev) interval, for each neutron emitted [13]. This expression applies to fission of both uranium-235 and plutonium-239; a similar equation, but with slightly different constants, is applicable to uranium-233. In Fig. 2.23 the values of $n(E)$ given by equation (2.52) are plotted as a function of E. It is seen that, although most of the prompt neutrons have energies between 1 and 2 Mev, there are nevertheless some with energies in excess of 10 Mev. These high-energy neutrons must be taken into consideration in the design of reactor shielding.

2.167. The gamma radiation accompanying fission is sometimes classified as prompt and delayed, as with neutron emission. The prompt gamma rays are generally defined, somewhat arbitrarily, as those produced within 0.1 micro-second of fission. They consist partly of the radiation emitted at about the same time as the prompt neutrons and partly of gamma rays from fission prod-ucts of very short half-life. The delayed gamma rays, on the other hand, are

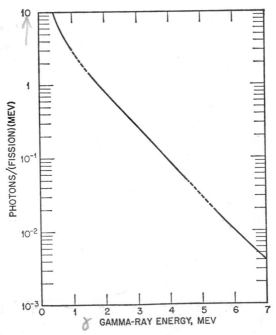

FIG. 2.24. Prompt fission gamma-ray energy spectrum

those given off over an extended period by the fission products still remaining after 0.1 microsecond.

2.168. The energy distribution of the prompt gamma rays accompanying fission is shown in Fig. 2.24 [14]. The spectrum is approximately exponential in the range from 0.2 to 4 Mev only, and can be represented within these limits by

$$n(E) = 10e^{-1.15E},$$

where $n(E)$ is the number of gamma-ray photons of energy E Mev per unit energy (Mev) interval. The total energy of the prompt gamma rays is about 7 Mev and the average energy per photon is roughly 1 Mev.

DELAYED NEUTRONS

2.169. Experimental studies of the rate of emission of the delayed neutrons have shown that these neutrons fall into six groups, each characterized by a definite exponential decay rate. It is thus possible to associate a specific half-life with each group. The average numbers of delayed neutrons per fission, in each group, and the total fraction of fission neutrons that are delayed are presented in Table 2.10 for the thermal fission of uranium-233, uranium-235,

TABLE 2.10.   CHARACTERISTICS OF DELAYED FISSION NEUTRONS IN THERMAL FISSION

Approximate Half-life (sec)	Number of Fission Neutrons Delayed per Fission			
	U-233	U-235	Pu-239	Energy (Mev)
56.....................	5.7×10^{-4}	5.2×10^{-4}	2.1×10^{-4}	0.25
23.....................	19.7	34.6	18.2	0.46
6.2....................	16.6	31.0	12.9	0.41
2.3....................	18.4	62.4	19.9	0.45
0.61...................	3.4	18.2	5.2	0.41
0.23...................	2.2	6.6	2.7	
Total Delayed...........	0.0066	0.0158	0.0061	
Total Fission Neutrons...	2.50	2.43	2.90	
Fraction Delayed........	0.0026	0.0065	0.0020	

and plutonium-239 [15]. The corresponding data for fission by moderately fast neutrons are not appreciably different. It should be noted that the approximate energies of the delayed neutrons, given in the last column of the table, are lower than for the great majority of prompt neutrons.

2.170. Most of the neutron-rich fission products undergo beta decay (§ 1.42); in a few cases, however, the daughter is produced in an excited state with sufficient energy to make possible the emission of a neutron. It is in this manner that the delayed neutrons arise, the characteristic half-life being determined by that of the parent (or *precursor*) of the actual neutron emitter. The 56-sec and 23-sec groups have been definitely associated with bromine-87 and iodine-137 as precursors, respectively, and krypton-87 and xenon-137 as the corresponding emitters. It is of interest to note that, in each case, the number of neutrons in the nucleus of the emitting species exceeds a magic number by unity, viz., 51 and 83, respectively. This means that the last neutron has a low binding energy and is, consequently, easily emitted.

2.171. The mechanism proposed to account for the 56-sec group of delayed neutrons is shown in Fig. 2.25. As the fission product bromine-87 continues

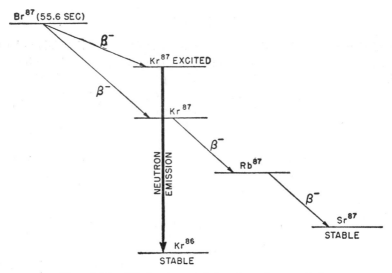

FIG. 2.25. Mechanism of delayed-neutron emission

to decay after fission, it produces krypton-87 in an excited state with sufficient excess energy (about 6 Mev) to permit the loosely bound neutron to be expelled. The observed rate of emission of the delayed neutrons is determined by the rate of formation of the neutron emitter, krypton-87, and this is dependent on the rate of decay of the precursor, bromine-87.

2.172. The precursor of the 23-sec delayed neutron group is probably iodine-137; this has a half-life of 22.0 sec, forming excited xenon-137 which instantaneously expels a neutron. Here again, the emission of the neutron is delayed because the rate of formation of the xenon-137, from which it originates, depends on the rate of decay of the precursor, iodine-137. The other five groups of delayed neutrons are probably produced in an analogous manner, although the precursors in these cases have not been definitely identified.

FISSION PRODUCTS

2.173. A detailed study of the slow-neutron fission of uranium-235 has shown that the compound nucleus splits up in more than 40 different ways, yielding over 80 primary fission products (or fission fragments). The range of mass numbers of the products is from 72, probably an isotope of zinc (atomic number 30) to 160, possibly an isotope of terbium (atomic number 65). In Fig. 2.26

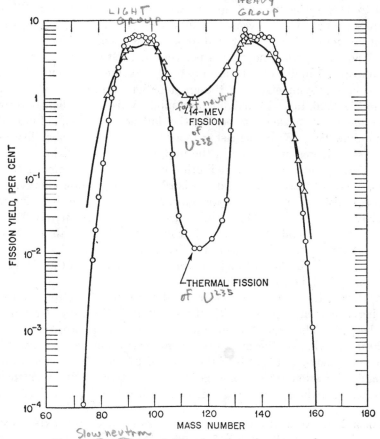

Fig. 2.26. Fission yield as function of mass number

the mass numbers of the products of thermal fission of uranium-235 and of the fast-neutron fission of uranium-238 are plotted against the corresponding *fission yields;* the fission yield is defined as the proportion (or percentage) of the total nuclear fissions that form products of a given mass number. Since the observed fission yields range from 10^{-5} to over 6 per cent, they are plotted on a logarithmic scale [16]. It should be noted that, as two nuclei result from each act of fission, the total yield for all mass numbers adds up to 200 per cent. Incidentally, the reason why mass numbers, rather than atomic numbers, are considered is that most of the fission fragments are radioactive, decaying by the loss of a negative-beta particle. The atomic numbers, consequently, change with time, but the mass numbers are unaffected by beta decay (§ 1.27).

2.174. An examination of Fig. 2.26 shows that the masses of nearly all the fission products fall into two broad groups, a "light" group, with mass numbers from 80 to 110, and a "heavy" group, with mass numbers from 125 to 155. There are some products between and outside these ranges, but altogether they represent no more than a few per cent or so of the fissions. The most probable type of fission, comprising nearly 6.4 per cent of the total, gives products with mass numbers 95 and 139. It is apparent that the slow-neutron fission of uranium-235 is unsymmetrical in the great majority of cases. Curves similar to that in Fig. 2.26, but with maxima and minima displaced slightly from those for uranium-235, have been obtained for the slow-neutron fission of uranium-233 and plutonium-239. With increasing neutron energy the probability of symmetrical fission increases, as seen in the upper curve of Fig. 2.26.

2.175. Corresponding to the distribution of mass numbers among the fission products, there has been observed a distribution of kinetic energy. Consequently, two distinct kinetic-energy groups, analogous to the two mass-number groups, have been detected; the kinetic energies are approximately 67 Mev for the most abundant member of the heavy group and 98 Mev for the corresponding one in the light group. The ratio of 98 to 67 is about 1.46, and this is very close, as it should be if momentum is conserved, to the ratio of the mass numbers for maximum yield, i.e., 139 to 95 or 1.46.

2.176. During the fission process many of the orbital electrons of the atom undergoing fission are ejected, with the result that the fission fragments are highly charged. The lighter fragments carry an average positive charge of about 20 units, whereas the heavy fragments carry some 22 positive charges. Such particles, moving at speeds of the order of 10^9 cm/sec, are able to produce considerable ionization in their passage through matter. Because of their large mass and charge, the specific ionization is high, and their range is, therefore,

TABLE 2.11.   APPROXIMATE RANGES OF GROSS FISSION FRAGMENTS FROM
THERMAL FISSION OF URANIUM-235

Material	Range $(10^{-3}$ cm$)$	Areal Density (mg/cm^2)
Aluminum	1.4	3.7
Copper	0.59	5.2
Silver	0.53	6.1
Gold	0.59	11.1
Uranium	0.66	12.6
Uranium oxide (U_3O_8)	1.4	10.0

relatively short (§ 2.28).   Thus, the ranges of the light and heavy groups of
fission fragments have been found to be about 2.5 cm (1 in.) and 1.9 cm (0.8 in.),
respectively, in air.   These are similar to the ranges of alpha particles from
radioactive sources.

2.177. The range of the fission fragments in various materials is important
for reactor design, since it is necessary to prevent their escape from the fuel
element.   The actual ranges in a number of materials and the corresponding
thickness (or areal) densities are given in Table 2.11 [17].   As a working
approximation, the range of fission fragments in any medium may be taken to be
the same as that of 4-Mev alpha particles.

RADIOACTIVITY AND DECAY OF FISSION PRODUCTS

2.178. Because they have neutron/proton ratios that are above the stability
range, nearly all (if not all) of the fission fragments are negative-beta emitters.
This may be seen from Fig. 2.27, which shows the relationship between the
fission fragments and the "stability curve" for nuclei (see also Fig. 1.2).   The
immediate decay products are also usually radioactive and, although some decay

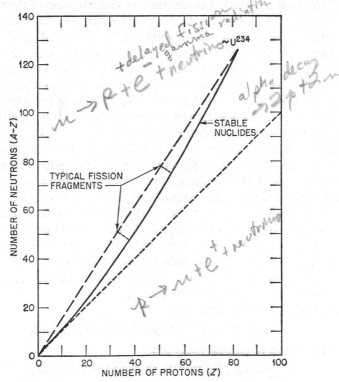

Fig. 2.27.   Instability of fission fragments

chains are longer and some shorter, each fragment is followed, on the average, by three stages of decay before a stable species is formed.   Since there are some 80 different radioisotopes, produced in fission and each is, on the average, the precursor of two others, there are over 200 radioactive species present among the fission products after a short time.

2.179. A large proportion of the radioactive fission products emit gamma rays, in addition to beta particles.   These represent the so-called delayed fission gamma radiation.   Most of the photons are of moderate energy, less than about 2 Mev, but a few of the fission products expel photons of higher energy.   The latter are of interest for some shielding problems and for reactor control.   The total energy of the delayed gamma radiation amounts to approximately 7 Mev per fission.

2.180. From the nuclear reactor standpoint, there are two important aspects of the radioactivity of fission products requiring consideration.   The first is to estimate the rate of energy release (as heat) from the fission product mixture due to the beta particles and gamma rays.   Even after a reactor has been shut down, large amounts of heat continue to be developed in the fuel because of the presence of fission products.   Steps must be taken to ensure removal of this heat, especially if the shutdown is a consequence of interruption of the reactor coolant flow; otherwise the fuel elements may suffer damage.   The second aspect of fission product radioactivity is concerned with the handling of the spent fuel after its removal from a reactor.   Initially, the activity is so high that it represents a serious hazard to personnel; furthermore, the beta and gamma radiations can bring about chemical decomposition of the fuel-processing solutions.   The fuel must therefore be set aside for a "cooling" period to permit the fission product activity to decay to a sufficient extent to make further treatment possible.

2.181. In principle, it is possible to express the rate of decay of the complex fission product mixture in terms of the fission yields and radioactive decay constants of the various nuclides present.   But this is quite impractical because of the complexity of the resulting expression.   Fortunately, a much simpler, if somewhat approximate, approach to the problem is available.   The half-lives of the fission products range from a small fraction of a second up to a million years.   Nevertheless, it has been found that the rate of emission of beta particles and of gamma-ray photons can be stated by means of a simple empirical expression which is probably accurate within a factor of two or less [18].   From about 10 sec up to several weeks after fission has taken place, these rates (per fission) are given approximately as follows:

Rate of emission of beta radiation

$$\approx 3.8 \times 10^{-6}t^{-1.2} \text{ particles/(sec)(fission)}, \tag{2.53}$$

Rate of emission of gamma radiation

$$\approx 1.9 \times 10^{-6}t^{-1.2} \text{ photons/(sec)(fission)}, \tag{2.54}$$

where t is the time after the fission event in *days*.   The mean energy of the beta particles from the fission products is about 0.4 Mev and that of the gamma-ray photons is about 0.7 Mev, so that

Rate of emission of beta and gamma energy ——— *days*

$$\approx 2.8 \times 10^{-6} t^{-1.2} \text{ Mev/(sec)(fission)}. \tag{2.55}$$

2.182. Suppose a reactor has been operating at a constant power of P_0 watts for a period of T_0 days, and that it is required to determine the rate of emission of beta and gamma energy from the fission products after shutdown. Consider a small interval dT at a time T days after startup (Fig. 2.28). The rate of

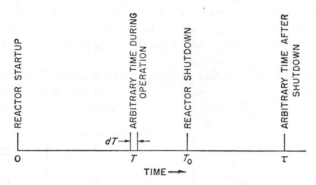

Fig. 2.28. Calculation of fission product decay power

emission of beta and gamma energy, observed at a time τ with reference to the initial startup, due to the particular fissions occurring during the interval dT is then given by equation (2.55) as $2.8 \times 10^{-6} (\tau - T)^{-1.2}$ Mev/(sec)(fission). Since 3.1×10^{10} fissions per sec or 2.68×10^{15} fissions per day yield 1 watt of power, the number of fissions occurring in the reactor (operating at P_0 watts) during the time interval dT days is $2.68 \times 10^{15} P_0 \, dT$. It follows, therefore, that

Rate of emission of beta and gamma energy at time τ, due to fissions in interval dT

$$\approx (2.8 \times 10^{-6})(2.68 \times 10^{15}) P_0 (\tau - T)^{-1.2} \, dT \text{ Mev/sec}$$

$$\approx 7.5 \times 10^{9} P_0 (\tau - T)^{-1.2} \, dT \text{ Mev/sec}.$$

2.183. Integration of this relation over the whole irradiation period, i.e., from zero time to T_0, gives the total beta and gamma energy emission rate from all fission products created during this period; thus, within a factor of two or so, the rate of emission of beta and gamma energy at time τ

$$\approx 7.5 \times 10^{9} P_0 \int_{T_0}^{0} (\tau - T)^{-1.2} dT$$

$$\approx 3.8 \times 10^{10} P_0 [(\tau - T_0)^{-0.2} - \tau^{-0.2}] \text{ Mev/sec}.$$

2.184. The rate of emission of beta and gamma energy is often called the *decay heat power*, since the decay energy appears in the form of heat. If this power is expressed in watts, by using the conversion factor 1 Mev $= 1.6 \times 10^{-13}$ watt-sec (§ 1.45), and is represented by P, it follows that

$$\frac{P}{P_0} = 6.1 \times 10^{-3}[(\tau - T_0)^{-0.2} - \tau^{-0.2}], \qquad (2.56)$$

where P and P_0 can be expressed in any convenient, but identical, units, since the ratio is dimensionless. In equation (2.56), $\tau - T_0$ is the time in days after shutdown, i.e., the cooling period. For long irradiation periods ($>$ about 1 year) and short cooling times ($<$ about 10 days), the approximate equation (2.56) can be simplified by neglecting the second term in the brackets. The rate of release of the decay energy is then dependent only on the cooling time, $\tau - T_0$, and not on the operating time.

2.185. The results of equation (2.56) can be represented in graphical form, as in Fig. 2.29. This gives the ratio of the beta and gamma decay heat power to the reactor operating power (P/P_0) as a function of time after shutdown

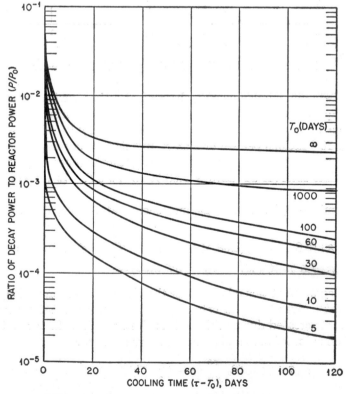

Fig. 2.29. Decay power as function of time after shutdown

for several operating periods. Since the curves represent the decay heat power after shutdown as a fraction of the reactor power during operation, the actual rate of energy release is obtained upon multiplying by P_0, the reactor operating power. It will be observed from the figure that the decay power decreases very rapidly in the period immediately following shutdown. Furthermore, as expected, the results are then essentially independent of the operating period (or irradiation time) of the reactor, provided this exceeds a week or so.

2.186. Although the times in equation (2.56) are in units of days, this equation (or the equivalent Fig. 2.29) can be used to determine the rate of decay heat generation in reactor fuel for periods as short as 10 sec after shutdown. However, a somewhat more accurate representation of the decay power, especially for short cooling times, is given by the empirical equation

$$\frac{P}{P_0} = 0.1[(\tau - T_0 + 10)^{-0.2} - 0.87(\tau - T_0 + 2 \times 10^7)^{-0.2}]$$
$$- 0.1[(\tau + 10)^{-0.2} - 0.87(\tau + 2 \times 10^7)^{-0.2}], \qquad (2.57)$$

where all times are now in *seconds* [19]. This expression, which is believed to be accurate within ±50 per cent for periods of 1 to 100 sec, includes an allowance for the heat produced by the beta decay of uranium-239 and neptunium-239 resulting from radiative capture of neutrons by uranium-238.

2.187. Some values of decay heat power, as a fraction of the operating power, calculated from equation (2.57) are quoted in Table 2.12. It is seen that at 10

TABLE 2.12. RATE OF PRODUCTION OF DECAY HEAT AFTER SHUTDOWN

	Time After Shutdown				
	1 sec	10 sec	1 min	1 hour	1 day
Operating Time	Fraction of Reactor Operating Power				
1 week.........	0.055	0.048	0.036	0.012	0.0035
30 days........	0.057	0.050	0.038	0.014	0.0050
1 year.........	0.058	0.051	0.039	0.016	0.0066
Infinite........	0.059	0.052	0.040	0.016	0.0073

sec after shutdown the heat liberated by the decay of fission products can be as large as 5 per cent of the power of the reactor during operation. Reactors cooled by liquid metal or gas are particularly vulnerable to damage from decay heat in the event of a failure in the coolant pumping system. In water-cooled reactors, boiling of the water acts as a partial safeguard, at least. It will be noted, too, from Table 2.12, that for operating times of a week or more, the decay power fraction is essentially independent of the operating period, for shutdown times up to almost an hour. At 1 day after shutdown, however, differences become apparent.

2.188. In connection with the handling of spent fuel after removal from a reactor, it is sometimes desirable to know the activity of the fission products in curies (§ 2.15). An approximate estimate can be made from equation (2.53), giving the rate of emission of beta particles, by utilizing the fact that every particle corresponds to one nuclear disintegration. Since 1 curie is equivalent to 3.7×10^{10} dis per sec, it follows that

Beta activity of fission products $\approx 1.03 \times 10^{-16} t^{-1.2}$ curie/fission,

where t is the time in days after fission. By means of arguments similar to those in the preceding paragraphs, it is found that, for a reactor which has been operating at a power of P_0 watts for T_0 days,

Beta activity of fission products at time τ
$$\approx 1.4 P_0 [(\tau - T_0)^{0.2} - \tau^{-0.2}] \text{ curies.} \qquad (2.58)$$

It is seen that if a reactor has been operating for some time at a power of the order of megawatts, the beta activity of the fission products will, soon after shutdown, be of the order of megacuries. Since equations (2.56) and (2.58) are similar in form but differ in the numerical factors, Fig. 2.29 can be used to give the activity in beta-curies; all that is necessary is to multiply the total decay power in watts by a factor of $1.4/(6.1 \times 10^{-3})$, i.e., by 230. Comparison of equations (2.53) and (2.54) shows that the gamma activity, expressed in gamma-curies, is half that of the beta activity.

ACCUMULATION OF SPECIFIC FISSION PRODUCTS

2.189. For certain purposes, e.g., the study of reactor poisoning by fission products (see Chapter 5) or the extraction of particular products, it is required to know how specific fission products accumulate during reactor operation and decay after shutdown. In the reactor a radioactive fission fragment A will be formed directly by fission and will decay to form the product B; the latter will decay in turn to yield C, and so on; thus,

$$\text{Fission} \rightarrow A \xrightarrow{\beta^-} B \xrightarrow{\beta^-} C \xrightarrow{\beta^-} \cdots$$

In addition to undergoing radioactive decay, the substances A, B, C, etc., may capture neutrons while the reactor is operating. By taking into account the various processes, it is possible to set up differential equations for the rate of change of concentration of the specific products, A, B, C, etc., and these can be solved relatively simply in certain cases. In the following treatment it is supposed that the neutrons either are monoenergetic or are in thermal equilibrium with their environment. Alternatively, it may be assumed that proper average values for the cross sections and neutron flux are used.

2.190. The concentration of the fission fragment A at any time may be obtained from the general balance equation

$$\begin{bmatrix} \text{Net rate of accumu-} \\ \text{lation of A per unit} \\ \text{volume} \end{bmatrix} = \begin{bmatrix} \text{Rate of formation of} \\ \text{A per unit volume} \\ \text{by fission} \end{bmatrix} - \begin{bmatrix} \text{Rate of loss of A per unit} \\ \text{volume by } (a) \text{ radioactive} \\ \text{decay, } (b) \text{ neutron cap-} \\ \text{ture.} \end{bmatrix}$$

If γ_A is the fission yield of A, expressed as a fraction of the total number of fissions in which the nuclide A is produced, then the rate of formation of A by fission is equal to $\gamma_A \Sigma_f \phi$ nuclei/(cm³)(sec), where $\Sigma_f \phi$ is the number of fissions/(cm³)(sec). If A is the number of nuclei/cm³ of A at any instant, and λ_A is its radioactive decay constant, rate of loss by radioactive decay will be $\lambda_A A$ nuclei/(cm³)(sec). In addition, A nuclei will be removed from the system as a result of neutron capture at a rate equal to $\sigma_A \phi A$ nuclei/(cm³)(sec), where σ_A is the capture cross section. The balance equation can thus be written as

$$\frac{dA}{dT} = \gamma_A \Sigma_f \phi - \lambda_A A - \sigma_A \phi A \qquad (2.59)$$

for the net rate of increase of A nuclei per unit volume with time T in the reactor.

2.191. An expression for $A(T)$, the concentration of A nuclei at any time T, can be obtained by rearranging equation (2.59) as

$$\frac{dA}{dT} + (\lambda_A + \sigma_A \phi)A = \gamma_A \Sigma_f \phi,$$

or

$$\frac{dA}{dT} + \lambda_A^* A = \gamma_A \Sigma_f \phi, \qquad (2.60)$$

where λ_A^*, defined by

$$\lambda_A^* \equiv \lambda_A + \sigma_A \phi, \qquad (2.61)$$

may be regarded as the effective removal (decay and capture) constant of A in the reactor, provided the neutron flux remains constant. The equilibrium amount of A, which is obtained by setting dA/dT in equation (2.60) equal to zero, is seen to depend on the neutron flux in the reactor. By multiplying both sides of equation (2.60) by $e^{\lambda_A^* T}$, this equation may be solved in a manner analogous to that used for equation (2.6). If $\Sigma_f \phi$ is constant, the result obtained upon integration is

$$A(t) = \frac{\gamma_A \Sigma_f \phi}{\lambda_A^*} (1 - e^{-\lambda_A^* T}) + A(0)e^{-\lambda_A^* T}, \qquad (2.62)$$

where $A(T)$ and $A(0)$ represent the concentration of A at times T and zero, respectively, in the reactor.

2.192. The general procedure described above can be extended to determine the concentration at any time of the decay products of A, when one or more members of the decay chain may be subject to simultaneous removal by neutron capture. In writing the balance equation for such decay products, e.g., B, a term for the rate of its formation by radioactive decay, e.g., from A, must be included, in addition to the rate of formation, if any, directly by fission.   The relations obtained in this manner, which are sometimes quite complicated, may be found in the literature [20].

SYMBOLS USED IN CHAPTER 2

A	mass number
$A(t)$, $A(T)$	activity (or number of nuclei) at time t, T
A_0	saturation activity
E, E'	energy (in ev or Mev)
E_m	maximum energy of beta particles
E_γ	gamma-ray energy
g	weight in grams
$g(T)$	correction for departure from $1/v$ behavior at temperature T
I	radiation (or neutron) intensity
k	Boltzmann constant (8.6×10^{-5} ev/°C or 4.8×10^{-5} ev/°F)
N	number of nuclei or atoms (or per cm³)
N_a	Avogadro number (0.602×10^{24})
n	neutron
n	neutron density (neutrons/cm³)
$n(E)$	number of neutrons of energy E per unit energy interval
nv	neutron flux (neutrons/(cm²)(sec))
P	reactor power (watts)
R	range of particles
R	nuclear radius
T	time of foil exposure
T	time after reactor startup
T	time in reactor
T_0	reactor operating period
T	absolute temperature
T_m	moderator temperature
T_n	effective neutron temperature
T_K	absolute temperature on Kelvin scale
T_R	absolute temperature on Rankine scale
t	time
t	time after fission
t	time after shutdown or after removal from reactor
$t_{1/2}$	half-life of radioisotope
t_m	mean life of radioisotope
V	volume
v	velocity (cm/sec)
w	weight in pounds
x	thickness of absorber

x	distance
Z	atomic number
α	alpha particle
γ	gamma-ray photon
ϕ	neutron flux (neutrons/(cm^2)(sec))
λ	radioactive decay constant
λ^*	effective decay constant $(\lambda + \sigma\phi)$
λ	mean free path
μ	linear attenuation coefficient (cm^{-1})
ν	number of atoms of given kind in molecule
ν	fission neutrons per fission (average)
ρ	density (g/cm^3)
σ	microscopic cross section (cm^2/nucleus or barns)
σ_a	absorption cross section (fission + capture)
σ_f	fission cross section
σ_s	scattering cross section
σ_t	total cross section
$\bar{\sigma}_{\text{th}}(T)$	average (Maxwellian) thermal neutron cross section at temperature T
$\sigma(kT),\ \sigma(T)$	cross section of neutrons having energy kT
Σ	macroscopic cross section (cm^{-1})
$\Sigma(E)$	macroscopic cross section at energy E
Σ_a	macroscopic absorption cross section
Σ_f	macroscopic fission cross section
Σ_s	macroscopic scattering cross section
Σ_t	macroscopic total cross section
θ	Compton scattering angle
τ	time during reactor shutdown (with reference to startup time)

REFERENCES FOR CHAPTER 2

1. O. Klein and Y. Nishina, Z. *Physik*, **52**, 853 (1929); see also W. Heitler, "The Quantum Theory of Radiation," Oxford University Press, Inc., New York, 3rd Ed., 1954, pp. 215 *et seq.*
2. Adapted from G. W. Grodstein, *NBS Circular 583* (1957).
3. R. R. Coveyou, R. R. Bate, and R. K. Osborn, *J. Nuc. Energy*, **2**, 153 (1956); M. J. Poole, *ibid.*, **5**, 325 (1957); for discussion, see A. M. Weinberg and E. P. Wigner, "The Physical Theory of Neutron Chain Reactions," University of Chicago Press, 1958, pp. 332 *et seq.*
4. C. H. Westcott, AECL (Canada) Report-670 (1958); C. H. Westcott and D. A. Ray, AECL (Canada) Report-869 (1959).
5. C. H. Westcott, *Nucleonics*, **16**, No. 10, 108 (1958).
6. E. P. Wigner and J. E. Wilkins, U. S. AEC Report AECD-2275 (1944); see also, H. J. Amster, *Nuc. Sci. and Eng.*, **2**, 394 (1957).
7. H. J. Amster, U. S. AEC Report WAPD-185 (1958).

8. G. Breit and E. P. Wigner, *Phys. Rev.*, **49**, 519 (1936); see also, A. M. Weinberg and E. P. Wigner, Ref. 3, Ch. III.
9. A. M. Weinberg and E. P. Wigner, Ref. 3, pp. 161 *et seq.*
10. U. S. AEC Report BNL-325 (1958) and subsequent revisions.
11. N. Bohr and J. A. Wheeler, *Phys. Rev.*, **56**, 426 (1939); S. Frankel and N. Metropolis, *ibid.*, **72**, 914 (1947).
12. "Reactor Physics Constants," 2nd Ed., U. S. AEC Report ANL-5800 (1962), Ch. 1.
13. B. E. Watt, *Phys. Rev.*, **87**, 1037 (1952); L. Cranberg, *ibid.*, **103**, 662 (1956).
14. F. C. Maienschein, *et al.*, *Proc. Second U. N. Conf. Geneva*, **15**, 366 (1958).
15. G. R. Keepin, T. F. Wimett, and R. Ziegler, *Phys. Rev.*, **107**, 1044 (1957); *J. Nuc. Energy*, **6**, 1 (1957).
16. For review, see S. Katcoff, *Nucleonics*, **18**, No. 11, 201 (1960).
17. E. Segrè and C. Wiegand, *Phys. Rev.*, **70**, 808 (1946); C. B. Fulmer, *ibid.*, **108**, 1113 (1957).
18. K. Way and E. P. Wigner, *Phys. Rev.*, **73**, 1318 (1948); see also J. F. Perkins and R. W. King, *Nuc. Sci. and Eng.*, **3**, 726 (1958); K. Shure, *Bettis Tech. Review, Reactor Technology*, U. S. AEC Report WAPD-PT-24 (Dec. 1961), p. 1.
19. S. Untermyer and J. T. Weills, U. S. AEC Report ANL-4790 (1952).
20. See, for example, M. Benedict and T. H. Pigford, "Nuclear Chemical Engineering," McGraw-Hill Book Co., Inc., New York, 1957, pp. 49 *et seq.*; H. Etherington (Ed.), "Nuclear Engineering Handbook," McGraw-Hill Book Co., Inc., New York, 1958, p. 1-38.

PROBLEMS

1. The radioactive decay constant of antimony-124 is given as 0.0115 day⁻¹. How many years would be required for the activity of this isotope to decay to 0.1 per cent of the initial value? How many half-lives does this represent?

2. A sample of radioactive iodine has an activity of 1 mc, due to iodine-126, half-life 13.3 days. What weight of this radioisotope is present?

3. Prior to 1951, all carbon compounds in nature emitted beta particles at an average rate of 15.3 per min per gram of carbon, due to the presence of carbon-14, with a half-life of 5760 years. (The rate has increased since 1951 as a result of the testing of thermonuclear weapons.) Determine the weight percentage of carbon-14 present in the carbon in nature.

4. Barium-140, a fission product (half-life 12.8 days), decays to lanthanum-140 (half-life 40.5 hr), and the latter forms stable cerium-140. Starting with 0.001 g of freshly purified barium-140, plot (a) the amount of each isotope and (b) the total activity in curies, as functions of time. (Be sure to extend the plots beyond any maxima in the curves.)

5. Gold consists of 100 per cent of the isotope Au^{197} which captures thermal neutrons to form radioactive Au^{198} (half-life 2.7 days); the activation cross section for 2240-meters/sec neutrons is 96 barns. A gold foil, weighing 0.05 g, is exposed for 30 min in a reactor and 2 hr after removal is found to emit beta particles at a rate of 2×10^4 per min. Calculate the flux of thermal neutrons in the reactor at 20°C.

6. If the fission product cesium-141 has an atomic mass of 140.9472 and that of its beta-decay product (barium-141) is 140.9438, estimate the approximate range of the beta particles, emitted by the former, in (a) air, (b) concrete of density 2.7 g/cm³, assuming there are no gamma rays.

7. The effect of a material in absorbing gamma rays is sometimes defined in terms of

the "tenth-value thickness," i.e., the thickness of the absorber which will decrease the radiation to one-tenth of its original value. Derive a general expression relating the tenth-value thickness for gamma rays to the linear attenuation coefficient of a material. What thickness of concrete of density 3.5 g/cm^3 would be required to attenuate 4-Mev gamma radiation by a factor of 10^6?

8. In traversing a thickness of 3.5 cm of material of atomic weight 122 and density 6.6 g/cm^3, the intensity of a narrow collimated beam of neutrons is decreased to 0.1 of its initial value. Calculate the total microscopic cross section in barns and estimate the absorption cross section.

9. A bismuth (density 10 g/cm^3) "window" is used in some experimental reactors to attenuate the gamma radiation without appreciably affecting the thermal neutrons. Compare the ratio of 4-Mev gamma rays to thermal neutrons incident upon an 8.5-in. thick window to the emergent ratio.

10. The most probable mode of fission of uranium-235 leads to nuclei with mass numbers 95 and 139. Identify the particular isotopes formed, assuming that the neutron/proton ratio is the same in both fragments. By comparing the mass numbers with those of stable isotopes of the same elements, account for the beta activity of the fission products.

11. If neutrons can be regarded as a monatomic gas, estimate the "neutron gas pressure" (in atmospheres) equivalent to a thermal flux of 10^{12} neutrons/(cm^2)(sec) at 25°C (77°F). Note that at this temperature and 1 atm pressure, 1 g atomic weight of a monatomic gas would occupy 22.6 liters.

12. The capture cross section of sulfur for neutrons of 0.025 ev (actual) energy is 0.52 barn. Calculate the average cross section for thermal neutrons at 80°C, assuming the $1/v$ law to be obeyed.

13. Determine the effective thermal average absorption (microscopic) cross section of uranium-235 for neutrons completely thermalized in a moderator at 500°F.

14. The core of the SM-1 (Army Package Power) reactor contains approximately 22.50 kg of uranium-235, 0.019 kg of boron-10, stainless steel equivalent to 210 kg of iron, and 111 kg of water. What is the thermal average fission cross section of uranium-235 in this system at 68°F? Estimate the total macroscopic absorption cross section for thermal neutrons in the core at the same temperature. (The absorption cross section of boron-10 for 2200-meters/sec neutrons is 4×10^3 barns; the density is 2.2 g/cm^3.)

15. Estimate the average thermal neutron flux in the reactor of Problem 14 when operating at a thermal power of 10 megawatts and a moderator temperature of 440°F.

16. A reactor operates at 100 megawatts for 100 days and is then shut down. What is the rate of energy release due to fission-product decay after a cooling period of 30 days? If the reactor contains 20 tons of slightly enriched uranium, what will be the gamma activity in curies of a fuel element weighing 1.57 lb?

17. How long would it require after removal of a slightly enriched uranium fuel element from a reactor for 99 per cent of the uranium-239 still present to be converted to plutonium-239?

18. Cobalt-60, half-life 5.3 yr, is made by exposing a 100-g block of cobalt-59 (density 8.9 g/cm^3), capture cross section 36 barns, to thermal neutrons at a flux of 10^{12} neutrons/(cm^2)(sec). What length of exposure, in days, would be necessary to attain an activity of 1 curie? Determine the maximum (or equilibrium) activity of the cobalt-60 under the given conditions. The thermal-neutron absorption cross section of the latter may be taken to be 6 barns. (An excited state of cobalt-60 is formed to some extent, but as it decays to the ground state with a relatively short half-life, it may be ignored.)

19. Tellurium-135 (half-life 2 min) is formed in a fractional yield of 0.061 in the fission of uranium-235; this decays to iodine-135 (half-life 6.7 hr) and the latter decays to

xenon-135 (half-life 9.2 hr), which is also formed directly to the extent of 0.002 in fission. Trace the growth of iodine-135 and xenon-135 with time in a reactor operating at a thermal flux of 2×10^{14} neutrons/(cm^2)(sec) and determine the equilibrium amounts relative to the number of fissile nuclei present, the latter being assumed to remain essentially constant. The capture cross section for iodine-135 is 7 barns and that for xenon-135 is 2.8×10^6 barns. (Note that because of its relatively short half-life the tellurium-135 may be ignored and the assumption made that iodine-135 is produced directly to the extent of 0.061 in fission.)

Chapter 3

DIFFUSION AND SLOWING DOWN OF NEUTRONS

INTRODUCTION

NEUTRON BALANCE

3.1. For a nuclear reactor to be capable of maintaining a fission chain reaction, the volume or mass of fissile material must exceed a certain critical value; this value depends on a variety of conditions, as seen in § 1.50. The determination of critical size is based on a consideration of the conservation (or balance) of neutrons in the reactor system. On the one hand, neutrons are produced in fission reactions, whereas, on the other hand, they are lost as a result of escape (or *leakage*) from the system as well as by absorption in fission and in various nonfission processes. The general form of the neutron balance equation is then

$$\begin{bmatrix} \text{Net rate of gain of} \\ \text{neutrons per unit} \\ \text{volume} \end{bmatrix} = \begin{bmatrix} \text{Rate of production of} \\ \text{neutrons by fission per} \\ \text{unit volume} \end{bmatrix} - \begin{bmatrix} \text{Rate of loss of neutrons} \\ \text{per unit volume by } (a) \\ \text{leakage, } (b) \text{ absorption.} \end{bmatrix}$$

In simpler form, this may be written as

$$\frac{\partial n}{\partial t} = \text{Production} - \text{Leakage} - \text{Absorption}, \tag{3.1}$$

where n is the neutron density, i.e., neutrons per unit volume, and $\partial n/\partial t$ is its time rate of change; the terms "production," "leakage," and "absorption" refer to the rates of these processes per unit volume of the system. When the system is in a steady state, $\partial n/\partial t$ is zero, so that the steady-state neutron balance equation is

$$\text{Production} = \text{Absorption} + \text{Leakage}. \tag{3.2}$$

In the critical system, the rate of production of neutrons by fission just balances the rate of loss in various ways, and thus equation (3.2) is the most general form of the critical equation for a reactor.

3.2. The problem of obtaining the physical conditions for reactor criticality

109

thus reduces itself to that of deriving expressions for the respective rates, i.e., of production, absorption, and leakage, in equation (3.2). Both the rates of neutron absorption and of the production of neutrons by fission depend upon various cross sections which must be determined experimentally. Hence, if the nuclear properties of the neutron multiplying system are known, the calculation of the production and absorption terms does not present any special difficulties.

3.3. The determination of leakage, on the other hand, is not quite such a simple matter; it is of interest to note, however, that the problem is one of classical mechanics rather than of nuclear physics. Leakage arises from the fact that neutrons are in motion. Consequently more neutrons will, on the average, leave the reactor, where the neutron density (or concentration) is high, than will return to it from the surrounding space, where the density is low. In actual practice the neutron motion is affected by collisions with atomic nuclei which cause scattering to occur; this plays a major role in neutron diffusion upon which the calculation of leakage is based [1].

NEUTRON DIFFUSION

3.4. Neutrons undergo elastic scattering collisions with all nuclei and, as a result, a typical neutron trajectory consists of a number of straight path elements (Fig. 3.1) joining the points (nuclei) where the neutron suffered scattering

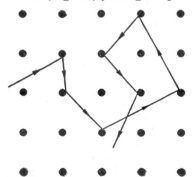

collisions. These are the scattering free paths of which the average is the scattering mean free path (cf. § 2.123). After a scattering collision the direction in which the neutron travels is not known exactly, but it can be given in terms of a probability distribution. However, when a large number of neutrons are considered, there is always a net movement of neutrons from regions of higher to those of lower neutron density. The rate at which this motion, referred to as *diffusion*, occurs can then be determined by means of simple and familiar laws, at least to a first approximation.

Fig. 3.1. Elastic scattering of neutrons by nuclei in a solid medium

3.5. In a reactor, the neutrons produced in fission are fast, and in an intermediate or thermal reactor they are subsequently slowed down by collisions with the nuclei of the moderator. During the slowing down process the neutrons may leak out of the system. Consequently, the complete neutron balance relations must involve both energy and space dependencies. After the neutrons have become thermalized, they will diffuse for a time until they are either absorbed or escape.   For purposes of theoretical treatment it is convenient to divide the problem into a number of parts.

3.6. The diffusion of monoenergetic neutrons, for which elastic scattering occurs without net energy loss, will be discussed first.   This represents a good approximation to the behavior of thermal neutrons in a weakly absorbing medium, where the Maxwell-Boltzmann distribution is maintained (§ 2.100).   Next a general outline of the mechanics of elastic collisions will be given, so that the energy loss suffered in a scattering collision may be determined.   Then the problem of slowing down in a medium in which fast neutrons are being produced continuously will be divided into two parts.   In the first place, consideration will be given to the distribution of neutrons as a function of energy, irrespective of their position.   Subsequently, the spatial distribution of the neutrons, as a result of diffusion during slowing down, will be considered.   By combining the results, the characteristic equation for a critical reactor will be derived in the next chapter.

THE DIFFUSION OF NEUTRONS

TRANSPORT THEORY

3.7. Neutron diffusion is, in some aspects, simpler than gaseous diffusion. Because of the extremely small density of neutrons in nuclear reactors, collisions between neutrons are rare.   Virtually all collisions leading to diffusion are conquently those between neutrons and the essentially stationary nuclei of the medium.   In this respect, therefore, the phenomenon of neutron diffusion bears a greater resemblance to the diffusion of electrons in a metal than to gaseous diffusion.   The rigorous treatment of neutron diffusion is, nevertheless, quite analogous to that used in the classical studies of gaseous diffusion.

3.8. The method of approach is to consider a small volume element located at a certain point in the system, and to derive expressions for the various ways in which neutrons having a given velocity vector enter and leave this volume element.   For the steady state, the algebraic sum of these expressions is equated to zero, and there is then obtained a complicated integro-differential equation, usually referred to as the (Boltzmann) transport equation.   Even if the detailed cross sections appearing in this equation were available, a complete solution would be prohibitively difficult and tedious in most cases.   Therefore, certain approximations are made which lead to considerable simplification.

3.9. A reactor is a complex system, involving fuel material, which may consist only partly of fissile nuclides, cladding of the fuel elements, coolant, structure, etc.   It is thus almost impossible to describe the nuclear and other properties completely.   Calculations cannot be relied upon, therefore, to give more than an approximate indication of the behavior of neutrons in such a system.

DIFFUSION THEORY APPROXIMATION

3.10. In the treatment of the transport equation, a simplification is achieved by supposing that all the neutrons have the same velocity (or energy) and that scattering collisions with nuclei do not involve any change of energy. These conditions apply, with a good degree of approximation, to neutrons in thermal equilibrium with a weakly absorbing medium, provided the cross sections used are properly averaged values (§ 2.131). In general, the neutron flux, which is defined as the product of the neutron density and velocity, is both spatially and velocity dependent. Since it is here postulated that the velocity is constant, the effect of spatial variation only need be considered.

3.11. In regions that are more than about two scattering mean free paths distant from neutron sources, strong absorbers, or boundaries, the distribution of the neutron velocity vectors is probably independent of direction. In these circumstances, the transport equation for monoenergetic neutrons, which are scattered without energy loss, reduces to the so-called *diffusion theory approximation*. The results are widely used in the theoretical treatment of thermal reactors.

3.12. In order to apply diffusion theory, it is necessary that there should be a gradient, i.e., a change with distance, of some function of the neutron density. The quantity whose gradient determines the net rate of neutron flow in a given direction, from higher to lower density, is taken to be the neutron density itself. The familiar Fick's law of diffusion, for monoenergetic neutrons, can consequently be written as

$$J_z = -D_0 \frac{\partial n}{\partial z},\qquad(3.3)$$

where n is the number of neutrons per unit volume, so that $\partial n/\partial z$ is the density gradient in the z direction; J_z is the *neutron current*, i.e., the net number of neutrons crossing unit surface per unit time, in the z direction, and D_0 is the appropriate *diffusion coefficient*. The latter is defined by equation (3.3) and has the dimensions of (length)(velocity) or (length)^2/time.

3.13. Since neutron diffusion in all directions must be taken into consideration, equation (3.3) is written in the general form

$$\mathbf{J} = -D_0 \operatorname{grad} n,\qquad(3.4)$$

where the neutron current vector $\mathbf{J}$ is the net number of neutrons flowing in a given direction in unit time through a unit area normal to the direction of flow. For the subsequent treatment, however, it is more convenient to use the neutron flux rather than the neutron density. Since the flux ϕ is equal to nv, where v is the (constant) neutron velocity, equation (3.4) can be modified to give

$$\mathbf{J} = -D \operatorname{grad} \phi,\qquad(3.5)$$

where D is equal to D_0/v.   The diffusion coefficient D for neutron flux, generally referred to merely as the *diffusion coefficient*, has the dimension of length; this follows immediately from the fact that $\mathbf{J}$ and ϕ have the same dimensions.*

3.14. If the conditions are such that the diffusion theory approximation is valid, the value of the diffusion coefficient is given by transport theory as

$$D = \frac{1}{3(\Sigma_t - \Sigma_s \bar{\mu}_0)}, \tag{3.6}$$

where Σ_t is the total macroscopic cross section of the particular medium for the monoenergetic neutrons and $\bar{\mu}_0$ is the average cosine of the neutron scattering angle per collision in the laboratory system (see § 3.49 *et seq.*).   If absorption of neutrons in the medium is relatively weak, Σ_t may be replaced by Σ_s, the macroscopic scattering cross section, so that

$$D = \frac{1}{3\Sigma_s(1 - \bar{\mu}_0)} = \frac{\lambda_s}{3(1 - \bar{\mu}_0)}, \tag{3.7}$$

where $\lambda_s(=1/\Sigma_s)$ is the scattering mean free path.   The quantity $1/\Sigma_s(1 - \bar{\mu}_0)$ or $\lambda_s/(1 - \bar{\mu}_0)$ is called the *transport mean free path* and is denoted by λ_{tr}; thus, provided neutron absorption is small,

$$\lambda_{tr} \equiv \frac{1}{\Sigma_s(1 - \bar{\mu}_0)} = \frac{\lambda_s}{1 - \bar{\mu}_0}, \tag{3.8}$$

and, from equation (3.7),

$$D = \tfrac{1}{3} \lambda_{tr}. \tag{3.9}$$

This result is applicable also to neutron absorbing media, but in view of equation (3.6), Σ_{tr} is now equal to $1/(\Sigma_t - \Sigma_s \bar{\mu}_0)$ which reduces to equation (3.8) when absorption is small, i.e., when $\Sigma_t \approx \Sigma_s$.

3.15. It may be noted that, if the scattering were isotropic, i.e., spherically symmetric, in the laboratory system, the average cosine, $\bar{\mu}_0$, of the scattering angle would be zero, and λ_s would be identical with λ_{tr}.   The factor $1/(1 - \bar{\mu}_0)$, which may be regarded as the correction for anisotropic scattering, represents the preferential scattering in the forward direction, i.e., in the initial direction of motion of the neutron before collision with a nucleus.   In general, $\bar{\mu}_0$ decreases with increasing mass of the struck nucleus, so that the tendency toward anisotropic scattering decreases correspondingly.   Hence, for heavy nuclei, the factor $1/(1 - \bar{\mu}_0)$ approaches unity and the total and transport mean free paths (and cross sections) are not very different.

---

* Both neutron current ($\mathbf{J}$) and the flux (ϕ) are usually expressed as neutrons per cm² per sec; it should be noted, however, that the current is a vector, i.e., it involves direction, whereas the flux is a scalar quantity. The situation would perhaps be less confusing if the flux were represented, as it should be strictly speaking, in terms of (neutrons/cm³)(cm/sec).

CALCULATION OF NEUTRON LEAKAGE

3.16. The rate at which neutrons leak out of a specified volume element can be calculated from the expressions for the current density, provided the neutron flux, $\phi(x, y, z)$, is known as a function of the spatial coordinates. Let a rectangular volume element dV, with dimensions dx, dy, dz, be located at a point whose coordinates are x, y, z (Fig. 3.2). Consider the two faces of area $dx\,dy$ which lie

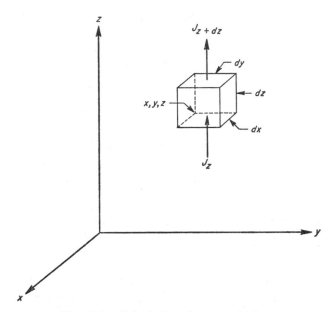

Fig. 3.2. Calculation of neutron leakage

parallel to the x, y plane. The number of neutrons entering the lower face per second is $J_z\,dx\,dy$, where J_z is the net neutron current in the z direction; similarly, the number of neutrons leaving the upper face is $J_{z+dz}\,dx\,dy$.

3.17. The currents in the z direction are given by expressions analogous to equation (3.3) with the neutron flux instead of the density and D in place of D_0. Thus, if D is constant, the net rate of flow of neutrons out of the given volume element through the faces parallel to the x, y plane is

$$(J_{z+dz} - J_z)\,dx\,dy = -D\left[\left(\frac{\partial\phi}{\partial z}\right)_{z+dz} - \left(\frac{\partial\phi}{\partial z}\right)_z\right]dx\,dy$$

$$= -D\,\frac{\partial^2\phi}{\partial z^2}\,dx\,dy\,dz$$

$$= -D\,\frac{\partial^2\phi}{\partial z^2}\,dV.$$

3.18. Similarly, the net rates of loss from the faces parallel to the y, z and x, z planes are

$$-D\,\frac{\partial^2 \phi}{\partial x^2}\,dV \quad\text{and}\quad -D\,\frac{\partial^2 \phi}{\partial y^2}\,dV,$$

respectively, and the total rate at which neutrons leak out of the volume element dV cm^3 is given by the sum of these three terms. The leakage rate per cm^3 may then be obtained upon dividing through by dV, so that

$$\text{Neutron leakage per cm}^3\text{ per sec} = -D\left(\frac{\partial^2 \phi}{\partial x^2} + \frac{\partial^2 \phi}{\partial y^2} + \frac{\partial^2 \phi}{\partial z^2}\right)$$

$$= -D\nabla^2 \phi, \tag{3.10}$$

where ∇^2 is the symbol used for the Laplacian operator.

THE DIFFUSION EQUATION

3.19. For convenience, equation (3.1) may be rewritten as

$$-\text{Leakage} - \text{Absorption} + \text{Production} = \frac{\partial n}{\partial t}, \tag{3.11}$$

where the various terms on the left refer to rates per cm^3 per sec, and n is the neutron density expressed as neutrons per cm^3. The first term on the left of equation (3.11) is given by equation (3.10) as $D\nabla^2\phi$, the two negative signs leading to a positive sign. The second term can be derived from equation (2.34) and is accordingly $\Sigma_a\phi$, where Σ_a is the macroscopic absorption cross section for the monoenergetic (or thermal) neutrons. Upon making these substitutions, equation (3.11) becomes

$$D\nabla^2 \phi - \Sigma_a \phi + S = \frac{\partial n}{\partial t}, \tag{3.12}$$

where S, called the *source term*, is the rate of production of neutrons per cm^3 per sec. This expression, generally referred to as the *diffusion equation*, is basic to thermal reactor theory. In view of its derivation, it is strictly applicable only to monoenergetic neutrons and then only at distances greater than two or three mean free paths from strong sources, absorbers, or boundaries (cf. § 3.11). It will be applied here to thermal neutrons in particular; this is justifiable, provided the medium does not absorb neutrons strongly.

3.20. For a system in the steady state $\partial n/\partial t$ is zero, and so the diffusion equation then becomes

$$D\nabla^2 \phi - \Sigma_a \phi + S = 0. \tag{3.13}$$

In a nonmultiplying medium, i.e., one containing no fissile material, the source term is zero everywhere except at the position of the neutron source. The differential equation (3.13) can be solved, with $S = 0$ for points outside the source region. The proper boundary conditions are then applied at the source, as will be illustrated below. If S is set equal to zero, the steady-state diffusion equation (3.13) reduces to the homogeneous form applicable anywhere except at the source; thus,

$$D\nabla^2\phi - \Sigma_a\phi = 0 \tag{3.14}$$

or

$$\nabla^2\phi - \kappa^2\phi = 0, \tag{3.15}$$

where

$$\kappa^2 \equiv \frac{\Sigma_a}{D}. \tag{3.16}$$

Since Σ_a has the dimensions of reciprocal length (§ 2.111) and D of length, κ is a reciprocal length.

SOLUTION OF THE DIFFUSION EQUATION: BOUNDARY CONDITIONS

3.21. The solution of the diffusion equation (3.15) will give the space distribution of the neutron flux at a steady state in a nonmultiplying medium at points other than the source. Since the diffusion equation is a differential equation, it does not provide a complete representation of the physical situation. The general solution will contain arbitrary constants of integration. In order to determine the proper values for these constants, it is necessary to apply restrictions upon the permitted solutions in the form of boundary conditions derived from the physical nature of the problem. The number of these boundary conditions must be sufficient to provide a unique solution, with no arbitrary constants. Some of the boundary conditions which are frequently used in the solution of neutron distribution problems are examined in the following paragraphs.

3.22. It is to be expected that at the boundary between two different media there should be continuity of both neutron flux and neutron current. Thus, assuming monoenergetic (or thermal) neutrons, one simple boundary condition is the following: *At an interface between two media, with different diffusion properties, the neutron flux will be the same in both media.* If the media are indicated by the symbols A and B, then

$$(\phi_A)_0 = (\phi_B)_0, \tag{3.17}$$

the zero subscript being used to indicate that the values apply at the interface.

3.23. The neutron current is a vector and the continuity condition requires continuity of the component normal to the boundary. The current boundary condition thus takes the form: *At a plane interface between two media with different diffusion properties, the net neutron currents, in the direction normal to the interface, are equal.* If the x direction is taken to be that normal to the boundary, the neutron current in this direction is given by equation (3.3) as

$$J_x = -D\frac{d\phi}{dx}. \tag{3.18}$$

Therefore the condition for continuity of current at the boundary may be written as

$$-D_A \left(\frac{d\phi_A}{dx}\right)_0 = -D_B \left(\frac{d\phi_B}{dx}\right)_0. \tag{3.19}$$

This result will hold at a spherical or infinite cylindrical interface as well as at a plane.

3.24. It should be noted that equation (3.19) implies that diffusion theory can be used at the boundary between the two media. Although this is actually not so, it has nevertheless been shown from transport theory that the equation represents a reasonably good approximation to the exact boundary condition.

THE LINEAR EXTRAPOLATION DISTANCE

3.25. At the boundary between a diffusion medium and a vacuum, different circumstances exist, since there is no scattering back of neutrons from the vacuum into the diffusion medium. In other words, there is a flow of neutrons in one direction only. The boundary condition is now as follows: *Near the boundary between a diffusion medium and a vacuum the neutron flux varies in such a manner that linear extrapolation would require the flux to vanish at a given distance beyond the boundary.* In general, the net neutron current represents the balance between currents in the positive and negative directions; according to diffusion theory these are given by

$$J_+ = \frac{\phi}{4} - \frac{D}{2} \cdot \frac{d\phi}{dx}$$

and

$$J_- = \frac{\phi}{4} + \frac{D}{2} \cdot \frac{d\phi}{dx}$$

for the components in the x direction [2]. The difference between these two components gives the net current, in agreement with equation (3.18). In the case under consideration, i.e., at the medium-vacuum boundary, the negative component of the neutron current is zero, so that

$$J_- = \frac{\phi_0}{4} + \frac{D}{2} \cdot \frac{d\phi_0}{dx} = 0. \tag{3.20}$$

The flux ϕ_0 at the boundary is positive, and so it follows from equation (3.20) that the slope $d\phi_0/dx$ of the flux distribution must be negative at the boundary, as indicated schematically in Fig. 3.3.

3.26. If the neutron flux is extrapolated into the vacuum, using a straight line with the same slope as at the boundary, i.e., $d\phi_0/dx$, the flux would vanish at a distance d given by

$$-\frac{\phi_0}{d} = \frac{d\phi_0}{dx}.$$

Consequently, by equation (3.20),

$$d = 2D \tag{3.21}$$

or, since $D = \frac{1}{3} \lambda_{tr}$, by equation (3.9),

$$d = \frac{2}{3} \lambda_{tr}. \tag{3.22}$$

This distance is called the *linear extrapolation distance* or, sometimes, the *augmentation distance.* On the basis of linear extrapolation, therefore, the neutron

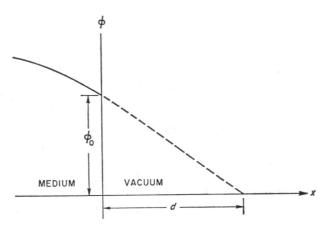

Fig. 3.3.   Extrapolation of neutron flux at plane boundary

flux should vanish at a distance of $\frac{2}{3} \lambda_{tr}$ beyond the plane boundary between a diffusion medium and a vacuum. This boundary condition is sometimes stated in the form: *the neutron flux vanishes at the extrapolated boundary*, which lies $\frac{2}{3} \lambda_{tr}$ (or $2D$) beyond the physical boundary in the case of a plane interface.

3.27. Since the diffusion theory approximation breaks down near a boundary, it is to be expected that the linear extrapolation distance obtained above will be incorrect. According to the more exact transport theory treatment, the linear extrapolation distance at a plane surface of a weakly absorbing medium is close to $0.71\lambda_{tr}$, rather than $\frac{2}{3} \lambda_{tr}$. Furthermore, the extrapolation distance is somewhat greater for curved than for plane surfaces. For a boundary of infinite curvature, i.e., of zero radius, this distance attains its maximum value of $\frac{4}{3} \lambda_{tr}$.

3.28. In postulating that the neutron flux vanishes at the extrapolated boundary, there is no implication, either from transport theory or diffusion theory, that the flux is actually zero there. The concept of the hypothetical boundary where the flux vanishes, as a result of linear extrapolation, is merely a convenient mathematical device used to obtain a simple boundary condition. Some indication of the actual state of affairs may be obtained with the aid of Fig. 3.4, which represents qualitatively the flux distribution in the x direction near a medium-vacuum interface.

3.29. It is seen that the actual (or transport theory) flux undergoes a sharp decrease within a mean free path or so of the interface. Hence, in order that the boundary condition under consideration may be used in conjunction with diffusion theory to determine the neutron flux distribution at *reasonable distances from the interface,* a so-called asymptotic linear extrapolation of the trans-

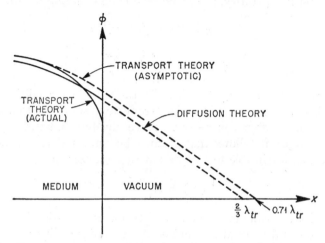

Fig. 3.4. Transport theory and diffusion theory extrapolations at a boundary

port theory distribution is made from the values not too near the boundary. The use of the resulting extrapolation distance of $0.71\lambda_{tr}$ (or about $2.1D$) for a plane interface, together with diffusion theory, provides a satisfactory approximation for the behavior at points in the medium more distant than a transport mean free path from the interface.

DIFFUSION OF MONOENERGETIC NEUTRONS FROM A POINT SOURCE

3.30. The procedure for solving the diffusion equation in a simple case will be illustrated by reference to the diffusion of monoenergetic neutrons from a point source into an infinite nonmultiplying medium. The result is of interest because it provides a physical interpretation of the quantity κ in equation (3.15) or, rather, of its reciprocal called the diffusion length (§ 3.36).

3.31. For convenience, the point source of neutrons, in the infinite medium, is assumed to be located at the origin of the coordinate system. The neutron distribution may then be regarded as having spherical symmetry, and the Laplacian operator in equation (3.15) may thus be expressed in spherical coordinates. The source-free diffusion equation for the flux distribution then takes the form

$$\frac{d^2\phi}{dr^2} + \frac{2}{r}\cdot\frac{d\phi}{dr} - \kappa^2\phi = 0, \tag{3.23}$$

where r is the distance from the point source.   This equation may be readily solved by setting $\phi \equiv y/r$, when it reduces to

$$\frac{d^2y}{dr^2} - \kappa^2 y = 0.$$

Since κ^2 is a positive quantity, as can be seen from equation (3.16), the general solution is

$$y = Ae^{-\kappa r} + Ce^{\kappa r},$$

and hence

$$\phi(r) = A\frac{e^{-\kappa r}}{r} + C\frac{e^{\kappa r}}{r}, \tag{3.24}$$

the constants A and C being determined by the boundary conditions.*

3.32. If C is not zero, then it is evident from equation (3.24) that the neutron flux would become infinite as $r \to \infty$.   Since the flux must always be finite, except at the origin, it follows that C must be zero, and the general solution of equation (3.23) applicable to the present case reduces to

$$\phi(r) = A\frac{e^{-\kappa r}}{r}. \tag{3.25}$$

3.33. In order to evaluate A the so-called source condition is used.   If J is the neutron current density at the surface of a sphere of radius r, with the source at the center, then the total number of neutrons passing through the whole of the surface per second will be $4\pi r^2 J$.   The limiting value of this number as r approaches zero will be equal to the source strength Q neutrons per sec, i.e., to the number of neutrons emitted by the point source in all directions per second.   The neutron current at a point distant r from the source is given by equations (3.3) and (3.25) as

$$J = -D\frac{d\phi}{dr} = DAe^{-\kappa r}\left(\frac{1 + \kappa r}{r^2}\right).$$

Consequently, from the foregoing argument,

$$Q = \lim_{r \to 0} 4\pi r^2 J$$

$$= \lim_{r \to 0} 4\pi r^2 DAe^{-\kappa r}\left(\frac{1 + \kappa r}{r^2}\right),$$

and hence

$$A = \frac{Q}{4\pi D}.$$

Equation (3.25) for the flux distribution accordingly becomes

$$\phi(r) = \frac{Q}{4\pi D} \cdot \frac{e^{-\kappa r}}{r}, \tag{3.26}$$

which is the required solution of the diffusion equation.

* Because B has a special significance in reactor theory (§ 4.17), it is not used here as an arbitrary constant.

Example 3.1. A hypothetical point source of thermal neutrons emits 10^6 neutrons per sec into a surrounding "infinite" graphite block. Determine the neutron flux at distances of 27, 54, and 108 cm from this source. (For graphite κ ($= 1/L$) is 0.0185 cm^{-1} and D is 0.94 cm.)

By equation (3.26),

$$\phi(27) = \frac{10^6}{(4\pi)(0.94)} \cdot \frac{e^{-(0.0185)(27)}}{27}$$

$$= 8.7 \times 10^4 \frac{e^{-0.5}}{27} = 1.9 \times 10^3 \text{ neutrons/(cm}^2)(\text{sec})$$

$$\phi(54) = 8.7 \times 10^4 \frac{e^{-1}}{54} = 5.9 \times 10^2$$

$$\phi(108) = 8.7 \times 10^4 \frac{e^{-2}}{108} = 1.1 \times 10^2.$$

THE DIFFUSION LENGTH

3.34. From equation (3.26) it is possible to derive an expression for the mean distance between the point at which a neutron originates, i.e., its source, to the point at which it is absorbed. However, instead of treating the first spatial moment of the flux, as would be required in this case, it is preferable, in view of later developments, to consider the second spatial moment. The volume of a spherical shell element of radius r and thickness dr surrounding the point source is $4\pi r^2\, dr$. Since $\Sigma_a\phi$ gives the number of neutrons absorbed per cm^3 per sec, in a region of flux ϕ and macroscopic absorption cross section Σ_a, it follows that the rate of neutron absorption in the spherical shell is $4\pi r^2\, dr\, \Sigma_a\phi$ neutrons per sec. This is a measure of the probability that a neutron leaving the source, at the center of the shell, will be absorbed within the element dr at a distance r from the source.

3.35. Hence, the mean square (net vector) distance, $\overline{r^2}$, from the neutron source to the point where it is absorbed, is given by

$$\overline{r^2} = \frac{\int_0^\infty r^2(4\pi r^2\Sigma_a\phi)\, dr}{\int_0^\infty 4\pi r^2\Sigma_a\phi\, dr}.$$

Upon substituting the value for ϕ from equation (3.26), this becomes

$$\overline{r^2} = \frac{\int_0^\infty r^3 e^{-\kappa r}\, dr}{\int_0^\infty r e^{-\kappa r}\, dr} = \frac{6/\kappa^4}{1/\kappa^2} = \frac{6}{\kappa^2}. \tag{3.27}$$

3.36. A quantity called the *diffusion length*, represented by L, is defined by

$$L \equiv \sqrt{\frac{D}{\Sigma_a}} = \frac{1}{\kappa}, \tag{3.28}$$

bearing in mind the definition of κ as given by equation (3.16). It follows, therefore, from equation (3.27) that

$$L^2 = \tfrac{1}{6}\overline{r^2}, \tag{3.29}$$

and so the square of the diffusion length of a monoenergetic neutron is equal to one-sixth of the mean square (net vector) distance from the source to the point at which the neutron is absorbed by a nucleus.*

3.37. Some alternative relationships between the diffusion length and other properties of the medium are obtained by replacing D in equation (3.28) by $\tfrac{1}{3}\lambda_{tr}$, as given by equation (3.9), and Σ_a by $1/\lambda_a$; the result is

$$L = \sqrt{\tfrac{1}{3}\,\lambda_{tr}\lambda_a}$$

or, if the macroscopic *transport cross section*, Σ_{tr}, is defined as $1/\lambda_{tr}$,

$$L = \frac{1}{\sqrt{3\Sigma_{tr}\Sigma_a}}.$$

For substances of high mass number Σ_{tr} is not very different from Σ_t, as stated in § 3.15; then

$$L \approx \frac{1}{\sqrt{3\Sigma_t\Sigma_a}}.$$

For a weakly absorbing medium, Σ_t may be replaced by Σ_s.

DETERMINATION OF DIFFUSION LENGTH

3.38. The variation of neutron flux with distance in a direction perpendicular to an infinite plane source in an infinite medium of a given material is represented by

$$\phi(z) = Ae^{-z/L},$$

where z is the distance from the plane and A is a constant. Although it is impossible to construct an infinite medium, the diffusion length of neutrons in a given medium can be determined by measuring the flux at various distances from a plane source in a finite rectangular parallelepiped of the diffusion material, shown in Fig. 3.5. The source is in the xy plane and the flux variation along any line parallel to the z axis, i.e., for any constant x and y, is found to be expressed by†

$$\phi(z) = \text{constant} \times e^{-\gamma z},$$

or

$$\frac{d\ln\phi(z)}{dz} = -\gamma,$$

---

* The net vector (or "crow-flight") distance considered here differs from the absorption mean free path of the neutrons in the respect that the latter is the corresponding total (or scalar) distance traveled by the neutron.

† For details of the theory and calculations, see Ref. 3; for experimental techniques, see Ref. 4.

at distances that are not too close to the source plane or to the upper end of the block of material.

3.39. The value of γ can be determined by measuring the neutron flux (or a quantity proportional to the flux) at various distances z from the source using the foil activation method described in § 2.119. If $\ln \phi(z)$ is plotted against z the slope of the linear portion is equal to $-\gamma$. A theoretical treatment shows that

$$\frac{1}{L^2} = \gamma^2 - \left(\frac{\pi}{a}\right)^2 - \left(\frac{\pi}{b}\right)^2,$$

where a and b are the dimensions of the rectangular block, including the extrapolation distance, in the x and y directions (Fig. 3.5). Since a, b, and γ are now known, the diffusion length L can be calculated. It is of interest to note that the terms $(\pi/a)^2$ and $(\pi/b)^2$ allow for leakage of neutrons from the sides of the parallelepiped with finite dimensions a and b.

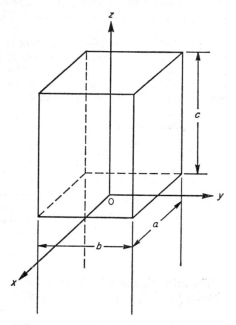

FIG. 3.5. Determination of thermal diffusion length

3.40. A compilation of measured diffusion lengths of thermal neutrons in various moderators is given in the Appendix; however, in order to indicate the general magnitude of the values, some data are quoted in Table 3.1 for four

TABLE 3.1. DIFFUSION PROPERTIES OF MODERATORS FOR THERMAL NEUTRONS

Moderator	Density (g/cm³)	L (cm)	Σ_a (cm⁻¹)	D (cm)
Water..........................	1.00	2.76	2.2×10^{-2}	0.17
Heavy water (99.75% D₂O).......	1.10	100	8.5×10^{-5}	0.85
Beryllium......................	1.84	21	1.2×10^{-3}	0.54
Carbon (graphite)*.............	1.70	54.2	3.2×10^{-4}	0.94

* High-purity commercial (reactor grade) product.

important moderating materials, namely, ordinary water, heavy water,* beryllium, and carbon (graphite). The thermal-neutron macroscopic absorption

* It should be noted that the properties of heavy water given here and later in this chapter refer to the commercial product containing 99.75 per cent of deuterium oxide (D_2O).

cross sections, Σ_a, for these substances are also given, as well as the diffusion coefficient, D, obtained from L and Σ_a by equation (3.28). The diffusion properties of a given material depend on its density and so this is quoted in each case. The transport mean free path λ_{tr} is equal to $3D$, by equation (3.9), and the absorption mean free path λ_a is equal to $1/\Sigma_a$.

Example 3.2. The diffusion length of beryllium metal is determined by measuring the flux distribution in an assembly 100 cm × 100 cm × 65 cm high, at the base of which is a plane source of thermal neutrons. The corrected saturation activities (in counts per minute) of foils located at various vertical distances above the base are given below. Similar measurements in a horizontal direction gave a value of 2.5 cm as the extrapolation distance, i.e., the distance from the assembly at which the neutron flux extrapolated to zero. What is the diffusion length of thermal neutrons in beryllium?

Vertical Distance (z cm)	*Saturation Activity* (c/m)
10	310
20	170
30	90
40	48
50	24
60	12

The saturation activity is proportional to the thermal neutron flux (§ 2.119); hence, the slope of the plot in Fig. 3.6 is equal to $d \ln \phi(z)/dz$, which is equal to $-\gamma$. The value

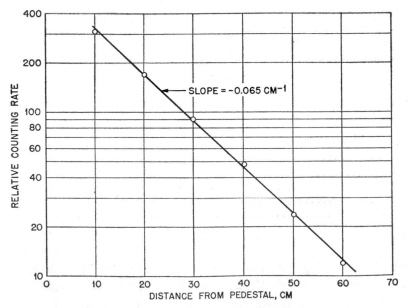

Fig. 3.6. Data for evaluation of diffusion length of beryllium

of γ is thus found to be 0.065 cm^{-1}.   The actual dimensions of the beryllium assembly in the x and y directions are both 100 cm; to these must be added $2 \times 2.5 = 5$ cm for the extrapolation distances at the two sides.   Hence, $a = b = 105$ cm, and

$$\frac{1}{L^2} = \gamma^2 - \left(\frac{\pi}{a}\right)^2 - \left(\frac{\pi}{b}\right)^2 = (0.065)^2 - 2\left(\frac{\pi}{105}\right)^2$$

$$= 2.5 \times 10^{-3} \text{ cm}^{-2}$$

$$L = 20 \text{ cm}.$$

(The value given in the literature is 21 cm.)

INFINITE PLANE SOURCE IN MEDIUM OF FINITE THICKNESS

3.41. Another case in which the solution of the diffusion equation is of some interest is that in which monoenergetic (or thermal) neutrons from an infinite plane source diffuse into a nonmultiplying medium consisting of a slab of infinite extent but of finite thickness (Fig. 3.7).   The situation here has some similarities

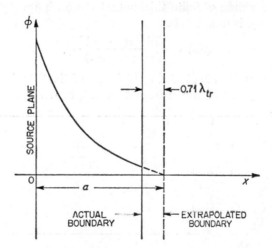

FIG. 3.7.   Neutron flux distribution in finite slab from infinite plane source

to that of a reactor (neutron source) surrounded by a reflector (diffusion medium).

3.42. The coordinate system is chosen so that the source plane coincides with the plane for which $x = 0$ at all points.   Since the source is assumed to be infinite in extent, it is evident that, for any given value of x, the neutron flux will be independent of y and z.   The Laplacian operator in rectangular coordinates is now merely d^2/dx^2, and so the diffusion equation (3.15) in the present case is

$$\frac{d^2\phi}{dx^2} - \kappa^2\phi = 0, \tag{3.30}$$

where, as before, κ is the reciprocal of the diffusion length in the medium. The general solution of equation (3.30), since κ^2 is a positive quantity, is

$$\phi(x) = Ae^{-\kappa x} + Ce^{\kappa x}, \tag{3.31}$$

where A and C are constants to be determined by the boundary conditions.

3.43. In order to evaluate C, use is made of the boundary condition that the flux shall vanish at the hypothetical extrapolated boundary (§ 3.26); thus,

$$\phi(a) = Ae^{-\kappa a} + Ce^{\kappa a} = 0$$
$$C = -Ae^{-2\kappa a},$$

where a is the thickness of the medium, *including the extrapolation distance*. The constant A is derived from the source condition by a procedure analogous to that described in § 3.33; the result is

$$A = \frac{Q}{2\kappa D(1 + e^{-2\kappa a})}.$$

Upon substituting these expressions for A and C into equation (3.31), the flux distribution in a medium of finite thickness, from an infinite plane source of Q neutrons/(cm²)(sec), is found to be

$$\phi(x) = \frac{\sinh\left[\kappa(a - x)\right]}{2\kappa D \cosh(\kappa a)} Q. \tag{3.32}$$

3.44. Some idea of the effect of the thickness of the diffusion medium on the flux distribution may be obtained from Fig. 3.8, in which ϕ/Q is plotted as a

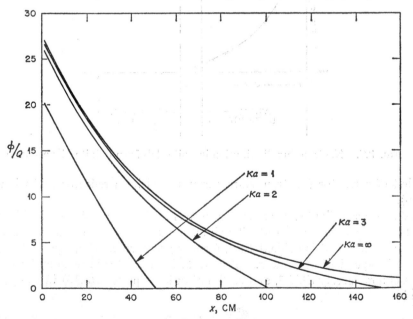

Fig. 3.8. Neutron flux distributions in media of different thicknesses

function of the distance x from a plane source in various thicknesses of graphite, namely, κa equal to 1, 2, 3, and infinity, respectively.* The value of $\kappa(=1/L)$ is 0.0185 cm^{-1} and D is 0.94 cm (Table 3.1). Since κa is equal to a/L, it gives the thickness of the medium in terms of the diffusion length; thus the curves in Fig. 3.8 are for thicknesses of one, two, and three diffusion lengths and infinity.

3.45. It is seen that when κa is 3, i.e., when the extrapolated thickness of the medium is three times the diffusion length for the particular monoenergetic (or thermal) neutrons, the flux distribution of these neutrons is not greatly different from that in an infinite medium, except near the boundary. Even when κa is 2, the difference from the infinite medium distribution is not large at distances greater than a diffusion length from the boundary, i.e., for $x < 50$ cm in Fig. 3.8. These results are quite general, for κa, i.e., the thickness of the medium in terms of the diffusion length, is the important property in determining how the neutron flux falls off with distance from a plane source. Thus, when the thickness of a medium is two or more times the diffusion length, it may be treated as an infinite medium, at distances greater than a diffusion length or so from the boundary.

3.46. A consideration of the physical interpretation of these results is of interest. In the infinite medium there is no loss of neutrons, but in slabs of finite thickness neutrons will leak out at the boundary. If the thickness is three or more times the diffusion length, most of the neutrons are scattered back before they reach the boundary and the leakage is very small. For slabs of lesser thickness there is a greater loss of neutrons and consequently a more rapid falling off of the flux near the boundary. This effect is very marked when the thickness (including the extrapolation distance) is equal to the diffusion length. i.e., when $\kappa a = 1$, or less.

THE SLOWING DOWN OF NEUTRONS

ELASTIC SCATTERING

3.47. The process of slowing down the fast neutrons resulting from fission, known as moderation, may involve both inelastic and elastic collisions of the neutrons with various nuclei. In a reactor, inelastic scattering is usually of importance only when collisions occur with certain nuclei of fairly high mass number, e.g., iron and uranium. Even in these cases, the neutron must have at least 0.1 Mev energy. Neutrons with less energy are unable to cause excitation of the nuclei and so they cannot be slowed down by inelastic collisions. In a reactor in which neutrons of energy below 0.1 Mev are to produce most of the fissions, it is necessary, therefore, to introduce a material of low mass number to act as moderator (§ 1.51). The nuclei then serve as targets for elastic col-

---

* Since equation (3.32) does not apply at the source itself, the curves are not extended to $x = 0$.

lisions with the neutrons; the lighter the nucleus the larger the fraction of energy lost by the neutron in such a collision.

3.48. Since most of the slowing down interactions of the neutrons in a thermal reactor usually occur at energies below 0.1 Mev, it follows that the elastic scattering processes predominate. An adequate mathematical picture of the slowing down of neutrons is thus usually based only on such collisions. These can be treated by the methods of classical mechanics, assuming the neutrons and scattering nuclei to behave as perfectly elastic spheres. By applying the principles of conservation of momentum and of energy, it is possible to derive a relationship between the scattering angle and the energy of the neutron before and after collision with a nucleus. Upon introducing an appropriate scattering law, various useful results can be obtained.

3.49. In considering scattering collisions, two convenient frames of reference are used; these are the laboratory (L) system and the center of mass (C) system. In the former, the scattering nucleus is assumed to be at rest before the collision, whereas in the latter the center of mass of the neutron + nucleus is taken to be stationary in the collision. In the L system the viewpoint is essentially that of an external observer, but in the C system it is that of an observer who is located at the center of mass of the neutron + nucleus. For purposes of theoretical treatment the C system is simpler, although actual measurements are made in the L system.

3.50. The conditions before and after a collision, as represented in the two systems, are shown in Fig. 3.9. The neutron, having a mass of approximately

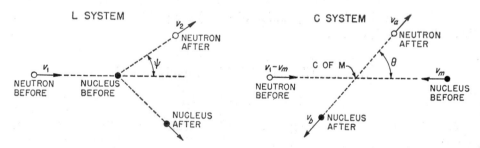

FIG. 3.9. Neutron scattering in laboratory (L) and center of mass (C) systems

unity on the conventional atomic mass scale, moves with a speed v_1 relative to the nucleus of mass number A. Since the target nucleus is stationary in the L system, the momentum v_1 of the neutron is also the total momentum. The total mass of the colliding particles is $A + 1$, and consequently the speed v_m of the center of mass in the L system is given by

$$v_m = \frac{v_1}{A + 1}. \tag{3.33}$$

In the C system, the velocity of the neutron before collision is $v_1 - v_m$, and, using the value of v_m just derived, it follows that

$$v_1 - v_m = \frac{Av_1}{A+1}. \tag{3.34}$$

3.51. It is seen, therefore, that in the C system the neutron and the scattering nucleus appear to be moving toward one another with velocities of $Av_1/(A+1)$ and $v_1/(A+1)$, respectively. The momentum of the neutron (mass unity) is consequently $Av_1/(A+1)$, while that of the nucleus (mass A) is also $Av_1/(A+1)$, but in the opposite direction. The total momentum before collision, with respect to the center of mass, is then zero, and by the principle of the conservation of momentum it must also be zero after the collision.

3.52. Following the collision, the neutron in the C system leaves in the direction making an angle θ with its original direction; this is the scattering angle in the C system. The recoil nucleus must then move in the opposite direction, since the center of mass is always on the line joining the two particles. If v_a is the speed of the neutron and v_b that of the nucleus after collision in the C system, then the requirement that the total momentum shall be zero is expressed by

$$v_a = Av_b. \tag{3.35}$$

3.53. The speeds of the nucleus and neutron before collision in the C system are given by equations (3.33) and (3.34), respectively, as seen above. The conservation of energy condition may therefore be written as

$$\frac{1}{2}\left(\frac{Av_1}{A+1}\right)^2 + \frac{1}{2}A\left(\frac{v_1}{A+1}\right)^2 = \frac{1}{2}v_a^2 + \frac{1}{2}Av_b^2, \tag{3.36}$$

where the left side gives the total kinetic energy before collision and the right side is that after collision. Upon solving equations (3.35) and (3.36) for v_a and v_b, it is found that

$$v_a = \frac{Av_1}{A+1} \quad \text{and} \quad v_b = \frac{v_1}{A+1}. \tag{3.37}$$

Comparison of these values with equations (3.34) and (3.33), respectively, shows that, in the C system, the speeds of the neutron and nucleus after collision are the same as the respective values before collision.

3.54. In order to determine the loss of kinetic energy of the neutron upon collision with an essentially stationary nucleus, it is necessary to transform the results obtained for the C system back to the L system. To perform this transformation, use is made of the fact that the two systems must always move relative to one another with the velocity (v_m) of the center of mass in the L system. Hence, the final velocity of the neutron after collision, in the L system, is obtained by adding the vector v_m, representing the motion of the center of mass in the L system, to the vector v_a, which is the velocity of the neutron after collision in the C system. This vector addition is shown in Fig. 3.10, the angle between the vectors being θ, the scattering angle in the C system.

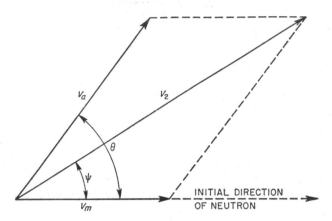

FIG. 3.10.   Transformation from C system to L system

3.55. If v_2 is the velocity of the neutron after collision in the L system, then by the law of cosines

$$v_2^2 = v_1^2 \frac{A^2 + 2A \cos \theta + 1}{(A + 1)^2}. \tag{3.38}$$

The angle ψ between the vectors v_2 and v_m is the scattering angle in the L system. It is the average value of the cosine of this angle for a large number of scatterings, i.e., $\overline{\cos \psi}$, which was represented by $\bar{\mu}_0$ in § 3.14.

ENERGY CHANGE IN SCATTERING

3.56. The kinetic energy E_1 of the neutron before a collision is $\frac{1}{2} mv_1^2$, and the energy E_2 after the collision is $\frac{1}{2} mv_2^2$; hence, the ratio of the neutron energy after collision to that before collision is given by equation (3.38) as

$$\frac{E_2}{E_1} = \frac{v_2^2}{v_1^2} = \frac{A^2 + 2A \cos \theta + 1}{(A + 1)^2}. \tag{3.39}$$

The assumption is made here, and subsequently, that the laboratory system, with an essentially stationary nucleus, represents the actual scattering conditions. If a quantity α, which is a property of the scattering nucleus related to its mass, is defined by

$$\alpha \equiv \left(\frac{A - 1}{A + 1}\right)^2, \tag{3.40}$$

then equation (3.39) becomes

$$\frac{E_2}{E_1} = \tfrac{1}{2}[(1 + \alpha) + (1 - \alpha) \cos \theta]. \tag{3.41}$$

3.57. The maximum value of the ratio E_2/E_1, i.e., the minimum loss of energy in a collision, occurs when $\theta = 0$, i.e., for a glancing collision; then $\cos \theta$ is unity, and equation (3.41) gives

$$\frac{E_{max}}{E_1} = 1 \quad \text{or} \quad E_{max} = E_1. \tag{3.42}$$

In this event, the energies of the neutron before and after scattering are equal, and the neutron suffers no energy loss in the collision. The minimum value of E_2/E_1, i.e., the maximum possible energy transfer, occurs, on the other hand, when $\theta = \pi$, i.e., in a head-on collision. The value of $\cos \theta$ is now -1, and from equation (3.41),

$$\frac{E_{min}}{E_1} = \alpha \quad \text{or} \quad E_{min} = \alpha E_1. \tag{3.43}$$

The minimum value of the energy to which a neutron can be reduced in an elastic scattering collision is therefore αE_1, where E_1 is the energy before the collision. The maximum fractional loss of energy in a collision is given by

$$\frac{E_1 - E_{min}}{E_1} = 1 - \alpha, \tag{3.44}$$

and the actual maximum possible energy loss in a collision is $E_1(1 - \alpha)$.

3.58. Since, by equation (3.40), the quantity α is related to the mass number of the target nucleus, it is evident that the loss of energy suffered by a neutron in a collision will also depend on the mass number. For hydrogen, $A = 1$ and so $\alpha = 0$; it is, consequently, possible for a neutron to lose all of its kinetic energy in one collision with a hydrogen nucleus. This arises, of course, because the masses of the neutron and the hydrogen nucleus (proton) are essentially equal. For carbon, $A = 12$ and $\alpha = 0.716$; hence, the maximum possible fractional loss of energy of a neutron in a collision with a carbon nucleus is $1 - 0.716 = 0.284$.

EMPIRICAL SCATTERING LAW

3.59. Equation (3.41) gives the relationship between the energy ratio E_2/E_1, and the scattering angle θ in the C system, but derivation of an expression for the *average* energy change in a collision requires the introduction of a scattering law, i.e., a postulate concerning the probability of various scattering angles. It appears that the assumption of spherically symmetric, i.e., isotropic, scattering of neutrons in the C system provides a reasonably satisfactory description of elastic scattering for neutrons with energies less than a few Mev.* In other

---

* The assumption also breaks down at low (thermal) energies, where the vibrational energy of a nucleus is significant in comparison with its kinetic energy.

words, it may be postulated that all values of $\cos \theta$ from -1 to $+1$ are equally probable. This empirical scattering law will be postulated as the basis of the subsequent treatment.

3.60. Although scattering may be spherically symmetric in the C system, it will not be so in the L system unless the mass of the scattering nucleus is large compared to the mass of the neutron. If this is the case, the center of mass of the system is located very close to the nucleus, and the L system approximates the C system. The same conclusion may be reached in another manner. It can be readily shown from Fig. 3.10 that

$$\cos \psi = \frac{A \cos \theta + 1}{\sqrt{A^2 + 2A \cos \theta + 1}}, \tag{3.45}$$

where ψ and θ are the scattering angles in the L and C systems, respectively. For a heavy nucleus, $A \gg 1$, and then $\cos \psi \to \cos \theta$, so that the scattering angle in the L system approaches that in the C system. Therefore, if scattering is spherically symmetric in the latter, it will also be so in the former.

3.61. The average cosine of the scattering angle in the L system, i.e., $\bar{\mu}_0$, assuming isotropic scattering in the C system, is given by

$$\overline{\cos \psi} \equiv \bar{\mu}_0 = \frac{\int_0^{4\pi} \cos \psi \, d\Omega}{\int_0^{4\pi} d\Omega},$$

where $d\Omega$ is an element of solid angle. Transforming the variable Ω to θ, i.e., $d\Omega = 2\pi \sin \theta \, d\theta$, and introducing equation (3.45) for $\cos \psi$, it is found that

$$\bar{\mu}_0 = \frac{2}{3A}. \tag{3.46}$$

Thus $\bar{\mu}_0$ decreases with increasing mass of the scattering nucleus as stated earlier.

THE AVERAGE LOGARITHMIC ENERGY DECREMENT

3.62. Instead of calculating the average energy change in an elastic scattering collision, a more useful quantity in the study of the slowing down of neutrons is the average value of the decrease in the natural logarithm of the neutron energy per collision, or the *average logarithmic energy decrement per collision*. It is the average for all collisions of $\ln E_1 - \ln E_2$, i.e., of $\ln (E_1/E_2)$, where E_1 is the energy of the neutron before and E_2 is that after a collision. If this quantity is represented by the symbol ξ then, taking into consideration the equal probability of all values of $\cos \theta$ from -1 to $+1$, it follows that

$$\xi \equiv \overline{\ln \frac{E_1}{E_2}} = \frac{\int_{-1}^{1} \ln \frac{E_1}{E_2} \, d(\cos \theta)}{\int_{-1}^{1} d(\cos \theta)},$$

where the integration limits refer to $\cos \theta$. Upon substituting the expression for E_2/E_1 from equation (3.39), it is found that

$$\xi = 1 + \frac{(A-1)^2}{2A} \ln \frac{A-1}{A+1}$$

or, using the definition of α from equation (3.40),

$$\xi = 1 + \frac{\alpha}{1-\alpha} \ln \alpha. \tag{3.47}$$

For values of A in excess of 10, a good approximation to equation (3.47) may be written as

$$\xi \approx \frac{2}{A + \frac{2}{3}}. \tag{3.48}$$

Even for $A = 2$, the error of equation (3.48) is only 3.3 per cent.

3.63. It will be seen from equation (3.47) that the value of ξ is independent of the initial energy of the neutron, provided scattering is symmetrical in the C system. In other words, in any collision with a given scattering nucleus, a neutron loses, on the average, the same fraction of the energy it had before collision.* This fraction decreases with increasing mass of the nucleus. It may be repeated here that the foregoing results involve the assumption of an essentially stationary nucleus prior to collision; this is justified only if the vibrational energy of the scattering nucleus is small compared to the kinetic energy of the neutron. Such is the case for neutron energies in excess of the thermal value.

3.64. The values of α and ξ for a number of elements, especially those of low mass number, are given in Table 3.2. The average number of collisions with nuclei of a particular moderator required to decrease the energy of fission neutron, with initial energy of, say, 2 Mev, to the thermal value at ordinary temperatures, i.e., 0.025 ev, is obtained upon dividing $\ln (2 \times 10^6 / 0.025)$ by ξ for the given moderator; thus,

$$\text{Average number of collisions to thermalize (2 Mev to 0.025 ev)} = \frac{\ln \left(\dfrac{2 \times 10^6}{0.025} \right)}{\xi} = \frac{18.2}{\xi}. \tag{3.49}$$

Some of the results obtained from this equation are included in Table 3.2.

TABLE 3.2. SCATTERING PROPERTIES OF NUCLEI

Element	Mass No.	α	ξ	Collisions to Thermalize
Hydrogen	1	0	1.000	18
Deuterium	2	0.111	0.725	25
Helium	4	0.360	0.425	43
Beryllium	9	0.640	0.206	86
Carbon	12	0.716	0.158	114
Uranium	238	0.983	0.000838	2172

* The arithmetic average value of E_1/E_2 per collision, which is equal to $\frac{1}{2}(\alpha + 1)$, is also independent of the initial neutron energy.

3.65. If the moderator is not a single element, but a compound containing n nuclei, the effective (or mean) value of ξ is given by

$$\bar{\xi} = \frac{\sigma_{s1}\xi_1 + \sigma_{s2}\xi_2 + \cdots + \sigma_{sn}\xi_n}{\sigma_{s1} + \sigma_{s2} + \cdots + \sigma_{sn}}, \tag{3.50}$$

where σ_s is the microscopic scattering cross section; the subscripts $1, 2, \cdots, n$ refer to the respective nuclei. For water, for example, containing two hydrogen nuclei and one oxygen nucleus, the appropriate expression for $\bar{\xi}_{H_2O}$ is

$$\bar{\xi}_{H_2O} = \frac{2\sigma_{s(H)} + \sigma_{s(O)}\xi_{(O)}}{2\sigma_{s(H)} + \sigma_{s(O)}},$$

since ξ_H is unity. Similar expressions are applicable to mixtures of elements or compounds.

SLOWING DOWN POWER AND MODERATING RATIO

3.66. According to equation (3.49), the value of ξ is inversely proportional to the number of scattering collisions required to slow down a neutron from fission energy to thermal values. It is thus a partial measure of the moderating ability of the scattering material. A good moderator is one in which there is a considerable decrease of neutron energy per collision on the average and, hence, it is desirable that ξ should be as large as possible. However, a large ξ is of little significance unless the probability of scattering, as indicated by the scattering cross section for neutrons with energy above thermal, is also large. The product $\xi\Sigma_s$, where Σ_s is the macroscopic scattering cross section of the moderator for epithermal neutrons, is called the *slowing down power*. For a compound, the slowing down power is given by

$$\overline{\xi\Sigma}_s = N(\nu_1\sigma_{s1}\xi_1 + \nu_2\sigma_{s2}\xi_2 + \cdots \nu_i\sigma_{si}\xi_i + \cdots \nu_n\sigma_{sn}\xi_n),$$

where ν_i is the number of atoms of the ith kind in the molecule and N is the number of molecules per unit volume of the moderator. The slowing down power is a better measure than ξ alone of the efficiency of a moderator, for it is equal to ξ/λ_s, and so represents the average decrease in the logarithm of the

TABLE 3.3. SLOWING DOWN PROPERTIES OF MODERATORS

Moderator	Slowing Down Power	Moderating Ratio
Water	1.28	58
Heavy water	0.18	21,000
Helium*	10^{-5}	45
Beryllium	0.16	130
Graphite	0.065	200

* At atmospheric pressure and temperature.

neutron energy per cm of path.   The slowing down powers of a number of materials consisting of (or containing) elements of low mass number are recorded in Table 3.3 and other data are given in the Appendix; the scattering cross sections are assumed to be constant in the energy range from 1 to 10^5 ev (cf. § 2.143).

3.67. Although the slowing down power gives a satisfactory indication of the ability of a material to slow down neutrons, it does not take into account the possibility that the substance may be a strong absorber of neutrons.   The slowing down power of boron, for example, is better than that of carbon, but boron would be useless as a moderator because of its high cross section for neutron absorption (§ 2.142).   In fact, any substance with appreciable absorption would be useless as a moderator.   It is for this reason that boron and lithium are omitted from the tables given above; their slowing down powers for neutrons are of no practical interest.

3.68. The ratio of the slowing down power, as defined above, to the macroscopic absorption cross section Σ_a, i.e., $\xi\Sigma_s/\Sigma_a$, called the *moderating ratio*, is consequently an important quantity, from the theoretical standpoint, in expressing the effectiveness of a moderator.   Some approximate values of this ratio, using absorption cross sections for thermal neutrons, are given in Table 3.3 [5].   It is seen that heavy water should be an excellent moderator.   Where the employment of a liquid is not convenient, beryllium and carbon are evidently possible, but less efficient, alternatives to heavy water.   Ordinary water can be used in certain circumstances, e.g., with enriched fuel, and has some advantages, especially the possibility of being able to design reactors of small size. Helium under high pressure, like water, could be used as both a moderator and heat-transfer medium.   The choice of the moderator in any particular circumstances will thus depend on other factors, in addition to slowing down power and moderating ratio.

LETHARGY

3.69. For many purposes it is convenient to express the energy E of a neutron in a logarithmic, dimensionless form by defining a quantity u, called the *lethargy* or *logarithmic energy decrement* [6]; thus,

$$u \equiv \ln \frac{E_0}{E}, \tag{3.51}$$

where E_0 is an arbitrary reference energy corresponding to zero lethargy.   This energy is usually taken as a high value, e.g., 10 Mev, so that essentially all the neutrons in a reactor have lower energies and hence positive lethargies.   It is seen from equation (3.51) that the lethargy of a neutron increases as it is slowed down.

3.70. If u_1 is the lethargy corresponding to E_1, the energy of a neutron before a scattering collision, and u_2 is the lethargy equivalent to E_2, the energy after a collision, then the lethargy change $u_2 - u_1$ is given by

$$u_2 - u_1 = \ln \frac{E_1}{E_2}.$$

Since the quantity ξ defined in § 3.62 is the average value of $\ln (E_1/E_2)$, it is evident that ξ may also be regarded as the average change in lethargy of a neutron per collision. Hence, the reciprocal, $1/\xi$, is equal to the average number of collisions per unit lethargy change. As stated in § 3.63, for spherically symmetric scattering in the C system, ξ (and hence $1/\xi$) is independent of the initial neutron energy. Thus, regardless of its energy, a neutron must, on the average, suffer the same number of collisions in a given medium to increase its lethargy by a specified amount. This fact represents one of the advantages of the use of the lethargy variable.

3.71. According to equation (3.51),

$$E = E_0 e^{-u},$$

so that the plot of E against u, as seen in Fig. 3.11, is exponential in nature.

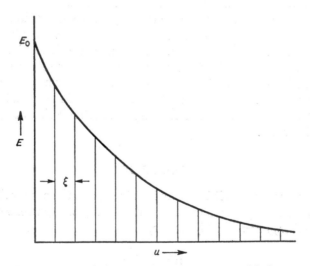

FIG. 3.11. Relationship between energy and lethargy

If a series of vertical lines are drawn at distances ξ apart, then, in view of the result derived in the preceding paragraph, the heights will represent the average values of the neutron energy for successive collisions. It is seen, therefore, that a neutron loses, on the average, considerably more energy in earlier scattering collisions than it does in the later ones.

SLOWING DOWN IN INFINITE MEDIA

ENERGY DISTRIBUTION WITHOUT ABSORPTION

3.72. Suppose that fast neutrons are being produced from a source, e.g., by fission, at a uniformly definite rate throughout the volume of a particular moderator. As the neutrons collide with the nuclei of the moderator, they steadily lose energy until they become thermal. But since fast neutrons are continuously being generated, a steady-state distribution of neutron energies will soon be attained. This energy distribution will depend on the extent to which neutrons are absorbed in the system and on their escape from it while being slowed down. For the present, however, it will be assumed that the moderator is infinite in extent, so that there is no loss of neutrons by leakage, and also that there is no absorption while slowing down. The ultimate fate of the thermal neutrons is of no consequence at the moment.

3.73. An expression for the energy distribution of the neutrons while being slowed down may be readily derived by considering the neutron balance during the slowing down process in an infinitesimal element dE in energy space, analogous to an infinitesimal element in volume space. Let $\phi(E)$ be the neutron flux of energy E *per unit energy interval*; the macroscopic scattering cross section at the same energy is $\Sigma_s(E)$. Then the number of neutrons per cm³ per sec leaving the element dE as a result of scattering collisions is $\Sigma_s(E)\phi(E)\,dE$.

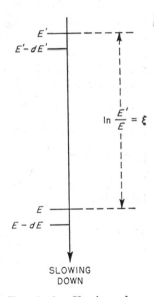

FIG. 3.12. Slowing down of neutrons

3.74. The rate at which neutrons are scattered into the energy interval dE while being slowed down can be calculated from the *slowing down density*, represented by the symbol q and defined as *the number of neutrons per cm³ per sec that slow down past a given energy E*. Consider an energy value E' such that $\ln (E'/E)$ is equal to ξ for the medium in which the neutrons are slowing down. Hence, *on the average*, a neutron scattered at E' will have its energy reduced to E (Fig. 3.12). Since it has been postulated that there is no loss of neutrons by absorption or leakage, the number of neutrons entering the energy interval dE between E and $E - dE$ will then be equal, on the average, to the number scattered in the interval from E' to $E' - dE'$. In view of the definition of ξ, the probability that a neutron of energy E' will be scattered in this interval is $[\ln E' - \ln (E' - dE')]/\xi$, i.e., $dE'/E'\xi$. If $q(E')$ is the slowing down density at E', then scattering occurs at the rate of $q(E')dE'/E'\xi$ neutrons per cm³ per sec in the interval dE', and this is the rate at which neutrons enter the energy interval dE. Because there are no losses, this rate may be written as $q(E)\ dE/E\xi$.

3.75. In the steady state, the neutron balance requires that the rate of scattering out of the interval dE shall be equal to the rate of scattering into this energy element, since no neutrons are lost by absorption or escape; hence,

$$\Sigma_s(E)\phi(E)\ dE = \frac{q(E)\ dE}{E\xi},$$

so that

$$\phi(E) = \frac{q(E)}{E\xi\Sigma_s(E)}. \tag{3.52}$$

A more rigorous treatment shows that equation (3.52) is satisfactory provided the neutron energy, E, is less than $\alpha^2 E_i$, where E_i is the initial (fission) energy [7]. It is thus not applicable to slowing down in a hydrogenous moderator, since α for hydrogen is zero. However, it is a good approximation over most of the slowing down energy range for moderators of larger mass number, even for deuterium.

3.76. Since there are no losses, the number of neutrons slowing down past any energy is constant; in other words, q is independent of energy in the case under consideration. Furthermore, Σ_s does not vary greatly with energy in the region just above thermal. Consequently, equation (3.52) implies that $\phi(E)$ is approximately proportional to $1/E$; this is the basis of the $1/E$ variation of the energy (or velocity) distribution for epithermal neutrons referred to in § 2.100.

Example 3.3. A reactor core contains enriched uranium as fuel and beryllium oxide as moderator ($\Sigma_s = 0.64$ cm^{-1} and $\xi = 0.17$ at a neutron energy of 7 ev). The thermal neutron flux is 2×10^{12} neutrons/(cm^2)(sec) and Σ_a for these neutrons in the fuel is 0.005 cm^{-1}; for each thermal neutron absorbed, 1.7 fission neutrons are produced. Neglecting all absorption during slowing down, estimate the epithermal neutron flux (at 7 ev) per unit *lethargy* interval.

If $\phi(u)$ is the neutron flux per unit lethargy interval and $\phi(E)$ is the flux per unit energy interval, then

$$\phi(u)\ du = -\phi(E)\ dE,$$

where u is the lethargy corresponding to the energy E; the minus sign is introduced because the lethargy increases as the energy decreases (see Fig. 3.11). In accordance with the definition of lethargy, e.g., equation (3.51),

$$du = -d \ln E = -\frac{dE}{E}$$

and so

$$\phi(u) = E\phi(E).$$

Hence, from equation (3.52),

$$\phi(u) = \frac{q(E)}{\xi\Sigma_s(E)},$$

where in the present case E is 0.7 ev.

The rate of fission neutron production is $1.7\Sigma_a\phi_{th} = (1.7)(0.005)(2 \times 10^{12}) = 1.7 \times 10^{10}$ neutrons/(cm^3)(sec). Since it is postulated that there is no absorption during slowing down, this represents the slowing down density at all energies.   Hence, $q(E)$ is 1.7×10^{10}, so that

$$\phi(u) = \frac{1.7 \times 10^{10}}{(0.17)(0.64)} = 1.6 \times 10^{11} \text{ neutrons/(cm}^2)(\text{sec) per unit lethargy interval.}$$

ENERGY DISTRIBUTION WITH ABSORPTION

3.77. When neutrons slow down in the presence of an absorber, the energy distribution differs from that derived above.   The slowing down density q is now no longer independent of the energy because neutrons are being lost by absorption while they slow down.   The change in the slowing down density across the energy interval dE arises from the absorption of neutrons in this interval; the rate of absorption is equal to $\Sigma_a(E)\phi(E) \, dE$, where $\Sigma_a(E)$ is the macroscopic absorption cross section for neutrons of energy E, so that

$$\frac{\partial q(E)}{\partial E} \, dE = \Sigma_a(E)\phi(E) \, dE$$

or

$$\frac{\partial q(E)}{\partial E} = \Sigma_a(E)\phi(E). \tag{3.53}$$

3.78. In considering the steady-state neutron balance in the energy element dE, it must be realized that neutrons are now removed by both scattering and absorption.   The total rate at which neutrons leave the interval dE is consequently $[\Sigma_s(E) + \Sigma_a(E)]\phi(E) \, dE$.   However, the rate at which neutrons are scattered into the interval while slowing down is the same as when there is no absorption, i.e., $q(E') \, dE'/E'\xi$, where E' has the same significance as before.   Provided the absorption between E' and E is not great, this is approximately equal to $q(E) \, dE/E\xi$.   Hence, in the steady state,

$$(\Sigma_s + \Sigma_a)\phi(E) \, dE \approx \frac{q(E)}{E\xi} \, dE,$$

or

$$\phi(E) \approx \frac{q(E)}{E\xi(\Sigma_s + \Sigma_a)}, \tag{3.54}$$

the energy argument following the cross sections having been omitted for simplicity of representation.   Upon substituting the value of $\phi(E)$ from equation (3.54) into (3.53) and integrating, the result is

$$q(E) = q_0 \exp\left[-\frac{1}{\xi} \int_E^{E_0} \frac{\Sigma_a}{\Sigma_a + \Sigma_s} \cdot \frac{dE}{E}\right], \tag{3.55}$$

where the subscript zero refers to the initial (or source) neutrons.

3.79. In Chapter 4, it will be seen that an important quantity in connection with the derivation of the conditions for reactor criticality is the *resonance escape probability*.   It is represented by $p(E)$ and is defined as *the fraction of* *source neutrons which escape capture while being slowed down to a particular energy* E. For a thermal-neutron reactor, the energy E of interest is, of course, the thermal value. The term "resonance" is used in describing the quantity $p(E)$ because most of the neutron captures will occur in the resonance region of the absorber, e.g., uranium-238. The fraction of neutrons which succeed in reaching an energy E is equal to the ratio of the slowing down density $q(E)$ at this energy to the initial value q_0. The latter is equal to the slowing down density at any energy in a medium in which there is no absorption or leakage; hence, $p(E)$ is equal to $q(E)/q_0$. It is apparent, therefore, from equation (3.55) that

$$p(E) = \exp\left[-\frac{1}{\xi} \int_E^{E_0} \frac{\Sigma_a}{\Sigma_a + \Sigma_s} \cdot \frac{dE}{E} \right]. \tag{3.56}$$

Some illustrations of the use of this expression for the resonance escape probability will be given in the next chapter.

SPATIAL DISTRIBUTION OF SLOWED DOWN NEUTRONS

FERMI AGE (CONTINUOUS SLOWING DOWN) MODEL

3.80. The next step in the discussion of the slowing down of neutrons is to investigate the spatial distribution of neutrons of various energies, resulting from diffusion during the slowing down process. The problem is important because the average (net vector) distance between the point at which a neutron originates and that at which it becomes thermal, as a result of slowing down, determines the neutron leakage in a finite medium. Consequently, it has a direct bearing on the critical size of a thermal reactor. If this average distance is large, then the reactor will have to be large, so as to reduce the probability of escape of neutrons before they have been slowed down to thermal energies. Thus the distribution of neutrons in space while slowing down determines their loss by leakage from a medium of finite dimensions.

3.81. A relatively simple, but approximate, approach to the problem is based on the *continuous slowing down model*, sometimes called the *Fermi age model* [8]. It is not applicable, however, to moderators containing very light nuclei, especially hydrogen. Consider a neutron produced at the fission (or source) energy E_0; it will travel for a certain time with this energy before it collides with a nucleus. After the collision its energy will be decreased, and it will then diffuse at this constant lower energy until it meets another nucleus. Since the neutron is now moving with a smaller speed, the time between collisions is, on the average, increased. This process of alternate diffusion at constant energy followed by a collision in which the energy is lowered will continue until the neutron is thermalized.

3.82. Since the average change in the logarithm of the neutron energy per collision, i.e., ξ, is independent of the energy, a plot of $\ln E$ of a neutron against time, t, from fission energy (E_0) down to thermal energy (E_{th}) should be of the form of Fig. 3.13. This consists of a series of steps of approximately equal height ξ, but of gradually increasing length. The horizontal lines represent the constant energy of the neutron while it is diffusing between collisions, and the lengths of these lines give the times elapsing between successive collisions.

3.83. Because individual neutrons behave differently, even if originating with the same energy and diffusing in the same medium, the plot of $\ln E$ against t, as in Fig. 3.13, will vary from one neutron to another. If the medium consists

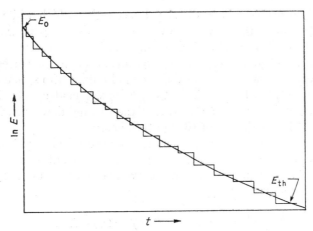

Fig. 3.13. Continuous slowing down approximation

of nuclei of moderate or high mass number, the spread of energies among the neutrons of the same generation is relatively small, and it is possible to represent the behavior of a large number of individual neutrons by means of an average. In these circumstances, too, the logarithmic change of energy per collision is small; that is to say, the height of the steps in Fig. 3.13 will be small. As a result, the series of steps may be replaced, without serious error, by a continuous curve, such as that shown in the diagram. This is the basis of the continuous slowing down (Fermi age) model. It is assumed that, during slowing down, neutrons lose energy continuously rather than discontinuously as is actually the case.

3.84. It should be emphasized that the continuous slowing down model can be valid, even as an approximation, only for moderators consisting of elements of fairly high mass number. It should not apply to hydrogen, deuterium, or to moderators containing these atoms, e.g., ordinary water or heavy water.* It is possible, as already seen, for a neutron to lose all its energy in a single collision with a hydrogen nucleus. There is, consequently, a considerable spread of ener-

---

* In practice, the Fermi age model turns out to be not too bad an approximation for heavy water (cf. § 3.75).

gies and it is not justifiable to represent the slowing down characteristics of a large number of neutrons by an average behavior. Furthermore, because of the relatively large changes in neutron energy per collision, and hence the small number of collisions required to reduce the energy to thermal values (see Table 3.2), the representation of the slowing down as continuous is completely invalid.

THE AGE EQUATION

3.85. The first problem is to develop an equation which will represent the spatial distribution of the slowing down density according to the continuous slowing down model in a nonabsorbing medium of finite dimensions. The following derivation, although not completely rigorous, leads to the correct result in a relatively simple manner. In a medium of finite dimensions, neutrons will inevitably escape at the boundaries. Consequently, neutron density (or flux) gradients will exist in the moderator, and the slowing down density will be a function of position as well as of the neutron energy.

3.86. By assuming that no neutrons are removed by absorption, the change in the slowing down density over an energy interval dE, at a given point in the medium, may be set equal to the rate at which neutrons of energy E diffuse away from that point; thus,

$$\frac{\partial q(E)}{\partial E}\, dE = -D(E)\nabla^2\phi(E)\, dE, \tag{3.57}$$

or

$$\frac{\partial q(E)}{\partial E} = -D(E)\nabla^2\phi(E), \tag{3.58}$$

where the right side of equation (3.57) is the rate of neutron diffusion per cm³ per sec, as expressed by equation (3.10), bearing in mind the definition of $\phi(E)$ given in § 3.73.

3.87. It was stated earlier that equation (3.52) is satisfactory, provided there is no absorption, in moderators of fairly high mass number. It is just to such media that the continuous slowing down model is applicable. Consequently, the value for $\phi(E)$ given by this equation may be substituted into equation (3.58), so that

$$\frac{\partial q}{\partial E} = -\frac{D}{\xi\Sigma_s E}\,\nabla^2 q,$$

or

$$\nabla^2 q = -\frac{\xi\Sigma_s E}{D}\cdot\frac{\partial q}{\partial E}, \tag{3.59}$$

the energy arguments being omitted from q, D, and Σ_s, although these quantities are all functions of the neutron energy.

3.88. A new variable $\tau(E)$ at energy E is now defined by

$$\tau(E) \equiv \int_{E_0}^{E} \frac{D}{\xi \Sigma_s E} \, dE, \tag{3.60}$$

where E_0 is the energy of the source neutrons. Upon making the transformation, whereby the variable E is replaced by τ, equation (3.59) reduces to

$$\nabla^2 q = \frac{\partial q}{\partial \tau}, \tag{3.61}$$

and this is known as the *Fermi age equation.* The quantity $\tau(E)$ is called the *Fermi age* or the *symbolic age.* It should be noted, however, that it is not a unit of time, but rather a length squared. This can be readily seen from equation (3.61), when it is realized that the Laplacian operator implies differentiation twice with respect to distance. Thus equation (3.61) represents the spatial distribution of the slowing down density in a nonabsorbing medium.

3.89. Although the age does not represent an elapsed time, it is nevertheless related to the chronological age of the neutrons, i.e., the time elapsing between formation as a source (fission) neutron and the attainment of a given energy. This can be seen, in a general way, from equation (3.60) which defines the age. When $E = E_0$, i.e., for neutrons of source energy, τ is zero. As E decreases, however, it is seen that $\tau(E)$ increases correspondingly. In other words, as the neutron slows down its age increases.

3.90. The foregoing results apply to a medium in which there is no absorption of neutrons, but it has been shown that, with a slight modification, the age equation can be used for a weakly absorbing medium. If $q(E)$ is a solution of equation (3.61) for the case of no absorption, then the corresponding slowing down density for a medium in which there is (weak) absorption is equal to $p(E)q(E)$, where $p(E)$ is the resonance escape probability for neutrons of energy E. This conclusion is seen to be in general agreement with the arguments in § 3.79.

SOLUTION OF THE AGE EQUATION: SIGNIFICANCE OF AGE

3.91. The Fermi age equation (3.61) is a form that is familiar to engineers; it is similar, for example, to the equation representing the unsteady-state conduction of heat in a continuous solid medium containing no sources or sinks.*

---

* The general form of the heat-conduction equation is

$$\nabla^2 T = \frac{1}{\alpha} \cdot \frac{\partial T}{\partial t},$$

where T is the temperature, t the time, and α is the thermal diffusivity (see W. H. McAdams, "Heat Transmission," 3rd Ed., McGraw-Hill Book Co. Inc., New York, 1954, p. 34). It will be noted that τ in the Fermi age equation actually corresponds to a time in the heat-conduction equation, the difference in dimensions being accounted for by α which has the dimensions of (length)^2/time.

Solutions for various forms are thus to be found in the literature. A case of immediate interest is that of a point source of fast neutrons, assumed to be mono-energetic, undergoing continuous slowing down in a nonabsorbing medium. The solution of the age equation is then

$$q(r, \tau) = \frac{e^{-r^2/4\tau}}{(4\pi\tau)^{3/2}}, \tag{3.62}$$

where $q(r, \tau)$ is the slowing down density for neutrons of age τ at a distance r from a point source of 1 neutron per sec. This expression is that for a Gauss error curve, and consequently the distribution of slowing down density for a given age is sometimes referred to as Gaussian.

3.92. The shape of the curve representing the variation of $q(r, \tau)$ with r, for a given value of τ, is shown in Fig. 3.14. It is seen that when τ is small, i.e., the

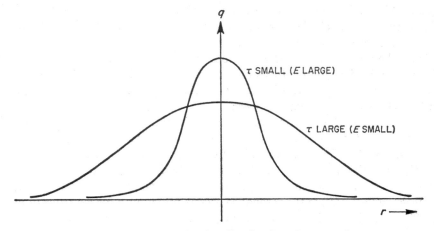

Fɪɢ. 3.14. Slowing down density distribution about a point source

neutron is near the source energy, the curve is high and narrow. For τ large, i.e., for neutrons of lower energy, however, the curve is lower and more spread out. This is to be expected on physical grounds. A small value of the age means that the neutrons have suffered little slowing down, and so they have not diffused far from the source. Consequently most of the neutrons will have high energies and be in the neighborhood of the source; this corresponds to a high and narrow distribution curve. On the other hand, when τ is large, the neutrons will have undergone considerable slowing down and will have diffused appreciable distances from the source. The distribution curve of the slowing down density will thus be low and broad.

3.93. A more precise physical significance can be associated with the Fermi age by calculating the mean square (net vector) slowing down distance, denoted by $\overline{r_s^2}$, about a point source, in a manner similar to that used in connection with the diffusion length (§ 3.35).   Thus,

$$\overline{r_s^2} = \frac{\int_0^\infty r^2 [4\pi r^2 q(r, \tau)] \, dr}{\int_0^\infty 4\pi r^2 q(r, \tau) \, dr}, \tag{3.63}$$

where $q(r, \tau)$ is given by equation (3.62), so that

$$\overline{r_s^2} = \frac{\int_0^\infty r^4 e^{-r^2/4\tau} \, dr}{\int_0^\infty r^2 e^{-r^2/4\tau} \, dr} = 6\tau$$

or

$$\tau = \frac{1}{6} \overline{r_s^2}. \tag{3.64}$$

The age τ is consequently one-sixth of the mean square (net vector) distance between the point of origin of the neutron, where its age is zero, to the point at which its age is τ.*

3.94. The age of neutrons of specified energy in a given medium can be determined experimentally, provided a material is available which strongly absorbs neutrons of that energy [9].   For example, the slowing down density can be measured at indium resonance, i.e., energy about 1.4 ev, at various distances from a point source of fast neutrons.   If an indium foil is covered with cadmium to prevent thermal neutrons from reaching the indium, the saturation activity, A_0 (§ 2.119), of the latter may be taken to be proportional to the neutron slowing down density at 1.4 ev.   Hence, from equation (3.63),

$$\overline{r_s^2} = \frac{\int_0^\infty A_0 r^4 \, dr}{\int_0^\infty A_0 r^2 \, dr} = 6\tau,$$

and $\overline{r_s^2}$ can be obtained by graphical integration, using measurements of the saturation activity at various distances in the moderator from a point source of fast, e.g., fission, neutrons.   In order to extend the integration to infinity, the variation of A_0 at large distances is taken to be proportional to $e^{-r/\lambda}/r^2$, i.e., an exponential absorption combined with an inverse square attenuation (cf. § 9.38).   The values of λ and the proportionality constant are derived from actual measurements of A_0 at moderately large distances from the source.   The procedure described above gives $\overline{r_s^2}$ for indium resonance neutrons and a correction is applied to obtain that for thermal neutrons [10].

Example 3.4.   An experiment was performed to measure the Fermi age of fission neutrons slowing down to indium resonance in an assembly of beryllium oxide blocks, using a fission plate as the neutron source. The corrected indium foil saturation activities,

---

* Attention may be called to the analogy between the neutron age and the square of the diffusion length as defined by equation (3.29).

normalized to 1000 at the source plate, are given in the accompanying table. Determine the neutron age at the indium resonance energy.

Distance from Fission Plate (r cm)	Relative Activity (A_0)	$A_0 r^2$	$A_0 r^4$
0	1000	–	–
4.0	954	1.53×10^4	2.44×10^5
7.9	828	5.16×10^4	5.16×10^6
11.7	660	9.04×10^4	1.24×10^7
19.35.........	303	1.13×10^5	4.24×10^7
27.15.........	103	7.59×10^4	5.59×10^7
34.8	29.3	3.55×10^4	4.3×10^7
42.7	7.2	1.31×10^4	2.4×10^7
50.5	1.6	4.08×10^3	1.04×10^7
62.2	0.45	1.74×10^3	6.7×10^6
77.8	0.15	9.08×10^2	5.5×10^6

According to the equation in § 3.94, it is necessary to determine the integrals of $A_0 r^2\, dr$ and $A_0 r^4\, dr$ over all values of r from zero to infinity. The integrations are performed graphically by plotting the values of $A_0 r^2$ and $A_0 r^4$, calculated above, and determining the areas under the curves; the curves are shown in Fig. 3.15. Summation of the ordinates

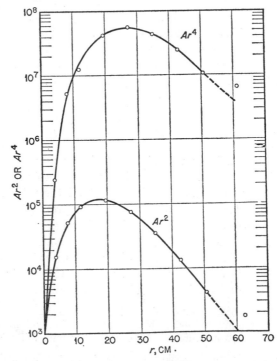

FIG. 3.15. Data for evaluation of Fermi age in beryllium oxide

gives the ratio of the areas as 560; this is equal to 6τ. Hence, τ is 93 cm²; this is the age of fission neutrons at indium resonance in beryllium oxide.

3.95. Some approximate values of the age of thermal neutrons at 20°C from fission sources in four important moderating media are given in Table 3.4 [11]. For ordinary water, at least, this is not the true Fermi age, since the continuous slowing down model cannot be applied to moderators containing hydrogen. The result is actually $\overline{r_s^2}/6$ for thermal neutrons about a point source of fission neutrons.

TABLE 3.4.   AGE OF THERMAL NEUTRONS
AT 20°C FROM FISSION SOURCE

Moderator	Age (cm²)
Water	31
Heavy water	120
Beryllium	85
Graphite	350

SLOWING DOWN AND MIGRATION LENGTHS

3.96. As indicated above, the Fermi age of neutrons is related to the mean square distance traveled while slowing down. The square root of the age is thus called the *slowing down length*. For thermal neutrons of age τ_{th}, for example, $\sqrt{\tau_{th}}$ is a measure of the net vector (or crow-flight) distance traveled from their formation as fission neutrons to their attainment of thermal energy.

3.97. The sum of the square of the diffusion length of slow (thermal) neutrons, i.e., L^2, and of the age, τ, is called the *migration area*, M^2, i.e.,

$$M^2 \equiv L^2 + \tau, \qquad (3.65)$$

and the square root is referred to as the *migration length*, M, i.e.,

$$M \equiv \sqrt{L^2 + \tau}. \qquad (3.66)$$

3.98. For thermal neutrons, L^2 is proportional to the mean square (net vector) distance the neutron travels from the point at which it becomes thermal to that at which it is captured. It is evident, therefore, that the migration length for thermal neutrons is related to the total (net vector) distance covered between

TABLE 3.5.   MIGRATION LENGTH OF THERMAL NEUTRONS

Moderator	Diffusion Length (cm)	Slowing Down Length (cm)	Migration Length (cm)
Water	2.76	5.6	6.2
Heavy water	100	11.0	101
Beryllium	21	9.2	23
Graphite	64.2	18.7	57

birth as a fission neutron and capture as a thermal neutron. The migration
lengths for thermal neutrons in four common moderators, at ordinary temperatures, are recorded in Table 3.5. The diffusion and slowing down lengths are
also included. For ordinary water and heavy water, M^2 is defined by

$$M^2 \equiv L^2 + \tfrac{1}{6}\overline{r_s^2}, \tag{3.67}$$

where $\overline{r_s^2}$ is the mean square slowing down distance from fission to thermal energies in these media. In fact, equation (3.67) may be regarded as the exact
definition, applicable to all moderators, with equation (3.65) representing an
alternative definition based on the continuous slowing down model. The migration length has a direct bearing on the size of a critical reactor, as will be seen
in the next chapter.

SYMBOLS USED IN CHAPTER 3

A	atomic weight (or mass number)
A	arbitrary constant
A_0	saturation activity
a	thickness of diffusion medium
C	center of mass system
C	arbitrary constant
cosh	hyperbolic cosine; $\cosh x = \tfrac{1}{2}(e^x + e^{-x})$
D	diffusion coefficient (of neutron flux)
D_0	diffusion coefficient (of neutron density)
E	neutron energy
E_0	initial (fission) neutron energy
E_1	neutron energy before collision
E_2	neutron energy after collision
$\mathbf{J}$	neutron current vector, neutrons/(cm²)(sec)
J_x	neutron current in x direction
J_+, J_-	neutron currents in opposite directions
L	laboratory system
L	diffusion length
M	migration length
M^2	migration area
m	neutron mass
n	neutron density, neutrons/cm³
$p, p(E)$	resonance escape probability; at energy E
Q	neutron source strength
q	slowing down density, neutrons/(cm³)(sec)
q_0	initial (fission energy) slowing down density
$q(E)$	slowing down density at energy E
$q(\tau)$	slowing down density at age τ
r	radial distance

$\overline{r^2}$	mean square distance traveled by thermal neutron
$\overline{r_s^2}$	mean square distance traveled by neutron while slowing down
S	source term in diffusion equation
sinh	hyperbolic sine; $\sinh x = \frac{1}{2}(e^x - e^{-x})$
t	time
u	lethargy
v	neutron velocity
x, y, z	coordinates
α	$(A - 1)^2/(A + 1)^2$
$-\gamma$	slope of plot
θ	scattering angle in C system
κ	Σ_a/D, reciprocal of diffusion length
λ_s	scattering mean free path
λ_{tr}	transport mean free path
$\bar{\mu}_0$	average cosine of neutron scattering angle
ξ	average logarithmic energy decrement per collision
$\Sigma_a; \Sigma_a(E)$	macroscopic absorption cross section; at energy E
$\Sigma_s; \Sigma_s(E)$	macroscopic scattering cross section; at energy E
Σ_t	total macroscopic cross section
Σ_{tr}	macroscopic transport cross section
σ_s	microscopic scattering cross section
$\tau; \tau(E)$	Fermi age, cm²; at energy E
ϕ	neutron flux, neutrons/(cm²)(sec)
$\phi(E)$	neutron flux at energy E per unit energy interval
$\phi(x)$	neutron flux at point having coordinate x
ψ	scattering angle in L system
∇^2	Laplacian operator

REFERENCES FOR CHAPTER 3

1. For a more detailed discussion of the topics in this chapter, see S. Glasstone and M. C. Edlund, "The Elements of Nuclear Reactor Theory," D. Van Nostrand Co., Inc., Princeton, N. J., 1952, Chs. V and VI.
2. Ref. 1, pp. 92 *et seq.*
3. Ref. 1, pp. 116 *et seq.*
4. D. J. Hughes, "Pile Neutron Research," Addison-Wesley Publishing Co., Inc., Reading, Mass., 1953, p. 219; J. B. Hoag, "Nuclear Reactor Experiments," D. Van Nostrand Co., Inc., Princeton, N. J., 1958, p. 77; W. J. Sturm (Ed.), "Reactor Laboratory Experiments," U. S. AEC Report ANL-6410 (1961), p. 299.
5. For additional data, see J. L. Feuerbacher, U. S. AEC Report GAT-T-673 (1959); T. S. Lundy and E. E. Gross, U. S. AEC Report ORNL-2891 (1960).
6. Ref. 1, p. 146.
7. R. V. Meghreblian and D. K. Holmes, "Reactor Analysis," McGraw-Hill Book Co., Inc., New York, 1960, p. 96; see also, A. M. Weinberg and E. P. Wigner, "The Physical Theory of Neutron Chain Reactors," University of Chicago Press, 1958, p. 287.

8. Ref. 1, pp. 172 *et seq.*
9. D. J. Hughes, Ref. 4, p. 123; J. B. Hoag, Ref. 4, p. 82; W. J. Sturm, Ref. 4, p. 317.
10. A. M. Weinberg and E. P. Wigner, Ref. 7, p. 331.
11. For additional data, see H. Goldstein, *et al.*, U. S. AEC Report ORNL-2639 (1961); T. S. Lundy and E. E. Gross, Ref. 5.

PROBLEMS

1. Calculate (a) the diffusion coefficient, (b) the extrapolation distance, and (c) the diffusion length, of thermal neutrons in reactor-grade graphite. The microscopic scattering and absorption cross sections are 4.8 barns and 0.0037 barn, respectively, and the density is 1.70 g/cm³.

2. A point source emits 10^5 monoenergetic neutrons per second into an infinite (nonmultiplying) medium. Plot on the same graph the variation of the flux with distance from the source up to 1 meter for (a) ordinary water, (b) heavy water, and (c) graphite.

3. Repeat the exercise in Problem 2 assuming a finite spherical medium with a radius of 1 meter. Compare and discuss the results.

4. By treating an infinite plane source of monoenergetic neutrons as made up of an infinite number of point sources spread over the plane, use equation (3.26) to show that the flux distribution in the x direction perpendicular to the plane source, emitting Q neutrons/(cm²)(sec) in an infinite, nonabsorbing medium, is represented by $Qe^{-\kappa x}/2\kappa D$.

5. Derive an expression for the neutron flux distribution in the radial direction about a line source emitting Q monoenergetic neutrons/(cm)(sec) in an infinite medium.

6. An infinite slab has a plane thermal-neutron source at the boundary. Derive in detail an expression for the net vector (crow-flight) distance a neutron travels before it is absorbed.

7. Show how the curves in Fig. 3.8 would be affected if the medium were (a) ordinary water, (b) heavy water, instead of graphite.

8. Neutrons of 1.5-Mev energy are introduced at the rate of 2×10^5 neutrons/(cm³)(sec) into an infinite slab of graphite. Calculate the number of elastic scattering collisions occurring per second per cm³ in the energy interval from 0.5 to 0.3 Mev.

9. Consider the scattering of a neutron by a deuteron in the laboratory system. What is the largest angle of scattering and what is the *maximum* possible fractional energy loss per collision? Derive a general relationship for the *average* fractional energy loss per collision.

10. Compare the properties of liquid diphenyl ($C_{12}H_{10}$, density 1.0 g/cm³) as a moderator with those of ordinary water, heavy water, and graphite. The following epithermal scattering cross sections are to be used: H, 20; D, 3.4; C, 4.7; and O, 3.8 barns. The thermal absorption cross sections are given in Table 2.6.

11. Indium foils were suspended at various distances, r, above a fission plate neutron source in a 900-gal tank of water. Activation measurements gave the following results:

r (cm)	$A_0 r^2 \times 10^6$	r (cm)	$A_0 r^2 \times 10^6$
4.23	556.7	21.21	137.3
5.73	653.8	23.81	84.9
7.37	783.3	26.41	59.4
9.12	761.2	29.15	44.6
10.88	566.1	32.11	26.5
14.08	410.6	37.10	14.7
16.38	264.7	42.57	6.3
18.77	187.1	50.28	2.4

A_0 is the corrected saturation activity in counts per minute of the foil at the indium resonance (1.44 ev) energy.   Determine the Fermi age of fission neutrons at 1.44 ev.

12. Consider a point source of fast neutrons and derive an expression for the actual mean net vector (crow-flight) distance traveled by a neutron while slowing down.   Compare the result with the so-called slowing down length defined in the text.

Chapter 4

REACTOR THEORY: THE STEADY STATE

THE CRITICALITY CONDITION

MULTIPLICATION FACTORS

4.1. In the discussion of the diffusion and slowing down of neutrons in the preceding chapter, no special attention was paid to the nature of the neutron source nor to the fate of the slowed down (or thermal) neutrons. However, when considering the neutron balance in a chain reacting system in order to determine its critical size, it is necessary to allow for the absorption of the thermal neutrons and for the production of fission neutrons, as well as for neutrons lost by escaping from the system, i.e., for leakage (§ 3.2). If a self-sustaining chain reaction is to be possible in a reactor of finite size, at least as many neutrons must be produced by fission as are absorbed or lost in one way or another.

4.2. One of the basic properties of a multiplying system in which neutrons are being produced by fission is the *infinite multiplication factor*. It is defined as *the ratio of the number of neutrons resulting from fission in each generation to the number absorbed in the preceding generation in a system of infinite size.* In an infinitely large system there is no loss of neutrons by escape (or leakage) and it is only by absorption in fuel, moderator, etc., that neutrons would be removed from the system. The condition for criticality, i.e., for a self-sustaining fission chain to be just possible, is that the infinite multiplication factor, represented by k_∞, should be unity; the number of neutrons produced in each generation would then be exactly equal to the number lost. Thus a steady state would be maintained with a constant neutron density and the chain reaction would proceed at a definite constant rate.

4.3. In a system of finite size, however, some neutrons are lost by leaking out; the criticality condition is then described by means of the *effective multiplication factor*, k_{eff}, defined as *the ratio of the number of neutrons resulting from fission in each generation to the total number lost by both absorption and leakage in the preceding generation.* The requirement for criticality in a finite system is thus

$k_{eff} = 1$; in these circumstances a steady-state fission chain would be possible. If the conditions are such that $k_{eff} < 1$, the chain would be convergent and would gradually die out. In each generation more neutrons would be lost in one way or another than are produced by fission, and so the neutron density, and hence the fission rate, would decrease steadily. Such a system is said to be *subcritical*. Finally, if $k_{eff} > 1$, the chain is divergent and the system is described as *supercritical*. More neutrons are produced than are lost in each generation, so that both the neutron population and fission rate increase continuously.

4.4. From the definitions of the infinite and effective multiplication factors given above, it is seen that

$$\frac{k_{eff}}{k_\infty} = \frac{\text{Neutrons absorbed}}{\text{Neutrons absorbed + neutrons leaking out}} \equiv P.$$

The fraction defining P is thus a measure of the probability that neutrons will not leak out of the finite system, but will remain until absorbed.* Hence, it is possible to write

$$k_{eff} = k_\infty P,$$

where P is called the *nonleakage probability* of the system. The problem of deriving the condition for criticality, i.e., $k_{eff} = 1$, thus falls into two parts. The first is the evaluation of the infinite multiplication factor which is a function of the materials, e.g., fuel, moderator, coolant, structure, etc., of the reactor. In a heterogeneous (or lattice) system of fuel and moderator, the physical arrangement of the materials is also important (§ 4.92 *et seq.*). The second involves the determination of the nonleakage probability, which is dependent partly on the materials but also largely on the geometry, i.e., size and shape, of the system. The nonleakage probability increases with size, for a reactor of given composition, and approaches unity for a system of infinite size, i.e., when k and k_{eff} are identical.

THE FOUR-FACTOR FORMULA

4.5. In order to determine the infinite multiplication factor for a given system, it is convenient to divide the factor into four parts. The following treatment refers more specifically to a thermal reactor in which most of the fissions arise from the capture of slow neutrons. Before proceeding with the derivation, however, it is necessary to define the quantity η as the *average number of neutrons liberated directly by fission for every thermal neutron absorbed in the fuel*. The term "fuel" as used here refers to the reactor material containing the fissile species. It may consist of the pure fissile nuclide, or fertile material may be present in

---

* The tacit assumption involved here is that the energy distribution of the neutrons will be the same in a system of finite size as in one that is infinitely large. This is essentially the case for large and moderately large reactors.

addition. The quantity η must be distinguished from ν which was defined in § 2.164 as the number of neutrons liberated for every neutron absorbed in a *fission reaction*. From the two definitions, it follows that

$$\eta = \nu \times \frac{\text{Neutrons absorbed in fission reactions}}{\text{Total number of neutrons absorbed in fuel}}. \tag{4.1}$$

This is the upper limit of the infinite multiplication factor for a system employing the given fuel.

4.6. An alternative way of expressing the relationship between η and ν is to write equation (4.1) as

$$\eta = \nu \frac{\Sigma_f}{\Sigma_a},$$

where Σ_f is the macroscopic fission cross section for the fissile nuclide and Σ_a is the total absorption cross section, for fission and nonfission processes, in the fuel material. For example, in a mixture of uranium-235 and uranium-238,

$$\eta = \nu \frac{\Sigma_f^{235}}{\Sigma_a^{235} + \Sigma_a^{238}}$$

$$= \nu \frac{N^{235}\sigma_f^{235}}{N^{235}\sigma_f^{235} + N^{235}\sigma_c^{235} + N^{238}\sigma_c^{238}}$$

$$= \nu \frac{\sigma_f^{235}}{\sigma_f^{235} + \sigma_c^{235} + \sigma_c^{238}/r},$$

where the N's represent the numbers of the indicated nuclei per unit volume and r, equal to N^{235}/N^{238}, is the ratio of the two isotopes in the fuel; σ_f and σ_c represent the microscopic fission and capture (nonfission) cross sections, respectively. Some values of η for the three fissile nuclides and for natural uranium are given in Table 4.1. It is because η is not significantly greater than 2

TABLE 4.1. AVERAGE NUMBER OF NEUTRONS LIBERATED
PER NEUTRON ABSORBED IN FUEL

	U-233	U-235	Pu-239	Natural U
Thermal (2200 m/sec) neutrons	2.27	2.06	2.10	1.33
Fast neutrons	2.60	2.18	2.74	1.09

for thermal-neutron fission of both uranium-235 and plutonium-239 that neither breeding nor pseudobreeding is possible with these nuclides, as stated in § 1.63. Only when η is appreciably greater than 2, to allow for inevitable loss of neutrons by leakage and parasitic capture in coolant, structure, etc., can breeding be realized.

4.7. As fairly typical of a thermal reactor system, the fuel is assumed to consist either of natural uranium or uranium that has been enriched to the extent of a few per cent in uranium-235. The arguments to be presented below are, however, applicable to other fuels.   Suppose that at a given instant, representing the initiation of a generation, there are available n thermal neutrons which are absorbed in fuel; then, as a result, $n\eta$ fast neutrons will be produced by fission. Before these fast neutrons have slowed down appreciably some will be captured by, and cause fission of, uranium-235 and uranium-238 nuclei.   At neutron energies greater than about 1 Mev, most of the fast-neutron fissions in the fuel will be of uranium-238 because of its greater relative macroscopic cross section. Since more than one neutron is produced, on the average, in each fast-neutron fission there will be an increase in the number of neutrons available.   Allowance for this effect may be made by introducing the *fast-fission factor*, denoted by ϵ; it is defined as *the ratio of the total number of (fast) neutrons slowing down past the fission threshold of uranium-238 to the number produced by thermal-neutron fissions.* Consequently, the capture of n thermal neutrons in fuel will result in $n\eta\epsilon$ fast neutrons slowing down past the fission threshold of uranium-238, i.e., about 1 Mev.

4.8. In the course of further slowing down some of the neutrons are captured in nonfission processes, and the fraction escaping capture is p, the *resonance escape probability* discussed in § 3.79.   Hence, the number of neutrons which reach thermal energies is $n\eta\epsilon p$.   According to equation (3.56), the resonance escape probability depends on the macroscopic absorption and scattering cross sections of the fuel-moderator system.   It is thus determined both by the nature of the materials and their proportions.   Some values of p for different systems will be given later.

4.9. When the neutrons have become thermalized, they will diffuse for a time in the infinite system until they are ultimately absorbed by fuel, by moderator, or by such impurities, often referred to as "poisons," as may be present (§ 1.55). Of the thermal neutrons, therefore, a fraction f, called the *thermal utilization*, will be absorbed in fuel material; the value of f is thus represented by

$$f = \frac{\text{Thermal neutrons absorbed in fuel}}{\text{Total thermal neutrons absorbed}}.$$

Like the resonance escape probability, f depends on the nature of the fuel and moderator and on their relative amounts.

4.10. According to equation (2.34), the rate at which thermal neutrons are absorbed in unit volume per unit time is equal to $\Sigma_a\phi$, where Σ_a is the appropriate macroscopic cross section for the absorption of thermal neutrons and ϕ is the thermal flux.   The rate of absorption in a volume V will then be $V\Sigma_a\phi$ per unit time.   The general expression for the thermal utilization is thus

$$f = \frac{V_u\Sigma_{au}\phi_u}{V_u\Sigma_{au}\phi_u + V_m\Sigma_{am}\phi_m + V_i\Sigma_{ai}\phi_i}, \tag{4.2}$$

where the subscripts u, m, and i indicate fuel (uranium), moderator, and impurities, respectively.   The manner in which this equation is used depends upon whether the system under consideration is homogeneous or heterogeneous.

4.11. For a *homogeneous system*, i.e., one in which fuel and impurities are uniformly dispersed throughout the moderator, V_u, V_m, and V_i are identical, each being equal to the volume of the whole system. Furthermore, the neutron flux is the same in each constituent, so that equation (4.2) reduces to the form

$$f(\text{homo}) = \frac{\Sigma_{au}}{\Sigma_{au} + \Sigma_{am} + \Sigma_{ai}}. \tag{4.3}$$

The macroscopic cross sections are the homogeneous system values determined in the manner described in § 2.112, based on the fact that each substance is uniformly distributed throughout the whole volume.

4.12. For a *heterogeneous system*, i.e., one in which the fuel, in particular, is present in the moderator in the form of discrete pieces, e.g., rods or plates, V_u, V_m, and V_i are the respective volumes of the individual fuel, moderator, and impurity materials, and not the volume of the whole system. Similarly, the macroscopic cross sections are determined on this basis, so that the values are the same as for the materials in separate form. An alternative version of equation (4.2) is useful for some purposes. It can readily be shown that $V\Sigma_a$ for any species in a heterogeneous system is equal to $V_t\bar{\Sigma}_a$, where V_t is the total volume of the system and $\bar{\Sigma}_a$ is the macroscopic cross section calculated by assuming a uniform distribution of all the materials at the same overall composition as in the heterogeneous system; hence, equation (4.2) can be written as

$$f(\text{hetero}) = \frac{\bar{\Sigma}_{au}\phi_u}{\bar{\Sigma}_{au}\phi_u + \bar{\Sigma}_{am}\phi_m + \bar{\Sigma}_{ai}\phi_i}. \tag{4.4}$$

4.13. Since all the $n\eta\epsilon p$ neutrons reaching thermal energies in the infinite system are absorbed, the number absorbed in fuel in each generation will be $n\eta\epsilon pf$. For the present purpose, it is convenient to define the infinite multiplication factor k_∞ for a thermal reactor as the ratio of the number of thermal neutrons produced (and hence absorbed) in one generation to the number of thermal neutrons produced (or absorbed) in the preceding generation, in an infinite medium; consequently,

$$k_\infty = \frac{n\eta\epsilon pf}{n} = \eta\epsilon pf. \tag{4.5}$$

This is sometimes referred to as the *four-factor formula*. The division of the infinite multiplication factor into four parts simplifies the procedure for evaluating this important property of a multiplying system. The methods used, which depend on whether the system is homogeneous or heterogeneous, will be described later.

Example 4.1. Determine the infinite multiplication factor of a uniform mixture of uranium-235 and beryllium oxide in the atomic (or molecular) ratio of 1 to 10,000. The value of σ_a for beryllium oxide is 0.010 barn. The resonance escape probability and the fast-fission factor may be taken to be unity.

Since p and ϵ are unity, $k_\infty = \eta f$ by equation (4.5). From Table 4.1, η is 2.06 for uranium-235, and for a uniform mixture f can be obtained by means of equation (4.3), with the impurity term omitted. Thus,

$$f = \frac{\Sigma_U}{\Sigma_U + \Sigma_{BeO}} = \frac{N_U\sigma_U}{N_U\sigma_U + N_{BeO}\sigma_{BeO}} = \frac{\sigma_U}{\sigma_U + (N_{BeO}/N_U)\sigma_{BeO}}$$

$$= \frac{683}{683 + (10^4)(0.010)} = 0.87,$$

where σ_U is obtained from Table 2.8. Hence,

$$k_\infty = 2.06 \times 0.87 = 1.79.$$

THE ONE-GROUP CRITICAL EQUATION

4.14. The criticality condition for a reactor of finite size, as seen above, is that $k_{\text{eff}} = 1$, or, in other words, that $k_\infty P = 1$. In order to make this result useful for calculating the critical size of a finite reactor, it is necessary to express the nonleakage probability, P, in terms of the physical properties of the fuel-moderator system. Since neutrons of all velocities will escape, at rates determined by the diffusion coefficients and the concentration gradients at all velocities, it is apparent that a reasonably complete calculation of leakage would be a very complex problem. In order to simplify the treatment, a number of approximate methods have been proposed and some of them will be considered here. The following discussion is applicable to *bare reactors*, i.e., reactors without reflectors (§ 1.52). The effect of a reflector is treated in § 4.46 *et seq.*

4.15. The simplest approach to the development of a reactor critical equation is to consider all the neutrons as having the same energy. This is called the "one-group" diffusion method, in which it is supposed that all production, diffusion-leakage, and absorption of neutrons occurs at a single energy. The approximation that results is more suited to a fast-reactor system, in which there is relatively little slowing down of the neutrons, than it is to a thermal reactor, in which neutrons having a wide range of energies are present.

4.16. When a reactor is critical, a steady-state neutron density is maintained without any extraneous (or primary) source; hence, in the steady-state diffusion equation (3.13), i.e.,

$$D\nabla^2\phi - \Sigma_a\phi + S = 0, \tag{4.6}$$

the source term S is derived entirely from fission neutrons produced from the fuel material in the system. As already seen, for each neutron absorbed in fuel, k_∞ fission neutrons are produced; in the case under consideration, the rate of neutron absorption per unit volume is $\Sigma_a\phi$ and so S is equal to $k_\infty\Sigma_a\phi$, since it is postulated that all neutrons have the same energy. Consequently, for a critical reactor system, equation (4.6) takes the form

$$D\nabla^2\phi - \Sigma_a\phi + k_\infty\Sigma_a\phi = 0$$

or, upon rearrangement,

$$\nabla^2\phi + \left[\frac{(k_\infty - 1)\Sigma_a}{D}\right]\phi = 0. \qquad (4.7)$$

Because all the neutrons are assumed to have the same energy, D/Σ_a is equal to L^2, the square of the diffusion length for that energy (§ 3.36); hence, equation (4.7) becomes

$$\nabla^2\phi + \frac{(k_\infty - 1)}{L^2}\phi = 0. \qquad (4.8)$$

4.17. The distribution of the neutron flux as a function of position in any multiplying system, not necessarily critical, can be represented (cf. § 3.20), with appropriate boundary conditions, by

$$\nabla^2\phi + B^2\phi = 0, \qquad (4.9)$$

where B^2 is called the *buckling* of the system because it is a measure of the bending (or buckling) of the spatial distribution of the neutron flux. It follows from equations (4.8) and (4.9) that the criticality condition for a bare reactor according to the one-group treatment can be written as

$$\frac{k_\infty - 1}{L^2} = B_c^2, \qquad (4.10)$$

where B_c^2 is the *critical buckling*, i.e., the value of the buckling in the critical system. It is seen from equation (4.10) that the critical buckling is determined, in the one-group method, by the quantities k_∞ and L^2 which are properties of the fuel-moderator (multiplying) system. For this reason, the quantity defined by equation (4.10) is sometimes called the *material buckling*. It is possible, as will be shown in § 4.29 *et seq.*, to solve equation (4.9) for a specified geometry, with appropriate boundary conditions, and so obtain a general expression for B^2 in terms of the dimensions of the system. If this result is set equal to the critical (or material) buckling, as given by equation (4.10), the size of the critical system for the given geometry can be readily obtained.

4.18. Upon rearrangement, equation (4.10) can be written in the form

$$\frac{k_\infty}{1 + L^2B_c^2} = 1, \qquad (4.11)$$

which is the one-group critical equation. Since the general condition for criticality is that $k_\infty P = 1$, it follows that

$$k_\infty\left(\frac{1}{1 + L^2B_c^2}\right) = 1 = k_\infty P.$$

Consequently, the factor $1/(1 + L^2B_c^2)$ represents the one-group nonleakage probability for the critical system.

THE TWO-GROUP CRITICAL EQUATION

4.19. In the two-group approximation, which is reasonably good for large thermal reactors, the neutrons are considered to be either fast or thermal. Although there are, strictly speaking, no neutrons of intermediate energy, allowance is made for the fact that there is an equivalent of resonance absorption. That is to say, not all the neutrons leaving the fast group enter the thermal group, since some are lost by capture in nonfission reactions. It will be shown in § 4.26 that k_∞/p fast (fission) neutrons are produced for each thermal neutron that is absorbed. Hence, the rate of production of fast neutrons by fission per unit volume is $(k_\infty/p)\Sigma_2\phi_2$, where Σ_2 is the macroscopic absorption cross section and ϕ_2 is the flux of thermal neutrons. The steady-state diffusion equation (4.6) for *fast neutrons* thus takes the form

$$D_1\nabla^2\phi_1 - \Sigma_1\phi_1 + (k_\infty/p)\Sigma_2\phi_2 = 0, \tag{4.12}$$

where the subscripts 1 refer to fast neutrons and 2 to thermal neutrons. The quantity Σ_1 is the macroscopic cross section for the slowing down of neutrons from the fast to the thermal group. If there had been no resonance capture, the rate at which fast neutrons are transferred to the thermal group would be $\Sigma_1\phi_1$, but since the probability that a fast neutron will be thermalized is p, the thermal-neutron source term is actually $p\Sigma_1\phi_1$. The steady-state diffusion equation for *thermal neutrons* is thus

$$D_2\nabla^2\phi_2 - \Sigma_2\phi_2 + p\Sigma_1\phi_1 = 0. \tag{4.13}$$

4.20. In order to obtain solutions of equations (4.12) and (4.13) the spatial distributions of the fast- and thermal-neutron fluxes are expressed by means of the equations

$$\nabla^2\phi_1 + B^2\phi_1 = 0 \tag{4.14}$$

and

$$\nabla^2\phi_2 + B^2\phi_2 = 0, \tag{4.15}$$

with appropriate boundary conditions. It can be shown that B^2 must then have the same value for both neutron groups [1]. Upon substituting $-B_c^2\phi_1$ for $\nabla^2\phi_1$ in equation (4.12) and $-B_c^2\phi_2$ for $\nabla^2\phi_2$ in equation (4.13), the critical buckling, B_c^2, being used since equations (4.12) and (4.13) apply to a steady-state, i.e., critical, system, it is seen that

$$-(D_1B_c^2 + \Sigma_1)\phi_1 + (k_\infty/p)\Sigma_2\phi_2 = 0$$

and

$$p\Sigma_1\phi_1 - (D_2B_c^2 + \Sigma_2)\phi_2 = 0.$$

4.21. The condition that these simultaneous equations shall have a nontrivial solution is that the determinant of the coefficients shall vanish (Cramer's rule), i.e.,

$$(D_1B_c^2 + \Sigma_1)(D_2B_c^2 + \Sigma_2) - k_\infty\Sigma_1\Sigma_2 = 0.$$

Upon dividing through by $\Sigma_1\Sigma_2$, and substituting L_1^2 for D_1/Σ_1 and L_2^2 for D_2/Σ_2, it follows that

$$\frac{k_\infty}{(1 + L_1^2 B_c^2)(1 + L_2^2 B_c^2)} = 1, \tag{4.16}$$

which is the critical equation for a bare reactor in the two-group diffusion approximation.   This is seen to be similar to the one-group equation (4.11), except that there are two factors of the form $1/(1 + L^2 B_c^2)$, representing the nonleakage probabilities from the critical system for neutrons in each of the two groups.

4.22. Since Σ_1 is physically a slowing down cross section, it can be shown, by a procedure similar to that used in § 3.34, that L_1^2 is equal to $\frac{1}{6}\overline{r_s^2}$, where $\overline{r_s^2}$ is the mean square distance traveled by a neutron in the fast group before it is transferred to the thermal group.   Consequently, according to equation (3.64), L_1^2 in the two-group treatment has the same general significance as the Fermi age, τ, in the continuous slowing down model, although the values are not necessarily those given in Table 3.4.   Upon replacing L_1^2 by τ, equation (4.16) may be written as

$$\frac{k_\infty}{(1 + \tau B_c^2)(1 + L_2^2 B_c^2)} = 1. \tag{4.17}$$

THE AGE-DIFFUSION METHOD

4.23. In the age-diffusion approximation, which leads to a critical equation that has been widely used, the leakage of neutrons while slowing down from fission to thermal energies is estimated by making use of the continuous slowing down concept, described in § 3.80 *et seq.*, to determine the source term in the steady-state diffusion equation.   By definition, the slowing down density at any energy E is equal to the number of neutrons per cm³ per sec slowing down past this energy.   If there were no resonance capture of neutrons while slowing down, the slowing down density could be taken as equal to the value of $q(E)$ which is the appropriate solution of the Fermi age equation (3.61).   As seen in § 3.79, for the case of weak absorption, the slowing down density is given by $p(E)q(E)$, where $p(E)$ is the resonance escape probability for neutrons of energy E.   The slowing down density for thermal neutrons with absorption is therefore pq_{th}, where p is the overall resonance escape probability of thermal neutrons, as defined in § 3.79, and q_{th} is the solution of the Fermi equation for neutrons of thermal energy.   Hence, pq_{th}, i.e., the number of neutrons per cm³ per sec which become thermal, is the source term to be used in the thermal diffusion equation.

4.24. In order to evaluate q_{th}, the Fermi equation (3.61) is written in the form

$$\nabla^2 q(\tau, r) = \frac{\partial q(\tau, r)}{\partial \tau}, \tag{4.18}$$

indicating that q is a function of both neutron energy (or age) and position

(§ 3.85).   A solution of this equation may be sought by separating the age and space coordinates by writing

$$q(\tau, r) = T(\tau)R(r),$$ (4.19)

where $T(\tau)$ is a function of age (or energy) only and $R(r)$ is a function of position only.   Substitution of equation (4.19) into (4.18) then gives

$$\frac{\nabla^2 R(r)}{R(r)} = \frac{1}{T(\tau)} \cdot \frac{dT(\tau)}{d\tau}.$$

Since each side involves a single variable only, it may be set equal to a constant;* thus,

$$\frac{\nabla^2 R(r)}{R(r)} = -B^2$$

or

$$\nabla^2 R(r) + B^2 R(r) = 0$$ (4.20)

and

$$\frac{1}{T(\tau)} \cdot \frac{dT(\tau)}{d\tau} = -B^2.$$ (4.21)

The solution of equation (4.21) is

$$T(\tau) = Ae^{-B^2\tau},$$

where A is a constant which is to be determined.   Hence, from equation (4.19),

$$q(\tau, r) = AR(r)e^{-B^2\tau},$$ (4.22)

where B^2 must be a real, positive quantity, in order to satisfy the physical requirement that the slowing down density cannot increase with the age.

4.25. A simple way to evaluate A is to consider the slowing down density of fission neutrons which have an age of zero; according to equation (4.22) this quantity can be represented by

$$q(0, r) = AR(r),$$ (4.23)

since τ is zero.   The rate at which neutrons slow down past the fission energy must be equal to the total number of fission neutrons generated per cm³ per sec in the given multiplying medium, and this can be calculated in the following manner.

4.26. If n thermal neutrons are absorbed, then nf is the number absorbed in fuel.   Using the arguments developed in § 4.7 *et seq.*, it can readily be shown that this will lead to the production of $nf\eta\epsilon$ fission neutrons, i.e., $f\eta\epsilon$ per thermal neutron absorbed.   Since the infinite multiplication factor k_∞ is equal to $f\eta\epsilon p$, where p is the resonance escape probability, it follows that k_∞/p fission neutrons are produced for each thermal neutron that is absorbed.   The number of thermal neutrons absorbed per cm³ per sec is $\Sigma_a\phi$, where ϕ is the thermal-neutron flux, and so the total number of fission neutrons generated, per cm³ per sec, which is

---

* The symbol B^2 is used for this constant because it will later be identified with the buckling, but for the present it must be regarded as an arbitrary constant.

also the slowing down density of fission neutrons, is $(k_\infty/p)\Sigma_a\phi$. It then follows from equation (4.23) that

$$AR(r) = \frac{k_\infty}{p}\Sigma_a\phi(r) \qquad (4.24)$$

and by equation (4.22)

$$q(\tau, r) = \frac{k_\infty}{p}\Sigma_a\phi(r)e^{-B^2\tau}. \qquad (4.25)$$

4.27. From the arguments presented above, the thermal-neutron source term is equivalent to $pq(\tau_{th}, r)$ and so, omitting the space coordinate r for simplicity, the steady-state (or critical) diffusion equation (4.6) for thermal neutrons becomes

$$D\nabla^2\phi - \Sigma_a\phi + k_\infty\Sigma_a\phi e^{-B_c^2\tau_{th}} = 0,$$

where B_c^2 is the value of B^2 in the critical system. Making use of the fact that D/Σ_a is equal to L^2, the square of the thermal diffusion length, it follows that

$$\nabla^2\phi + \frac{k_\infty e^{-B_c^2\tau_{th}} - 1}{L^2}\phi = 0 \qquad (4.26)$$

is the condition for criticality.

4.28. According to equation (4.24), $R(r)$ is proportional to $\phi(r)$, and so equation (4.20) may be written as

$$\nabla^2\phi + B^2\phi = 0, \qquad (4.27)$$

which is the familiar equation (4.9) giving the spatial distribution of the neutron flux, in this case that of thermal neutrons, in terms of the buckling. This means that the constant B_c^2 which appears in the critical equation (4.26) is actually the critical (or material) buckling. Comparison of equations (4.26) and (4.27) then shows that

$$B_c^2 = \frac{k_\infty e^{-B_c^2\tau_{th}} - 1}{L^2} \qquad (4.28)$$

or

$$\frac{k_\infty e^{-B_c^2\tau_{th}}}{1 + L^2B_c^2} = 1, \qquad (4.29)$$

which is the age-diffusion critical equation for a bare reactor. In the foregoing treatment it has been tacitly assumed that the source neutrons, i.e., the fission neutrons, all have the same energy. However, equation (4.29) is applicable to the general case, provided τ_{th} is taken as the average value of the Fermi age of thermal neutrons properly weighted according to the fission spectrum (§ 2.166).

THE CRITICAL SIZE

4.29. As stated in § 4.17, it is possible to solve equation (4.27) for a given geometry, with the appropriate boundary conditions, and thereby obtain a general equation for B^2 in terms of the dimensions of the system. The critical (or material) buckling, as defined by equation (4.28) in terms of certain properties of the reactor materials, can then be used to evaluate the critical size for a prescribed shape. The general treatment will be illustrated here by reference to a spherical reactor, but the results for other geometries will also be given.

4.30. The general solution of equation (4.27) for a sphere, with the center taken to be the origin of the radial coordinate r, is

$$\phi(r) = \frac{C \sin Br}{r} + \frac{C' \cos Br}{r},$$

where C and C' are arbitrary constants. The fact that the flux is finite at the center of the sphere, i.e., when r is zero, eliminates the second term on the right, so that a permissible solution is

$$\phi(r) = \frac{C \sin Br}{r}. \tag{4.30}$$

4.31. If R is the radius of the spherical reactor, including the extrapolation distance (§ 3.26), then the boundary condition requires that $\phi(r)$ shall be zero when $r = R$; hence, from equation (4.30),

$$\frac{C \sin BR}{R} = 0.$$

Since C and R are not zero, this means that $\sin BR = 0$, and consequently

$$BR = n\pi,$$

so that

$$B^2 = \left(\frac{n\pi}{R}\right)^2, \tag{4.31}$$

where n is an integer, i.e., 1, 2, 3, etc. Strictly speaking, $n = 0$ gives a permissible solution, but it is not included since it makes the neutron flux zero.

4.32. Although the various integral values of n give solutions (eigenvalues) of equation (4.27) or (4.9) that satisfy the mathematical requirements, there is only one, i.e., $n = 1$, which can give a steady-state solution for the neutron flux. Consequently, the required solution is

$$B^2 = \left(\frac{\pi}{R}\right)^2. \tag{4.32}$$

If the buckling has the critical value, B_c^2, as given by the appropriate critical equation, e.g., equation (4.28), the critical radius, R_c, of a bare spherical reactor is then given by

$$R_c = \frac{\pi}{B_c}. \tag{4.33}$$

4.33. The flux distribution in the critical reactor is obtained by inserting the value for B_c from equation (4.33) into (4.30); thus,

$$\phi(r) = \frac{C}{r} \sin \frac{\pi r}{R_c}. \tag{4.34}$$

The arbitrary constant C in equation (4.34) is determined by the power level of the reactor, since this is proportional to the neutron flux (§ 2.162). For a given flux (or flux distribution), the power output of a reactor is, in principle, independent of its size, provided it is critical.

4.34. In order to determine the critical size of a bare reactor of a specified composition, the procedure is to derive the values of k, L^2, and τ for the given multiplying medium. Then these are used to obtain the critical (or material) buckling B_c^2 by means of equation (4.28). Insertion of the corresponding (positive) value of B_c into equation (4.33) gives the critical radius of the bare sphere. The same general procedure is employed if either the one-group or two-group approximations is used, the appropriate values of the critical buckling being obtained by means of equation (4.10) or (4.16), respectively. The relationship between the buckling and the radius is given by equation (4.33) in every case, since this merely represents the appropriate solution of equation (4.27).

REACTORS OF VARIOUS SHAPES

4.35. By solving equation (4.27) for reactors of various shapes, expressions can be determined for the buckling in terms of the applicable spatial dimensions [2]. The flux distribution in the critical reactor can then be derived in a manner similar to that described above for a sphere. The results are summarized in Table 4.2 for a sphere, a rectangular parallelepiped, and a finite cylinder, using the dimensions, which include the extrapolation distances, shown in Fig. 4.1.

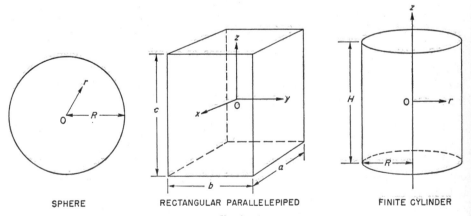

SPHERE   RECTANGULAR PARALLELEPIPED   FINITE CYLINDER

Fig. 4.1.   Dimensions and coordinates for reactors of three geometries

The function J_0 is the Bessel function of the first kind of zero order. It will be noted that the buckling always has the dimensions of (length)$^{-2}$.

4.36. For a given value of the buckling, i.e., for a system of given composition, there will be a certain minimum volume for each kind of geometry which can be

TABLE 4.2. BUCKLING AND FLUX DISTRIBUTION IN BARE REACTORS

Geometry	Buckling	Critical Flux Distribution	Minimum Critical Volume
Sphere	$\left(\dfrac{\pi}{R}\right)^2$	$\dfrac{A}{r}\sin\dfrac{\pi r}{R_c}$	$\dfrac{130}{B_c^3}$
Rectangular parallelepiped	$\left(\dfrac{\pi}{a}\right)^2+\left(\dfrac{\pi}{b}\right)^2+\left(\dfrac{\pi}{c}\right)^2$	$A\cos\dfrac{\pi x}{a_c}\cos\dfrac{\pi y}{b_c}\cos\dfrac{\pi z}{c_c}$	$\dfrac{161}{B_c^3}$ (for $a = b = c$)
Finite cylinder	$\left(\dfrac{2.405}{R}\right)^2+\left(\dfrac{\pi}{H}\right)^2$	$A J_0\left(\dfrac{2.405 r}{R_c}\right)\cos\dfrac{\pi z}{H_c}$	$\dfrac{148}{B_c^3}$ (for $H = 1.847R$)

readily calculated from the appropriate buckling equations. The expressions for these minimum volumes are recorded in the last column of Table 4.2; in the case of the rectangular parallelepiped, the critical volume is least when the three sides are equal, i.e., for a cube. It is evident that, for a specified multiplying medium, the critical volume (and mass) of a spherical reactor will be less than for any of the other shapes. The reason is that the sphere has the smallest area-to-volume ratio. Neutron production takes place throughout the volume of the system, but leakage occurs only at the surface. Hence, for a given composition, the geometry with the smallest ratio of surface to volume will have the smallest critical mass.

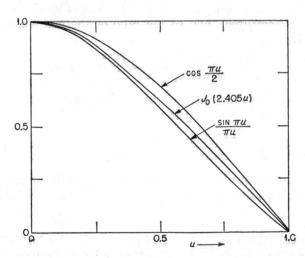

FIG. 4.2. Functions for determination of neutron flux distribution

4.37. An examination of the expressions for the critical flux distribution in Table 4.2 shows that each is determined by one of three functions, namely, $(1/r) \sin (\pi r/R)$, $\cos (\pi x/a)$ or similar, and $J_0(2.405r/R)$. These quantities are plotted in Fig. 4.2 versus the coordinate u, which represents r/R, $2x/a$ (or similar), and r/R, respectively. It is seen that, as a first approximation, the neutron flux in the critical reactor may be taken to have a cosine distribution about any of the axes in Fig. 4.1, with a relative value of unity at the axis and zero at the extrapolated boundary.

NONLEAKAGE PROBABILITIES

4.38. Before considering the applications of the age-diffusion critical equation (4.29), it is instructive to examine the significance of the terms involved. If a reactor were infinite in size, so that no neutrons were lost by leakage, the critical condition would be simply that the infinite multiplication factor k_∞ should be unity. Then, for every neutron absorbed, another equivalent neutron would be produced. The use of the continuous slowing down model introduces two factors, namely, $e^{-B^2\tau}$ and $1/(1 + L^2B^2)$,* which take into account the fact that the reactor has finite dimensions and that neutron leakage takes place.

4.39. Since the slowing down density of source (fission) neutrons is $(k_\infty/p)\Sigma_a\phi$, and the slowing down density of thermal neutrons is $(k_\infty/p)\Sigma_a\phi e^{-B^2\tau}$, it follows that $e^{-B^2\tau}$ is the fraction of the source neutrons that remain in the reactor and do not escape. In other words, $e^{-B^2\tau}$ represents the nonleakage probability of the neutrons during the course of slowing down from source to age τ.

4.40. The rate at which neutrons leak away from a specified point in the reactor is given by diffusion theory as $-D\nabla^2\phi$, and by equation (4.27) this may be taken as equal to $DB^2\phi$ neutrons/(cm³)(sec). The rate of absorption is $\Sigma_a\phi$ thermal neutrons/(cm³)(sec), and so the ratio of thermal leakage to thermal absorption is given by

$$\frac{\text{Thermal leakage}}{\text{Thermal absorption}} = \frac{DB^2}{\Sigma_a} = L^2B^2,$$

since D/Σ_a is equal to L^2, the square of the thermal diffusion length. By adding unity to each side and inverting, it is seen that

$$\frac{\text{Thermal absorption}}{\text{Thermal absorption} + \text{thermal leakage}} = \frac{1}{1 + L^2B^2}. \tag{4.35}$$

Since the denominator on the left side represents the total number of neutrons thermalized, it follows that $1/(1 + L^2B^2)$ is the nonleakage probability for thermal neutrons.

---

* Since it is understood that τ refers to thermal neutrons, the subscript "th" will be omitted for simplicity. Moreover, the buckling is not necessarily the critical value, since the conclusions derived below are of general applicability.

4.41. It is of interest to consider the physical significance of the quantities appearing in the nonleakage probability factors. In view of the inverse relationship between B^2 and the dimensions of the reactor (see Table 4.2), B^2 will be large when the reactor is small, and vice versa. A large value of B^2 will mean that both $e^{-B^2\tau}$ and $1/(1 + L^2B^2)$ are relatively small. Thus, as is to be expected, the neutron leakage probability, both during slowing down and while thermal, will be large for a small reactor. Next, consider the age τ, which is related to the average (net vector) distance traversed by a neutron while slowing down. It is to be anticipated, therefore, that for a reactor of given size, a large τ will mean a small nonleakage probability; this is in harmony with the results obtained above, for $e^{-B^2\tau}$ is then small. Similar considerations apply to the diffusion length, L, which is a measure of the average (net vector) distance which the thermal neutron traverses before capture. A large value of L should mean a large probability of leakage from the reactor, and this is in agreement with the thermal nonleakage probability represented by $1/(1 + L^2B^2)$.

4.42. Since B^2 is inversely related to the reactor size, it is evident that for a *large reactor* the exponential $e^{-B^2\tau}$ can be expanded in series form and all terms beyond the second neglected, so that

$$e^{-B^2\tau} \approx 1 - B^2\tau \approx (1 + B^2\tau)^{-1},$$

it being understood that τ refers to thermal neutrons. In these circumstances the critical equation (4.29) becomes

$$\frac{k_\infty}{(1 + \tau B_c^2)(1 + L^2B_c^2)} = 1, \tag{4.36}$$

which is identical with the two-group equation (4.17). If terms in B_c^4 are neglected, since they are very small for a large reactor, equation (4.36) reduces to

$$\frac{k_\infty}{1 + B_c^2(L^2 + \tau)} = 1. \tag{4.37}$$

As seen in § 3.98, the sum $L^2 + \tau$, or $L^2 + \frac{1}{6}\overline{r_s^2}$, in general, is defined as the migration area M^2 for thermal neutrons, so that the critical equation (4.37) for a large reactor is

$$\frac{k_\infty}{1 + M^2B_c^2} = 1. \tag{4.38}$$

According to this result, the total nonleakage probability is given by $1/(1 + M^2B_c^2)$; the greater the value of M^2 the smaller is the nonleakage probability, i.e., the greater the neutron loss by leakage, for a reactor of specified size. Since the migration length (§ 3.98) includes both slowing down and thermal diffusion lengths, this result is to be expected.

Example 4.2. The infinite multiplication factor for a heterogeneous lattice of natural uranium fuel in heavy water (D_2O) moderator is 1.28. In this particular lattice L^2 for thermal neutrons is 175 cm^2 and τ is 120 cm^2. Determine the critical buckling using

(a) the one-group (thermal) critical equation, (b) equation (4.38) for a large reactor, (c) the age-diffusion equation, and (d) equation (4.36), which is an approximation of the age-diffusion equation for a large reactor.

(a) $B_c^2 = \dfrac{k_\infty - 1}{L^2} = \dfrac{0.28}{175} = 1.6 \times 10^{-3} \text{ cm}^{-2}.$

(b) $B_c^2 = \dfrac{k_\infty - 1}{M^2} = \dfrac{k_\infty - 1}{L^2 + \tau} = \dfrac{0.28}{175 + 120} = 9.5 \times 10^{-4} \text{ cm}^{-2}.$

(c) The age-diffusion equation is transcendental and in order to obtain a solution it is written in the form

$$e^{-B_c^2\tau} = \frac{1 + L^2 B_c^2}{k_\infty}.$$

This is solved by iteration in the following manner. The value 9.5×10^{-4} is inserted for B_c^2 on the right side; this gives $e^{-B_c^2\tau}$ as 0.910, so that $B_c^2\tau$ is 0.094 and B_c^2 is 7.8×10^{-4}. The trial value was obviously too large, and so a smaller value, e.g., 8.5×10^{-4}, is chosen; this is found to be too small. Finally, B_c^2 is found to be $9.0 \times 10^{-4} \text{ cm}^{-2}$.

(d) Equation (4.36) can be written as

$$\tau L^2 B_c^4 + (L^2 + \tau) B_c^2 + (1 - k_\infty) = 0$$

which is a quadratic in B_c^2. The solution is $B_c^2 = 9.0 \times 10^{-4}$, in agreement with that obtained in (c).

Example 4.3. The system to which Example 4.2 refers consists of 1-in. diameter rods of natural uranium metal arranged in a square lattice with a pitch of 6 in., suspended in a cylindrical vessel containing the heavy water moderator. The ratio of the height to diameter of the bare core is 1.2. Estimate the mass of uranium which will make this bare reactor just critical.

The first step is to determine the dimensions of the critical system, taking B_c^2 to be $9.0 \times 10^{-4} \text{ cm}^{-2}$, as obtained in Example 4.2. For a cylinder (see Table 4.2),

$$B_c^2 = \left(\frac{2.405}{R_c}\right)^2 + \left(\frac{\pi}{H_c}\right)^2,$$

but since $H_c/2R_c = 1.2$, i.e., $H_c = 2.4 R_c$,

$$B_c^2 = \left(\frac{2.405}{R_c}\right)^2 + \left(\frac{\pi}{2.4 R_c}\right)^2 = \frac{7.50}{R_c^2};$$

Hence,

$$R_c^2 = \frac{7.50}{9.0 \times 10^{-4}}$$

$$R_c = 91 \text{ cm} \quad\text{and}\quad H_c = 218 \text{ cm}.$$

The fuel rods are in a square lattice with a pitch of $6.0 \times 2.54 = 15.2$ cm, so that the effective area occupied per rod is $(15.2)^2 = 230 \text{ cm}^2$. The cross-sectional area of the core is $\pi R_c^2 = (3.14)(91)^2 = 26{,}000 \text{ cm}^2$, and so the number of fuel rods is $26{,}000/230 \approx 110$. Each rod has a diameter of 1 in., i.e., 2.54 cm, and the length of the core, i.e., 218 cm; hence, the volume of the 110 rods is

$$\text{Total volume of fuel} = (\tfrac{1}{4})(\pi)(2.54)^2(218)(110)$$
$$= 1.53 \times 10^5 \text{ cm}^3.$$

1. Must have a $k_\infty > 1$ in the material
2. Eigen values.
3. Match Buckling to go critical

The density of uranium metal is 19 g/cm³, and so the total weight of uranium in the bare critical reactor is

$$\text{Weight of fuel} = (1.53 \times 10^5 \times 19) = 2.91 \times 10^6 \text{ grams}$$
$$= 2.9 \text{ tonnes.}$$

THE EFFECTIVE MULTIPLICATION FACTOR

4.43. The product of the two terms $e^{-B^2\tau}$ and $1/(1 + L^2B^2)$ is equal to the total nonleakage probability of neutrons in a reactor of finite size, as derived from age-diffusion theory. This is the quantity represented by P in § 4.4; hence, $k_\infty e^{-B^2\tau}/(1 + L^2B^2)$ must, by definition, be equal to the effective multiplication factor, i.e.,

$$\frac{k_\infty e^{-B_c^2\tau}}{1 + L^2B_c^2} = k_{\text{eff}}. \qquad (4.39)$$

For a critical reactor, of finite size, k_{eff} is equal to unity, and equation (4.39) then becomes identical with the critical equation (4.29).

4.44. If the reactor is larger than the critical size for the given materials and composition, i.e., the reactor is supercritical, then the value of B^2, as determined from the geometry of the system, will be less than for the critical reactor, because B^2 is related inversely to the reactor dimensions, as seen above. The left side of equation (4.39) will then exceed unity, so that $k_{\text{eff}} > 1$, as required for a

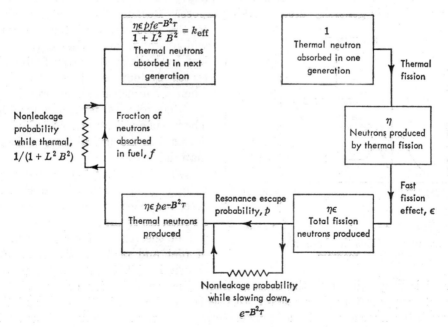

FIG. 4.3.   Neutron cycle in a thermal fission system

supercritical system (§ 4.3). In this case, the two nonleakage probabilities, $e^{-B^2\tau}$ and $1/(1 + L^2B^2)$, will both be greater than required to make the reactor critical. If the reactor is subcritical, so that B^2 is greater than the critical value, k_{eff} as given by equation (4.39) will be less than unity, because the nonleakage probabilities are then smaller than necessary for the given system to become critical.

4.45. Combination of the arguments in § 4.5 *et seq.*, leading to the four-factor formula, with the foregoing discussion of nonleakage probabilities, leads to the neutron balance representation in Fig. 4.3. The cycle starts with the absorption of a single thermal neutron in fuel in one generation, in the box at the top right, which is followed by the absorption of k_{eff} thermal neutrons in fuel in the next generation, as shown in the box at the top left. If the quantities in the two boxes are equal, the system is just critical. On the other hand, if they are different, the neutron density will either increase ($k_{eff} > 1$) for a supercritical system or decrease ($k_{eff} < 1$) for a subcritical one.

REFLECTED REACTORS

EFFECTS OF REFLECTOR

4.46. The *core* of a reactor, i.e., the region containing the fuel, is usually surrounded by a neutron *reflector* consisting of a material having good neutron scattering properties. For a thermal reactor, heavy water, beryllium (or its oxide), or graphite is generally used for this purpose. Because the reflector returns to the core many of the neutrons which would otherwise have been lost by leakage, the given fuel-moderator system can become critical when the dimensions are appreciably less than required for a bare reactor. The use of a reflector thus results in a decrease in the critical mass of fissile material.

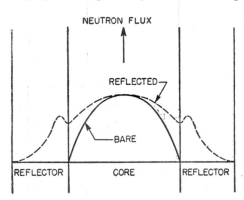

FIG. 4.4. Approximate thermal-neutron flux distributions in bare and reflected reactors

4.47. In addition to decreasing the volume (and mass) of the critical core, the reflector can make possible an increase in the average power output of the reactor for a given mass of fuel and peak neutron flux. As shown in § 2.161 *et seq.*, the power level is proportional to the average thermal-neutron density or flux in the core. Suppose that, in a reactor with reflector, the flux in the center of the core is the same as that for a bare reactor. Because of the return of neutrons to the core by the reflector, the flux near the boundaries will be much greater than in the absence of a reflector (Fig. 4.4). The average thermal-neutron flux over the whole core, and hence the average power output for a given peak power (or flux), will then be increased by the reflector. The multiplying properties of the outermost regions of the core are used more effectively when a reflector is employed.

4.48. In the presence of a reflector, the treatment of the neutron balance, by means of the steady-state equation, is more difficult than for a bare reactor. For a thermal reactor, many of the fast neutrons escaping from the core will be slowed down and will return with reduced energies. Since the absorption and diffusion properties of the reflector will undoubtedly be different from those of the core, calculation of neutron losses, both by absorption and by leakage, in the course of slowing down to thermal energies becomes very complicated. Furthermore, as can be seen from Fig. 4.4, the reflector has a marked effect on the spatial distribution of neutrons.

THE ONE-GROUP METHOD

4.49. A solution of the age-diffusion equation for a reflected reactor can, in general, be obtained only by numerical procedures, because of the complication mentioned above. An alternative approach is to make use of the group treatment outlined earlier, and the simplest approximation is to postulate that all the neutrons are in a single group. Although the results obtained are by no means exact quantitatively, the general conclusions that will be reached below are in complete agreement with those arrived at by more highly involved methods.

4.50. In the following treatment, the properties of the reflector are indicated by the subscript r; symbols without a special subscript refer to properties of the core, as in the preceding discussion. The one-group steady-state diffusion equation *for the core* (§ 4.16) is then

$$D\nabla^2\phi - \Sigma_a\phi + k_\infty\Sigma_a\phi = 0 \qquad (4.40)$$

and *for the reflector*, which contains no fuel material and is nonmultiplying (§ 3.20), it is

$$D_r\nabla^2\phi_r - \Sigma_{ar}\phi_r = 0 \qquad \text{or} \qquad \nabla^2\phi_r - \kappa_r^2\phi_r = 0, \qquad (4.41)$$

where $\kappa_r^2 = \Sigma_{ar}/D_r$, as defined in equation (3.16). By solving these equations and applying the appropriate boundary conditions, the effect of the reflector on the critical size of the core can be determined.

4.51. The procedure will be illustrated by considering the simple case of a spherical reactor surrounded by an infinitely thick reflector. The flux distribution in the critical core, which is the solution of equation (4.40), is given by equation (4.30) as

$$\phi(r) = \frac{C\sin B_c r}{r}, \qquad (4.42)$$

where B_c^2 is the one-group critical buckling in the core, which can be obtained from equation (4.10). Furthermore, the solution of equation (4.41) in a non-multiplying medium of infinite thickness is given by equation (3.25); thus, writing L_r for $1/\kappa_r$, in accordance with equation (3.28),

$$\phi_r(r) = \frac{Ae^{-r/L_r}}{r}. \tag{4.43}$$

4.52. The arbitrary constants A and C can now be eliminated by applying the boundary conditions that the neutron flux and current density shall be continuous at the core reflector interface (§ 3.22), i.e., when $r = R_c$, where R_c is the critical radius of the core. As far as the flux is concerned, it follows, therefore, from equations (4.42) and (4.43), that

$$\frac{C \sin B_c R_c}{R_c} = \frac{Ae^{-R_c/L_r}}{R_c}. \tag{4.44}$$

The current is equal to $-D(d\phi/dr)$, and so the continuity condition, after some simplification, leads to

$$CD(\sin B_c R_c - B_c R_c \cos B_c R_c) = AD_r e^{-R_c/L_r}\left(1 + \frac{R_c}{L_r}\right). \tag{4.45}$$

Upon dividing equation (4.45) by (4.44), it is found that

$$D(1 - B_c R_c \cot B_c R_c) = D_r\left(1 + \frac{R_c}{L_r}\right)$$

or

$$B_c R_c \cot B_c R_c = -\frac{D_r}{D}\left(1 + \frac{R_c}{L_r}\right) + 1. \tag{4.46}$$

This is the one-group critical equation for a spherical core with an infinitely thick reflector.

4.53. If the core radius of the reflected reactor is relatively large in comparison with the diffusion length of neutrons in the reflector, i.e., R_c/L_r, is large, the right side of equation (4.46) is large and negative. Consequently, $\cot B_c R_c$ must also have a large negative value, and this means that $B_c R_c$ must approach π in magnitude. For any angle θ radians which is not very different from π, it is known that $\tan \theta \approx \theta - \pi$ or $\cot \theta \approx 1/(\theta - \pi)$; hence, in the present circumstances,

$$\cot B_c R_c \approx \frac{1}{B_c R_c - \pi}.$$

Moreover, since it has been postulated that R_c is large, unity can be neglected both inside and outside the parentheses in equation (4.46), and so the latter can be reduced to yield

$$R_c \approx \frac{\pi}{B_c} - \frac{D}{D_r} L_r. \tag{4.47}$$

This gives the value of the critical radius of the core of the reflected reactor.

REFLECTOR SAVINGS

4.54. The decrease in critical dimensions of a reactor core due to a reflector is expressed by means of the *reflector savings*, δ, defined in the case of a spherical reactor by

$$\delta \equiv R_0 - R_c,$$

where R_0 is the critical radius of the unreflected reactor. For a sphere, R_0 is given by equation (4.33) as π/B_c; hence, the reflector savings in the present case, i.e., for an infinitely thick reflector, is obtained from equation (4.47) as

$$\delta(\text{thick reflector}) \approx \frac{D}{D_r} L_r.$$

It is of interest to mention that exactly the same result is obtained by applying the one-group method to a cubic (or parallelepiped) reactor with an infinite reflector. In this case, the dimension to which the reflector savings refers is half the edge length.

4.55. The general procedure described above has been applied to the situation in which the reflector has a finite, rather than an infinite, thickness. For a large reactor, the result obtained by the one-group treatment [3] is given by

$$\delta \approx \frac{D}{D_r} L_r \tanh \frac{T}{L_r}, \tag{4.48}$$

where T is the reflector thickness; "tanh" represents the hyperbolic tangent. When T/L_r is large, i.e., for a "thick" reflector, tanh T/L_r approaches unity and the reflector savings then has the same value as that derived in § 4.54. On the other hand, when T/L_r is small, i.e., for a "thin" reflector, tanh T/L_r may be replaced by T/L_r so that

$$\delta(\text{thin reflector}) \approx \frac{D}{D_r} T.$$

A fairly good approximation to the one-group equation (4.48) is to write

$$\delta \approx \frac{D}{D_r} L_r(1 - e^{-T/L_r}),$$

which reduces to the same values as does equation (4.48) for thick and thin reflectors.

4.56. If the reflector material is the same as the moderator in the core of a thermal reactor, D and D_r are not very different. The reflector savings for a thick reflector is then seen to be approximately equal to the diffusion length of thermal neutrons in the reflector. An examination of the data in Table 3.1 shows that, except for water, this can be quite substantial. For reactors moderated and reflected by water, the reflector savings obtained from equation (4.48) is not a good approximation and better results are obtained from an empirical expression given in § 4.141.

Example 4.4. A bare natural uranium-graphite reactor in the form of a cube is critical when the edge length is 560 cm; it then contains 50 tonnes of fuel. Estimate the amount of fuel required for criticality when the core is completely surrounded by a graphite reflector 100 cm thick.

The migration length of thermal neutrons in graphite is about 57 cm (Table 3.4), and so the reflector may be regarded as "thick." Since the moderator and reflector are of the same material, D/D_r will not be very different from unity and the reflector savings will be approximately equal to the diffusion length of thermal neutrons in the reflector, i.e., 54 cm in graphite. This reflector savings will apply to each face of the cube, and so the edge length of the critical core is reduced by $2 \times 54 = 108$ cm. The edge length of the reflected critical core is thus $560 - 108 = 452$ cm. The mass of uranium required to make the reflected core just critical is therefore $(50)(452/560)^3 = 26$ tonnes.

4.57. For a thin reflector, which is of the same material as the core moderator, the reflector savings is seen to be approximately equal to the reflector thickness itself. This result implies that fuel may be removed from the outer layers of a bare reactor without appreciably affecting the criticality, provided the thickness of the layer is small compared with the diffusion length of thermal neutrons in the moderator [4].

Example 4.5. The cylindrical lattice referred to in Example 4.3 is surrounded, at the sides only, by a 25-cm thick reflector of heavy water. What is the mass of uranium required to make the reflected reactor just critical?

The diffusion length of thermal neutrons in heavy water is 100 cm (Table 3.1), and so the 25-cm thick reflector may be regarded as being "thin" in the sense referred to in § 4.55. Since the reflector and moderator are of the same material, the reflector savings

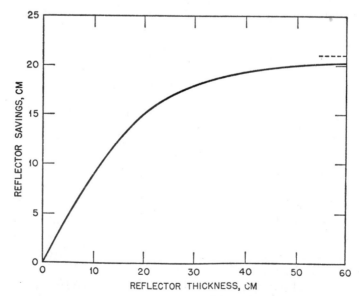

Fig. 4.5. Calculated (one-group) reflector savings as function of beryllium reflector thickness

is equal to the reflector thickness, i.e., 25 cm. The radius of the reflected critical core is thus $91 - 25 = 66$ cm. There is no reflector at the top and bottom of the core and so the length will remain unchanged. The reflected core volume is thus $(66/91)^2$, i.e., 0.53, of the bare core. The minimum mass of uranium required for criticality is thus $2.9 \times 0.53 = 1.54$ tonnes, for the reflected core.

4.58. A plot of the reflector savings as a function of the reflector thickness, derived from equation (4.48) for a reactor moderated and reflected by beryllium, is shown in Fig. 4.5; the broken line at the extreme right indicates the reflector savings for an infinitely thick reflector. It is seen that beyond a certain reflector thickness—about 40 cm in Fig. 4.5—further increase has very little effect on the reflector savings. According to the one-group treatment, a reflector having a thickness of about two thermal diffusion lengths, i.e., 2×21 cm for beryllium, behaves effectively as one of infinite thickness. More exact calculations, which take into account the fact that the neutrons do not all have the same energy, show that a reflector thickness of 1.5 to 2 migration lengths (§ 3.98) is almost equivalent to an infinite reflector as far as the effect on the critical size (and mass) of the core is concerned. These conclusions are in general agreement with those stated in § 3.45, namely, that for the diffusion of monoenergetic neutrons from a plane source, a thickness of two diffusion lengths behaves essentially like an infinite medium.

4.59. The reflector savings concept is useful in providing a method for making an approximate estimate of the critical size of a reflected reactor. From the known properties of the core, the critical buckling is first determined for a bare reactor by means of a critical equation, such as the one based on age-diffusion theory. Then, by using the appropriate relationship between the buckling and the dimensions for the required shape (Table 4.2), the critical dimensions of the bare reactor are calculated. The corresponding (approximate) values for the reflected core are then obtained by subtracting the respective reflector savings.

MULTIGROUP METHODS

4.60. The one-group treatment is highly approximate, but it is very simple and leads to conclusions which are qualitatively correct. However, the reflector savings calculated on the basis of one group of neutrons are too low for at least two reasons. In the first place, since the one-group method assumes all neutrons to be thermal, no allowance is made for the fact that fast neutrons escaping from the core make more collisions in the reflector than do thermal neutrons and so they have a higher probability of being scattered back into the core. Secondly, neutrons entering the reflector with energies above the resonance level are moderated in the reflector; they then return to the core as thermal neutrons, having completely avoided the resonance capture to which they would have been liable if they had reached thermal energies in the core. Both these factors tend to increase the effectiveness of the reflector, as compared with that to be expected if all the neutrons had thermal energies.*

4.61. A second approximation to the treatment of reflected reactors is to use the two-group method. It is then postulated that all the neutrons can be divided into two energy groups, namely, fast (fission) and slow (thermal) neutrons. There are now four differential equations; two are similar to equation (4.40), one for each neutron group in the core, and two others, like equation (4.41), give the fast and slow flux distributions in the reflector. In solving these equations, six boundary conditions are used: continuity of fast- and slow-neutron flux and current density at the core-reflector boundary, and zero fast- and slow-neutron flux at the (same) extrapolated boundary of the reflector. The actual solution is somewhat involved and will not be given here [5].

4.62. There is one result of the two- (and higher-) group treatment which is worth mentioning; it refers to the spatial distribution of the neutron flux. For convenience, and in accordance with the two-group method, the neutrons are divided into two groups: thermal neutrons and all other (fast) neutrons, respectively. The general nature of the variation of the flux with distance from the

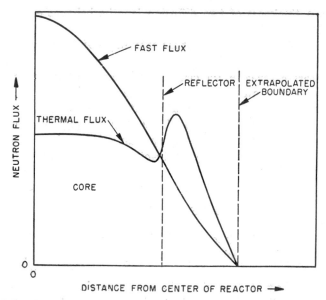

FIG. 4.6. Thermal- and fast-neutron flux distributions (qualitative) in reflected reactor

center of the reactor core is shown in Fig. 4.6. It will be observed that the thermal-neutron flux exhibits a maximum within the reflector at a short distance from the core-reflector interface. This may be ascribed to the fact that thermal

* A reasonably good empirical rule is that the reflector savings is roughly 1.2 times that calculated by the one-group method.

neutrons are produced in the reflector by the slowing down of fast ones, but they are absorbed to a smaller extent in the reflector than in the core. Farther from the interface the thermal flux, like the fast-neutron flux, falls to zero at the extrapolated boundary of the reflector. Since the thermal-neutron flux within a good reflector may be as high as that in the core, some thermal reactors are controlled by means of an absorber inserted into the reflector instead of into the core.

4.63. The derivation of the critical condition for any system is based on the development of equations which express the neutron distribution as a function of space and energy in the steady state. In the treatment presented earlier, this has been achieved by combining the diffusion equation either with the concept of one or two energy groups or with the continuous slowing down (Fermi age) model. The results thus obtained can be used, under appropriate conditions, to determine the approximate critical sizes of thermal reactors, both bare and reflected. In connection with problems of reactor design, especially when an appreciable proportion of the fissions result from the absorption of neutrons with energies in excess of the thermal values, e.g., in intermediate and fast reactors, the equations derived above are not adequate. One reason for this is the approximate nature of diffusion theory which, at best, can be applied accurately only to systems of large size.

4.64. An exact mathematical description of the space and energy distribution of neutron density (or flux) is provided by transport theory but, as stated in § 3.8, this leads to a complex integro-differential (Boltzmann) equation which can be solved in only a very few highly restricted cases. For practical purposes, therefore, it is necessary to make certain simplifications and approximations. For example, it may be postulated that the neutrons all have the same energy (or velocity); this leads to the one-velocity (or one-group) transport equation. One method of solution is then to express the neutron source and flux distributions by a convergent series of spherical harmonics ($l = 0, 1, 2, \cdots, \infty$), thereby approximating the integro-differential equation by a set of coupled differential equations having no angular dependence. By assuming that the contributions of all the terms beyond $l = 1$ are negligible, and making certain simplifications, the resulting equation, called the P_1 approximation, is equivalent to diffusion theory for monoenergetic neutrons. The same procedure may be extended to several groups of neutrons by using multigroup techniques. A considerable improvement in accuracy, accompanied by an increase in complexity, is obtained by using the P_3 approximation, which includes two more spherical harmonic terms and neglects only those for which l is greater than 3.

4.65. Several other methods of simplifying (and approximating) the transport equation have been proposed and used in various cases [6]. One in particular will be mentioned since it has the merit of being easily adaptable to digital machine calculations. In this procedure, called the S_n method, the directional distribution of the flux is expressed by dividing the range of directions, from

$\cos \theta = +1$ to $\cos \theta = -1$, into n intervals and then assuming that the flux varies linearly in each interval. The result is that the Boltzmann (transport) equation is reduced for each neutron group to a set of n equations which can be solved numerically. The accuracy and complexity of the S_n method may be expected to increase with the number of intervals, but for most purposes $n = 4$, i.e., the S_4 method, is adequate. The procedure has been found especially useful for the treatment of fast-reactor systems.

4.66. For the accurate calculation of critical size and spatial distribution of the neutron flux, a multigroup approach, sometimes involving either a few or as many as 30 or more energy groups, is employed. The neutron steady state in each group is expressed by means of some approximation of the one-velocity transport equation. Solution of the resulting set of coupled differential equations is a major problem and can be achieved only by the use of high-speed digital computers. An essential requirement for the application of these multigroup methods is a knowledge of detailed absorption, scattering, and fission cross sections for each energy group for the fuel isotopes, moderator, and structural material; this information is becoming increasingly available. In addition, a considerable number of "codes" have been devised for the solution of nuclear reactor problems by means of electronic computers [7].

HOMOGENEOUS REACTOR SYSTEMS

INFINITE MULTIPLICATION FACTOR

4.67. The treatment of homogeneous reactors, in which the fuel is dispersed uniformly throughout the moderator (§ 4.11), is generally less involved than that of heterogeneous reactors in which the fuel is arranged in a definite geometrical pattern (or lattice) within the moderator mass. In fact, as will be seen shortly, the calculations for some lattice systems, such as those of slightly enriched uranium and uranium dioxide in water as moderator, are often made as though they were homogeneous. The procedure is thus greatly simplified and the results, although not exact, are an adequate approximation for their intended purpose.

4.68. In evaluating the properties of a homogeneous multiplying system, consideration will first be given to the calculation of the resonance escape probability and the thermal utilization. In equation (3.56), which can be used to express the resonance escape probability in a homogeneous system, Σ_a may be assumed to refer to the fuel only, since neutron absorption in the moderator is, in general, relatively small. Hence Σ_a may be replaced by $N_u\sigma_{au}$, where N_u is the number of fuel atoms (or nuclei) per cm³ of the system, and σ_{au} is the absorption cross section for neutrons of energy E. It is then possible to write

$$\frac{\Sigma_a}{\Sigma_a + \Sigma_s} = \frac{N_u\sigma_{au}}{\Sigma_s} \cdot \frac{\Sigma_s}{\Sigma_a + \Sigma_s}.$$

Since, in the resonance region, the scattering cross section may be taken to be independent of neutron energy, equation (3.56) becomes

$$p(E) = \exp\left[-\frac{N_u}{\xi\Sigma_s} \int_E^{E_0} \left(\frac{\Sigma_s}{\Sigma_a + \Sigma_s} \sigma_{au} \right) \frac{dE}{E} \right].$$ (4.49)

The integral in this equation is called the *effective resonance integral*,* and it can be represented by

$$\text{Effective resonance integral} = \int (\sigma_{au})_{\text{eff}} \frac{dE}{E},$$ (4.50)

where

$$(\sigma_{au})_{\text{eff}} = \frac{\Sigma_s}{\Sigma_a + \Sigma_s} \sigma_{au},$$

so that

$$p(E) = \exp\left[-\frac{N_u}{\xi\Sigma_s} \int (\sigma_{au})_{\text{eff}} \frac{dE}{E} \right],$$ (4.51)

the integration being carried over the resonance energy range.

4.69. Since Σ_s is equal to $N_m\sigma_{sm} + N_u\sigma_{su}$,† it follows that

$$\frac{\Sigma_s}{N_u} = \frac{N_m}{N_u} \sigma_{sm} + \sigma_{su} \approx \frac{N_m}{N_u} \sigma_{sm},$$

because in the great majority of cases of interest σ_{su}, the scattering cross section of the fuel nuclei, can be neglected in comparison with the other term. Hence, equation (4.51) can be written as

$$p(E) \approx \exp\left(-\frac{N_u}{N_m} \cdot \frac{I}{\sigma_{sm}\xi} \right)$$ (4.52)

where I is the effective resonance integral. This form of the expression of the resonance escape probability shows more clearly how this quantity is affected by the physical properties and composition of the fuel-moderator system.

4.70. Values of the effective resonance integral, for various mixtures of natural uranium and several different moderators, have been determined experimentally. The results for a temperature of 300°K fall on (or close to) the line shown in Fig. 4.7, in which the effective resonance integral is plotted as ordinate against the total scattering cross section per fuel atom, i.e., Σ_s/N_u, as abscissa [8]. Provided Σ_s/N_u is less than about 400 barns, a good approximation for natural uranium is

$$I = 3.06 \left(\frac{\Sigma_s}{N_u} \right)^{0.472} \text{ barns,}$$ (4.53)

* The actual resonance integral for any material is the integral of $\sigma_a dE/E$ over the resonance energy range; it is equal to the extrapolated value of the effective resonance integral as Σ_s/N_u approaches infinity, i.e., for infinite dilution of fuel in moderator.

† The N values here, as for homogeneous systems in general, refer to the numbers of nuclei (or atoms) per unit volume of the whole system. However, since the essential equations always involve the ratio N_u/N_m (or N_m/N_u), the actual numbers are not needed.

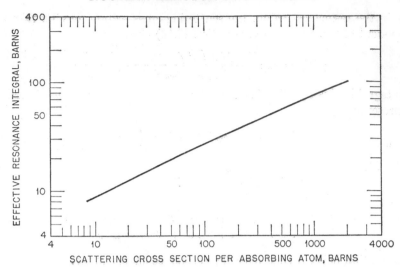

FIG. 4.7. Effective resonance integral of natural uranium at 300°K [8]

but for larger values of Σ_s/N_u the results obtained from this expression are too high, so that the calculated resonance escape probabilities are too small.

Example 4.6. The resonance escape probability in a uniform mixture of graphite and natural uranium in the atomic ratio (N_m/N_u) of 400 is 0.74. Compare this with the value given by the empirical relationship for the resonance integral in § 4.70; σ_{sm} for resonance neutrons in graphite may be taken to be 4.7 barns.

By combining equations (4.52) and (4.53), the result is

$$p(E) \approx \exp\left[-\frac{3.06}{\xi}\left(\frac{N_u}{N_m\sigma_{sm}}\right)^{0.528}\right].$$

Because of the large proportion of graphite, ξ may be set equal to the value for carbon, i.e., 0.158; hence,

$$p(E) \approx \exp\left[-\frac{3.06}{0.158}\left(\frac{1}{400 \times 4.7}\right)^{0.528}\right] = e^{-0.28}$$

$$= 0.76.$$

It will be noted that $\Sigma_s/N_u = \sigma_s(N_m/N_u) = 1880$, so that equation (4.53) is not really applicable to this case; however, the discrepancy between the estimated value of 0.76 and that (0.74) based on the measured resonance integral is not large.

4.71. For a homogeneous system, in the absence of impurities, the thermal utilization can be obtained from equation (4.3); thus,

$$f = \frac{\Sigma_{au}}{\Sigma_{au} + \Sigma_{am}}$$

$$= \frac{N_u\sigma_{au}}{N_u\sigma_{au} + N_m\sigma_{am}} = \frac{\sigma_{au}}{\sigma_{au} + \sigma_{am}(N_m/N_u)}, \tag{4.54}$$

where the σ_a's are now the absorption cross sections for *thermal* neutrons. Comparison of equation (4.54) with equation (4.52) shows that, whereas the resonance escape probability decreases with increasing proportion of fuel to moderator, i.e., as N_u/N_m increases, the reverse is true for the thermal utilization.

4.72. For natural uranium as fuel, the value of η is 1.33 (§ 4.6) and the fast fission factor ϵ is approximately unity for a homogeneous mixture. These data, together with the values of p and f calculated from equations (4.52) and (4.54), using the effective resonance integrals from Fig. 4.7 and the known cross sections, are combined in Table 4.3 to obtain the infinite multiplication factors. The

TABLE 4.3.   INFINITE MULTIPLICATION FACTOR FOR UNIFORM
MIXTURES OF NATURAL URANIUM AND GRAPHITE

N_m/N_u	p	f	k_∞
200	0.64	0.92	0.79
300	0.70	0.89	0.83
400	0.74	0.86	0.85
500	0.77	0.82	0.85
600	0.78	0.80	0.84

results are given for various proportions of natural uranium as fuel and graphite as moderator. Since p increases while f decreases, the product pf, and consequently the infinite multiplication factor, passes through a broad maximum in the region in which N_m/N_u is about 400 to 500. However, it is seen that the highest possible value of k_∞ for a homogeneous natural uranium-graphite mixture is 0.85. Since this is less than unity, it is clearly impossible to establish a self-sustaining chain reaction in this system. In other words, no matter how large a reactor were constructed, it would never become critical.

4.73. A similar situation arises in connection with homogeneous systems of natural uranium with either ordinary water or beryllium as the moderator, but when heavy water is the moderator, a critical system is possible for certain ratios of fuel to moderator. The optimum ratio gives a maximum value of about 1.03 for the infinite multiplication factor, so that the critical system would probably be too large to be practical. As will be shown below, reactors of reasonable size with heavy water (or other material) as moderator are obtained by using the fuel in the form of lumps or rods.

4.74. There are two reasons why the behavior of natural uranium fuel in heavy water in a homogeneous system is different from that in the other moderators. In the first place, the absorption cross section for thermal neutrons is smaller so that the thermal utilization is larger. Second, and more important, is the increase in the resonance escape probability for a given fuel to moderator ratio; this arises from the higher values of both σ_{sm} and ξ in equation (4.52). It is true that for ordinary water the value of $\sigma_{sm}\xi$ is even larger than for heavy water, so that the resonance escape probability is greater. However, there is a marked decrease in the thermal utilization, because of the relatively large absorption cross section of hydrogen for thermal neutrons, and so the product pf is much less than in heavy water.

4.75. The problem of obtaining a critical system with a uranium fuel and graphite or beryllium as moderator can be solved in two ways. The same procedures make it possible to design a heavy-water moderated system of practical size. One is to use enriched fuel, i.e., fuel containing an appreciably greater proportion of the fissile uranium-235 than does natural uranium. The main effect of such enrichment is to increase the factor η because of the increase in Σ_f (§ 4.6). If the maximum value of pf is taken to be about 0.64, as in Table 4.3, it is seen that η must be at least 1.55 if k_∞ is to equal (or exceed) unity; this requires an enrichment of at least 1.3 atomic per cent of uranium-235. Roughly the same composition would be the minimum necessary for a homogeneous critical system with ordinary water as moderator. It should be noted that, for a practical reactor, the infinite multiplication factor would have to be about 1.05 to 1.2, to allow for the consumption of fuel, accumulation of poisons, etc., during operation. The actual enrichment would thus have to be appreciably larger than the minimum value.

4.76. The second method for attaining criticality with uranium as the fuel is to take advantage of the fact that the resonance escape probability, for a given ratio of fuel to moderator, can be increased by using a heterogeneous lattice system consisting of lumps (or rods) of natural uranium arranged in a matrix of moderator. However, with ordinary water as moderator, it is apparently impossible to achieve a critical system with natural uranium as fuel under any circumstances. Consequently, even with a lattice system, it is still necessary to use enriched uranium fuel when water is the moderator. With graphite, beryllium, and, of course, heavy water it is possible to attain criticality in heterogeneous reactors using natural uranium as the fuel. Such systems are discussed in § 4.92 *et seq.*

4.77. A thermal reactor employing highly enriched material can be made critical with a very low concentration of fuel. Because of the large ratio of moderator to fuel, the fraction of neutrons captured in fuel while slowing down is small, since the probability of scattering is large compared to absorption. The resonance escape probability is then not much less than unity, and as ϵ is slightly greater than unity, it is a good approximation to write

$$k_\infty = \eta f. \tag{4.55}$$

4.78. If the fuel material contains about 20 per cent or more of uranium-235, η is approximately 2.06. The thermal utilization could then be as low as 0.5, i.e., $k = 1.03$, and the system still be capable of becoming critical. For a homogeneous mixture of fuel and moderator, in the absence of impurities, f is given by equation (4.54) and the system would be capable of maintaining a chain reaction even when $N_u \sigma_{au} = N_m \sigma_{am}$, so that f is 0.5. For uranium-235, σ_{au} is $577 + 106 =$

683 barns, and if graphite is the moderator, σ_{am} is 0.0034 barn.   Hence, in these circumstances, N_m/N_u would be 2.0×10^5.   This means that a homogeneous mixture containing 1 atom of uranium-235 to roughly 200,000 atoms of graphite could be used in a thermal reactor.   Since k is only 1.03, the critical reactor would, of course, be large, so that the nonleakage probability exceeds 0.97.   If the ratio of uranium-235 to moderator is increased, the thermal utilization is increased, and so also is the infinite multiplication factor.   Criticality could then be attained in a homogeneous reactor of smaller size.

CALCULATION OF CRITICAL SIZE

4.79. After the infinite multiplication factor has been determined, in the manner outlined above, the next step in the evaluation of the critical size of the homogeneous reactor is to calculate the critical buckling, e.g., by means of equation (4.28) based on the age-diffusion model.   In general, the proportion of fuel in a thermal reactor is relatively low so that in a homogeneous system the diffusion coefficient and age of thermal neutrons will be essentially the same as in pure moderator.   However, the diffusion length in the reactor core is generally quite different from that in the moderator alone.   By equation (3.28), L^2 is equal to D/Σ_a, where all the properties apply to the fuel-moderator mixture.   As stated above, D may be taken to be the diffusion coefficient of the pure moderator, i.e., D_m, and so

$$L^2 = \frac{D_m}{\Sigma_{am} + \Sigma_{au}}.$$

If this result is combined with equation (4.54), which defines f, it is found that

$$L^2 = \frac{D_m}{\Sigma_{am}} (1 - f) = L_m^2(1 - f). \tag{4.56}$$

4.80. Since k, τ, and L^2 for the given reactor composition are now known, it is possible, in principle, to determine B_c^2 from equation (4.28).   It should be noted, however, that this is a transcendental equation and its solution can be performed either graphically [9] or by a trial and error (iteration) procedure.   When the critical buckling has been evaluated, the dimensions of the bare critical reactor are then obtained from the expression in Table 4.2 appropriate to the specified geometry, e.g., sphere, cube, or cylinder.

Example 4.7.   Calculate the critical radius (at ordinary temperature) of a spherical reactor containing a uniform mixture of uranium-235 and beryllium in the atomic ratio of 1 to 10,000 (see Example 4.1), having a 25-cm thick beryllium reflector.

From Example 4.1, the value of f is 0.87, and from Table 3.1, L_m, the diffusion length of thermal neutrons in beryllium is 21 cm; hence, in the reactor, using equation (4.56),

$$L^2 = L_m^2(1 - f) = (21)^2(0.13) = 57 \text{ cm}^2.$$

To determine the critical buckling of the bare reactor, a preliminary value is first obtained by means of equation (4.38). From Example 4.1, k_∞ is 1.79 and from Table 3.4, τ is 85 cm; hence,

$$B_c^2 = \frac{k_\infty - 1}{M^2} = \frac{0.79}{57 + 85} = 5.5 \times 10^{-3} \text{ cm}^{-2}.$$

Starting with this value and using the iteration procedure described in Example 4.2(c), the age-diffusion critical equation (4.28) can be solved to give the critical buckling of the bare reactor as

$$B_c^2 = 4.3 \times 10^{-3} \text{ cm}^{-2}.$$

From Table 4.2, the critical radius of the bare spherical reactor would be given by

$$R_c^2 = \left(\frac{\pi}{B_c}\right)^2 = \frac{(3.14)^2}{4.3 \times 10^{-3}}$$

$$R_c = 48 \text{ cm}.$$

(It will be noted that for this relatively small reactor, equation (4.38) is not as good an approximation to the age-diffusion equation as it is for the larger reactor in Example 4.2.)

The next step is to determine the reactor savings. The reflector thickness is 25 cm and, since the thermal diffusion length in beryllium is 21 cm, this reflector is neither "thick" nor "thin." For a thick reflector, the savings in this case would be given by the equation in § 4.54 as roughly equal to L_r, which is 21 cm. On the other hand, for a thin reflector, the savings would be approximately the same as the reflector thickness (§ 4.57), i.e., 25 cm. These values are not very different, and so an average may be taken in this case, namely, 23 cm. The critical radius of the reflected reactor is thus $48 - 23 = 25$ cm.

4.81. For highly enriched fuel, k_∞ and L^2 in the critical equation (4.29) may be replaced by the corresponding values from equations (4.55) and (4.56). After some rearrangement, it is readily found that

$$\frac{N_m}{N_u} = \frac{\eta e^{-B_c^2 \tau} - 1}{1 + L_m^2 B_c^2} \cdot \frac{\sigma_{au}}{\sigma_{am}}, \tag{4.57}$$

where τ is essentially the age of thermal neutrons in the moderator. By means of this equation it is possible to calculate the atomic ratio of moderator to fuel required to make a homogeneous reactor of specific composition and size just critical.

4.82. In order to indicate the general relationship between the core composition and the critical buckling for various moderators, equation (4.57) has been used to determine N_m/N_u as a function of B_c^2 for pure uranium-235 as fuel and either heavy water, beryllium, or carbon (graphite) as moderator. The results are plotted in Fig. 4.8. (A curve for ordinary water as moderator is also given, but this is derived in a different manner as will be seen later.) In the case of heavy water as moderator, N_m refers to the number of molecules of this substance; for the other moderators it represents the number of atoms. If the composition of the fuel-moderator system, i.e., N_m/N_u, is specified, then B_c^2 can be obtained from the figure; the critical dimensions for a given shape can then be calculated. On the other hand, if the geometry is specified, B_c^2 is known and the composition required to make the system critical can be read from the curves.

4.83. It is seen from Fig. 4.8 that as the ratio of moderator to fuel is increased, the dimensions of the critical reactor must increase. However, by varying the composition of the mixture, i.e., by changing the moderator-to-fuel ratio, it is

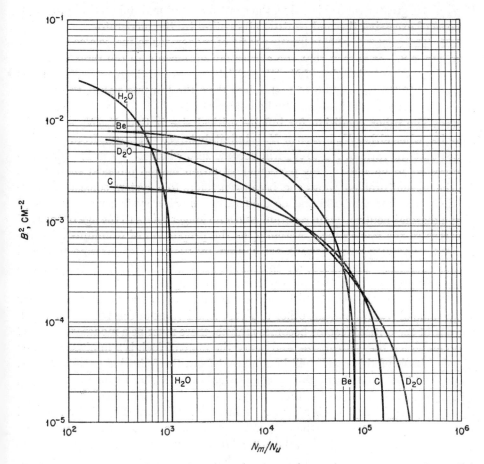

FIG. 4.8. Critical buckling as function of composition of homogeneous systems with pure uranium-235 as fuel

possible to obtain a system which will become critical with a minimum mass of fuel. This is related to the fact that two extreme situations are theoretically possible in which the critical size is infinite.

4.84. In the first place, when k_∞ is unity, B_c^2 in the critical equation (4.29) must be zero and hence the critical reactor would be infinitely large. For a highly enriched fuel system this requires that f be equal to $1/\eta$, by equation (4.55). The proportion of moderator is then so large that sufficient neutrons are absorbed to make the product $f\eta$ equal to unity.   From equation (4.57), it is readily shown that the critical reactor would be infinitely large when

$$\frac{N_m}{N_u} = \frac{\sigma_{au}(\eta - 1)}{\sigma_{am}}. \tag{4.58}$$

If N_m/N_u is greater than this value, the system cannot maintain a chain reaction because of excessive thermal neutron absorption by the moderator.

4.85. The existence of this limiting ratio of moderator to fuel is also apparent from Fig. 4.8.   It is seen that the curves fall off steeply to the right and tend toward an asymptotic value of N_m/N_u as B_c^2 decreases and the reactor dimensions increase.   If the proportion of fuel is less than the limiting value, the reactor cannot be made critical, no matter how large it is.

4.86. The other condition for infinite critical size is that N_m/N_u should be zero, i.e., when either N_m is zero or N_u is infinitely large.   It is seen from equation (4.57) that the ratio of N_m to N_u will become zero when $\eta e^{-B_c^2 \tau} = 1$, which also requires B_c^2 to be zero.   This particular situation has no physical significance for the present problem, however, because it refers to a system consisting of essentially pure uranium-235 and no moderator.   Such a system would not support a thermal-neutron chain reaction, and τ no longer refers to the age of thermal neutrons in the moderator.   Actually, as the proportion of uranium-235 is increased, relatively more fissions are caused by neutrons of intermediate and higher energies.   The theory of thermal-neutron chain reactions then breaks down and predicts critical masses which are too large.

4.87. Between the two hypothetical conditions—one corresponding to a very large value of N_m/N_u and the other to a very small value of this ratio—for which the size of the critical reactor becomes infinite, there must be a moderator-to-fuel ratio for which the critical size, and hence the mass of fuel, is a minimum.   The physical basis of this result will be apparent from the following considerations. When N_m/N_u is zero, the infinite multiplication factor k_∞ has its maximum value, which is equal to η, since f is then unity.   As N_m/N_u is increased, however, k_∞ does not change very much at first.   Thus, for a mixture of uranium-235 and graphite, k_∞ is 0.99η even when N_m/N_u is as high as 1.4×10^3.   This means, then, when N_m/N_u is 1.4×10^3, the critical reactor will have the same nonleakage probability within about 1 per cent, and hence almost the same size, as when N_m/N_u is almost zero.*   The greatly increased moderator-to-fuel ratio in essentially the same total critical volume obviously means a large decrease in the critical mass of fuel as N_m/N_u increases.

4.88. As the ratio N_m/N_u increases still further, k_∞ begins to fall off rapidly. Consequently a large increase is necessary in the nonleakage probability, and hence in the reactor size, if the system is to be critical.   Beyond a certain point, the critical volume increases more rapidly than N_m/N_u, and then the critical

---

* It is assumed that L^2 and τ are roughly the same in the two cases, although this is not strictly true.

mass of fuel begins to increase with increasing moderator-to-fuel ratio. It is evident, therefore, that there must be certain ratio of the given moderator to fuel for which the critical mass of fuel is a minimum.

4.89. If the reactor composition and geometry, i.e., N_m/N_u and B_c^2, respectively, are known, it is not a difficult matter to calculate the critical mass of fuel, using the densities of the fuel and moderator. This has been done for systems consisting of uranium-235 as fuel and either heavy water, beryllium, or graphite as moderator, the required data, except for the densities, being obtained from Fig. 4.8. The results plotted in Fig. 4.9 show the dependence of the critical

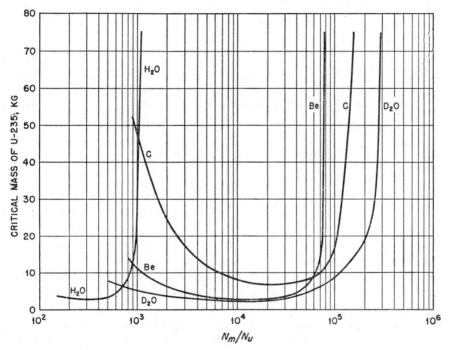

FIG. 4.9. Critical mass as function of composition of homogeneous systems with pure uranium-235 as fuel

mass of uranium-235 in a bare (spherical) reactor, i.e., one without a reflector, upon the composition, as expressed by the ratio N_m/N_u. The existence of a minimum critical mass is apparent in each case.

4.90. When water is the moderator, the continuous slowing down (or Fermi age) model is not very satisfactory, as stated in § 3.84. A somewhat better approach for homogeneous, aqueous reactor systems is to use a multigroup critical equation which is, in effect, an extension of equation (4.16), namely,

$$\frac{k_\infty}{(1 + L_1^2 B_c^2)^2(1 + L_2^2 B_c^2)(1 + L^2 B_c^2)} = 1, \tag{4.59}$$

where L^2 is the square of the thermal diffusion length in the reactor system, obtained from equation (4.56), and L_1^2 and L_2^2 are empirical constants derived from measurements on the slowing down of fast neutrons in water. Appropriate values are 2.625 and 26.539 cm^2, respectively. Several other methods have been proposed for the treatments of homogeneous systems with water as moderator, but they are more complicated than the one given here, which was used to derive the appropriate curves in Figs. 4.8 and 4.9.

4.91. Although some of the arguments presented above have been based on the supposition that pure uranium-235 is the fuel, the general conclusions are applicable to thermal reactors of all types. No matter what the nature of the fuel or the shape of the reactor, there will, in principle, be a particular moderator-to-fuel ratio for which the critical mass of fuel is a minimum. It should be noted that the data in Figs. 4.8 and 4.9 refer to bare reactors, without reflectors. For reflected systems the method mentioned in § 4.59 may be employed. The critical dimensions are first determined from B_c^2 for a bare reactor and then the appropriate reflector savings are deducted to give the size of the reflected core.

HETEROGENEOUS REACTOR SYSTEMS

INTRODUCTION

4.92. By arranging the fuel in the form of a lattice of lumps, rods, or plates in the moderator matrix, it is possible to achieve an increase in the infinite multiplication factor, largely as a result of the gain in the resonance escape probability. For example, in a system consisting of natural uranium and graphite, the value of k_∞ is increased from a maximum of 0.85 in a homogeneous mixture (Table 4.3) to about 1.08 in an optimized heterogeneous (or lattice) system. Consequently, criticality can be attained in the latter case, although it is impossible, with exactly the same materials, in the former circumstances. It should be mentioned, too, that the use of rod or plate fuel elements in a reactor provides a convenient design to permit removal of the heat generated in fission by circulating a coolant through the core (see Chapter 6).

4.93. Consider an arrangement of uranium fuel rods in a moderator. Fast neutrons formed by fission in the rods enter the moderator and are gradually slowed down. When in the vicinity of a rod, epithermal neutrons having energies near the resonance values (cf. Fig. 2.18) will be absorbed by uranium-238 nuclei, principally in the outer layers of the rod. Few neutrons will, therefore, be absorbed in the interior of the rod. Consequently, of the epithermal neutrons entering a natural uranium rod, only those with energies at (or very close to) the resonance peaks will be captured. All others will have a high probability of passing through the fuel rod, since they suffer little energy loss in collisions with the heavy uranium nuclei.　Upon re-entering the moderator, the neutrons will undergo further slowing down, and they may pass through several resonances before being captured as thermal neutrons.　The resonance escape probability will thus be greater than in a homogeneous system of the same composition. The larger the fuel-rod radius the greater the proportion of fuel protected from resonance neutrons, since the latter are nearly all absorbed in the outer layers. Hence, other things being equal, e.g., for a given fuel-moderator ratio, the resonance escape probability will increase with the rod radius.

4.94. In a heterogeneous system there is also an increase in the fast-fission factor, as compared with a uniform mixture of the same composition.　The fission neutrons produced within the uranium rod suffer very little slowing down therein.　Absorption of these fast neutrons by uranium-238 nuclei, and the resulting fissions, will therefore be greater than in a homogeneous system where the neutrons are rapidly slowed down in the moderator to energies at which uranium-238 fission is no longer possible.　Incidentally, some of the fast neutrons liberated in the fission of uranium-238 nuclei in a fuel rod will produce more fissions of the same type before escaping into the moderator, thereby adding to the fast-fission factor.

4.95. The one factor which is less in a lattice than in a homogeneous mixture of the same composition is the thermal utilization.　This is because the thermal neutron flux in the moderator is larger than in the interior of the fuel rod (Fig. 4.10), since absorption is greater in the latter.　The effect is to reduce the ther-

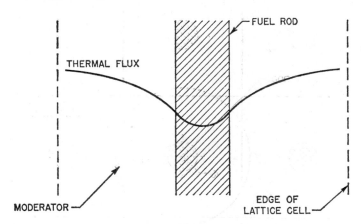

FIG. 4.10.　Thermal-neutron flux distribution in fuel and moderator

mal utilization.　However, the decrease is usually more than counterbalanced by the decrease in the effective resonance integral, and consequent increase in the resonance escape probability, resulting from the self-protection of the uranium rod against absorption of resonance neutrons.

CALCULATION OF THERMAL UTILIZATION

4.96. Disregarding absorption of neutrons by impurities, the thermal utilization is given by equation (4.2) as

$$f = \frac{V_u \Sigma_{au} \phi_u}{V_u \Sigma_{au} \phi_u + V_m \Sigma_{am} \phi_m} = \frac{1}{1 + \dfrac{\Sigma_{am}}{\Sigma_{au}} \cdot \dfrac{V_m \phi_m}{V_u \phi_u}} \tag{4.60}$$

or

$$\frac{1}{f} = 1 + \frac{\Sigma_{am}}{\Sigma_{au}} \cdot \frac{V_m \phi_m}{V_u \phi_u}, \tag{4.61}$$

where ϕ_m and ϕ_u are the average values of the thermal-neutron flux in moderator and fuel rod (or lump), respectively. The ratio of the fluxes in the two materials, i.e., ϕ_m/ϕ_u, is called the *thermal disadvantage factor*. Since the absorption cross sections for thermal neutrons are usually available, the calculation of the thermal utilization thus becomes a matter of finding the thermal-neutron flux distribution in the fuel and moderator. It should be noted that the macroscopic cross sections in equations (4.60) and (4.61) are the heterogeneous values, and are those for the separate (pure) fuel and moderator (§ 4.12).

4.97. In the treatment of uniform heterogeneous systems, it is convenient to divide the lattice up into a number of identical unit cells and to suppose that a square cross section can be replaced by a cylindrical cross section of the same

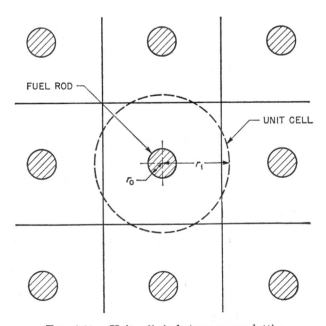

FIG. 4.11. Unit cells in heterogeneous lattice

area.   If the reactor consists of cylindrical rods of uranium arranged in the manner shown in Fig. 4.11, it may be divided up into a number of parallelepipeds of square cross section.   The equivalent unit cell of circular cross section, indicated by the broken-line circle, will then be a long cylinder with the fuel rod at the center.   The radius is such that the circle has the same area as the unit square upon which the lattice is based.   Let r_0 be the radius of the fuel rod and r_1 that of the unit cell; then πr_1^2 is the area of the unit square and V_m/V_u is equal to $(r_1^2 - r_0^2)/r_0^2$.

4.98. For the determination of the neutron distribution in the unit cell, two basic approximations are made.   First, it is supposed for simplicity that simple diffusion theory is applicable.   Transport theory methods could be used but the treatment would be more complicated.   Diffusion theory is adequate if the dimensions of the system are large compared with the scattering mean free path of neutrons, as is generally the case for thermal reactors using natural or slightly enriched uranium as fuel.   Furthermore, it is necessary that there should be only relatively weak absorption and that no neutron sources should be present. The second postulate is that the slowing down density of neutrons is constant throughout the moderator and zero in the uranium.   This condition will hold provided the separation of the fuel rods is not too large in comparison with the slowing down distance.

4.99. As a consequence of the assumption just mentioned, the neutrons in the fuel may be treated as monoenergetic and the equation of the one-group model can be employed.   Thus, the thermal diffusion equation for the fuel rod is

$$D_0 \nabla^2 \phi_0 - \Sigma_{a0} \phi_0 = 0, \tag{4.62}$$

or, upon dividing through by D_0 and replacing Σ_{a0}/D_0 by κ_0^2,

$$\nabla^2 \phi_0 - \kappa_0^2 \phi_0 = 0, \tag{4.63}$$

where the zero subscripts refer to thermal neutrons in the fuel.   It will be noted that there is no thermal-neutron source term in equation (4.62) or (4.63), since it has been postulated that there is no slowing down in the uranium.

4.100. For the moderator, indicated by the subscript 1, the thermal diffusion equation is

$$D_1 \nabla^2 \phi_1 - \Sigma_{a1} \phi_1 + q = 0 \tag{4.64}$$

or

$$\nabla^2 \phi_1 - \kappa_1^2 \phi_1 + \frac{q}{D_1} = 0, \tag{4.65}$$

where $\kappa_1^2 (= \Sigma_{a1}/D_1)$ is the reciprocal of the square of the thermal diffusion length in the moderator.   The source term, q, is equal to the number of neutrons becoming thermal in the moderator per cm³ per sec, and is constant because of the postulate that the slowing down density is constant.

4.101. By utilizing the appropriate boundary conditions, namely, that there is continuity of neutron flux and current density at the fuel rod-moderator interface, and that there is no net flow of neutrons at the outer boundary of the unit cell, since just as many neutrons diffuse into any given cell as diffuse out of it, it is possible to solve equations (4.63) and (4.65). The result is conveniently expressed as the reciprocal of the thermal utilization; thus,

$$\frac{1}{f} = 1 + \frac{V_1 \Sigma_{a1}}{V_0 \Sigma_{a0}} F + (E - 1), \tag{4.66}$$

where the quantities F and E for a cylindrical fuel rod are given by*

$$F = \frac{\kappa_0 r_0}{2} \cdot \frac{I_0(\kappa_0 r_0)}{I_1(\kappa_0 r_0)}$$

and

$$E = \frac{\kappa_1(r_1^2 - r_0^2)}{2r_0} \left[\frac{I_0(\kappa_1 r_0) K_1(\kappa_1 r_1) + K_0(\kappa_1 r_0) I_1(\kappa_1 r_1)}{I_1(\kappa_1 r_1) K_1(\kappa_1 r_0) - K_1(\kappa_1 r_1) I_1(\kappa_1 r_0)} \right],$$

I_0 and K_0 being zero-order modified Bessel functions of the first and second kinds, respectively, and I_1 and K_1 are the corresponding first-order functions.

4.102. The modified Bessel functions may be expanded in series form for cases where the ratio of moderator to fuel is fairly high and absorption in the fuel is weak. When $\kappa_0 r_0$ is less than 1 and $\kappa_1 r_1$ is less than 0.75, the results may be approximated [12] to give

$$F \approx 1 + \frac{(\kappa_0 r_0)^2}{8} - \frac{(\kappa_0 r_0)^4}{192}$$

and

$$E \approx 1 + \frac{(\kappa_1 r_1)^2}{2} \left[\frac{r_1^2}{r_1^2 - r_0^2} \ln \frac{r_1}{r_0} + \frac{1}{4} \left(\frac{r_0}{r_1} \right)^2 - \frac{3}{4} \right].$$

4.103. It can be shown that F is physically equivalent to the ratio of the thermal-neutron flux at the fuel-moderator interface to the average flux in the interior of the fuel; consequently, the second term on the right of equation (4.66) is sometimes referred to as the *relative absorption*. The quantity $E - 1$, i.e., the last term in equation (4.66), is called the *excess absorption*; it arises from the fact that the average thermal neutron flux in the moderator is greater than that at the fuel-moderator interface (cf. Fig. 4.10).

Example 4.8. A reactor consists of natural uranium rods, of 1 in. diameter, arranged in a square lattice with a pitch of 6 in. in heavy water (see Example 4.3). Calculate the thermal utilization for this lattice. The diffusion length of thermal neutrons in natural uranium metal is 1.55 cm; the effective macroscopic absorption cross section is 0.314 cm^{-1}.

The value of f is obtained from equation (4.66) and it may be assumed that E and F are given by the expressions in § 4.102. For the specified lattice, the rod diameter, r_0, is 0.5 in. = 1.27 cm, and $\kappa_0 = 1/L$ for uranium, i.e., $1/1.55 = 0.65$ cm^{-1}. Hence,

$$F \approx 1 + \frac{(\kappa_0 r_0)^2}{8} - \frac{(\kappa_0 r_0)^4}{192}$$

$$= 1 + \frac{[(0.65)(1.27)]^2}{8} - \frac{[(0.65)(1.27)]^4}{192} = 1.083.$$

* The values are different for other shapes; see, for example, Refs. 10 and 11.

The area of a unit cell is 36 in.^{2}, i.e., 232 cm^2, and this is equal to πr_1^2, so that $r_1 = 8.60$ cm. For heavy water, L is 100 cm (Table 3.1), and $\kappa_1 = 1/L_1 = 0.010$ cm^{-1}. Hence,

$$E \approx 1 + \frac{(\kappa_1 r_1)^2}{2}\left[\frac{r_1^2}{r_1^2 - r_0^2}\ln\frac{r_1}{r_0} + \frac{1}{4}\left(\frac{r_0}{r_1}\right)^2 - \frac{3}{4}\right]$$

$$= 1 + \frac{[(0.010)(8.60)]^2}{2}\left[\frac{(8.60)^2}{(8.60)^2 - (1.27)^2}\ln\frac{8.60}{1.27} + \frac{1}{4}\left(\frac{1.27}{8.60}\right)^2 - 0.75\right]$$

$$= 1.0044.$$

It may be noted that in this and many other cases, the expression for E could have been simplified to

$$E \approx 1 + \frac{(\kappa_1 r_1)^2}{2}\left[\ln\frac{r_1}{r_0} - \frac{3}{4}\right].$$

In order to obtain f from equation (4.66), the values of V_1/V_0, Σ_{a1}, and Σ_{a0} are still required. As stated in § 4.97 (see Fig. 4.11),

$$\frac{V_1}{V_0} = \frac{r_1^2 - r_0^2}{r_0^2} = \left(\frac{r_1}{r_0}\right)^2 - 1$$

$$= \left(\frac{8.60}{1.27}\right)^2 - 1 = 44.9.$$

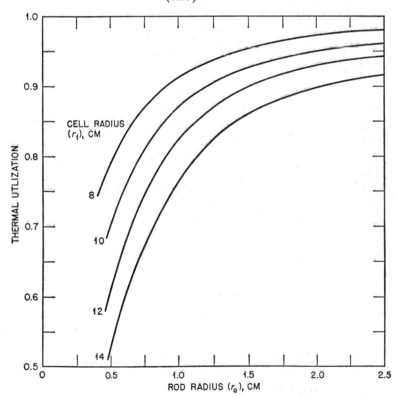

FIG. 4.12.   Thermal utilization in natural uranium-graphite lattice

For the heavy-water moderator, Σ_{a1} is obtained from Table 3.1 as 8.5×10^{-5}, and Σ_{a0} is given as 0.314 cm^{-1}; hence

$$\frac{1}{f} = 1 + \frac{V_1}{V_0} \cdot \frac{\Sigma_{a1}}{\Sigma_{a0}} F + (E - 1)$$

$$= 1 + \frac{(44.9)(8.5 \times 10^{-5})}{0.314} (1.083) + 0.0044 = 1.017.$$

$$f = 0.983.$$

4.104. For a given ratio of moderator to fuel, i.e., V_m/V_u (or V_1/V_0) constant, the thermal utilization is found to increase as the rod (or lump) radius decreases. For rods (or lumps) of given radius, the thermal utilization decreases as V_m/V_u (or V_1/V_0) increases. It is seen, therefore, that the thermal utilization increases (1) with decreasing radius of the rods at constant reactor composition, and (2) with decreasing ratio of moderator to fuel volume at constant rod radius. The results of some calculations for a square lattice of natural uranium rods in a graphite matrix are shown in Fig. 4.12.

CALCULATION OF RESONANCE ESCAPE PROBABILITY

4.105. Because the neutron flux is different in the fuel and the moderator, equation (4.51) for the resonance escape probability, which is applicable to a homogeneous system, must be modified and takes the approximate form

$$p(E) \approx \exp\left[-\frac{N_u V_u \phi_u}{\xi \Sigma_{sm} V_m \phi_m} \int (\sigma_{au})_{\text{eff}} \frac{dE}{E} \right], \tag{4.67}$$

where V_m and V_u are the volumes of moderator and fuel in the lattice, and ϕ_m and ϕ_u are the average values of the *resonance* flux in the moderator and in the interior of the fuel lump, respectively. The ratio ϕ_m/ϕ_u is here the disadvantage factor for resonance neutrons.

4.106. A quantity f_r, called the *resonance utilization*, is now defined by analogy with the thermal utilization as the ratio of the neutrons absorbed by fuel nuclei in the resonance region to the total number of resonance neutrons produced. The reciprocal of f_r can be represented by equations similar to (4.61) or (4.66); thus, using the subscripts 0 and 1 for fuel and moderator, respectively,

$$\frac{1}{f_r} = 1 + \frac{\Sigma_{a1}}{\Sigma_{a0}} \cdot \frac{V_1 \phi_1}{V_2 \phi_2} \tag{4.68}$$

or

$$\frac{1}{f_r} = 1 + \frac{V_1 \Sigma_{a1}}{V_0 \Sigma_{a0}} F_r + (E_r - 1), \tag{4.69}$$

where Σ_{a1} and Σ_{a0} now refer to resonance neutron absorption; F_r is the ratio of the resonance neutron flux at the fuel-moderator interface to the average in the fuel and $E_r - 1$ is the excess slowing down of neutrons out of the resonance region.

4.107. The quantity Σ_{a1} in equations (4.68) and (4.69) is not strictly an absorption cross section but rather a cross section for the removal of resonance neutrons by slowing down in the moderator, similar to that described in the two-group treatment in § 4.19.   It can be shown that

$$\Sigma_{a1} = \frac{\xi \Sigma_{s1}}{\ln (E_1/E_2)}, \tag{4.70}$$

where Σ_{s1} is the macroscopic scattering cross section for resonance neutrons and E_1 and E_2 represent the upper and lower energy limits of the resonance interval; for both uranium metal and uranium dioxide as fuel, it is a good approximation to take $\ln (E_1/E_2)$ to be equal to 3.0.   However, for the present purpose the value of $\ln E_1/E_2$ is immaterial since it cancels out, as will be seen below.

4.108. The absorption cross section Σ_{a0} of resonance neutrons in the fuel can be expressed in an analogous manner by

$$\Sigma_{a0} = N_0 \frac{\int (\sigma_{a0})_{\text{eff}} \dfrac{dE}{E}}{\ln (E_1/E_2)}, \tag{4.71}$$

the integral in the numerator being the appropriate effective resonance integral for the heterogeneous system.   Upon combining equations (4.68), (4.69), (4.70), and (4.71) with equation (4.67), it is found that the resonance escape probability can be written in the simple form

$$p(E) \approx \exp \left(-\frac{f_r}{1 - f_r} \right). \tag{4.72}$$

The problem of evaluating p is thus reduced to the determination of f_r, and this involves the properties κ and Σ_a for both fuel and moderator.

4.109. The expressions for evaluating F_r and E_r are exactly the same as those given above for the thermal utilization, except that κ_0 and κ_1 now refer to resonance neutrons.   Appropriate data for κ_0 appear to be as follows:

Uranium metal $\kappa_0 = 0.44$ cm^{-1}
Uranium dioxide $\kappa_0 = 0.31$ cm^{-1},

and recommended values of κ_1, which may be used for both uranium metal and the oxide, are given in Table 4.4 [13].   The evaluation of Σ_{a0} by means of equation (4.71) requires a knowledge of the effective resonance integral which is derived from experimental measurements.

TABLE 4.4.   RESONANCE NEUTRON DATA FOR MODERATORS

Moderator	κ_1(cm^{-1})
Water	0.885
Heavy water	0.22
Beryllium	0.325
Beryllium oxide	0.19
Graphite	0.145

4.110. By postulating that the effective absorption cross section for a hetero-geneous system can be separated into two parts, namely, volume absorption and surface absorption, it has been found that the effective resonance integral can be expressed in the form

$$\int (\sigma_{au})_{\text{eff}} \frac{dE}{E} = a + b \frac{S}{M} \text{ barns,} \qquad (4.73)$$

where S is the surface area of the uranium fuel elements in cm^2 and M is the mass in grams. Values of a and b for uranium metal and uranium dioxide based on experimental measurements are as follows [14]:

	a	b
Uranium metal....................	8.0	27.5
Uranium dioxide...............	11.0	24.5

A number of modifications of equation (4.73) and different values of a and b are to be found in the literature, but the data given here are adequate for most calculations of the resonance escape probability.*

Example 4.9. Calculate the resonance escape probability for the natural uranium-heavy water lattice in Example 4.8.

The value of f_r, as defined by equation (4.69), is first determined. Following the procedure in Example 4.8, with $\kappa_0 = 0.44$ cm^{-1} (for uranium metal) and $\kappa_1 = 0.22$ cm^{-1} (from Table 4.4), for resonance neutrons,

$$F_r \approx 1 + \frac{[(0.44)(1.27)]^2}{8} - \frac{[(0.44)(1.27)]^4}{192}$$

$$= 1.039.$$

For E_r the simplified expression in Example 4.8 may be used; thus,

$$E_r \approx 1 + \frac{[(0.22)(8.61)]^2}{2} \left(\ln \frac{8.61}{1.27} - 0.75 \right)$$

$$= 3.17.$$

To evaluate Σ_{a1} for resonance neutrons, equation (4.70) is used; the numerator is equal to the slowing down power of the heavy water moderator, which is 0.18 (Table 3.3). Hence,

$$\Sigma_{a1} = \frac{0.18}{\ln (E_1/E_2)},$$

where $\ln (E_1/E_2)$ does not need to be evaluated. To obtain Σ_{a0}, equations (4.71) and (4.73) may be combined, so that

$$\Sigma_{a0} = \frac{N_0(a + bS/M)}{\ln (E_1/E_2)}.$$

---

* An alternative expression for the effective resonance integral which is sometimes used is of the form $a + b(S/M)^{1/2}$, the a and b values being different from those given here [15].

Uranium metal contains 0.048×10^{24} nuclei/cm³, for a density of 19 g/cm³, and so N_0 is 0.048 if the microscopic cross section is in barns; hence,

$$\Sigma_{a0} = \frac{(0.048)(8.0 + 27.5S/M)}{\ln{(E_1/E_2)}}.$$

For a cylindrical rod,

$$\frac{S}{M} = \frac{2\pi r_0}{\pi r_0^2 \rho} = \frac{2}{(1.27)(19)} = 0.084,$$

so that

$$\Sigma_{a0} = \frac{0.513}{\ln{(E_1/E_2)}}.$$

Insertion of the appropriate values into equation (4.69) then gives

$$\frac{1}{f_r} = 1 + \frac{(44.9)(0.18)}{0.513}(1.039) + 2.17 = 19.6$$

$$f_r = 0.051.$$

Finally, by equation (4.72),

$$p \approx \exp\left(-\frac{0.051}{0.949}\right) = 0.947.$$

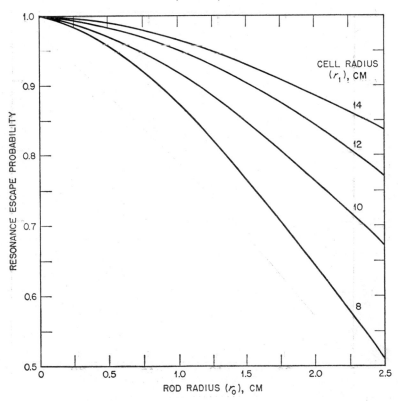

FIG. 4.13.   Resonance escape probability in natural uranium-graphite lattice

4.111. The calculated resonance escape probabilities for various rod and cell radii in a heterogeneous natural uranium-graphite system with a square lattice are given in Fig. 4.13. The effects of changing the fuel rod (or lump) radii and of the composition are the reverse of those on the thermal utilization given in § 4.104. Thus, the resonance escape probability increases (1) with increasing radius of the fuel rod at constant composition, and (2) with increasing moderator-to-fuel volume ratio at constant rod radius.

FAST-FISSION FACTOR

4.112. Since η may be regarded as known for natural uranium (§ 4.6), there remains only the fast-fission factor ϵ to be determined in order that the infinite multiplication factor may be evaluated for a given lattice. The method of calculation is based on probability theory and requires a knowledge of fission, non-fission capture, and inelastic and elastic scattering cross sections for fast neutrons in the fuel [16]. The values of ϵ increase with the probability that a primary

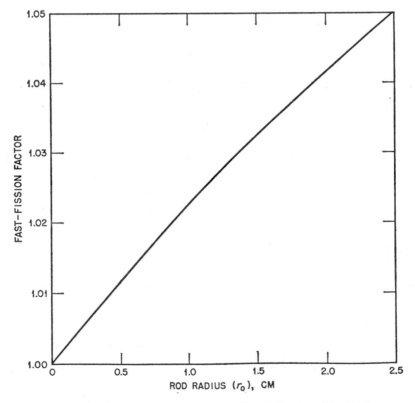

FIG. 4.14. Fast-fission factor in natural uranium-graphite lattice

fission neutron will make a collision of some kind inside the fuel rod (or lump) in which it was created. As may be expected, this probability increases with the total neutron cross section of the fuel material and with the radius of the rod, and consequently so also does the fast-fission factor. Lumping of the fuel thus provides some advantage, although a relatively small one, because of the increase in the fast-fission factor. The dependence of ϵ upon the radius of cylindrical fuel rods of natural uranium is indicated in Fig. 4.14. In several thermal reactors, moderated by graphite or heavy water, the radius of the uranium metal fuel rods is close to 1.4 cm, so that ϵ is about 1.03.

4.113. In calculating the fast-fission factors given in Fig. 4.14, it was assumed that none of the neutrons leaving the fuel rod and entering the moderator will return to the fuel element with sufficient energy to produce fission of the uranium-238. Although this is reasonably true for lattices in heavy water, beryllium, or graphite, it is not the case for water-moderated (slightly enriched) systems. The reason is that in the latter the fuel elements are close together, because the ratio of moderator to fuel in a critical system is generally much less than for other moderators (see, for example, Fig. 4.8). Consequently, some of the neutrons leaving a given fuel element can pass through the moderator and enter an adjacent fuel element with sufficient energy to cause fast-neutron fission. It has been found that for fuel rods of 0.76-cm radius, of uranium dioxide containing 1.3 per cent of uranium-235, the value of ϵ varied from 1.04 when the volume ratio of water to uranium was 4 to about 1.10 when the ratio was 1, i.e., for much closer rod spacing. Thus, the *interaction effect*, as it is called, increases as the volume ratio of water moderator to fuel is decreased. A common value of this ratio in slightly enriched fuel reactors with water as moderator is about 3 or 4, so that the fast-fission factor is approximately 1.04. These systems present some problems of special interest and so they are treated more fully later.

Example 4.10. Using the values of f and p obtained in Examples 4.8 and 4.9, calculate the infinite multiplication factor for the specified natural uranium-heavy water lattice.

From Fig. 4.14, the value of ϵ for a rod radius of 1.27 cm is seen to be close to 1.03, and from Table 4.1, η in natural uranium is 1.33; hence, by equation (4.5),

$$k_\infty = \eta \epsilon p f$$
$$= (1.33)(1.03)(0.947)(0.983)$$
$$= 1.28.$$

This is the value which was used in Example 4.2.

DETERMINATION OF OPTIMUM LATTICE

4.114. From the statements made earlier or from an examination of Figs. 4.12 and 4.13, it is evident that the conditions which favor an increase in the resonance escape probability are just those which cause the thermal utilization to decrease, and vice versa. The problem in the construction of a heterogeneous lattice is to determine the particular arrangement of fuel and moderator that gives a maximum value of the product pf. The procedure used is to consider a number of fuel-moderator lattices, with various rod radii and lattice spacings (cell radii), and to calculate the values of p, f, and ϵ for each lattice. Then, using the known value of η, the corresponding infinite multiplication factors, k_∞, can be determined.

4.115. Some indication of the results obtained from a series of such calculations may be obtained from the data in Table 4.5 [17]. These refer to a natural

TABLE 4.5. PROPERTIES OF HETEROGENEOUS NATURAL
URANIUM-GRAPHITE LATTICES

Cell Radius (cm)	f	p	k_∞
10	0.907	0.866	1.073
11	0.888	0.890	1.082
12	0.867	0.908	1.076
13	0.846	0.923	1.066

uranium-graphite lattice, with cylindrical fuel rods of 1.25-cm radius located in cells of various radii. The appropriate values of η and ϵ are thus 1.33 (Table 4.1) and 1.027 (Fig. 4.14), respectively. Allowance is made in each case for an aluminum cladding 0.115 cm thick and a 2-cm air space surrounding each fuel rod. It is apparent that, for a specified rod radius, the value of k_∞ passes through a maximum as the cell radius, i.e., the pitch of a square lattice, is increased. This maximum generally occurs when the values of p and f, one of which decreases regularly while the other increases, are almost equal.

4.116. If a particular cell radius is chosen and the calculations performed for different radii of the fuel rods, somewhat similar results are obtained. With increasing rod radius the values of f increase whereas those of p decrease, so that k_∞ has a maximum for a particular fuel-rod radius. By combining the various data it is thus possible to choose the best dimensions for the reactor. It was on the basis of such calculations that the lattices of the natural uranium-graphite reactors were designed, e.g., rod radius 1.4 cm (about 1.1 in. diameter) and 11.5-cm cell radius (about 8 in. pitch in a square lattice).

WATER-MODERATED, SLIGHTLY ENRICHED LATTICE SYSTEMS

INTRODUCTION

4.117. Criticality calculations for systems using slightly enriched uranium as fuel and water as moderator have proved to be quite complicated because such reactors are not truly thermal. As seen above, the ratio of moderator to fuel is fairly low in a critical system, with the result that in these close-packed (or "tight") lattices thermalization of the neutrons is not complete. An important consequence is that there is a significant contribution to the multiplication from fissions in uranium-235 caused by the absorption of resonance neutrons. Hence, consideration must be given to resonance capture and fission in uranium-235, as well as to neutron leakage and to resonance capture in uranium-238, while the neutrons are slowing down.

4.118. Although large digital computers are in general use for making detailed critical calculations (§ 4.66), a need exists in conceptual design surveys for simpler, more approximate procedures that do not require the use of calculating machines. Unfortunately, most such approximate procedures, although quite adequate for slightly enriched lattices with heavy water, beryllium, or graphite moderators, and for water-moderated, highly enriched systems, have given poor results for water lattices with slightly enriched fuels. Although some errors can be tolerated in a simplified method of calculation, the application of the age-diffusion method or the ordinary two-group model yields results which are too inaccurate for most purposes.

4.119. The semi-empirical method to be described here, however, has proved to be simple and sufficiently accurate for its intended purpose [18]. It is similar to the four-factor approach except that it is modified to include the effect of epithermal fissions in uranium-235. By recognizing that the design parameters for practical reactors using water (under pressure) as moderator (see Chapter 12) cover a fairly limited range, some of the empirical relations and cross sections can be assumed to remain constant in situations of interest. Because of the importance of pressurized-water systems and the usefulness of the procedure, it will be described in some detail.

MODIFIED FOUR-FACTOR MODEL

4.120. In order to establish a physical picture of the factors included, a neutron cycle will be developed to fit the present situation. Let the point A in Fig. 4.15 be the starting point of the cycle where one thermal neutron is absorbed. As a result, ηf fission neutrons will be produced as shown by the line AB, the quantities η and f having the same significance as before. To these neutrons, however, there will be added x fission neutrons formed by the epithermal fission of uranium-235, as indicated by the line DB, so that the total number of neutrons at B is $\eta f + x$. As a consequence of the fast-fission effect mainly in uranium-238, the total number of fission neutrons available for slowing down is $(\eta f + x)\epsilon$, where ϵ is the appropriate fast-fission factor.

4.121. These neutrons are now considered to slow down from C to E, the process being arbitrarily divided into two parts, first from C to the intermediate point D, taken to be 0.625 ev, where the radiative capture in uranium-235 is significant, and then from D to E, where the energy is thermal. The number of neutrons reaching D is equal to $(\eta f + x)\epsilon P_1 p'_{28}$, where P_1 is the nonleakage probability of neutrons with energies above D and p'_{28} is the neutron resonance

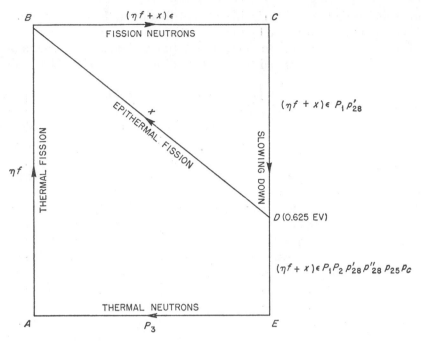

Fig. 4.15. Neutron cycle in water-moderated system

escape probability for uranium-238 in the same energy range. In slowing down from D to E, the following additional factors are introduced: P_2, the nonleakage probability for neutrons in the energy range from D to E; p_{28}'', the capture escape probability for uranium-238 in this energy range; p_{25}, the capture escape probability for uranium-235; and p_c, the probability that neutrons while slowing down will escape capture in various materials other than fuel present in the reactor.* Finally, introduction of P_3, the nonleakage probability of thermal neutrons, completes the cycle, so that the effective multiplication factor is given by

$$k_{\text{eff}} = P_1 P_2 P_3 (\eta f + x) \epsilon p_{28} p_{25} p_c, \tag{4.74}$$

where p_{28}' and p_{28}'' have been combined to give p_{28}, the overall capture escape probability for uranium-238.

4.122. Since p_{25} is the capture escape probability for uranium-235, the fraction of the neutrons reaching the point D which undergo epithermal absorption in this isotope is $1 - p_{25}$. Consequently, the number x of fission neutrons produced by this process is

$$x = \eta_{25}(1 - p_{25})[(\eta f + x)\epsilon P_1 p_{28}'],$$

---

* The symbols "25" and "28," representing a combination of the final integers of the atomic and mass numbers, are frequently used to represent uranium-238 and uranium-235, respectively.

where η_{25} is the number of neutrons produced per epithermal neutron absorbed in uranium-235. From this relationship, it follows that

$$(\eta f + x) = \eta f[1 - \epsilon P_1 p'_{28} \eta_{25}(1 - p_{25})]^{-1} \qquad (4.75)$$

and substitution into equation (4.74) yields

$$k_{\text{eff}} = P_1 P_2 P_3 \eta f \epsilon p_{28} p_c \beta, \qquad (4.76)$$

where

$$\beta \equiv \frac{p_{25}}{1 - \epsilon P_1 p'_{28} \eta_{25}(1 - p_{25})}. \qquad (4.77)$$

Comparison of equation (4.76) with the conventional four-factor formula shows that β accounts for the contribution to the neutron multiplication of resonance fissions in uranium-235.

CALCULATION OF f AND η

4.123. The calculational procedure will be illustrated by evaluating the critical buckling for a particular lattice which has been the subject of experimental study in connection with the development of the Yankee Atomic Electric Company reactor. The core, in one particular experiment, utilized fuel rods containing sintered pellets of uranium dioxide (density 10.2 g/cm³) enriched to 2.70 per cent in uranium-235. The rod diameter was 0.300 in. (0.762 cm), clad with a Type 304 stainless steel tube of 0.306 in. (0.778 cm) inner diameter and 0.016 in. (0.041 cm) wall thickness. The active length of the fuel rods was 48 in. and they were arranged in a square lattice with a pitch of 0.470 in. A lattice parameter that is commonly stated is the ratio of the volume of water in a unit cell to the volume that the uranium in the fuel would occupy in the form of metal. In the present case, this water to uranium volume ratio is 3.9 to 1.

4.124. The number density, N, of the various elements present, expressed for convenience in units of 10^{24} atoms per cm³, have been evaluated on the assumption of a uniform distribution throughout the core. The results are given in Table 4.6, together with the microscopic thermal (2200 m/sec) neutron cross

TABLE 4.6.   BASIC DATA FOR WATER-MODERATED REACTOR CORE

Element (or isotope)	N (10^{24} atoms/cm³)	σ_a (barns)	g	Σ_a (cm^{-1})
Uranium-235.....	0.000198	683	0.974	0.132
Uranium-238.....	0.00707	2.71	1.002	0.0192
Hydrogen........	0.0396	0.332	—	0.0131
Oxygen..........	0.0343	—	—	—
Iron.............	0.0045	2.6	—	⎫
Chromium.......	0.00125	2.9	—	⎬ 0.0180
Nickel..........	0.00058	4.5	—	⎭

sections, and the factor g which allows for departure from $1/v$-behavior (§ 2.132). The last column gives the products of these quantities, so that the values are the macroscopic absorption cross sections for the homogenized system to be used in the calculations.*

4.125. For the calculation of the thermal utilization it is convenient to use equation (4.4) which may be written in the form

$$f = \frac{\bar{\Sigma}_{au}}{\bar{\Sigma}_{au} + \bar{\Sigma}_{am}(\phi_m/\phi_u) + \bar{\Sigma}_{ac}(\phi_c/\phi_u)}, \qquad (4.78)$$

where the subscripts u, m, and c, refer to fuel, moderator (water), and cladding (stainless steel), respectively; the macroscopic cross sections are those based on the assumption of a uniform distribution and hence are the values in Table 4.6. Because the radius of the fuel rods (r_0) is not very different from that of the unit cell (r_1) in a closely packed, water-moderated lattice, the procedure for obtaining the disadvantage factor described in § 4.101 *et seq.* is not satisfactory. The moderator-fuel (ϕ_m/ϕ_u) and the cladding-fuel (ϕ_c/ϕ_u) disadvantage factors can be calculated by means of transport theory [19], but for the present purpose use may be made of the experimental observation [20] that ϕ_m/ϕ_u is close to 1.12 for water to uranium volume ratios ranging from about 2 to 4. Hence the moderator-fuel disadvantage factor may be taken to be 1.12 in the present calculations. The cladding-fuel disadvantage factor is not known, but it is apparent that no serious error will be introduced by assuming that the neutron flux in the cladding is midway between the values in the moderator and in the fuel, so that ϕ_c/ϕ_u may be set equal to 1.06. Upon inserting these results into equation (4.78), together with the macroscopic cross sections from Table 4.6, it is found that

$$f = \frac{0.1512}{0.1512 + (0.0131)(1.12) + (0.0180)(1.06)}$$

$$= 0.817.$$

4.126. The value of η is obtained from the equations in § 4.6; with ν equal to 2.43 (Table 2.9) and 577 barns as the fission cross section of uranium-235, the result is

$$\eta = \nu \frac{\Sigma_f^{235}}{\Sigma_a^{235} + \Sigma_a^{238}}$$

$$= (2.43) \frac{(0.000198)(577)(0.974)}{(0.0192 + 0.132)} = 1.79,$$

since g for uranium-235 is 0.974 (Table 4.6). Hence the product ηf is (0.817)(1.79), i.e., 1.46.

---

* Strictly speaking, it would be desirable to use the average thermal cross sections which allow for incomplete thermalization, as described in § 2.135. However, since most of the calculations involve cross-section ratios, the values in Table 4.6 are adequate for preliminary calculations.

4.127. The value of ηf obtained above requires some correction because the temperature of the fuel exceeds that of the moderator when the reactor is in operation. This tends to decrease Σ_{au} relative to Σ_{am}, so that the net result is a decrease in the thermal utilization, in particular [21]. For the purpose of the approximate calculations described here, it is probably adequate to decrease ηf by about 1 per cent; hence,

$$\eta f = \frac{1.46}{1.01} = 1.45.$$

NEUTRON CAPTURE AND ESCAPE PROBABILITIES

4.128. The calculation of the capture and escape probabilities, p_{28}, and p_c, respectively, requires a knowledge of scattering cross sections, logarithmic energy decrements, and epithermal capture integrals (in impurities). The data for uranium, hydrogen, oxygen, and stainless steel are given in Table 4.7; values for aluminum are included to permit calculations to be made for cases in which this metal is used as cladding.

TABLE 4.7.   EPITHERMAL NEUTRON CONSTANTS

Element	σ_s (barns)	$\xi\sigma_s$ (barns)	Epithermal Capture Integral (barns)
Uranium.........	9.5	0.080	—
Hydrogen........	20.1	20.1	0.132
Oxygen..........	3.8	0.460	0.088
Stainless steel.....	12.0	0.420	2.53
Aluminum........	1.4	0.100	0.18

4.129. In a closely packed lattice, the ratio $V_u\phi_u/V_m\phi_m$ in equation (4.67) is not very different from unity. It is, therefore, a good approximation to estimate the resonance escape probability by utilizing equation (4.51), to which equation (4.67) reduces in the circumstances, with the effective resonance integral evaluated by means of equation (4.73). Hence, in this case

$$p_{28} = \exp\left[-\frac{N_{28}I_{28}}{\xi\Sigma_s}\right], \tag{4.79}$$

where I_{28} is the effective resonance integral of uranium-238 in the dioxide; $\xi\Sigma_s$ refers to the sum over all the nuclei present in the reactor core. From the number densities in Table 4.6 and the $\xi\sigma_s$ values in Table 4.7, it is found that $\xi\Sigma_s$ is 0.810 cm^{-1}; N_{28} is 0.00707×10^{24} atoms per cm^3. The effective resonance integral is given by equation (4.73) with a equal to 11.0 and b to 24.5.

4.130. For a cylindrical fuel element, S/M is equal to $2/r_0\rho$, where r_0 is the radius (0.381 cm) and ρ the density of the uranium dioxide (10.2 g/cm³), i.e., 0.514 cm² per g. However, when the fuel rods are closely spaced, the surface term in equation (4.73) must be modified to allow for "shadowing," i.e., the effect of one rod in depleting the epithermal neutron flux incident on adjacent rods. For the present situation, the correction term for shadowing is found to be 0.97 [21], so that the expression for the capture escape probability becomes

$$p_{28} = \exp\left\{-\left(\frac{0.00707}{0.810}\right)[11.0 + (24.5)(0.514)(0.97)]\right\}$$

$$= \exp(-0.203) = 0.816.$$

4.131. The factor p_{28} is the product of p'_{28} and p''_{28}, the former of which must be known to determine β by equation (4.77). It is possible to write $p'_{28} = p_{28}{}^c$, where c is an empirically derived constant; it is found that the best fit to the data is obtained by taking c as 0.5, so that p'_{28} is the square root of p_{28}. Hence, in the present situation,

$$p'_{28} = 0.903.$$

4.132. The capture escape probability, p_c, in materials other than fuel is obtained from an expression similar to equation (4.79), except that the numerator of the exponential term involves the sum of the NI terms for hydrogen, oxygen, and the elements in the stainless steel. Using the number densities in Table 4.6 and the resonance capture integrals in the last column of Table 4.7, it is found that

$$p_c = 0.970.$$

4.133. Finally, the capture escape probability p_{25} in uranium-235 is obtained from equation (4.79) with $N_{28}I_{28}$ replaced by $N_{25}I_{25}$. Experimental values of the effective capture integral, for a homogeneous system, have been plotted as a function of the total scattering cross section per atom of uranium-235, i.e., Σ_s/N_{25} [23]. In the region of interest, the capture integral does not vary greatly and a constant value of 290 barns may be assumed, in order to simplify the treatment, without introducing any serious errors. In these circumstances,

$$p_{25} = \exp\left[-\left(\frac{0.000198}{0.810}\right)(290)\right]$$

$$= \exp(-0.0709) = 0.931.$$

FAST-FISSION FACTOR

4.134. The fast-fission factor in a system of slightly enriched uranium dioxide in water can be represented by*

$$\epsilon = 1 + \frac{0.156}{1 + 0.62\rho_w(V_w/V_u) + 0.288(V_c/V_u)},$$

* This expression is a modified form of that given in Ref. 24.

where ρ_w is the density of the water in grams per cm³, V_w/V_u is the volume ratio of the water to uranium, and V_c/V_u is that of cladding to uranium. From the data given in § 4.123, it is found that V_w/V_u is close to 3.9 whereas V_c/V_u is 0.59; the density of water at ordinary temperatures is about 1.0, and so

$$\epsilon = 1.044.$$

NONLEAKAGE PROBABILITIES

4.135. To simplify the calculation of the nonleakage probabilities, a two-group (fast and thermal) model is used, with empirical fast-group constants which are known to be in agreement with experimental measurements of slowing down of neutrons in various elements and simple mixtures. The "slowing down area" for the fast neutrons, which is equivalent to the Fermi age, is represented (cf. § 3.37) by

$$\tau = \frac{1}{3\Sigma_{tr}\Sigma_{sl}},$$

where Σ_{tr} and Σ_{sl} are the appropriate macroscopic transport and slowing down cross sections for the fast-neutron group. The values of the corresponding microscopic cross sections for elements of interest are given in Table 4.8. From these cross sections and the number densities in Table 4.6, it is found that

TABLE 4.8.   FAST-GROUP CROSS SECTIONS FOR CALCULATION OF AGE

Element	σ_{tr} (barns)	σ_{sl} (barns)
Uranium	9.00	0.800
Hydrogen	1.85	0.655
Oxygen	3.31	0.027
Stainless steel	4.38	0.064
Aluminum	2.25	0.012

for the reactor system under discussion Σ_{tr} is 0.280 cm⁻¹ and Σ_{sl} is 0.0331 cm⁻¹, so that

$$\tau = \frac{1}{(3)(0.280)(0.0331)} = 36.1 \text{ cm}^2.$$

4.136. For the thermal diffusion area, L^2, equation (4.56), which is applicable to homogeneous reactors, is modified slightly and takes the form

$$L^2 = L_m^2(1 - f)(V/V_m),$$

where V/V_m is the ratio of the total core volume to that of the moderator. For moderators other than water, V/V_m is close to unity and can usually be ignored. In the present case, L_m is 2.76 (Table 3.1), f has been found above to be 0.817, and V/V_m is 1.68; hence,

$$L^2 = (2.76)^2(0.183)(1.68)$$
$$= 2.37 \text{ cm}^2.$$

The total nonleakage probability for fast neutrons, i.e., P_1P_2, is $1/(1 + \tau B^2)$; however, for the determination of β, defined by equation (4.77), it is necessary to know P_1, i.e., the nonleakage probability for neutrons with energies in excess of the arbitrary value 0.625 ev. The procedure used is to define P_1 as $1/(1 + b\tau B^2)$, where b is an empirically fitted constant equal to 0.86. The thermal nonleakage probability, P_3, is given in the usual manner by $1/(1 + L^2B^2)$.

INFINITE MULTIPLICATION FACTOR AND BUCKLING

4.137. The evaluation of the various nonleakage probabilities requires a knowledge of B^2, and this is obtained by a series of approximations. If $P_1P_2P_3$ is taken as unity, equation (4.76) gives the infinite multiplication factor; thus,

$$k_\infty = \eta f \epsilon p_{28} p_c \beta.$$

As a first approximation β is set equal to 1.00, so that

$$k_\infty \approx (1.45)(1.044)(0.816)(0.970)(1.00)$$
$$= 1.20.$$

A rough value of B^2 may now be obtained from equation (4.37) in the form

$$B^2 \approx \frac{k_\infty - 1}{L^2 + \tau} = \frac{1.20 - 1.00}{2.37 + 36.1}$$

$$= 0.0052 \text{ cm}^{-2},$$

from which it follows that

$$P_1 = \frac{1}{1 + b\tau B^2} \approx \frac{1}{1 + (0.86)(36.1)(0.0052)}$$

$$= 0.861.$$

4.138. The number of neutrons released by the resonance fission of uranium-235 per neutron absorbed, i.e., η_{25}, varies somewhat with the quantity Σ_s/N_{25}, but there is little error involved in taking it to be constant at 1.70. Hence, a first approximation to β, using equation (4.77), is found to be

$$\beta \approx \frac{0.931}{1 - (1.044)(0.861)(0.903)(1.70)(0.069)} = 1.03.$$

4.139. A better estimate of the infinite multiplication factor can now be made by taking β as 1.03; the result is $k_\infty \approx 1.24$. A recalculation of B^2 by the same procedure as before now gives B^2 as 0.0061 cm^{-2}; from this it is found that

$$P_1P_2 = \frac{1}{1 + \tau B^2} = \frac{1}{1 + (36.1)(0.0061)} = 0.820$$

$$P_3 = \frac{1}{1 + L^2B^2} = \frac{1}{1 + (2.37)(0.0061)} = 0.986.$$

Consequently,

$$k_{\text{eff}} = P_1P_2P_3k_\infty = (0.820)(0.986)(1.24) = 1.00.$$

Since k_{eff} is 1.00, it means that, as a rough approximation, the buckling used in this calculation is equal to the critical value; consequently, the estimated critical buckling is

$$B_c^2 = 0.0061 \text{ cm}^{-2},$$

which compares favorably with the experimental value of 0.00633 cm^{-2}.

4.140. If k_{eff} had been appreciably different from unity, it would have been necessary to repeat the iteration process by calculating β with B^2 equal to 0.0061 cm^{-2}, and then redetermining k_∞, and so on.

4.141. The reflector savings δ of a water reflector is a function of the leakage of neutrons from the core and so should be related to the migration area M^2, i.e., $L^2 + \tau$. An expression which appears to be consistent with calculations and with experimental measurements [20] is

$$\delta(\text{cm}) = 7.2 + 0.10(M^2 - 40.0).$$

For the reactor system considered above, M^2 is $2.37 + 36.1 = 38.5$, and so δ is approximately 7 cm. For a cylindrical core, therefore, the critical radius when reflected would be 7 cm less and the overall (axial) length 14 cm less than for a bare reactor.

FAST REACTORS*

INTRODUCTION

4.142. A so-called fast reactor is one in which there is very little moderator or other material, such as coolant, of low mass number to slow down the neutrons by elastic collisions. Some reduction in energy of the neutrons does occur, however, mainly as a result of inelastic collisions with nuclei of the fuel and structural materials. Most of the fissions result from the absorption of neutrons of high energy, e.g., greater than 0.1 Mev (100 kev), but there are contributions from neutrons of lower energy, down to 1 kev or less.

4.143. The great interest in fast reactors for power production lies in the fact that they offer the prospect of the most efficient conversion of uranium-238

TABLE 4.9.   FISSION PARAMETERS FOR FAST AND THERMAL REACTORS

Parameter	Uranium-235		Uranium-238		Plutonium-239	
	Fast	Thermal	Fast	Thermal	Fast	Thermal
σ_f (barns)	1.44	577	0.112	—	1.78	746.0
ν	2.52	2.43	2.61	—	2.98	2.90
α	0.152	0.172	1.44	—	0.086	0.38
$\eta - 1$	1.18	1.06	0.07	—	1.74	1.10

* Some of the material in this section was contributed by R. M. Kiehn.

into fissile plutonium-239 (§ 1.58). The situation may be understood by reference to Table 4.9 which shows some of the parameters for fast-neutron and thermal-neutron fission of uranium-235, uranium-238, and plutonium-239. The fast-neutron data apply to a typical neutron energy distribution (or spectrum) such as might exist in a fast reactor [25]. The quantities ν and η are defined in § 4.5, and α is the ratio of the cross section for nonfission capture to that for fission, i.e., σ_c/σ_f. For the pure fissile material,

$$\eta = \nu \frac{\sigma_f}{\sigma_f + \sigma_c} = \frac{\nu}{1 + \alpha}.$$

4.144. Since η fission neutrons are produced per neutron absorbed in the fuel, and one neutron is necessary to maintain the chain reaction, $\eta - 1$ is the number of neutrons remaining which can escape from the reactor, be parasitically captured in reactor materials and poisons, or be absorbed by fertile materials, e.g., uranium-238, leading to the production of new fissile material. For thermal fission, the values of $\eta - 1$ are so close to unity that, after allowing for neutron leakage and parasitic capture, less than one neutron is available for conversion of fertile to fissile nuclei per neutron absorbed. Consequently, breeding or its conversion equivalent, i.e., the production of more fissile material than is consumed, cannot be achieved in a thermal reactor using uranium-235 or plutonium-239 as fuel.*

4.145. In a fast reactor $\eta - 1$ is larger than for the same fuel in a thermal system; furthermore, the parasitic capture of neutrons in nonfuel materials is generally less in the former case because of the smaller macroscopic cross sections. There is then a possibility for a conversion ratio exceeding unity in a fast reactor fueled by uranium-235 and a breeding ratio, i.e., ratio of fissile nuclei produced to fissile nuclei consumed, approaching 1.7 when plutonium-239 is the fuel. The actual breeding (or conversion) ratio that can be achieved may exceed these values because neutrons produced by fission of uranium-238 can make a significant contribution to the neutron balance in a fast reactor. A typical fast breeder could consist of a core, containing a fuel of about 25 per cent fissile material plus 75 per cent uranium-238, surrounded by a blanket of natural or depleted uranium. The blanket serves both as a neutron reflector and a region in which neutrons escaping from the core are captured by uranium-238 to produce plutonium-239. As a general rule, about half the core volume consists of fuel and the remainder of coolant and structural materials.

MULTIGROUP EQUATIONS

4.146. With a thermal reactor the estimation of the rough critical size is simplified by the approximately Maxwellian neutron energy spectrum and by the

---

* Breeding is theoretically possible in a thermal reactor based on uranium-233 as the fissile material, with thorium-232 as the fertile species (see Table 4.1).

$1/v$-dependence of the cross sections.   With appropriate corrections, for "hardening" and for departure from $1/v$-behavior, therefore, properly averaged cross sections are adequate for preliminary calculations, at least.   The situation in a fast reactor is, however, much more complicated.   In the first place, the neutron spectrum varies considerably with the composition of the core.   The data for three systems are depicted in Fig. 4.16, in which the ordinates represent the

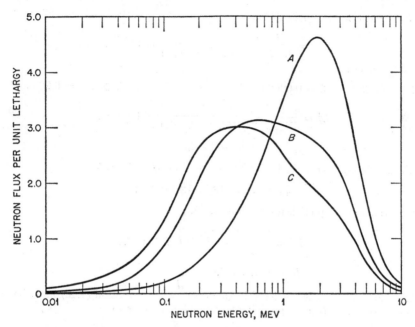

FIG. 4.16.   Fast-neutron spectra in three different systems [25]

fluxes (per unit lethargy interval) for a range of neutron energies from 0.01 to 10 Mev [25].   Curve A refers to a bare core consisting of essentially uranium-235 alone; curve B to a uranium-235 core surrounded by a 9-in. thick blanket of natural uranium; and curve C to a typical reflected fast breeder reactor in which the fuel is 25 per cent uranium-235 and 75 per cent uranium-238.   Second, the cross sections, for fission, capture, and inelastic and elastic scattering, required in the criticality calculations are highly dependent on the neutron energy. Consequently, although very rough estimates of critical size can sometimes be made by the one-group method, employing empirical averaged fast-neutron parameters, the most satisfactory treatment is based on the use of multigroup methods.

Example 4.11.   By means of the one-group treatment, estimate the critical radius of a spherical core of uranium-235 surrounded by a thick reflector of uranium-238.   The cross sections, etc., may be taken as the average values for neutrons in the energy range of 0.4 to 1.35 Mev (see Example 4.12); thus,

Uranium-235: $\sigma_f = 1.27,$ $\sigma_a = 1.40,$ $\sigma_{tr} = 5.7$ barns; $\nu = 2.52$
Uranium-238: $\sigma_a = 0.13,$ $\sigma_{tr} = 5.8$ barns.

The density of uranium is approximately 19 g/cm³, so that N for both uranium-235 and -238 is close to 0.048×10^{24} nuclei per cm³.

For the *core*,

$$\Sigma_a = 0.048 \times 1.40 = 0.067 \text{ cm}^{-1} \quad \text{and} \quad \Sigma_{tr} = 0.048 \times 5.7 = 0.27 \text{ cm}^{-1}$$

$$L^2 = \frac{1}{3\Sigma_{tr}\Sigma_a} = \frac{1}{(3)(0.27)(0.067)} = 18 \text{ cm}^2.$$

For the *reflector*,

$$\Sigma_a = 0.048 \times 0.13 = 0.0062 \text{ cm}^{-1} \quad \text{and} \quad \Sigma_{tr} = 0.048 \times 5.8 = 0.28 \text{ cm}^{-1}$$

$$L_r^2 = \frac{1}{3\Sigma_{tr}\Sigma_a} = \frac{1}{(3)(0.28)(0.0062)} = 190 \text{ cm}^2.$$

$$L_r = 14 \text{ cm.}$$

In the uranium-235 core, k_∞ is equal to η, i.e., to $\nu\sigma_f/\sigma_a$, so that

$$k_\infty = (2.52)(1.27)/1.40 = 2.3;$$

hence, by the one-group critical equation (4.10),

$$B_c^2 = \frac{k_\infty - 1}{L^2} = \frac{1.3}{18} = 0.072 \text{ cm}^{-2}.$$

$$B_c = 0.27 \text{ cm}^{-1}.$$

It is found that R_c/L_r is not large in this case, so that equation (4.46) must be used; however, since D and D_r are approximately equal, it reduces to

$$\tan B_c R_c = -B_c L_r.$$

Hence, $\tan (0.27R_c) = -(0.27)(14) = -3.8$, and R_c is 6.8 cm. (The experimental value is close to 6.0 cm.)

4.147. A brief description will be given here of a relatively simple multigroup procedure which can be applied, with the aid of a desk calculator, to determine the critical size of a fast reactor. The same general method, with a large number of neutron groups, can be used for more accurate machine calculations.

4.148. For a single neutron group, a so-called asymptotic solution of the integral form of the transport theory equation is given by [26]

$$B = \beta \tan^{-1}\left(\frac{B}{\alpha}\right), \tag{4.80}$$

for $B \leq \alpha$, where B^2 is the buckling, α is the total macroscopic removal cross section (Σ_t), i.e., the sum over all the neutron events, including inelastic and elastic collisions, capture, and fission, and β is the macroscopic cross section for "secondary" neutron production; thus,

$$\alpha = N(\sigma_{\text{elastic}} + \sigma_{\text{inelastic}} + \sigma_{\text{capture}} + \sigma_{\text{fission}}) = \Sigma_t,$$
$$\beta = N(\sigma_{\text{elastic}} + \sigma_{\text{inelastic}} + \nu\sigma_{\text{fission}}).$$

The solution given above is applicable provided β/α is equal to or greater than unity, as is the case for an infinite critical or supercritical system. In order to correct for anisotropic scattering, transport cross sections should be used where possible in place of ordinary scattering cross sections.

4.149. When B^2 is determined by equation (4.80), the integral form of the one-group transport equation is satisfied, in regions some distance from the boundaries of a finite reactor, by values $\phi(r)$ of the flux which are solutions of the wave equation

$$\nabla^2\phi(r) + B^2\phi(r) = 0$$

for the flux distribution. Consequently, the B^2 defined by equation (4.80) is identical with the familiar buckling and when the system is just critical, the value of B_c^2 can be used to calculate the critical size of the reactor.

4.150. In extending the treatment to several neutron groups, in which the fluxes are represented by $\phi_1(r)$, $\phi_2(r)$, $\phi_3(r)$, etc., it is first assumed that

$$\nabla^2 F(r) + B^2 F(r) = 0,$$

where $F(r)$ is a factor, depending on position in the reactor, which is independent of energy, that is to say, it has the same value for all neutron groups. In other words, it is assumed that the neutron energy spectrum is the same at all positions in the core; this cannot be rigorously true, even for the diffusion theory approximation, but it is adequate for the present purpose. It is this assumption which makes possible the solution of fast reactor problems involving as many as ten groups using only a desk calculating machine.

4.151. It is now possible to express the group fluxes in terms of $F(r)$; thus,

$$\phi_1(r) = \psi_1 F(r), \qquad \phi_2(r) = \psi_2 F(r), \qquad \phi_3(r) = \psi_3 F(r), \text{etc.}$$

The solution of the integral transport equation equivalent to equation (4.80) can then be expressed in the general form, for the jth group, by

$$\psi_j B = (\text{neutrons entering } j\text{th group}) \tan^{-1}\left(\frac{B}{\Sigma_t^j}\right), \tag{4.81}$$

where Σ_t^j is the total removal cross section for the group. The quantity referred to as "neutrons entering jth group" consists of the neutrons scattered into this group, both elastically and inelastically, from all higher energy groups, as well as those scattered within the jth group, plus that fraction of the fission neutrons having energies falling into the range of this group.

4.152. In group notation, equation (4.81) may be written as

$$\psi_j B = \left[\sum_{i=1}^{j} \Sigma_{ij}\psi_i + \chi_j \sum_{j=1}^{n} (\nu\Sigma_f)_j\psi_j \right] \tan^{-1}\left(\frac{B}{\Sigma_t^j}\right). \qquad (4.82)$$

The first term in the brackets on the right is the sum of all the inelastic and elastic (transport) scatterings into the jth group from all the groups $1, 2, 3, \cdots j$, i.e., from all groups of higher energy and from the jth group itself. The symbol χ_j represents the fraction of the total fission neutrons born with energy in the jth group and the summation in the second term gives the total numbers of fission neutrons for all n energy groups. It is convenient to set this total fission source equal to unity, i.e.,

$$\sum_{j=1}^{n} (\nu\Sigma_f)_j\psi_j \equiv 1,$$

which is the relation whereby the various ψ's, i.e., $\psi_1, \psi_2, \cdots \psi_j, \cdots \psi_n$, are defined.

4.153. Furthermore, scatterings taking place within the jth group may be separated from the others, so that equation (4.82) becomes

$$\psi_j B = \left[\sum_{i=1}^{j-1} \Sigma_{ij}\psi_i + \Sigma_{jj}\psi_j + \chi_j \right] \tan^{-1}\left(\frac{B}{\Sigma_t^j}\right),$$

and solving for ψ_j, the result is

$$\psi_j = \frac{\left[\sum_{i=1}^{j-1} \Sigma_{ij}\psi_i + \chi_j \right] \tan^{-1}\left(\frac{B}{\Sigma_t^j}\right)}{B - \Sigma_{jj} \tan^{-1}\left(\frac{B}{\Sigma_t^j}\right)}. \qquad (4.83)$$

This expression represents a set of n equations, one for each neutron energy group, which must be solved to determine B, and hence the buckling of the critical system.

EVALUATION OF BUCKLING

4.154. In general, the choice of the number of groups is determined by the capacity of the calculator or computer to be used. Another consideration is the availability of the group cross sections, such as those given in the Appendix and in the literature.* The energy (or lethargy) width of each group is arbitrary, although a good choice can be made as the result of experience. In particular, the average cross section for each group will be determined by the neutron energy variation within the group. It is desirable, to choose the groups, as far as possible, so that there is relatively little variation of cross section within each

---

* See, for example, Ref. 27 for 6, 10, and 11-group constants, and Ref. 28 for 16-group constants.

group.   In other words, group boundaries should be selected to correspond to neutron energies at which fission and capture cross sections, in particular, undergo a marked change.   The procedure of assessing group cross sections, especially for inelastic scattering, is made difficult by the lack of experimental data.   The values in the literature are generally based on a combination of experimental results, theoretical analysis, and critical measurements.

4.155. It was stated in § 4.148 that the asymptotic solution, given by equation (4.80), is applicable only for a system which would be supercritical (or critical) when infinite.   Unless it is evident from comparisons with other reactors that this is indeed the case, it is desirable to apply a test before assuming that equation (4.83) is valid.   For an infinite medium, B is zero, and the asymptotic transport theory development leads to

$$\psi_j = \frac{\sum_{i=1}^{j-1} \Sigma_{ij}\psi_i + \chi_j}{\Sigma_t^j - \Sigma_{jj}}. \tag{4.84}$$

Starting with the highest energy group, and noting that the summation in the numerator of equation (4.84) is zero when determining ψ_j for $j = 1$, the various group ψ's can be calculated.   The corresponding values of $(\nu\Sigma_f)_j\psi_j$ are then computed, and if the sum for all n groups is equal to or greater than unity, the infinite system is critical or supercritical.   In these circumstances, equation (4.83) may be used for the finite system.

4.156. The group constants and all the macroscopic cross sections, for the specified amounts of fuel and structural materials, as well as the appropriate values of ν_i and χ_j, may be regarded as known.   The next step is to estimate a preliminary trial value for the critical buckling.   This can be obtained by using the one-group equation (4.10), replacing k_∞ by η, i.e., by $\nu\Sigma_f/\Sigma_a$ (§ 4.143), and L^2 by $1/3\Sigma_{tr}\Sigma_a$ (§ 3.37), so that

$$B_c^2 \approx 3\Sigma_{tr}(\nu\Sigma_f - \Sigma_a). \tag{4.85}$$

The cross sections to be used in this expression may be either one-group averaged values or those for a group of intermediate energy.   The particular group may be selected on the basis of the following working rules.   Species present in the core that have high capture cross sections or high atomic numbers (above about 60) may be considered to be equivalent to uranium-238.   Elements with small cross sections or atomic number less than 60 may be regarded as diluents.   If there are up to 2 atoms of uranium-238 (or equivalent) plus less than 8 atoms of diluent per atom of fissile species, then the spectrum may be considered to be "very fast" and the cross sections to be used in equation (4.84) are for the range from 0.35 to 1 Mev.   On the other hand, a "fast" spectrum will correspond to 3 to 6 atoms of uranium-238 plus 8 to 35 atoms of diluent per fissile atom; the energy range for the cross section is then 0.05 to 0.35 Mev.

4.157. The trial value of B obtained in this manner is inserted in each of the n

equations represented by equation (4.83) and the corresponding value of ψ is determined. From each of the n ψ's, the corresponding $\nu\Sigma_f\psi$ is calculated, since ν and Σ_f for each group are known. The values of $\nu\Sigma_f\psi$ are then added; if the B chosen is a correct solution, then the total will be equal to unity, as postulated above. If it is greater than unity, a slightly larger value of B is selected for the second trial. The procedure is repeated until a consistent solution is obtained. From the critical buckling derived in this manner, the dimensions of the bare reactor can be calculated. The neutron flux spectrum in the given system is derived from the values of ψ which give the correct solution to the group equations.

4.158. A preliminary value of the radius of the reflected core, with a thick blanket, can be derived from the one-group equation (4.47); recalling that $D = 1/3\Sigma_{tr}$ by equation (3.9) and $L^2 = 1/3\Sigma_{tr}\Sigma_a$, it follows that

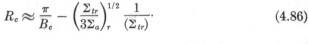

$$R_c \approx \frac{\pi}{B_c} - \left(\frac{\Sigma_{tr}}{3\Sigma_a}\right)_r^{1/2} \frac{1}{(\Sigma_{tr})}. \qquad (4.86)$$

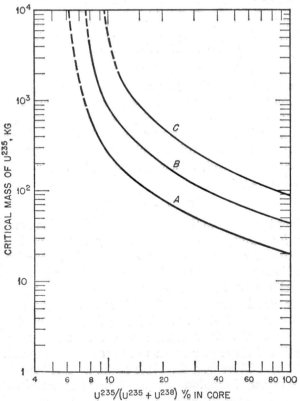

Fig. 4.17. Critical mass of uranium-235 in fast reactors [25]

Utilizing the one-group cross section, an approximate estimate of R_c can be made. This value of R_c can then be used, by a process of iteration, to obtain one which satisfies equation (4.46).

CORE COMPOSITION AND CRITICAL MASS

4.159. The effect of core composition on the calculated critical masses of fast reflected reactors containing various volume percentages of either uranium-235 or plutonium-239 in the fuel, the remainder being uranium-238 in each case, is shown in Figs. 4.17 and 4.18, respectively [25].   Curve A refers to a core consisting of fuel material only; curve B applies for 50 volume per cent of fuel, $33\frac{1}{3}$ per cent of sodium (coolant), and $16\frac{2}{3}$ per cent of stainless steel (structural material); and curve C to 25 volume per cent of fuel, 50 per cent of sodium, and 25 per cent of steel.   The reflector in each case is a 20-in. thick blanket consisting of 40 volume per cent depleted uranium, 40 per cent sodium, and 20 per cent steel.   The critical mass of fissile material is seen to increase rapidly as the

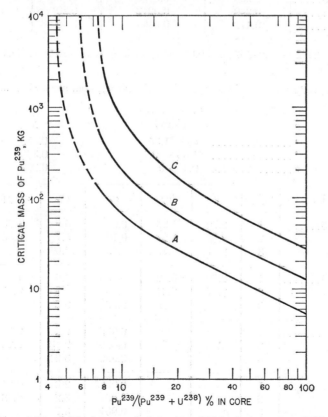

FIG. 4.18. Critical mass of plutonium-239 in fast reactors [25]

percentage in the fuel decreases, especially at the lower concentrations. Since large inventories of uranium-235 or plutonium-239 are likely to be uneconomical, a practical fast reactor, the composition of which would approximate that of curve C, may be expected to contain at least 20 per cent of fissile material in the fuel. It may be noted, incidentally, that the critical masses of fast reactors are considerably greater than for thermal reactors employing the same fuel.

Example 4.12. Make a preliminary estimate, using a three-group procedure, of the core volume of a fast reactor with core and blanket compositions similar to those specified for the Enrico Fermi Atomic Power Plant (§ 13.91) as given below. Assume a spherical geometry and a homogeneous core and blanket.

Component	Core Volume Fraction	Core N (10^{24} atoms/cm^3)	Blanket Volume Fraction	Blanket N (10^{24} atoms/cm^3)
Uranium-235.....	0.071	0.0034	—	—
Uranium-238.....	0.208	0.0100	0.40	0.0192
Iron*...........	0.249	0.0212	0.20	0.0170
Sodium.........	0.472	0.0104	0.40	0.0088

* Includes 0.142 Fe, 0.049 Zr, and 0.058 Mo; the Zr and Mo are treated as Fe to simplify the calculation.

The three-group microscopic cross sections (in barns), etc., are as follows:

j	Energy range (Mev)	χ
1..............	$\infty - 1.35$	0.575
2..............	$1.35 - 0.4$	0.326
3..............	$0.4 - 0$	0.099

Material	j	ν	σ_f	σ_c	σ_{tr}	$\sigma_{j\to j+1}$	$\sigma_{j\to j+2}$
Uranium-235.......	1	2.70	1.29	0.08	4.5	1.00	0.50
	2	2.53	1.27	0.13	5.7	0.50	—
	3	2.47	1.77	0.49	10.0	—	—
Uranium-238.......	1	2.60	0.524	0.036	4.6	1.41	0.64
	2	2.47	0.01	0.13	5.8	0.25	—
	3	—	—	0.26	9.6	—	—
Iron..............	1			0.005	2.0	0.60	0.10
	2			0.006	2.13	0.22	—
	3			0.006	3.24	—	—
Sodium...........	1			0.0005	2.0	0.24	0.06
	2			0.001	3.2	0.18	—
	3			0.001	3.7	—	—

The values derived from the foregoing data for the *core* are as follows:

	$j = 1$	$j = 2$	$j = 3$
$\nu\Sigma_f$.............	0.0254	0.0111	0.0149
$\Sigma_{tr} \approx \Sigma_t^*$........	0.1245	0.1558	0.2372
$\Sigma_{j\rightarrow j+1}$..........	0.0327	0.0107	—
$\Sigma_{j\rightarrow j+2}$..........	0.0108	—	—
$\Sigma_a = \Sigma_f + \Sigma_c$....	0.0104	0.0063	0.0104
Σ_{jj}†.............	0.0706	0.1388	0.2268

* $\Sigma_{tr} \approx \Sigma_t$ because most of the core components have high mass numbers (§ 3.15).
† $\Sigma_{jj} = \Sigma_t - \Sigma_a - \Sigma_{j\rightarrow j+1} - \Sigma_{j\rightarrow j+2}$.

Start with equation (4.84)

$$\psi_j = \frac{\sum_{i=1}^{j-1} \Sigma_{ij}\psi_i + \chi_j}{\Sigma_t^j - \Sigma_{jj}},$$

and evaluate the ψ_j's; thus,

$$\underset{(j=1,\, i=0)}{\psi_1} = \frac{0.575}{0.1245 - 0.0706} = 10.7$$

$$\underset{(j=2,\, i=0,\, 1)}{\psi_2} = \frac{(0.0327)(10.7) + 0.326}{0.1558 - 0.1388} = 39.7$$

$$\underset{(j=3,\, i=0,\, 1,\, 2)}{\psi_3} = \frac{(0.0108)(10.7) + (0.0107)(39.7) + 0.099}{0.2372 - 0.2268} = 61.5$$

$$\sum_j (\nu\Sigma_f)\psi = (0.0254)(10.7) + (0.0111)(39.7) + (0.0149)(61.5)$$
$$= 1.63.$$

Since this is greater than unity, the specified core can become critical.

The next step is to make a rough guess at B, using equation (4.85). Assume that the various cross sections are weighted according to the ψ's derived above. The sum of $\psi_1 + \psi_2 + \psi_3$ is 112; consequently,

$$\Sigma_{tr} = 0.1245 \left(\frac{10.7}{112}\right) + 0.1558 \left(\frac{39.7}{112}\right) + 0.2372 \left(\frac{61.5}{112}\right) = 0.194$$

$$\nu\Sigma_f = 0.0254 \left(\frac{10.7}{112}\right) + 0.0111 \left(\frac{39.7}{112}\right) + 0.0149 \left(\frac{61.5}{112}\right) = 0.0145$$

$$\Sigma_a = 0.0104 \left(\frac{10.7}{112}\right) + 0.0063 \left(\frac{39.7}{112}\right) + 0.0104 \left(\frac{61.5}{112}\right) = 0.0089.$$

Hence, from equation (4.85),

$$B = [3\Sigma_{tr}(\nu\Sigma_f - \Sigma_a)]^{1/2} = [(3)(0.194)(0.0145 - 0.0089)]^{1/2}$$
$$= 0.057 \text{ cm}^{-1}.$$

Take B as 0.0570 cm^{-1} and now determine the ψ_j's from equation (4.83). For the first group,

$$\frac{B}{\Sigma_t^1} = \frac{0.0570}{0.1245} = 0.458; \qquad \tan^{-1} 0.458 = 0.430$$

$$\underset{(j=1,\, i=0)}{\psi_1} = \frac{(0.575)(0.430)}{0.0570 - (0.0706)(0.430)} = 9.25.$$

For the second group,

$$\frac{B}{\Sigma_t^2} = \frac{0.0570}{0.1558} = 0.360; \qquad \tan^{-1} 0.360 = 0.351$$

$$\underset{(j=2,\, i=1,\, 0)}{\psi_2} = \frac{[(0.0327)(9.25) + 0.326](0.351)}{0.0570 - (0.1388)(0.351)} = 26.6.$$

For the third group,

$$\frac{B}{\Sigma_t^3} = \frac{0.0570}{0.2372} = 0.240; \qquad \tan^{-1} 0.240 = 0.236$$

$$\underset{(j=3,\, i=2,\, 1,\, 0)}{\psi_3} = \frac{[(0.0108)(9.25) + (0.0107)(26.6) + 0.099](0.236)}{0.0570 - (0.2268)(0.236)} = 32.6$$

$$\sum_j (\nu\Sigma_f)\psi = (0.0254)(9.25) + (0.0111)(26.6) + (0.0149)(32.6)$$
$$= 1.02.$$

This result is close enough to unity to make a further trial unnecessary. However, if the sum of the $(\nu\Sigma_f)\psi$ terms had been appreciably greater than unity, it would be necessary to repeat the calculations given above with a larger value of B; had the sum been smaller than unity, the next trial value of B would be smaller than the first. In the present case, the critical value, B_c, may be taken to be 0.057 cm^{-1}, so that the critical radius of the bare spherical reactor is given by

$$R_c = \frac{\pi}{B_c} = 55 \text{ cm,}$$

and the volume is 700 liters.

The one-group critical condition for a reactor with a thick, but finite, reflector (or blanket) is

$$z \cot z - 1 \approx -\frac{\Sigma_{tr}}{\Sigma_{tr(r)}}\left(1 + \frac{z}{BL_r}\right),$$

where

$$z \equiv B_c R_c,$$

the subscript r referring to the reflector (blanket); R_c is here the radius of the reflected critical core, which is determined by solving the critical equation. For this type of preliminary calculation, average properties of the blanket may be assumed to be those of the group lower in energy to the one corresponding to the maximum core flux. This assumption is satisfactory for a ten-group calculation, but it cannot be applied in the three-group treatment given here. Consequently, the second-group properties are taken to be representative of the core, whereas the third (lowest energy) group is assumed to be applicable to the blanket. Hence, using the N values given at the beginning and the appropriate group microscopic cross sections, the following results are obtained:

Core: $\Sigma_{tr} = 0.156$
Blanket: $\Sigma_a = 0.0025$ $\Sigma_{tr} = 0.176.$

If the blanket is regarded as nonmultiplying, $1/L_r$ may be represented by

$$1/L_r = (3\Sigma_{tr}\Sigma_a)^{1/2}$$
$$= [(3)(0.176)(0.0025)]^{1/2} = 0.0363.$$

Since B_c has been found to be 0.057, the reflected (one-group) critical equation becomes

$$z \cot z - 1 \approx -\frac{0.156}{0.176}\left(1 + \frac{0.0363}{0.057}z\right)$$

so that

$$z \cot z \approx 0.114 - 0.55z.$$

Inspection of this transcendental equation shows that $\cot z$ is probably negative, so that z may lie between $\pi/2$ and π, i.e., between 1.57 and 3.14. By trying $z = 1.8$, 2.0, and 2.2 as possible solutions, it is found that the correct value is close to 2.0; hence,

$$z = B_c R_c = 2.0$$

$$R_c = \frac{2.0}{0.057} = 35 \text{ cm.}$$

The critical volume of the core with a blanket is thus roughly 200 liters.

(A ten-group calculation based on the same procedure gives a critical volume of 340 liters, which may be compared with 335 liters for the cylindrical core of the Fermi reactor. Cross sections and other data for the ten-group treatment are given in the Appendix.)

EXPERIMENTAL DETERMINATION OF CRITICAL SIZE

THE CRITICAL ASSEMBLY

4.160. Confirmation of the calculation of the critical size of a reactor, either homogeneous or heterogeneous, is obtained by experiment [29]. The method of the critical assembly can be used for reactors of all sizes, but it is more suited to reasonably small (enriched) reactors. A system consisting of fuel and moderator, in the desired proportions and shape, is gradually built up until it approaches the critical dimensions. At the center of the assembly is placed a source, emitting about 10^7 neutrons per sec; this is called the primary source. As a result of the fissions induced in the fuel material, there is a multiplication of the primary source neutrons. Even though the reactor is less than the critical size, a steady state of the neutron flux will be attained, provided the source is present. The (subcritical) multiplication is then defined as the ratio of the total neutron flux, due to both the primary source and the fissions, to the flux due to the source only, i.e., in the same medium in the absence of fissile material.

4.161. If Q neutrons are obtained from the source, these will produce Qk_{eff} neutrons at the end of one generation, Qk_{eff}^2 at the end of two generations, and so on. The effective multiplication factor, k_{eff}, for the particular assembly under consideration depends on its size as well as its composition and arrangement.   The steady-state multiplication of the neutrons in the system can then be represented by

$$\frac{Q + Qk_{eff} + Qk_{eff}^2 + \cdots}{Q} = \frac{1}{1 - k_{eff}}, \tag{4.87}$$

provided $k_{eff} < 1$, i.e., the assembly is subcritical.   As the size of the assembly is increased and approaches critical, k_{eff} tends to unity, and the multiplication becomes infinite.

4.162. The multiplication can be determined experimentally by measuring the neutron flux within the assembly at a certain distance from the source.   The measurement is first made in the absence of the fuel, i.e., with source and moderator (if any) only, and then repeated at the same position with fuel present. The ratio of the two values gives the required multiplication for the particular size of the assembly.   The observations are repeated for assemblies of various sizes approaching the critical.

4.163. In order to determine the actual critical size the reciprocal of the multiplication is plotted against the mass of the fuel, as in Fig. 4.19.   The plot is then extrapolated to zero value of the reciprocal multiplication, i.e., infinite multiplication, to give the critical mass of the system.   By making observations at two (or more) different positions, two (or more) sets of data are obtained, and the curves should extrapolate to the same point, as shown in the figure.   The size of the critical reactor can thus be determined from measurements with subcritical assemblies.   A final check can then be made by constructing the critical reactor and showing that a chain reaction can be maintained in the absence of an extraneous source.

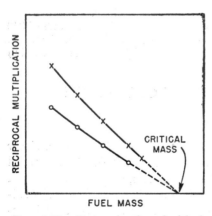

FIG. 4.19.   Determination of critical mass

4.164. In view of the statement based on equation (4.87) that the multiplication is infinite when k_{eff} is unity, it might appear that an exactly critical reactor, i.e., one for which $k_{eff} = 1$, would be unstable.   In a sense, this is true if a primary source is present.   For stability, k_{eff} must be less than unity, but the difference from unity may be only infinitesimal, as will be seen from the following calculations.

4.165. A good primary source might be one emitting 10^7 neutrons/sec; in a reactor volume of 10^6 cm^3, this would represent 10 neutrons/(cm^3)(sec).   The thermal-neutron flux in a critical reactor might be 2×10^{13} neutrons/(cm^2) (sec), and, taking the thermal velocity as 2×10^5 cm/sec (§ 2.98), the neutron density would be 10^8 neutrons/cm^3. The average lifetime of a thermal neutron, e.g., in a natural uranium-graphite system, is about 10^{-3} sec. Hence neutrons are produced in the reactor at the rate of 10^{11} neutrons/(cm^3) (sec). In other words, the 10 primary source neutrons are multiplied up to 10^{11} as a result of the fission chain in the fuel. The neutron multiplication, in this case, is consequently 10^{10}.

4.166. If this result is inserted into equation (4.87), it is seen that k_{eff} is $1 - 10^{-10}$, which is indistinguishable from unity for all practical purposes. However, it is true that when a primary neutron source is present,* as it frequently is, k_{eff} cannot actually be unity. If the reactor is to be stable, the departure from unity is so small as to be negligible when the reactor is operating at an appreciable power level. In the complete absence of any primary source, a steady state could be maintained only if k_{eff} were precisely unity.

THE EXPONENTIAL EXPERIMENT

4.167. In an exponential (pile) experiment [31], a subcritical assembly is constructed having exactly the same composition and structure as the proposed (large) reactor, but with linear dimensions about one-third or one-fourth of those expected for the critical system. If a primary source is present, a steady state can be attained in the subcritical assembly, as described above. The flux distribution in such an assembly is not the same as that in a critical reactor. Nevertheless, it can be shown that if the subcritical system is relatively large, as would be the case for a natural-uranium or slightly enriched fuel assembly, the overall (or large scale) thermal-neutron flux distribution, at a distance from boundaries and from the source, can be represented fairly well by

$$\nabla^2\phi + B_c^2\phi = 0, \tag{4.88}$$

where B_c^2 is the critical value of the buckling for the given fuel-moderator system. The value of B_c^2 is determined from measurements of the thermal flux distribution in the regions of the large subcritical assembly where equation (4.88) may be expected to apply.

4.168. The subcritical assembly, e.g., in the form of a rectangular parallelepiped or a cylinder, is constructed upon a base from which emerges a constant supply of thermal neutrons. The thermal-neutron source may consist of a structure of graphite and uranium within which is placed an ordinary neutron source, or it may be part of an existing thermal reactor. If z is the vertical distance of a point from the bottom of the assembly, then theoretical considerations show that the thermal-neutron flux in the z direction is given by

$$\phi(z) = Ce^{-\gamma z}, \tag{4.89}$$

where C is a constant. This relationship, which holds at appreciable distances

* Such a source is used in some reactors to simplify startup (§ 5.261).

from the source and boundaries, means that, in the experimental arrangement under consideration, the neutron flux along any line parallel to the (vertical) z axis falls off in an exponential manner. It is for this reason that the system has been called an "exponential pile."

4.169. According to equation (4.89) the plot of $\ln \phi$ against z should be a straight line of slope $-\gamma$, so that the latter can be determined from experimental measurements of the thermal flux at various points along a z axis, not too near the boundaries of the assembly. The importance of this result lies in the fact that $-\gamma^2$ constitutes one of the terms in the expression for the buckling (Table 4.2). Thus, for an exponential assembly having the form of a rectangular parallelepiped, with dimensions a and b in the x and y directions, respectively,

$$B_c^2 = \left(\frac{\pi}{a}\right)^2 + \left(\frac{\pi}{b}\right)^2 - \gamma^2.$$

For an exponential assembly cylinder of radius R,

$$B_c^2 = \left(\frac{2.405}{R}\right)^2 - \gamma^2,$$

the value of γ being obtained from the flux variation in the axial direction. Hence, if γ is determined, as just described, the value of B_c^2 can be obtained. The dimensions of the critical reactor, having the same structure and composition as the experimental assembly, can then be calculated.

SYMBOLS USED IN CHAPTER 4

A	arbitrary constant
a, b	constants in equation for effective resonance integral
a, b, c	dimensions of a parallelepiped
B^2	buckling
C	arbitrary constant
D	diffusion coefficient
D_c, D_m, D_r	diffusion coefficient of core, moderator, reflector
E	neutron energy
E	quantity in lattice calculations
E_1	upper energy of resonance region
E_2	lower energy of resonance region
F	quantity in lattice calculations
$F(r)$	factor in fast-reactor calculations
f	thermal utilization
f_r	resonance utilization
H	height of cylinder
I	effective resonance integral
I_0, I_1	Bessel functions

i, j	energy groups in fast-reactor calculations
K_0, K_1	Bessel functions
k_∞	infinite multiplication factor
L	diffusion length
L_c, L_m, L_r	diffusion length in core, moderator, reflector
M	mass (of fuel lump)
N	atoms (or nuclei)/cm^3
N_m, N_u	atoms/cm^3 of moderator, fuel
N_0, N_1	atoms/cm^3 of fuel, moderator
n	neutron density, neutrons/cm^3
n	number of neutrons
n	integer
P	nonleakage probability
P_1, P_2, P_3	nonleakage probabilities
$p, p(E)$	resonance escape probability, at energy E
p_c	epithermal capture escape probability in nonfuel materials
p_{25}	epithermal capture escape probability in U^{235} at 0.625 ev
p_{28}	overall epithermal capture escape probability in U^{238}
p'_{28}	epithermal capture escape probability in U^{238} above 0.625 ev
p''_{28}	epithermal capture escape probability in U^{238} below 0.625 ev
$q, q(E)$	slowing down density, at energy E
q_{th}	slowing down density of thermal neutrons
$q(\tau)$	slowing down density of neutrons at age τ
$q(\tau, r)$	slowing down density of neutrons at age τ and position r
R	radius of sphere or cylinder
$R(r)$	space function of neutron age
r	radial distance
r_0	radius of fuel rod
r_1	radius of unit cell in lattice
S	source term in diffusion equation
S	surface area of fuel lump
T	reflector thickness
$T(\tau)$	time function of neutron age
V_0, V_u	volume of fuel
V_1, V_m	volume of moderator
V_i	volume of impurity
α	ratio of capture to fission cross sections
$-\gamma$	slope of plot in exponential experiment
δ	reflector savings
ϵ	fast-fission factor
η	fission neutrons per neutron absorbed in fuel (average)
η^r_{25}	fission neutrons per resonance neutron absorbed in U^{235}
κ	reciprocal of diffusion length

κ_0, κ_1	values of κ in fuel, moderator
κ_r	value of κ in reflector
ν	fission neutrons per fission (average)
ξ	average logarithmic energy decrement per collision
ρ	density
Σ_a	macroscopic absorption cross section
Σ_{ac}, Σ_{ar}	absorption cross section of core, reflector
$\Sigma_{am}, \Sigma_m, \Sigma_{a1}$	absorption cross section of moderator
$\Sigma_{au}, \Sigma_u, \Sigma_{a0}$	absorption cross section of fuel
Σ_f	fission cross section
Σ_i	absorption cross section of impurities
Σ_{sm}, Σ_{s1}	scattering cross section of moderator
Σ_{su}, Σ_{s0}	scattering cross section of fuel
Σ_{sl}	slowing down cross section
Σ_t	total cross section
Σ_{tr}	transport cross section
σ_a	microscopic absorption cross section
σ_{am}, σ_{a1}	absorption cross section of moderator
σ_{au}, σ_{a0}	absorption cross section of fuel
σ_c	capture cross section
σ_f	fission cross section
σ_{sl}	slowing down cross section
σ_{sm}, σ_{s1}	scattering cross section of moderator
σ_{su}, σ_{s0}	scattering cross section of fuel
σ_{tr}	transport cross section
τ	Fermi age of neutrons
τ_{th}	Fermi age of thermal neutrons
ϕ	neutron flux, neutrons $/(\text{cm}^2)(\text{sec})$
$\phi(r)$	neutron flux at point r
ϕ_i, ϕ_m, ϕ_u	neutron flux in impurity, moderator, fuel
ϕ_c, ϕ_r	neutron flux in core, reflector
χ_j	fission neutrons born with energy in the jth group
ψ_j	quantity related to flux in jth group
∇^2	Laplacian operator

REFERENCES FOR CHAPTER 4

1. S. Glasstone and M. C. Edlund, "The Elements of Nuclear Reactor Theory," D. Van Nostrand Co., Inc., Princeton, N. J., 1952, p. 241.
2. Ref. 1, pp. 203 et seq.
3. Ref. 1, p. 235.

4. A. M. Weinberg and E. P. Wigner, "The Physical Theory of Neutron Chain Reactors," University of Chicago Press, 1958, p. 498.
5. See, for example, H. Etherington (Ed.), "Nuclear Engineering Handbook," McGraw-Hill Book Co., Inc., New York, 1958, pp. 6–60 *et seq.;* R. V. Meghreblian and D. K. Holmes, "Reactor Analysis," McGraw-Hill Book Co., Inc., New York, 1960, pp. 456 *et seq.*
6. H. Etherington, Ref. 5, pp. 6–10 *et seq.;* R. V. Meghreblian and D. K. Holmes, Ref. 5, Ch. 7; A. M. Weinberg and E. P. Wigner, Ref. 4, Ch. IX; "Reactor Handbook," 2nd Ed., Interscience Publishers, Inc., New York, 1962, Vol. III, Part A, Ch. 3; W. S. Sangren, "Digital Computers and Nuclear Reactor Calculations," John Wiley and Sons, Inc., New York, 1960.
7. W. S. Sangren, Ref. 6; V. Nather and W. S. Sangren, *Nucleonics,* **19**, No. 11, 154 (1961); "Reactor Handbook," Ref. 6, Ch. 4; *Proceedings of the Seminar on Codes for Reactor Calculations,* International Atomic Energy Agency, 1961; "Reactor Physics Constants," U. S. AEC Report ANL-5800, 2nd Ed., 1962, Section 10.
8. L. Dresner, *Nuc. Sci. and Eng.,* **1**, 68 (1956).
9. F. T. Miles and H. Soodak, *Nucleonics,* **11**, No. 1, 66 (1953).
10. Ref. 1, p. 272; Ref. 4, p. 624.
11. Ref. 4, Chs. XVIII and XIX; J. Chernick, U. S. AEC Report BNL-622 (1960).
12. Ref. 4, p. 635.
13. Ref. 4, p. 669.
14. Ref. 4, p. 662.
15. E. Hellstrand, *J. Appl. Phys.,* **28**, 1493 (1957); I. E. Dayton and W. G. Pettus, *Nuc. Sci. and Eng.,* **3**, 286 (1958).
16. Ref. 1, pp. 276 *et seq.;* Ref. 4, Ch. XX; B. I. Spinrad, *Nuc. Sci. and Eng.,* **1**, 455 (1956); M. R. Fleishman and H. Soodak, *ibid.,* **7**, 217 (1960).
17. E. A. Guggenheim and M. H. L. Pryce, *Nucleonics,* **11**, No. 2, 50 (1953); see also, I. Kaplan and J. Chernick, *Proc. (First) U. N. Conf. Geneva,* **5**, 295 (1956).
18. W. H. Arnold, Jr., U. S. AEC Report YAEC-152 (1960).
19. "Reactor Physics Constants," Ref. 7, Section 4.1.4.
20. R. W. Deutsch, U. S. AEC Report GNEC-133 (1960).
21. H. Takahashi, *Nuc. Sci. and Eng.,* **5**, 338 (1959).
22. Ref. 19, Section 4.2.3.
23. Ref. 18, p. 22.
24. R. Gardner and B. A. Lee, M. S. Thesis, Mass. Inst. of Technology, Cambridge, 1957.
25. L. J. Koch and H. C. Paxton, *Am. Rev. Nuc. Sci.,* **9**, 437 (1959).
26. See J. Codd, L. R. Shephard, and J. H. Tait, *Progress in Nuclear Energy,* Series I (Physics and Mathematics), Vol. 1, McGraw-Hill Book Co., Inc. (and Pergamon Press), New York, 1956, Ch. 9.
27. Ref. 19, Section 7.
28. S. Yiftah, D. Okrent, and P. A. Moldauer, "Fast Reactor Cross Sections," Pergamon Press, Inc., New York, 1960.
29. "Reactor Handbook," Ref. 6, Section 2.6; "Critical Facilities in Nuclear Technology," *Nucleonics,* **16**, No. 12, pp. 39–54 (1958); see also, P. W. Davison, *et al.,* U. S. AEC Report YAEC-94 (1959).
30. For theory, see Ref. 1, pp. 281 *et seq.;* Ref. 4, pp. 428 *et seq.*
31. "Reactor Handbook," Ref. 6, Section 2.5.6; "Nuclear Engineering Handbook," Ref. 5, p. 5–117; J. B. Hoag (Ed.), "Nuclear Reactor Experiments," D. Van Nostrand Co., Inc., Princeton, N. J., 1958, Ch. 4; R. W. Campbell and R. K. Paschall, U. S. AEC Report NAA-SR-5409 (1960); see also, C. N. Kelber, U. S. AEC Report ANL-6487 (1962).

PROBLEMS

1. Calculate the thermal utilization, f, for a homogeneous mixture of (a) uranium-235 and graphite and (b) natural uranium and graphite, with $N_m/N_u = 1000$ in each case. What are the respective values of ηf?

2. For a large bare reactor, the infinite multiplication factor is 1.03 and the neutron migration length is 50 cm. Determine the critical volume of a spherical system.

3. Calculate the critical mass of uranium-235 in a bare, spherical thermal homogeneous reactor consisting of uranium-235 and graphite with $N_m/N_u = 2 \times 10^3$.

4. Using the expression in Table 4.1 for the buckling of a finite cylinder, show that there is a limiting radius below which a cylindrical thermal reactor cannot become critical, for a given fuel-moderator ratio.

5. The SUPO version of the Los Alamos Water Boiler contains 870 g of uranium-235 (as uranyl nitrate) in 13 liters of aqueous solution. A proposal has been made to store this solution in a cylindrical tank 7.88 in. in diameter and 36 in. high. Is the solution likely to go critical in this tank?

6. Determine the minimum mass ratio of plutonium-239 to water that will just sustain a chain reaction in an infinitely large homogeneous system.

7. Calculate the critical volume at ordinary temperatures of a cubical thermal reactor containing heavy water as moderator and pure uranium-235 as fuel in the atomic (or molecular) ratio of 10^5 to 1, assuming the continuous slowing down model to be applicable. Estimate the critical mass of uranium-235 for this reactor.

8. In the preceding example, what fraction of the neutrons produced in fission escape (a) while slowing down, (b) while at thermal energies?

9. Determine the size of a bare spherical reactor (Water Boiler) containing uranium-235 and ordinary water in the weight ratio of 1 to 15, using equation (4.59). Compare the result with that given by the use of the continuous slowing model.

10. Estimate the error in the preceding problem resulting from the neglect of the neutron capture by sulfur and oxygen if the uranium were present in solution as uranyl sulfate (UO_2SO_4).

11. What would be the approximate critical mass of uranium-235 for the reactor referred to in Problem 9 if the sphere were surrounded by a thick graphite reflector?

12. Consider a homogeneous solution of uranium-235 in ordinary water, in which there are 200 moles of H_2O to each atom of U^{235}. Calculate the critical radius of a bare sphere (a) assuming the volume to be large, (b) using Fermi age theory, (c) by means of equation (4.38), and (d) from the critical equation [A. M. Weinberg, *Am. J. Phys.*, **20**, 401 (1952)] $k_\infty \Sigma_0 \tan^{-1}(B/\Sigma_0) = B(1 + L^2B^2)$, where Σ_0^2 may be identified with $1/3\tau$, which is particularly applicable to hydrogen-moderated systems.

13. Determine the values of the infinite multiplication factor in homogeneous mixtures of heavy water and natural uranium in the atomic (or molecular) ratios of 100, 200, 300, and 400 to 1, respectively. Hence estimate the maximum value of this factor, if parasitic capture of neutrons is neglected.

14. The infinite multiplication factor of a reactor constructed of natural uranium rods of 1 in. diameter arranged in a square lattice with a pitch of 6 in. in heavy water as moderator was found in Example 4.10 to be 1.28. Compare this with the value to be expected for a homogeneous system of the same overall composition.

15. The British BEPO reactor consists of 0.90-in. diameter natural uranium rods in a square lattice, with a spacing of 7.25 in., with graphite as moderator. Neglecting the aluminum fuel element cladding and the air (coolant) space, calculate the infinite multiplication factor for this heterogeneous system.

16. The reactor core referred to in the preceding problem is cylindrical with the height and diameter approximately equal. Assuming that a 3-ft thickness of graphite reflector surrounds the core, at top and bottom as well as at the sides, estimate the mass of natural uranium metal required for criticality. (The actual mass is about 28 tonnes.)

17. Assuming that all neutrons captured in uranium-238, either while slowing down or when thermal, produce plutonium-239, derive an expression for the ratio of plutonium formed to uranium-235 consumed in a reactor, in terms of p, f, and the uranium cross sections. Discuss the design conditions which might be used in a production reactor to obtain a high yield of plutonium-239. (Assume that neutron leakage while slowing down is negligible.)

18. Critical measurements have been made [V. E. Grob, et al., Nuc. Sci. and Eng., **7**, 481 (1960)] of a lattice system consisting of uranium dioxide (4.48 per cent uranium-235 enrichment) in ordinary water. The uranium dioxide (density 10.2 g/cm³) pellets have a diameter of 0.300 in. and are clad with stainless steel 0.305 in. i.d. and 0.347 in. o.d. The square lattice has a pitch of 0.470 in. Calculate the buckling, using the method described in the text for partially enriched, water-moderated lattices.

19. Using a three-group treatment, compute the critical mass of uranium-235 required for a spherical fast reactor of which the core and 50-cm thick blanket (reflector) have the following compositioms:

Material	Core Volume %	Blanket Volume %
Uranium-235................	8	—
Uranium-238................	42	40
Sodium....................	33	40
Iron......................	17	20

(A calculation by the S_4 method gave a critical mass of 300 kg of uranium-235.)

20. Calculate the critical mass of plutonium-239 in a reactor similar to that in the preceding problem except that the core contains 15 volume per cent of plutonium-239 and 35 volume per cent of uranium-238. All other compositions are unchanged. (The value obtained by an S_4 calculation is 40 kg.) The following three-group constants, for the same energy ranges as in Example 4.12, are to be used for plutonium-239:

Group	ν	σ_f	σ_c	σ_{tr}	$\sigma_{j \to j+1}$	$\sigma_{j \to j+2}$
1......	3.10	1.95	0.10	4.6	0.45	0.45
2......	2.98	1.75	0.12	6.0	0.50	—
3......	2.91	1.80	0.35	10.0	—	—

21. The Los Alamos Godiva I was a bare, spherical fast reactor consisting of essentially pure uranium-235 (94 per cent enriched). Using the ten-group data in the Appendix, calculate the critical radius of this reactor. (The actual value was 8.7 cm.)

22. What would be the critical radius of the core in the preceding example if it were surrounded by a thick reflector of uranium-238? (The reflected fast reactor Topsy had a critical radius of 6.04 cm.)

Chapter 5

CONTROL OF NUCLEAR REACTORS

REACTOR KINETICS

INTRODUCTION

5.1. The operating power of a nuclear reactor is dependent on the mass of fissile material present, the microscopic fission cross section, and the neutron flux (§ 2.163). Of these factors, it is clear that the quantity which lends itself most readily to the control of the power is the flux or the neutron density. Later in this chapter the means by which this control can be achieved will be described. As a first step in the development of the subject of reactor control, however, consideration will be given to the manner in which the neutron flux (or density) responds to various situations, especially changes in the effective multiplication factor (or reactivity), in the temperature, and as a result of the accumulation of fission product poisons.

5.2. In the preceding chapter the steady-state conditions of a nuclear reactor were examined in order to derive the equations for criticality. The neutron density was then constant and independent of time. It is now required to consider the transient changes resulting from a departure from the critical state. Such situations arise particularly during the startup and shutdown of a reactor and also as a result of accidental disturbances in the course of what is intended to be steady-state operation. The theoretical treatment to be given here of the time-dependent behavior (or kinetics) of a reactor, due to changes in the effective multiplication factor, refers strictly to bare, homogeneous thermal reactors, but the general conclusions are applicable to reactors of all types.

NEUTRON LIFETIME

5.3. An important quantity in reactor kinetics is the *neutron lifetime*, i.e., the average time elapsing between the release of a neutron in a fission reaction and its loss from the system by absorption or escape. For convenience of calculation, the lifetime in a thermal reactor may be divided into two parts, namely: (a) the slowing down time, i.e., the mean time required for the fission neutrons to slow down to thermal energies, and (b) the thermal neutron lifetime (or *diffusion time*), i.e., the average time the thermal neutrons diffuse before being lost in some way.

5.4. Suppose that the neutrons are slowing down in a nonabsorbing medium and that after a time t they have a velocity v. Let λ_s be the scattering mean free path, i.e., the average total distance a neutron travels between successive slowing down collisions with nuclei. The average number of collisions a neutron suffers during the subsequent time interval dt is then $v\,dt/\lambda_s$. Since ξ is the average logarithmic energy change per collision (§ 3.62), it follows that the average decrease in the logarithm of the energy, i.e., $-d \ln E$, in the time dt is equal to $\xi v\,dt/\lambda_s$; hence,

$$-d \ln E = -\frac{dE}{E} = \frac{v\xi}{\lambda_s}\,dt,$$

so that

$$dt = -\frac{1}{v\xi\Sigma_s} \cdot \frac{dE}{E}, \tag{5.1}$$

where $1/\lambda_s$ has been replaced by its equivalent Σ_s, the macroscopic scattering cross section. If E_0 and E_{th} represent the average energies of fission and thermal neutrons, respectively, the slowing down time, t, is obtained upon integrating equation (5.1) between these limits; thus,

$$t = -\int_{E_0}^{E_{th}} \frac{1}{v\xi\Sigma_s} \cdot \frac{dE}{E}. \tag{5.2}$$

5.5. The energy of a neutron is kinetic, provided the velocity does not approach that of light, and so $E = \frac{1}{2}mv^2$, where m is the actual mass of a neutron; hence v may be replaced by $\sqrt{2E/m}$. Furthermore, assuming an average value of Σ_s, represented by $\bar{\Sigma}_s$, equation (5.2) becomes

$$t = \frac{\sqrt{2m}}{\xi\bar{\Sigma}_s}\left(\frac{1}{\sqrt{E_{th}}} - \frac{1}{\sqrt{E_0}}\right).$$

In order to evaluate t in sec, $\bar{\Sigma}_s$ must be in cm^{-1}, as usual, m in grams (1.67

TABLE 5.1.   SLOWING DOWN AND DIFFUSION TIMES FOR
THERMAL NEUTRONS IN AN INFINITE MEDIUM

Moderator	$\bar{\Sigma}_s$(cm^{-1})	Slowing Down Time (sec)	Diffusion Time (sec)
Water..........	1.40	7.1×10^{-6}	2.4×10^{-4}
Heavy water.....	0.35	5.0×10^{-5}	6.0×10^{-2}
Beryllium.......	0.75	5.7×10^{-5}	4.2×10^{-3}
Graphite........	0.41	1.4×10^{-4}	1.6×10^{-2}

$\times 10^{-24}$ g), and E_0 and E_{th} in ergs (1 Mev = 1.6×10^{-6} erg). The calculated slowing down times from $E_0 = 2$ Mev to $E_{th} = 0.025$ ev, for four moderators, are given in Table 5.1.

5.6. In an infinite medium, thermal neutrons are lost by absorption only; the thermal lifetime, l, is then equal to the absorption mean free path, λ_a, divided by the average velocity, v, of the thermal neutrons, so that

$$l = \frac{\lambda_a}{v} = \frac{1}{v\Sigma_a}, \tag{5.3}$$

where Σ_a is the total macroscopic absorption cross section for thermal neutrons. The mean velocity of these neutrons at ordinary temperatures is taken as 2.2 $\times 10^5$ cm/sec, and the values of the thermal lifetime (or diffusion time) are quoted in Table 5.1. It is seen that the average slowing down time in a moderator is usually much less than the thermal-neutron diffusion time. For this reason, it is common practice to neglect the slowing down time and to refer to l as the neutron lifetime in an infinite medium. In order to distinguish this from the effective lifetime, which takes into account the delayed fission neutrons (§ 5.32), it is often called the *prompt* neutron lifetime.

5.7. It should be noted that the data in Table 5.1 refer to the lifetimes in the moderator. In a reactor core, Σ_a would be equal to the total absorption cross section of fuel, moderator, and impurities. It can then readily be shown that Σ_a in equation (5.3) is equivalent to $\Sigma_{am}/(1 - f)$, where Σ_{am} refers to the moderator and f is the thermal utilization, as defined in § 4.9.

5.8. In a system of finite size, the average lifetime is decreased because some neutrons are lost by leakage. The ratio of thermal neutrons absorbed to the sum of those absorbed and leaking out is equal to the thermal nonleakage probability. It was shown in Chapter 4 that this is given by diffusion theory as $1/(1 + L^2B^2)$, and so the prompt thermal lifetime in a finite system is $l/(1 + L^2B^2)$, where L is the diffusion length of thermal neutrons in the reactor core.

Example 5.1. Determine the thermal lifetime (or diffusion time) of neutrons in the bare critical reactor referred to in Example 4.7, consisting of beryllium and uranium-235 in the atomic ratio of 10^4 to 1.

It was found in Example 4.7 that, for this system, B^2 is 0.0051 cm^{-2} and L^2 is 57.3 cm^2, so that L^2B^2 is 0.29 and $1 + L^2B^2$ is 1.29. The diffusion time l in beryllium alone is 4.2×10^{-3}; hence, in the given critical reactor it is $4.2 \times 10^{-3}/1.29 = 3.2 \times 10^{-3}$ sec. (In a larger reactor, the difference would be less.)

KINETIC EQUATIONS OF BARE REACTOR: ONE-GROUP MODEL

5.9. In the subsequent development of the time-dependent behavior of a reactor, a one-group model will be used; in other words, it is postulated that production, diffusion, and absorption of neutrons occur at a single energy. This will make it possible to neglect the slowing down time of the fission neutrons and also their leakage and resonance capture while slowing down. The treatment is thereby greatly simplified, but the final results are essentially identical with those obtained by following more complicated procedures.

5.10. In the absence of a primary source, a noncritical system is not in a steady state and the neutron density is changing with time;* hence, the diffusion equation takes the general form of equation (3.12), namely,

$$D\nabla^2\phi - \Sigma_a\phi + S = \frac{dn}{dt},\qquad(5.4)$$

where, in the present case, all the quantities refer to neutrons of a single energy. According to the arguments in § 4.26, the source term is equal to $k_\infty\Sigma_a\phi$, where k_∞ is the infinite multiplication factor. However, because time-varying quantities are involved, it is necessary to distinguish between the contribution of the prompt and delayed neutrons produced in fission (§ 2.168).

5.11. Let β represent the fraction of the fission neutrons which are delayed; then $1 - \beta$ is the prompt neutron fraction, and the rate of production of these neutrons is $k_\infty\Sigma_a\phi(1 - \beta)$. This is the prompt neutron contribution to the source term in equation (5.4). As stated in § 2.168, there are six groups of delayed neutrons and the rate of formation in any one group, designated by i, is equal to the rate of radioactive decay of the precursor, i.e., to λ_iC_i neutrons per cm³ per sec, where C_i nuclei per cm³ is the concentration of these precursors at any given time and λ_i is the appropriate radioactive decay constant (§ 2.4). For the six groups of delayed neutrons, therefore,

$$\text{Total rate of formation of delayed neutrons} = \sum_{i=1}^{6} \lambda_iC_i \text{ neutrons}/(\text{cm}^3)(\text{sec}),\qquad(5.5)$$

the summation being taken over all the groups; this sum represents the delayed neutron source term.

5.12. Upon inserting the prompt and delayed neutron source terms into equation (5.4), the result is

$$D\nabla^2\phi - \Sigma_a\phi + k_\infty\Sigma_a\phi(1 - \beta) + \sum_{i=1}^{6} \lambda_iC_i = \frac{dn}{dt},\qquad(5.6)$$

where n is the neutron density.† In this expression, ϕ (and n) and C_i are functions of both space and time in a finite reactor; however, as a general rule, these

---

* A subcritical system can be in a steady state if an extraneous (or primary) source is present.

† Equation (5.8) is applicable to the one-group model in which it is postulated that all the neutrons have the same energy. However, since the delayed fission neutrons have lower energies than do the prompt neutrons, they escape less readily while slowing down, so that they are more effective in the chain-reacting system. This difference in behavior depends on the nature of the reactor, and for a given reactor type it is more significant for a small than for a large core, because of the greater importance of leakage in the former case. The greater effectiveness of the delayed neutrons can be allowed for either by using a "β effective" which is larger than the experimental value or by including a factor greater than unity in front of the sum term in equation (5.6). The difference ranges from a few per cent in a large reactor to about 25 per cent in a small one and will be ignored in the treatment given here.

variables are separable and then $\nabla^2\phi$ can be replaced by $-B^2\phi$ (cf. § 4.17), provided the system is not far from critical. Recalling that D/Σ_a is equal to L^2 and that ϕ is defined as nv (§ 2.118), equation (5.6) can be readily transformed into

$$k_\infty v\Sigma_a \left[(1 - \beta) - \frac{1 + L^2B^2}{k_\infty} \right] n + \sum_{i=1}^{6} \lambda_i C_i = \frac{dn}{dt}. \qquad (5.7)$$

For the one-group model, it follows from equation (4.39) that the effective multiplication factor, k_{eff}, is equal to $k_\infty/(1 + L^2B^2)$, and so equation (5.7) becomes

$$k_\infty v\Sigma_a \left[\left(1 - \frac{1}{k_{\text{eff}}}\right) - \beta \right] n + \sum_{i=1}^{6} \lambda_i C_i = \frac{dn}{dt}. \qquad (5.8)$$

5.13. By equation (5.3), $1/v\Sigma_a$ is equal to l, the infinite medium neutron lifetime; hence,

$$\frac{1}{k_\infty v\Sigma_a} = \frac{l}{k_\infty} \equiv l^*, \qquad (5.9)$$

where l^*, defined as l/k_∞, is called the prompt *neutron generation time*. This quantity is dependent only on the composition and design of the reactor, since it is determined by k_∞, v, and Σ_a, and is independent of its size. The factor at the extreme left of equation (5.8) may thus be replaced by $1/l^*$, which is a characteristic property of the system.

5.14. A further simplification of equation (5.8) is possible by introducing the concept of *reactivity*; this quantity, represented by ρ, is defined† by

$$\rho \equiv \frac{k_{\text{eff}} - 1}{k_{\text{eff}}} = 1 - \frac{1}{k_{\text{eff}}}. \qquad (5.10)$$

Consequently, substituting l^* and ρ for the equivalent terms on the left of equation (5.8), the latter reduces to

$$\frac{dn}{dt} = \frac{\rho - \beta}{l^*} n + \sum_{i=1}^{6} \lambda_i C_i. \qquad (5.11)$$

5.15. To solve this equation it is necessary to have an expression for the variation of the concentrations C_i of the delayed neutron precursors with time. The rate of formation of the precursor of the ith group is equal to $\beta_i k_\infty \Sigma_a \phi$, where $k_\infty \Sigma_a \phi$, as seen above, is the total rate of production of fission neutrons and β_i is the fraction belonging to the ith delayed group. However, radioactive decay occurs simultaneously at the rate $\lambda_i C_i$, and so the net rate of formation of the delayed-neutron precursors of the ith group is given by

$$\frac{dC_i}{dt} = \beta_i k_\infty \Sigma_a \phi - \lambda_i C_i.$$

† It is common to use $\Delta k/k$ or $\delta k/k$ (or even just δk) as essentially equivalent to the more precise definition of reactivity given here.

Since ϕ is equal to nv, and $k_\infty v \Sigma_a$ to $1/l^*$, it follows that

$$\frac{dC_i}{dt} = \frac{\beta_i}{l^*} n - \lambda_i C_i. \tag{5.12}$$

5.16. The differential equations (5.11) and (5.12) are basic to the study of reactor kinetics; both of these equations are linear and first order. The rate at which the neutron density n increases is determined by the fission rate and this is proportional to n. Since the space and time variables are separable, it appears that the solution to equation (5.11) will be of the form

$$n(t) = n_0 e^{t\omega}, \tag{5.13}$$

where ω is a parameter, having the dimensions of reciprocal time, which is to be evaluated. Furthermore, since C_i must be proportional to the neutron density, the solution to equation (5.12) can be written in the analogous form

$$C_i(t) = C_{i0} e^{t\omega}. \tag{5.14}$$

5.17. In these equations, n_0 and C_{i0} are the neutron density and the concentration of the precursors of the ith group of delayed neutrons, respectively, at time $t = 0$; prior to this time it may be supposed that the reactor was in a stationary state. At $t = 0$, the reactor suffered a sudden (step) increase in its effective multiplication factor (or reactivity) so that the neutron density commenced to change with time.

5.18. From equations (5.12), (5.13), and (5.14), it is found that

$$C_{i0} = \frac{\beta_i}{l^*(\omega + \lambda_i)} n_0, \tag{5.15}$$

and from this result, together with equations (5.13) and (5.14), equation (5.11) can be transformed into

$$\omega l^* = \rho + \sum_{i=1}^{6} \left(\frac{\lambda_i \beta_i}{\omega + \lambda_i} - \beta_i \right),$$

where β has been replaced by the sum of the β_i terms. It follows, therefore, that

$$\rho = \omega l^* + \sum_{i=1}^{6} \frac{\omega \beta_i}{\omega + \lambda_i}. \tag{5.16}$$

5.19. Equation (5.16) is the characteristic equation which relates the parameters ω to the properties of the materials present in the reactor, as represented by the quantities ρ, l^*, β_i, and λ_i. It is seen to be an algebraic equation of the seventh degree in ω, so that there are seven values of ω corresponding to each value of the reactivity ρ. The variation of the neutron density with time may then be expressed by a linear combination of seven terms of the form of equation (5.13), i.e.,

$$n(t) = A_0 e^{t\omega_0} + A_1 e^{t\omega_1} + \cdots + A_6 e^{t\omega_6}, \tag{5.17}$$

where $\omega_0, \omega_1, \cdots, \omega_6$ are the seven roots of equation (5.16). The A_0, A_1, etc., are constants determined by the initial conditions of the reactor in the steady state, as will be seen shortly.

STABLE REACTOR PERIOD

5.20. If the reactivity ρ is positive, i.e., $k_{\text{eff}} > 1$, it is found that six roots are negative and one is positive; thus, in equation (5.17) ω_0 may be taken as positive, and ω_1, ω_2, etc., as negative. Numerically, each of the six negative values of ω is of the same order as one of the six decay constants, λ_i, of the delayed-neutron precursors. As the time t, i.e., the time interval following the step change in the reactivity, increases, the contributions of all terms beyond the first on the right of equation (5.17) decrease rapidly to zero, because of the negative ω values. In other words, these terms make a transient contribution to the neutron flux; they soon become negligible compared to the first term, which increases with time because of the positive ω value. Hence, after a very short time interval, equation (5.17) reduces to the first term only, i.e.,

$$n(t) = A_0 e^{t\omega_0}. \tag{5.18}$$

5.21. The *reactor period*, T_p, or e-folding time, is defined as the time required for the neutron density (or flux) to change by a factor e, so that

$$n = n_0 e^{t/T_p}, \tag{5.19}$$

or

$$\frac{1}{n} \cdot \frac{dn}{dt} \equiv \frac{1}{T_p}. \tag{5.20}$$

Since the parameters ω have the dimensions of reciprocal time, comparison of equations (5.18) and (5.19) shows that $1/\omega_0$ is the reactor period after the lapse of sufficient time to permit the contributions of the transient terms to damp out. Consequently T_p, defined by

$$T_p \equiv \frac{1}{\omega_0}, \tag{5.21}$$

is called the *stable reactor period*. The quantities of $1/\omega_1$, $1/\omega_2$, etc., are sometimes referred to as *transient periods*, but they are negative and have no physical significance as reactor periods in the sense that $1/\omega_0$ does. The ω_1, ω_2, etc., are merely values of the parameter ω which satisfy the characteristic equation of the problem considered here. They affect the rate at which the neutron flux changes for a short time only after the effective multiplication factor has been increased (or decreased).

5.22. If ρ is negative, i.e., $k_{\text{eff}} < 1$, all seven values of ω, which are solutions of equation (5.16) for a particular ρ, are negative. The expression for the flux as a function of time, as given by equation (5.17), will then consist of seven terms with negative exponents. The magnitude of each term will thus decrease with time.   The first term, with the smallest numerical value of ω_0, will decrease at a slower rate than the others, and will ultimately yield a stable (negative) period equal to $1/\omega_0$.

ONE GROUP OF DELAYED NEUTRONS [1]

5.23. Some indication of the relative importance of the stable and transient terms in equation (5.17) may be obtained by considering the simple case in which there is assumed to be only one group of delayed neutrons with a decay constant, λ, equal to the properly weighted average of the six actual groups.* It will be seen later that this average is given by

$$\lambda = \frac{\beta}{\sum\limits_{i=1}^{6} \frac{\beta_i}{\lambda_i}}. \tag{5.22}$$

The values of λ_i and β_i for the six groups of delayed neutrons from uranium-235, derived from the data in Table 2.10, are given in Table 5.2;† it will be recalled

TABLE 5.2.   DECAY CONSTANTS AND YIELDS OF DELAYED-NEUTRON
PRECURSORS IN THERMAL FISSION OF URANIUM-235

$t_{1/2}$ (sec)	$\lambda_i(\text{sec}^{-1})$	β_i	β_i/λ_i (sec)
55.7	0.0124	0.00021	0.0171
22.7	0.0305	0.00141	0.0463
6.22	0.111	0.00127	0.0114
2.30	0.301	0.00255	0.0085
0.61	1.1	0.00074	0.0007
0.23	3.0	0.00027	0.0001
Total		0.0065	0.084

from equation (2.4) that λ for any radioactive species is equal to $0.693/t_{1/2}$, where $t_{1/2}$ is the half-life.   The average decay constant, as defined by equation (5.22), is thus $0.0065/0.084$, i.e., approximately 0.08 sec^{-1}.

5.24. For a single group of delayed neutrons, equation (5.16) is

$$\rho = \omega l^* + \frac{\omega\beta}{\omega + \lambda}, \tag{5.23}$$

which is a quadratic that can be solved for ω.   Provided $(\beta - \rho + \lambda l^*)^2 \gg$

---

* For a method of complete solution, taking into account all the delayed-neutron groups, using the Laplace transform (§ 5.173 *et seq.*), see Ref. 2.   Alternatively, an electrical simulator may be employed.

† The corresponding values for uranium-233 and plutonium-239 can be readily calculated, if required, from the data in Table 2.10.

$2|\lambda \rho l^*|$, as will be the case for thermal reactors with positive reactivities of about 0.0025 or less, and for all negative reactivities, i.e., $\rho \ll \beta$, the two solutions of equation (5.23) are

$$\omega_0 \approx \frac{\lambda \rho}{\beta - \rho} \quad \text{and} \quad \omega_1 \approx \frac{-(\beta - \rho)}{l^*}. \tag{5.24}$$

Since these solutions hold when $\beta - \rho$ is positive, it is evident that ω_0 is positive for positive reactivities whereas ω_1 is always negative; hence, for positive reactivities $1/\omega_0$ is equal to the stable period. It will be seen later that this is also true for negative reactivities.

5.25. For a single (average) group of delayed neutrons, the variation of the neutron density with time is given by equation (5.17) as

$$n = A_0 e^{t\omega_0} + A_1 e^{t\omega_1}, \tag{5.25}$$

and the corresponding expression for the delayed-neutron precursor is

$$C = B_0 e^{t\omega_0} + B_1 e^{t\omega_1}. \tag{5.26}$$

The problem of immediate interest is to determine the constants A_0, A_1, B_0, and B_1. This is achieved by writing equation (5.12) for the single precursor, i.e.,

$$\frac{\partial C}{\partial t} = \frac{\beta}{l^*} n - \lambda C, \tag{5.27}$$

and substituting equations (5.25) and (5.26) for n and C, respectively. Another expression for $\partial C/\partial t$ can be obtained by differentiating equation (5.26) and, upon comparing the coefficients, expressions are obtained relating A_0 to B_0 and A_1 to B_1.

5.26. If n_0 is the neutron density in the steady state, before the step change in the reactivity, i.e., at $t = 0$, it follows from equation (5.25) that it is equal to $A_0 + A_1$. Similarly, the steady-state value of the precursor concentration, C_0, is equal to $B_0 + B_1$, by equation (5.26). Furthermore, in the steady state $\partial C/\partial t$ is zero, and so a relationship between C_0 and n_0 can be derived from equation (5.27). With the information now available, and using equation (5.24) for ω_0 and ω_1, it is found, after neglecting certain terms of small magnitude, that

$$A_0 \approx \frac{\beta}{\beta - \rho} n_0 \quad \text{and} \quad A_1 \approx - \frac{\rho}{\beta - \rho} n_0. \tag{5.28}$$

The variation of the neutron density with time is consequently given by equations (5.24), (5.25), and (5.28) as

$$n \approx n_0 \left[\frac{\beta}{\beta - \rho} e^{\frac{\lambda \rho t}{\beta - \rho}} - \frac{\rho}{\beta - \rho} e^{- \frac{(\beta - \rho)t}{l^*}} \right], \tag{5.29}$$

where n is the neutron density at time t after the step change in reactivity.

5.27. If the reactivity is positive, then $\beta - \rho$ will be positive under the conditions postulated in the foregoing derivation, i.e., $\rho \leq 0.0025$, and so the first term in the brackets in equation (5.29) has a positive coefficient and the second a negative coefficient. The neutron density is thus the difference of two terms, of which one (the first) increases while the other decreases with increasing time after the step change ρ in the reactivity. It should be mentioned that, since the neutron velocity has been assumed to be independent of time, equation (5.29) also relates the neutron flux ϕ at any time t to the value, ϕ_0, in the steady state prior to the reactivity change.

5.28. The second (or transient) term, with the negative exponent, dies out in a very short time. Subsequently the flux variation is given by the first term of equation (5.29), namely,

$$n \approx n_0 \frac{\beta}{\beta - \rho} e^{\frac{\lambda \rho t}{\beta - \rho}}, \tag{5.30}$$

so that the stable reactor period, assuming one (average) group of delayed neutrons, is

$$T_p \approx \frac{\beta - \rho}{\lambda \rho}. \tag{5.31}$$

Example 5.2. The reactivity in a steady-state reactor, in which the neutron generation time is 10^{-3} sec, is suddenly made 0.0022 positive; assuming one group of delayed neutrons, determine the subsequent change of neutron flux with time. Compare the stable period with that which would result if there had been no delayed neutrons.

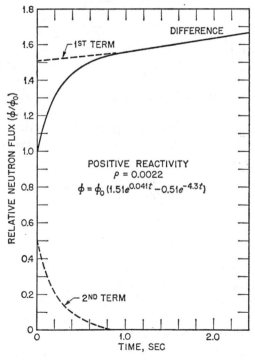

FIG. 5.1. Change of neutron flux with time for positive reactivity

For one (average) group of delayed neutrons λ is 0.08 sec^{-1}; ρ is given as 0.0022 and l^* as 10^{-3} sec; β for uranium-235 is 0.0065. Consequently, by equation (5.29),

$$\phi = \phi_0(1.51e^{0.041t} - 0.51e^{-4.3t}).$$

The numerical values of each term in the parentheses and of their difference are plotted in Fig. 5.1. The second (transient) term is seen to die out within about 0.8 sec.

By equation (5.31) the stable period is $(0.0065 - 0.0022)/(0.08)(0.0022) = 24$ sec, assuming one group of delayed neutrons. If there had been no delayed neutrons, the period would have been given by equation (5.33) as $10^{-3}/0.0022 = 0.45$ sec. The effect of the delayed neutrons in increasing the reactor period, i.e., in decreasing the growth rate of the neutron flux, after a change in multiplication is very marked.

5.29. A more complete calculation, taking into account all the groups of delayed neutrons, gives a stable period of 14 sec for a positive reactivity of 0.0022. The difference between this value and that obtained in Example 5.2 (24 sec) arises largely from the approximations made in deriving equation (5.29). These approximations are much less serious for small reactivities, e.g., 0.001 or less, and then the results obtained using one group of delayed neutrons are reasonably good.

5.30. For very small reactivities, i.e., for very long periods, ρ in equation (5.31) may be neglected in comparison with β; this equation then becomes

$$T_p \approx \frac{\beta}{\lambda\rho}. \tag{5.32}$$

Since β/λ has a definite value for a given fissile material, e.g., 0.084 for uranium-235 (see Table 5.1), it is seen that at low reactivities the stable reactor period is inversely proportional to the reactivity. As will be mentioned later, this fact has been utilized in connection with large reactors; for such reactors the effective multiplication factor is never much greater than unity and the reactivities are usually small.

5.31. It is of interest at this point to call attention to the effect of the delayed neutrons on the reactor period. If there were no delayed neutrons, equation (5.11) would reduce to

$$\frac{dn}{dt} = \frac{\rho}{l^*}\,n,$$

so that the reactor period would be

$$T_p = \frac{l^*}{\rho}. \tag{5.33}$$

In Example 5.2, in which $l^* = 10^{-3}$ sec and $\rho = 0.0022$, the reactor period in the absence of delayed neutrons would be 0.45 sec, which is very much less than the actual stable period of 14 sec. In other words, provided the reactivity is not too large (§ 5.53), the effect of the delayed neutrons is to prevent the neutron density from rising so rapidly when the reactivity suffers a sudden increase. The reason is that, because of the delay in the release of a proportion of the fission neutrons, the average *effective* generation time is greater than l^* by an amount depending on the extent to which the delayed neutrons contribute to the fission chain.

5.32. A reasonable approximation, based on one average group of delayed neutrons, is to write the *effective generation time*, $\bar{l}$, as

$$\bar{l} \approx l^* + (\beta - \rho)/\lambda, \tag{5.34}$$

so that $\bar{l}$ becomes equal to l^* when the reactivity is equal to or greater than β, i.e., 0.0065, the fraction of delayed neutrons. In these circumstances (§ 5.54) the stable reactor period is given by equation (5.33). On the other hand, when ρ is small, l^* can be neglected in comparison with $(\beta - \rho)/\lambda$ and equation (5.34) becomes

$$\bar{l} \approx (\beta - \rho)/\lambda.$$

Replacing l^* in equation (5.33) by this value of $\bar{l}$, the stable reactor period becomes identical with that in equation (5.31), or with equation (5.32) when ρ is very small.

5.33. It will be noted from Fig. 5.1 that immediately after the step change in the multiplication factor, i.e., at times very soon after $t = 0$, the neutron flux increases rapidly; this is sometimes referred to as the initial prompt rise (or "prompt jump") of the flux. The rate of the initial rise may be calculated by first differentiating equation (5.25) with respect to time and setting $t = 0$, i.e.,

$$\left(\frac{\partial n}{\partial t}\right)_{t=0} = A_0\omega_0 + A_1\omega_1.$$

Upon introducing the values of A_0, A_1, ω_0, and ω_1 derived above, it is found that

$$\frac{1}{n_0} \cdot \frac{dn_0}{dt} \approx \frac{\lambda\rho\beta}{(\beta - \rho)^2} + \frac{\rho}{l^*}.$$

According to equation (5.20), the left side is equivalent to T_0, the initial reactor period, i.e., at $t = 0$; hence,

$$\frac{1}{T_0} \approx \frac{\lambda\rho\beta}{(\beta - \rho)^2} + \frac{\rho}{l^*}. \tag{5.35}$$

5.34. For moderately small values of ρ, the first term on the right side of equation (5.35) may be neglected, and then

$$T_0 \approx \frac{l^*}{\rho}. \tag{5.36}$$

5.35. It is seen from equation (5.36) that the initial reactor period, T_0, is identical with that given by equation (5.33) for the case where there are no delayed neutrons. Thus, for small reactivities, the initial rate of growth of the neutron flux, following a step change, is almost the same as if all the fission neutrons were prompt.†

---

† It should be understood that T_0 does not represent a stable reactor period. An examination of the derivation given above shows that it is a purely transient period arising from the second (transient) term on the right of equation (5.25).

5.36. The physical interpretation of this situation is as follows.   The number of delayed neutrons contributed to the fission chain reaction at any instant is proportional to the thermal-neutron flux at some previous time, but the number of delayed-neutron precursors produced by fission, i.e., the number of neutrons being delayed, is proportional to the flux at the given instant.   For a short time after the change in the effective multiplication factor the neutron flux will not change very greatly, provided the reactivity is small.   The number of delayed neutrons contributed to the system will then not be appreciably less than those being delayed.   Hence, for a very short time during the initial prompt rise, the reactor behaves as if all the fission neutrons were prompt, and the flux increases with a period of approximately l^*/ρ.   However, as the neutron flux grows with time, fewer delayed neutrons enter the chain-reacting system than are being delayed, and the rate of increase of the flux gradually falls off.   Ultimately, the rate of growth is determined by the stable period, equal to $1/\omega_0$.

ONE GROUP OF DELAYED NEUTRONS: NEGATIVE REACTIVITY

5.37. The approximations made in deriving equation (5.29) are applicable for almost any negative values of ρ, especially for large negative values of the reactivity.   This equation may thus be used to provide an indication of the manner in which the neutron flux decreases as the result of a negative step change in the effective multiplication factor.   For purposes of comparison with the results obtained in the example given above, it will be supposed that ρ is -0.0022;

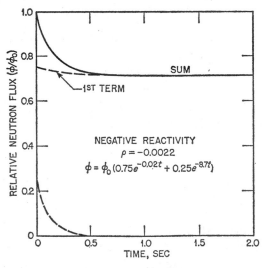

Fig. 5.2.   Change of neutron flux with time for negative reactivity

taking the values of l^*, λ, and β to be the same as before, equation (5.29) now becomes

$$n = n_0(0.75e^{-0.02t} + 0.25e^{-8.7t}).$$

The values of the two terms and their sum are plotted in Fig. 5.2. It will be noted that the coefficients of both terms are now positive and the exponents are both negative. However, the large negative exponent in the second term on the right makes this term damp out fairly rapidly, in about 0.5 sec, as seen in Fig. 5.2. When the second term ceases to be significant, the stable period appears to be $-1/0.02$, i.e., -50 sec, compared to roughly 24 sec when ρ is 0.0022 positive (§ 5.29). It will be seen later that when all six groups of delayed neutrons are taken into account, the stable negative period is probably even larger numerically than that derived here.

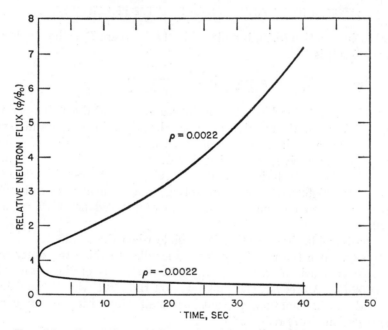

Fig. 5.3. Comparison of effects of positive and negative reactivities

5.38. If all the fission neutrons were prompt, the reactor periods would be -0.45 sec for $\rho = -0.0022$ and 0.45 sec for $\rho = 0.0022$. It is evident, therefore, that, whereas the presence of delayed neutrons slows down the rate of increase of flux when the reactivity is positive, it slows down the rate of decrease to an even greater extent when the reactivity is negative. This may be seen more clearly from Fig. 5.3 which covers a considerably longer period of time than the two preceding figures.† It will be shown below that, for a reactor operating in a steady state, it is not possible to reduce the neutron flux in a reactor more rapidly than is permitted by the most-delayed neutron group (§ 5.47).

5.39. Since the arguments leading to equation (5.36) apply irrespective of the sign of the reactivity, it follows that, immediately after the multiplication factor of the reactor undergoes a step decrease, the initial period is equal to l^*/ρ, where ρ is now negative. The rate of decrease of the neutron flux is then the same as if all the neutrons were prompt. Thus, for very small times after the change in the multiplication factor, the two curves in Fig. 5.3 are seen to be symmetrical. The physical basis of this result is the same as given above for the case of positive reactivity. After a short time, however, the delayed neutron effect causes a considerable slowing down in the rate of decrease of flux.

REACTIVITY AND PERIOD: POSITIVE REACTIVITIES

5.40. If ω in equation (5.16) is replaced by $1/T_p$, where T_p is the stable reactor period, the result is

$$\rho = \frac{l^*}{T_p} + \sum_{i=1}^{6} \frac{\beta_i}{1 + \lambda_i T_p}, \tag{5.37}$$

which provides a relationship between the reactivity and the stable period. If the period T_p is given, the reactivity can be calculated, since l^*, β_i, and λ_i may be regarded as known. The reverse procedure, i.e., the calculation of the stable reactor period for a given reactivity, involves solution of the seventh-order equation. Methods of solution have been developed, but for positive reactivities that are not too high, results in reasonably good agreement can be obtained by the procedure based on one average group of delayed neutrons, as already described.

5.41. Considerable simplification is possible when the stable period is large, that is to say, when the reactivity is very small. In this case, unity may be neglected in comparison with each $\lambda_i T_p$ in the terms of the summation in equation (5.37). Moreover, the first term on the right will be small in comparison with the summation, so that for small reactivities (or large reactor periods) equation (5.37) reduces to

$$\rho \approx \frac{1}{T_p} \sum_{i=1}^{6} \frac{\beta_i}{\lambda_i}$$

or

$$T_p \approx \frac{1}{\rho} \sum_{i=1}^{6} \frac{\beta_i}{\lambda_i}. \tag{5.38}$$

† These curves are based on calculations using one (average) group of delayed neutrons. For the particular value of the positive reactivity chosen, i.e., 0.0022, the results, except at small times, are too low (see Ref. 2), although they are qualitatively correct. Better agreement can be obtained by means of a two-group treatment (Ref. 3).

Since all the β's and λ's have definite values for a particular fissile material, it is apparent that the reactivity and stable reactor period are inversely proportional to each other. The proportionality constant should be the same for all thermal reactors using a given fissile material, and should be independent of the reactor size and structure, provided the reactivity is small.

5.42. It will be noted that equation (5.38) becomes identical with equation (5.32), based on a single group of delayed neutrons, provided the average decay constant λ is defined by

$$\lambda \equiv \frac{\beta}{\sum\limits_{i=1}^{6} \dfrac{\beta_i}{\lambda_i}}.$$

This is, in fact, the definition used in § 5.23, so as to make the single group roughly equivalent to the six actual groups of delayed neutrons.

5.43. For small reactivities, the stable period for a given reactivity is seen to be independent of the neutron generation time, l^*, because the first term on the right of equation (5.37) is then small in comparison with the second. However, for larger values of ρ, i.e., for shorter reactor periods, the effect of the generation time becomes evident. The relationship between the reactivity and the stable

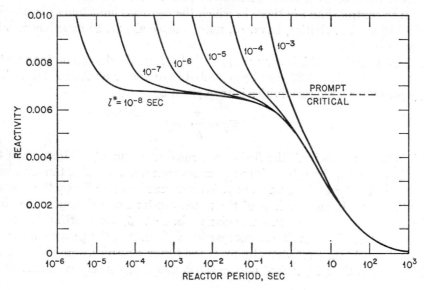

FIG. 5.4. Relationship between reactor period and reactivity for various neutron generation times [4]

period, is shown in Fig. 5.4 for neutron generation times of 10^{-3} to 10^{-8} sec, with uranium-235 as the fissile species [4]. The stable periods for a few reactivity values are also given in Table 5.3, for three generation times.

TABLE 5.3. REACTOR PERIOD AND REACTIVITY FOR URANIUM-235

	Neutron Generation Time (sec)		
	10^{-3}	10^{-4}	10^{-5}
Reactivity	Reactor Period (sec)		
0.001........	50	50	50
0.003........	7.1	6.4	6.4
0.004........	3.5	3.0	2.8
0.007........	0.6	0.11	0.016
0.010........	0.3	0.03	0.003

5.44. At low reactivities, all the curves coincide, so that for a specified reactivity the period is independent of the neutron generation time, as stated above. For higher reactivities, e.g., greater than about 0.003, the stable period corresponding to a given reactivity decreases as the generation time decreases. This means that for higher reactivities the neutron flux will increase at a more rapid rate the smaller the neutron generation time. At such reactivities, reactors with short generation times, e.g., fast reactors, are more difficult to control than are those in which the neutrons have longer generation times. At low reactivities, however, no difference should be experienced with respect to control.

5.45. At sufficiently high values of ρ, so that T is small, the second term on the right side of equation (5.37) can be neglected in comparison with the first. The stable period for large reactivities is then given by

$$T_p = \frac{l^*}{\rho},$$

which is the same as if all the fission neutrons were prompt. Thus, when the reactivity is small, the rate of increase of neutron density (after the transients have died out), for a given value of ρ, is determined essentially by the neutron delay time, and is independent of the much shorter neutron generation time. At high reactivities, however, particularly when $\rho > \beta$, the contribution of the delayed neutrons is negligible in comparison with that of the prompt neutrons. The reactor behaves as if virtually all of the fission neutrons were prompt; the stable period is then dependent on the generation time, as shown earlier.

REACTIVITY AND PERIOD: NEGATIVE REACTIVITIES

5.46. For numerically small negative values of the reactivity, the approximations made in deriving equation (5.32) are valid. The negative stable period is then inversely proportional to the negative reactivity. However, for large negative reactivities, the stable periods calculated on the basis of one average group of delayed neutrons, which are fairly satisfactory for positive reactivities, are too small. This is because, after the transients have died out, the rate of change of neutron density, especially for moderate and large negative reactivities, is determined by the most-delayed neutrons, and not by the average, as is essentially true for positive reactivities.

5.47. In physical terms, it may be said that when the reactivity is made negative, as in reactor shutdown, delayed neutrons resulting from fissions occurring before shutdown continue to be released. These prevent the neutron density from decreasing as rapidly as would have been the case if there were no delayed neutrons. After a little time, the shorter-lived precursors will have decayed almost completely, and the delayed neutrons remaining are essentially those belonging to the group having the precursor of longest life, i.e., the one for which λ is 0.0124 sec^{-1} (see Table 5.2). In these circumstances, equation (5.37) simplifies to

$$\rho \approx \frac{\beta}{1 + \lambda_1 T_p},$$

where λ_1 has the value just given. As ρ increases numerically, it is evident that $(1 + \lambda_1 T_p) \to 0$, so that, for large negative reactivities, the stable period, T_p, approaches $-1/\lambda_1$, i.e., $-1/0.0124 = -80$ sec.

5.48. Observations have shown that thermal reactors containing carbon (or graphite) or ordinary water as moderator do, in fact, attain a stable period of about -80 sec when shut down. When the moderator is beryllium or heavy water (deuterium oxide), the period is even longer because the gamma rays from the fission products continue to release neutrons as a result of the photoneutron reaction with the moderator (§ 2.74). Allowance for the effect of the photoneutrons can be made by introducing appropriate values for β and λ into the kinetic equations [5].

THE INHOUR FORMULA

5.49. The reactivity is sometimes expressed in terms of the inverse hour or "inhour" unit, defined as the reactivity which will make the stable reactor period equal to 1 hour, i.e., 3600 sec [6]. Thus, the value of the inhour unit in terms of reactivity is obtained by setting T_p in equation (5.37) equal to 3600 sec, since the λ_i are usually expressed in sec^{-1}. The reactivity of a reactor in inhours, represented by Ih, is then obtained upon dividing equation (5.37) by the corresponding value of the inhour unit; thus,

$$\text{Ih} = \frac{\dfrac{l^*}{T_p} + \displaystyle\sum_{i=1}^{6} \dfrac{\beta_i}{1 + \lambda_i T_p}}{\dfrac{l^*}{3600} + \displaystyle\sum_{i=1}^{6} \dfrac{\beta_i}{1 + 3600\lambda_i}}. \tag{5.39}$$

This expression is the correct form of the inhour formula, although the related equations (5.16) and (5.37) are sometimes referred to by this name.

5.50. It can be seen from equation (5.39) that, in general, the reactivity given in inhours does not define the stable reactor period in a simple manner. Nevertheless, for large reactors, such as the natural uranium-graphite reactors, the reactivity is frequently expressed in inhours. The reason is that for small reactivities, i.e., for long reactor periods, unity may be neglected in comparison with $\lambda_i T_p$ and $3600\lambda_i$, and equation (5.39) then reduces to

$$\text{Ih} \approx \frac{3600}{T_p} \quad \text{or} \quad T_p \approx \frac{3600}{\text{Ih}}, \tag{5.40}$$

so that the reactivity in inhours is inversely related to the stable reactor period in a simple manner. In fact, if the reactor period is expressed in hours, it is equal to the reciprocal of the reactivity in inhours.

5.51. By comparing equations (5.38) and (5.40), a relationship between ρ and Ih may be obtained, namely,

$$\frac{1}{\rho} \sum_{i=1}^{6} \frac{\beta_i}{\lambda_i} \approx \frac{3600}{\text{Ih}}.$$

For uranium-235, the sum of the β_i/λ_i values is 0.084 (Table 5.2), and so

$$\rho \approx 2.3 \times 10^{-5} \text{ Ih}.$$

Although this result is strictly applicable only for long reactor periods, it is sometimes combined with equation (5.37) to give the relationship

$$\text{Ih} \approx \frac{1}{2.3 \times 10^{-5}} \left(\frac{l^*}{T_p} + \sum_{i=1}^{6} \frac{\beta_i}{1 + \lambda_i T_p} \right).$$

The times l^* and T_p are usually expressed in seconds, and so the λ_i's are in reciprocal seconds. Upon inserting the known values of β_i and λ_i, from Table 5.2, and that for l^* for thermal neutrons in the given reactor, an expression is obtained relating the reactivity in inhours to the reactor period.

THE PROMPT-CRITICAL CONDITION

5.52. When a reactor is critical on prompt neutrons alone, it is said to be *prompt critical*. The required condition for prompt criticality can be obtained from equation (5.11). If the delayed-neutron source term is ignored, so that prompt fission neutrons only are considered, this equation becomes

$$\frac{dn}{dt} = \frac{\rho - \beta}{l^*} n.$$

For the reactor to be critical, dn/dt must be zero; since n and l^* are not zero, it follows that

$$\rho = \beta,$$

i.e., a reactor becomes prompt critical when the reactivity is equal to the fraction of delayed neutrons. A thermal reactor in which the fissile material is uranium-235 thus becomes prompt critical when ρ is 0.0065, so that k_{eff}, which is equal to $1/(1 - \rho)$ by equation (5.10), is then close to 1.0065.

5.53. When the effective multiplication factor is greater than the prompt-critical value, the chain reaction can be maintained on the prompt neutrons alone. The reactor is then said to "outrun" the delayed neutrons. If the delayed neutrons made no contribution at all, it can be readily shown that the stable reactor period would be given by

$$T_p = \frac{l^*}{\rho(\text{prompt})}, \tag{5.41}$$

where $\rho(\text{prompt})$ is the amount by which the actual reactivity ρ exceeds the prompt-critical value, i.e.,

$$\rho(\text{prompt}) = \rho - \beta.$$

The stable period for reactivities *in excess* of the prompt-critical value is then given by equation (5.41) as

$$T_p = \frac{l^*}{\rho - \beta}. \tag{5.42}\dagger$$

This result is applicable only when $\rho > \beta$; it does not hold at prompt critical for then $\rho - \beta$ is zero, and the stable period can be obtained only by solving equation (5.37).

5.54. If ρ is large enough for β to be neglected in comparison with it, equation (5.42) reduces to

$$T_p \approx \frac{l^*}{\rho},$$

as stated in § 5.32. The reactor period is then approximately equal to that which would result if all the neutrons were prompt. The contribution of the delayed fission neutrons is negligible in comparison with that of the prompt neutrons.

5.55. Some indication of the rate of increase in the neutron density as a result of fission of uranium-235, for various reactivities, will be found in Fig. 5.5; the values of ρ are 0.001, 0.0065, i.e., prompt critical, and 0.014, respectively, the neutron generation time l^* being taken as 10^{-3} sec in each case. It should be noted that in order to plot the three curves on one figure, the time scale has been made logarithmic; the change in behavior as ρ is increased is thus more marked than would appear at first sight. An examination of the curves shows that when ρ is 0.001 it takes more than 100 sec for the neutron density to increase tenfold, but when ρ is 0.0065, the same increase occurs in about 1 sec, and when

---

† The same expression can be obtained directly from equation (5.37) by neglecting $\lambda_i T_p$ in comparison with unity, since the reactor period is short. The treatment given above, however, brings out the physical significance of the result.

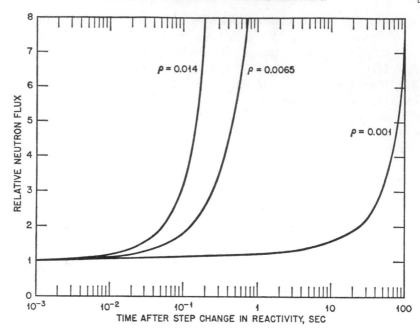

FIG. 5.5. Rate of increase of neutron flux following a step increase in reactivity

ρ is 0.014, it would require about 0.2 sec. The rapid increase in the neutron density when the reactivity exceeds the prompt-critical value would make the reactor difficult to control, and special precautions must be taken to make sure that this condition does not arise in reactor operation.

5.56. On account of its special significance, the prompt-critical condition is sometimes used as the basis for a reactivity unit, especially for reactors having considerable latent (or built-in) excess multiplication (§ 5.149). The unit, called "the dollar," is defined by [7]

$$\text{Reactivity in dollars} \equiv \frac{\rho}{\beta}.$$

A "cent" represents a one-hundredth part of a dollar. When prompt critical, a reactor has a reactivity of exactly one dollar. For a thermal reactor using uranium-235 as the fissile material, for example, a reactivity of \$1.50 would imply a value ρ of 1.50 × 0.0065, i.e., 0.0097, and an effective multiplication factor of about 1.0098.

Example 5.3. The prompt neutron lifetime in a reactor moderated by heavy water is 5.7 × 10^{-4} sec. For a reactivity of 0.00065, express the reactor period in (a) seconds, (b) inhours. What is the reactivity in dollar units?

Since the reactivity is quite small, the reactor period can be determined from equation (5.32); thus, since β/λ is 0.084 (Table 5.2),

$$T_p = \frac{\beta}{\lambda \rho} = \frac{0.084}{0.00065} = 130 \text{ sec.}$$

The period in inhours is then obtained from equation (5.40) as

$$\text{Ih} = \frac{3600}{T_p} = \frac{3600}{130} = 28 \text{ inhours.}$$

By the definition in § 5.56, since β is 0.0065,

$$\text{Reactivity in dollars} = \frac{0.00065}{0.0065} = \$0.10 \text{ (or 10 cents).}$$

EFFECT OF TEMPERATURE ON REACTIVITY

GENERAL CONSIDERATIONS

5.57. An important cause of transient changes in the effective multiplication factor in an operating reactor is variation in the temperature of the system. Such variations may be localized, e.g., due to nonuniformity of structure affecting the flow of coolant at particular points, or they may affect the reactor as a whole. A change in the rate of flow, for example, would gradually alter the temperature of the reactor. A similar effect would result from a change in the power demand, i.e., the heat removed from the coolant, so that the temperature at which the latter enters the reactor would be changed. Whether such temperature transients cause the multiplication factor of the reactor to increase or decrease and how fast these changes manifest themselves are of great practical significance.

5.58. The effective multiplication factor may be expressed as the product of the infinite multiplication factor, k_∞, and the nonleakage probablility, P, as stated in § 4.4; hence,

$$k_{\text{eff}} = k_\infty P.$$

Differentiation with respect to temperature, T, then gives

$$\frac{1}{k_{\text{eff}}} \cdot \frac{dk_{\text{eff}}}{dT} = \frac{1}{k_\infty} \cdot \frac{dk_\infty}{dT} + \frac{1}{P} \cdot \frac{dP}{dT}. \tag{5.43}$$

Since k_{eff} is not very different from unity and bearing in mind the definition of the reactivity in equation (5.10), it is apparent that the left side of equation (5.43) is approximately the same as $d\rho/dT$, so that it represents the temperature coefficient of reactivity; hence,

$$\frac{d\rho}{dT} \approx \frac{1}{k_\infty} \cdot \frac{dk_\infty}{dT} + \frac{1}{P} \cdot \frac{dP}{dT}. \tag{5.44}$$

Thus, the general magnitude and sign of the temperature coefficient of reactivity can be studied by considering the effect of temperature on (a) the infinite multiplication factor and (b) the nonleakage probability.

5.59. It was shown in § 4.13 that the infinite multiplication factor in a thermal reactor may be treated as the product of four terms, namely, ϵ, η, p, and f. Of these, the fast-fission factor, ϵ, may be omitted from consideration here since the effect of temperature on it is probably not very significant. The influence of temperature on the other three factors will be considered in turn. Because there are important differences between the temperature effects on homogeneous and heterogeneous reactors, these two types of system will be treated separately [8].

HOMOGENEOUS SYSTEMS

5.60. It is seen from the definition of η (§ 4.5) that the effect of temperature is determined essentially by the changes in ν, the average number of neutrons liberated per fission, and the cross sections. If the capture and fission cross sections in the thermal region all obey the $1/v$ law, the variation of η with temperature is dependent only on the changes in ν. With increasing temperature, and hence increasing neutron energy, there will be some increase in ν (Table 2.9), but the effect is small within the temperature range of interest. It is to be expected, therefore, that ν will not be very sensitive to temperature changes during normal reactor operation. Since the cross sections exhibit deviation from $1/v$ behavior, varying with the neutron process and the nuclide involved in the reaction, the actual temperature dependence of η is not as simple as implied above. For natural uranium, for example, η is almost constant up to 300°C (570°F), but at higher temperatures the value decreases. For the present purpose, however, it is sufficient to assume that the variation of η with temperature is small. This conclusion applies to both homogeneous and heterogeneous systems.

5.61. Examination of equation (4.52) for the resonance escape probability in a homogeneous reactor shows that the change in I/σ_{sm} will determine the effect of temperature. In general, σ_{sm} is essentially independent of neutron energy in the epithermal region and so the effective resonance integral, I, is the significant factor. In this connection consideration must be given to the so-called neutron *Doppler effect*, which is a broadening of the resonance peaks (§ 2.138) with increasing temperature due to the increased random motion of the reacting nuclei [9]. For natural or moderately-enriched uranium fuel, it is to be expected that the effective resonance integral will increase somewhat with temperature as a result of the Doppler broadening. It follows, therefore, that the resonance escape probability will decrease to a certain extent with increasing temperature. However, the effect is relatively small and is less the smaller the ratio of fuel to moderator atoms, N_u/N_m.

5.62. The influence of temperature on the thermal utilization in a homogeneous system depends on relative changes in the macroscopic cross sections, as indicated by equation (4.3). Since all the materials occupy the same volume, the effect of a change in density on the numbers of nuclei of each kind present in unit volume will cancel out. Hence, the temperature coefficient of the thermal utilization is determined by changes in the microscopic absorption cross sections. If these obey the $1/v$ law, however, there will be no net change with temperature. There are, of course, deviations from $1/v$ behavior, but the overall effect on the thermal utilization is minor (see, however, § 5.68).

5.63. A review of the foregoing discussion leads to the conclusion that, in a homogeneous reactor, temperature causes a relatively small change in the infinite multiplication factor. The main effect arises from the decrease in the resonance escape probability as the temperature rises. Hence, the infinite multiplication factor will have a negative temperature coefficient, but its value will be small in magnitude.

HETEROGENEOUS SYSTEMS

5.64. For a homogeneous reactor there is only one overall (or total) temperature coefficient of the infinite multiplication factor, because the fuel and moderator are intimately mixed. With a heterogeneous system, however, it is desirable to distinguish between the *fuel (or metal) temperature coefficient** and the *moderator temperature coefficient*, which give the effects of changes in the temperature of the fuel and moderator, respectively, on that part of the reactivity which is determined by the infinite multiplication factor. These coefficients will, in general, be different in magnitude, and also in sign, because they are dependent upon entirely different factors. In addition, the time constants, i.e., the times required for a temperature change to produce an appreciable change in the reactivity, are usually quite different for the fuel and the moderator. The fact that the fuel temperature coefficient has a short time constant compared with that of the moderator makes it much more significant in reactor operation.

5.65. An important effect of temperature on the infinite multiplication factor is associated with the resonance escape probability in a heterogeneous reactor. This arises from the increase in temperature of the effective resonance integral, defined by equation (4.73). In a heterogeneous system the relatively large resonance escape probability achieved by "lumping" results from the shielding of the inner regions of the fuel lump (or rod) by resonance capture in the outer layers (§ 4.93). Since this self-shielding is diminished, because of the increased absorption of neutrons in the outer layers arising from the Doppler broadening, there will be a decrease in the resonance escape probability with increasing temperature. The effect is much more marked than in a homogeneous system because in the latter there is no self-shielding. Other factors make some contribution to the change in the resonance escape probability with temperature, but these are small in comparison with the effect of the Doppler broadening.

---

* In some circumstances, especially for fast reactors, it is convenient to divide the effect of temperature on fuel reactivity into two parts: (a) the thermal base effect, resulting from the influence of temperature on the neutron cross sections, and (b) the Doppler effect due to the broadening of resonances. The former is generally negative but small, whereas the latter may be either positive or negative depending on the composition of the fuel.

5.66. It is seen that in a heterogeneous reactor the resonance escape probability can make a significant negative contribution to the temperature coefficient of the infinite multiplication factor. Since the effect is determined by the temperature of the fuel, and is independent of the moderator temperature, it is apparent that the fuel temperature coefficient may be expected to have an appreciable negative value, especially in reactors employing natural or moderately enriched uranium as fuel. It is in such reactors that the role of neutron capture by uranium-238 in the resonance region is important.

5.67. Neglecting the effects of cladding and other impurities, equation (4.2) for the thermal utilization can be written, for a heterogeneous reactor, as

$$f = \frac{V_u N_u \sigma_{au}}{V_u N_u \sigma_{au} + V_m N_m \sigma_{am}(\phi_m/\phi_u)}$$

$$= \frac{N_{tu}\sigma_{au}}{N_{tu}\sigma_{au} + N_{tm}\sigma_{am}(\phi_m/\phi_u)}, \tag{5.45}$$

where the symbol N_t represents the total number of nuclei of the indicated species present in the reactor. If the numbers of fuel and moderator nuclei (or their ratio) remain constant with changes of temperature, and the absorption cross sections exhibit $1/v$ behavior, it follows from equation (5.45) that the effect of temperature on the thermal utilization will depend on changes in the disadvantage factor, ϕ_m/ϕ_u. Comparison of equations (4.61) and (4.65) shows that the disadvantage factor is determined by the quantities F and $E - 1$. It is found that both of these, especially the latter which is determined by the properties of the moderator, decrease as the temperature rises. This means that the disadvantage factor decreases and approaches unity with increasing temperature. It follows, therefore, that in a heterogeneous reactor the thermal utilization should have a positive temperature coefficient, provided the absorption cross sections of fuel and moderator are inversely proportional to the neutron velocity. However, for uranium-235 this cross section falls off somewhat more rapidly, and for plutonium-239 less rapidly, with increasing temperature, than required for $1/v$ behavior. It will be seen from equation (5.45) that the former will consequently make a negative contribution and the latter a positive contribution to the thermal utilization temperature coefficient.

5.68. Since it is the moderator, rather than the fuel, temperature which determines the disadvantage factor and the neutron energy, upon which the cross sections depend, the variation of the thermal utilization with temperature provides a good indication of the moderator temperature coefficient of reactivity. It is evident therefore that, as far as the nuclear aspects are concerned, this coefficient may be positive or negative, depending upon various circumstances including the nature of the fissile species. It should be mentioned that the overall effect of the moderator temperature increase on reactivity includes the change in nonleakage probability, as will be seen shortly, and this is generally negative.

5.69. If the moderator is a liquid, increase of temperature will cause it to expand, and if the reactor has a constant volume, N_{tm} in equation (5.45) will decrease. Even if the volume is not constant, differences in the coefficients of expansion of the fuel and moderator may result in the inclusion of additional fuel in the core, so that there is an overall decrease in N_{tm} relative to N_{tu}. This effect will cause an additional increase in the positive temperature coefficient of the thermal utilization. The physical significance of this result is that, as the temperature rises, relatively more neutrons are absorbed in fuel than in the moderator. A similar situation occurs in connection with the coolant, assuming it to be different from the moderator. If the coolant is a fairly good neutron absorber, as is the case with ordinary water or sodium, for example, an increase of temperature will result in decreased absorption due to the decrease in density of the coolant. The latter will thus make a positive contribution to the temperature coefficient of thermal utilization.

5.70. Because the resonance escape probability decreases with temperature whereas the thermal utilization frequently increases, the net temperature coefficient of the infinite multiplication factor, which is the *sum* of the individual coefficients, may be positive or negative. As a general rule, the resonance escape contribution is numerically the larger of the two, so that the net coefficient is small but negative. However, this is not necessarily always the case, as will be seen later.

NONLEAKAGE PROBABILITIES

5.71. The nonleakage probability may be divided into two factors, namely, P_{sl}, the nonleakage probability while slowing down, which is equal to $e^{-B^2\tau}$ according to the age-diffusion treatment, and P_{th}, the nonleakage probability for thermal neutrons, which is $1/(1 + L^2B^2)$, as explained in § 4.38 *et seq.* Hence, the contribution of the nonleakage probability to the temperature coefficient of reactivity is

$$\frac{1}{P} \cdot \frac{dP}{dT} = \frac{1}{P_{sl}} \cdot \frac{dP_{sl}}{dT} + \frac{1}{P_{th}} \cdot \frac{dP_{th}}{dT},$$

where

$$\frac{1}{P_{sl}} \cdot \frac{dP_{sl}}{dT} = -B^2\tau \left(\frac{1}{\tau} \cdot \frac{d\tau}{dT} + \frac{1}{B^2} \cdot \frac{dB^2}{dT} \right) \tag{5.46}$$

and

$$\frac{1}{P_{th}} \cdot \frac{dP_{th}}{dT} = -\frac{L^2B^2}{1 + L^2B^2} \left(\frac{1}{L^2} \cdot \frac{dL^2}{dT} + \frac{1}{B^2} \cdot \frac{dB^2}{dT} \right). \tag{5.47}$$

The effect of the nonleakage probability on the temperature coefficient of reactivity is thus dependent on the temperature coefficients of the Fermi age, τ, of the square of the diffusion length of thermal neutrons, L^2, and of the reactor buckling, B^2.

5.72. According to the definition of the Fermi age, by equation (3.60), the temperature dependent quantity, disregarding the small change in the energy range, is D/Σ_s. By equation (3.6), D is inversely proportional to $1/\Sigma_{tr}$, and so

$$\frac{1}{\tau} \cdot \frac{d\tau}{dT} = \Sigma_s \Sigma_{tr} \frac{d(1/\Sigma_s \Sigma_{tr})}{dT}.$$

Since Σ_s and Σ_{tr} are determined almost entirely by the moderator, $\Sigma_s = N_m \sigma_s$ and $\Sigma_{tr} = N_m \sigma_{tr}$. Both σ_s and σ_{tr} are essentially independent of temperature in the range of interest; hence,

$$\frac{1}{\tau} \cdot \frac{d\tau}{dT} = 2N_m \frac{d(1/N_m)}{dT},$$

where N_m is the number of moderator nuclei per unit volume. The mass of moderator, and hence the total number of nuclei, is taken to be constant, so that N_m is inversely proportional to the volume; consequently,

$$\frac{1}{\tau} \cdot \frac{d\tau}{dT} = \frac{2}{V} \cdot \frac{dV}{dT}.$$

If the temperature coefficient of volume expansion of the moderator is represented by α_m, i.e.,

$$\alpha_m = \frac{1}{V} \cdot \frac{dV}{dT},$$

it follows that

$$\frac{1}{\tau} \cdot \frac{d\tau}{dT} = 2\alpha_m. \tag{5.48}$$

5.73. The square of the thermal diffusion length of the moderator, L_m^2, is equal to $1/3\Sigma_s\Sigma_a$, i.e., $1/3(N_m\sigma_{sm})(N_m\sigma_{am})$, as seen in § 3.37. Assuming σ_s to be independent of temperature and σ_a to have a $1/v$ variation, it is readily found by using the same procedure as given above that

$$\frac{1}{L_m^2} \cdot \frac{dL_m^2}{dT} = 2\alpha_m + \frac{1}{2T},$$

remembering that $1/v$ is proportional to $1/E^{1/2}$ and hence to $1/T^{1/2}$.

5.74. In a reactor, the square of the diffusion length, L^2, is related approximately to the value in the moderator, L_m^2, by equation (4.56), i.e.,

$$L^2 = L_m^2(1 - f),$$

so that

$$\frac{1}{L^2} \cdot \frac{dL^2}{dT} = 2\alpha_m + \frac{1}{2T} - \frac{1}{1-f} \cdot \frac{df}{dT}. \tag{5.49}$$

In a homogeneous reactor the additional term can be neglected, since df/dT is small, as noted in § 5.62, but in a heterogeneous system df/dT has a significant positive value. The temperature coefficient of L^2 is then less than in a homogeneous reactor with the same moderator.

5.75. It is seen from Table 4.1 that B^2 is inversely related to the square of one or more of the reactor's linear dimensions; in general, therefore, B^2 is proportional to $1/u^2$, where u has the dimensions of length. Hence,

$$\frac{1}{B^2} \cdot \frac{dB^2}{dT} = -\frac{2}{u} \cdot \frac{du}{dT}$$

$$= -\frac{2}{3}\alpha_r, \tag{5.50}$$

where α_r is the coefficient of cubical expansion of the reactor as a whole; for many cases this will be almost the same as the coefficient of expansion of the moderator.

5.76. By inserting the results of equations (5.48), (5.49), and (5.50) into equations (5.46) and (5.47), it is found that

$$\frac{1}{P_{\rm sl}} \cdot \frac{dP_{\rm sl}}{dT} = -B^2\tau\left(2\alpha_m - \frac{2}{3}\alpha_r\right) \tag{5.51}$$

and

$$\frac{1}{P_{\rm th}} \cdot \frac{dP_{\rm th}}{dT} = -\frac{L^2B^2}{1 + L^2B^2}\left(2\alpha_m + \frac{1}{2T} - \frac{2}{3}\alpha_r - \frac{1}{1-f} \cdot \frac{df}{dt}\right). \tag{5.52}$$

If the reactor core volume is independent of temperature, B^2 will be constant and the terms involving α_r can be omitted. Apart from the effect of the thermal utilization in a heterogeneous reactor, which is generally small in comparison with the other terms in equation (5.52), it is evident that the nonleakage probability factors both make negative contributions to the temperature coefficient of reactivity. A large temperature coefficient of thermal expansion of the moderator is advantageous in this respect, as also is a large value for the buckling, i.e., a small reactor core.

REACTOR TEMPERATURE COEFFICIENTS

5.77. Since the factors which determine reactivity temperature coefficients are themselves dependent on the temperature, the coefficients will also vary with temperature to some extent. However, for comparative purposes, at least, it is convenient to list the values at ordinary temperatures, as in Table 5.4 which

TABLE 5.4.   REACTIVITY TEMPERATURE COEFFICIENTS

Reactor	Temperature Coefficient Per °C	Per °F
Brookhaven (graphite moderated): Fuel.......	-2×10^{-5}	-1×10^{-5}
Overall.....	-4×10^{-5}	-2×10^{-5}
SRE (sodium-graphite): Fuel.................	-1.4×10^{-5}	-7.9×10^{-6}
Overall..............	$+1.2 \times 10^{-5}$	$+6.7 \times 10^{-6}$
Calder Hall (graphite, gas-cooled).............	-6×10^{-5}	-3×10^{-5}
OMRE (organic moderated and cooled)........	$+3.5 \times 10^{-4}$	$+1.8 \times 10^{-4}$
Shippingport (pressurized water)..............	-5.5×10^{-4}	-3.0×10^{-4}
Argonne CP-5 (heavy-water moderated).......	-4×10^{-4}	-2.2×10^{-4}
Water Boiler (homogeneous)...................	-3.0×10^{-4}	-1.7×10^{-4}
EBR-I (fast)................................	-3.5×10^{-5}	-1.9×10^{-5}
Enrico Fermi (fast).........................	-1.8×10^{-5}	-1.0×10^{-5}

refers to reactors of a number of different types. For the original Brookhaven (natural uranium-graphite) reactor and for the SRE, values are given for the fuel, as well as for the overall coefficient. It should be noted that for the SRE, which is a heterogeneous reactor with a graphite moderator and sodium coolant, the overall coefficient of reactivity is positive, at ordinary temperatures, but the fuel coefficient is negative. This means that the positive contribution to the reactivity temperature coefficient made by the moderator exceeds the negative contribution of the fuel in this reactor. The same is probably true of the OMRE, which has an overall positive temperature coefficient at ordinary temperatures. At higher temperatures, however, both the SRE and OMRE have negative coefficients, probably because of the increased effect of neutron leakage [10].

5.78. A negative temperature coefficient of reactivity is desirable since it tends to counteract the effects of transient temperature changes during reactor operation. Actually, it is the fuel coefficient which is significant in this respect, and this aspect of the problem will be discussed below. If, for any reason, the effective multiplication factor of a steady-state reactor suffers a transient increase, there will be a corresponding increase in the neutron density and, hence, in the fission rate. As a result, the temperature of the reactor will tend to rise. But if the (fuel) temperature coefficient of reactivity is negative, this will cause the effective multiplication factor to decrease, thus offsetting the original increase. If the (fuel) reactivity temperature coefficient were positive, any disturbance producing an increase in reactor temperature would result in an increase in fission rate and a continuously rising temperature. Such a reactor would be unstable to temperature changes and its operation would be unsafe.

5.79. It is seen from Table 5.4 that the sodium-graphite (SRE) reactor has a positive overall temperature coefficient of reactivity, yet this reactor is completely stable against temperature transients. This is because the fuel has a negative temperature coefficient and the time constant, i.e., the response time, of the fuel is considerably less than for the moderator. When an increase takes place in the fission rate, the temperature of the fuel will rise immediately, but that in the moderator will be delayed because of its large heat capacity and moderate thermal conductivity. In the SRE, for example, the time constant for the fuel temperature coefficient is about 2 sec, whereas that for the reactor is approximately 3 min [11]. Hence, a transient increase in temperature of the fuel will be counteracted promptly by the effect of the negative temperature coefficient, long before the moderator can respond with its positive coefficient. The reactor is thus inherently stable against transient temperature changes.

5.80. The foregoing situation would be applicable, in general, to any heterogeneous reactor having a fairly large mass of moderator. A special situation could arise, however, in a reactor in which ordinary water or sodium is the coolant within the reactor core while neutron moderation takes place in another medium, e.g., heavy water, beryllium, or graphite. As seen in § 5.69, the water or sodium can make a positive contribution to the reactivity temperature coefficient. Because of the relatively small mass of coolant and its proximity to the fuel, the time constant of this coefficient would be fairly small. Care must be taken, therefore, in the design of such reactors to ensure that the negative coefficient of the fuel exceeds the positive coefficient of the coolant.

5.81. During the operation of a reactor, the temperature coefficient of reactivity can undergo significant changes. In a reactor utilizing natural or slightly enriched uranium as fuel, both uranium-235 and uranium-238 are consumed but some plutonium-239 is produced. The alteration in the composition of the fuel material will affect the temperature coefficient in various ways. The most important is the change in the moderator coefficient resulting from the partial replacement of uranium-235 by plutonium-239 as the fissile material. It was stated in § 5.67 that the contribution of the latter is positive, whereas that of the former is negative. Consequently, it has been observed that in reactors employing natural or very slightly enriched uranium as fuel, the temperature coefficient of the moderator, which may be initially negative, can become positive after a period of operation. This change has an insignificant effect on the operational safety, because of the relatively long time constant of the moderator response. By using moderately (or highly) enriched uranium as fuel, sufficient uranium-235 will always be present to prevent the moderator temperature coefficient from becoming positive during the lifetime of the fuel.

5.82. In some water-moderated and cooled thermal reactors a neutron poison, such as boric acid, is added to the water to compensate for the excess reactivity of the freshly charged reactor (§ 5.149). There are then two additional effects which make positive contributions to the temperature coefficient. First, an increase of temperature will cause the solution to expand, thereby decreasing the amount of poison within the reactor, and second, the absorption cross section of the poison decreases. The net effect is markedly to decrease the effectiveness of the poison with increasing temperature, so that the reactivity increases correspondingly.

5.83. Because the great majority of reactors have overall negative temperature coefficients, the effective multiplication factor will be less at higher temperatures than at ordinary temperatures. During startup, the temperature of a reactor—especially one designed for producing useful power—inevitably increases, so that k_{eff} decreases. In order that the effective multiplication factor may exceed unity at the operating temperature, it is necessary, therefore, to include additional reactivity in the cold reactor. In the Yankee Atomic Power (pressurized water) reactor, for example, the average temperature coefficient of reactivity, over the range between ordinary and operating temperatures, is estimated to be -1.6×10^{-4} per °F. The difference between these two temperatures has an average value of about 450°F, and so the additional reactivity required is $(450)(1.6 \times 10^{-4})$, i.e., 0.072.

5.84. It should be pointed out, however, that this estimate is based on the isothermal temperature coefficient, as given in Table 5.4, involving the assumption that the fuel and moderator are at the same temperature. This is referred to as the "zero power" condition. When operating at appreciable power, the temperature of the fuel in a heterogeneous reactor is invariably greater than that of the moderator, because of the inevitable delay in the transfer of heat from the fuel, where it is generated, to the moderator. Since the fuel temperature coefficient is negative, additional reactivity is required in the cold reactor to permit operation at power. The negative fuel coefficient arises almost entirely from the Doppler broadening of the capture resonances in uranium-238, and so the extra reactivity is ascribed to what is called the power Doppler effect. In a pressurized water reactor, such as was referred to above, this may amount to about 0.03 or so.

5.85. The preceding discussion has had particular reference to thermal reactors. In a fast reactor there is no moderator, and so the overall coefficient is essentially that of the fuel, apart from the effect of changes in the nonleakage probability. The fuel contains both fissile (uranium-235 or plutonium-239) and fertile (uranium-238) species and the net reactivity temperature coefficient depends largely on the opposing effects of Doppler broadening on neutron absorption in the two nuclides. An increase of temperature will result in enhanced fission absorption in the fissile species, leading to a positive coefficient, and increased nonfission capture in the fertile material, accompanied by a negative reactivity coefficient. Consequently, in order to obtain the necessary negative temperature coefficient in the fuel, there must be an upper limit to the ratio of fissile to fertile nuclei. It has been estimated that, for safe operation, the proportion of fissile nuclides in the fuel should not exceed about 50 per cent.* In actual fast reactors, especially those designed for breeding and the production of economical power, it is desirable to keep the proportion of fissile material in the fuel at a low value, in any event, e.g., below 25 per cent. Hence, the temperature coefficients of reactivity are negative, although relatively small, as seen in Table 5.4.

FISSION PRODUCT POISONING

EFFECT OF POISONS ON REACTIVITY

5.86. During the course of operation of a nuclear reactor, the fission fragments and their many decay products accumulate. Among these substances there are some, xenon-135 and samarium-149, in particular, which have large cross sections for thermal-neutron absorption. These nuclei, therefore, act as reactor poisons and affect the multiplication factor, chiefly by decreasing the thermal utilization.

---

* This proportion, which applies to the uranium-238 and plutonium-239 system, is not known precisely [12]. In any event, it depends on the fast-neutron spectrum in the reactor and so varies with the composition of the core (cf. § 4.149).

The concentration of fission product poisons in a reactor is related to the thermal-neutron flux (or density). Consequently, when the reactivity is changed, so that there is an accompanying change in the neutron density, the concentration of fission product poisons will be affected and this will, in turn, influence the reactivity. It would appear, therefore, that some allowance for the effect of poisoning should be included in the reactor kinetics equations considered in the early sections of this chapter. However, the rate of change of fission product concentration with time is, in general, small compared to that of the neutron density. Hence, the equations giving the time rate of change of neutron density may be treated quite independently of those for the fission product poisons.

5.87. Although fission products thus have little direct effect on reactor kinetics, they may have an important influence on the reactivity, and this must be taken into consideration in designing both the reactor core and the control system. A specific fission product, such as xenon-135, can be formed both directly and indirectly in fission and be removed by radioactive decay and by neutron capture. As a result of the two opposing types of reactions, the concentration will attain an equilibrium value while the reactor operates at a specific power level. When the reactor is shut down, however, indirect formation of xenon-135, by the decay of its parent (iodine-135), will continue, but the main method for its removal, by neutron capture, will be greatly decreased because of the small neutron flux. As a result, the concentration of the fission product may increase to a maximum before finally decreasing. Both the equilibrium amount during operation and the maximum after shutdown are determined by the neutron flux (or power output) of the reactor. The effect on the reactivity of a particular fission product poison will thus depend on the conditions of operation of the reactor as well as on the nature of the poison [13].

5.88. Of the four factors which make up the infinite multiplication factor (§ 4.13), the thermal utilization is essentially the only one affected by the poison. Let f be the thermal utilization in a reactor without poison and let f' be the value with poison; assuming, for simplicity, that the system is homogeneous, then

$$f = \frac{\Sigma_u}{\Sigma_u + \Sigma_m} \quad \text{and} \quad f' = \frac{\Sigma_u}{\Sigma_u + \Sigma_m + \Sigma_p},$$

where Σ_u, Σ_m, and Σ_p are the macroscopic cross sections for the absorption of thermal neutrons in fuel, moderator, and poison, respectively. The *poisoning*, ψ, of a reactor has been defined as the ratio of the number of thermal neutrons absorbed by the poison to those absorbed in the fuel; hence, in the present case,

$$\psi \equiv \frac{\Sigma_p}{\Sigma_u}. \tag{5.53}$$

5.89. Because of the change in the thermal utilization, the poison will affect the thermal diffusion length [cf. equation (4.56)] and, consequently, the nonleakage probability of thermal neutrons. The effect is small, however, and for the present purpose it is reasonable to make the approximation of taking the effective multiplication factor to be proportional to the thermal utilization, with or without poison. Let the effective multiplication factor without poison be k_{eff} and that with poison k'_{eff}; then

$$\frac{k'_{eff} - k_{eff}}{k'_{eff}} = \frac{f' - f}{f'} = -\frac{\psi z}{1 + z}, \tag{5.54}$$

where z is used to represent the ratio Σ_u/Σ_m. If the reactor without poison is just critical, so that k_{eff} is unity, the left side of equation (5.54) becomes equivalent to the (negative) reactivity change due to the poison; thus,

$$\rho = -\frac{\psi z}{1 + z}.$$

In an enriched reactor, in particular, z is large in comparison with unity, and so the reactivity decrease resulting from the presence of a poison is roughly equal to the poisoning, ψ, i.e., to the ratio of Σ_p to Σ_u.

XENON POISONING DURING OPERATION

5.90. The most important fission product poison is xenon-135 because of its exceptionally large capture cross section—about 3.0×10^6 barns—for thermal neutrons. This isotope is formed to a small extent (about 0.2 per cent) as a direct product of fission, but the main proportion in a reactor originates from the radioactive decay of tellurium-135 and iodine-135, produced in 6.1 per cent of the slow-neutron fissions of uranium-235. The negative beta decay stages are as follows:

$$\text{Te}^{135} \xrightarrow{<1m} \text{I}^{135} \xrightarrow{6.7\,h} \text{Xe}^{135} \xrightarrow{9.2\,h} \text{Cs}^{135} \xrightarrow{2 \times 10^6\,y} \text{Ba}^{135} \text{ (stable)}.$$

5.91. Because tellurium-135 has such a short half-life, the analytical treatment of xenon poisoning may be simplified, without introducing appreciable error, by assuming that iodine-135 is produced directly in fission with a fractional yield of 0.061. According to the argument in § 2.190, the net rate of formation of iodine-135 may thus be represented by

$$\frac{dI}{dt} = -\lambda_I I - \sigma_I \phi I + \gamma_I \Sigma_f \phi, \tag{5.55}$$

where I is the concentration of the iodine-135, λ_I is its radioactive decay constant, σ_I its thermal-neutron absorption cross section, and γ_I is the fission yield, i.e., 0.061; Σ_f is the macroscopic thermal fission cross section of the fuel material in the reactor and ϕ is the thermal-neutron flux. The first two terms on the right of equation (5.55) give the rate of removal of iodine-135, by radioactive decay and neutron capture, respectively, whereas the third term represents its rate of formation by fission. The equilibrium concentration, I_0, attained after the reactor has been operating for some time, is obtained by setting dI/dt equal to zero, so that

$$I_0 = \frac{\gamma_I \Sigma_f \phi}{\lambda_I + \sigma_I \phi} \approx \frac{\gamma_I \Sigma_f \phi}{\lambda_I}, \tag{5.56}$$

since σ_I is quite small.

5.92. The rate of increase of the xenon-135 concentration is given by an expression similar to equation (5.55) with the addition of a term on the right side to express the rate of formation by the decay of iodine-135; thus,

$$\frac{dX}{dt} = -\lambda_X X - \sigma_X \phi X + \lambda_I I + \gamma_X \Sigma_f \phi, \tag{5.57}$$

where the symbol X is used to represent properties associated with xenon-135. The direct fission yield of xenon-135, represented by γ_X, is 0.002. The equilibrium concentration, X_0, of xenon is obtained from equation (5.57) as

$$X_0 = \frac{\lambda_I I_0 + \gamma_X \Sigma_f \phi}{\lambda_X^*} = \frac{(\gamma_I + \gamma_X)\Sigma_f \phi}{\lambda_X^*}, \tag{5.58}$$

where $\lambda_X^* \equiv \lambda_X + \sigma_X \phi$; because σ_X is so high, λ_X^* is appreciably greater than λ_X.

5.93. The corresponding equilibrium value of the poisoning is then

$$\psi_0 = \frac{X_0 \sigma_X}{\Sigma_u} = \frac{\sigma_X(\gamma_I + \gamma_X)\Sigma_f \phi}{\lambda_X^* \Sigma_u}. \tag{5.59}$$

The ratio Σ_f/Σ_u is equal to $\sigma_f/(\sigma_f + \sigma_c)$, where σ_f and σ_c are the fission and capture (nonfission) cross sections, respectively; for uranium-235, these are given in Table 2.8 as 577 and 106 barns, so that Σ_f/Σ_u is 0.83.† As seen above, $\gamma_I + \gamma_X$ is $0.061 + 0.002$, i.e., 0.063; σ_X is 3.0×10^6 barns, i.e., 3.0×10^{-18} cm²; and λ_X, the radioactive decay constant of xenon-135, is 2.1×10^{-5} sec⁻¹. Equation (5.59) can thus be written as

$$\psi_0 = \frac{1.57 \times 10^{-19}\phi}{(2.1 \times 10^{-5}) + (3.0 \times 10^{-18}\phi)}, \tag{5.60}$$

recalling that $\lambda_X^* = \lambda_X + \sigma_X \phi$. Some values of ψ_0, calculated from equation (5.60), for various steady-state neutron fluxes, are given in Table 5.5 to two

TABLE 5.5. EQUILIBRIUM VALUES OF XENON POISONING
DURING OPERATION OF REACTOR

Thermal Flux (ϕ)	Poisoning (ψ_0)
10^{12}	0.0065
10^{13}	0.031
10^{14}	0.049
10^{15}	0.052

† The 2200-meters/sec cross sections are used here, since a ratio is involved. The xenon cross section, σ_X, however, is a thermal average value.

significant figures. It is seen that the equilibrium poisoning is small for a thermal flux of 10^{12} neutrons/(cm²)(sec), but increases rapidly at higher fluxes and approaches a limiting value.

5.94. If ϕ is about 10^{11} (or less) neutrons/(cm²)(sec), the second term in the denominator of equation (5.60) may be neglected in comparison with the first; then

$$\psi_0 \approx 8 \times 10^{-15}\phi,$$

and the poisoning will be negligible, i.e., 8×10^{-4} or less. Even for a flux of 10^{12} neutrons/(cm²)(sec), the poisoning is only 0.0065, so that about 0.65 per cent of the thermal neutrons are absorbed by the equilibrium amount of xenon. However, for values of ϕ greater than 10^{12}, the poisoning increases rapidly, at first (Fig. 5.6). If the flux is 10^{15} (or more) neutrons/(cm²)(sec), λ_X can be

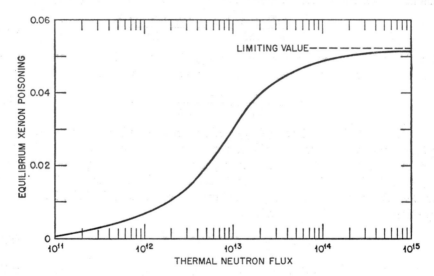

Fig. 5.6. Equilibrium xenon poisoning during reactor operation

neglected in comparison with $\sigma_X\phi$, so that $\lambda_X^* \approx \sigma_X\phi$; the poisoning, as given by equation (5.59), then reaches a limiting value expressed by

$$\psi_{\text{lim}} \approx (\gamma_I + \gamma_X)\frac{\Sigma_f}{\Sigma_u} = (0.063)(0.83) = 0.052.$$

This is the maximum value of the poisoning *during operation* for thermal reactors, with uranium-235 as fissile material, no matter how high the flux. In view of the relationship between the poisoning and reactivity derived in § 5.89, it is seen that the maximum decrease in reactivity due to xenon poisoning during reactor operation is about 0.052. The maximum after shutdown can, however, be many times this value.

XENON POISONING AFTER SHUTDOWN

5.95. Because iodine-135 has a shorter half-life than xenon-135, i.e., $\lambda_I > \lambda_X$, the conditions are suitable for the xenon concentration to increase to a maximum after the reactor is shut down completely. This occurs because the iodine-135 present at shutdown forms xenon-135 by radioactive decay at a rate that is initially greater than the rate of decay of the xenon-135, almost none of the latter now being lost by neutron capture. If equilibrium has been attained before shutdown, the concentration of iodine-135 at any time t_s after shutdown is given by

$$I = I_0 e^{-\lambda_I t_s}. \tag{5.61}$$

If the neutron flux after shutdown is assumed to drop to zero, the terms containing ϕ in equation (5.57) are eliminated and substitution of equation (5.61) for I then yields

$$\frac{dX}{dt} = -\lambda_X X + \lambda_I I_0 e^{-\lambda_I t_s}. \tag{5.62}$$

Upon multiplying through by the integrating factor $e^{\lambda_X t_s}\, dt$, equation (5.62) can be solved in the manner described in § 2.191; the result is

$$X(t_s) = \frac{\lambda_I}{\lambda_X - \lambda_I}\, I_0(e^{-\lambda_I t_s} - e^{-\lambda_X t_s}) + X_0 e^{-\lambda_X t_s}, \tag{5.63}$$

where $X(t_s)$ is the xenon-135 concentration at time t_s after shutdown. The values of I_0 and X_0, attained at equilibrium before shutdown, are expressed by equations (5.56) and (5.58), respectively. The first term on the right of equation (5.63) gives the xenon concentration due to the decay of the iodine-135 followed by decay of the xenon-135 thus formed after shutdown. The second term, on the other hand, represents the decay of the xenon present in the reactor prior to shutdown. Since I_0 and X_0 depend on the neutron flux during operation, it is evident that this will also affect the xenon "buildup," as it is called, after shutdown.

5.96. In the derivation of equation (5.63), it was assumed that the flux dropped to zero immediately when the reactor was shut down. Although this is not strictly true for a reactor, it is an adequate approximation for the purpose at hand. Taking into consideration the initial rapid drop and the subsequent stable negative period of 80 sec due to delayed neutrons (§ 5.47), it is found that a reactor can be shut down from a flux of 10^{12} to 10^3 neutrons/(cm²)(sec) in about 20 min. This time is short compared with the several hours during which the xenon concentration builds up, and so very little error results from the assumption that the neutron flux drops immediately to zero when the reactor is shut down.

5.97. The time to attain the maximum concentration of xenon after shutdown, mentioned in § 5.87, can be found by differentiating equation (5.63); the result is found to be

$$t_{max} = \frac{1}{\lambda_X - \lambda_I} \ln \frac{\lambda_X}{\lambda_I} \left(1 - \frac{\lambda_X - \lambda_I}{\lambda_I} \cdot \frac{X_0}{I_0}\right). \tag{5.64}$$

If the value of t_{max} obtained from this equation is inserted in equation (5.63), the maximum xenon poisoning after shutdown can be calculated for any given operating flux. The procedure is illustrated in the following example.

Example 5.4. A thermal reactor using uranium-235 as fissile material has been operating for some time at an average flux of 2×10^{14} neutrons/(cm²)(sec); how long after shutdown will the xenon poisoning reach a maximum and what is the poisoning at this time?

The time to attain the poisoning maximum is given by equation (5.64), with λ_I equal to 2.9×10^{-5} sec^{-1} (derived from the half-life of 6.7 hr) and λ_X given above as 2.1×10^{-5} sec^{-1}. The values of I_0 and X_0 can be obtained from equations (5.56) and (5.58), respectively; however, the ratio X_0/I_0 which appears in equation (5.64) is readily shown to be so small, for an operating flux of 2×10^{14} neutrons/(cm²)(sec), that equation (5.64) reduces to

$$t_{max} \approx \frac{1}{\lambda_X - \lambda_I} \ln \frac{\lambda_X}{\lambda_I} = \frac{1}{0.8 \times 10^{-5}} \ln \frac{2.9}{2.1}$$

$$= 4 \times 10^4 \text{ sec}$$
$$= 11 \text{ hr.}$$

The poisoning is found by writing equation (5.53) in the general form

$$\psi(t) = \frac{X(t)\sigma_X}{\Sigma_u},$$

with $X(t)$ given by equation (5.63). However, some simplification in the latter is possible because, as seen above, X_0 is small in comparison with I_0 at the specified flux; hence, the last term on the right of equation (5.63) can be neglected. The quantities λ_I, λ_X, $t_s(4 \times 10^4$ sec), and σ_X are known, but I_0 (or rather I_0/Σ_u) must be evaluated from equation (5.56); for this purpose γ_I is taken as 0.061, ϕ as 2×10^{14}, and Σ_f/Σ_u as 0.83. Upon substituting into the expression given above, the poisoning at 11 hr after shutdown is found to be 0.46.

5.98. The xenon poisoning, $\psi(t_s)$, at time t_s after shutdown is given by equation (5.53) as $X(t_s)\sigma_X/\Sigma_u$, where $X(t_s)$ can be obtained from equation (5.63). It is thus possible to calculate $\psi(t_s)$ at various times after shutdown for different values of the neutron flux before shutdown. The results for thermal fluxes of 10^{13}, 10^{14}, and 2×10^{14} neutrons/(cm²)(sec) are indicated by the curves in Fig. 5.7. For fluxes of the order of 10^{13} (or less) neutrons/(cm²)(sec), the increase in poisoning after shutdown is seen to be negligible, but it assumes greater significance for higher operating fluxes. Thus, for a flux of 2×10^{14} neutrons/(cm²) (sec), the poisoning maximum, attained about 11 hours after shutdown is 0.46; this may be compared with the equilibrium value of less than 0.05 when the reactor was in operation. For higher fluxes, the maximum poisoning after shutdown is approximately proportional to the flux.

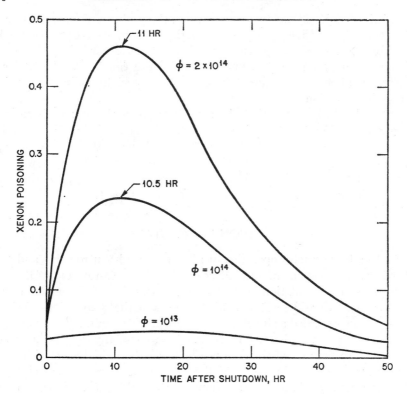

FIG. 5.7. Xenon poisoning after shutdown for various operating fluxes

5.99. In view of the approximate equality of reactivity decrease and the poisoning, it is evident that if a high-flux reactor is to be capable of being re-started at any time after shutdown, it must have an appreciable amount of excess multiplication (or reactivity), in the form of additional fuel, built into it. In order that a thermal reactor, normally operating at a flux of 2×10^{14} neutrons/(cm²)(sec), shall be able to "override" the maximum xenon concentration after shutdown, it must have about 0.46 of excess reactivity available.

5.100. As an alternative to the capability of overriding the xenon poisoning at any time, a reactor may be designed with sufficient reactivity available to override the xenon within a certain limited time only, e.g., within 1 hour, after shutdown.   For a flux of 2×10^{14} neutrons/(cm²)(sec), this is considerably less than the maximum of 0.46 referred to above.   However, should the delay in restarting be greater than the specified time, it may be necessary to wait a day or more before the reactor can be started up again.   As soon as the flux becomes appreciable after startup, the xenon concentration will decrease rapidly toward the equilibrium value for the particular steady-state flux.   Because of the rapid increase in the xenon poisoning and the high maximum for thermal fluxes in excess of about 2×10^{14} neutrons/(cm²)(sec), the problem of startup after shutdown becomes increasingly difficult. For this reason, an average thermal flux of roughly 2×10^{14} neutrons/(cm²)(sec) is generally regarded as the practical maximum for a power reactor, unless a procedure can be developed for continuous removal of the progenitors of xenon-135, especially iodine-135, while the reactor is in operation.

5.101. It should be noted that the equations given above have involved the assumption of uniform distribution of neutron flux, as well as of iodine-135 and xenon-135, throughout the reactor. If ϕ is taken as the average flux, the resulting error in the value of the poisoning will be no more than a few per cent. A more exact treatment is possible, using perturbation theory, but the method used here will suffice for all preliminary purposes, at least.

XENON INSTABILITY

5.102. In a large reactor operating at a high thermal-neutron flux and at constant power, the phenomenon called *xenon instability* can occur [14]. This is apparent as a slow oscillation in the spatial distribution of the neutron flux (or reactor power). Consider a reactor which is so large that two regions, designated I and II, can be regarded as functioning as independent units; that is to say, the neutrons produced by fission in one region do not cause any significant fissions in the other region because of the distance between them. This condition would be realized if a linear dimension of the core, e.g., diameter or height, were several times greater than the migration length (Table 3.5). Assume that the reactor has been operating for some time so that the xenon concentration has attained equilibrium, and suppose that the neutron flux in region I undergoes a small increase.

5.103. As a result of the increase in flux, the rate of consumption of xenon-135 in region I increases promptly because of increased neutron capture. The rate of formation of iodine-135, the parent of the xenon, increases at the same time, but since the former has a half-life of 6.7 hours, there is a considerable delay between the increase in the flux in region I and the associated increase in the rate of xenon-135 formation. Consequently, the net prompt result is that the xenon concentration decreases. This has the effect of decreasing the neutron capture in region I, so that the flux increases still further. The result is a continued decrease in the xenon concentration and a steady increase in the thermal flux until the delayed production of xenon, by the decay of the increasing concentration of iodine-135, brings about an increase in the amount of xenon-135. Then the neutron flux in region I will start to decrease.

5.104. In the meantime, if the reactor power is being maintained constant, the increase in flux in region I will be offset by a decrease in region II. This decrease will continue until it is reversed, as described above, by the delayed production of xenon-135 in region I. Then the neutron flux in region II will commence to increase and will go through the same phases as did region I when its flux was increasing. In due course, therefore, delayed production of xenon by the decay of iodine-135 in region II will cause another reversal of the flux, and so on. Consequently, a continuous series of thermal-neutron flux (and reactor power) oscillations, having a period of about a day, will occur between regions I and II.

5.105. The reactor oscillations arising from xenon-135 are not a nuclear hazard, in the sense that there is no danger that the reactor will become super-critical. The main problem is that a local increase in the neutron flux means that fission heat is being generated more rapidly than is expected, and provided for, in the reactor design. There is thus a possibility of damage resulting from some of the fuel elements becoming overheated. In addition to the requirement of a large reactor and constant operating power, xenon instability can occur only if the neutron flux is sufficiently high to make the rate of consumption of xenon-135 by neutron capture large in comparison with the rate of decay; this condition requires the thermal flux to be appreciably greater than 10^{13} neutrons/(cm^2)(sec). A large negative temperature coefficient of reactivity could overcome xenon instability because the changes in the temperature would oppose the flux changes due to the xenon. Apart from this self-regulating possibility, xenon oscillations can be avoided by having neutron flux measuring instruments distributed over the reactor so that local changes could be detected and compensated by the movement of control rods.

SAMARIUM POISONING

5.106. Next to xenon-135, the most important fission product poison is samarium-149. This is a stable isotope with a capture cross section of 5×10^4 barns for thermal neutrons. It is the end product of the decay chain

$$\mathrm{Nd}^{149} \xrightarrow{\ 2.0\mathrm{h}\ } \mathrm{Pm}^{149} \xrightarrow{\ 53\mathrm{h}\ } \mathrm{Sm}^{149}\ (\text{stable}),$$

which occurs in about 1.1 per cent of uranium-235 fissions by slow neutrons. Since the half-life (2.0 hours) of neodymium-149 is short in comparison with that of promethium-149 (53 hours), it may be supposed that the latter is a direct product of fission, its fractional yield being 0.01. The problem of the formation of promethium and of samarium in a reactor is thus analogous to that considered above for iodine and xenon, respectively. As before, the capture cross section of promethium-149 is so small that $\sigma\phi$ can be neglected in comparison with λ. Because the samarium-149 is not radioactive, it is found that the equilibrium concentration and the poisoning during reactor operation are independent of the neutron flux.

5.107. Taking γ for promethium as 0.01 and for samarium as zero, the values of λ as 3.6×10^{-6} sec^{-1} and zero, respectively, since samarium-149 is a stable isotope, and σ for samarium-149 as 5×10^{-20} cm^2, the equilibrium value of the poisoning is calculated to be about 0.01. The maximum change in reactivity due to samarium in an operating reactor is thus -0.01, regardless of the thermal-neutron flux.

5.108. After shutdown, the samarium concentration does not pass through a maximum, as does that of xenon, but it increases toward an asymptotic value, ψ_∞, represented by

$$\psi_\infty = \frac{\Sigma_f}{\Sigma_u}\left(\frac{\sigma_S \phi \gamma_P}{\lambda_P} + \gamma_P\right),$$

where the subscripts P and S stand for promethium and samarium, respectively; ϕ is the steady-state flux prior to instantaneous shutdown. Inserting the known values for Σ_f, Σ_u, σ_S, γ_P, and λ_P, it is found that

$$\psi_\infty = 1.3 \times 10^{-16}\phi + 0.01.$$

Thus, for a steady-state flux of 2×10^{14} neutrons/(cm²)(sec), which appears to be the practical limit for a thermal power reactor, the samarium poisoning after instantaneous shutdown increases to 0.036. The presence of samarium-149 in a reactor thus requires, at most, an addition of 0.04 to the reactivity, so that samarium poisoning is a minor problem compared with that of xenon poisoning. It should be noted that the results in this section, like those for xenon poisoning, refer to uranium-235 as the fissile species.

OTHER POISONS PRODUCED BY FISSION

5.109. Although xenon makes the major contribution to fission product poisoning of a reactor operating at a flux of 10^{13} neutrons/(cm²)(sec) or more, there are many other nuclides which accumulate during reactor operation and cause a decrease in the reactivity [15]. The method for determining the poisoning of each nuclide is the same as that described earlier for iodine-135 and xenon-135, starting with balance equations of the form of equations (5.55) and (5.57). Since there are some 200 species present among the fission products after a short time, the summation of the effects of all these substances would be very tedious, even if all the fission yields, capture cross sections, and radioactive decay constants were known. However, a selection has been made of about 100 nuclides which represent the most significant poisons, based on their high fission yield, large capture cross section, or relatively long half-life. The poisoning caused by these nuclides has been determined for a completely thermal neutron flux and for various partially thermalized distributions.

5.110. In addition to fission products, the buildup of higher isotopes, as they are called, of uranium, neptunium, and plutonium has a significant effect on the reactivity of the fuel in a reactor during operation. These isotopes are formed by various neutron reactions, e.g., (n, γ) and $(n, 2n)$, starting with uranium-235 and -238, sometimes followed by beta decay. Although the higher isotopes are not fission products in the strict sense, they are produced in a fission reactor and contribute to the overall poisoning by absorbing neutrons. Consequently, it is appropriate to consider them here. The most important of these nuclides of high mass number as far as poisoning is concerned are uranium-236, neptunium-237, and plutonium-239, -240, -241, and -242. Although plutonium-239 and -241 are fissile and thus add to the net reactivity, both of these isotopes have large nonfission capture cross sections so that they are also reactor poisons.

5.111. The results of calculations of poisoning in a completely thermalized reactor are shown in Fig. 5.8 [16]. The poisoning, defined in accordance with

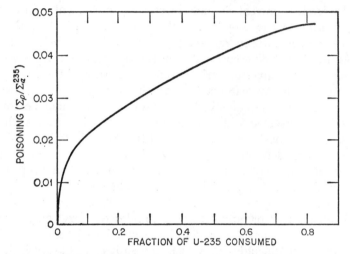

FIG. 5.8.   Poisoning due to fission products other than xenon-135

equation (5.53) as Σ_p/Σ_u, where Σ_u refers to the initial amount of fuel, is plotted against the fraction of uranium-235 consumed in the reactor. The poisoning effect of samarium-149 as well as that of the heavier isotopes is included, but the xenon-135 has been excluded. In order to simplify the computations, radioactive decay of the poisons is neglected. The probable error resulting from this neglect and from uncertainties in fission yields and cross sections is estimated to be not more than 10 to 15 per cent. For incompletely thermalized systems, the poisoning for any given degree of uranium-235 consumption is larger than shown in Fig. 5.8 because of the increased neutron capture in the resonance region. In a power reactor, the fraction of the fissile material used up prior to renewal of the fuel will generally not exceed about 0.2. The corresponding poisoning, apart from that due to xenon-135, will thus have a maximum of less than 0.03 in a completely thermal reactor. It might increase to 0.04 if there is a larger degree of neutron capture by epithermal neutrons.

GENERAL FEATURES OF REACTOR CONTROL [17]

BASIC PRINCIPLES OF CONTROL

5.112. The basic purpose of a reactor control system is to provide smooth and steady operation in all situations for which the device was designed. Methods must be available for starting the reactor, for bringing the power output to the desired level, to maintain it at that level, and then to shut down the reactor when necessary. In addition, as with other energy converters, safety devices must be provided to prevent damage in case of an accident. The control (or operation) of a nuclear reactor is thus the same, in principle, as the operation of other means for making energy available from a fuel, such as coal or oil.

5.113. An important difference between the control of a reactor and the control of other energy converters arises from the nature of the fuel. With but few exceptions, in the nuclear reactors which have so far been constructed or designed, the fuel supply, namely, the fissile material, cannot be continuously replaced as it is consumed. It is consequently necessary that such a reactor have included in its construction all the fuel that is likely to be required for the production of a predetermined quantity of energy. In the case of mobile reactors, such as for use on shipboard or for propulsion of an aircraft, this may be quite considerable, since sufficient fuel should be included to permit continuous operation over extended periods.

5.114. A nuclear reactor cannot be used to release energy continuously unless the critical mass for the particular fuel, shape, etc., is exceeded. For this reason a power reactor must include extra fuel, and so it must obviously have excess (or built-in) reactivity to an appreciable extent at the time it commences operation. In addition, other factors, such as the need to override the effect of fission product poisons and of temperature, require that additional reactivity be built into the reactor. Because of the high reactivity, the neutron flux (or density) and, hence, the fission rate and the power are capable of an extremely rapid, and dangerous, increase in a very short time. An essential aspect of reactor control thus lies in the precautions to be taken to prevent an excessive rate of increase in neutron flux when the power level is being raised.

5.115. Since the power level of a reactor is virtually proportional to the neutron flux, the obvious basis for reactor control is to vary the effective multiplication or reactivity. If the effective multiplication factor is greater than unity, the reactor is supercritical, and the power level will increase continuously. Upon decreasing the factor to unity, so that the reactor is just critical, the power output will remain constant (apart from transient changes) at the level attained at that time. Finally, by making the reactor subcritical, i.e., by reducing the effective multiplication factor below unity, the power level will be decreased.

METHODS OF CONTROL

5.116. There are four general methods which can be used to vary the effective multiplication factor (or reactivity); they involve addition or removal of (1) fuel, (2) moderator, (3) reflector, or (4) a neutron absorber. Each one of these methods, or a combination of them, has been used or proposed for the control of nuclear reactors already in existence or under construction.

5.117. The control procedure which has hitherto been widely employed in thermal reactors, because of its simplicity and convenience, is the insertion or withdrawal of a material, such as boron or cadmium, which has a large capture cross section for thermal neutrons. Rods or strips of boron steel or of Boral (boron carbide and aluminum) or of cadmium have been widely used for reactor control. The control rods may be located within the core or in the reflector close to the core where the thermal-neutron flux is high (§ 4.62). Because of its high melting point and other useful properties, the element hafnium is employed in water-cooled reactors. There is interest in the application of certain rare-earth oxides in reactors operating at very high temperatures.

5.118. The chief disadvantage of control by a strong absorber is the resulting loss of neutrons. This is not serious in systems having small excess reactivity and which have a low power output, for during operation the control rods are almost entirely withdrawn. However, such would not be the case for reactors having considerable built-in (excess) reactivity to take care of fuel depletion, fission product poisoning, etc., and which have high neutron fluxes. An improvement in neutron economy can be achieved by combining motion of the core material, i.e., fuel and moderator, with that of an absorber. The lower portion of a coarse-control rod may be constructed of the same material as the reactor core, whereas the upper portion contains cadmium. When such a rod is lowered, so that the neutron absorber is inserted, some of the core is removed at the same time, thereby bringing about a further decrease of reactivity.

5.119. Instead of using a wasteful absorber, such as cadmium or boron, there is a possibility that the control rods could be made of an absorber, such as uranium-238 (or natural uranium) or thorium, which would produce a fissile material—plutonium-239 or uranium-233, respectively—as the result of neutron capture and subsequent beta decay. Alternatively, the absorption of neutrons by the control rod could be used for the production of certain desired isotopes, such as cobalt-60. It is to be expected that, ultimately, control by absorption in nonproductive nuclei will be eliminated.

5.120. In a fast-neutron reactor, control by means of a neutron absorber is not generally satisfactory because of the low capture cross sections for neutrons of high energy. The reactivity can, however, be changed by removal of fuel material from (or by addition to) the core or by movement of part of the reflector. It should be noted that, when fuel or reflector is used for control purposes, its direction of motion will be opposite to that of an absorber. In order to decrease the reactivity, fuel or reflector must be removed, whereas an absorber, on the other hand, must be inserted. This suggests the possibility of control by combining insertion of an absorber with removal of part of the reflector in a manner similar to that described for the core in § 5.118.

CONTROL MATERIALS

5.121. It was stated above that the neutron absorbers (or poisons) most widely used for thermal reactor control are boron, cadmium, hafnium, and some rare-earth (samarium, europium, and gadolinium) oxides [18]. In addition, silver and indium, which have large resonances, have been employed as alloying

TABLE 5.6. PROPERTIES OF CONTROL MATERIALS

Material	Abundance (per cent)	Thermal σ_a (barns)	Thermal Σ_a (cm^{-1})	Major Resonances	
				Energy (ev)	σ_a (barns)
Boron...............		755	107	—	—
Boron-10...........	20	3800	—	None	
Silver...............		62	3.64	—	—
Silver-107..........	51.3	31	—	16.6	630
Silver-109..........	48.7	87	—	5.1	12,500
Cadmium............		2450	113	—	—
Cadmium-113......	12.3	20,000	—	0.18	7200
Indium..............		190	7.3	—	—
Indium-113........	4.2	58	—		
Indium-115........	95.8	197	—	1.46	30,000
Samarium...........		5600	155		
Samarium-149......	13.8	41,000	—	0.096	16,000
Samarium-152......	26.6	225	—	8.2	15,000
Europium...........		4300	90		
Europium-151......	47.8	7700	—	0.46	11,000
Europium-153......	52.2	450	—	2.46	3000
Gadolinium.........		46,000	1400		
Gadolinium-155.....	14.7	61,000	—	2.6	1400*
Gadolinium-157.....	15.7	240,000	—	17	1000*
Hafnium............		105	4.71		
Hafnium-177.......	18.4	380	—	2.36	6000†
Hafnium-178.......	27.1	75	—	7.8	10,000
Hafnium-179.......	13.8	65	—	5.69	1100†
Hafnium-180.......	35.4	14	—	74	130

* Gd^{155} and Gd^{157} have several important resonances in the energy range from 2 to 17 ev.

† Hf^{177} and Hf^{179} have several important resonances in the energy range from 1.1 to 50 ev (and smaller ones up to about 100 ev).

materials with cadmium. The essential nuclear properties of these elements and of their strongly absorbing isotopes are summarized in Table 5.6. With the exception of boron-10, where the neutrons are absorbed in (n, α) reactions, radiative capture is responsible for the absorption of the thermal and resonance neutrons.

5.122. The materials used in control rods must be able to withstand very severe conditions. Absorption of neutrons and exposure to other radiations should not result in dimensional changes, loss of strength, and decrease in corrosion resistance (cf. § 7.5 et seq.). Since heat is generated in the control rod, as a result of the neutron reactions, a high thermal conductivity is desirable.

5.123. Boron, either natural or enriched in boron-10, is generally employed as an alloy or as a dispersion (cermet) in stainless steel. It is also used as the carbide (B_4C), either alone in the form of a ceramic compress, or as a dispersion in a metal; in Boral, for example, the metal is aluminum [19]. Dispersions of boron in iron containing some aluminum have been used in gas-cooled reactors [20]. Borides of chromium, iron, nickel, titanium, or zirconium have also been mixed with cadmium. In dispersed systems, the material generally has the essential properties, e.g., corrosion resistance and thermal conductivity, of the matrix metal. A problem associated with the use of boron for reactor control is the liberation of helium gas by the (n, α) reaction; this can cause swelling and cracking of the material. Nevertheless, because of other desirable properties, control rods containing boron have probably been used more extensively than any other type.

5.124. The melting point of cadmium is 321°C (610°F) and so the metal is useful as a control material for low-temperature applications only. It is extremely ductile and can be fabricated with ease. However, it does not have good corrosion resistance and so it is clad with either aluminum or stainless steel. The capture cross section drops off rapidly at energies above the resonance at 0.18 ev, and this is a drawback to its use in water-moderated power reactors where, as seen in § 4.117, a considerable proportion of the neutrons are in the epithermal (or resonance) region. In order to overcome this situation, an alloy of silver with 5 per cent cadmium and 15 per cent indium has been proposed for use in pressurized-water reactors [21].

5.125. It is largely because hafnium is so effective for capturing neutrons in the epithermal (resonance) energy region that it is regarded as being the best control material for water-cooled (and moderated) power reactors. The fact that hafnium has three isotopes with large and one with moderately large cross sections (Table 5.6) means that it remains effective for a long time. The capture of neutrons by the isotope of mass number 177 leads to the formation of mass number 178, and so on. The chemical properties of hafnium are similar to those of zirconium, its homologue in the periodic system of elements. It can be fabricated by the same methods as are used for zirconium and it is also quite resistant to attack by water at high temperature (cf. § 7.59 et seq.). Its mechanical properties are quite adequate for control rod purposes. Hafnium has been employed as the control material in pressurized- and boiling-water reactors.

5.126. The rare-earth oxides are of interest because they have large thermal cross sections and high resonances in the epithermal region. The use of these materials, in the form of a dispersion in stainless steel, has been somewhat limited. Because of their ceramic nature and ability to withstand high temperatures they may find application for the control of advanced, gas-cooled power reactors (§ 13.62).

CONTROL LOOPS

5.127. Although manual control is feasible in reactors operating at very low powers, it is not always possible, neither is it advisable, to rely entirely on human operation. Emergency safety action, for example, should be made automatic. Thus, if an instrument indicates that the neutron flux is rising at a dangerously rapid rate, or if the power level is appreciably higher than that for which the reactor was designed, an operator may not be able to act quickly enough to decrease the reactivity in time to avoid an accident. In circumstances such as these, control rods should be moved automatically to rectify the situation. Ideally, too, it would be advantageous if automatic devices were available for raising the power output of the reactor to the required level and for maintaining it there.

5.128. A block diagram depicting some of the general aspects of reactor control is given in Fig. 5.9. Arrowheads indicate the direction of flow of information, signal, or effect; broken lines imply flow of information while solid lines represent power. Following the lines representing the "operator loop," it is seen that the reactor affects certain instruments which pass information to the operator. The latter receives the information and, in turn, exerts appropriate

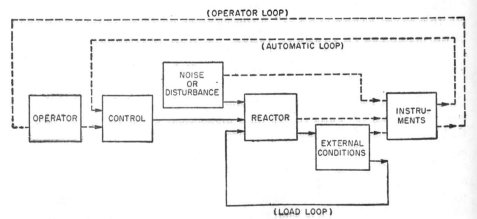

Fig. 5.9. General block diagram of reactor control system

action on, or conveys a signal to, the control block. It is then the function of the control block to exert the desired control action on the reactor, usually with the expenditure of power from an external source. In the "automatic loop" the information received from the reactor by the instruments is fed directly into the control block, the operator being by-passed.

5.129. A third loop, shown in Fig. 5.9 as the "load loop," is intended to represent the interaction of the nuclear aspects of the reactor, i.e., the fuel, moderator, and reflector, with the external conditions, or nonnuclear aspects. Among the latter may be included the status of equipment, such as a power plant, designed to be closely associated with the reactor, and of process facilities, such as water, air, and electricity, that may be essential to its operation. The external conditions may be regarded as constituting, in a general sense, the "load" on the reactor. The closed cycle in the block diagram implies that the load can affect the reactor, which in turn influences the load. The changes in the external conditions are indicated on suitable instruments, and information is fed back to the control block via operator and automatic loops, so that the necessary action may be taken.

5.130. An important aspect of the load loop is found in the interaction of the reactor and its cooling system. The heat released inside a reactor goes partly into heating the reactor itself and partly into the thermal load. In an equilibrium condition, all the heat goes to the load. Depending on the power level and, especially, on the power density of the reactor, the thermal load is a more or less elaborate cooling system designed to permit rapid transfer of heat from reactor to coolant. The coolant flow may be open-ended, as the flow of a gas into the atmosphere, e.g., in the Oak Ridge and Brookhaven graphite reactors, or of water into a river, e.g., the Hanford reactors; or it may be a closed circulating flow of liquid, e.g., when heavy water, liquid metal, or pressurized water is the coolant, or of a gas, e.g., carbon dioxide under pressure in several British power reactors. In any case, the heat exchange between coolant and reactor is an element of the control problem, as was indicated earlier.

5.131. In addition to the three closed loops described above, it will be observed that Fig. 5.9 includes a block labeled "noise or disturbance." These two terms as used here are essentially synonymous; the only reason for giving both is because of their more limited customary implications. "Noise" has been widely used beyond its acoustical meaning to signify relatively fine-grained, random fluctuations in any physical quantity. The term "disturbance," on the other hand, usually implies relatively gross, unexpected, or undesired effects. The label "noise or disturbance" is thus intended to include all undesirable effects on the reactor. As indicated on the block diagram, one method of counteracting some kinds of noise or disturbance is to use instruments to detect them as soon as possible and to initiate corrective control action.

5.132. The operator, automatic, and load loops have been shown as closed, but they are capable of modification to some extent. For example, the operator brings into the control pattern a factor not shown on the block diagram: this is his concept of how the reactor should be operated. Depending on how the total responsibility is divided between the automatic loop and the operator loop, the operator is more or less free to put his concept into practice. In any case, the operator's concept, together with the concepts built into the automatic loop by the designer, determine the overall performance of the total control system. It is clear, therefore, that the system is not altogether isolated, since information, in the form of instructions or criteria of performance, enters from outside.

5.133. Information also leaves the system in various ways. For example, the instruments may furnish information about the reactor or the load to persons or places other than the operator or the control block. This aspect of the subject, however, has no immediate bearing on the control of a reactor and so it will not be considered here. In the same category is the information obtained from health physics instruments used for the monitoring of maintenance and other operations connected with the reactor.

5.134. The reactor system is also open with regard to energy; this leaves the system mainly via the load. In several isotope-producing and experimental reactors most of the energy is carried off by the coolant, such as air or water, which flows away. In a power-generating reactor, a part of the energy is delivered to the useful load. There is also energy input, mainly in the form of the fuel going into the reactor itself. In addition, as implied by the solid line representing the output of the control block in Fig. 5.9, an external source of power is required to amplify the weak signals from operator or instruments so as to cause effective action in the reactor. There are similar, though less important, energy inputs to other blocks in the diagram, e.g., to the instruments.

DESIGN OF THE CONTROL SYSTEM

SPECIFICATION OF THE CONTROL SYSTEM

5.135. An important factor—perhaps the most important factor—underlying the whole treatment of control is the purpose and design of the reactor to be controlled. Some typical purposes for which reactors are designed are as follows: research and teaching, isotope and plutonium production, generation of electric power in a stationary plant, aircraft or ship propulsion, and breeding of fissile material. A given reactor may, of course, serve more than one purpose, but no matter what its function, this must motivate control design just as it motivates other features of reactor design. The controls for any reactor system cannot be properly designed until the required purposes and, therefore, the criteria of performance of the system are clearly understood.

5.136. Although the details of the control design will vary from one type of reactor to another, there are nevertheless certain general requirements which all control systems should meet. In any given reactor the control action should have sufficient range to cover all contingencies and sufficient delicacy of action to permit fine adjustment. It should respond quickly to activating signals, without itself creating undue disturbance. The control system should be simple and foolproof in action, thus assuring a maximum of reliability. It should be designed so that there will be adequate safeguards in the event that any part should fail.* Ideally, too, it should be as economical as possible of neutrons.

5.137. Specification of a complete control system can thus be discussed in terms of three main requirements: range, accuracy, and efficiency. *Range* is concerned with the total amount of reactivity which must be at the command of the system; this obviously affects both operation and safety. *Accuracy* of operation control refers to the closeness with which the actual behavior of the reactor responds to the wishes of the operator or to the formula of the control system. In safety action, the term "accuracy" may be used to imply the promptness with which safety action follows upon the need for such action. Reliability may be considered as an aspect of accuracy. *Efficiency* is an overall, combined measure of neutron economy, power economy, and economy of operation. Qualitatively, it may be said that an efficient control system works unobtrusively, using this word in its widest sense.

CONTROL ROD FUNCTIONS†

5.138. Before considering these specifications in further detail, it may be noted that to achieve both range and accuracy might prove difficult if a single type of control rod were used. Consequently, most reactors have rods with different functions or serving different purposes. There are, first, the so-called *shim rods* for coarse control. They are used to bring the reactor approximately to the desired power level when the system is started up. The shim rods must, therefore, have a fairly large reactivity equivalent, the actual amount depending on the particular reactor, as will be seen later. Because the rate of increase of reactivity must be closely regulated, the driving mechanism must be such that the shim rods cannot move at high speed during startup.

5.139. When the desired power level of the reactor is approached, the shim rods are returned to such a position as to make the reactivity very small, that is to say, the effective multiplication factor is reduced almost to unity. The task of bringing the reactor up to the operating level and of maintaining the power essentially constant, by counteracting the effects of rapid, transient changes, then passes to the *regulating rods*. The reactivity equivalent of these

---

* The dictum that component parts should be designed to "fail safe" applies not merely to controls, but to all aspects of reactor design, in so far as possible.

† The term "control rod," as employed here and subsequently, applies quite generally to either fuel, moderator, reflector, or absorber, the movement of which is accompanied by a change in the reactivity of the reactor.

rods may be quite small, but in general they should be capable of moving rapidly so as to provide quick response to changes affecting the reactor. By keeping the reactivity equivalent small and, in some reactors, by limiting their speed and the distance over which they can move, the regulating rods are unable to produce any dangerous increase of reactivity.

5.140. As a general rule, the reactivity equivalent of a regulating rod should not exceed the fraction of delayed neutrons, e.g., 0.0065 (or 0.65 per cent) when uranium-235 is the fissile material. Then, if the reactor is critical, complete withdrawal of the regulating rod, due to operator error or to failure of the automatic control system, would not permit the system to become prompt critical.

5.141. If a high-flux reactor operates continuously at constant power, a gradual withdrawal of the regulating rod is necessary to compensate for the decrease in reactivity due to fuel depletion, accumulation of poisons, etc. When the rod reaches the end of its range, it must be reinserted into the reactor while a shim rod is withdrawn to an equivalent extent. Therefore in some reactors the shim rods and regulating rods are interlocked. An alternative to this procedure is to make use of the idea of shim control by suitable distribution of a "burnable" poison throughout the reactor core (§ 5.153). If the consumption of the poison, due to neutron capture, occurs at a rate which largely compensates for the fuel depletion, the gradual withdrawal of the regulating (or shim) rods can be almost entirely avoided.

5.142. Finally, there are the *safety rods*, the purpose of which is to shut down the reactor quickly in the event of an emergency; this is generally referred to as "scram." Safety rods must, obviously, be capable of moving very rapidly, and their reactivity equivalent must be appreciably greater than the maximum excess of reactivity built into the reactor. In some reactors the same rods are used as shim rods and safety rods. Although both types have high reactivity equivalents, it has been seen that the former must move slowly while the latter must move rapidly. These apparently contradictory requirements can be met by attaching the rods to the drive mechanism by means of magnetic clutches. When the clutches are energized the rods act as shim rods and can move only at a predetermined slow rate. In the event of an emergency, however, the clutches are de-energized and the rods fall rapidly, under the influence of gravity (and sometimes of the pressure of the cooling water), to fulfill their safety function.

5.143. In addition to the control rods described above, most reactors, especially those of large dimensions, have what is called a "backup" safety device. This is intended for use in an extreme emergency, such as an earthquake, when distortion of the reactor system would prevent normal motion of the safety rods. For example, the backup safety may consist of boron-steel shot which, in the event of an emergency, can be run into voids in the reactor. Several water-cooled, power reactors have a provision for the rapid injection of a solution of boric acid (or a borate) to be used in the event of failure of the safety rods. Homogeneous reactors using a liquid medium, e.g., one form of the Water Boiler, can be fitted with "dump" valves so that fuel solution can be rapidly removed from the core into vessels having noncritical geometry. For reactors operating at low power level, the employment of "nuclear fuses" has been proposed; these would permit injection of a neutron poison into the core when the temperature exceeds a prescribed value (§ 9.113).

RANGE OF CONTROL SYSTEM

5.144. The range requirements or, in other words, the total reactivity equivalence of various means of control, especially safety rods and backup safety devices, are fixed almost entirely by design and operating criteria of the reactor. Important factors in establishing reactivity needs are temperature coefficient effects, degree of fuel depletion to be allowed, cumulative fission product (and other) poisoning, isotope production, experiments, etc. These may demand a total of 1 or 2 per cent of additional reactivity (or $\Delta k/k$) for a reactor of low power, or they may need as much as 20 per cent or more for reactors of high flux and high specific power.

TEMPERATURE EFFECT

5.145. Since most reactors have overall negative temperature coefficients, the reactivity decreases when the temperature rises, as it inevitably will when the reactor operates at appreciable power. A reactor which is just critical when cold will thus become subcritical when the temperature increases. A certain amount of excess reactivity, i.e., additional fuel, must therefore be included in the reactor to overcome the negative effect of temperature during operation. It was seen in § 5.83 that in the Yankee (pressurized water) reactor, the reactivity decreases by 0.072 in going from the cold condition to the operating temperature at zero power. The corresponding reactivity allowance is thus about 7 per cent (Table 5.7). To this must be added the amount (2.5 per cent) required to overcome the Doppler power effect, although it is sometimes included with the operational allowance for the reactor.

FUEL DEPLETION

5.146. In order to avoid the necessity for shutting down the reactor at frequent intervals to replace the fuel that has been consumed, additional fuel, over and above the essential requirements, is included at the time of initial startup. The allowance for fuel depletion is greatly dependent on the design of the reactor, as may be seen from the data in Table 5.7. In a thermal reactor having natural (or partially enriched) uranium as fuel, the allowance will depend on the extent to which uranium-238 is converted into plutonium-239. If the conversion ratio (§ 12.110) is large, the amount of additional fuel required to maintain criticality will be small. In a fast (breeder) reactor the fuel depletion allowance may be small if the reactor is designed to produce fissile material in the core to replace that which is consumed.

FISSION PRODUCT POISONING

5.147. The most important fission product poisons are xenon-135 and samarium-149, and the excess reactivity required to compensate for the effects of these isotopes can be estimated from the equations given in § 5.90 *et seq*. In most power reactors, the average thermal-neutron flux is in the range of 10^{13} to 10^{14} neutrons/(cm^2)(sec), and so the equilibrium poisoning is roughly 0.03 to 0.045; the reactivity requirement is thus about 3 or 4 per cent. This will compensate for xenon and samarium poisoning during normal operation of the reactor but not for the large increase that may occur within a few hours after shutdown (Fig. 5.7). If it is considered desirable that the reactor have the capability of being started up under such circumstances, additional reactivity must be included to override the maximum possible xenon poisoning.

5.148. Other fission products and higher isotopes also cause some poisoning, as was seen in § 5.109 *et seq*. The necessary reactivity allowance is small, since it rarely exceeds 2 per cent if the contribution of samarium-149 is omitted, and is generally included with the allowance for fuel depletion.

REACTIVITY SPECIFICATIONS

5.149. The reactivity specifications for a number of power reactors of different types are summarized in Table 5.7. In addition to the allowances for temperature, fuel depletion, and fission product poisoning, some excess reactivity is included for operation and other effects. The purpose of this added fuel is to provide a control margin toward the end of the core life, to allow for uncer-

TABLE 5.7. REACTIVITY SPECIFICATIONS FOR POWER REACTORS

Specification	Reactivity (or $\Delta k/k$) Requirements in Per Cent					
	Yankee (Pressurized Water)	Calder Hall (Gas Cooled)	Hallam (Sodium-Graphite)	Piqua (Organic)	Dresden (Boiling Water)	Fermi (Fast Breeder)
Temperature.........	7	1.6	1.5	1.4	3	0.14
Fuel depletion.......	7	—	2.5	1.0	6	0.28
Xe and Sm..........	3.3	2.5	2.5	3.1	4	0.014
Operation, etc........	2.7	1.2	0.5	1.0	—	0.21
Total (per cent)....	20	5.3	7.0	6.5	13	0.644

tainties in the analysis upon which the reactivity estimate is based, and to provide for unforeseen or special circumstances which might arise while the reactor is operating.

5.150. In the design of the control system for a reactor, allowance must be made for the possibility that it may have the maximum fuel loading; in addition, the control rods should include a margin, generally about 50 per cent, for control flexibility and to act as a safety factor. The actual number of rods used will depend on the power, size, and nature of the reactor, but a reactivity equivalent of 0.05 is common for a single safety rod.

5.151. The reactivity range of the shim rods is also determined largely by the reactor design; this is especially true, of course, when the same rods are used for coarse control and for safety purposes. Regulating rod range, on the other hand, is left largely to the control designer, since the function of the rod is to permit fine operating control. For the reason given in § 5.55, the reactivity equivalent of a regulating rod is kept below 0.0065, a common value being 0.005 (or 0.5 per cent). As a general rule only one regulating rod of this reactivity would be used, although another might be available as a stand-by.

5.152. The presence of as much as 20 per cent excess fuel in a reactor in its initial state has some disadvantages. The control rods must be inserted to compensate for the additional reactivity and then withdrawn during startup and as the fuel is consumed. If the rods are accidentally withdrawn too rapidly during startup, the large excess reactivity could produce a dangerous rate of increase in the power level. In any case, the absorption of neutrons by the control rods is wasteful, unless the product has some value. Furthermore, the insertion of the rods causes a distortion of the neutron flux (and power) distribution which changes in the course of reactor operation. This leads to complication in the problem of heat removal.

5.153. One way in which the difficulty arising from changes in the distribution of neutron flux, and hence of heat generation, is to make use of a "burnable poison" in the moderator or coolant [22]. For example, a compound of boron or of lithium can be added to the coolant, or boron may be included within, or as an aluminum-boron strip adjacent to, the fuel element. The poison captures neutrons and is thereby consumed (or "burned") during operation of the reactor; consequently, the effectiveness of the added poison steadily decreases as the fission product poisons accumulate and the fuel is consumed. As a result, largely automatic compensation of the decreasing reactivity is provided with the minimum use of control rods. A drawback to the use of a soluble burnable poison is that it makes a significant positive contribution to the reactivity temperature coefficient (§ 5.82).

5.154. Various suggestions have been made to overcome the wasteful loss of neutrons to a burnable poison. For example, if the latter were lithium-6, the hydrogen isotope tritium could be produced by the (n, α) reaction; or plutonium-240, which has a thermal capture cross section of about 300 barns, could be converted into the fissile plutonium-241. An alternative possibility is provided by the "spectral shift control" concept [23]. The operation of a close-packed lattice of slightly enriched uranium, such as is used in a pressurized-water reactor (§ 4.123), is commenced with a large proportion of heavy water in the moderator. Since heavy water is much less effective in slowing down neutrons than is ordinary water, there is an increase in the resonance flux and a consequent increase in resonance capture of neutrons and conversion of uranium-238 into fissile plutonium-239. With continued operation, the loss of fuel reactivity is compensated by introducing ordinary water into the moderator. This has the effect of increasing the resonance escape probability and the reactivity of the close-packed lattice system.

EFFECTIVENESS OF CONTROL RODS

5.155. The theoretical calculation of the reactivity equivalent of a control rod is a difficult problem, although it has been solved for a number of special cases. For preliminary design purposes, however, the one-group treatment is usually adequate to calculate the rod effectiveness in terms of known properties of the reactor. It is assumed that a cylindrical "black" (perfect neutron absorber) rod is inserted fully into the center of a cylindrical, bare homogeneous reactor and that it expels an equivalent cylinder of core. The result, after making a correction to allow for fast neutrons in addition to the one-group (thermal) neutrons, is

$$k_{\text{eff}} - 1 \approx \frac{7.5L^2}{R^2}\left(0.116 + \ln\frac{R}{2.4r_{\text{eff}}}\right)^{-1},$$

where L is the neutron thermal diffusion length in the reactor, R is the cylindrical core radius, and r_{eff} is the rod radius minus the linear extrapolation distance (§ 3.26); $k_{\text{eff}} - 1$ represents the rod "worth," i.e., the excess multiplication (or reactivity) that can be controlled by the rod when it is fully inserted [24].

5.156. The effectiveness of a noncylindrical rod can be determined by utilizing an equivalent radius for a cylindrical rod. For example, for a single rod with a symmetrical cruciform section, of the type in common use in water-moderated power reactors, the equivalent cylindrical radius is $a/\sqrt{2}$, where a is the distance from the central axis of the rod to the end of an arm. Values for other shapes have been calculated and are available in the literature [25].

5.157. Most reactors have control rods which are off-center and, although theoretical methods for treating such situations have been developed, the results are too involved to be given here. A good approximation to the worth of a noncentral rod can be obtained from the central-rod value by taking the effectiveness to be proportional to the square of the neutron flux at the rod location. Assuming a cosine (or similar) distribution of the thermal flux in a bare, cylindrical reactor (§ 4.37), a vertical rod located halfway along the radius would be about half as effective as the same rod at the center. For a reflected reactor, the ratio would be somewhat larger, because of the higher neutron flux away from the center.

5.158. The size of the region over which a control rod can influence reactivity is roughly equal to the diffusion length in the reactor. Consequently, in order to avoid large variations in the neutron flux when the rod is inserted, it is the practice to use several rods distributed throughout the reactor. The total effectiveness (or worth) of two or more rods, however, is not necessarily equal to the sum of the values for the individual rods. The overall effect is a function of the spacing between the rods, as well as of their location in the reactor. When two rods are close together, their total worth is less than the sum of their individual worths. The phenomenon is referred to as control rod "shadowing" and arises from the distortion of the flux distribution accompanying the insertion of a rod into the reactor. The situation is shown qualitatively in Fig. 5.10

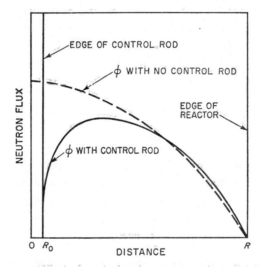

Fig. 5.10. Effect of control rod on neutron flux distribution

for a bare cylindrical reactor with a central control rod. Because of the decrease in the flux in the vicinity of this rod, a second rod adjacent to it will be less effective than in the absence of the first rod. On the other hand, the central rod may cause the neutron flux to increase some distance away, as indicated in the figure. A second rod inserted in this region will then have a greater worth than it would if used alone in the same location.

5.159. The small value of the diffusion length (about 2 cm) in a water-moderated reactor and the close-packed nature of the lattice present a special problem in control. In order to minimize the neutron flux distortion, a relatively large number of rods, with a cruciform geometrical cross section, are distributed more or less uniformly throughout the core. For arrangement of this kind, an approximate estimate of the reactivity worth of the whole system of rods can be made by the "poison cell" technique. In this method, a control rod (or portion of a rod) is associated with an appropriate cell of the reactor core. The fraction of the neutrons in a single cell absorbed by the rod, which is assumed to be "black," is then calculated. If the entire core is occupied by poison cells of the same type, the worth of the whole system of rods in the reactor is equal to this fraction [26].

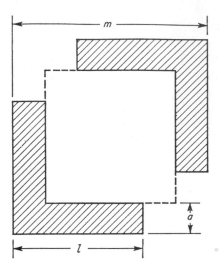

FIG. 5.11.   Poison cell for control rod calculation

5.160. Suppose that the rods are distributed uniformly through the reactor core in the form of a square array with an additional rod at the center of each square. This is a common arrangement in water-moderated power reactors. A single poison cell will then include a quarter of each of two cruciform control elements, as shown in Fig. 5.11; l is the length and a is the half-thickness of each arm of the cross, and m is the cell dimension. If q is the rate at which neutrons appear per unit cell, then

$$\frac{\text{Fraction of neutrons}}{\text{absorbed in rod}} = \frac{\text{Net leakage rate across surface of rod}}{\text{Net production rate in cell}}$$

$$= \frac{\int_S D\nabla^2\phi \, dS}{\int_V q \, dV},$$

where S and V signify surface and volume, respectively.

5.161. If the leakage at the cell boundary is assumed to be zero, the neutron flux may be determined from the diffusion equation,

$$D\nabla^2\phi - \Sigma_a\phi + q = 0,$$

where Σ_a is the macroscopic absorption cross section of the core material in the cell. Upon using slab geometry, it is found that

$$D\nabla^2\phi \text{ (at rod surface)} = \frac{qL}{\dfrac{hD}{L} + \coth\left(\dfrac{m - 2a}{2L}\right)},$$

where L is the thermal diffusion length of the neutrons in the cell medium and h is the linear extrapolation distance into the rod (§ 3.26) divided by the diffusion coefficient; for a black rod h is 3×0.71. If q is taken to be constant outside the rod and zero inside, then the fraction of neutrons absorbed in a single rod, which may be taken as the total worth of all the rods, as stated above, is given by

$$\delta k = \frac{(\text{Exposed perimeter of rod}) \, D\nabla^2\phi \, (\text{at rod surface})}{q \, (\text{Area of source region})}$$

$$\approx \frac{4l}{(m - 2a)^2} \cdot \frac{1}{h\Sigma_a + \dfrac{1}{L} \coth\left(\dfrac{m - 2a}{2L}\right)},$$

noting that D/L^2 has been replaced by Σ_a.

Example 5.5. In a water-moderated reactor, in which 32 cruciform control rods are distributed evenly throughout the core (§ 12.33), the following dimensions are applicable: $l = 10.0$ cm, $a = 0.336$ cm, and $m = 19.5$ cm. The thermal-neutron diffusion length in the core is 1.8 cm, and Σ_a, determined by the method described in § 2.135, is 0.114 cm^{-1}. Estimate the total worth of the system of control rods.

Substitution of the data in the equation given above yields

$$\delta k \approx \frac{(4)(10)}{[19.5 - (2)(0.336)]^2} \cdot \frac{1}{(2.13)(0.114) + \dfrac{1}{1.8} \coth \dfrac{(19.5 - 0.67)}{(2)(1.8)}}.$$

The hyperbolic cotangent term is essentially unity; hence

$$\delta k \approx 0.14,$$

so that the rods can control about 14 per cent of the reactivity.

CALIBRATION OF CONTROL RODS

5.162. Although a rough idea of the reactivity equivalent of a control rod may be obtained by calculation, as indicated above, the values must be checked by direct measurement. In any event, in addition to knowing the total effectiveness of the rod when fully inserted, it is necessary to calibrate it, so that its reactivity equivalent is known for various degrees of insertion. The required data can be obtained only by direct measurement on the reactor itself or, preferably, on a low-power mockup or critical assembly [27].

5.163. One method used is the following. The reactor is made critical by means of other control rods with the rod to be calibrated, hereafter called the test rod, in its fully inserted position. The test rod is then withdrawn a small distance x_1 so that the reactor is slightly supercritical and the neutron flux starts to increase. After allowing a minute or two for the transients to die out (§ 5.20), the reactor period is determined from the rate of increase of the flux, e.g., by means of a period meter (§ 5.232). From the observed period it is possible to compute the corresponding reactivity, utilizing equation (5.37) in the manner to be described below. Let the reactivity corresponding to the rod insertion x_1 be ρ_1. One of the other control rods is now inserted into the reactor to bring it back to critical, and the test rod is withdrawn further to a total distance x_2. The period is again determined and the reactivity calculated from it is ρ_2. The total reactivity change corresponding to a rod withdrawal of x_2 is thus $\rho_1 + \rho_2$. The procedure is repeated until the test rod has been calibrated along its whole length.

5.164. From equation (5.37) and the known delayed-neutron characteristics (Table 5.2), it is possible to express the relationship between the reactivity and the stable period, T_p, for a particular value of l^*, the neutron generation time. Thus, if uranium-235 is the fissile material,

$$\rho = \frac{l^*}{T_p} + \frac{0.0171}{80.0 + T_p} + \frac{0.0463}{32.8 + T_p} + \frac{0.0114}{9.0 + T_p}$$
$$+ \frac{0.0085}{3.3 + T_p} + \frac{0.0007}{0.09 + T_p} + \frac{0.0001}{0.33 + T_p},$$

with T_p in seconds. If the stable period is large, as it usually is in control rod calibration measurements, the first term on the right of this equation can be neglected, so that the value of l^* need not be known. Consequently, from a measurement of the reactor period, the corresponding reactivity required for the control rod calibration can be readily evaluated.

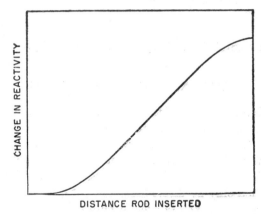

FIG. 5.12. Change in reactivity as function of insertion of control rod

5.165. In the rod-drop method for calibrating a control rod, the reactor is brought to steady-state operation with the rod withdrawn and the neutron flux, ϕ_0, is measured. The rod is then suddenly dropped into the reactor to a known extent so as to provide what is essentially a step decrease in reactivity. The neutron flux (or density) will then change somewhat in the manner shown by the full curve in Fig. 5.2. If that portion of the curve corresponding to the stable period is extrapolated back to zero time, along the broken line representing the first term in equation (5.29), the value of the flux, ϕ, so obtained will be equivalent to $\phi_0\beta/\beta - \rho)$, assuming one (averaged) group of delayed neutrons. It follows, therefore, that

$$\frac{\phi_0}{\phi} = \frac{\beta - \rho}{\beta}.$$

Since β is known, e.g., 0.0065 for uranium-235, the reactivity change ρ produced by inserting the control rod a known distance can be determined from the neutron flux measurements. The observations are repeated for various insertions of the rod.

5.166. The general nature of the results obtained in the calibration of a control rod is indicated by Fig. 5.12; the change in reactivity is shown as a function of the distance the rod is inserted. The S-shaped curve is quite characteristic for reactors of most types. When the rod is inserted to a small extent, it will affect the reactivity only slightly. Beyond a certain point, the change in reactivity is linear, that is to say, it is directly proportional to the extent of insertion of the rod. Finally, when the rod is almost completely inserted, further insertion has relatively little effect. The small changes in reactivity at the beginning and end of the rod insertion can be accounted for by the lower neutron flux near the edges of the reactor as compared with that closer to the center.

DANGER COEFFICIENT AND PILE OSCILLATOR

5.167. There are two matters of general interest, related somewhat indirectly to reactor control, which merit brief consideration here. The poisoning effect of a neutron absorber in a reactor is expressed by means of its *danger coefficient*; this is the change in reactivity resulting from the insertion of unit mass of the material into the reactor core. The measurement of the danger coefficient is performed most simply with the aid of a calibrated control rod. The rod is first adjusted so that the reactor is exactly critical, that is to say, the neutron flux does not change with time. The absorber is then inserted and the rod is withdrawn so as to establish the critical condition. From the extent of withdrawal of the calibrated rod, the change in reactivity produced by the inserted material can be determined. The actual value of the danger coefficient is proportional to the square of the neutron flux at the point where the absorber is located (cf. § 5.157). By utilizing a substance of known cross section for comparison, the danger coeffiicent method can be used to derive the neutron absorption cross section of the material being studied [28].

5.168. The *pile oscillator* has been employed to determine either the reactor poisoning effect or the absorption cross section of a given substance. If a sample of the absorber is moved in and out of the reactor, by means of a mechanical oscillating device, the reactor power (and neutron flux) will oscillate accordingly. It has been shown that the amplitude of the neutron flux oscillations, as measured by a suitable instrument placed close to the sample, is proportional to the neutron absorption cross section of the material. Absolute values can be obtained by calibration with a material of known cross section [29].

REACTOR SYSTEM ANALYSIS [30]

INTRODUCTION

5.169. An important aspect of reactor system design, especially in relation to the control elements, is an examination of the response to changes resulting from external disturbances, e.g., movement of a control rod, change in flow of coolant, change in power demand, etc., of a reactor operating in the steady state. It is necessary to know (a) whether following such a change the reactor reaches another steady state, so that the system is stable, and (b) if it is stable, how the reactor behaves in the course of the transition from the initial to the final steady state. If the system is unstable, then obviously steps must be taken to overcome the causes of instability. However, a reactor system may be stable, in the sense defined above, but its transient behavior may not be satisfactory; for example, the power may undergo a number of oscillations of such magnitude, in the transition from one state to another, that safe operation would not be possible. For the present purpose, therefore, a system is regarded as stable if the transient behavior between two steady states is represented by oscillations of a minor nature which damp out quite rapidly (cf. § 5.203).

5.170. The effect of external changes on a steady-state system can be determined from the resulting transient response and this can be studied by system analysis. The approach may be by experiment, by calculation, or by a combination of both. In the SPERT (Special Power Excursion Reactor Tests) experiments, for example, it was found that under certain conditions an increase in reactivity, resulting from the withdrawal of a control rod, was accompanied by violent power oscillations [31]. On the other hand, application of theoretical analysis to the observations made on the EBWR (Experimental Boiling Water Reactor) showed that its normal operating power level could be increased from 20 to 50 megawatts (Mw) with complete safety [32].

5.171. A nuclear reactor is intrinsically unstable in the sense that, when in the steady state, an increase in reactivity will cause the neutron flux (and power) to rise indefinitely without reaching a second steady state. However, there are generally some self-limiting characteristics of a reactor, such as a negative temperature coefficient, which may serve as a stabilizing influence. These characteristics provide what is called *negative feedback*. The term "feedback" is used to describe the behavior of any controlled system when the system responds to a change (or error) in such a manner as either to enhance (positive feedback) or to counteract (negative feedback) the error. For example, an increase in reactivity of a steady-state reactor, without any change in rate of flow of the coolant, will be accompanied by an increase in temperature. If the system has a negative temperature coefficient, the reactivity will decrease to such an extent that a new steady state, at a higher temperature, may be attained. Certain fission products, although not xenon and samarium because they are formed only after a considerable delay, can also contribute negative feedback, since the poisoning increases, and the reactivity decreases, with increasing neutron flux.

5.172. In considering the reactivity (or effective multiplication) feedback, it should be realized that it is the change in the output of the reactor which determines the magnitude of the feedback. The reactor output is its power and the *power coefficient* of reactivity is defined as the change in reactivity per unit increase in the power. There are several factors which may contribute to this coefficient; two obvious ones which will be considered here are temperature and fission product poisoning. Since the temperature and poisoning feedbacks are essentially independent of one another, the reactor system may be represented by a block diagram of the type shown in Fig. 5.13. The feedback loops are

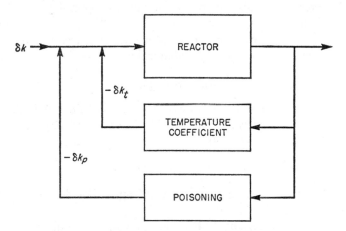

FIG. 5.13. Feedback due to temperature coefficient and poisoning

shown in parallel with the reactor system. Suppose that, as a result of a power demand, the effective multiplication (or reactivity) is increased by δk, e.g., by moving a control rod, then the accompanying temperature change provides a negative reactivity feedback of $-\delta k_t$ and the poisoning contributes $-\delta k_p$. The actual (or net) change δk_n in the effective multiplication factor is thus given by

$$\delta k_n = \delta k - \delta k_t - \delta k_p,$$

so that the negative feedback tends to counteract the reactivity change resulting from the power demand. Thus, in the event of an accidental increase in the power, the negative contributions to the power coefficient would help to stabilize the reactor system. In order for a given contribution to the feedback to be effective, however, the response time (or time constant) must be short. This will be apparent from the remarks in § 5.79, where it was seen that because of the delay in heating the moderator, the transient behavior of the reactor was affected mainly by the fuel temperature. It is of interest to mention, too, that a negative temperature coefficient of reactivity can cause power oscillations, and contribute to instability, if it has a suitably long time constant.

LAPLACE TRANSFORM REPRESENTATION

5.173. In the theoretical analysis of a physical system, the first step is to develop a differential equation which represents the behavior of the system as the result of a small external change (or error). A simple example is provided by equation (5.11) which gives the rate of change in neutron density following upon a change in reactivity of a reactor operating in the steady state. By solving the differential equation it is possible to obtain information concerning both the transient behavior and the stability of the system.

5.174. In systems analysis it is often convenient to employ a Laplace transform representation of the differential equation [33]. In the present connection, the Laplace transformation converts a differential equation expressed in terms of the time variable, t, into an algebraic equation in terms of a complex variable, s, where

$$s = a + jb,$$

in which a and b are real quantities and j is $\sqrt{-1}$. Within certain mathematical conditions, which are generally satisfied in situations of present interest, the Laplace transformation of the function $f(t)$ is described by

$$\mathcal{L}[f(t)] \equiv \int_0^\infty f(t)e^{-s}t\, dt \equiv F(s). \tag{5.65}$$

By this operation the function is transformed from a time domain to a frequency domain.

5.175. After the transformation has been made, the resulting equations can be treated like ordinary algebraic equations. When the required manipulations have been completed, the algebraic equation may be converted back into the time domain to give the solution of the original differential equation. More commonly, however, in systems analysis the required information can be obtained directly from the algebraic equation by the use of graphical methods.

5.176. In making a Laplace transformation of a differential equation, it is usually not necessary to perform the operation indicated by equation (5.65).

TABLE 5.8. LAPLACE TRANSFORMS

$f(t)$	$F(s)$	$f(t)$	$F(s)$
a	a/s	t^n	n/s^{n+1}
t	$1/s^2$	e^{at}	$1/(s-a)$
at	a/s^2	$d^n f(t)/dt^n$	$s^n F(s)*$

* Provided the system is in a steady state when t is zero and $f(0)$ is zero.

The general practice is to employ tables of Laplace transforms, some examples of which are given in Table 5.8. By utilizing standard and well established techniques, the required transformations of complex differential equations can be performed without serious difficulty.

FUEL-MODERATOR TIME CONSTANT

5.177. As a simple example of the application of the Laplace transformation, an expression will be derived for the relationship between the temperature of the fuel and that of the moderator in a heterogeneous reactor following a disturbance. Suppose the system has been operating in the steady state and the reactivity is then increased in an arbitrary manner; the temperature T_f of the fuel will increase correspondingly, but that of the moderator, T_m, will lag behind to some extent. At any time t, a reasonable expression relating $T_f(t)$ and $T_m(t)$ is

$$T_f(t) - T_m(t) = \tau \frac{dT_m}{dt}, \tag{5.66}$$

where τ is the time constant. Writing $T_f(s)$ and $T_m(s)$ for the Laplace transforms of the fuel and moderator temperatures, respectively, it follows, from Table 5.8, that

$$T_f(s) - T_m(s) = \tau s T_m(s)$$

or

$$\frac{T_m(s)}{T_f(s)} = \frac{1}{\tau s + 1} = \frac{1/\tau}{s + 1/\tau}. \tag{5.67}$$

This result is quite general, regardless of the manner in which the temperature of the fuel increases, provided $T_f(0)$ is zero or T_f and T_m are initially the same (§ 5.178).

5.178. A simple solution is possible in the case of a step change in reactivity and fuel temperature. If the latter suddenly increases and then remains constant, the Laplace transform of the fuel temperature is given by $T_f(s) = T_f(t)/s$, so that, by equation (5.67),

$$T_m(s) = T_f(t) \frac{1/\tau}{s(s + 1/\tau)}.$$

Upon taking the inverse Laplace transforms, it is found that

$$T_m(t) = T_f(1 - e^{-t/\tau}).$$

Instead of regarding the initial fuel temperature as being zero prior to the step change, the initial temperature of the fuel and moderator may be taken to be the same, namely, T_0; if this is added to both sides, the result is

$$T_m(t) = T_0 + T_f(1 - e^{-t/\tau}),$$

where T_f is the magnitude of the step increase in the fuel temperature, so that the actual fuel temperature is $T_0 + T_f$. It is seen that the moderator temperature will approach the fuel temperature asymptotically at a rate dependent upon the time constant. The magnitude of the time constant is determined by the rate of heat transfer from the fuel to the moderator and the thermal conductivity and heat capacity of the moderator. If the time constant is large, the moderator temperature will lag considerably behind the fuel temperature (§ 5.79).

TRANSFER FUNCTIONS

5.179. The use of the Laplace transform representation lends itself particularly to the development of the important property of a system called the *transfer function* [34]. In the broadest sense, the transfer function is a mathematical expression which describes the effect of a physical system on the information (or signal) transferred through it. It may be defined by

$$\text{Transfer function} = \frac{\text{Laplace transform of the output signal}}{\text{Laplace transform of the input signal}}. \qquad (5.68)$$

The transfer function is a characteristic property of the system for the signal or disturbance which is transferred; for example, a reactor has a definite transfer function for the response of the neutron flux to a change in reactivity of a steady-state system.

5.180. If the input signal has a sinusoidal variation with time, then when the system reaches a steady state, the output signal is also sinusoidal, differing from the input only in amplitude and phase, provided the system under consideration is linear. Thus, if the input signal (or *forcing function*) is represented by

$$\theta_i(t) = A \sin \omega t,$$

where A is the amplitude of the sine wave and ω is the angular frequency in radians per second, i.e., $2\pi f$, where f is the actual frequency in cycles per second, then the steady-state output signal for a linear system is

$$\theta_o(t) = B \sin (\omega t + \phi),$$

where B is the amplitude of the output signal and ϕ is the phase angle between $\theta_i(t)$ and $\theta_o(t)$.

5.181. It can be shown that, in these circumstances, the Laplace transform variable s is equal to $j\omega$, and the transfer function defined by equation (5.68) may be written as

$$\text{Transfer function} = \frac{\theta_o(j\omega)}{\theta_i(j\omega)} = KG(j\omega), \qquad (5.69)$$

i.e., the product of two quantities K and $G(j\omega)$. The factor K, called the *gain*, is independent of frequency, whereas $G(j\omega)$ describes the magnitude and phase relationship of the transfer function in terms of complex numbers. In general, $G(j\omega)$ may be written as the sum of a real and an imaginary term; thus,

$$G(j\omega) = R$$

$$= |G($$

where $|G(j\omega)|$ is the real part (or modulus) of
Furthermore, the value of the phase angle ϕ is and $e^{j\phi}$ is t

$$\phi = \tan^{-1}\frac{y}{x} = \tan^{-1}\frac{\text{Imaginary part}}{\text{Real part of } G(j\omega)}.$$

Consequently, if the transfer function is expressed, eithe
by mathematical analysis, in terms of
the frequency ω of the sinusoidal forcing function, it is generally possible to
evaluate the two quantities $K|G(j\omega)|$
and ϕ. Another interpretation of
$KG(j\omega)$ is as a vector with a magnitude equal to the amplitude ratio
$K|G(j\omega)|$ and direction determined by

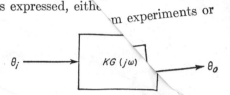

FIG. 5.14.   Transfer function for single
unit in open loop system

the phase angle ϕ.   These are the basic components of the transfer function of
the system from which its behavior can be predicted.

5.182. One of the advantages of the transfer function concept is that the
representation of the function for a complete system consisting of several dependent or independent units is a relatively simple matter.   For this purpose,
a block diagram is useful.   Each individual unit may be represented as in Fig.
5.14, and an open system consisting of a series (or cascade) of noninteracting

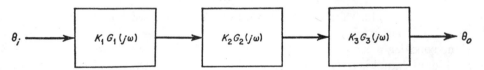

FIG. 5.15.   Transfer function for cascade of units in open loop system

units by Fig. 5.15.   It can then be readily shown that the overall transfer function of the cascade system is given by

$$KG(\text{overall}) = K_1G_1(j\omega) \times K_2G_2(j\omega) \times K_3G_3(j\omega)$$

$$= (K_1K_2K_3)[G_1(j\omega)G_2(j\omega)G_3(j\omega)]. \tag{5.70}$$

5.183. A simple system with feedback, generally called a *closed loop system*, is
shown in Fig. 5.16, where K_fG_f is the feedback transfer function.*   Let θ_i be
the input signal, θ_s the output of the block marked "system" without feedback,
and θ_o the output signal of the feedback; then the error signal ϵ applied to the
system is equal to $\theta_i - \theta_o$.   For the system without feedback,

---

* Since the variable $j\omega$ is used in nearly all the subsequent treatment, it will be omitted from
now on.

NUCLEAᵣ $$\frac{\theta_s}{\theta_i - \theta_o} = K_s G_s,$$

and for the feedback cₑ ..tion

$$\frac{\theta_o}{\theta_s} = K_f G_f.$$

The transfer fun⌐ of the system including the effect of feedback is then

$$\frac{\cdot}{\theta_i} = \frac{K_s G_s}{1 + (K_s G_s)(K_f G_f)} = \frac{K_s G_s}{1 + (K_s K_f)(G_s G_f)}. \tag{5.71}$$

5.184. ₚ utilizing these general principles, it is possible to express the overall response ̶ a system consisting of a number of dependent units. As an illustra-

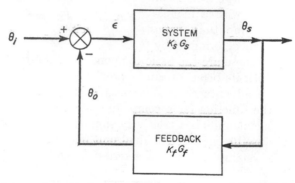

Fɪɢ. 5.16. Transfer function with feedback in closed loop system

tion, consider a simple reactor system with feedback indicated in Fig. 5.17. The input signal is a change δk in the effective multiplication which causes the

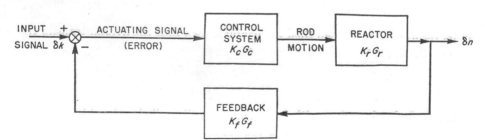

Fɪɢ. 5.17. Transfer function for simple reactor system

control rods to move, thereby increasing the steady-state neutron density by ₷n; $K_c G_c$ is the transfer function of the control system, including rods, actuating ₑchanism, etc., $K_r G_r$ is the transfer function of the reactor without feedback, and $K_f G_f$ is the feedback contribution, e.g., from temperature poisoning. It follows, therefore, that

$$\frac{\delta n}{\delta k} = \frac{(K_c K_r)(G_c G_r)}{1 + (K_c K_r K_f)(G_c G_r G_f)}.$$

5.185. In addition to the advantage of simple algebraic manipulation by the use of block diagrams, there are some other aspects of the transfer function representation which should be mentioned [35]. In the first place, transfer functions are well suited to experimental measurement by well known frequency response techniques. For example, reactor transfer functions have been determined by utilizing an oscillating control rod to produce a sinusoidal time variation of the reactivity and observing the corresponding changes in the neutron density. Furthermore, experimental reactor transfer functions may be combined with those derived analytically from the appropriate differential equations. In this manner it is possible to calculate transfer functions which are not readily obtainable either analytically or by direct measurement. Finally, by utilizing transfer functions the analysis of the complete system by means of an analog computer is greatly simplified.

REACTOR KINETICS TRANSFER FUNCTION

5.186. In deriving the transfer function for the kinetics of a reactor, a single average group of delayed neutrons will be assumed for simplicity and the modification required to allow for the actual six groups will be indicated later. For a single group of delayed neutrons, equations (5.11) and (5.12) become

$$\frac{dn}{dt} = \frac{\delta k - \beta}{l^*} n + \lambda C \tag{5.73}$$

and

$$\frac{dC}{dt} = \frac{\beta}{l^*} n - \lambda C, \tag{5.74}$$

where δk, equal to $k_{\text{eff}} - 1$, has been written in place of the reactivity ρ, since only small changes in k_{eff} are to be considered. Addition of equations (5.73) and (5.74) then gives

$$\frac{dn}{dt} = \frac{\delta k}{l^*} n - \frac{dC}{dt}. \tag{5.75}$$

The requirement of a small change in k_{eff} is valid for most purposes. It is introduced here, however, in order to achieve linearization of equation (5.75), as will be seen below, since this is an essential step in the development of the transfer function.

5.187. Suppose the reactor is initially in the steady state in which the neutron density and concentration of delayed-neutron precursors are n_0 and C_0, respec-

…ve multiplication factor is then made to undergo a sinusoidal

98 , and at any instant

tiv.
$$n = n_0 + \delta n \quad \text{and} \quad C = C_0 + \delta C.$$

…nce, equation (5.75) may be written as

$$\frac{dn}{dt} = \frac{d(\delta n)}{dt} = \frac{\delta k}{l^*} n_0 + \frac{\delta k}{l^*} \delta n - \frac{d(\delta C)}{dt}. \tag{5.76}$$

Provided the change δk is small, δn will also be small, so that the term $\delta k \, \delta n/l^*$ can be neglected, leaving equation (5.76) in the required linearized form

$$\frac{dn}{dt} = \frac{\delta k}{l^*} n_0 - \frac{d(\delta C)}{dt}. \tag{5.77}$$

Similarly, equation (5.74) becomes

$$\frac{dC}{dt} = \frac{d(\delta C)}{dt} = \frac{\beta}{l^*} (n_0 + \delta n) - \lambda(C_0 + \delta C). \tag{5.78}$$

In the steady state, $dC/dt = 0$, and also $\delta n = 0$ and $\delta C = 0$; hence,

$$\frac{\beta}{l^*} n_0 = \lambda C_0,$$

and combination of this result with equation (5.78) yields

$$\frac{d(\delta C)}{dt} = \frac{\beta}{l^*} \delta n - \lambda \delta C. \tag{5.79}$$

5.188. Upon applying the Laplace transformation to equations (5.77) and (5.79) and writing $j\omega$ for the Laplace transform variable, since a sinusoidal time change is postulated, the results are

$$j\omega \delta n(j\omega) = \frac{n_0}{l^*} \delta k(j\omega) - j\omega \delta C(j\omega)$$

and

$$j\omega \delta C(j\omega) = \frac{\beta}{l^*} \delta n(j\omega) - \lambda \delta C(j\omega),$$

because the system has been assumed to be initially in the steady state. From these two algebraic equations it is found that

$$\frac{\delta n(j\omega)}{n_0 \delta k(j\omega)} = \frac{1}{j\omega l^* + \dfrac{j\omega \beta}{j\omega + \lambda}},$$

and the corresponding expression for six groups of delayed neutrons is

$$\frac{\delta n(j\omega)}{n_0 \delta k(j\omega)} = \frac{1}{j\omega l^* + \displaystyle\sum_{i=1}^{6} \dfrac{j\omega \beta_i}{j\omega + \lambda_i}}. \tag{5.80}$$

This relationship gives the kinetic transfer function of the reactor;
in such a manner as to make it independent of the neutron densit
the relative change in neutron density $\delta n/n_0$, rather than δn, to be ~~the response~~
to the reactivity change δk.

5.189. There are various alternative ways of writing equation (5.80) but it can
be seen, in general, that since all the β_i's and λ_i's are known, it is possible to calculate the transfer function for a range of frequencies, using a given value for l^*,
the generation time. By employing the standard methods of complex number

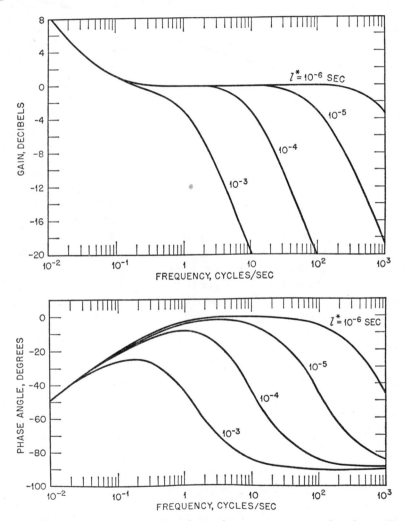

Fig. 5.18. Gain and phase angle plots for various neutron generation times. (Redrawn
by permission from "Control of Nuclear Reactors and Power Plants," by M. A. Schultz.
Copyright, 1961: McGraw-Hill Book Co., Inc.)

algebra, it is possible to break up the transfer function into its real and complex parts. The former gives what is variously called the "gain," "amplitude ratio," or "modulus," and the latter determines the phase angle, for each frequency. In the *Bode diagram*, the logarithm of the gain and the phase angle in degrees are plotted against the logarithm of the frequency. The gain is generally expressed in *decibels*, defined by

$$\text{Gain in decibels} = 20 \log K|G| = 20 \log K + 20 \log |G|,$$

the values being sometimes normalized, so that the gain is zero decibels when the frequency (either angular or in cycles per second) is unity for some particular condition.

5.190. The Bode diagram corresponding to equation (5.80) for uranium-235 and various values of l^* is depicted in Fig. 5.18; the curves for the gain have been normalized so that the value of 0 decibels occurs at 1 cycle per second when l^* is 10^{-4} sec [36]. Although more detailed methods of analysis are generally required, as explained below, it is at once evident from the gain curves that the reactor (without feedback) is unstable because the gain tends to become infinite as the frequency becomes smaller. The same conclusion can, of course, be reached directly from equation (5.80).

NEGATIVE TEMPERATURE COEFFICIENT FEEDBACK

5.191. It was seen earlier that the temperature effect is one of the factors contributing to the power coefficient of reactivity. A negative temperature coefficient, with a short time constant, can limit the rate of increase in the power level of a reactor and thus make it easier to control. Since the reactor power is proportional to neutron density, the power coefficient of reactivity can be represented by Cdk/dn, where dk is the change in effective multiplication (or reactivity) accompanying a change dn in the neutron density and C is a constant. The total change in reactivity is made up of various contributions, and the portion of the power coefficient which depends on the temperature may be written as Cdk_t/dn. In practice, it is convenient to split this temperature dependent part of the power coefficient of reactivity into two parts, viz.,

$$C \frac{dk_t}{dn} = \frac{dk_t}{dT} \left(C \frac{dT}{dn} \right),$$

where the first factor on the right is essentially equal to the ordinary temperature coefficient of reactivity considered earlier in this chapter. The second factor represents the rate of change of temperature, generally that of the fuel since it responds rapidly, with the reactor power. For simplicity, it will be taken to be constant in the treatment below.

5.192. The block diagram for a reactor with temperature coefficient feedback is equivalent to that in Fig. 5.16, and the overall system transfer coefficient is given by equation (5.71) as

$$KG = \frac{K_r G_r}{1 + (K_r G_r)(K_t G_t)},\tag{5.81}$$

where the subscripts r refer to the reactor and t to the temperature feedback. The transfer coefficient $K_r G_r$ may be taken to be that given by equation (5.80). It is required, therefore, to determine $K_t G_t$. This may be achieved in the following somewhat oversimplified manner.

5.193. It will be assumed that it is the fuel temperature only which affects the reactivity in a transient change. This is a good approximation, as was seen earlier, because the time constant for the fuel in a heterogeneous reactor is usually much less than for the moderator. However, the treatment can be generalized using an overall temperature change, rather than that of the fuel. Suppose an increase δn in the neutron density produces a total change of δT_f in the fuel temperature. It will be postulated that δT_f is directly proportional to δn, so that

$$\delta T_f = A \, \delta n \qquad \text{or} \qquad \delta n = \delta T_f / A,$$

where A is a constant which depends upon the conditions in the reactor.

5.194. The response time of the temperature coefficient feedback due to the change in the neutron density is determined by the rate at which the temperature of the fuel element increases by the amount δT_f. If δT is the temperature increase at any time t after the increase in the neutron density, a reasonable expression for the change in δT with time is given by

$$\delta T_f - \delta T = \tau \frac{d(\delta T)}{dT},$$

which implies that δT approaches δT_f asymptotically in an exponential manner, with τ as the time constant. This equation is seen to be analogous to equation (5.66), so that the Laplace transformation gives, as in equation (5.67),

$$\frac{\delta T(j\omega)}{\delta T_f(j\omega)} = \frac{1}{j\omega\tau + 1}.$$

5.195. The power coefficient of reactivity, as defined in § 5.172, is equal to αAC, where α is the ordinary (negative) temperature coefficient of reactivity. The feedback transfer function is then

$$\frac{\delta k(j\omega)}{\delta n(j\omega)} = \alpha AC \frac{\delta T(j\omega)}{\delta T_f(j\omega)} = \frac{K_t}{j\omega\tau + 1},$$

where K_t has been written for αAC. Hence, from equation (5.81),

$$KG = \frac{K_r G_r}{1 + K_r G_r \left[K_t / (j\omega\tau + 1) \right]}.\tag{5.82}$$

The gain and phase angle plots for $\tau = 0.159$ sec and l^* in equation (5.80) for $K_r G_r$ equal to 10^{-4} sec, derived from equation (5.82), are shown in Fig. 5.19 for several arbitrary values of K_t, corresponding to negative temperature coeffi-

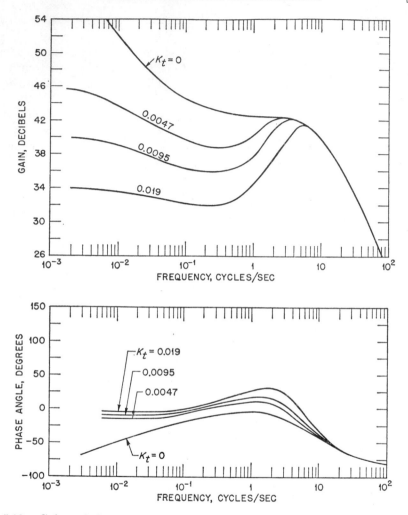

Fig. 5.19. Gain and phase angle plots for various temperature coefficients. (Redrawn by permission from "Control of Nuclear Reactors and Power Plants," by M. A. Schultz. Copyright, 1961: McGraw-Hill Book Co., Inc.)

cients [37]. Comparison with Fig. 5.18 indicates immediately the difference in the gain curve at the lower frequencies. The gain now remains finite as the frequency approaches zero and so stability is possible.

COOLANT LOOP TRANSFER FUNCTIONS

5.196. A nuclear plant system consists of a number of components in addition to the reactor core which must be taken into consideration in the analysis of the transient behavior.   These components include the coolant loop and the power-producing equipment.   Some aspects of the coolant loop are susceptible of simple analysis and these will be described here.   Consider, for example, the transfer of the coolant from reactor to heat exchanger, and vice versa (Fig. 5.20).

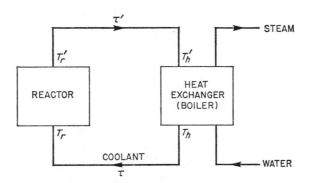

FIG. 5.20.   Transfer of coolant between reactor and heat exchanger

Suppose that there is no loss of heat in the piping but there is a simple time delay before a temperature change occurring at the outlet of the heat exchanger is apparent at the entrance to the reactor; this delay time is the average time required for the coolant to traverse the pipe length.   The relationship between the temperature T_h at the heat exchanger exit and T_r at the reactor entrance following a temperature change is

$$T_r(t) = T_h(t - \tau),$$

so that the temperature at the reactor lags by the time τ.   The Laplace transformation of this expression is

$$T_r(s) = T_h(s)e^{-s\tau}$$

and the transfer function is

$$\frac{T_r(s)}{T_h(s)} = e^{-s\tau}.$$

For a sinusoidal temperature change the variable s would be replaced by $j\omega$. Similarly, if τ' is the time lag between the reactor exit and the entrance to the heat exchanger, the transfer function is

$$\frac{T_h'(s)}{T_r'(s)} = e^{-s\tau'},$$

where the symbols have the significance shown in Fig. 5.20.

5.197. Another case is the delay associated with coolant mixing at a location where there is a change in the cross-sectional area of the flow channel.   Examples are the mixing occurring at the inlet header of the reactor or heat exchanger. Suppose a coolant flowing in a channel at temperature T_i and volume flow rate

F enters a wider channel of volume V where the temperature is T_o, and consider a time interval Δt.   In this time a volume $F\Delta t$ of coolant will have entered the wider channel and changed the temperature by ΔT_o, assuming complete mixing. A heat and volume balance then requires that

$$V(T_o + \Delta T_o) = F\Delta t + (V - F\Delta t)T_o.$$

For a small time interval, Δt becomes dt and ΔT_o is correspondingly dT_o, so that upon rearrangement it is seen that

$$\frac{dT_o}{dt} + \frac{F}{V}\, T_o = \frac{F}{V}\, T_i$$

or in Laplacian notation

$$sT_o(s) + \frac{F}{V}\, T_o(s) = \frac{F}{V}\, T_i(s),$$

provided the temperature was the same initially in both channels.   The transfer function is then

$$\frac{T_o(s)}{T_i(s)} = \frac{1}{(V/F)s + 1} = \frac{1}{\tau s + 1},$$

where V/F has the dimensions of time and is equivalent to the time constant τ. This result is similar to equation (5.67) and implies that the temperature in the wider channel approaches the temperature of the entering coolant in an asymptotic exponential manner, the time constant being determined by the ratio of the volume of the wider channel to the coolant flow rate.

5.198. The various transfer functions associated with the coolant loop are part of the temperature feedback loop.   The latter is thus seen to depend upon the heat transfer and other physical characteristics of the plant external to the reactor core.   To represent the transfer function of the reactor system as a whole it is necessary, therefore, to consider in detail the characteristics of each contributing component.

STABILITY ANALYSIS

5.199. One of the most important characteristics of any feedback control system is its stability.   A system may be considered to be stable if an excitation or disturbance results in transients which die out with increasing time.   It should be understood that the stability depends upon the system and not upon the nature of the disturbance or driving function.   Any excitation will cause an unstable system to oscillate in a divergent manner, a characteristic which would render any feedback control system useless.   The determination of stability conditions is thus a basic objective in control system analysis and design.   By the use of transfer functions and the algebraic methods devised for combining terms, complicated systems may be expressed by means of a single function.   The analysis of such functions, which provides an indication of performance characteristics, can be carried out in various ways. Graphical methods, such as those employing Bode (§ 5.190) or Nyquist plots, are often used [38].

TRANSIENT BEHAVIOR DURING NORMAL OPERATION

5.200. It should be recognized that in the development of the transfer function representation of a complicated system, such as a reactor plant, it is convenient to consider only small deviations from normal operating conditions. The control system equation can then be expressed in linear form. For large deviations, however, a separate analysis may be required. The main treatment here is concerned with small disturbances, but reference will be made later to large changes in operating conditions arising from accidental circumstances.

5.201. The Laplace transform changes the representation of system behavior from the time domain to the frequency domain. In principle, it is possible to convert the transfer function in the Laplace transform representation back to the time domain and thus determine the transient behavior of a stable system. Although this procedure is sometimes carried out, it will not be described here, but reference will be made to some qualitative rules relating to the behavior of a system in passing from one steady state to another.

5.202. It is seen that in Fig. 5.19 the logarithmic plots of gain versus frequency, for systems with appreciable negative temperature coefficients, exhibit maxima at particular frequencies; these are referred to as *resonances*. The existence of a resonance implies that the system can undergo oscillations with a frequency roughly equal to the resonance value. When the resonance peak is high, the system is underdamped, so that the oscillations may be quite large and there may be several cycles of overshooting and undershooting, before the steady state is attained. On the other hand, when there is no peak, the system is overdamped and it will change to the new state very slowly. Neither of these extreme conditions is satisfactory. In the design of the servomechanism for a reactor control system, the objective is to achieve optimum damping for the system as a whole, including the servomechanism, control rod actuators, temperature coefficient, poisoning feedback, and all other components contributing to the overall transfer function.

5.203. In general, a desirable situation is one in which the resonant frequency is high but the resonance peak is moderate; such a system will have a rapid response with very few oscillations which are damped out in a short time. A well-damped system of this kind will have a minimum response time about twice the reciprocal of the resonant angular frequency. It is of interest to mention that the frequency and height of the resonant peak, as derived from an analysis of transfer functions of experiments made at lower powers, led to the conclusion that the power of the EBWR could be safely increased to 50 Mw (§ 5.170). The peak height at 50 Mw was only about 2 decibels higher than at 20 Mw and the frequency was also slightly larger at 10 radians per sec.

REACTOR SIMULATORS

5.204. The analysis of reactor systems can be greatly facilitated by the use of electrical simulators [39]. A simulator, in general, is a "real time" analogue computer; that is to say, it is a device which displays the analytical behavior of a system on the same time scale as would apply to the system itself. Each component of the block diagram of the reactor system, such as that in Fig. 5.17, is represented by the appropriate electrical simulation. The simulator can then be used to determine the transfer function of the system and to study its stability. The effect of changes in the individual components (or blocks) can be investigated and valuable information obtained for utilization in the design of the reactor control system. In addition, simulators can be constructed to include as components the actual servomechanisms which drive the control rods, thus making possible the testing of equipment before installation in the reactor.

5.205. With the correct conversion factors, a simulator will act in a manner that is essentially identical with the actual system. There are, however, certain limitations upon the accuracy of the analysis performed with the aid of a simulator. In the first place, even if the simulator can be designed to obey a set of differential equations, it is doubtful whether the behavior of the simulated system is represented exactly by these equations. For example, an electrical simulator can be designed to satisfy the reactor kinetics equations obtained earlier in this chapter, but various approximations were made in deriving these equations, e.g., the assumption of linear behavior. Consequently, the simulator may not provide a complete representation of the actual kinetic behavior. Furthermore, it is not possible to take into account many transient or spatial variations of a relatively minor character that must inevitably occur within an active system, such as a power reactor. Bearing in mind these obvious limitations, it is still true that simulators can give results in a much shorter time that are at least as good as could be obtained in a more laborious manner by theoretical analysis.

TRANSIENT BEHAVIOR DURING ABNORMAL OPERATION

5.206. A somewhat different transient behavior to that discussed in the preceding sections is concerned with the response of the reactor system to large changes in operating conditions resulting from emergencies. Linearizing assumptions may not be permissible so that the transfer function analysis is unreliable. Behavior of the reactor system following some potentially dangerous occurrence can be described by a set of differential equations which must be solved simultaneously, usually by machine computation.

5.207. An example of the kind of behavior requiring analysis is that which would occur as a result of a complete loss of the primary coolant flow; this might be regarded as the "maximum credible accident" for many reactor systems (§ 9.121). The events which must be considered in the analysis of such abnormal behavior may be quite complicated and not at all the same as those treated earlier when there are only small perturbations from normal operating conditions. For instance, as a result of a large power increase or reduction in coolant flow, a sodium coolant could increase in temperature to such an extent that boiling would result. Pumping and flow characteristics then become very different for the consequent two-phase system than was previously the case. An unstable situation would then probably arise in a system which is normally stable.

REACTOR CONTROL INSTRUMENTATION [40]

INTRODUCTION

5.208. In addition to what might be called conventional instruments, such as are used for measuring temperatures, coolant flow rates, pressure drops, etc., the control and operation of a nuclear reactor require devices of a different type. The average power output of a reactor can be obtained from the product of the temperature rise of the coolant and its mass flow rate. However, for control purposes the determination of the power in this manner is too slow for safe operation to be possible. As already seen, the power level of a reactor, at any instant and at any location, is proportional to the neutron flux. Instruments are available that measure this flux (or density) with an almost instantaneous response; they are particularly suitable for indicating power levels and for providing signals to automatic control and safety mechanisms.

PRODUCTION OF ION-PAIRS

5.209. The property of neutrons most widely used for the measurement of flux is their ability to cause ionization indirectly in suitable circumstances. The problem of determining neutron flux is thus reduced to one of devising methods for the observation of the resulting ion-pairs. Both alpha and beta particles can form ion-pairs when they interact with atoms and molecules, especially in gases, and so they produce ionization directly. Gamma rays, on the other hand, cause most of their ionization indirectly. Each of the three major types of interaction of gamma rays with matter, i.e., the Compton effect, photoelectric effect, and pair production (see § 2.48 *et seq.*), results in the liberation of free (recoil) electrons. The energy of these electrons depends on the circumstances, but it is usually high enough for them to be capable of producing considerable ionization. Since all the nuclear radiations lead to the formation of ion-pairs, either directly or indirectly, they can be detected and measured by procedures depending on the same basic principles. Consequently, much of the subsequent treatment will be applicable to alpha and beta particles and gamma rays as well as to neutrons.

5.210. In order for neutrons to cause ionization, they must first take part in certain nuclear interactions which are accompanied by the emission of ionizing particles. Three main types of reactions have been used: (1) elastic scattering by hydrogen nuclei; (2) the (n, α) reaction with boron; and (3) fission. The first of these processes is applicable only to fast neutrons, of energy about 0.1 Mev or more; in their passage through hydrogen gas, or a compound rich in hydrogen, e.g., a hydrocarbon, the fast neutrons can transfer a large proportion of their energy to hydrogen nuclei by elastic scattering (§ 3.63). The resulting recoil protons, moving with high speed, produce considerable ionization in their passage through a gas.

5.211. The second and third reactions referred to above are used for the detection of slow neutrons. The reaction of such neutrons with boron-10 produces a helium nucleus (alpha particle) and a lithium nucleus (§ 2.85), both of which have relatively high energy and are capable of causing ionization.* Slow neutrons are able to induce fission of uranium-235 and the resulting fission fragments can produce large numbers of ion-pairs in their paths.

BEHAVIOR OF ION-PAIRS IN ELECTRIC FIELD

5.212. Many instruments for the detection of nuclear radiations are dependent upon the behavior in an electric field of the ion-pairs formed by the ionizing particles in their passage through a gas. This behavior varies with the potential gradient of the electric field and can best be explained by considering two electrodes, e.g., a metal cylinder as one and a wire running along its axis as the other (Fig. 5.21), placed in a vessel of gas and a potential difference established across them. As a general rule, the central wire is the positive electrode (anode) and the outer cylinder the negative electrode (cathode), so that (negative) electrons are attracted to the former and positive ions to the latter.

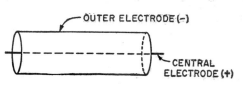

Fig. 5.21.   Electrical field for study of behavior of ion-pairs

5.213. Suppose a burst (or pulse) of ionizing radiation enters the gas between the electrodes so that a definite number of ion-pairs are formed. Under the influence of the electric field the positive ions will move toward the negatively charged electrode and the electrons will migrate toward the positive electrode. The magnitude of the charge collected on the electrodes will depend on the ap-

* The Li⁷ nucleus is usually in an excited state; the kinetic energy of the alpha particle is then 1.5 Mev (range 0.7 cm in air) and that of the lithium nucleus is 0.85 Mev (range 0.4 cm in air).

plied voltage, and the general nature of the variation is indicated in Fig. 5.22, in which the charge collected is shown (on a logarithmic scale) as a function of the voltage. Curve *A* refers to a case in which the radiation pulse produces 10 ion-pairs and curve *B* to a larger pulse producing 1000 ion-pairs. It can be seen that the curves can be divided into six fairly distinct regions, marked I to VI. Three of these regions, namely, II, III, and V, are used in various types of nuclear radiation instruments, and these will be considered in turn [41].

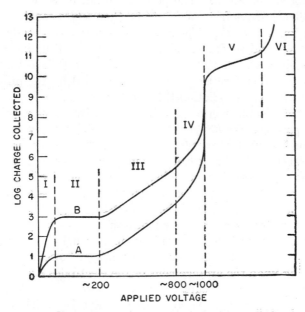

FIG. 5.22. Variation of charge collected with applied voltage

5.214. At very low voltages the charge collected at first increases with the applied voltage (region I) and then attains a constant value (region II). At very low potentials, the ions and electrons move slowly in the electrical field, with the result that many of them recombine before they reach the electrodes. With increasing voltage, the charged particles migrate more rapidly toward the respective electrodes; the extent of recombination is thus decreased and the charge collected on the electrodes increases. This accounts for the behavior in region I.

5.215. Above a certain voltage, the ions and electrons move so fast that the recombination is virtually zero, and essentially every charged particle produced by the nuclear radiation reaches the electrode. A further increase in the applied voltage will not result in any increase in the number of ion-pairs collected. Hence, for region II, referred to as the *ionization chamber* (or *ion chamber*) region, the charge remains constant as the voltage is increased. The charge collected is then equal to the total charge carried by the ion-pairs produced in the ionization chamber gas by the nuclear radiation. This is 10 unit charges (log 10 = 1) for case A and 1000 unit charges (log 1000 = 3) for case B, in Fig. 5.22.

5.216. The ionization chamber region has a limited range, and as the potential between the electrodes is increased, a new phenomenon appears. In region III, the quantity of charge collected for a given amount of radiation is larger than, but proportional to, the original amount of ionization; it has consequently been called the *proportional region*. The potential gradient is now so high that the (primary) electrons move fast enough to be themselves capable of causing secondary ionization in the gas. The secondary electrons so produced may cause further ionization and so on. A single ion-pair can thus initiate a chain or "avalanche" of ionizations. The charge collected on the electrodes in the proportional region is therefore larger than that carried by the primary ion-pairs formed by the passage of radiation through the gas.

5.217. The total number of ion-pairs produced by a single, primary ion-pair is called the *gas-amplification factor*. This factor is unity in the ionization chamber region (II), but it may become as large as 10^4 or more in the proportional region (III). The charge collected in both of these regions is equal to the original number of ion-pairs multiplied by the amplification factor. Consequently in Fig. 5.22 the curves A and B remain a constant distance apart, because of the logarithmic scale of the ordinates, in regions II and III.

5.218. It will be apparent from the figure that in the proportional region the gas-amplification factor increases rapidly, in a roughly exponential manner, with the applied voltage. For any given voltage, however, it has a definite value, depending upon the arrangement of the electrodes and the nature of the gas in which the ionization occurs.

5.219. As the voltage is increased, other factors arise which limit the production of secondary ion-pairs, so that the increase in the gas-amplification factor does not continue indefinitely. The atmosphere of positive ions around the central (positive) electrode gives rise to a space charge which causes a decrease in the electric field, thus tending to decrease the extent of gas amplification. At sufficiently high voltage, e.g., in region IV, the amplification factor consequently approaches a limit and the charge collected is no longer proportional to the initial ionization. Region IV is therefore referred to as the limited proportional region.

5.220. In region V, known as the *Geiger region*, the ionization avalanche, which has hitherto been confined to a point or small region of the (positive) electrode, spreads over the whole of its length. The fact that the curves A and B in Fig. 5.22 now coincide means that the total amount of charge collected is independent of the extent of the primary ionization, i.e., of the number of ion-pairs produced by the pulse of nuclear radiation. Thus, although the gas-amplification factor is very large, on the order of 10^8, it does not have a definite value, as it does in the proportional region. The charge collected increases with the applied voltage, but its amount does not depend on the nature or energy of the radiation causing the initial ionization. In the Geiger region (V), therefore, an ionizing particle giving rise to a large number of primary ion-pairs will cause essentially the same amount of charge to be collected on the electrodes as will a particle producing a small number.

5.221. If the applied voltage is increased beyond the Geiger region, there is a rapid increase in the charge collected; this represents region VI of Fig. 5.22. The potential is then so high that, once a discharge is initiated, others follow in rapid succession, and there is effectively a continuous discharge. This region is not used for the detection or measurement of ionizing radiations.

IONIZATION CHAMBERS

5.222. A number of different instruments for the measurement of nuclear radiations operate in the ionization-chamber region, i.e., where the charge collected on the electrodes is equal to that carried by the primary ion-pairs. Although there are marked differences even among the instruments in each category, ionization-chamber measurements fall into two types, namely, integrating and nonintegrating. In the integrating instruments, which will be considered first, the total quantity of charge carried by a number of ionizing particles is collected over a period of time. In the nonintegrating (or counting) devices, on the other hand, each particle capable of causing ionization is recorded separately.

5.223. Integrating devices also can be divided into two classes, which may be referred to as electrostatic and electrodynamic (or current indicating), respectively. Those of the electrostatic type are often called *electroscopes*, since they operate on the same principle as the familiar gold-leaf electroscope, which was used in some of the earliest measurements of radioactivity. Since electroscopes are not employed in reactor control, but in surveying (or monitoring) to protect personnel from the harmful effects of radiation, they will be discussed in Chapter 9.*

5.224. In the second type of integrating ionization chamber, a constant potential is maintained between the electrodes by means of a battery (Fig. 5.23). If ionizing radiation enters the chamber at a sufficiently high rate, the ions and electrons produced are swept continuously to the respective electrodes, and a steady current flows. The strength of this ion current is a direct measure of the rate of entry of the ionizing particles and hence of the radiation intensity. The ion current may be measured directly, or it may be determined by means of a high-impedance voltmeter V connected across a resistance R through which the current flows. The reading on the voltmeter is then proportional to the ion current and consequently to the radiation intensity.

5.225. Ionization chambers with current-measuring devices have been designed for many different purposes. In the parallel-plate chamber, as its name

---

* Other monitoring instruments, such as Geiger counters, are also described in Chapter 9.

implies, the electrodes are two parallel plates, which may be flat or curved (§ 5.230). Some chambers have central electrodes consisting of a single wire or a network of wires, insulated from the container, e.g., a cylindrical vessel, which serves as the other electrode. The electrode to which the measuring instrument

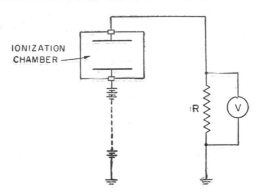

IONIZATION
CHAMBER

R

V

FIG. 5.23.   Integrating type of ionization chamber

is attached is called the collecting electrode. This is frequently maintained at (or close to) ground potential, while the other (high-voltage) electrode is kept at the required voltage to permit operation in the ionization-chamber region. The gas filling the chamber may be air, carbon dioxide, nitrogen, argon, or methane, among others.

5.226. For the detection of fast neutrons the ionization chamber is filled with hydrogen or with a hydrocarbon gas; alternatively, or in addition, the interior walls may be lined with a hydrogenous substance, such as paraffin or polyethylene. If exposed to fast neutrons, recoil protons of high energy will be liberated, as indicated above. The ion current due to the ionization produced by the recoil protons is then a measure of the neutron density (or flux).

5.227. In principle, the adaptation of an ionization chamber to the measurement of slow neutrons is a simple matter. All that is necessary is to coat one or both electrodes with boron, preferably enriched in the boron-10 isotope, since this is the one that has a high cross section for the reaction with slow neutrons. Alternatively, the chamber may be filled with gaseous boron trifluoride. Integrating ionization chambers for neutron detection have also been made by coating the electrodes with a uranium-235 compound.

5.228. For use in reactor control there are certain special requirements. The chamber must be small, since many are needed in a reactor, yet it must deliver a reasonably large ion current in order to provide rapid and smooth response. These two conditions mean that the chamber should be placed in a region of high neutron flux. As a result, problems of radiation damage (see Chapter 7), particularly to the insulators, become acute. Moreover, the chamber materials become highly radioactive; this not only means that repair is difficult but, more serious, the beta particles emitted by these materials interfere with the neutron measurement.   In some reactors, placing the chamber in a high neutron flux means that it is subjected to high temperatures, and this further complicates the problem of construction.

5.229. In order to overcome some of these difficulties, ionization chambers have been designed to operate in regions of relatively low flux.   In some instruments the electrodes are parallel and cup-shaped, thus providing a relatively large area within a small volume.   The same objective has been achieved in an ionization chamber using a number of flat, circular, parallel (boron-coated) plate electrodes, connected alternately.   This instrument, called the *PCP* (*parallel circular-plate*) *chamber*, was designed to respond quickly to a change in the neutron flux in the reactor core.

5.230. Since the gamma radiation, from fission and fission products, in a reactor can be very intense, instruments are often required for measuring neutron flux in the presence of gamma rays.   The *compensated ionization chamber* has been designed for this purpose.   It consists of three parallel electrodes, *A*, *B*, and *C*; these are shown in Fig. 5.24 as flat plates, but they are usually cup-shaped

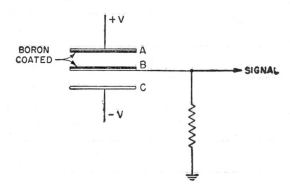

FIG. 5.24.   Compensated ion chamber for neutron detection

in order to increase the exposed areas.   The electrode *B* is the collecting electrode, which is grounded through the measuring circuit, and *A* and *C* are the accelerating electrodes, whose potentials are opposite with respect to ground. If the position of *B* is such that the volume *AB* and *BC* are roughly equal, the amounts of ionization produced in them by gamma rays will be approximately the same. The net ion current to the collecting electrode will then be almost zero.   If boron is coated on the surfaces of one volume of the chamber, e.g., *AB*, the net ion current will be a measure of the neutron flux.

5.231. By suitable modification of the output circuit an integrating ionization chamber may be used to indicate the logarithm of the neutron flux (or density). This type of instrument, called a *log n meter*, is particularly useful in the startup of a reactor when the neutron density covers a very wide range.   Its operation is based on the fact that, within certain current limits, the voltage across a thermionic or semiconductor diode is proportional to the logarithm of the current flowing through it. The current from a neutron-sensitive ionization chamber is taken directly to the plate of the diode and the voltage, after amplification, is indicated on a recorder. The readings are then directly related to the logarithm of the neutron density.

5.232. If the output from a log n meter is connected to an RC differentiating circuit, the current in the latter is a measure of the derivative of log n with respect to time, i.e., $d \log n/dt$. As seen in § 5.21, the reciprocal of the reactor period is equal to $d \ln n/dt$, and so the combination of a log n meter and an RC circuit gives an output which is proportional to $1/T_p$. When attached to a suitable indicator calibrated to read in seconds, the system constitutes a reactor *period meter*.

5.233. The recorder chart tracing of the output of a log n meter in a reactor is shown in Fig. 5.25 and that from an attached differentiating circuit in Fig. 5.26.

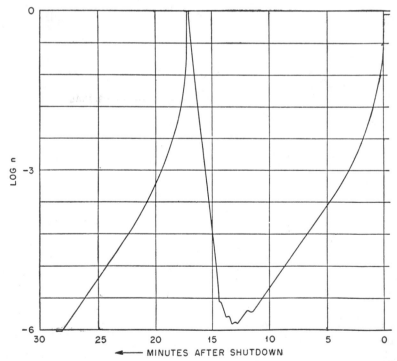

FIG. 5.25.   Recorder tracing accompanying insertion and withdrawal of control rods

The time scale runs from right to left in each case, and in Fig. 5.26 the record is given in two parts for convenience of presentation. The period during the first shutdown is seen to attain a stable value of about -80 sec (§ 5.47). When the reactor is restarted about 14 min after shutdown, there is a transient stage followed by a stable period of approximately 15 sec. A subsequent shutdown occurs at 17 sec.

5.234. If the output circuit of an ionization chamber has a time constant which is small in comparison with the average time between the arrival of successive ionizing particles in the chamber, the current is not continuous but is in the form of a series of pulses. Each pulse represents the entry of a particle into the ionization chamber. By connecting the resulting *pulse chamber* to an amplifier and a mechanical recorder, it is possible to count the individual pulses and hence the number of particles.

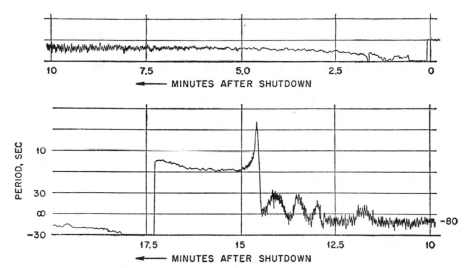

FIG. 5.26. Tracing of period indicator based on differentiation of log n curve

5.235. One of the important advantages of the pulse chamber is that it can be used to determine neutron densities in the presence of considerable intensities of beta particles and gamma rays such as exist in an operating reactor. If the ionization chamber is fairly small, e.g., a few centimeters between the electrodes, then the number of ion-pairs produced (indirectly) by a neutron will greatly exceed that due to entry of a beta particle or gamma-ray photon. The voltage pulses will, accordingly, be much greater with the strongly ionizing particles. It will be seen below that the output pulses from the chamber are fed into an amplifier followed by a counting-rate meter; if a discriminator circuit is included, the meter will respond only to the large pulses produced by the neutrons and not at all to the weak ones.

5.236. For detecting neutrons, the pulse chamber may be lined with boron or filled with boron trifluoride gas. More generally, however, the interior of the chamber (or the electrode surface) is coated with a uranium compound enriched with uranium-235; it is then called a *fission chamber*. The entry of a slow neutron causes fission of the uranium, and the resulting fragments produce considerable ionization in the gas. Fission chambers have been found particularly useful for measuring neutron flux during the early stages of reactor startup and shutdown. The pulse due to the fission fragments is so large that there is no difficulty in discriminating even against "pile-up" pulses from gamma rays. This is, of course, very important soon after shutdown when the gamma-ray intensity is considerable.

5.237. For use in reactor control, fission chambers are generally employed in association with a *counting-rate meter*, since the rate at which neutrons enter the chamber is a measure of the local neutron density. Insertion of a thermionic or semiconductor diode in the output of the fission chamber gives a *log count-rate meter* (or *log CRM*) which can be calibrated to indicate the neutron flux. If the voltage from the logarithmic circuit is applied to a differentiating system, the output can be used to indicate reactor periods, as described above.

PROPORTIONAL COUNTERS

5.238. The use of the proportional region III of Fig. 5.22 for the counting of ionizing particles has certain advantages. In the first place, the large internal (gas) amplification means that less external amplification is required to make the voltage pulses large enough to actuate a mechanical counter or scaler. In the second place, a properly designed proportional counter has a very short recovery time, and so very high counting rates are possible. Since the extent of gas amplification increases with the voltage between the electrodes, it is possible, by operating at moderate voltages, to count highly ionizing particles in the presence of those having a low specific ionization.

5.239. The electrodes in the proportional counter consist of a cylinder and a central wire, as in Fig. 5.21. The central wire is invariably made the anode (positive electrode) and this is usually the collecting electrode connected to the amplifier. The electric field between a cylinder and a central wire electrode is nonuniform, since it varies in proportion to $1/r$, where r is the distance from the wire. The field strength near the central electrode is therefore very large and a pulse of appreciable size, due to gas amplification, can result from the entry of an ionizing particle. In principle, the proportional region could be used in an integrating instrument in which the ion current is measured. But this would require such exact control of the electric field between the electrodes as not to be practical. Consequently, pulse counting is always employed, and this accounts for the name "proportional counter."

5.240. For the determination of neutron densities, the tube containing the electrodes is filled with boron (preferably enriched in boron-10) trifluoride gas for slow neutrons and with hydrogen for fast neutrons. By applying a suitable voltage bias to the counting instrument, e.g., a counting-rate meter, weak pulses produced by the beta particles and gamma radiation in a reactor are not recorded. In nuclear reactor work, proportional counters containing boron-10 trifluoride, often referred to as *boron counters*, are employed.   Because of their large internal amplification, they are especially useful in the early stages of startup when the neutron flux is very low.

REACTOR OPERATION

INSTRUMENTS FOR VARIOUS RANGES

5.241. Normal operation of a reactor may be divided broadly into three phases: startup, power operation, and shutdown.   In each phase, special instruments are required in order to indicate the power level at all times to permit safe control.   The neutron flux at any point in the reactor is proportional to the fission rate and hence also to the power level in the region where the flux is measured.   Because instruments can be made to respond very rapidly to changes in the flux, they provide a virtually immediate indication of local power variations.   For control purposes, therefore, they have, in certain respects, advantages over more conventional devices for indicating power levels.   During the early stages in the startup of a reactor, the power output is so small that instruments of the latter type would not register.   Under these conditions recourse must be had to ionization chambers and similar neutron detectors, in any event.

5.242. The safe operation of a reactor requires that the neutron flux (or fission rate) be known over a very large range; for example, the flux may vary by a factor 10^{10} or more between startup and operating level.   During startup it is

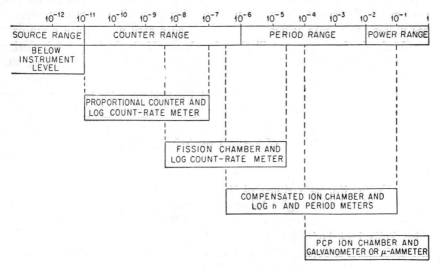

Fig. 5.27.   Reactor startup and operating ranges with appropriate instruments

quite possible for the reactor to become critical and for the reactor period to be very short while the neutron flux is still low.   If this occurred, the reactor power might increase so rapidly that dangerous levels would be reached before steps could be taken to reduce the fission rate.   Short periods should be avoided for safety reasons, and so the fission rate must be measured from the initial level at startup to the operating level, if possible.

5.243. No single instrument, nor even any one type of instrument, can be used for satisfactory measurement of neutron flux in a reactor over the whole possible range; as a result, it has been found convenient to divide the range into four regions.   Although these regions are not exactly defined, and will vary to some extent from one reactor to another, they are approximately as indicated in Fig. 5.27.   The right end of the scale, indicated by unity, represents full power of the reactor and the various negative powers of ten are the respective fractions of full power.   In the *power range*, the power (or flux) is from full power to 10^{-2} of full power; the *period range* covers from 10^{-2} to 10^{-6} of full power; the *counter range* is from 10^{-6} to 10^{-11}; while the *source range* is that below 10^{-11} of full power.   No suitable method is available for indicating the neutron flux in the source range, but none is needed because a source is usually present which brings the flux up into the counter range.

5.244. The ideal instrument for reactor control should be capable of detecting neutrons in the presence of strong gamma radiation; in addition, it should be accurate and respond rapidly to changes in neutron flux.   These requirements are sometimes in conflict.   When the neutron flux is low but the gamma radiation high, an ionization chamber or proportional counter in conjunction with a pulse-rate counting system is desirable because it is capable of discriminating between the large pulses due to neutrons and the small pulses produced by gamma rays.   However, since the neutron flux is small, the rate of arrival of pulses is low and a long integrating time is necessary if accurate readings are to be obtained; the response will then be slow.   If the instrument is designed to respond rapidly, the readings will be inaccurate.   At higher neutron fluxes the response of a counting-rate meter would be more rapid, for a given accuracy, but a limit is set by the resolving time of the system.   Fortunately, for large rates of arrival of pulses a current-measuring (or integrating) type of instrument is capable of accuracy and rapid response.   The effect of gamma radiation can then be overcome by the use of a compensated chamber.

Counter Range

5.245. During the early stages of the restart of a reactor after shutdown, the gamma radiation field is high while the neutron flux is relatively low.   A pulse-type ionization chamber (or proportional counter), with appropriate bias on the counter, is the best device for measuring the neutron flux under these conditions. A fission chamber is better in this connection than one containing boron, because the ionization pulses are much larger in the former case, thus facilitating discrimination against gamma radiation.   Since the neutron flux is small, the number of ionizing events in the chamber will be small and the accuracy of the reading will be low.   However, this does not matter in the circumstances, and so, at the commencement of the startup of a reactor, approximate values of the neutron flux are usually indicated by means of a pulse-type fission chamber or a proportional counter with boron trifluoride, combined with a biased counting-rate meter.   It is for this reason that the description "counter range" has been applied to the region in which the flux ranges from about 10^{-11} to 10^{-6} of full power.

5.246. Since the neutron flux during startup may vary over several powers of ten, in the counter range, it is convenient to use a meter that indicates the neutron counting rate on a logarithmic scale.   Such an instrument is the log count-rate meter (§ 5.237); meters reading from 1 count per sec to 10,000 per sec, i.e., from zero to four on the logarithmic scale, are used with many reactors. As the power level increases, the measuring range of the fission chamber or detector can be extended by moving it into a region of the reactor core where the flux is lower.   The withdrawal of the chamber from the high-flux region is an advantage, in any event, since the problems of radiation damage and induced radioactivity of the chamber materials are thereby decreased.

5.247. Because of the resolving time of the pulse chamber, the maximum rate of arrival of pulses, for which satisfactory readings are obtainable, is about 10^4 or 10^5 per sec.   When this condition is reached, the neutron flux is large enough to make possible the use of an integrating ionization chamber, thus taking advantage of its more rapid response.   Satisfactory operation of such a chamber commences when the ion current is about 10^{-10} amp, corresponding to about 10^4 particles (or pulses) per sec, and this is approaching the point where the pulse chamber ceases to be effective.   Since the number of ionizing events in the chamber is still relatively small, the accuracy and steadiness of the current will not be great at first, but both increase with increasing neutron flux.

Period Range

5.248. In the period range the main concern is the rate at which the neutron flux is increasing, and so the reactor period is determined by means of a meter which indicates the time derivative of the logarithm of the neutron flux (§ 5.232). The magnitude of the flux is recorded simultaneously by means of a log n meter. Since it is desirable to know the reactor period at as low a power (or neutron) level as possible, so that dangerously short periods may be detected, a period meter is used in conjunction with a compensated ionization chamber, such as the one described in § 5.230.   In this way the effect of the large gamma-ray intensity is eliminated.

Power Range

5.249. In the power range, i.e., from about 10^{-2} of full power to full power, accuracy and sensitivity are the essential requirements of the indicating device, especially if automatic control is employed. For this purpose an ionization chamber is used, but, instead of logarithmic amplification, the ion current, after necessary linear amplification, is recorded directly on a galvanometer or micro-ammeter appropriately calibrated to read neutron flux or reactor power. The output of such an instrument may be matched against a prescribed value, representing the desired full power of the reactor, and the difference applied to the automatic control-rod actuating mechanism. As long as any difference exists, the neutron flux (and power level) is permitted to increase at a controlled rate but when full power is reached the increase is halted automatically. The reactor is then maintained at the desired power level in an analogous manner; transient changes in flux, as indicated by means of the ion chamber, are compensated for by prompt, automatic movement of a regulating rod.

5.250. As it has a very rapid response, the parallel-plate (PCP) ionization chamber referred to in § 5.229 is often used for automatic control. It is true that the ion current includes some contribution from gamma radiation, but this is not important since it can be allowed for in the reference voltage (or current) against which the output of the chamber is matched. For the same reason as given above, the parallel-plate chamber is recommended for the actuation of safety circuits.

5.251. In the foregoing discussion it has been assumed, for convenience, that the counter, period, and power ranges are more or less distinct. This is, of course, not the case, and the various instruments used overlap the boundaries, as indicated in Fig. 5.27, thus providing continuity of indication of neutron flux or power level. The log CRM instrument, for example, has a range of 10^4, part of which is covered by the log n meter with a range of about 10^6; finally, the microammeter or galvanometer used in the power region is usually provided with shunts which give an operating range of 10^4 or 10^5.

Operating Level

5.252. When a reactor has reached its normal operating level, the *neutron thermopile*, which is a sturdy and reliable instrument not much larger than a fountain pen, may be useful. This device contains some 40 thermocouples, alternate junctions being coated with boron. The heat generated when the boron interacts with neutrons is capable of producing an appreciable voltage. Thus, in a thermal flux of 10^{12} neutrons/$(cm^2)(sec)$, the output is approximately 10 millivolts. Since thermocouples have a low impedance, it is not necessary to maintain high insulation resistances, such as are required for ionization chambers. Their chief disadvantage is a relatively long response time, of the order of seconds, so that they are not recommended for control purposes. Nevertheless, neutron thermopiles can play a part in reactor instrumentation.

5.253. A single neutron detector placed at a particular point in a reactor indicates only the flux (or power level) in its immediate vicinity, but does not give a good idea of the average power level of the reactor as a whole. The flux in the region of the detector may be considerably influenced by absorbers in the vicinity and by the position of the rods used to control the reactor. A much more satisfactory indication of the overall power level would be provided by indication of the neutron flux averaged over the reactor volume. This can be approximated by placing a number of detectors at various points in the reactor and connecting their output terminals in an appropriate manner. The relatively large volume occupied by an ionization chamber makes its use as a detector impractical in this connection, but boron thermopiles are small enough to be very suitable for the purpose of recording the average reactor power level.

5.254. Boron thermopiles can be used as a part of the safety system of a reactor even though their response is slow. Certain corrective actions of the control rods can be initiated automatically when the average power level, as indicated by the thermopiles, exceeds the permitted full operating level of the reactor by a specified amount. In addition, boron thermopiles have been employed for monitoring neutron flux at a given point in the reactor, for example, where an experiment is in progress. Knowing that the reactor power has been held constant is not sufficient for certain experiments, because of the local "shadowing" effects, etc.

5.255. In reactors using a coolant which contains oxygen, e.g., water, air, or carbon dioxide, the activity of the gamma radiation from nitrogen-16, produced in the coolant by the action of fast neutrons on oxygen nuclei, can be employed as a power indicator [42]. A gamma-ray detector located in the coolant outlet will respond to changes in the average power of the reactor. A single detector of this kind will provide the same type of information as is obtained from a large number of boron thermopiles distributed throughout the reactor core.

Summary

5.256. Some of the general conclusions based on the foregoing discussion are summarized in Table 5.9 which lists the types of neutron measuring instruments in most common use for various aspects of reactor control and operation.

TABLE 5.9.   REACTOR INSTRUMENTATION FOR NEUTRON MEASUREMENT

Purpose	Detector	Indicating Meter
Power indication (Counter range)	Fission (ion) chamber or BF_3 proportional counter	Counting-rate meter (logarithmic scale)
Power indication (Period range)	Compensated ion chamber	Log n meter
Period indication (Period range)	Compensated ion chamber	Period meter
Power indication (Power range)	Compensated ion chamber	Galvanometer or micromicroammeter
Automatic control (Operating level)	Parallel-plate (PCP) chamber	(To servomechanism)
Power indication (Operating level)	Boron thermopile	Voltage indicator

INITIAL STARTUP

5.257. It is during the startup of a reactor that the problem of control requires careful attention in order to avoid the possibility of an accident. In a reactor designed for high neutron flux and having considerable excess (built-in) reactivity, the reactor period can become quite short during startup. If, as a result of operator or mechanical failure, this is allowed to continue through the prompt-critical stage, the power will very rapidly exceed the normal operating level and the consequences could be serious [43]. Actually, three different types of startup may be distinguished. There is, first, what may be called "initial startup," frequently referred to as startup from "cold," which applies to the starting of a new reactor, one which has been refueled, or one which has been shut down for several days. Second, there is startup within a short time after accidental or deliberate shutdown; this is sometimes called "startup after scram." Finally, it may be necessary to start up the reactor after a reduction in power, e.g., to about 0.1 or 1 per cent of full operating power; this has been designated as "power range startup."

5.258. All startup operations have the same objective, namely, bringing the reactor to critical and then increasing the neutron flux (and power output) in a carefully controlled manner until the desired level is reached. Initial startup requires special attention because the neutron level may be so low that measurements with the usual instruments are not very reliable. In any event, it is of the utmost importance that no control rod be moved until the neutron flux is large enough to be detected, on a counting-rate meter at least. This detectable

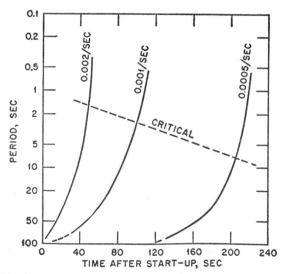

FIG. 5.28. Reactor period as function of time for various rates of increase of reactivity

flux may arise from spontaneous fission or it may result from the inclusion of a special neutron source in the reactor, as will be explained below.

5.259. Even in these circumstances, there is a possibility that the reactor may go through prompt critical on a short period; this situation can be avoided by careful regulation of the rate of increase of reactivity, i.e., of the rate of withdrawal of the control rods, during initial startup. An examination of Fig. 5.28 will prove instructive in this connection. It gives the reactor period* as a function of the time after startup for various rates of increase of reactivity. The initial effective multiplication factor was taken to be 0.9 in each case, and the neutron generation time in the reactor was 10^{-4} sec. It is seen that when the rate of change of reactivity is 0.002 per sec the reactor becomes critical after 50 sec and the period is then only about 1.5 sec. However, by increasing the reactivity more slowly, e.g., 0.0005 per sec, the period when the reactor becomes critical is some 7 sec. Whereas the former period is dangerously short, the latter is relatively safe. As a precautionary measure, a reactor may have a safety circuit associated with the period meter which prevents any further insertion of the control rods when the period reaches a predetermined small value.

5.260. For large natural uranium reactors the chances of going into a short period during startup are less than in other systems. The prompt neutron lifetime in reactors moderated by heavy water or graphite are relatively long (cf. Table 5.1). Consequently, a given reactivity greater than 0.005 will correspond to a considerably longer period than would be the case if the generation time were shorter (§ 5.45). Furthermore, spontaneous fission of uranium-238 yields about 15 neutrons per sec per kilogram of natural uranium, so that the resulting neutron density may be sufficiently large to detect. In this event, the neutron level can be readily followed by instruments during the startup operation.

5.261. Because of the small total mass of fuel in an enriched reactor, as compared with one employing natural uranium as fuel, and the small amount of uranium-238 present, spontaneous fission alone will not produce neutrons at a sufficient rate for their detection to be possible. It is consequently necessary that an additional neutron source, e.g., one consisting of polonium and beryllium (§ 2.72), be built into the reactor, for initial startup at least. Sources such as this yielding 10^6 to 10^7 neutrons per sec are readily available, and so an adequate indication on a fission counter or a proportional counter containing boron is possible before the control rods are moved. It is thus feasible to follow the neutron level by means of suitable instruments over the entire range, from startup to the operating power level. The possibility of a very rapid rise in the neutron flux, i.e., a very short period, without the operator being aware of the situation, is thus very greatly decreased.

5.262. Because instrumentation at low neutron levels, at the lower end of the

---

* The term "period," as used here, is the reciprocal of $d \ln n/dt$ at a particular instant; it is not a true (stable) reactor period.

counter range, is very unreliable, it is sometimes inadvisable to withdraw the control rods continuously. Intermittent withdrawal, e.g., 0.5 sec on and 4.5 sec off, can then provide an important element of safety. As long as the reactor is subcritical, the neutron flux will increase during the "on" period, due to subcritical multiplication, but will tend to become constant (or decrease) during the "off" period. When the reactor passes through the critical state, the flux continues to increase even during the "off" period. In this way, the operator will become aware that the system has gone critical, and he can then take precautions to prevent the reactor "running away."

STARTUP AFTER SCRAM

5.263. Startup after a brief shutdown is a much less hazardous process than initial startup. If the delay has not been too great, the neutron level will probably be high enough to give good instrumental information. There are several factors contributing to this situation. One is the limiting effect of the delayed neutrons, so that even 10 min after shutdown the neutron flux will have decreased by a ratio of only about 10^5 and it will still be in the instrument range. Another factor is the formation of neutrons by the action of the fission product gamma radiation upon various reactor materials. This will be particularly important when heavy water or beryllium is employed as moderator or reflector. For some time after shutdown the subcritical reactor will thus behave as if it contained a large primary source. When the reactor is started up again, this large source and the subcritical multiplication will give a neutron flux for which instrument response will be possible even prior to withdrawal of the control rods. Period information, however, may be unreliable because of the high gamma-ray background.

5.264. Because of the growth of the xenon concentration after shutdown in a high-flux thermal reactor, it may be desirable to start up again as soon as possible. If this is done within a short time, the startup would be carried out as described above. However, if the delay is long enough to permit the xenon concentration to grow to such an extent that there is insufficient built-in reactivity available to override the poisoning, then it will be necessary either to replace some of the fuel elements or else to allow the reactor to stand until the xenon has decayed. In this case, as in other instances where the shutdown has been prolonged, the neutron level will have decreased greatly and the startup procedure will be essentially the same as for initial startup.

POWER RANGE STARTUP

5.265. In certain circumstances the power level of a reactor will be reduced to a small fraction, e.g., 0.1 to 1 per cent, of its full operating power. While this is quite small from the power standpoint, the neutron flux is so high that reliable information is available from both period and neutron level instruments. Startup under these conditions presents no special problem.   It can be performed manually, if desired, or it can be automatically controlled by a servo-system. All that is necessary is to set a period that will bring the reactor up to its normal power at a reasonably fast rate compatible with safety.

NORMAL OPERATION OF REACTOR

5.266. During normal operation of a reactor at the required power level the automatic control system compensates for transient changes in reactivity, such as might result from minor variations in coolant flow, as well as for steady changes due to fuel depletion and the accumulation of poisons.   An error signal, representing the difference between the demand power (or neutron density) and the actual power, is fed to a servomechanism which then causes the regulating rod to be moved in such a manner as to reduce the error to zero.   From the treatment in § 5.188 it can be shown that $\delta n/\delta k$ is proportional to n_0, the neutron density corresponding to the demand power level.   In order that the servo-mechanism may be suitable for operation over a range of neutron densities (or power levels), the automatic control system is designed so as to respond to an error of $\delta n/n_0$ rather than δn.

SHUTDOWN OF REACTOR

5.267. There are, in general, two ways in which a reactor may be shut down: controlled (or deliberate) and enforced (or "scram"). If the controlled shutdown is to be of relatively long duration, e.g., for refueling or for maintenance purposes, then the procedure adopted is immaterial. All the control rods may be inserted as rapidly as desired. On the other hand, if the shutdown is expected to be temporary, while an external repair is performed, then the neutron level should be decreased, by gradual insertion of the shim rods, until the power output is the maximum that can be tolerated with safety. This will not only minimize the xenon buildup, but will also simplify the restart operation. In deciding upon a safe degree of shutdown, it must be remembered that considerable heat will continue to be liberated as a result of radioactive decay of the fission products.

5.268. In order to ensure continued operation of the reactor, enforced shutdown (or scram) should be avoided as far as possible. Small fluctuations in power, which may be caused by variation in the coolant, etc., or apparent fluctuations, which may be due to changes in the controlling instruments and their associated circuits, should not be able to produce a scram. Thus, the safety system of a reactor can be designed so that in the event of a lesser emergency there is a time delay before the scram is initiated; this is referred to as a slow scram. Only when the power gets dangerously high, e.g., 50 per cent above the normal level, or the period becomes dangerously low, e.g., 1 sec, will a fast scram occur automatically.

5.269. One of the most potentially dangerous situations which could arise in a reactor is the too rapid withdrawal of the control rods during startup, leading to what is called a startup accident (§ 5.251). It is in these circumstances, in particular, that a fast scram is essential. There is invariably some delay time between the receipt of a scram signal and the insertion of the safety rods into the core. In order to avoid damage to the reactor, this delay should be as short as possible, generally 30 to 50 millisec.

SYMBOLS USED IN CHAPTER 5

A	constant
$A_0, A_1, \ldots$	constant coefficients
a	half-thickness of cruciform control rod
B^2	buckling
$B_0, B_1, \ldots$	constant coefficients
C	constant
C	precursor nuclei/cm³ (weighted average)
C_0	steady-state concentration of precursor nuclei
C_i	precursor nuclei of ith kind/cm³
C_{i0}	C_i at $t = 0$
D	diffusion coefficient
E	energy
E_0	fission energy
E_{th}	thermal energy
e	base of natural logarithms
f	thermal utilization
f'	thermal utilization with poiso
$G(j\omega)$	complex quantity in transfer function
$|G(j\omega)|$	real part of $G(j\omega)$
h	linear extrapolation distance
I	effective resonance integral
I	iodine concentration (nuclei/cm³)
I_0	iodine concentration at equilibrium
Ih	inhour unit
j	square root of -1
K	gain factor in transfer function
KG	transfer function
k_∞	infinite multiplication factor
k_{eff}	effective multiplication factor
k'_{eff}	k_{eff} with poison
L	diffusion length

l	half-length of arm of cruciform rod
l	neutron lifetime
l^*	neutron generation time
$\bar{l}$	effective generation time
m	mass
m	poison cell dimension
N	nuclei/cm³ (of indicated species)
N_t	total number of nuclei (of indicated species)
n	neutron density (neutrons/cm³)
n_0	steady-state neutron density
P	nonleakage probability
P_{th}	nonleakage probability of thermal neutrons
P_{sl}	nonleakage probability while slowing down
p	resonance escape probability
q	rate of formation of neutrons
R	core radius
S	neutron source strength
s	complex variable
T	temperature
T_f	fuel temperature
T_m	moderator temperature
T_p	reactor period
T_0	initial reactor period
t	time
t_s	time after shutdown
t_{max}	time for maximum poisoning after shutdown
V	Volume
v	velocity
X	xenon concentration (nuclei/cm³)
X_0	xenon concentration at equilibrium
$X(t_s)$	xenon concentration at time t_s after shutdown
z	ratio Σ_u/Σ_m
α	temperature coefficient of reactivity
α_m	coefficient of volume expansion of moderator
α_r	coefficient of volume expansion of reactor
β	total fraction of delayed neutrons
β_i	fraction of delayed neutrons of ith kind
γ_I	fission yield of iodine-135
γ_P	fission yield of promethium-149
γ_X	fission yield of xenon-135
η	fission neutrons per neutron absorbed in fuel
θ_i	input signal
θ_o	output signal

λ	radioactive decay constant
λ_i	precursor decay constant (ith kind)
λ^*	effective decay constant ($\lambda + \sigma\phi$)
λ_I	decay constant of iodine-135
λ_P	decay constant of promethium-149
λ_X	decay constant of xenon-135
ν	fission neutron per neutron absorbed in fission
ρ	reactivity
Σ_a	macroscopic absorption cross section
Σ_f	macroscopic fission cross section
Σ_m	moderator macroscopic absorption cross section
Σ_p	poison macroscopic absorption cross section
Σ_s	macroscopic scattering cross section
Σ_{tr}	macroscopic transport cross section
Σ_u	fuel macroscopic absorption cross section
σ_a	absorption cross section
σ_s	scattering cross section
σ_I	absorption cross section of iodine-135
σ_S	absorption cross section of samarium-149
σ_X	absorption cross section of xenon-135
τ	time constant
τ	Fermi age
ϕ	phase angle
ϕ	neutron flux
ϕ_0	neutron flux at $t = 0$
ϕ_m	neutron flux in moderator
ϕ_u	neutron flux in fuel
ω	angular frequency
ω	parameter (reciprocal time)
$\omega_0, \omega_1, \ldots$	values of parameter ω (roots of equation 5.16)
ω_0	reciprocal of stable period
∇^2	Laplacian operator

REFERENCES FOR CHAPTER 5

1. S. Glasstone and M. C. Edlund, "The Elements of Nuclear Reactor Theory," D. Van Nostrand Co., Inc., Princeton, N. J., 1952, pp. 30 *et seq.;* R. V. Meghreblian and D. K. Holmes, "Reactor Analysis," McGraw-Hill Book Co., Inc., New York, 1960, pp. 573 *et seq.*
2. H. S. Isbin and G. W. Gorman, *Nucleonics,* **10**, No. 11, 68 (1952); for a graphical procedure, see H. Smets, *Nucleonics,* **17**, No. 4, 104 (1959).
3. R. E. Skinner and E. R. Cohen, *Nuc. Sci. and Eng.,* **5**, 291 (1959).
4. G. R. Keepin and T. F. Wimett, *Nucleonics,* **16**, No. 10, 86 (1958).
5. A. Lundby, *Nucleonics,* **11**, No. 4, 16 (1953).

6. E. Fermi, *Phys. Rev.*, **72**, 16 (1947).
7. J. P. Franz, U. S. AEC Report AECD-3260 (1951); see also, A. M. Weinberg and E. P. Wigner, "The Physical Theory of Neutron Chain Reactions," University of Chicago Press, 1958, p. 595.
8. A. M. Weinberg and E. P. Wigner, Ref. 7, pp. 482, 635, 680; R. V. Meghreblian and D. K. Holmes, Ref. 1, p. 308.
9. E. Creutz, *et al.*, *J. Appl. Phys.*, **26**, 257 (1955); R. M. Pearce, *J. Nuc. Energy*, **A**, **13**, 150 (1961).
10. C. Starr and R. W. Dickinson, "Sodium Graphite Reactors," Addison-Wesley Publishing Co., Inc., Reading, Mass., 1958, pp. 52, 239; J. R. Dietrich and W. H. Zinn (Eds.), "Solid Fuel Reactors," Addison-Wesley Publishing Co., Inc., Reading, Mass., 1958, p. 721.
11. C. Starr and R. W. Dickinson, Ref. 10, p. 241.
12. W. J. McCarthy, *et al.*, *Proc. Second U. N. Conf. Geneva*, **12**, 212 (1958); *Power Reactor Technology*, **4**, No. 4, 14 (1961).
13. S. Glasstone and M. C. Edlund, Ref. 1, p. 329; R. V. Meghreblian and D. K. Holmes, Ref. 1, p. 610; A. M. Weinberg and E. P. Wigner, Ref. 7, p. 599.
14. *Power Reactor Technology*, **1**, No. 2, 12 (1958); D. Randall and D. S. St. John, *Nucleonics*, **16**, No. 3, 82 (1958); G. C. Fullmer, *Nuc. Sci. and Eng.*, **9**, 93 (1961).
15. A. M. Weinberg and E. P. Wigner, Ref. 7, p. 83; R. W. Deutsch, *Nucleonics*, **14**, No. 9, 89 (1956); J. D. Garrison and B. W. Roos, *Nuc. Sci. and Eng.*, **12**, 115 (1962).
16. C. R. Greenhow and E. C. Hansen, U. S. AEC Report KAPL-2172 (1961).
17. M. A. Schultz, "Control of Nuclear Reactors and Power Plants," McGraw-Hill Book Co., Inc., New York, 2nd Ed., 1961; "Power Reactor Control," *Nucleonics*, **16**, No. 5, 61–92 (1958).
18. "Symposium on Reactor Control Materials," *Nuc. Sci. and Eng.*, **4**, 357–494 (1958); W. K. Anderson, *Nucleonics*, **15**, No. 1, 44 (1957); H. P. Iskenderian, *ibid.*, **15**, No. 11, 150 (1957); J. A. Ransohoff, *ibid.*, **17**, No. 7, 80 (1959); J. L. Russell, Jr., *et al.*, *ibid.*, **18**, No. 12, 88, 94 (1960); *Power Reactor Technology*, **3**, No. 3, 26 (1960); **4**, No. 4, 39 (1961).
19. V. L. McKinney and T. Rockwell, U. S. AEC Report AECD-3625 (1949).
20. T. H. Middleham, *et al.*, *J. Iron and Steel Inst.*, **187**, 1 (Sept. 1957).
21. I. Cohen, *et al.*, *Nucleonics*, **16**, No. 8, 122 (1958).
22. A. Radkowsky, *Proc. Second U. N. Conf. Geneva*, **13**, 426 (1958); R. V. Meghreblian and D. K. Holmes, Ref. 1, p. 621.
23. M. C. Edlund and G. K. Rhode, *Nucleonics*, **16**, No. 5, 80 (1958).
24. S. Glasstone and M. C. Edlund, Ref. 1, p. 318; A. M. Weinberg and E. P. Wigner, Ref. 7, p. 754.
25. H. Etherington (Ed.), "Nuclear Engineering Handbook," McGraw-Hill Book Co., Inc., New York, 1958, p. 8–16.
26. A. F. Henry, U. S. AEC Report TID-7532, Part 1 (1957), p. 3; U. S. AEC Report WAPD-218 (1959); see also, W. H. Arnold, Jr., U. S. AEC Report YAEC-62 (1959).
27. H. Etherington, Ref. 25, p. 5–114; F. J. Jankowski, *et al.*, *Nuc. Sci. and Eng.*, **2**, 288 (1957); W. R. Kimel, *et al.*, *ibid.*, **6**, 233 (1959); W. J. Sturm (Ed.), "Reactor Laboratory Experiments," U. S. AEC Report ANL-6410 (1961), p. 33.
28. D. J. Hughes, "Pile Neutron Research," Addison-Wesley Publishing Co., Inc., Reading, Mass., 1953, p. 196; W. J. Sturm, Ref. 27, p. 59.
29. A. M. Weinberg and H. C. Schweinler, *Phys. Rev.*, **74**, 851 (1948); J. I. Hoover, *et al.*, *ibid.*, **74**, 864 (1948); W. J. Sturm, Ref. 27, p. 66.
30. M. A. Schultz, Ref. 17, Chs. 5, etc.; A. Hitchcock, "Nuclear Reactor Stability," Simmons-Boardman Books, New York, 1960.

31. W. E. Nyev and S. G. Forbes, U. S. AEC Report IDO-16701 (1961). See also, *Power Reactor Technology*, **5**, No. 1, 22 (1961); this quarterly journal frequently reviews work on reactor stability.
32. A. W. Kramer, "Boiling Water Reactors," Addison-Wesley Publishing Co., Inc., Reading, Mass., 1958, pp. 358 *et seq.*
33. See R. V. Churchill, "Modern Operational Mathematics in Engineering," McGraw-Hill Book Co., Inc., New York, 1944.
34. The subject of transfer functions is developed in textbooks on servomechanisms and related topics; see, for example, J. D. Trimmer, "Response of Physical Systems," John Wiley and Sons, Inc., New York, 1950; G. S. Brown and D. P. Campbell, "Principles of Servomechanisms," John Wiley and Sons, Inc., New York, 1951; H. Chestnut and R. W. Mayer, "Servomechanisms and Regulating System Design," John Wiley and Sons, Inc., New York, 1951; G. J. Thaler and R. G. Brown, "Servomechanism Analysis," McGraw-Hill Book Co., Inc., New York, 1953; and others.
35. J. M. Harrer, *et al.*, *Nucleonics*, **10**, No. 8, 32 (1952); "Transfer Function Measurement and Reactor Stability Analysis," U. S. AEC Report ANL-6205 (1960).
36. M. A. Schultz, Ref. 17, p. 115.
37. M. A. Schultz, Ref. 17, p. 141.
38. Refs. 30, 34, and 35.
39. M. A. Schultz, Ref. 17, Ch. 14; H. Etherington, Ref. 25, pp. **8**–89 *et seq.*
40. H. Etherington, Rev. 25, pp. **8**–49 *et seq.; Power Reactor Technology*, **5**, No. 1, 37 (1961).
41. D. H. Wilkinson, "Ionization Chambers and Counters," Cambridge University Press, New York, 1950; E. Segrè (Ed.), "Experimental Nuclear Physics," John Wiley and Sons, Inc., New York, 1953, Vol. 1, Part I; S. A. Korff, "Electron and Nuclear Counters," D. Van Nostrand Co., Inc., Princeton, N. J., 1955; W. J. Price, "Nuclear Radiation Detection," McGraw-Hill Book Co., Inc., New York, 1958.
42. E. E. Drucker and W. D. Wallace, *Nuc. Sci. and Eng.*, **3**, 215 (1958).
43. H. Hurwitz, Jr., *Nucleonics*, **12**, No. 3, 57 (1954); *Nuc. Sci. and Eng.*, **6**, 11 (1959).

PROBLEMS

1. Compare the prompt thermal-neutron lifetime in plutonium-239 with that in uranium-235.

2. Determine the weighted average (one-group) decay constant for the delayed fission neutrons from (a) uranium-233, and (b) plutonium-239.

3. From the results of Problem 2, compare qualitatively the effect of a given step change in the effective multiplication factor on the neutron flux in thermal reactors employing (a) uranium-233, (b) uranium-235, and (c) plutonium-239 as fuel.

4. A reactor with uranium-235 as fissile material undergoes a step increase in reactivity of 0.005 after operating in the steady state; the neutron generation time is 10^{-4} sec. Plot $\phi(t)/\phi_0$ on a semilogarithmic scale against the time t in seconds, assuming a single average group of delayed neutrons.

5. Assuming one average group of delayed neutrons, plot the variation of thermal-neutron flux with time (for the first 2 sec) after a natural-uranium reactor undergoes a sudden increase of 0.0015 in reactivity. The neutron generation time in the reactor is 10^{-3} sec.

6. For the conditions of the preceding problem, estimate the time required for the transient contribution to the neutron flux to decrease to about 1 per cent of the total. Determine (a) the initial and (b) the stable value of the reactor period.

7. An air-cooled, graphite-natural uranium reactor, in which the thermal-neutron generation time is 10^{-3} sec, is operating at a steady power level. A small cylinder of slightly enriched uranium is inserted rapidly in the reactor and the neutron flux is observed to increase with a stable period of 8.33 min. Calculate the reactivity in (a) ordinary units, (b) in inhours, and (c) in dollars.

8. From the data in Table 5.1 and equation (5.37), plot a curve relating the reactivity to the stable period for a reactor in which the neutron generation time is 10^{-3} sec. (Determine ρ for T values of 1, 5, 10, 25, and 50 sec and plot ρ versus $1/T$.) At what period does the magnitude of the neutron lifetime become significant?

9. Compare the results of taking into account all six groups of delayed neutrons, as in the preceding problem, by plotting ρ against $1/T$ for the case of one average group of delayed neutrons.

10. A Water Boiler reactor has a fuel solution consisting of 12.7 liters of aqueous (enriched) uranyl nitrate of density 1.1 g/cm^3 contained in a sphere 12 in. in diameter. It is operating at a steady power level of 40 kw when a step increase of 10 cents in reactivity occurs. Develop the information needed to plot (a) the neutron flux and (b) the temperature as a function of time. Indicate, in a semiquantitative manner, the variation of these two quantities on the same graph.

11. The reactor in a nuclear rocket must reach full power in 30 sec after startup. (a) If the power is to increase by a factor of 10^5 during this time, estimate the stable period required. (b) What is the corresponding reactivity if a step increase is assumed?

12. Use the data in Example 4.2 to estimate the reactivity temperature coefficient of the heavy-water, natural-uranium reactor due to neutron leakage at a temperature of 27°C. The coefficient of volume expansion of heavy water is 3.0×10^{-4} per °C. The volume of the reactor may be taken to remain constant.

13. After operating at a thermal flux of 10^{14} neutrons/(cm²)(sec) for a long time, a reactor is shut down completely. How long will it take the xenon concentration to rise to a maximum? What is the value of the xenon poisoning at that time?

14. Repeat the preceding problem assuming a thermal flux of 10^{15} neutrons/(cm²)(sec). From these results and those in the text, draw general conclusions as to the effect of the neutron flux on (a) the time required to attain the maximum xenon poisoning after shutdown, and (b) the value of the maximum poisoning.

15. Calculate the xenon poisoning for a reactor operating at a thermal flux of 10^{14} neutrons/(cm²)(sec) at the following times after initial startup: (a) 20, (b) 30, and (c) 50 hr; and at the following times after complete shutdown: (a) 5, (b) 15, (c) 25, (d) 35, and (e) 50 hr. Combine the results with those from Problem 13 and plot a curve showing the variation of xenon poisoning, for the given flux, with time during 50 hr of operation followed by 50 hr of shutdown.

16. A reactor operating at a thermal flux of 10^{14} neutrons/(cm²)(sec) has excess reactivity of 0.10 available for overriding xenon poisoning. Suppose the reactor has been operating for some time and is then completely shut down. During what time interval can the reactor be restarted? If this time has elapsed, how long will it be necessary to wait before startup is again possible, without refueling? (Use the results of Problem 15.)

17. The aqueous fuel solution of a homogeneous reactor has a concentration of 120 g of uranyl sulfate (UO_2SO_4), enriched to the extent of 90 per cent in uranium-235, in a volume of 1 liter. Estimate the concentration of lithium sulfate (Li_2SO_4) which would be required to compensate for burnup of the uranium-235, so as to minimize continuous adjustment of the regulating rod during reactor operation. (Use cross section data in the Appendix.)

18. Verify the statement in § 5.108 that the maximum poisoning in an operating reactor due to samarium-149 is about 0.01, regardless of the neutron flux.

Chapter 6

ENERGY REMOVAL

THERMAL PROBLEMS IN REACTOR DESIGN

INTRODUCTION

6.1. One of the unusual features of a nuclear reactor is that there is no theoretical upper limit to the rate of energy release, i.e., of power production, due to fission. In practice, however, the maximum power level of a reactor is frequently determined by the rate at which the energy (heat) can be removed. Thus, in nuclear reactors operating at high neutron fluxes, such as those intended for central-station power, for ship or aircraft propulsion, or for plutonium production, the design of the core may depend just as much on the heat-removal aspects as on nuclear considerations. The transfer of heat from fuel elements to the coolant is facilitated by increasing both the contact area and the coolant-channel volume. However, such increases generally result in a decrease of the multiplication factor of the system, so that additional fissile material is required to make the reactor critical. The heat-transfer and nuclear requirements may consequently be in conflict, and the actual design may represent a compromise between the opposing factors (see Chapter 12).

6.2. Most existing reactor-design concepts represent particular solutions of the heat-transfer problem in which the choice of coolant, the arrangement of fuel and coolant, and the method of heat removal are the primary considerations. Although each type of reactor has its own specific thermal problems, the solution of these problems can be approached in all cases by the application of conventional engineering principles of heat transfer, hydrodynamics, and thermal stress. The equations needed to analyze the thermal behavior of a reactor are, therefore, those concerned with temperature distribution and thermal stress in solid components and with the flow of reactor coolants or liquid fuels. It is the purpose of this chapter to show how some of these equations are derived and to indicate how they are applied to the design of the various components of the heat-removal system of a reactor.

6.3. Before considering the individual problems, however, mention should be made of some unusual thermal aspects of nuclear reactor design. Since heat is being continuously produced by fission and other nuclear processes, a path must be provided for its transmission to a sink to prevent a steady temperature increase. Unlike a conventional power plant, where the temperature is limited to that resulting from the combustion of coal, oil, or gas, the temperature in a nuclear reactor could increase continuously until the reactor is destroyed if the rate of heat removal were less than the rate of heat generation. An example of this type of behavior would be an uncooled fuel element in which the heat from a sustained fission reaction raises the temperature to the point where the element melts or vaporizes and destroys itself and its container. The rates of heat generation and of heat removal must therefore be properly balanced in an operating reactor.

6.4. For a reactor of given design, the maximum operating power is limited by some temperature in the system. This limit may be set by the phase stability of the fuel elements, the coolant, or the moderator, by allowable thermal stresses in some parts of the system, by the influence of temperature on corrosion, or by other thermal effects. The maximum permissible temperature must be definitely established to make sure that the cooling system is adequate under all anticipated operating conditions. This is done by, first, estimating the magnitude and distribution of the heat sources in the reactor and, second, determining the temperature differences along the various paths of heat flow in the system. These two interrelated areas of investigation form the basis for both the concept of a new reactor system and the thermal analysis of a specific design.

GENERATION AND DISPOSAL OF HEAT IN REACTOR SYSTEMS

6.5. The determination of the manner in which the heat sources are distributed in the reactor core and structure is complicated by the fact that the heat is generated in a number of different ways. The major portion arises from the kinetic energy of the fission fragments which usually manifests itself as heat released entirely within the fuel elements, but heat is also produced from the slowing down of neutrons and beta particles, and the absorption of various gamma radiations. Since the neutrons and gamma rays are not uniformly distributed throughout the reactor core and structure, the associated heat-source distribution will also be nonuniform. An accurate knowledge of neutron and gamma-ray fluxes is therefore necessary for a complete solution of the reactor cooling problem. These fluxes are difficult to estimate with sufficient accuracy for thermal calculations, and so, in the final analysis, the reactor designer must frequently rely upon data obtained from experiments or from operating reactors.

6.6. Once the heat-source distribution is known, the determination of the temperature distribution in the reactor and of the temperatures of the various coolant circuits leading to the heat sink can be made in a straightforward manner.

Here the main objective is to follow the heat flow from the fuel element to the coolant and from the coolant to a sink. Although a heat sink must be provided in all cases, the heat-flow path will vary with the nature of the reactor system. In some cases the heat is discharged directly from the primary coolant to a sink, e.g., a river or the atmosphere. In other instances, a secondary or even a tertiary coolant may be used to transfer heat from the primary coolant to the sink. Further reference to this aspect of the design of the thermal system of a reactor will be made shortly.

6.7. The temperature at any point in a reactor will be greater than that of the sink by an amount equal to the sum of all the temperature drops along the heat-flow path. The temperature drops will usually be proportional to the amount of heat flowing and, hence, to the reactor power. Thus, the steady-state temperature in any component is generally determined by the sink temperature, the reactor power level, the total power generated in the component, and the effective thermal resistance between the component and the sink.

SPECIAL THERMAL PROBLEMS

6.8. In addition to the manner in which heat is generated and disposed of in a reactor, there are other special problems unique to reactor design which must be considered. There is, in particular, the matter of selecting reactor materials and coolants. In a conventional heat engine or power plant, materials of construction are chosen on the basis of mechanical performance, but in a reactor system the choice is dictated, to a considerable extent, by nuclear properties. This often results in the selection of unconventional, high-cost materials with their attendant problems. Such materials do not always have the most desirable thermal, physical, or mechanical properties, and frequently these properties are not well known. Beryllium metal, for example, is an excellent material for use as moderator and reflector, but it is relatively brittle (§ 7.78).

6.9. Another factor which adds to the problem of heat removal from a reactor is that the volumetric heat-release rates may be higher than for any other thermal systems designed for continuous operation. There are three reasons for this: (1) the necessity for keeping the size of the reactor small, e.g., in fast reactors and in all types of mobile reactors, (2) the desire to increase the power output per unit mass of expensive fuel material in order to reduce ultimate power costs (Chapter 14), and (3) the need to keep the core volume small so as to decrease costs by requiring a smaller container and less extensive biological shielding. Some typical volumetric heat-release rates (or power densities), as well as specific powers, i.e., power per unit mass of fuel, are listed in Table 6.1. The data are in two parts: one is for typical civilian power reactors and the other for various other systems of interest, including experimental reactors.

6.10. The combination of high operating temperatures (for maximum thermal output of the reactor and maximum thermodynamic efficiency for ultimate conversion of heat into power) and large power densities makes special demands upon the design of the reactor heat-removal system. Not only do the large densities cause stresses due to temperature differences within core components

TABLE 6.1. POWER DENSITIES AND SPECIFIC POWERS

Reactors	Power Density		Specific Power
	kw (thermal)/ft³	10⁴ Btu/(hr)(ft³)	kw (thermal)/kg U
Power Reactor Type			
Gas cooled–natural U.....	15	5.1	3.0
High-temperature gas.....	220	73	80
Sodium–graphite..........	290	99	13
Organic cooled...........	390	130	6.3
Heavy water–natural U...	510	170	32
Boiling water............	820	280	13
Pressurized water........	1550	530	20
Sodium cooled fast breeder...............	21,500	7300	900
Other Thermal Sources			
Brookhaven (natural U) reactor (28 Mw)........	1.7	0.58	—
Steam boiler (natural convection)............	14	4.8	—
Steam boiler (forced convection)............	280	97	—
Homogeneous reactor (1 Mw)..............	570	195	—
Aircraft gas-turbine.......	1250	400	—
Experimental breeder reactor (1 Mw).........	6800	2300	—
Materials testing reactor (30 Mw)........	8500	2900	—
V-2 rocket..............	566,000	193,000	—
Nuclear rocket reactor [1].............	280,000	95,000	—

to increase rapidly as their thicknesses increase, but the nuclear radiations at these power densities, e.g., high neutron fluxes, may have adverse effects on the thermal conductivity and other properties of reactor materials (see Chapter 7). In general, the major advances in reactor design have come from material and fabrication developments, but it is only by optimizing the reactor systems from the heat-transfer standpoint that their full advantage can be realized.

DESIGN OF COOLING SYSTEM

COOLANT CIRCUITS

6.11. The simplest solution of the reactor heat-removal problem is to employ a direct-cycle cooling system in which the coolant is passed once through the reactor and then to a sink.   In the Hanford reactors, for example, river water is the coolant and after passage through the reactor it returns to the river, which acts as the sink.   The Oak Ridge (X-10) and Brookhaven uranium–graphite reactors, and the British (Windscale) production reactors* use air as coolant and the atmosphere as the sink.   These systems, although simple in principle, are not without problems.   River water must be purified before it is acceptable as a reactor coolant, and provision must be made for preventing the discharge of radioactive air or water to the surroundings.

6.12. The use of a secondary coolant system in conjunction with a closed primary coolant circuit has the advantage of better control over the escape of radioactivity.   Such systems are considerably more flexible than those using a direct cycle, particularly with respect to location of the reactor and choice of primary coolant.   Heavy water, for example, is a desirable coolant from the standpoint of its nuclear properties, but its cost necessitates a closed primary circuit, so that a secondary coolant, e.g., ordinary water, is required to transfer the heat to the sink.   Cooling systems of this type, i.e., heavy water as primary coolant in a closed circuit and ordinary water in an open circuit, are used in the CP-5 reactor at the Argonne National Laboratory and in other reactors of similar design.

6.13. Even when the coolant is ordinary water, but highly purified, as in the Materials Testing Reactor (MTR), and in the Engineering Test Reactor (ETR), it may be desirable for reasons of economy to use a second cooling system.   The secondary coolant may be either air or water and the heat sink will be either the atmosphere or a river.

6.14. In the systems considered above, little or no use is made of the heat generated by fission.   In reactors designed for power production, however, three successive separate coolant circuits may be necessary.   In most reactor power system designs the primary coolant is used to boil water in a secondary circuit and thus produce steam.   The steam then drives a turbine, the water from which is condensed by means of a third coolant circuit which discharges heat to the sink.   A boiling-water power reactor, however, does not require the three coolant circuits, since the steam from the primary circuit is used directly in the turbine, so that the secondary system is eliminated.

* These reactors were closed down in October 1957 following an accident to one of them.

6.15. In a reactor using sodium (or sodium-potassium alloy) as coolant it may be desirable to interpose a liquid, such as mercury, molten lead, or an organic compound, between the sodium and boiling-water (steam) loops to avoid the possibility of chemical reaction between these two materials. The use of an intermediate system of this type can also serve the purpose of restricting the volume that has to be shielded because of the gamma radiations from radioactive sodium-24 produced by neutron capture in the reactor. In some recent designs, the heat is transferred from the radioactive sodium to a nonradioactive system, then to boiling water through a heat exchanger designed especially to minimize the effects of a possible chemical reaction. A sodium-cooled power reactor may thus have four coolant circuits, e.g., radioactive sodium, nonradioactive sodium, boiling water, and condenser water.

6.16. The use of a circulating liquid fuel, in the form of an aqueous solution or a molten metallic alloy, simplifies the heat-transfer problem in some respects, because the heat produced by fission is removed outside, rather than inside, the reactor. However, circulating liquid-fuel systems have their own particular problems, such as the design of the primary heat exchanger in which heat is transferred from the liquid fuel to water for the production of steam in the secondary circuit. Because of the intense radioactivity of the circulating fuel, due to the presence of fission products, the heat exchanger must be constructed from corrosion-resistant materials and be absolutely reliable and leakproof. It should also have a small volume in order to minimize the hold-up (and inventory) of fuel solution. If the liquid fuel is a molten metal or a fused salt, the construction specifications of the heat exchanger become even more stringent.

HEAT-SOURCE DISTRIBUTION

6.17. In a reactor using solid fuel, the design of the fuel elements and of the spacing of coolant passages through the various solid components is influenced by the temperature distribution within these components. The first step in the determination of this distribution is to establish the maximum volumetric heat-release rate in each of the components being considered. The procedure for doing this will be discussed later. Although the heat-source distribution depends upon the particular reactor design, there are certain generalizations which can be used for preliminary calculations, at least.

6.18. It may be assumed that, in a thermal reactor, 90 per cent of the total heat is liberated in the fuel elements and that this heat source will be distributed in a manner proportional to the neutron flux. About half of the remaining heat, i.e., 5 per cent of the total, will be released in the moderator, and the remaining 5 per cent in the reflector and shields. The moderator heat-source distribution will be somewhere between a uniform distribution and one proportional to the thermal-neutron flux. In many reactor designs, each cooling channel takes up heat from both the fuel element and the surrounding moderator, and in these cases the two heat sources are combined.

6.19. The determination of the heat-source distribution in the reflector and shields is somewhat complicated, because it involves an estimate of the neutron leakage from the core and of the gamma-radiation flux as a function of the energy spectrum. Since gamma rays arise from a number of sources, e.g., fission, fission product decay, and radiative capture (cf. § 10.14, *et seq.*), the details of each originating nuclear process must be known and the subsequent behavior of the radiations calculated. Although only 5 per cent of the total heat is released in the reflector and shields, including the pressure vessel if one is used, the cooling of these components may be a significant problem in a reactor operating at high power.

TEMPERATURE DISTRIBUTION

6.20. The second step in analyzing or designing the cooling system of a reactor is to calculate the temperature distribution in the solid components from their known heat-source distributions, thermal conductivities, and configurations, using appropriate heat-flow equations. Since fuel elements are usually in the form of cylinders, hollow rods, or parallel plates, the heat flow and temperature distribution in these cases can be determined in a relatively simple manner, as will be seen subsequently.

6.21. Solid moderator and reflector components, on the other hand, do not always have simple configurations; moreover, the spacing of cooling channels in the shields may not be uniform, because the heat liberated falls off with increasing distance from the core. Such cases can often be treated by simplifying the geometry to a form, e.g., a cylinder, which permits the use of an established heat-flow equation. Alternatively, a point-to-point analysis can be made using a "relaxation" technique (§ 6.88).

6.22. The design of a reactor cooling system next requires the determination of the flow characteristics and heat-transfer coefficients of the coolant. This fixes the temperature drop between the surface of the solid being cooled and that of the bulk of the coolant, as well as the temperature rise in the coolant stream. The increase in temperature of the coolant flowing through the hottest channel is of special significance, since it establishes the maximum temperature of the system.

6.23. Finally, it is necessary to consider the transfer of heat, in stages, from the primary coolant to the sink. Such transfer may involve heat exchange from the primary coolant to a steam system, or from the primary coolant to a secondary coolant and then to a steam system, and ultimately to a heat sink. This aspect of the subject is treated along more or less conventional lines.

AUXILIARY THERMAL PROBLEMS

6.24. In addition to the matters dealt with above, there are a number of auxiliary problems, related to the thermal characteristics of a reactor, that have a bearing on the design of the cooling system. One of the most important of these is the problem of stress in solid components due to temperature differences. For example, problems of stress may arise in the design of thermal and biological shields, because of the considerable differences between average and surface temperatures. Furthermore, in pressurized-water reactors, thermal stresses in the thick-walled pressure vessel become very significant. This subject will be treated in Chapter 11.

6.25. The coolant itself often presents special problems which will be considered in a later section of this chapter. However, mention may be made here of such considerations as purification, induced radioactivity resulting from neutron capture, decomposition by radiation, and corrosion (or erosion), including mass transfer. These are highly important aspects of the design of the thermal system of a reactor, but they represent problems in materials and equipment rather than of heat transfer.

6.26. In summarizing the foregoing discussion it can be seen that, apart from problems associated with thermal stresses and the choice of coolant, the design (or analysis) of the cooling system of a reactor involves the following topics:

(1) Heat sources and their distribution.
(2) Heat conduction through solid components and to the primary coolant (or heat convection within liquid fuels).
(3) Heat transport along reactor channels due to flow of fluid.
(4) Heat exchange from primary to secondary coolant and to the steam-generating system (or heat sink).

These are the main subjects which will be considered in the succeeding sections of this chapter.

HEAT SOURCES IN REACTOR SYSTEMS

FISSION ENERGY

6.27. The energy released during the fission process appears in various forms, but mainly as the kinetic energy of the fission fragments, the fission neutrons, and the beta particles resulting from radioactive decay of the fission products. The fission fragments are usually stopped within the fuel elements themselves; the small fraction that escapes into the cladding penetrates only about 0.01 mm

(0.0004 in.).   The beta particles of high energy may travel up to 4 mm (0.16 in.) in a cladding material such as aluminum, and so a large fraction of these particles may escape from the fuel element into the moderator or coolant, but they will not get out of the reactor core.   The fission neutrons lose most of their energy in the first few collisions with moderator atoms, and they travel distances ranging from some centimeters to a few feet.   It is seen, therefore, that most of the heat from the three sources under consideration, comprising about 90 per cent or more of the total energy generated, will be released within the reactor core.*

6.28. The remaining 10 per cent (or less) of the energy released in fission appears as gamma radiations from the several sources mentioned earlier.   The manner in which these are distributed throughout the reactor core, reflector, and shields depends upon the materials present and the configurations of the various components.   The distribution will consequently depend upon the reactor type.   A complete analysis of the heat generation due to gamma radiation is complicated; it requires a knowledge of the energy spectrum and spatial distribution of each of the gamma-ray sources, as well as the absorption coefficients of the materials in the system for radiations of different energies.

6.29. In addition to the spatial distribution of heat sources, the time dependence must also be considered, since an appreciable proportion of the energy of fission is released over a period of time.   For example, the fission fragments may lose essentially all of their kinetic energy in a few microseconds, whereas the energy accompanying the radioactive decay of the fission products is released over many years.   Actually, the approach to equilibrium thermal conditions after the startup of a reactor is very rapid, and any departure from this equilibrium can usually be neglected.

6.30. However, provision must be made for cooling the fuel elements and other reactor components for some time after shutdown of the reactor because of the heat generated in the decay of the fission products (cf. § 2.180).   The source distribution will then be different from that when the reactor is in operation. One hour after shutdown, for example, the heat generation rate in the fuel elements will be about 1 per cent of the operating value, whereas in the reflector and shield it will be approximately 10 per cent of the rate during operation. This difference arises because a large fraction of the heat released after shutdown is due to absorption of gamma rays from fission product decay.   Although this is a small proportion of the heat generated in the core during operation, it makes a considerable contribution to the heat liberated in other parts of the reactor system.

6.31. It was seen in Chapter 1 that the total energy released in fission, and which ultimately appears as heat, is made up of contributions from a number

---

* According to the statement in § 6.18, the heat released in the core is about 95 per cent (90 per cent in the fuel elements and 5 per cent in the moderator) of the total.   The additional 5 per cent or so arises from gamma-ray absorption in the core.

of sources.   In general, the total energy, exclusive of the neutrino energy which is lost to the reactor system, may be expressed by

$$E = 191 + E_c \text{ Mev},$$

where E_c is the energy liberated as a result of various parasitic neutron capture processes, e.g., nonfission capture in uranium-235 and uranium-238, and capture in moderator, coolant, structure, etc.; this includes the energy of the capture gamma radiations and the decay energies, i.e., the energies of the alpha and beta particles and gamma rays, of any radioactive species that are formed by parasitic neutron capture.*   Since the value of E_c will obviously depend upon the nature of the materials present in the reactor core, it is evident that the total amount of heat produced by fission will vary, to some extent, from one type of reactor to another.

Example 6.1.   Calculate the total energy release per thermal fission in a pressurized water reactor, using 3.5 per cent enriched uranium as fuel, based on the following data:

	U-238	U-235	Water	Iron
Volume per cent	33.78	0.22	53.0	12.0
$N \times 10^{24}$	0.0161	5.85×10^{-4}	0.178	0.0115
Σ_c (2200 m/sec), cm^{-1}	0.0438	0.062	0.0106	0.0291
Σ_f (2200 m/sec), cm^{-1}	—	0.338	—	—
Neutrons captured per fission	0.129	0.183	0.031	0.086
Total energy released per neutron captured (Mev)	6.8	6.8	2.2	6.0

The energy released as a result of (nonfission) capture is

$$E_c = (6.8 \times 0.129) + (6.8 \times 0.183) + (2.2 \times 0.031) + (6.0 \times 0.086)$$
$$= 2.7 \text{ Mev}.$$

The total energy release per fission is thus

$$E = 191 + 2.7 \approx 194 \text{ Mev}.$$

SPATIAL DISTRIBUTION OF ENERGY SOURCES IN REACTOR CORE

6.32.   As already noted, a large proportion of the energy from fission is available in the form of heat within a very short distance of the fission event.   The total rate of heat generation is proportional to the fission rate, i.e., to $\Sigma_f \phi$ or $N \sigma_f \phi$, where σ_f is the microscopic fission cross section, ϕ is the neutron flux, and N is the number of fissile nuclei per unit volume of fuel.   If N remains uniform throughout a particular reactor, it follows that the thermal source function, expressed as the power density, will have the same spatial dependence as the neutron flux.   Local perturbations of the flux, such as occur within a fuel ele-

---

* Although E_c should include the decay energies of all radioactive species formed as a result of nonfission neutron reactions, only those of relatively short half-life contribute to the energy release in the state of pseudo-equilibrium attained by the reactor soon after startup.

ment in a heterogeneous reactor, normally have little effect on the temperature distribution and so they can be ignored.

6.33. The overall flux distribution in the reactor core can be calculated by methods described in Chapter 4, and some of the results are tabulated in Table 4.2. For the common case of a reflected cylindrical reactor in which the fuel is distributed uniformly, the flux distribution in the core and reflector can be represented approximately by

$$\frac{\phi}{\phi_{max}} = J_0 \left(2.405 \, \frac{r}{R'} \right) \cos \frac{\pi z}{H'}, \tag{6.1}$$

where ϕ_{max} is the flux at the center of the core where it is assumed to have its maximum value; J_0 is the zero-order Bessel function of the first kind, r and z are the radial and vertical coordinates, as in Fig. 4.1, and R' and H' are the effective radius and height, respectively, of the reactor, including an allowance for the reflector.

6.34. The average neutron flux in the core is given by

$$\phi_{av} = \frac{1}{\pi R^2 H} \int_0^R \int_{-\frac{1}{2}H}^{\frac{1}{2}H} \phi 2\pi r \, dr \, dz, \tag{6.2}$$

where R and H are the actual radius and height of the fuel-containing region. Upon substituting the expression for ϕ from equation (6.1) and performing the integration, the result is

$$\frac{\phi_{av}}{\phi_{max}} = \frac{2RR'J_1 \left(2.405 \, \frac{R}{R'} \right)}{2.405 R^2} \cdot \frac{2H' \sin \left(\frac{\pi}{2} \cdot \frac{H}{H'} \right)}{\pi H}, \tag{6.3}$$

where J_1 is the first-order Bessel function.

6.35. Since the power density, P, is proportional to the neutron flux, it follows that the reciprocal of equation (6.3) is equal to P_{max}/P_{av}, i.e., the ratio of the maximum to the average power density. For a reflected cylindrical reactor, R/R' and H/H' may be taken as roughly 5/6, and then

$$\frac{P_{max}}{P_{av}} \text{ (reflected)} \approx 2.4.$$

For a bare reactor, i.e., one without a reflector, R/R' and H/H' are both unity; then equation (6.3), after inversion, yields

$$\frac{P_{max}}{P_{av}} \text{ (bare)} = \frac{2.405}{2J_1(2.405)} \cdot \frac{\pi}{2} = 3.64.$$

If in a bare cylindrical reactor the arrangement of fuel elements or control rods (or both) is such that the flux is uniform in the radial direction, the factor containing the Bessel function is unity, and P_{max}/P_{av} is $\pi/2$, i.e., 1.57. Similar calculations have been made for reactors of other shapes and the results for some cores are given in Table 6.2. It may be noted that, because of the arguments presented in § 4.37, based on the similarity of the curves in Fig. 4.2, a cosine distribution of the flux may be assumed in all cases, e.g., even in the radial direction of a cylindrical reactor.

TABLE 6.2.   RATIO OF MAXIMUM TO AVERAGE POWER
DENSITIES IN VARIOUS REACTORS

Core Geometry	P_{max}/P_{av}
Sphere (bare)	3.29
Rectangular parallelepiped (bare)	3.87
Cylinder (bare)	3.64
Cylinder (bare, flat radial flux)	1.57
Cylinder (reflected)	2.4
Pool type (water reflected)	2.6

AVERAGE AND MAXIMUM POWER IN SINGLE FUEL CHANNEL

6.36. In reactor cooling problems, it is often of interest to estimate the maximum heat-generation rate (or power density) at a point in a given fuel channel, rather than for the whole reactor core. In equation (6.1), the first (Bessel function) term gives the radial flux distribution, whereas the second (cosine) term represents the longitudinal distribution in a cylindrical reactor. In any specified longitudinal fuel channel, at a fixed radial distance, r, from the reactor center, the neutron flux distribution is

$$\phi_{\text{long}} = (\phi_{\text{max}})_{\text{long}} \cos \frac{\pi z}{H'}, \tag{6.4}$$

where $(\phi_{\text{max}})_{\text{long}}$, the maximum flux at the center of the given channel, is equal to $\phi_{\text{max}} J_0(2.405 r/R')$.

6.37. The average flux in a longitudinal fuel channel in a cylindrical core is given by

$$(\phi_{\text{av}})_{\text{long}} = (\phi_{\text{max}})_{\text{long}} \frac{1}{H} \int_{-\frac{1}{2}H}^{\frac{1}{2}H} \cos \frac{\pi z}{H'} \, dz$$

$$= (\phi_{\text{max}})_{\text{long}} \frac{2H' \sin\left(\frac{\pi}{2} \cdot \frac{H}{H'}\right)}{\pi H}.$$

Hence, assuming the power density to be proportional to the neutron flux, as before, it follows that

$$\left(\frac{P_{\text{max}}}{P_{\text{av}}}\right)_{\text{long}} = \frac{\pi H}{2H' \sin\left(\frac{\pi}{2} \cdot \frac{H}{H'}\right)}. \tag{6.5}$$

6.38. The result expressed by equation (6.5), or the simplification for the core without end reflectors, is also applicable to fuel channels in a rectangular parallelepiped reactor. For this geometry the flux distribution, in any direction parallel to any of three principal axes, is represented by an expression analogous to equation (6.4). The only change necessary is to replace H by the actual length of the reactor, in the given direction, and H' by the effective length including allowance for the reflector.

POWER AND FLUX FLATTENING

6.39. In many reactor designs the power output is limited by the maximum permissible temperature of the fuel elements, i.e., by the value of P_{max}. In some cases it is possible to increase the power output by decreasing P_{max}/P_{av}. In other words, the power output of the reactor can be increased by a more uniform distribution of the power density. This is sometimes referred to as a "flattening" of the power distribution or of the neutron flux.

6.40. In addition to the obvious use of a reflector (see Fig. 4.4), other techniques can be utilized in an effort to achieve a more uniform power distribution. One possibility is to vary the number of fissile atoms per unit volume in different parts of the core. For example, the central region, where the neutron flux is highest, may be loaded with a fuel of a lower enrichment than that utilized in the outer regions. Such "fuel management" schemes, described in § 12.114 *et seq.*, may also provide improved utilization of the available neutrons and thus decrease the cost of the power produced by the reactor.

6.41. Another approach to flux flattening is to vary the extent of insertion of the control rods, so that neutron capture is greatest in those regions in which the flux would otherwise be high. The same result may perhaps be achieved more effectively by using a neutron absorber (or poison) either incorporated in the fuel material or circulated as a solution to reduce the neutron flux in certain parts of the core.

6.42. As an alternative to varying the rate of heat production by changing the flux, the rate of heat removal may be varied, e.g., by adjusting the number or size of the coolant channels or the flow rate of the coolant. In a reactor having parallel channels, entry orifices of different sizes may be employed, so that the rate of coolant flow varies appropriately throughout the reactor. Ideally, the rate of flow through any channel should be proportional to the heat generated in that channel. A somewhat similar effect can result from the employment of multipass (or multiflow) systems. For example, the entering coolant may first flow through the center of the reactor, where the most efficient heat removal is required, then through the outer regions, where smaller temperature differences between a fuel element and the coolant are adequate.

HEAT GENERATION IN MODERATOR

6.43. The heat produced in the moderator is the result of the slowing down of fission neutrons, the stopping of beta particles from the fission products, and the absorption of gamma rays from various sources. As seen in Chapter 1, this constitutes about 5 per cent of the heat generated in the reactor system. Since most of the kinetic energy of the fission neutrons is lost during the first few collisions, the distribution of heat from this source will depend upon the mean free path of the fast neutrons. In reactors where this mean free path is short, e.g., in those having ordinary water as moderator, the heat distribution in the moderator is roughly the same as that of the thermal-neutron flux. If, however, the mean free path is long, the distribution of heat resulting from the slowing down of the fission neutrons will be more nearly uniform throughout the reactor core.

6.44. If the cooling of the moderator is treated as a separate problem from the cooling of the fuel elements, the heat distribution in the moderator must be determined independently. On the other hand, when the cooling of fuel and moderator is combined, the heat-generation distribution in the moderator may be supposed to be similar to the overall distribution in the fuel.

HEAT GENERATION IN REFLECTOR AND SHIELD

6.45. Most of the heat generated in the reflector and shield, amounting to roughly 5 per cent of the total heat produced (§ 6.19), results from the absorption of neutrons and gamma radiation escaping from the reactor core. A study of the heat-source distribution consequently involves consideration of the neutron and gamma-ray interactions. The calculations are similar to, and in many respects the same as, those concerned with the attenuation of the radiations for reasons of safety to operating personnel. Consequently, the subject of heat generation in shields, in particular, is treated as an aspect of reactor shielding.

6.46. Fast neutrons escaping from the reflector deposit essentially all their energy as heat as they are slowed down by inelastic and elastic scattering collisions in the shield. Similarly, the energy of the prompt gamma rays accompanying fission and the fission product decay gamma rays is converted into heat as the radiations from the core are absorbed in the shield. In addition, neutron capture gamma rays and gamma rays accompanying the decay of radioactive neutron capture products formed in the shield itself contribute considerably to the amount of heat generated. These matters will be considered more fully in Chapter 10.

HEAT TRANSMISSION PRINCIPLES [2]

CONDUCTION AND CONVECTION

6.47. There are three general mechanisms whereby heat is transferred from one point to another, namely: (1) conduction, (2) convection, and (3) radiation. These three methods for heat removal will be reviewed and their general characteristics outlined. The subsequent discussion, however, is devoted to the first two, since they are of major importance in existing reactor designs. The preliminary treatment will refer to heat transmission in systems without internal sources, and subsequently the problems associated with internal sources, such as exist in reactor fuel elements, will be taken up.

6.48. The term *conduction* refers to the transfer of heat by molecular (and sometimes electronic) interaction without any accompanying macroscopic displacement of matter. The flow of heat by conduction is governed by the Fourier equation

$$q = -kA \frac{dt}{dx}, \tag{6.6}$$

where q is the rate (per unit time) at which heat is conducted in the x direction through a plane of area A normal to this direction, at a point where the temperature gradient is dt/dx.* The quantity, k, defined by equation (6.6), is the *thermal conductivity*. The units employed for expressing k must, of course, be consistent with those used for the other variables in equation (6.6). Thus, if the heat is given in calories, time in seconds, length in centimeters, and temperature on the centigrade (or Celsius) scale, then the appropriate units of k are cal/(sec)(cm²)(°C/cm). On the other hand, if the units used are British thermal units, hours, feet, and degrees Fahrenheit, then k will be in Btu/(hr)(ft²)(°F/ft).†

6.49. The thermal conductivity, k, is a physical property of the medium through which the heat conduction occurs. For anisotropic substances, the value of k is a function of direction; although methods are available for making allowance for such variations, they are ignored in most analytical solutions of conduction problems. The thermal conductivity is also temperature dependent and can generally be expressed in a power series; thus,

$$k = c_0 + c_1 t + c_2 t^2 + \cdots,$$

which, in many cases, may be approximated to the simple linear form

$$k = k_0(1 + at).$$

Where considerable accuracy is desirable (and possible), allowance must be made for the variation of thermal conductivity with temperature. But very frequently, especially when the temperature range is not great, k is taken to be constant. However, when uranium oxide is used as fuel, rather than uranium metal, the temperature gradients are quite large and allowance must be made for the variation of the thermal conductivity with temperature (§ 6.80). Some values of k of interest in reactor design are given in the Appendix.

6.50. Upon integration of equation (6.6), it is found, for unidirectional heat flow by conduction in a slab of constant cross section, with k independent of temperature, that

---

* The quantity q/A, having the dimensions of heat/(time)(area), e.g., Btu/(hr)(ft²), is called the *heat flux*.

† For conversion factors, see Table A.1 in the Appendix.

$$q = -kA \frac{t_1 - t_2}{x_1 - x_2}, \tag{6.7}$$

where t_1 and t_2 are the temperatures at two points whose coordinates are x_1 and x_2. The result means that the temperature gradient at a point, i.e., dt/dx, in the Fourier equation (6.6) may be replaced by the average gradient over any distance, i.e., by $(t_1 - t_2)/(x_1 - x_2)$.

6.51. If $t_1 - t_2$ is replaced by Δt, the temperature difference, and $x_2 - x_1$ by L, the length of the heat-flow path, equation (6.7) upon rearrangement takes the form

$$q = \frac{\Delta t}{\dfrac{L}{kA}}. \tag{6.8}$$

This expression is analogous to Ohm's law, $I = E/R$; hence the quantity L/kA is often called the *thermal resistance* for a slab conductor. The analogy between the flow of heat by conduction and of electricity is the basis of the *thermal circuit* concept which is very useful in solving heat-transfer problems. In general, the rate of heat flow, q, is equivalent to the current, I; the temperature difference, Δt, is the analogue of the potential difference (or EMF), E; and the thermal resistance replaces the electrical resistance.

6.52. Heat transmission by *convection*, which is usually concerned with the transfer of heat across a solid-fluid interface, involves macroscopic motion of the fluid. In *free convection* the motion is a consequence of the buoyant forces generated in the fluid due to temperature differences within it. In *forced convection*, on the other hand, the fluid is displaced by mechanical means, e.g., by a pump. When applied to heat removal in reactors, two aspects of convective heat transmission must be considered; there is, first, transfer of heat from the material which is being cooled, e.g., the fuel element, to the coolant; and, second, the transport of this heat, usually in sensible form, by flow of the coolant from one point in the system to another. Fluid flow is thus an important problem, as will be seen later, in the study of reactor cooling systems.

6.53. The fundamental equation of convective heat transfer, for both free and forced motion of the fluid, is Newton's law of cooling, which may be written as

$$q = hA_h \, \Delta t, \tag{6.9}$$

where q is the rate of convective heat transfer to or from a surface of area A_h when the temperature difference is Δt. The quantity h, defined by equation (6.9), is commonly called the *heat-transfer coefficient*, but some authors prefer the term *unit thermal conductance*. For engineering calculations, it is usually expressed in units of Btu/(hr)(ft^2)(°F). It should be noted that equation (6.9) applies to convective heat-flow in either direction, i.e., from solid to fluid or from fluid to solid; the actual direction of the flow depends, of course, upon the sign of Δt.

6.54. Although the value of h is dependent upon the physical properties of the fluid medium, it is also a function of the shape and dimensions of the interface, and of the nature, direction, and velocity of the fluid flow. Thus, the heat-transfer coefficient is a property of the particular system under consideration. Another factor which determines h is the exact definition of Δt, i.e., the temperature difference between the surface and the fluid. Whereas the temperature of the solid is uniquely defined, that of the fluid is subject to several arbitrary definitions. The latter is usually taken as the "mixed-mean (or bulk) fluid temperature," t_m, given by

$$t_m \equiv \frac{\int_{A_f} \rho c_p v t \, dA_f}{\int_{A_f} \rho c_p v \, dA_f},$$

where ρ, c_p, and v are the density, specific heat, and flow velocity, respectively, of the fluid, and A_f is the cross-sectional (or transverse) flow area of the fluid. The value of the heat-transfer coefficient in any given circumstances can be determined experimentally, but for design purposes the general practice is to use results predicted by means of various theoretical and semiempirical expressions (see § 6.113 et seq.).

6.55. By slight rearrangement, equation (6.9) can be written in the Ohm's law form

$$q = \frac{\Delta t}{\dfrac{1}{hA_h}},$$

so that, as before, q and Δt are analogous to current and potential difference, respectivly, but $1/hA_h$ now represents the thermal resistance to convective heat transfer. Thermal circuits can thus be constructed with one or more stages of thermal conduction combined with convection for the solution of problems involving both types of heat transfer, as will be shown in the succeeding paragraphs.

HEAT CONDUCTION IN INFINITE SLAB WITH CONVECTION BOUNDARY CONDITIONS

6.56. Although thermal problems in reactors generally involve internal sources, the first (and simplest) case to be considered is that of an infinite slab of finite thickness with *no internal heat source*. It is postulated that heat is transferred at both faces by convection and that the heat-transfer coefficients are h_a and h_b, respectively. The idealization of an infinite slab is a good approximation to real (finite) systems in which the transverse dimension is large compared with the thickness, e.g., the cladding of a parallel-plate type fuel element or the flat plate of a heat exchanger. In the following treatment, steady-state conditions with the thermal conductance independent of temperature will be postulated.

6.57. The characteristics of the physical system are shown at the left in Fig. 6.1, and the equivalent thermal-circuit diagram is given at the right. The total

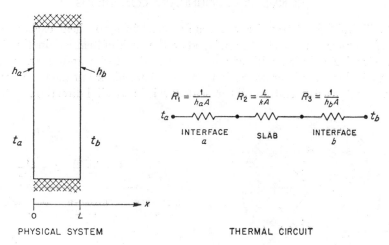

PHYSICAL SYSTEM   THERMAL CIRCUIT

FIG. 6.1.   Heat conduction in infinite slab with convection boundaries

thermal resistance is the sum of three terms, namely, two (R_1 and R_3) for the convective resistances and the third (R_2) for the conductive resistance of the slab; hence, if A is the heat-flow area, the total thermal resistance, R, is

$$R = \frac{1}{h_a A} + \frac{L}{kA} + \frac{1}{h_b A} = \frac{1}{UA}. \tag{6.10}$$

The relationship between the heat-flow rate and the temperature difference $t_a - t_b$, which is equal to Δt, is thus given by

$$q = \frac{t_a - t_b}{R} = UA(t_a - t_b)$$

or

$$t_a - t_b = \frac{q}{UA}.$$

The quantity U, defined by equation (6.10), is called the *overall coefficient of heat transfer*; it has the same dimensions as the heat-transfer coefficient.

6.58. The equations derived above may be used to determine the steady-state difference in temperature, $t_a - t_b$, between two fluids in a heat exchanger, if the heat-flow rate, q, is given. Alternatively, if $t_a - t_b$ is fixed, the corresponding value of q can be calculated for specified conditions. The individual temperature differences, i.e., the two fluid-solid differences and that between the two faces of the slab, can be determined by the use of equations (6.9) and (6.7), respectively.

HEAT CONDUCTION IN HOLLOW CYLINDER WITH CONVECTION BOUNDARY CONDITIONS

6.59. Tubular heat-transfer surfaces are used in many heat exchangers, and the performance of these can be analyzed on the basis of heat conduction through hollow circular cylinders with convection at both interior and exterior surfaces. The physical system in cross section and the thermal-circuit diagram are shown in Fig. 6.2; it is assumed that all the heat flow is in a radial direction. The inner

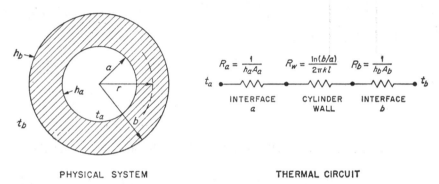

PHYSICAL SYSTEM   THERMAL CIRCUIT

Fig. 6.2.   Heat conduction in hollow cylinder with convection boundaries

and outer radii of the cylinder are a and b, respectively; h_a and h_b are the heat-transfer coefficients for the inner and outer fluids, and t_a and t_b are the mixed-mean temperatures of inner and outer fluids, respectively.

6.60. The convective thermal resistances are equal to $1/h_a A_a$ and $1/h_b A_b$, where A_a and A_b are the areas of the inner and outer surfaces. The thermal conductive resistance of the cylinder wall remains to be determined. If l is the length of the cylinder, then the heat-flow rate, expressed at an arbitrary radius r, due to conduction within the wall is given by equation (6.6) as

$$q = -k(2\pi r l)\frac{dt}{dr},$$

where k is the thermal conductivity and $2\pi r l$ is equivalent to A, the area through which heat flow occurs. This expression may be rearranged and integrated over the wall thickness, i.e., from $r = a$ to $r = b$; the result is

$$q = \frac{\Delta t}{\dfrac{\ln (b/a)}{2\pi k l}}, \tag{6.11}$$

where Δt is the difference in temperature across the cylinder wall.

6.61. It is evident from equation (6.11) that the thermal resistance, R_w, of the wall is given by

$$R_w = \frac{\ln{(b/a)}}{2\pi kl}. \tag{6.12}$$

The total thermal resistance of the system under consideration is consequently

$$R = \frac{1}{h_a A_a} + \frac{\ln{(b/a)}}{2\pi kl} + \frac{1}{h_b A_b} = \frac{1}{U_b A_b},$$

where U_b is the overall heat-transfer coefficient based on the outer surface of the cylinder. According to the thermal-circuit concept

$$q = \frac{t_a - t_b}{R} = U_b A_b (t_a - t_b)$$

or

$$t_a - t_b = \frac{q}{U_b A_b},$$

where

$$U_b = \frac{1}{\dfrac{b}{h_a a} + \dfrac{b \ln{(b/a)}}{k} + \dfrac{1}{h_b}}. \tag{6.13}$$

RADIATION HEAT TRANSFER

6.62. As a result of changes in the thermal motions of its constituent particles, which are a function of the temperature, every body emits energy in the form of electromagnetic radiations over a range of wave lengths; this is called *thermal radiation*. The amount of energy carried by the radiation is unchanged as it passes through a vacuum and is largely unaffected by dry air and many other gases, with the notable exceptions of carbon dioxide and water vapor. But when the radiation falls on a solid body or in its passage through the last-mentioned gases, part or all of the energy is absorbed. The fraction of the incident radiation that is absorbed is called the *absorptivity*. An ideal material which absorbs all the radiation falling upon it, and thus has an absorptivity of unity, is designated a *black body*. The emissive power or thermal radiation flux, i.e., the energy radiated in unit time per unit area of a black body, is given by the Stefan-Boltzmann law, namely,

$$\text{Thermal radiation flux} = \sigma T^4,$$

where the constant σ has the value 0.173×10^{-8} Btu/(hr)(ft²)(°R⁴), with the absolute temperature, T, in degrees Rankine, or 1.38×10^{-12} cal/(sec)(cm²)(°K⁴) if the temperature is in degrees Kelvin. The ratio of the emissive power of an actual surface to that from a black body is referred to as the *emissivity*, and for a body in thermal equilibrium the emissivity and absorptivity are equal (Kirchhoff's law).

6.63. The emissivity (and absorptivity) of surfaces vary with the nature of the material and its temperature, as well as with its physical condition, e.g., roughness, cleanliness, etc.   For metals, the emissivity is relatively low, the values generally ranging from about 0.05 or less for a highly polished surface to 0.2 or 0.3 for a roughened surface.   If the metal is covered with an oxide film the emissivity is greatly increased.   Nonmetals have fairly high emissivities, although, in contrast to metals, the values decrease with increasing temperature.   The emissivity of graphite, for example, approaches unity at temperatures up to about 1000°F, so that it approximates a black body.   It is therefore a good emitter and absorber of thermal radiation.   When radiation falls on a body, the proportion which is not absorbed is partly transmitted, e.g., through a "transparent" material such as air, and partly reflected, e.g., by a polished metal surface.

6.64. If two surfaces at different temperatures are separated by a nonabsorbing medium, there is an interchange of radiation between them since they both act as absorbers and emitters.   However, the net result is the transfer of energy from the hotter to the colder surface, the rate of energy transfer, e.g., in Btu/hr or cal/sec, being given by

$$q_r = A_1 \epsilon_{1,2} \sigma (T_1^4 - T_2^4),$$

where T_1 and T_2 are the absolute temperatures of the hotter and colder bodies, respectively, A_1 is the surface area of the former, and $\epsilon_{1,2}$ is an interchange factor which is related to the emissivities (or absorptivities) of the two surfaces; if both bodies were black, i.e., perfect absorbers and emitters, $\epsilon_{1,2}$ would be unity. Certain geometrical factors should be included in this expression, but they may be disregarded here.

6.65. A relationship of the same form is applicable when a hot surface radiates energy into an absorbing gas, such as carbon dioxide or water vapor, at a lower temperature.   Consequently, heat can be transferred by thermal radiation across a solid-fluid interface, in addition to the transfer taking place by convection, as already described.   It is convenient, therefore, to define a *radiative heat-transfer coefficient*, h_r, by an expression analogous to equation (6.9), i.e.,

$$q_r = h_r A_1 (T_1 - T_2),$$

so that, using the relationship for q_r given above,

$$h_r = \frac{q_r}{A_1(T_1 - T_2)} = \frac{\epsilon_{1,2}\sigma}{(T_1 - T_2)}(T_1^4 - T_2^4),$$

where T_2 is here the gas temperature.   For ease of calculation, this may be written in the form

$$h_r = \frac{0.173 \epsilon_{1,2}}{(T_1 - T_2)}\left[\left(\frac{T_1}{100}\right)^4 - \left(\frac{T_2}{100}\right)^4\right] \text{Btu/(hr)(ft}^2),$$

with the temperatures in °R (°R = °F + 460).

6.66. The heat-transfer coefficient defined in this manner differs from the heat-transfer coefficient for convection in the respect that the former is very highly temperature dependent; its value is not only determined by the actual temperatures T_1 and T_2 but also by their difference. For radiative heat transfer to be significant in comparison with convective heat transfer to a gaseous coolant in a reactor, h_r should be about 50 Btu/(hr)(ft²), and if the temperature difference between the fuel element and the gas, e.g., carbon dioxide or steam, is 100°F, it can be calculated, taking $\epsilon_{1,2}$ as unity, that the temperature of the surface must exceed about 1500°F. There are no reactors under consideration using carbon dioxide or water vapor (steam) as coolant in which these conditions would exist.

6.67. Even though radiation does not transfer any appreciable amount of energy directly to the coolant in a reactor, it may do so indirectly. In gas-cooled reactors operating at high temperatures, heat is transferred by convection from the fuel to the coolant, which is generally helium gas. The latter is transparent to thermal radiation and so does not absorb any radiant energy directly. However, radiation is transferred through the gas to the moderator (graphite) and this then loses energy by convection to the gaseous coolant. Thus, radiative transfer provides a means for conveying heat from the fuel element to the gas in an indirect manner. This process is of importance in the Ultra-High Temperature Reactor Experiment (§ 13.91) in which the fuel surface temperature is expected to reach about 2700°F. Radiation may also be significant in the transfer of heat from one reactor component, such as a fuel element, to another component, thus leading to thermal gradients which must be taken into account in stress analysis.

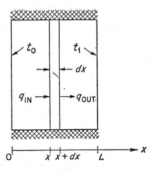

Fig. 6.3. Heat transmission in infinite slab with internal source

HEAT TRANSMISSION IN SYSTEMS WITH INTERNAL SOURCES

GENERALIZATION FROM ONE-DIMENSIONAL SLAB

6.68. The treatment of heat transmission in the steady state will now be extended to systems with internal sources and, for this purpose, the simple one-dimensional case of an infinite slab will be examined (Fig. 6.3). Consider a thin section of the slab of thickness dx at x; then in the steady state the heat rate balance requires that

Heat conducted out of $A\ dx$ − Heat conducted into $A\ dx$

$$= \text{Heat generated in } A\ dx,$$

where A is the heat-conduction area and $A\ dx$ is the volume of the section. By equation (6.6)

$$q_{in} = -kA \left(\frac{dt}{dx}\right)_x$$

and

$$q_{out} = -kA \left(\frac{dt}{dx}\right)_{x+dx},$$

assuming the heat flow is from left to right. Taking the difference of these terms and expanding the expression for q_{out}, it is found that

$$q_{out} - q_{in} = -kA \left\{ \left[\left(\frac{dt}{dx}\right)_x + \left(\frac{d^2t}{dx^2}\right)_x dx + \cdots \right] - \left(\frac{dt}{dx}\right)_x \right\}$$

$$= -kA \left(\frac{d^2t}{dx^2}\right) dx, \tag{6.14}$$

higher-order terms being neglected.

6.69. If $Q(x)$ represents the volumetric thermal source strength, expressed as heat generated per unit time per unit volume at x, e.g., in Btu/(hr)(ft^3), then $Q(x)A\,dx$ is the heat generated per unit time in the section of the slab under consideration. It follows, therefore, from the heat-balance equation and from equation (6.14) that

$$-kA \left(\frac{d^2t}{dx^2}\right) dx = Q(x)A\,dx$$

so that

$$-k\frac{d^2t}{dx^2} = Q(x), \tag{6.15}$$

where $Q(x)$ may or may not vary with x.

6.70. It can be shown that equation (6.15) is a particular form of the general equation for the steady state in a system with an internal source, namely,

$$-k\nabla^2 t = Q(r), \tag{6.16}$$

where the thermal conductivity k is the same in all directions, i.e., for a truly isotropic medium, and ∇^2 is the Laplacian operator (§ 3.18). It is apparent that in the one-dimensional case, in Cartesian coordinates, equation (6.16) reduces to equation (6.15). For conduction in a cylinder, in the radial direction, equation (6.16) becomes

$$-k \left(\frac{d^2t}{dr^2} + \frac{1}{r} \cdot \frac{dt}{dr}\right) = Q(r) \tag{6.17}$$

and for a sphere

$$-k \left(\frac{d^2t}{dr^2} + \frac{2}{r} \cdot \frac{dt}{dr}\right) = Q(r). \tag{6.18}$$

CONDUCTION IN INFINITE SLAB WITH UNIFORM SOURCE

6.71. Many reactor designs have plate-type fuel elements which can be idealized as an infinite slab with a uniformly distributed thermal source. In this case, equation (6.15) may be used with Q constant, independent of x. The general solution of this differential equation is then

$$t = -\frac{Qx^2}{2k} + C_1 x + C_2, \tag{6.19}$$

where the constants C_1 and C_2 are to be evaluated from the boundary conditions, which in this case (see Fig. 6.3) are

$$x = 0, \, t = t_0 \quad \text{and} \quad x = L, \, t = t_1.$$

It is thus found that

$$t = -\frac{Qx^2}{2k} + \left(\frac{t_1 - t_0}{L} + \frac{QL}{2k}\right) x + t_0, \tag{6.20}$$

which gives the steady-state temperature t at any point x in the slab. Differentiating equation (6.20) with respect to x to obtain dt/dx, and inserting the result in equation (6.6), it follows that

$$\frac{q}{A} = \frac{Q}{2} (2x - L) + \frac{k(t_0 - t_1)}{L}. \tag{6.21}$$

6.72. The slab of thickness L shown in Fig. 6.3 may be regarded as the right half of a plate-type fuel element, of thickness $2L$, in which both surface temperatures are the same. In this case $q = 0$ at $x = 0$, at the middle of the fuel element, since the temperature is a maximum and dt/dx is, consequently, zero at this point. If q is equal to zero for $x = 0$, equation (6.21) reduces to

$$t_0 - t_1 = \frac{QL^2}{2k}, \tag{6.22}$$

where $t_0 - t_1$ represents the temperature difference from the central plane to the outer surface of the fuel element (Fig. 6.4).

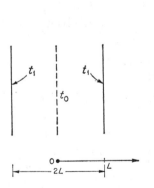

FIG. 6.4. Temperature in plate-type fuel element

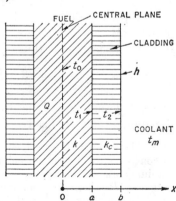

FIG. 6.5. Heat transmission in clad plate-type fuel element

HEAT TRANSMISSION IN CLAD PLATE-TYPE FUEL ELEMENT

6.73. The physical system for a plate-type fuel element clad on each face to form a "sandwich" is shown in Fig. 6.5. Here a is the half-thickness of the fuel portion and b is the half-thickness of the sandwich; t_0 is the temperature at the central plane of the fuel and t_1 is that at the fuel-cladding interface. The flow of heat is assumed to occur only in the x direction. The fuel portion of the sandwich is equivalent to the infinite slab treated above, with a replacing L as the half-thickness; hence, from equation (6.22),

$$t_0 - t_1 = \frac{Qa^2}{2k}. \tag{6.23}$$

6.74. The heat transmission through the cladding into the coolant fluid can be treated by the simple thermal-circuit method, with convection at one boundary;* thus, if t_m is the mixed-mean temperature of the fluid, then

$$t_1 - t_m = q(R_{\text{clad}} + R_{\text{conv}})$$

$$= \frac{q}{A}\left(\frac{b-a}{k_c} + \frac{1}{h}\right), \tag{6.24}$$

where $R_{\text{clad}} = (b - a)/k_c A$ is the thermal resistance of the cladding (thermal conductivity $= k_c$) and $R_{\text{conv}} = 1/hA$ is the convective thermal resistance at the cladding-fluid interface (heat-transfer coefficient $= h$); A is the area of the heat-transfer surface. From a heat-rate balance, q at the interface is found to be AQa. Since there is no thermal source in the cladding, this is the value of q in equation (6.24); hence

$$t_1 - t_m = Qa\left(\frac{b-a}{k_c} + \frac{1}{h}\right), \tag{6.25}$$

which gives the difference in temperature between the fuel-cladding interface and the coolant.

6.75. The temperature difference between the center of the fuel element and the fluid is now obtained by adding $t_0 - t_1$ and $t_1 - t_m$ as given by equations (6.23) and (6.25), respectively; thus,

$$t_0 - t_m = \frac{Qa^2}{2k} + Qa\left(\frac{b-a}{k_c} + \frac{1}{h}\right). \tag{6.26}$$

It should be noted that in deriving this result, the thermal resistance, if any, at the fuel-cladding interface has been neglected.

Example 6.2. Determine the steady-state temperature distribution in a plate-type fuel element consisting of 1 mm of a uranium-aluminum alloy, containing 18 per cent by weight of uranium enriched to 90 per cent in U-235, clad on each side with 0.5-mm thickness of aluminum. The thermal conductivities of the fuel region and the cladding may be taken to be the same, namely, 118 Btu/(hr)(ft²)(°F/ft). The thermal-neutron flux in the fuel region is 10^{14} neutrons/(cm²)(sec). The plate is cooled on both sides by water whose temperature is 80°F, and the heat-transfer coefficient for the given conditions is 7000 Btu/(hr)(ft²)(°F). (The thermal source in the fuel-bearing section may be assumed to be uniform and k may be taken as being independent of temperature.)

* The small quantity of heat liberated in the cladding from various sources is neglected in this calculation.

The first problem is to determine the thermal source strength, Q; this is equal to $\Sigma_f \phi E_f$, where Σ_f is the thermal macroscopic fission cross section in the fuel region, ϕ is the thermal-neutron flux (which is known), and E_f is the energy released in the fuel per fission (which may be taken as 200 Mev). Hence it is required to evaluate $\Sigma_f = N\sigma_f$, where σ_f is the effective thermal fission cross section of uranium-235 (about 470×10^{-24} cm^2 at $\sim 120°$F) and N is the number of these nuclei per cm^3 of the fuel.

Taking the density of uranium as 19 g/cm^3 and that of aluminum as 2.7 g/cm^3, the density of the fuel region (18% U-82% Al by weight) is found to be 3.2 g/cm^3. Since the uranium contains 90% of U-235, there are $(0.9)(0.18)$ g of U-235 per gram of fuel, or $(0.9)(0.18)(3.2)$ g of U-235 per cm^3 of fuel. Each 235 g of U-235 contains 0.602×10^{24} (the Avogadro number) nuclei, so that

$$\Sigma_f = N\sigma_f = \frac{(0.9)(0.18)(3.2)(0.602)(470)}{235}$$

$$= 0.62 \text{ cm}^{-1}.$$

The fission rate is $\Sigma_f \phi = 0.62 \times 10^{14}$ fissions/(sec)(cm^3), and so, at 200 Mev/fission, the rate of energy release is 1.24×10^{16} Mev/(sec)(cm^3); hence,

$$Q = 1.24 \times 10^{16} \frac{\text{Mev}}{(\text{sec})(\text{cm}^3)} \times 1.52 \times 10^{-16} \frac{\text{Btu}}{\text{Mev}} \times 3600 \frac{\text{sec}}{\text{hr}} \times 2.83 \times 10^4 \frac{\text{cm}^3}{\text{ft}^3}$$

$$= 1.9 \times 10^8 \text{ Btu/(hr)(ft}^3).$$

Noting that $a = 0.05$ cm $= (0.05)(3.28 \times 10^{-2}) = 1.64 \times 10^{-3}$ ft, $b = 0.1$ cm $= 3.28 \times 10^{-3}$ ft, $t_m = 80°$F, $k = k_c = 118$ Btu/(hr)(ft^2)(°F/ft), and $h = 7800$ Btu/(hr)(ft^2)(°F), the temperature t_0, at the central plane, where $x = 0$, is given by equation (6.26) as 127°F. Similarly, from equation (6.25), t_1 is found to be 124°F.

The temperature difference, $t_2 - t_m$, between the outside of the cladding and the coolant may be obtained from the convective term in equation (6.25); thus,

$$t_2 - t_m = \frac{Qa}{h} = \frac{(1.9 \times 10^8)(1.64 \times 10^{-3})}{7800}$$

$$= 40°\text{F},$$

so that t_2 is 120°F. The results are plotted in Fig. 6.6.

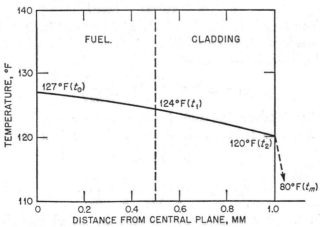

FIG. 6.6. Temperature distribution in clad plate-type fuel element

HEAT TRANSMISSION IN CLAD CYLINDRICAL FUEL ELEMENT

6.76. In many power reactors the fuel elements are in the form of cylinders, usually clad with stainless steel or a zirconium alloy, the heat being removed from the outer surface by convection. The determination of the steady-state temperature distribution is similar to that described for a plate-type fuel element. Consider first the fuel cylinder of radius a (Fig. 6.7), with uniform source Q. For heat conduction in the radial direction, equation (6.17) is applicable, the general solution being

$$t = -\frac{Qr^2}{4k} + C_1 \ln r + C_2, \qquad 0 \leqslant r \leqslant a. \tag{6.27}$$

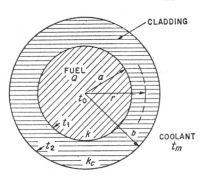

FIG. 6.7. Heat transmission in clad cylindrical fuel element

6.77. The boundary conditions in the present case are

$$\frac{dt}{dr} = 0, r = 0 \qquad \text{and} \qquad t = t_1, r = a,$$

so that the appropriate solution is

$$t - t_1 = \frac{Q}{4k}(a^2 - r^2), \qquad 0 \leqslant r \leqslant a,$$

where t is the temperature at the radial distance r. If t_0 is the temperature along the central axis, where $r = 0$, then

$$t_0 - t_1 = \frac{Qa^2}{4k} \tag{6.28}$$

for the temperature drop across the fuel itself.

6.78. For the transmission of heat through the cladding to the coolant,

$$t_1 - t_m = q(R_{\text{clad}} + R_{\text{conv}})$$

as before, where R_{clad} is identical with R_w of § 6.61 and is given by equation (6.12), and R_{conv} is $1/hA$, where A is the outside area of the cylinder of length l, i.e., $2\pi bl$. It follows, therefore, that

$$t_1 - t_m = \frac{q}{2\pi l}\left(\frac{1}{k_c} \ln \frac{b}{a} + \frac{1}{hb}\right).$$

At the interface between fuel and cladding, where $r = a$, $dt/dr = -Qa/2k$; since the surface area is here $2\pi al$, it follows from a heat-rate balance that

$$q = \pi a^2 l Q$$

and, hence,

$$t_1 - t_m = \frac{Qa^2}{2}\left(\frac{1}{k_c} \ln \frac{b}{a} + \frac{1}{hb}\right). \tag{6.29}$$

This gives the temperature difference between the fuel-cladding interface and the coolant.

6.79. Upon adding equations (6.28) and (6.29), the temperature drop from the center of the fuel element to the coolant is found to be

$$t_0 - t_m = \frac{Qa^2}{4k} + \frac{Qa^2}{2}\left(\frac{1}{k_c}\ln\frac{b}{a} + \frac{1}{hb}\right), \tag{6.30}$$

assuming negligible thermal resistance at the fuel-cladding interface. Equations for the temperature distribution in the cladding or for the difference between the temperature at any point in the fuel or the cladding and the coolant can be readily derived, if required, by a procedure similar to that used for plate-type elements.

Example 6.3. Determine the maximum specific power (in kw/kg) obtainable with a water-cooled, cylindrical fuel rod of natural uranium, 0.9 in. in diameter, having an aluminum cladding 50 mils (0.05 in.) thick, assuming that the maximum metal temperature does not exceed 1000°F and the maximum mean coolant temperature does not exceed 200°F. What is the temperature drop through (a) the fuel, (b) the cladding, (c) between the cladding and coolant, at maximum power? The following data are to be used: For uranium, $k = 14$, and for aluminum, $k = 118$ Btu/(hr)(ft²)(°F/ft); $h = 6000$ Btu/(hr)(ft²)(°F), a representative value for water flowing at about 30 ft/sec in a narrow channel.

The specific power is related to Q, the volumetric source strength, since this is a measure of the rate of heat release by fission per unit volume of fuel. Hence, the maximum value of Q must be determined; this can be done by means of equation (6.30), setting $t_0 = 1000°F$, and $t_m = 200°F$, so as to give the maximum permissible temperature drop. Then, with $a = 0.45$ in. $= 0.0375$ ft, and $b = 0.50$ in. $= 0.0416$ ft,

$$1000 - 200 = \frac{Q(0.0375)^2}{(4)(14)} + \frac{Q(0.0375)^2}{2}\left[\frac{1}{118}\ln\frac{0.50}{0.45} + \frac{1}{(6000)(0.0416)}\right]$$

$$= 2.85 \times 10^{-5}Q$$

$$Q = \frac{800}{2.85 \times 10^{-5}} = 2.80 \times 10^7 \text{ Btu/(hr)(ft}^3\text{)}.$$

Since 1 Btu/hr $= 2.93 \times 10^{-4}$ kw, $Q = 8.22 \times 10^3$ kw/ft³. The density of uranium is about 19 g/cm³, i.e., 0.019 kg/cm³, so that 1 ft³ $= 2.83 \times 10^4$ cm³ of uranium weighs $(2.83 \times 10^4)(0.019) = 5.37 \times 10^2$ kg. The required maximum specific power is consequently $8.22 \times 10^3/5.37 \times 10^2 = 15.4$ kw/kg of fuel. The specific power per kilogram of fissile material is 140 times this value, i.e., $(140)(1.45) = 2160$ kw/kg of U-235.

The temperature drop across the fuel is $t_0 - t_1$, and by equation (6.28), using Btu as the heat unit in Q and k, this is

$$t_0 - t_1 = \frac{Qa^2}{4k} = \frac{(2.80 \times 10^7)(0.0375)^2}{(4)(14)}$$

$$= 704°F.$$

The temperature drop across the cladding is given by the first term in the parentheses in equation (6.30); hence,

$$t_1 - t_2 = \frac{Qa^2}{2k_c}\ln\frac{b}{a} = \frac{(2.80 \times 10^7)(0.0375)^2}{(2)(118)}\ln\frac{0.50}{0.45}$$

$$= 17°F.$$

The temperature drop between the cladding and coolant is obtained from the second term in the parentheses in equation (6.30); thus,

$$t_2 - t_m = \frac{Qa^2}{2hb} = \frac{(2.80 \times 10^7)(0.0375)^2}{(2)(6000)(0.0416)}$$

$$= 79°F.$$

(It will be noted that the three temperatures add up to 800°F, as they should.)

6.80. In Example 6.3, the thermal conductivities were assumed to be independent of temperature; this is a satisfactory approximation for metallic fuel elements. With oxide fuels, however, the errors introduced in this manner may be significant. Allowance for the variation of k with temperature can be made by writing equations (6.23) and (6.28) in the form

$$\int_{t_1}^{t_0} k\,dt = \frac{Qa^2}{2} \text{ (for slab)} \quad \text{or} \quad \frac{Qa^2}{4} \text{ (for cylinder)}. \tag{6.31}$$

An illustration of the application of this type of equation is given below.

Example 6.4. Determine the maximum temperature in a cylindrical uranium oxide fuel element, having a radius of 0.02 ft and an outside temperature of 1000°F, when the volumetric heat generation rate is 3.50×10^7 Btu/(hr)(ft³). The heat conductivity of the oxide can be expressed approximately as a function of temperature by

$$k = \frac{3300}{T} \text{ Btu/(hr)(ft}^2)(°F/ft),$$

where T is the temperature in °R; note that °R = °F + 460. (This relationship is probably not applicable at temperatures above about 2400°F, but it is satisfactory for illustrative purposes.)

Equation (6.31) may be written in the form

$$Q = \frac{4}{a^2} \int_{1460}^{T_0} \frac{3300}{T} \, dT,$$

where T_0 is the required temperature; since Q is 3.50×10^7 Btu/(hr)(ft³) and a is 0.02 ft, it follows that

$$3.50 \times 10^7 = \frac{(4)(3300)}{(0.02)^2} \ln \frac{T_0}{1460},$$

from which it is found that $T_0 = 4220°R$ or $t_0 = 3760°F$.

If a constant value of k had been taken for a mean temperature of 2840°R, namely, 1.16 Btu/(hr)(ft²)(°F/ft), then t_0 is found to be 4020°F, from equation (6.28).

HEAT TRANSMISSION IN SHIELDS AND PRESSURE VESSELS: SLAB WITH EXPONENTIAL HEAT SOURCE

6.81. In all the preceding examples it has been postulated for simplicity, since it is a good approximation, that the heat source is uniformly distributed within a conductor. A situation will now be considered in which the heat source has an exponential distribution. This is sometimes the case in certain external reactor components, such as thermal shields and pressure vessels, where the heat generated is due to the absorption of gamma rays and the slowing down of neutrons. Although the heat generation processes are actually quite complex, it may happen that, as the result of a combination of circumstances, the volumetric heat source can be represented approximately by

$$Q = Q_0 e^{-\mu x},$$

where μ is the linear attenuation coefficient (or macroscopic cross section) of the radiations (cf. § 10.164).

6.82. If the shield or pressure vessel is a sphere or cylinder of large dimensions or is rectangular in shape, it may be treated as a slab; equation (6.15) then takes the form

$$-k \frac{d^2 t}{dx^2} = Q_0 e^{-\mu x},$$

of which the general solution, provided k is constant, is

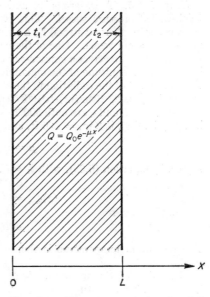

FIG. 6.8. Heat transmission in slab with exponential source

$$t = -\frac{Q_0}{k\mu^2} e^{-\mu x} + C_1 x + C_2.$$

The boundary conditions are (see Fig. 6.8),

$$t = t_1, x = 0 \qquad \text{and} \qquad t = t_2, x = L,$$

where L is the thickness of the slab. These lead to the solution

$$t - t_1 = (t_2 - t_1)\frac{x}{L} + \frac{Q_0}{k\mu^2}\left[\left(e^{-\mu L} - 1\right)\frac{x}{L} - e^{-\mu x} + 1\right] \qquad (6.32)$$

for the steady-state temperature distribution in the slab. It may be noted that the first term on the right represents the linear temperature distribution due only to the difference in temperature between the two faces of the slab, whereas the second gives the effect of the exponential heat source.

6.83. Under certain conditions the combination of the two terms just mentioned leads to a temperature maximum within the slab. The point at which this occurs may be obtained by equating to zero the derivative of equation (6.32) with respect to x. The result is

$$(x)_{t_{max}} = -\frac{1}{\mu} \ln\left[(t_1 - t_2)\frac{k\mu}{Q_0 L} + \frac{1 - e^{-\mu L}}{\mu L}\right]. \qquad (6.33)$$

The maximum temperature can be obtained by inserting this value for x into equation (6.32).

Example 6.5. A water-cooled and -moderated power reactor is contained within a thick-walled pressure vessel. This vessel is protected from excessive irradiation (and, thus, excessive thermal stress) by a series of steel thermal shields between the reactor core and the pressure vessel. One of these shields, 2 in. thick, whose surfaces are both maintained at 500°F, receives a gamma-ray energy flux of 10^{14} Mev/(cm²)(sec). Calculate the location and magnitude of the maximum temperature in this shield. The linear attenuation coefficient of the radiation in the steel may be taken to be 0.27 cm⁻¹, and the thermal conductivity as 23 Btu/(hr)(ft²)(°F/ft).

In this case, t_1 and t_2 are both 500°F, so that $t_1 - t_2$ is zero. The thickness of the shield, i.e., L (Fig. 6.8), is 2 in. = 1/6 ft, and μ in ft⁻¹ is given by

$$\mu = (0.27)(30.5) = 8.2.$$

Hence, by equation (6.33), with $t_1 = t_2$,

$$(x)_{t_{max}} = -\frac{1}{8.2} \ln \left(\frac{1 - e^{8.2/6.0}}{8.2/6.0} \right)$$

$$= 0.074 \text{ ft (or 0.89 in.).}$$

The maximum temperature is thus attained slightly inside from the midplane of the shield.

The value of the maximum temperature is calculated from equation (6.32) with $t_1 = 500°F$, $k = 23$ Btu/(hr)(ft²)(°F/ft), and $x = 0.074$ ft. The value of Q_0 is given in § 7.27 as $\phi E \mu_e$, where μ_e is the energy absorption coefficient, which is 0.164 cm⁻¹ for steel. Since ϕE is given as 10^{14} Mev/(cm²)(sec), it follows that

$$Q_0 = (10^{14})(0.164) = 1.64 \times 10^{13} \text{ Mev/(cm}^3)(\text{sec})$$
$$= 2.54 \times 10^5 \text{ Btu/(ft}^3)(\text{hr})$$

so that

$$t_{max} - 500 = \frac{(2.54 \times 10^5)}{(23)(8.2)^2} \left[(e^{-8.2/6.0} - 1) \frac{(0.074)}{(1/6.0)} - e^{-(8.2)(0.074)} + 1 \right]$$

$$= 21°F,$$

and t_{max} is consequently 521°F.

CONTACT RESISTANCES BETWEEN SOLIDS

6.84. In the calculation given above for clad fuel elements, it was assumed that the thermal contact resistance at the fuel-cladding interface was negligible because there is usually metallurgical bonding. The resistance to the flow of heat between two solid surfaces in simple mechanical contact is highly unpredictable. Estimates can be made by calculating the thermal conductance of a gas film of the thickness indicated by the roughness of the surface, but the results are very uncertain. Experimental data are somewhat more significant and it has been found that thermal conductances of the order of 5000 to 10,000 Btu/(hr)(ft²)(°F) exist at the surface of contact between uranium metal rods and various cladding materials [3]. Because uranium dioxide fuel cracks during use, the contact conductance appears to be on the order of only 1000 Btu/(hr)(ft²)(°F).

6.85. It should be noted, however, that thermal cycling, i.e., alternate increase and decrease of temperature, may misalign surfaces which had previously been in reasonably good contact, and thus decrease the conductance. For example, if a cylindrical cladding jacket is drawn onto a fuel slug (or element) by passing it through a die, the inner surface of the jacket will be pressed into irregularities of the slug surface. Heating of this assembly, particularly from the inside, e.g., by fission, may stretch the jacket so that upon cooling it fits rather loosely, and subsequent reheating may not result in perfect alignment between the fuel element and its cladding.

6.86. In metal oxide fuel elements there is generally a gas space between the oxide and the cladding to accommodate gaseous fission products and to allow for the different thermal expansions of the fuel material and the cladding. In such cases, therefore, the surfaces may not even be in contact, so that the thermal conductances will be quite low.

CONDUCTION IN IRREGULAR GEOMETRIES

6.87. The combination of nonuniform heat-source distribution and irregular geometries, which occur in many reactor components, complicates the calculation of thermal conduction rates and temperature distributions. The problem is an important one because high thermal stresses frequently result from large temperature gradients. A study of the temperature stresses in reactor components, which frequently have irregular shapes, depends upon a knowledge of the temperature pattern within the materials. An example of such a situation is found in a graphite-moderated reactor, consisting of massive graphite blocks pierced by long holes with cylindrical fuel elements and passages for coolant. Pressure vessels containing the reactor and thermal and biological shields frequently present similar difficulties in the determination of temperature distribution and thermal-stress analysis. For such cases a rigorous solution of the basic heat-conduction equation (6.16) is usually impossible. However, approximate methods have been developed in some instances; thus, a circular harmonics series solution has been proposed for the case of a massive graphite block treated as equivalent to a number of adjacent cylinders [4].

6.88. Temperature patterns for nonregular boundary conditions may be determined using numerical "relaxation" methods. The procedure is first to divide the irregular volume, so far as is possible, into a number of regular subvolumes. The heat-conduction equation may now be applied to these subvolumes in a systematic manner based on an assumed temperature distribution compatible with the boundary conditions. By carrying out a heat balance calculation for each of the subvolumes, the inaccuracy of the assumed temperature pattern is determined and expressed as a so-called "residual" for each sub-volume.   On the basis of these residuals, the assumed temperatures are changed in a systematic manner to reduce (or "relax") the residuals from point to point. The heat balance procedure and temperature adjustment are repeated until the residuals have been reduced to zero; the corresponding temperature distribution is then taken to be the correct one.   The lengthy, iterative calculations were originally performed by hand methods, but these are now obsolescent.   High-speed digital computer methods have been developed for the solution of complex heat-conduction problems, including radiation and both natural and forced convection effects, using numerical relaxation procedures.   These methods are applicable to transient as well as to steady-state conditions [5].

6.89. There are many cases for which thermal conduction problems cannot be solved by exact analytical procedures and for which approximate numerical solutions are too tedious.   For situations of this kind analog methods are frequently useful, the heat flow being simulated by an electric current, a flow of fluid, by ions in an electrolytic bath, or in other ways.   These rate processes are governed by mathematical relationships similar to those for conduction (cf. § 6.51 *et seq.*).   Direct-current electrical network analogs are particularly useful since an electrical resistance network can be easily assembled to simulate different geometries as needed.   The temperature distribution can then be determined from potentiometric measurements.

TEMPERATURE DISTRIBUTION ALONG PATH OF REACTOR COOLANT

GENERALIZED COOLANT CHANNEL

6.90. The foregoing treatment of temperature distribution has dealt exclusively with radial variations; consideration will now be given to the steady-state longitudinal temperature distribution along a fuel element, for example, resulting

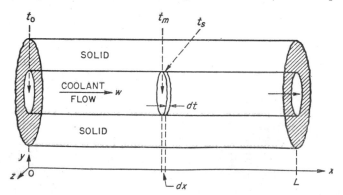

FIG. 6.9.   Generalized coolant-flow channel

from the flow of a coolant.   The most common type of reactor design involves regularly distributed fuel elements; these are cooled by the flow of some fluid through a number of parallel channels which traverse the entire reactor core. Associated with each channel is an effective area, transverse to the channel length, from which the coolant in the channel removes the heat.

6.91. A generalized representation of a coolant channel, with its associated effective heat-removal area, is shown in Fig. 6.9, in which the different variables are identified.   The x-coordinate is seen to represent the direction of coolant flow.   The rate, $dq(x)$, at which heat is added to the stream in a differential length, dx, of any coolant passage is then given by

$$dq(x) = wc_p \, dt, \tag{6.34}$$

where w is the mass flow-rate (mass/time) of the coolant in the channel, c_p is the specific heat, and dt is the differential temperature increase in the coolant across the length dx.   For simplicity it will be postulated that all heat flow within the solid is normal to the coolant stream, i.e., there is no heat conduction within the solid parallel to the coolant passage.*   Then $dq(x)$ is the rate of heat generation in a volume $A \, dx$, where A is the fuel area in the yz plane associated with the given coolant channel.   Since the volumetric heat source is, in general, a point function, $Q(x, y, z)$, it is possible to write

$$dq(x) = \left[\int Q(x, y, z) \, dy \, dz \right] dx,$$

the integration being carried over the fuel area.

6.92. If a local average heat source per unit volume, $Q(x)$, is defined by

$$Q(x) = \frac{1}{A} \int Q(x, y, z) \, dy \, dz,$$

it follows that

$$dq(x) = A Q(x) \, dx. \tag{6.35}$$

Hence, from equations (6.34) and (6.35), the longitudinal temperature distribution, in the direction of coolant flow, must satisfy the relation

$$\frac{dt}{dx} = \frac{AQ}{wc_p},$$

it being understood that Q is really $Q(x)$.   If the quantity A/wc_p is independent of x, the longitudinal distribution of the coolant mixed-mean temperature, t_m, will be given by

$$t_m - t_0 = \frac{A}{wc_p} \int_0^x Q \, dx, \tag{6.36}$$

where the fluid entrance temperature, t_0 at $x = 0$, is chosen as the datum.   This

---

* This postulate implies that the following analysis gives a good approximation provided $dt/dx \ll dt/dn$, where n is the coordinate normal to x.

expression gives the increase in temperature of the coolant due to the heat added as it flows along the channel.

6.93. By equation (6.9), the local (solid) surface temperature, t_s, is related to the mixed-mean coolant temperature, t_m, by

$$\frac{dq}{dA_h} = h(t_s - t_m), \tag{6.37}$$

where h is the heat-transfer coefficient for the given conditions, and dA_h is a small element of heat-transfer area. Although h will vary to some extent along the length of the channel, it will be assumed to be constant and independent of x, in order to simplify the treatment. If p is the passage perimeter, then $dA_h = p\,dx$, and equation (6.37) may be written as

$$dq = hp(t_s - t_m)\,dx. \tag{6.38}$$

From equations (6.35) and (6.38) it is seen that

$$t_s - t_m = \frac{AQ}{hp}, \tag{6.39}$$

which is the local temperature difference across the solid-coolant convective resistance. If numerator and denominator are multiplied by L, the total channel length, it is found that

$$t_s - t_m = \frac{QV}{hA_h}, \tag{6.40}$$

where A_h and V are the heat-transfer area and heat-generation volume, respectively, associated with the channel.

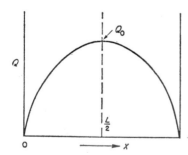

FIG. 6.10. Cosine distribution of volumetric heat source

COSINE LONGITUDINAL SOURCE DISTRIBUTION

6.94. In a bare cylinder or rectangular parallelepiped reactor, the neutron flux has a cosine distribution in the longitudinal direction (Table 4.1). The volumetric heat-source distribution in this direction will thus also be represented by a cosine function (§ 6.36). Previously the origin of the coordinates was chosen at the center of the reactor. For the present purpose, however, it is more convenient to take the origin at the reactor face (Fig. 6.10), so that the heat-source distribution along a particular fuel channel can be represented by

$$Q = Q_0 \sin \frac{\pi x}{L}, \tag{6.41}$$

where Q is the volumetric heat-release rate at the point x, and Q_0 is the maximum value at the center of the channel; L is the total channel length, assuming no reflector. Upon inserting this expression for Q into equation (6.36) and performing the integration, the result is

$$t_m - t_0 = \frac{Q_0 A L}{\pi w c_p} \left(1 - \cos \frac{\pi x}{L}\right)$$

$$= \frac{Q_0 V}{\pi w c_p} \left(1 - \cos \frac{\pi x}{L}\right), \qquad (6.42)$$

which gives the rise in temperature of the coolant as it flows through the channel.

6.95. The temperature of the coolant increases continuously, as shown in the bottom curve of Fig. 6.11; the overall (mixed-mean) temperature rise, represented by Δt_c, is thus seen to be

$$\Delta t_c = (t_m - t_0)_{x=L} = \frac{2 Q_0 V}{\pi w c_p}. \qquad (6.43)$$

It is of interest to note that this result can be derived in another manner. Since Q_0 is the maximum value of the heat-release rate per unit volume, the average is $2Q_0/\pi$ (cf. § 6.35), and the total heat-release rate for the given channel is then $2Q_0 V/\pi$. In the steady state this must, of course, be equal to the rate at which heat is removed by the coolant, i.e., $w c_p \Delta t_c$, so that equation (6.43) follows immediately.

6.96. The temperature difference between the solid and the fluid at any point is obtained from equations (6.40) and (6.41) as

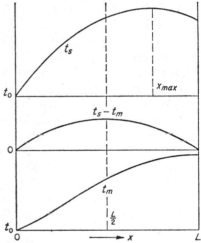

FIG. 6.11. Longitudinal variations of coolant and fuel-surface temperatures

$$t_s - t_m = \frac{Q_0 V}{h A_h} \sin \frac{\pi x}{L}, \qquad (6.44)$$

which is seen to be analogous to equation (6.41) for the source distribution. The value of $t_s - t_m$, consequently, passes through a maximum when $x = \frac{1}{2}L$, i.e., in the middle of the channel, as shown in the second curve of Fig. 6.11. This result is, of course, to be expected since at any point the fluid-solid temperature difference will be proportional to the radial heat-flow rate at that point, and this is a maximum in the center of the fuel channel.

6.97. The value of $t_s - t_0$, i.e., the fuel surface temperature at any point with respect to the value at the channel entrance, is obtained by adding equations (6.42) and (6.44), so that

$$t_s - t_0 = \frac{Q_0 V}{h A_h} \sin \frac{\pi x}{L} + \frac{Q_0 V}{\pi w c_p} \left(1 - \cos \frac{\pi x}{L}\right). \qquad (6.45)$$

It is apparent that $t_s - t_0$ must pass through a maximum at some point beyond the middle of the channel, as shown in the top curve of Fig. 6.11. This maximum represents the highest surface temperature of the fuel element. The point at which the maximum is attained is found by differentiating equation (6.45) and setting the result equal to zero; thus,

$$x_{max} = \frac{L}{\pi} \tan^{-1}\left(-\frac{\pi w c_p}{h A_h}\right). \tag{6.46}$$

From this expression it is seen that the position of the maximum value of t_s approaches the end of the channel, i.e., $x_{max} \to L$, as w and c_p decrease, and h and A_h increase.

6.98. The maximum surface temperature in the given channel can now be obtained by insertion of x_{max} into equation (6.45). Assuming the coolant flow rate to be the same in all channels, the maximum fuel-element surface temperature occurs in the channel where the neutron flux is a maximum. In this case Q_0 is the volumetric heat-release rate at the center of the reactor, and the corresponding maximum value of t_s will affect the maximum reactor power.

6.99. It should be noted that the surface temperature, t_s, considered above, is less than the temperature in the middle of the fuel element. The difference, represented by the temperature drop through the fuel region and the cladding, can be calculated by the methods described earlier (§ 6.73 *et seq.*). Because Q varies along the length of the fuel element, the point at which the surface temperature is a maximum will generally not be the same as that at which the interior temperature of the fuel has its maximum value.

6.100. For purposes of computation, it is convenient to represent $\pi x/L$ by the symbol α, i.e.,

$$\alpha \equiv \frac{\pi x}{L}.$$

It can then be found from equation (6.46) that

$$\tan \alpha_{max} = -\frac{\pi w c_p}{h A_h}, \tag{6.47}$$

where $\alpha_{max} = \pi x_{max}/L$. Utilizing this expression for $\tan \alpha_{max}$, equation (6.45) can be reduced to

$$(t_s)_{max} - t_0 = \frac{Q_0 V}{\pi w c_p} (1 - \sec \alpha_{max}). \tag{6.48}$$

With these two equations the value of the maximum surface temperature and its location can be readily evaluated.

Example 6.6. Flat fuel elements, 2.5 in. wide and 30 in. long, are arranged in an approximately cylindrical array in a water-cooled and -moderated reactor. To prevent the water from boiling, the maximum surface temperature of the elements is set at 200°F. The water enters at 100°F and flows upward at a rate of 3 ft/sec through cooling channels formed by spacing the fuel plates at 0.120 in. apart (Fig. 6.12). Assuming for simplicity that the water reflector has a negligible effect on the neutron flux in the reactor, determine (a) the minimum number of channels required if the reactor is to operate at 3000 kw, and (b) the average temperature of the exit water under these conditions. Given that at the metal-water interface $h = 1050$ Btu/(hr)(ft²)(°F), the density of the water is 61.8 lb/ft³, and the specific heat is 1.00 Btu/(lb)(°F). (The heating of the water in its capacity as moderator may be neglected.)

The mass rate of flow, w, of the coolant per channel is equal to $v\rho A_f$, where v is the velocity of the coolant (3 ft/sec = 10,800 ft/hr), ρ is the density of water (61.8 lb/ft³), and A_f is the cross-sectional (or flow) area of a coolant channel. In this case $A_f = (2.5)(0.12)/144 = 2.08 \times 10^{-3}$ ft²; hence,

$$w = v\rho A_f$$
$$= (10,800)(61.8)(2.08 \times 10^{-3})$$
$$= 1.39 \times 10^3 \text{ lb/hr.}$$

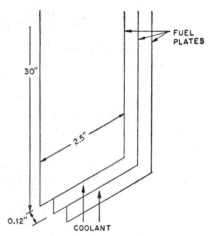

Fig. 6.12. Coolant channels between vertical fuel plates

In order to determine x_{max}, the value of A_h, the heat-transfer area per channel, is required. Since the coolant in each channel receives heat from two fuel-element surfaces, it follows that

$$A_h = \frac{(2)(2.5)(30)}{144} = 1.04 \text{ ft}^2.$$

Consequently, by equation (6.47),

$$\tan \alpha_{max} = -\frac{(3.14)(1.39 \times 10^3)(1.00)}{(1050)(1.04)} = -3.98$$

and, hence,

$$\sec \alpha_{max} = -4.11.$$

The problem stipulates that $(t_s)_{max}$ is 200°F and, since t_0 is given as 100°F, it follows from equation (6.48) that

$$Q_0 V = 100 \times \frac{(3.14)(1.39 \times 10^3)(1.00)}{5.11} = 8.55 \times 10^4 \text{ Btu/(hr)(channel).}$$

The value of Q_0, the volumetric heat source at the center of the core, may be taken to be 3.64 times the average; hence,

$$Q_{av} V = \frac{8.55 \times 10^4}{3.64} = 2.35 \times 10^4 \text{ Btu/(hr)(channel),}$$

which gives the average rate of heat generation per channel. The total power of the reactor is to be 3000 kw, i.e., 10.24×10^6 Btu/hr, and so the minimum number, n, of coolant channels required is

$$n = \frac{10.24 \times 10^6}{2.35 \times 10^4} = 435 \text{ channels.}$$

The average temperature rise of the coolant, Δt_c, is obtained by equating $Q_{av}V$ (heat generated per channel) to $wc_p \, \Delta t_c$ (heat taken up by coolant per channel); then

$$\Delta t_c = \frac{Q_{av}V}{wc_p} = \frac{2.35 \times 10^4}{(1.39 \times 10^3)(1.00)} = 16.9°F.$$

6.101. The treatment given above is strictly applicable to a bare reactor. For a reflected reactor, the flux (or power) distribution is flatter than in an equivalent bare core, and the ratio of maximum to average specific power (or volumetric heat source) is lower, as shown in § 6.35. One method of treating the problem of a reactor with a thick reflector is to assume a cosine source distribution, as in Fig. 6.10, but to suppose that the value of Q goes to zero at some distance, e.g., about one or two reflector diffusion lengths, beyond the boundary of the core. The integration of sin $\pi x/L$, referred to in § 6.94, then does not start from zero, but from x equal to roughly two diffusion lengths in the reflector.

HOT-CHANNEL FACTORS

6.102. There are several factors of a physical nature which affect the accuracy of temperature distribution calculations in the coolant channel of a reactor. For example, the coolant flow rate may not be the same in all channels, and the overall heat-transfer coefficient may vary, as a result of such causes as corrosion of surfaces, deposits of scale, etc. Furthermore, conduction along the axis of the fuel elements and the variation of the heat-transfer coefficient along the channel, which were neglected, may have an important effect on the temperature distribution. In order to allow for many of these variations from the ideal situation, the "hot-channel factor" concept has been introduced in connection with the thermal design of a reactor system. This subject will be described more fully and illustrated in Chapter 12.

HEAT-TRANSFER CHARACTERISTICS OF FLUIDS

LAMINAR AND TURBULENT FLOW

6.103. The study of the flow of fluids in connection with reactor systems is important for several reasons. As seen in earlier sections, all the heat generated in a reactor is removed and transferred from one medium to another either as sensible heat or (in a boiling reactor) as latent heat. These processes clearly involve the flow of one or more fluids. From the properties of these fluids and their flow characteristics, it is usually possible to make reasonable predictions concerning the heat-transfer coefficients for simple geometries (§ 6.113). For complex cases, however, the experimental determination of the heat-transfer coefficients is required.

6.104. In addition, a knowledge of the pressure drops in various parts of the cooling system is necessary in order to establish the overall pumping power requirements. The methods whereby these pressure drops are estimated in a number of cases will be described later (§ 6.153 *et seq.*).

6.105. Before proceeding with the more detailed aspects of the problems of interest, the essential nature of the two important types of fluid flow will be reviewed briefly. When a fluid flows through a straight pipe at low velocity, the particles of fluid move in straight lines parallel to the axis of the pipe, without any appreciable radial motion. This is described as laminar, streamline, or viscous flow, and the "velocity profile" is depicted diagrammatically in Fig. 6.13A; the lengths of the arrows indicate the relative magnitudes of the fluid

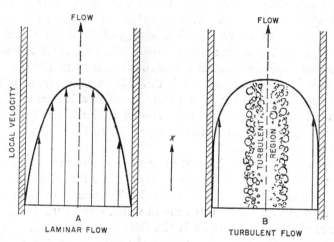

FIG. 6.13. A. Laminar flow B. Turbulent flow

velocity at various points across the pipe. The velocity distribution curve in a circular pipe is a parabola, and the average velocity is half the maximum at the center of the pipe.

6.106. One characteristic of streamline flow is that it obeys Newton's equation,*

$$F = \mu A \frac{du}{dy},\qquad(6.49)$$

where F is the shearing force (or fluid friction) over an area A between two parallel layers of fluid, moving with different velocities in a region where the velocity gradient is du/dy perpendicular to the direction of flow; the symbol μ represents the absolute (or dynamic) viscosity of the fluid, which is defined by equation (6.49). The dimensions of viscosity are seen to be mass/(length)(time); in the cgs system, the unit of viscosity, expressed in g/(cm)(sec), is called the poise, and it is in terms of this unit (or its hundredth part, called a centipoise)

---

* Certain fluids, e.g., suspensions, which do not obey the Newton equation, are referred to as non-Newtonian fluids.

that viscosity values are frequently calculated. In the English (or engineering) system of units, the dynamic viscosity is conveniently expressed as $lb/(ft)(hr)$; the values in this system are obtained upon multiplying the number of centipoises by 2.42.

6.107. Laminar flow may be distinguished from other types of flow by means of a dimensionless quantity called the *Reynolds number* (or *modulus*); this is represented by Re and defined by

$$\text{Re} \equiv \frac{Dv\rho}{\mu}, \tag{6.50}$$

where D is the pipe diameter, v is the mean velocity of the fluid, ρ is its density, and μ its viscosity. Experiments with many fluids have shown that, in general, flow is laminar, and Newton's equation is obeyed, in ducts of uniform cross section as long as Re is less than about 2100. The precise critical value of the Reynolds number depends, to some extent, on the flow conditions.

6.108. If the circumstances are such that Re exceeds about 4000 in such systems, the fluid motion is turbulent; this type of flow is characterized by the presence of numerous eddies which cause a radial motion of the fluid, i.e., motion across the stream, in addition to the flow parallel to the pipe axis. Newton's equation is then no longer valid.

6.109. The velocity profile for turbulent flow is shown in Fig. 6.13B. As indicated, three more-or-less distinct regions exist in a turbulent stream. First, there is a layer near the wall in which the flow is essentially laminar. This is followed by a transition (or buffer) zone in which some turbulence exists, and finally, around the pipe axis, there is the fully turbulent core. In the latter region turbulence mixes the fluid so that the velocity in the axial direction changes less rapidly with radial distance than it does under laminar flow conditions. As a result, there is a marked flattening of the velocity profile; this flattening becomes more pronounced with increasing Reynolds number. The ratio of the mean flow velocity to the maximum ranges from about 0.75 to 0.81, as Re increases from 5000 to 100,000.

6.110. In the range between the Re values of 2100 and 4000 there is generally a transition from purely laminar flow to complete turbulence. Laminar behavior sometimes persists in this region or turbulence may increase, depending, among other factors, on entrance conditions, the occurrence of upstream turbulence, the roughness of the pipe, the presence of obstacles, and on pulsations in flow rate produced by the pump or some other component in the system.

6.111. The Reynolds number, as given by equation (6.50), is applicable only to pipes of circular cross section; for noncircular channels, especially long rectangular ducts and annular spaces, a good approximation is obtained if D in equation (6.50) is replaced by the *equivalent* (or *hydraulic*) *diameter*, D_e, defined by

$$D_e = 4 \times \frac{\text{Cross section of stream}}{\text{Wetted perimeter of duct}}, \tag{6.51}$$

which becomes identical with the actual diameter for a circular pipe. The use of the equivalent diameter makes possible the application of tube relationships to the prediction of heat-transfer coefficients, pressure drops, and burnout heat fluxes for noncircular channels, as will be seen in due course (§ 6.116). For the channel between two parallel plates, as in reactors of the pool type, D_e is approximately twice the distance between the plates (see Example 6.7).

Example 6.7. Determine the Reynolds number for the cooling water in the reactor in Example 6.6.

The channel cross section is (2.5 in.)(0.12 in.), and 2.5 in. on each side plus 0.12 in. on each end are wetted; hence,

$$D_e = 4 \times \frac{(2.5)(0.12)}{(2)(2.5 + 0.12)} = 0.23 \text{ in.}$$

i.e., roughly twice the distance between the plates. The mean flow velocity may be taken as (3)(3600) = 10,800 ft/hr; the density of water is 61.8 lb/ft³; and the viscosity at 100°F is 1.66 lb/(hr)(ft); hence,

$$\text{Re} = \frac{(0.02)(10,800)(61.8)}{1.66} = 8050.$$

The flow is thus definitely in the turbulent region.

6.112. Although the use of the equivalent diameter is convenient for preliminary design calculations, certain limitations should be recognized. The theoretical validity of the equivalent diameter concept for generalized use has, in fact, been questioned. For example, for a duct having a cross section in the shape of an isosceles triangle, the turbulent flow friction for air was found to be 20 per cent lower than the value calculated using the equivalent diameter [6]. Heat-transfer coefficients also varied considerably around the circumference. Several correlation procedures have been proposed for ducts with noncircular cross sections, but additional experimental work is required for their verification.

PREDICTION OF HEAT-TRANSFER COEFFICIENT FOR TURBULENT FLOW OF ORDINARY FLUIDS

6.113. In essentially all cases of reactor cooling, where forced convection is used, the coolant is under turbulent flow conditions.* Satisfactory predictions of heat-transfer coefficients in long straight channels of uniform cross section can be made on the assumption that the only variables involved are the mean velocity of the fluid coolant, the diameter (or equivalent diameter) of the coolant channel, and the density, heat capacity, viscosity, and thermal conductivity of the coolant. From the fundamental differential equations or by the use of the methods of dimensional analysis,† it can be shown that heat-transfer coefficients

---

* Some reactors which operate at very low (almost zero) power, and reactors of the pool type at powers up to almost 1 megawatt (§ 13.153), make use of conduction and free convection for cooling.

† See standard texts on heat transfer, e.g., Ref. 2.

can be expressed in terms of three dimensionless moduli; one of these is the Reynolds number, already defined, and the others are the *Nusselt number* (Nu) and the *Prandtl number* (Pr), defined by

$$\text{Nu} \equiv \frac{hD}{k} \quad \text{and} \quad \text{Pr} \equiv \frac{c_p \mu}{k}.$$

6.114. As a result of numerous experimental studies of heat transfer, various expressions relating the three moduli have been proposed; one of these, for a coolant fluid in a long straight channel, is [7]

$$\frac{hD}{k} = 0.023 \left(\frac{Dv\rho}{\mu}\right)^{0.8} \left(\frac{c_p \mu}{k}\right)^{0.4} \tag{6.52}$$

or

$$\text{Nu} = 0.023 \, \text{Re}^{0.8} \text{Pr}^{0.4}, \tag{6.53}$$

with all the physical properties evaluated at the bulk temperature of the fluid. In the modified form [8]

$$\text{Nu} = 0.023 \, \text{Re}^{0.8} \text{Pr}^{0.33},$$

the physical properties, except the specific heat, are the values at the film temperature, i.e., in the laminar (film) layer of fluid adjacent to the surface. This is taken as the arithmetic mean of the wall (or surface) temperature and the bulk fluid temperature

6.115. It is seen from equation (6.52), that, if the viscosity, thermal conductivity, density, and specific heat of the coolant are known, the heat-transfer coefficient can be estimated for turbulent flow of given velocity in a pipe or channel of specified diameter (or equivalent diameter). The results appear to be satisfactory for values of Re in excess of about 10,000, and for Pr values of from 0.7 to 120. This range of Prandtl numbers includes gases and essentially all liquids, except liquid metals; the latter have very low Prandtl numbers, primarily because of their high thermal conductivity but also often because of their low viscosity and heat capacity. The correlations for liquid metals will, therefore, be considered separately (§ 6.121). For gases, Pr is approximately 0.75 and so equation (6.53) reduces to

$$\text{Nu} \approx 0.020 \, \text{Re}^{0.8} \text{ (for gases).}$$

6.116. Where large differences exist between the temperatures of the solid and of the fluid, the associated variations in the physical properties of the coolant can influence the heat transfer. As indicated above, such variation is most significant for the viscosity, and in order to make allowance for it the relationship

$$\text{Nu} = 0.027 \, \text{Re}^{0.8} \text{Pr}^{0.33} \left(\frac{\mu}{\mu_w}\right)^{0.14}$$

has been proposed, where μ is the viscosity at the bulk fluid temperature, and μ_w is that at the wall temperature [9]. This equation, used in conjunction with the equivalent diameter concept, has been found to be satisfactory for predicting local and average heat-transfer coefficients for ordinary fluids flowing through thin rectangular channels [10]. For gaseous coolants, a correlation based on temperature ratio has been proposed for use when the temperature differences are large [11].

6.117. For water as coolant, allowance for the variations in physical properties with temperature may be made by writing [12]

$$h \text{ Btu}/(\text{hr})(\text{ft}^2)(°F) = 0.148(1 + 10^{-2}t - 10^{-5}t^2)\frac{v^{0.8}}{D^{0.2}}, \qquad (6.54)$$

where v is in feet per hour and D in feet. The values for the heat-transfer coefficient given by this relationship are in good agreement with those obtained from equation (6.52) for temperatures up to 600°F.

Example 6.8. Calculate the heat-transfer coefficient for the water coolant in Example 6.6 using (a) equation (6.52), (b) equation (6.54).

The bulk temperature of the water is 100°F, so that $\rho = 61.8$ lb/ft³, $c_p = 1.00$ Btu/(lb)(°F), and μ is 1.66 lb/(ft)(hr). From Table A.6 in the Appendix, k for water is 0.364 Btu/(hr)(ft²)(°F/ft). The value of v is 3 ft/sec or 10,800 ft/hr and the equivalent diameter of the duct, as found in Example 6.7 is 0.23 in. or 0.020 ft. Hence, from equation (6.52)

$$h = 0.023 \times \frac{0.364}{0.020}\left[\frac{(0.020)(10,800)(61.8)}{1.66}\right]^{0.8}\left[\frac{(1.00)(1.66)}{0.364}\right]^{0.4}$$

$$= 1050 \text{ Btu}/(\text{hr})(\text{ft}^2)(°F).$$

By equation (6.54) for water,

$$h = (0.148)[1 + (10^{-2})(100) - (10^{-5})(100)^2]\frac{(10,800)^{0.8}}{(0.020)^{0.2}}$$

$$= 1040 \text{ Btu}/(\text{hr})(\text{ft}^2)(°F).$$

LIQUID-METAL HEAT-TRANSFER COEFFICIENT FOR TURBULENT FLOW

6.118. For ordinary fluids, the principal mechanism of heat transport is by the effect of turbulence, as a result of which a "parcel" of fluid is rapidly moved from a region close to the hot wall into the main body of fluid. In liquid metals, however, thermal transport occurs mainly by molecular conduction. Whereas this mechanism may provide 70 per cent of the heat transfer for a liquid metal, it contributes only about 0.2 per cent to heat transfer in water. This means that the laminar boundary thickness, which is important for ordinary liquids, is not significant for liquid metals and heat-transfer relationships applicable to gases and nonmetallic liquids cannot now be used.

6.119. The essential difference between the heat-transfer properties of liquid metals and ordinary fluids is brought out by the temperature profiles in a heated tube shown in Fig. 6.14 [13]. In these curves the approach of the fluid temperature to the tube-wall temperature is represented by the dimensionless quantity

$(t_w - t)/(t_w - t_0)$, where t_w and t_0 are the wall and centerline temperatures, respectively. The abscissa is the ratio y/r_0, where y is the distance from the wall at which the fluid temperature is t, and r_0 is the tube radius. The Prandtl numbers are the parameters for the various curves and the Reynolds number is 10^4 in all cases.

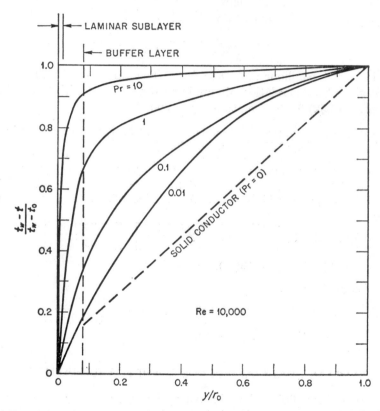

FIG. 6.14.   Dependence of temperature profile on Prandtl number [13]

6.120. For Pr = 1, the velocity and temperature profiles are identical; most of the resistance to heat transfer occurs in the laminar sublayer and in the buffer layer, and there is little further change in temperature as the center of the tube ($y/r_0 = 1$) is approached. For liquid metals (Pr ≪ 1), however, molecular conduction is so significant that there is a marked thermal gradient from the buffer layer boundary all the way to the center, much as would be observed for a solid rod (Pr = 0). The heat-transfer coefficient is normally based on a mixed mean temperature obtained by integration of the thermal profile (§ 6.54). It can be seen that for fluids of low Prandtl number this temperature may be quite different from that at the centerline.

6.121. Application of the momentum heat-transfer analogy to fluids having a high molecular conductivity has led to the correlation, for flow in long tubes [14],

$$\frac{hD}{k} = 7 + 0.025 \left(\frac{Dv\rho c_p}{k}\right)^{0.8} \tag{6.55}$$

or

$$Nu = 7 + 0.025 Pe^{0.8},$$

where Pe is the *Peclet number* defined by

$$Pe \equiv Re \times Pr \equiv \frac{Dv\rho c_p}{k}.$$

Although equation (6.55) has been used quite extensively, it gives values for the heat-transfer coefficient for many liquid metals that are too high. For design purposes, it is preferable, therefore, to use the more conservative relationship [15]

$$Nu = 0.625 \, Pe^{0.4}. \tag{6.56}$$

An alternative proposal [16], for Prandtl numbers between 0.005 and 0.05 and Reynolds numbers exceeding 10,000, leads to h values that are even lower than those derived from equation (6.56).

6.122. When conduction is the predominant heat-transfer mechanism, as is the case for liquid metals, the equivalent-diameter concept for noncircular channels, referred to in § 6.111, is not valid. A modified expression, which is effectively the result of multiplying the right side of equation (6.55) by 0.8, namely,

$$Nu = 5.8 + 0.020 \, Pe^{0.8},$$

has been proposed for flow in the channel between parallel plates with heat being removed uniformly from one side only [17]. The equivalent diameter used in computing Nu and Pe is taken as twice the distance between the plates (§ 6.111).

TABLE 6.3. VALUES OF SLUG NUSSELT NUMBERS FOR LIQUID METALS IN DUCTS

Geometry	Slug Nusselt Numbers	
	A*	B†
Circle......................	5.80	8.0
Square.....................	$\pi^2/2$ ($= 4.93$)	7.03
Equilateral triangle..........	—	6.67
90° Isosceles triangle.........	—	6.55
Infinite slot................	π^2 ($= 9.87$)	12
Infinite slot with one wall insulated.............	$\pi^2/2$ ($= 4.93$)	6

* A = constant wall temperature in flow direction around periphery.
† B = constant peripheral wall temperature and constant heat input per unit length.

6.123. Another approach to the problem of evaluating the Nusselt number for liquid metals in noncircular ducts is based on the use of the "slug" Nusselt number, which is evaluated for each specific cross section, plus a semi-empirical turbulent contribution. Thus, if Nu_s represents the slug Nusselt number, then

$$Nu = 0.67\ Nu_s + 0.015\ Pe^{0.8}, \qquad (6.57)$$

the conventional equivalent diameters being used to express the dimensionless ratios Nu and Pe. The slug Nusselt numbers of a number of geometries, for two different boundary conditions, are given in Table 6.3 [18].

6.124. For the flow of liquid metals in annular channels, with or without uniform heat flux, a good approximation, based on available reliable data, is [19]

$$Nu = 0.8 \left(\frac{r_o}{r_i}\right)^{0.3} (5.1 + 0.028\ Pe^{0.8}),$$

where r_o and r_i are the outer and inner radii, respectively, of the annulus, and the equivalent diameter is used to evaluate the dimensionless moduli.

6.125. An interesting fact, to which attention may be called here, is that, although the heat-transfer coefficients for both nonmetals and metals are increased by increasing the rate of flow of the coolant, the effect is relatively small for liquid metals. An examination of equation (6.52) shows that for a nonmetal h is proportional to $v^{0.8}$, whereas in some of the correlations for liquid metals it is seen that the term containing $v^{0.8}$ is added to a constant quantity which is independent of the flow velocity. A possible physical interpretation of this result is that, even under turbulent flow conditions, part of the heat transferred from a solid surface to a liquid metal is by molecular conduction and this is not affected by the flow rate of the coolant.

HEAT-TRANSFER COEFFICIENTS IN FREE CONVECTION

6.126. The prediction of heat-transfer coefficients associated with free (or natural) convection is much more uncertain than for forced convection. In some reactor systems, however, free convection cooling may become important, e.g., for the removal of fission product decay heat after shutdown when the forced flow of coolant is stopped. In free convection, heat is transferred from the surface of a heated solid to the adjacent fluid coolant layers. The density of the latter decreases, so that these layers rise and thereby create a flow which transports heat away from the solid.

6.127. For free convection heat-transfer correlations, another dimensionless modulus, the *Grashof number*, Gr, is used; this is defined by

$$Gr \equiv \frac{L^3 \rho^2 g \beta \Delta t}{\mu^2},$$

where L is an appropriate length (or diameter), g is the acceleration of gravity, β is the thermal coefficient of volume expansion of the fluid, and Δt is a prescribed temperature difference within the fluid; ρ and μ are, as before, the density and viscosity of the fluid. For laminar free convection from either a vertical plate or a large cylinder whose surface is at a uniform temperature and which is in an infinite body of fluid, the local Nusselt number is related to the product of the Grashof and Prandtl moduli, sometimes called the *Rayleigh number*, by

$$\text{Nu}_x = \frac{hx}{k} = 0.55(\text{Gr}_x\text{Pr})^{0.25}, \tag{6.58}$$

where the physical properties of the fluid are to be evaluated at the film temperature. This relationship is applicable for nonmetals in the Rayleigh number (laminar flow) range of 10^4 to 10^9. In the turbulent region of free convection flow, when the Rayleigh number is from 10^9 to 10^{12}, the proposed correlation is

$$\text{Nu}_x = 0.13(\text{Gr}_x\text{Pr})^{0.33}.$$

The subscript x in these equations implies that the heat-transfer coefficient is given at a point on the surface at a distance x from the bottom. In evaluating Nu_x and Gr_x, this distance is used for the length dimension, i.e., for D or L. The average value of Nu is obtained by integration over the total height of the plate or cylinder [20].

ENTRANCE EFFECTS

6.128. When a fluid enters a heat-transfer region both the flow velocity and temperature patterns begin to develop in accordance with the existing boundary layer conditions. In general, however, for turbulent flow a section length of the order of 50 diameters is required before the patterns are fully developed. The various equations given above for the prediction of heat-transfer coefficients in turbulent flow apply only beyond this entrance region. For tubes of sufficient length, the entrance effect will have a minor influence on the average value of the coefficient. For short tubes, however, the entrance region is significant and various methods for predicting its effect have been proposed [21]. For fluids with Reynolds number of about 10^4 in a heat exchanger 10 diameters in length, the mean coefficient will be 20 per cent greater than for a long exchanger (40 diameters). For higher Reynolds numbers the increase in the heat-transfer coefficient at the entrance will be smaller.

HEAT TRANSFER TO BOILING LIQUIDS [22]

SURFACE AND VOLUME BOILING

6.129. Boiling is of importance in nuclear reactor systems both as a means of achieving high heat-transfer rates and of generating steam. The use of a boiling coolant within the core is applicable to reactors employing either solid or liquid fuels. In a solid-fuel reactor, such as the various forms of boiling water reactors, heat is transferred by "surface boiling"; heat generated in the solid fuel must flow to the coolant across a solid-fluid interface at which bubble development occurs. Similar considerations apply to the generation of steam in a secondary coolant system. In a reactor with a boiling-liquid fuel, "volume boiling" occurs; fuel distributed throughout the coolant, usually as a solution, will cause bubble formation within the bulk of the coolant itself.

6.130. For the treatment of boiling heat transfer from a solid fuel element, it is convenient to consider a region of heated surface at temperature t_s, in contact with a liquid at an average (mixed-mean) temperature t_m. Suppose the temperature difference between the heated surface and the liquid bulk, i.e., $t_s - t_m$, is steadily increased; the corresponding variation in the heat flux, i.e., q/A, across the surface is then as shown in Fig. 6.15, in which the scales are both logarithmic.

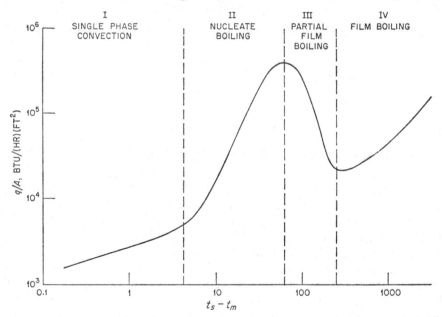

FIG. 6.15. Variation of heat flux with surface-liquid temperature difference in boiling system

Although the data in this figure are representative of natural convection boiling from a heated wire in a pool of water at atmospheric pressure and a liquid temperature of 212°F, some of the same general characteristics apply to forced convection boiling and other pressure and temperature conditions.

6.131. The curve can be divided into a number of regions, in each of which the mechanism of heat transfer is somewhat different from that in the others. Until the heated surface exceeds the saturation temperature* by a small amount, heat

---

* The saturation temperature is the temperature of the saturated vapor, i.e., saturated steam, at the existing pressure.

is transferred by single-phase convection; this occurs in region I. The system is heated by slightly superheated liquid rising to the liquid pool surface where evaporation occurs. In region II, vapor bubbles form at the heated surface; this is the *nucleate boiling* range in which formation of bubbles occurs upon nuclei, such as solid particles or gas adsorbed on the surface, or gas dissolved in the liquid. Nucleate boiling is a common phenomenon, since it is encountered in standard power-plant steam generators.

6.132. The steep slope of the heat flux curve in region II is a result of the mixing of the liquid caused by the motion of the vapor bubbles. A maximum flux is attained when the bubbles become so dense that they coalesce and form a vapor film over the heating surface. The heat must then pass through the vapor film by a combined mechanism of conduction and radiation, neither of which is particularly effective in this temperature range. Consequently, beyond the maximum, the heat flux decreases appreciably in spite of an increase of temperature.* In region III, the film is unstable; it spreads over a part of the heating surface and then breaks down. Under these conditions violent nucleate boiling occurs at some portions of the surface, while *film boiling*, due to heat transfer by conduction and radiation, occurs at other portions.

6.133. For sufficiently high values of $t_s - t_m$, as in region IV, the film becomes stable, and the entire heating surface is covered by a thin layer of vapor; boiling is then exclusively of the film type. If attempts are made to attain large heat fluxes with film boiling, as high as those possible with nucleate boiling, for example, the temperature of the heated surface may become so high as to result in destruction of the heater. This is called a "burnout" and is, of course, to be avoided. It may be noted, too, from Fig. 6.15, that if a system undergoing nucleate boiling is operating at conditions near the maximum of the curve, a slight increase in the heat flux will cause a sudden change to film boiling, which usually results in burnout.

6.134. *Subcooled (or surface) boiling* occurs when the bulk temperature of the liquid is below saturation but that of the surface is above saturation. Vapor bubbles form at the heating surface but condense in the cold liquid, so that no net generation of vapor is realized. Very high heat fluxes can be obtained under these conditions; values as large as 1.4×10^7 Btu/(hr)(ft²) have been reported in forced-convection heating of water, but 2×10^6 Btu/(hr)(ft²) appears more realistic if the surface temperatures are to be kept low enough to avoid burnout.

6.135. In *volume boiling* the heat is generated within the liquid and only ordinary nucleate boiling occurs. Hence, with volume boiling burnout problems are eliminated. The boiling liquid temperature is very slightly higher than the saturation value and is essentially independent of the rate of heat generation, i.e., of the power density.

---

* The maximum flux is sometimes referred to as the DNB (departure from nucleate boiling) value; it is a design limitation similar to the burnout flux (§ 6.141).

BOILING IN REACTOR SYSTEMS

6.136. Boiling of the coolant is possible in any reactor using a liquid for heat removal. In fact, boiling organic liquid, liquid sulfur, and liquid metal systems have been considered in addition to boiling-water reactors. However, since the latter are the only type which have been developed so far, the subsequent discussion will refer to boiling water in particular, although the general conclusions are applicable to any boiling-liquid coolant.

6.137. Suppose that water, below the saturation temperature, is forced through a channel between or around the solid fuel elements of a reactor; heat is then transferred from the solid surface to the water. As long as the fuel-wall temperature, which increases along its length in the manner described in § 6.90 *et seq.*, remains below the steam saturation temperature, single-phase heat transfer only will occur. In water-cooled power reactors the pressure on the cooling system is increased in order to raise the saturation temperature and thus prevent bulk boiling, but some local boiling is tolerated (§ 13.10). Pressurized-water and boiling-water reactors may therefore be regarded as having certain features in common.

6.138. The various boiling regimes along a typical reactor fuel-element channel are shown in Fig. 6.16. At first, the only effect is an increase in the sensible heat of the coolant, and this is followed by a region of subcooled boiling. After this there is the region in which ordinary nucleate boiling occurs ac-

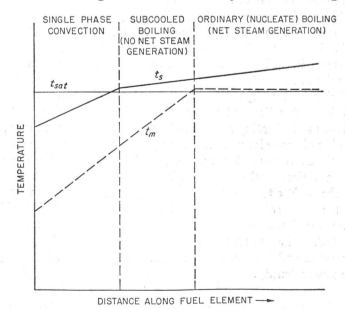

Fig. 6.16.   Variation of surface and liquid temperatures with distance along fuel element

companied by steam generation. In this section of the fuel element the heat-transfer rate is high and in the design of a reactor with a boiling coolant it is desirable to maximize the length of the nucleate boiling region.

6.139. If, as a result of local circumstances, the temperature difference between a point (or region) on the fuel-element surface and the water is increased, film boiling may take place. There is then a danger that burnout will occur. This is most likely to happen near the point in the coolant channel where the temperature of the metal surface is a maximum (Fig. 6.11). If two or more channels are connected in parallel, as is usually the case, the probability of burnout is greatly increased, for the following reason. The formation of a film or layer of steam on the metal in a particular channel will result in an increased flow resistance over the length of that channel. When several channels are fed from a common header, the flow of water in the channel under consideration will then decrease, since it offers the largest resistance. This will tend to increase steam-film formation, which will be accompanied by a further decrease in flow, and so on until burnout occurs.

6.140. Vapor generated in the boiling liquid in the interior of the reactor will rise as a result of its buoyancy. In addition to this natural rise of the vapor, the gross motion or circulation of the liquid can contribute to the transport of vapor. A natural circulation can be achieved if the flow can be directed to take advantage of the variation of fluid density within the core. The thermal and hydraulic design procedures for boiling-water reactors are therefore unique.

PREDICTION OF BURNOUT CONDITIONS

6.141. The heat flux at which burnout is likely to occur represents an important design consideration for the normal operation of both pressurized- and boiling-water reactors. A knowledge of burnout conditions is also necessary for possible accidental situations arising in nonboiling reactors as a result of loss of coolant flow or excessive fuel temperatures due to a power excursion. Consequently, methods for estimating the burnout flux in a reactor are required. Most of the information available is for pressurized-water reactors with forced circulation operating under conditions of local or subcooled boiling. Several empirical relationships have been proposed for the maximum attainable flux, $(q/A)_{max}$, before burnout occurs in water; one which has been widely used is [23]

$$\left(\frac{q}{A}\right)_{max} \approx 10^6 C \left(\frac{G}{10^6}\right)^m (\Delta t_{sub})^{0.22}, \qquad (6.59)$$

based on experimental data for water flowing upward in vertical tubes with inside diameters of 0.143 to 0.226 in. and L/D ratios of 21 to 168. The pressures ranged from 100 to 2000 lb/in.^{2} (abs.), water temperatures from 229 to 636°F, and mass velocities from 8×10^3 to 7.7×10^6 lb$_m$/(hr)(ft^2) of cross section. In equation (6.59), q/A is expressed in Btu/(hr)(ft²), G is equal to $v\rho$, i.e., the mass velocity of the water, in lb_m/(hr)(ft²), and Δt_{sub}, the amount of subcooling, is in °F; the constants C and m are functions of the pressure as follows:

Pressure	m	C
500psia............	0.16	0.817
1000..............	0.28	0.626
2000..............	0.50	0.445

Although equation (6.59) is probably not applicable in all cases, it is useful for design purposes. Where burnout occurs under conditions other than subcooled boiling, different relationships must be used [24].

6.142. In the region of nucleate boiling, when the bulk temperature of the coolant is above the saturation temperature, the maximum heat flux is primarily dependent on the fluid enthalpy, but it decreases as the mass velocity increases at constant burnout enthalpy. This trend is opposite to that for subcooled boiling, as is apparent from equation (6.59). The change in the effect of mass velocity on the burnout flux occurs when the bulk temperature is near to the saturation point. The channel length and its ratio to the distance between the heated surfaces appear to have little effect on the maximum flux. On the other hand, the heat flux distribution along the channel length does influence the results. For a cosine-shaped distribution in the axial direction, the burnout flux is less than for a uniformly heated channel under the same coolant conditions at the burnout location [25].

Example 6.9. In the Yankee pressurized-water reactor, the average outlet temperature of the coolant is 533°F and the pressure is 2000 psia. Determine the burnout flux for the subcooled (or local) boiling condition; G is 2.21×10^6 lb/(hr)(ft²), and the saturation temperature at 2000 psia is 636°F.

By equation (6.59), since $m = 0.50$ and $C = 0.445$ for 2000 psia,

$$\frac{q}{A} \text{ (burnout)} \approx (10^6)(0.445)(2.21)(636 - 533)^{0.22}$$

$$= 1.95 \times 10^6 \text{ Btu/(hr)(ft²)}.$$

This result, based on average conditions, is actually too optimistic for design purposes. A more realistic value is believed to be about 10^6 Btu/(hr)(ft²), as will be seen in § 12.25, i.e., roughly half the value obtained from equation (6.59).

BOILING HEAT-TRANSFER COEFFICIENTS

6.143. It is apparent from Fig. 6.15 that log (q/A) is a linear function of log $(t_s - t_m)$ over a considerable portion of the nucleate boiling range, so that the general expression

$$\frac{q}{A} = c(t_s - t_m)^n, \tag{6.60}$$

where c and n are constants, is applicable to a particular set of conditions. Since t_m is essentially the same as the saturation temperature (see Fig. 6.16),

$t_s - t_m$ is virtually equal to the difference between the temperature of the heated surface and the saturation temperature of the coolant; this difference may be represented by Δt_b. Hence equation (6.60) may be written as

$$\frac{q}{A} = c(\Delta t_b)^n. \tag{6.61}$$

If h_b represents the boiling heat-transfer coefficient, it may be defined by

$$\frac{q}{A} = h_b \Delta t_b, \tag{6.62}$$

so that

$$h_b = c(\Delta t_b)^{n-1}.$$

6.144. Curves showing h_b in Btu/(hr)(ft²)(°F) as a function of Δt_b in °F,

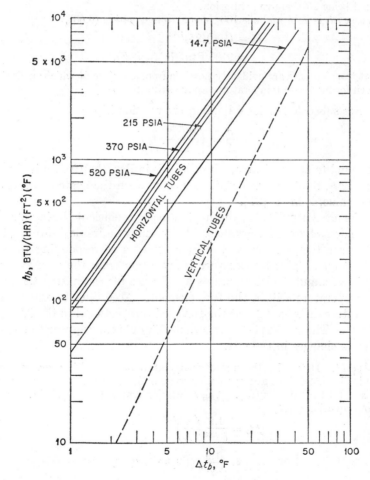

Fig. 6.17. Boiling heat-transfer coefficient

obtained from a variety of experimental studies of heat transfer to boiling water in a free convection system at several pressures, are given in Fig. 6.17.* The value of the exponent $n - 1$ is found to be 1.42 in all cases, but the factor c increases with the pressure; thus,

$$h_b = c(\Delta t_b)^{1.42}, \tag{6.63}$$

with the following values of c:

Pressure, psia	14.7 (1 atm)	215	370	520
c	42	82	90	100

These results are applicable over a range of Δt_b from about 1° to 45°F.

Example 6.10. Determine the boiling heat-transfer coefficient of water in a boiling-water reactor operating at a pressure of 520 psia, the temperature at the surface of the fuel element being 25°F above saturation.

In this case Δt_b is 25°F and c is 100; hence, from equation (6.63),

$$h_b = (100)(25)^{1.42}$$

$$= 9100 \text{ Btu}/(\text{hr})(\text{ft}^2)(°\text{F}).$$

(High heat-transfer rates are evidently possible from a surface at which nucleate boiling occurs, even for moderate temperatures above saturation.)

6.145. For subcooled or local boiling, the relationship

$$\frac{q}{A} = \left(\frac{e^{p/900}}{1.9} \Delta t_b\right)^4 \tag{6.64}$$

has been found to be applicable, where q/A is in Btu/(hr)(ft²), p is the pressure in psia, and Δt_b, the temperature difference between the heated surface and the saturation value, is in °F [27].

6.146. For design purposes, the heat-transfer coefficient during steady-state boiling is seldom of great importance. It is usually desirable, however, to evaluate the surface temperature of the fuel elements corresponding to a desired average heat flux. This temperature is important from the standpoint of corrosion resistance and it also serves as a reference for estimating the maximum internal temperature within the fuel element. According to equation (6.64), the temperature difference Δt_b at the surface is proportional to the 0.25 power of the heat flux. The surface temperature itself would thus appear to be relatively insensitive to changes in the flux.

Example 6.11. Determine the surface temperature of the fuel in a reactor core under subcooled boiling conditions at a system pressure of 600 psia when the average heat flux is (a) 10^5 and (b) 10^6 Btu/(hr)(ft²). The saturation temperature is 486°F.

(a) From equation (6.64)

$$\Delta t_b = \frac{1.9}{e^{p/900}} \left(\frac{q}{A}\right)^{0.25}$$

$$= \frac{1.9}{e^{600/900}} (10^5)^{0.25} = 17°\text{F}.$$

* Based upon data in the published literature; see, for example, Ref. 26.

The fuel surface temperature is consequently $486 + 17 = 503°F$.

(b) $$\Delta t_b = (17)(10)^{0.25} = 30°F,$$

and so the fuel surface temperature is $486 + 30 = 516°F$. (As stated in § 6.146, the surface temperature is not very sensitive to changes in the heat flux; the temperature increases from 503°F to only 516°F for a tenfold increase in heat flux.)

6.147. Heat-transfer coefficients to flowing saturated and superheated steam are much lower than boiling coefficients. Since the fluid is essentially a gas, the heat-transfer coefficient may be calculated from equation (6.53).

HEAT EXCHANGE TO STEAM SYSTEM

6.148. The design of the heat exchanger for transferring heat from the primary to a secondary coolant without boiling follows conventional engineering practice as far as the heat-transfer calculations are concerned. Because of induced radio-activity in most coolants, however, more rigid specifications with regard to construction, in order to avoid leakage and minimize corrosion and other factors causing maintenance problems, are imposed than is usual for commercial equip-ment. In addition, the need for locating the heat exchanger behind a shield may affect the design.

6.149. The operation of nuclear power plants, with the exception of those using boiling reactors, involves the production of steam by transfer of heat from the primary or secondary coolant to water. The design of the heat exchanger or steam generator involves an application of boiling heat-transfer data.

FLUIDS CONTAINING A VOLUME HEAT SOURCE

HEAT TRANSFER FROM CIRCULATING FUEL

6.150. Because reactors with circulating fluid fuel have certain advantages (see Chapter 13), they have attracted interest for possible use in power pro-duction. A consideration of heat transfer in fluid fuels is, therefore, an impor-tant aspect of the design of such reactors. Since the fluids (while in the reactor) contain heat sources due to fission, the determination of the temperature distri-bution in the flowing system and at the walls involves unusual principles of heat transfer and fluid flow [28].

6.151. If the fuel is forced through a channel or pipe in the reactor, the heat will be convected away relatively slowly by the slow-moving part of the stream close to the wall. As a result, there will be a temperature rise toward the wall of the channel, and a flow of heat from the slower-moving parts of the stream to the faster-moving parts near the middle of the stream. If the temperature of the wall is limited, some prediction of the difference between the wall temper-ature (t_w) and the mixed-mean (bulk) fluid temperature (t_m) is required.

6.152. The temperature difference obtained for the case of turbulent flow of a fluid containing a uniform heat source, but with no cooling at the pipe wall, is depicted in Fig. 6.18. Here the dimensionless "temperature difference,"

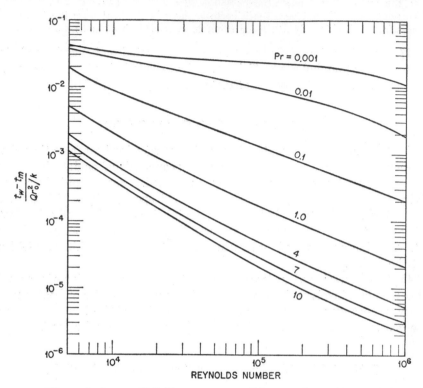

FIG. 6.18.   Dimensionless wall-fluid temperature difference for fluid containing a heat source [28]

$(t_w - t_m)/(Qr_0^2/k)$, where r_0 is the pipe radius, is shown as a function of the Reynolds number for certain specified values of the Prandtl modulus.   Experimental measurements have confirmed the reliability of the calculations in the range of Pr = 5 to 10 and Re = 750 to 14,000.

CORE HYDRAULICS

PRESSURE DROP DUE TO FRICTION

6.153. Pumping power requirements are determined by the pressure drop in the cooling system and the rate of flow of the coolant.   The total pressure drop is the sum of terms arising from (a) the hydrostatic pressure due to the increase in level, (b) the acceleration of the coolant fluid, and (c) friction in channels, pipes, etc., and in regions where there are changes in the cross-sectional area. Of these, the hydrostatic pressure is simply the integrated product of the increase in level, the density, and the acceleration of gravity. The kinetic energy term, for a given mass flow rate, depends on the difference in velocity between inlet and outlet. For an incompressible fluid this is usually negligible, and even for gases the contribution is very small for reactor operating conditions. The subsequent discussion will therefore be restricted to the effects of friction and flow area changes on the pressure drop. Although these topics are treated in standard texts on fluid dynamics, they will be reviewed briefly here [29].

6.154. The pressure drop, Δp_f, accompanying isothermal *laminar* flow at a mean velocity v in a circular pipe of diameter D and length L is given by the familiar Poiseuille equation

$$\Delta p_f = \frac{32\mu v L}{g_c D^2},$$

which may be written as

$$\Delta p_f = 64\,\frac{\mu}{Dv\rho} \cdot \frac{L}{D} \cdot \frac{\rho v^2}{2g_c}, \tag{6.65}$$

where μ is the viscosity and g_c the factor for relating mass and force. The subscript f is used in Δp_f to indicate that the pressure loss described is due to fluid friction. It will be observed that the factor $\mu/Dv\rho$ on the right of equation (6.65) is the reciprocal of the Reynolds number (§ 6.107), and so

$$\Delta p_f = \frac{64}{\mathrm{Re}} \cdot \frac{L}{D} \cdot \frac{\rho v^2}{2g_c}. \tag{6.66}$$

This gives the pressure drop due to fluid friction for laminar flow in a straight pipe of circular cross section.

6.155. For turbulent flow at a mean velocity v, the relationship corresponding to equation (6.66) is

$$\Delta p_f = 4f\,\frac{L}{D} \cdot \frac{\rho v^2}{2g_c} \tag{6.67}$$

known as the Fanning equation, in which the dimensionless quantity f is called the *friction factor*.* Comparison of equations (6.66) and (6.67) shows that for laminar flow $f = 16/\mathrm{Re}$, but the relationship between the friction factor and the Reynolds number in the case of turbulent flow is more complicated. Several empirical expressions have been developed; one of the simplest of these, which holds with a fair degree of accuracy for flow in smooth pipes at Reynolds numbers up to about 2×10^5, is the Blasius equation

$$f = 0.079\,\mathrm{Re}^{-0.25}. \tag{6.68}$$

This permits the evaluation of the turbulent-flow friction factor for a given Reynolds number.

---

* In the equivalent Darcy-Weisbach equation a friction factor equal to $4f$ is used.

6.156. For turbulent flow in a commercial rough pipe the friction factor is larger than for a smooth pipe at the same Reynolds number, as shown in Fig. 6.19. The deviation from the ideal behavior of a smooth pipe increases with the

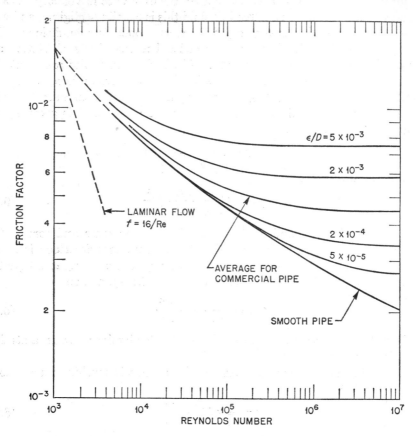

Fig. 6.19. Friction factor as a function of pipe roughness [30]

roughness of the pipe, especially at high Re values. The variation of the friction factor with Reynolds number has been determined experimentally for different degrees of roughness, expressed in terms of a dimensionless quantity ϵ/D, where ϵ is a measure of the size of the roughness projections and D is the pipe diameter [30].

6.157. For turbulent flow in noncircular channels of relatively simple form, the pressure drop due to friction may be calculated by substituting the equivalent diameter, defined in § 6.111, for D in equation (6.67). The same value is used in determining the Reynolds number. It may be mentioned that the replacement of D by D_e for noncircular channels is incorrect for laminar flow.

Example 6.12. Estimate the pressure in lb/in.² (psi) required to overcome friction for the turbulent flow of water in the channel between the fuel plates in the examples given earlier.

The value of Re was found in Example 6.7 to be 8050 and so f, assuming the plates to be moderately smooth, is seen from Fig. 6.19 to be about 0.008. Since ρ is 61.8 lb/ft³, v is 10,800 ft/hr, L is 2.5 ft, D_e is 0.02 ft, and g_c is 32.2 ft/sec² = (32.2)(3600)² ft/hr², it follows from equation (6.67) that

$$\Delta p_f = (4)(0.008)\,\frac{(61.8)(10,800)^2(2.5)}{2[(32.2)(3600)^2](0.0833)}$$

$$= 34.5 \text{ lb/ft}^2$$

$$= 0.24 \text{ lb/in.}^2 \text{ (psi).}$$

6.158. In a flowing fluid there will be changes in pressure due to changes in velocity resulting from gradual or abrupt changes in flow area. Such pressure changes are usually considered in terms of the "velocity head," defined as $v^2/2g_c$; the pressure corresponding to a head H is equal essentially to $H\rho$. Since v^2 is generally high for turbulent flow, the pressure losses accompanying changes in cross-sectional area of fluid conduits may be very significant. In general, for

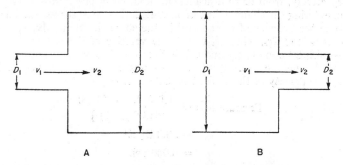

A   B

Fig. 6.20.   A. Abrupt expansion in fluid flow
B. Abrupt contraction in fluid flow

abrupt expansion (Fig. 6.20A) or contraction (Fig. 6.20B), the pressure change Δp due to the loss of head ΔH, which occurs in either case, can be represented by

$$\Delta p = \rho\Delta H \text{ (expansion)} = K_e\frac{\rho v_1^2}{2g_c}$$

$$\Delta p = \rho\Delta H \text{ (contraction)} = K_c\frac{\rho v_2^2}{2g_c},$$

where v_1 and v_2 are the upstream and downstream velocities, respectively; the value of K, the loss coefficient, depends upon the conditions, as indicated below.

6.159. For abrupt *expansion*,

$$K_e = \left(1 - \frac{D_1^2}{D_2^2}\right)^2,$$

where D_1 and D_2 are the pipe diameter upstream and downstream, respectively. For expansion into a large reservoir, D_2 is large, and K_e becomes virtually unity. The loss of head is then almost equal to $v_1^2/2g_c$.

6.160. For abrupt *contraction* the value of K_c varies with the ratio D_2/D_1 in the following manner:

D_2/D_1......	0.8	0.6	0.4	0.2	0
K_c.........	0.13	0.28	0.38	0.45	0.50

For the case of inlet from a very large reservoir, D_2/D_1 approaches zero, and K_c is then approximately 0.5; the corresponding loss of head is $v_2^2/4g_c$. It should be understood that the results given above apply only to cases of abrupt expansion or contraction. The loss of head decreases if the fluid exit is rounded, making the change less abrupt. Where the exit is tapered so that the included angle is 7° or less, the losses will usually be negligible.

Example 6.13. Estimate the pressure loss due to contraction and expansion of coolant entering or leaving a channel between two parallel-plate fuel elements considered previously.

The coolant undergoes sudden contraction and expansion as it enters and leaves the channel, respectively, from or to a header or manifold of large cross section. The loss of head upon entering the channel, i.e., upon contraction, may then be taken as $0.5v_2^2/2g_c$, where v_2 is here the velocity of the coolant in the channel. Similarly, upon leaving the channel, the loss is $v_1^2/2g_c$, where v_1 is also the velocity of the coolant in the channel. The total loss of head upon entry and exit is thus $1.5v^2/2g_c$, where v is the velocity of the coolant in the channel, i.e., 10,800 ft/hr. The pressure loss is equal to the loss of head multiplied by the density of the coolant (§ 6.158); hence,

$$\text{Pressure loss} = \frac{(1.5)(10,800)^2(61.8)}{2[(32)(3600)^2]}$$

$$= 13.0 \text{ lb/ft}^2$$

$$= 0.090 \text{ psi.}$$

6.161. Losses in pipe fittings, due to changes in direction, e.g., in elbows, curves, etc., or to contraction in valves, can also be expressed in terms of the velocity head, e.g., $K_1v^2/2g_c$. The factor K_1 may range from 0.25 or less for a gradual 90° curve or a fully opened gate valve to 1.0 for a standard screwed 90° elbow, or as high as 10 for a fully opened globe valve.*

TWO-PHASE PRESSURE DROP

6.162. In a channel containing boiling water there are two phases, namely, liquid water and steam. The problem of the two-phase pressure drop in a vertical channel is of particular importance in boiling-water reactor design. A

---

* For further information, standard reference books and manufacturers' catalogs should be consulted. The loss coefficients, K_1, are often expressed in terms of an "equivalent" length of straight pipe.

convenient treatment involves the use of a *friction factor multiplier*, R, defined as the ratio of the two-phase friction pressure gradient, $(dp/dL)_2$, to the single phase value for the liquid phase alone, $(dp/dL)_l$, i.e.,

$$R \equiv \frac{(dp/dL)_2}{(dp/dL)_l} \approx \frac{(\Delta p)_2}{(\Delta p)_l},$$

where $(\Delta p)_2$ is the two-phase pressure drop in a given channel in which $(\Delta p)_l$ would be the pressure drop for the liquid phase alone. Values of R, obtained by semiempirical correlation procedures and graphical integration of local values with respect to length, are given in Fig. 6.21 as a function of the system pressure

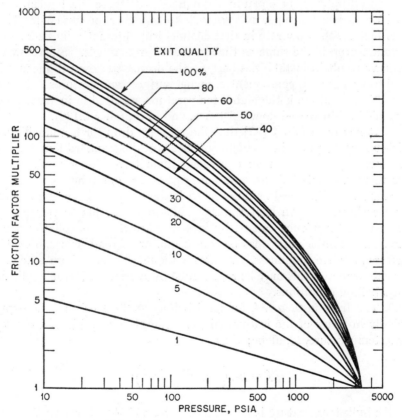

Fig. 6.21. Friction factor multiplier for two-phase flow [31]

for steam ranging in exit quality from 1 to 100 per cent [31]. Other curves have been determined for the situation in which the axial heat flux has a cosine-shaped distribution.

EFFECT OF LOCAL BOILING ON PRESSURE DROP

6.163. In the design of the first pressurized-water reactors, such as the PWR (§ 13.10), local (nucleate) boiling was considered to be undesirable and conditions were chosen to prevent its occurrence. In later designs, however, local boiling during steady-state operation became a design requirement. Local boiling will affect the pressure drop and, in an open lattice, will result in flow redistribution, as explained below, with the possibility of burnout. An understanding of the factors which affect the pressure drop is therefore essential.

6.164. Where the parallel flow channels have the same inlet and outlet headers, the pressure drop must be equal in all channels. If there is a tendency in any channel for the resistance per unit flow rate to be higher, because of local boiling or other factors, the flow rate in that channel will automatically decrease until the pressure drop is the same as that in the other channels. Such flow distribution may result in instabilities because the decreased flow will mean a lower thermal transport and, consequently, higher surface temperatures. This could lead to ultimate failure if the higher temperatures resulted in increased boiling, accompanied by increased flow resistance, and so on (cf. § 6.139).

6.165. It has been found [32] that the pressure drop for a nonboiling, open parallel-tube system may be predicted by means of the standard Fanning friction factor correlation based on an equivalent diameter. Under nonboiling conditions there are only minor variations caused by differences in bulk water temperatures. When local boiling occurs, however, there is an increase in momentum exchange and in heat transfer which may be due to the increase in hydrodynamic turbulence caused by repeated growth and collapse of steam bubbles. The bubbles may also be considered, in a broad sense, to contribute an increased surface roughness. Whatever the mechanism, there is an increase in pressure drop and this may be represented by a friction factor applicable to the local boiling conditions.

6.166. The friction factor, f, for local boiling at 2000 psi may be expressed empirically in terms of the isothermal (nonboiling) factor, f_{iso}, as a function of the bulk temperature, t_m, alone; thus,

$$f = f_{\text{iso}} \left(1 + \frac{t_m - 560}{126} \right),$$

where the bulk temperature is between 560 and 635°F (saturation temperature of water) and above the local boiling temperature. The pressure drop is equal to that for nonboiling until the bulk water temperature reaches 560°F. Alternative procedures for local boiling pressure-drop calculations have been based on two-phase flow predictions.

PRESSURE DROP AND FREE CONVECTION COOLING

6.167. Some information concerning the cooling of reactors of the pool type by free convection can be obtained from pressure drop considerations. The flow of water between the parallel-plate elements in these reactors is approximately laminar, because of the close spacing of the plates, the low heat flux, and the consequent small convective driving force. The flow rate and, hence, the ability to carry heat from the core by free convection alone may be estimated in the following manner.

6.168. For isothermal laminar flow between parallel plates, the appropriate form of the Poiseuille equation leads to the expression

$$\Delta p_f = \frac{48\mu vL}{g_c D_e^2} \qquad (6.69)$$

for the pressure drop through the entire length L of the fuel plate; as before, D_e is the equivalent diameter. For convective cooling of a vertical plate, v in equation (6.69) will then be the average upward velocity of the coolant when Δp_f is equal to the average pressure drop due to the difference in density at bottom and top. If ρ_1 and ρ_2 are the densities of the coolant at entrance and discharge, respectively, from the channel, the average pressure differential is approximately $\frac{1}{2}(\rho_1 - \rho_2)L$. If this is equated to the right side of equation (6.69), the average velocity of the fluid is given by

$$v_{av} = \frac{(\rho_1 - \rho_2)g_c D_e^2}{96\mu_{av}}, \qquad (6.70)$$

where μ_{av} is the average viscosity.

6.169. The total rate of heat production by the reactor, q_{tot}, is equal to the product of the total mass flow rate, w, of the coolant, its specific heat, c_p, and the average rise, Δt_c, in the mixed-mean (or bulk) coolant temperature, i.e.,

$$q_{tot} = wc_p\Delta t_c.$$

Since w may be replaced by $v\rho A_f$, where A_f is here the total flow area, it follows that

$$q_{tot} = (\rho c_p)vA_f\Delta t_c. \qquad (6.71)$$

In the present case, v is given by equation (6.70), ρ is the mean density, i.e., $\frac{1}{2}(\rho_1 + \rho_2)$, and Δt_c is $t_2 - t_1$, the temperature difference between the top and bottom of the vertical fuel plate. Hence,

$$q_{tot} = (\rho_1 + \rho_2)(\rho_1 - \rho_2)\frac{A_f g_c D_e^2 c_p(t_2 - t_1)}{192\,\mu_{av}},$$

which may be used to estimate the power output for specified lower and upper temperatures.

Example 6.14. The fuel plates in a water-cooled reactor are 0.01 ft apart and the total flow area for the water is 0.75 ft²; the temperature at the bottom of the plates is 80°F

Estimate the rate at which heat is removed by free convection when the upper temperature is (a) 100°F, (b) 120°F, using the following data:

t °F	ρ lb/ft³	μ lb/(ft)(hr)	c_p Btu/(lb)(°F)
80°	62.21	2.08	1.00
100°	62.00	1.66	1.00
120°	61.73	1.36	1.00

(a) Upper temperature = 100°F. Taking the average viscosity as the arithmetic mean for the lower and upper temperatures,

$$q_{tot} = \frac{(124.21)(0.21)(0.75)(32)(3600)^2(0.02)^2(1.00)(20)}{(192)[(\tfrac{1}{2})(2.08 + 1.66)]} \text{ Btu/hr}$$

$$= 53 \text{ kw.}$$

(b) Upper temperature = 120°F

$$q_{tot} = \frac{(123.94)(0.48)(0.75)(32)(3600)^2(0.02)^2(1.00)(40)}{(192)[(\tfrac{1}{2})(2.08 + 1.36)]} \text{ Btu/hr}$$

$$= 260 \text{ kw.}$$

6.170. In reactors of the parallel-plate type, with free convection cooling, higher power levels than those calculated here are possible by operating at higher exit-water temperatures. However, at power levels above about 200 kw other factors, such as radioactivity of the pool water and surface temperature of the fuel elements, must be taken into consideration. These may limit the total power of reactors of this type.

6.171. It should be pointed out that, in the foregoing analysis, no consideration has been given to the existence of the heated jet of liquid rising in a column above the reactor core. That an additional driving force due to this column of liquid does exist has been proved by experiment; it is much more important in promoting coolant flow through the reactor than the vertical driving force taken over the length of the fuel plates. Other effects which may be significant, including pressure losses at channel inlet and discharge, deviations from laminar flow, and nonuniform surface temperatures, have also been neglected.

PUMPING POWER IN COOLANT SYSTEMS

6.172. The power utilized in pumping the coolant through a reactor is given by

$$\text{Pumping power} = \text{Pressure drop} \times \text{Volume flow rate}$$

$$= (\Delta p)(A_f v), \tag{6.72}$$

where Δp is the total pressure drop due to all causes, and A_f is the total cross-sectional (or flow) area of all the coolant channels. Since flow is ordinarily turbulent, where forced convection heat transfer is employed, the pressure drop due to friction is given by equation (6.67); to this must be added the values for entrance and exit effects, given above, as well as for various fittings, etc., in the coolant circuit.

6.173. Attempts have been made to compare the pumping power, relative to the total reactor power, required to overcome the pressure drop due to friction for various coolants. This has been done by combining equations (6.9), (6.52), and (6.67), and the relationship of equation (6.71). In making such comparisons, however, it is necessary to postulate that various conditions in the reactor are independent of the nature of the coolant. Thus, it is supposed that, for a given reactor power, the temperature rise, Δt_c, of the coolant is always the same. But this may not necessarily be true in practice for optimum operation. For example, using a gaseous coolant, it may be desirable to have Δt_c much larger than for a liquid metal. Furthermore, the comparisons of pumping power are based on overcoming fluid friction in the reactor only and do not take into account the heat exchanger, as well as entrance and exit effects which may be considerable.

6.174. Without being quantitative, however, it is possible to draw some general conclusions with regard to pumping power. It is seen that the pressure losses due to fluid friction as well as to sudden changes in flow area all increase with v^2. Moreover, since the *volumetric heat capacity*, ρc_p, of a gaseous coolant is normally less than that of a liquid, it follows from equation (6.71) that the removal of a certain quantity of heat will usually require a very much higher coolant velocity for a gas than for a liquid, assuming that $A_f \Delta t_c$ is not greatly different in the two cases. Hence, the pumping power requirements for a gas will be much larger than for a liquid coolant. The discrepancy can be partially overcome by increasing the density of the gas, i.e., by operating at high pressures.

Example 6.15. Estimate the power required for pumping the cooling water through the core of the 3000-kw reactor considered in previous examples, assuming the minimum number of cooling channels to be 450.

The pressure drop due to fluid friction and entrance and exit effects has been found to be $34.5 + 13 = 47.5$ lb/ft². The total flow area of the 450 channels is

$$A_f = \frac{(450)(2.5)(0.12)}{144} = 0.94 \text{ ft}^2.$$

The velocity of the coolant is 10,800 ft/hr so that

$$\text{Pumping power} = (47.5)(0.94)(10,800)$$

$$= 4.82 \times 10^5 (\text{ft})(\text{lb})/\text{hr}.$$

Converting to theoretical horsepower, making no allowance for pump efficiency, gives

$$\text{Pumping power} = \frac{4.82 \times 10^5}{(33,000)(60)} = 0.243 \text{ hp}.$$

(The total pressure drop for the entire cooling system, including piping and heat exchangers, will be considerably larger than the value used here and thus the total pumping power will also be greater.)

REACTOR COOLANTS

GENERAL CHARACTERISTICS [33]

6.175. Since most of the energy liberated in fission appears as heat, a nuclear reactor must have an adequate cooling system which will prevent the attainment of undesirable temperatures within the reactor. Such temperatures may be determined by a property of the fuel, e.g., uranium metal undergoes a phase change at 665°C (1230°F) accompanied by a relatively large increase in volume (§ 8.37), or of the coolant, e.g., it may be undesirable in a water-cooled reactor for the water to boil. The cooling medium (or coolant) may be gaseous or liquid, and the specific requirements for this substance will depend largely upon the rate of heat production, i.e., the reactor power density, and the operating temperatures. For a reactor of low power output, in which no use is made of the heat generated, the cooling system may be relatively simple. On the other hand, for economic power production a reactor should operate at high temperatures and then careful consideration must be given to the choice of coolant.

6.176. In general a coolant should possess the following characteristics: (1) good thermal properties, e.g., high specific heat and thermal conductivity; (2) low power requirement for pumping; (3) high boiling point and low melting point; (4) stability to heat and radiation; (5) suitable corrosion characteristics in the given system; (6) small cross section for neutron capture; (7) nonhazardous, including low induced radioactivity; and (8) low cost. These properties will be reviewed briefly in the order given, although this must not be regarded as representing the order of their importance. In actual practice the coolant chosen for any particular application will represent a compromise among conflicting requirements.

6.177. Since the primary purpose of a coolant is the removal of heat from a reactor, it must possess good heat-transfer properties. For a given geometry and dynamics of flow this condition is best met by high thermal conductivity and high specific heat; for nonmetallic liquids in particular, low viscosity is also advantageous in this connection. The fraction of the power output required to pump the coolant through the reactor and heat exchanger should be small, and this requires a coolant with a high density and low viscosity. For use at high temperatures a liquid coolant should have a low vapor pressure, so that a strong pressurized system is not necessary. At the same time, the coolant should remain liquid down to low temperatures to avoid the possibility of solidification when the reactor is shut down. These two requirements are, to a certain extent, contradictory; the former implies a high boiling point and the latter a low melting point. Nevertheless, a number of substances are known, e.g., liquid metals and molten salts, that have a very wide temperature range for the liquid phase. The problem considered here does not arise, of course, when a gas is used as the coolant.

6.178. A coolant should be stable with respect both to high temperatures and to the action of nuclear radiation. If the reactor has a circulating fuel, so that the secondary coolant is not exposed to a very high radiation field, the problem of radiation decomposition (and of induced radioactivity) is not so serious as when the coolant actually flows through the reactor. Further, it is necessary that the fluid should not attack the materials, either inside or outside the reactor, with which it comes into contact. Since the coolant system is a dynamic one, mass transfer due to temperature gradients may be very significant. As far as physical and chemical attack are concerned, the coolant and containing materials must, of course, be considered together.

6.179. Because loss of neutrons must be compensated for by an increase in the fuel loading of the reactor, the coolant should consist of an element or elements having small cross sections for neutron capture. This restriction is less important for fast reactors, because the capture cross sections for high-energy neutrons are invariably small. If, as a result of neutron capture, the (primary) coolant acquires induced activity, this should be weak with little or no gamma radiation in order to avoid the necessity for shielding the pumps, heat exchanger, and piping. It is desirable, in any event, that the cooling medium not be toxic or otherwise hazardous, and that it does not become a hazard as a result of exposure to radiation. Finally, the coolant should be readily available at relatively low cost.

6.180. For a thermal reactor it is desirable, too, that the elements present in the coolant should have low atomic weights. In other words, it is an advantage if the coolant acts also as a moderator and makes some contribution to the slowing down of the neutrons. On the other hand, for a fast reactor, where slowing down is to be avoided, materials of low mass number should be eliminated, as far as possible, unless they happen to have very small scattering cross sections, e.g., lithium-7.

6.181. The general characteristics of a number of individual coolants will be reviewed below. It will be seen that there is no single substance (or mixture) which satisfies all the requirements laid down for a satisfactory coolant. It is for this reason that several different coolants, including water, organic liquids, liquid metals, and gases, are at present being tested in various reactor systems. The important physical properties of those coolants which appear to offer some prospect for use in reactors, now or in the future, are summarized in Table A.6 in the Appendix. Because a coolant may be used over a range of temperatures, values are given for several temperatures where possible.

COMPARISON OF HEAT-TRANSFER CHARACTERISTICS OF COOLANTS

6.182. A good coolant is, in general, one having a high heat-transfer coefficient (and a large volumetric heat capacity) under the conditions that exist (or can exist) in a reactor. Thus, if h is large, it is possible to realize a specified rate of heat removal from the fuel elements with a relatively small heat-transfer area. It is also possible to decrease the rate of flow of the coolant and, as a result, the power required for pumping. The decrease in the flow rate will decrease the heat-transfer coefficient, but if the latter is sufficiently high, the decrease may not be serious, especially for a liquid metal.

6.183. It can be seen from equations (6.52) and (6.55) that, for a given flow rate and coolant channel diameter, large heat-transfer coefficients are associated with high thermal conductivity, density, and heat capacity. In addition, for nonmetallic fluids, a low viscosity is advantageous. It is apparent, from general considerations, therefore, that water and many liquid metals may be expected to have good heat-transfer characteristics. Such materials will therefore permit considerable flexibility in the design of the heat-removal system of the reactor.

6.184. Gases have the disadvantage of low density and thermal conductivity, but some compensation can be achieved, as far as the former property is concerned, by operating at high pressures. The high flow rate required with a gaseous coolant (§ 6.174) also helps to increase the heat-transfer coefficient although at the cost of very significant increase in pumping power. In reactors designed for the production of useful power, the flow rate of liquid coolants is generally about 15 to 20 ft/sec, whereas in gas-cooled reactors it is approximately ten times as great.

6.185. Another basis for the comparison of coolants is the volumetric heat capacity, ρc_p generally expressed in $Btu/(ft^3)(°F)$, which appears in parentheses in equation (6.71). It is seen from this equation that for a given coolant flow rate, v, and total flow area, A_f, a large volumetric heat capacity will make possible a specified rate of heat removal without an excessive (average) rise in the coolant temperature. This is advantageous from the thermal stress standpoint. Although liquid metals may be somewhat superior to water with respect to heat transfer, the latter has a better volumetric heat (or heat-transport) capacity. Thus, at 400°F, ρc_p for water is 53.7, whereas that for liquid sodium is 17.1 $Btu/(ft^3)(°F)$. At 10-atm pressure the volumetric heat capacity of carbon dioxide gas is only 0.174 $Btu/(ft^3)(°F)$, so that very high flow rates are necessary with gaseous coolants.

Example 6.16. Compare the heat-transfer coefficient for turbulent flow at 400°F of (a) carbon dioxide at 10 atm and a velocity of 100 ft/sec, (b) water at 20 ft/sec, and (c) liquid sodium at 20 ft/sec, flowing through a long, straight, circular tube of diameter 0.10 ft.

The physical properties of the three coolants at 400°F are tabulated below:

	Carbon Dioxide	Water	Liquid Sodium
ρ, lb/ft^3	0.656	53.7	53.2
μ, $lb/(hr)(ft)$	0.0504	0.148	—
k, $Btu/(hr)(ft^2)(°F/ft)$	0.0184	0.380	47.1
c_p, $Btu/(lb)(°F)$	0.265	1.00	0.320

For *carbon dioxide*, using equation (6.52)

$$h = \frac{0.0184}{0.10} \times 0.023 \left[\frac{(0.1)(100 \times 3600)(0.656)}{0.0540} \right]^{0.8} \left[\frac{(0.265)(0.0540)}{0.0184} \right]^{0.4}$$

$$= 124 \text{ Btu}/(\text{hr})(\text{ft}^2)(°F).$$

For *water*, from the same equation (6.52)

$$h = \frac{0.380}{0.10} \times 0.023 \left[\frac{(0.1)(20 \times 3600)(53.7)}{0.148} \right]^{0.8} \left[\frac{(1.00)(0.148)}{0.380} \right]^{0.4}$$

$$= 8150 \text{ Btu}/(\text{hr})(\text{ft}^2)(°F).$$

For *liquid sodium*, however, equation (6.55) is used; thus,

$$h = \frac{47.1}{0.10} \left\{ 7 + 0.025 \left[\frac{(0.10)(20 \times 3600)(53.2)(0.320)}{47.1} \right]^{0.8} \right\}$$

$$= 9650 \text{ Btu}/(\text{hr})(\text{ft}^2)(°F).$$

WATER AS COOLANT

6.186. Water is one of the most important coolants for power reactors, partly because of its low cost but principally because it can act simultaneously as moderator and coolant. It has a high specific heat, although only a fair thermal conductivity as compared to liquid metals; it has a relatively low viscosity and is easily pumped. The pumping power for equivalent heat removal is roughly on the order of one-tenth of that required with a gaseous coolant at 10 atm pressure. The high latent heat of vaporization, combined with its other desirable properties, makes water attractive as a boiling coolant in various boiling-water reactors (§ 13.21 *et seq.*).

6.187. The use of water as coolant has, however, a number of drawbacks. Among these may be noted a fairly large neutron capture cross section, decomposition by radiation, corrosive action on metals, and low boiling point at normal pressures. The loss of neutrons can be compensated by loading additional (enriched) fuel into the reactor in order to attain criticality or by increasing the uranium-235 content of the same mass of fuel. The effect of radiation on water is fairly well understood and the extent of decomposition can be minimized by removing ionic impurities, a procedure which is desirable for other reasons. If decomposition is appreciable, a degasifier may be necessary to remove the oxygen and hydrogen gases formed.

6.188. For the efficient production of useful power, a reactor must be operated at temperatures well above the normal boiling point of water under atmospheric pressure. In these circumstances, high pressures must be used, which means the reactor must be enclosed in a pressure vessel. However, at high temperatures even pure water, demineralized and freed from oxygen, becomes quite corrosive. As a result of the corrosion problem and the need for high pressures, the use of water as coolant in a power reactor imposes limitations upon materials and equipment.   Among the materials which possess the necessary properties of strength and corrosion resistance are certain stainless steels, zirconium alloys, titanium (or its alloys), Inconel, and K Monel metal.

6.189. When using water as a coolant (or for any other purpose) consideration must be given to the emission of gamma rays and neutrons as a result of induced radioactivity; the appropriate information in this connection is summarized in Table 6.4.   Of special interest are the fast-neutron (n, p) reactions with oxygen-16 and oxygen-17.   The product in the former case is nitrogen-16 which emits gamma rays of very high energy, whereas in the latter case the nitrogen-17 formed is one of the few nuclides to emit neutrons.   In addition, the sodium present as an impurity yields the gamma-emitter sodium-24, as a result of radiative capture of thermal (and other) neutrons.

TABLE 6.4.   NEUTRON INDUCED ACTIVITY DATA

Target Nuclide	Isotopic Per Cent	Activation Cross Section (barns)	Radio-active Product	Half-life	Energy of Gamma Rays (Mev)	Gammas per Disinte-gration
O^{16}	99.8	2×10^{-5} *	N^{16}	7.4 sec	6.13, 7.10	0.76, 0.06
O^{17}	0.039	5×10^{-6} *	N^{17}	4.1 sec	Neutron	1 neutron
O^{18}	0.204	2×10^{-4} †	O^{19}	29.4 sec	1.6	0.7
Na^{23}	100	0.53†	Na^{24}	14.9 hr	2.75, 1.38	1, 1
K^{41}	6.8	1.15†	K^{42}	12.4 hr	1.51	0.25

* Fast (n, p) cross sections averaged over fission spectrum.
† 2200-meters/sec cross sections.

HEAVY WATER AS COOLANT

6.190. The only significant difference between ordinary water and heavy water as a reactor coolant (or coolant-moderator) lies in the appreciably smaller cross section of the latter for the capture of thermal neutrons.   The improvement in neutron economy means that less fissile material is required to make the reactor critical and natural uranium can be used as the fuel.   The additional cost of heavy water is not compensated by any gain in coolant properties.   It is therefore normally specified as a moderator (§ 7.90) and may incidentally also serve as a coolant.

LIQUID METALS AS COOLANTS [34]

6.191. For reactors with high thermal fluxes, operating at high temperatures, liquid metals have special interest as coolants.   They have excellent thermal properties, e.g., high thermal conductivity and low vapor pressure; metals of low atomic weight, e.g., lithium and sodium, also have relatively high specific heats and volumetric heat capacities.   In addition, liquid metals are stable at high temperatures and in intense radiation fields.   Their main disadvantage arises from the generally great chemical reactivity of liquid metals at high temperatures; they must be protected from oxidation, and suitable containing materials must be chosen to minimize corrosion.   Nevertheless, liquid sodium and sodium-potassium alloys have been used successfully as coolants in both fast and thermal reactors.

6.192. The heat-transfer characteristics of several liquid metals are superior to those of water, so that it is possible to achieve the same rate of heat removal with smaller areas of contact between fuel and coolant.   On the other hand, the volumetric heat capacities of liquid metals are less than that for water, so that a higher flow rate is necessary to remove the same amount of heat.   The pumping power depends on several factors, and it may be greater or less than for water; for the metals of low mass number it is generally somewhat less, but it is larger for elements of high mass number, e.g., bismuth and lead.   On the whole, pumping power would not be an important consideration in the choice of a coolant among various liquid metals and water.

6.193. Some indication of the metals which might be used as coolants in thermal reactors can be obtained by selecting elements having relatively low absorption cross sections for thermal neutrons and melting points below 500°C (930°F).   These are given in Table 6.5; they are divided into two classes,

TABLE 6.5.   METALS WITH MODERATE THERMAL-NEUTRON
CROSS SECTIONS AND LOW MELTING POINT

Small Thermal Cross Sections				Intermediate Thermal Cross Sections			
Metal	σ_a, barns	M.P. °C	M.P. °F	Metal	σ_a, barns	M.P. °C	M.P. °F
Bismuth............	0.032	271	520	Potassium........	2.0	62	145
Lithium-7.........	0.033	186	367	Gallium.........	2.7	30	86
Lead..............	0.17	327	621	Thallium........	3.3	302	576
Sodium...........	0.50	98	208				
Tin...............	0.65	232	450				

namely, those with absorption cross sections less than 1 barn, and those with cross sections from 1 to 10 barns.   Mercury is not included in this table because of its large cross section (340 barns) for the capture of thermal neutrons.

6.194. For fast neutrons the capture cross section is much less, and mercury was used as coolant in the Los Alamos Fast Reactor, which was dismantled in 1953.   However, the boiling point of mercury is 357°C (675°F), and its vapor pressure at moderate temperatures is relatively so high that it is considered less desirable than other possible coolants.

6.195. Of the elements in Table 6.5, potassium, gallium, and thallium do not present any notable advantages, especially as the two latter are expensive. Furthermore, liquid gallium is difficult to contain at high temperatures, and the same is true of tin. Bismuth and lead have satisfactorily low cross sections, but their melting points are fairly high. A eutectic alloy of these two elements, containing 44.5 per cent of lead by weight, melts at 125°C (257°F) and may be a possible reactor coolant; another possibility is a lead-magnesium eutectic, containing 97.5 per cent of lead, which melts at 250°C (482°F).

6.196. A significant drawback to the use of bismuth in a reactor is the fact that it undergoes the (n, γ) reaction to form bismuth-210 (radium-E), and this is a beta-particle emitter of 5.0 days half-life, decaying to yield polonium-210. The latter constitutes an exceptionally insidious hazard because it is very toxic and is difficult to contain. Nevertheless, the problem is not regarded as insoluble, and a reactor has been designed with a circulating fuel of uranium dissolved in molten bismuth [35]. The best containing materials for liquid bismuth, and also probably for lead and its alloys, are chrome steels, e.g., those containing $2\frac{1}{4}\%$ Cr-1% Mo or 5% Cr-1$\frac{1}{2}$% Si. Mass transfer, which becomes important at temperatures around 550°C (1020°F), can be controlled by addition of small amounts of zirconium and magnesium to the liquid metal.

6.197. Liquid sodium has attracted much attention as a reactor coolant, as will be seen shortly, as also has an alloy of 22 per cent (by weight) of sodium and 78 per cent of potassium, written NaK and generally referred to as "nack," which is liquid at room temperatures. Lithium-7 is included in Table 6.5 in spite of the large capture cross section of the natural form of the element for thermal neutrons; this is because separated lithium-7, although expensive and at present unavailable in quantity, may ultimately be important as a coolant.

6.198. The general opinion at present seems to be that sodium is, on the whole, the most suitable liquid metal generally available as the coolant for a reactor operating at high temperatures, particularly for fast reactors where moderating (low mass number) materials cannot be used. If free from oxygen, liquid sodium does not attack stainless steels, nickel, many nickel alloys, or beryllium at temperatures below 600°C (1110°F). At higher temperatures mass transfer can occur. Because of its fairly high melting point (98°C or 208°F), sodium may solidify in the reactor cooling system; this can be avoided by heating the piping and vessels either electrically or with hot oil or steam. In any event, after the reactor has been in operation at high power for some time, the heat which continues to be liberated after shutdown will keep the sodium in the liquid state for several days. One of the reasons why NaK, which is inferior to sodium as a heat-transfer medium, was chosen as coolant for the first Experimental Breeder Reactor (EBR-I) was the fact that it is liquid at ordinary temperatures, so that there is no danger of solidification either inside or outside of the reactor.

6.199. A drawback associated with the use of sodium as a reactor coolant, either alone or as an alloy, is the formation of sodium-24 as a result of neutron capture (see Table 6.4). This radioisotope has a half-life of nearly 15 hours and emits, in addition to beta particles, two gamma-ray photons of fairly high energy (1.38 and 2.75 Mev). Consequently some shielding is necessary for coolant tanks, piping, pumps, and heat exchanger, and maintenance problems are increased. In the Fermi fast reactor and in the Sodium Reactor Experiment (Chapter 13), the volume to be shielded is reduced by means of a secondary cooling circuit of sodium. The exchanger in which heat is transferred from the primary to the secondary coolant is shielded, but the remainder of the secondary circuit, including the heat exchanger (or boiler) in which steam is produced, is not radioactive and does not require shielding.

6.200. Sodium is chemically reactive and is easily oxidized by oxygen in the air or in water. At high temperatures, sodium exposed to air will burn with a low flame and the evolution of the oxide as smoke. The reaction of sodium (or NaK) with water produces hydrogen gas which, in proper concentration, can react explosively with atmospheric oxygen. Extensive studies, however, have established the fact that, by the exercise of proper precautions, the hazards can be greatly minimized.

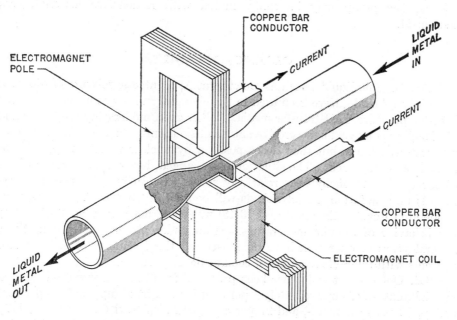

Fig. 6.22. Schematic representation of electromagnetic pump

6.201. Two different types of pumps are used for liquid metals: centrifugal pumps and electromagnetic pumps [36]. Centrifugal pumps have been built with capacities up to 13,000 gal per min of sodium and operating temperatures up to 540°C (950°F). A primary consideration in the design of a centrifugal pump is the method of sealing the shaft. The "free-surface" vertical shaft pump, in which a liquid-gas interface is maintained in the pump case, is the most widely used.

6.202. In the so-called "electromagnetic" pumps, advantage is taken of the electrical conductivity of a liquid metal to force it to flow in a pipe under the influence of a magnetic field. These pumps operate on the same principle as an electric motor. If a conductor in a magnetic field carries a current flowing at right angles to the direction of the field, a force is exerted on the conductor in a direction perpendicular to both the field and the current. In the electromagnetic pump the liquid metal is the conductor which passes through a duct located between the poles of an electromagnet (Fig. 6.22). A current from an external source is passed across the liquid metal in a direction perpendicular to both the magnetic field and the desired direction of flow of the liquid metal. The force exerted on the conductor by the magnetic field then causes flow to take place. One advantage of the electromagnetic pump is that it has no moving parts, so that, with proper design, maintenance is greatly simplified. This is important, not only because of the nature of the liquid being pumped but also because the pump may be contaminated with radioactive material, e.g., sodium-24.

ORGANIC COOLANTS

6.203. Organic liquids, such as Dowtherm A, a eutectic mixture of diphenyl and diphenyl oxide, have been used in the chemical industry for many years as high-temperature, heat-transfer media. Their good moderating properties and high boiling point and noncorrosive nature made them seem attractive as reactor moderator-coolants, but their decomposition by nuclear radiation appeared to be a serious drawback. The situation has changed, however, as a result of the discovery that terphenyls are relatively stable to both high temperature (pyrolysis) and radiation (radiolysis), as will be seen in Chapter 7. The Organic Moderated Reactor Experiment (§ 13.45) was constructed as a facility to study the behavior of organic moderator-coolants under power reactor conditions. Several years of operation have demonstrated the feasibility of such reactor concepts with tolerable amounts of coolant decomposition.

6.204. One coolant which has been proposed for power reactor use is a commercial mixture of approximately 1 per cent by weight of diphenyl, 12 per cent ortho-, 59 per cent meta-, and 27 per cent para-terphenyl (Santowax R) [37]. During operation, partial purification of the coolant is carried out by distillation under reduced pressure. About 30 per cent of high boiling point polymer products of radiolytic and pyrolytic decomposition are allowed to accumulate in the coolant for optimum operating conditions.   The vapor pressure of the coolant is then 34 psia at 800°F.   Since the organic material has a relatively low thermal conductivity, the heat-transfer properties of the system are improved by the use of fuel elements with lateral fins that increase the exposed area (cf. § 6.210).

6.205. For a 240-Mw (thermal) organic cooled and moderated reactor design, it has been estimated that the rate of radiolytic damage would be about 1.1 lb per Mw-hr at 675°F.   This would require a makeup of 260 lb per hr of fresh coolant in the specified reactor; and additional 10 lb per hr would be needed to allow for pyrolytic decomposition.   The cost of coolant makeup would be about 0.58 mill per kw-hr of electric power.

MOLTEN SALTS

6.206. Molten salts, especially fluorides, have been studied for use both as coolant and as the fuel-bearing medium in circulating fuel reactors for high-temperature (about 1200°F) operation.   Fluorine has a very low capture cross section for thermal neutrons and certain mixtures of fluorides of metals of low mass number have reasonable melting points, e.g., 350°C (660°F) for an equimolar mixture of lithium and beryllium fluorides.   The fluorides are stable at high temperatures and in intense radiation fields, and have low vapor pressures. Their volumetric heat capacities and thermal conductivities are not very different from those of water, but they have much higher viscosities at temperatures of interest.

6.207. A mixture of lithium-7, beryllium, and zirconium fluorides is being used as the circulating, fuel-bearing medium in the Molten-Salt Reactor Experiment (§ 13.112), with a lithium fluoride-beryllium fluoride mixture as the coolant in the primary heat exchanger.   A nickel (66 to 71 weight per cent)-molybdenum (15 to 18 per cent) alloy, containing a few per cent each of chromium and iron, known as INOR-8, has been found to be a satisfactory containing material for these molten fluorides at high temperatures.

GASEOUS COOLANTS

6.208. On the basis of general radiation and thermal stability, ease of handling, and absence of hazardous conditions, there is much to recommend the use of a gas as coolant.   Although the heat-transfer characteristics are not as good as those of water and liquid metals, it appears that, by suitable design, gas-cooled power reactors may be competitive with those using liquid coolants.

6.209. As a basis of comparison of gaseous coolants among themselves, a merit index has been proposed based on the ratio of the heat removed to the pump work expended.   The values of this index given in Table 6.6 were obtained on the assumption that all conditions, e.g., gas pressure, fuel and gas temperatures, were the same for all the gases, which might not necessarily be the case [38]. The results are applicable to a temperature of 212°F and 1 atm pressure; the index for hydrogen is taken to be 100.

TABLE 6.6.   HEAT-TRANSFER MERIT INDEX OF GASES

Gas	Merit Index
Hydrogen	100
Methane	29
Ammonia	22
Helium	18
Steam	11
Carbon dioxide	11
Air	7.5

6.210. In the earlier stages of power reactor development, it was thought that the relatively poor heat-transfer properties of gases would limit the attainable power density and fuel specific power, so that power costs would be higher than for reactors utilizing other coolants. However, certain design developments have resulted in an increased interest in gas-cooled power reactors. In the first place, improvements in thick-walled, pressure vessels have made it possible to use the coolant at high pressure, thereby increasing its heat-transfer properties. Second, the use of ceramic fuels, with high-temperature cladding or as unclad fuel elements, permit operation at higher temperatures than have been achieved with metallic fuels. Finally, the rate of heat removal can be increased by means of fuel elements having large heat-transfer areas. Simple examples of finned

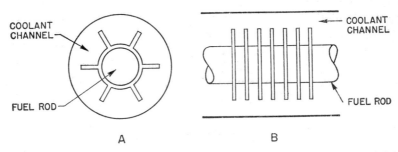

Fig. 6.23.   Simple finned fuel elements.   A. Longitudinal fins   B. Radial fins

fuel elements are shown in Figs. 6.23A and B; the former (longitudinal fin) type was originally employed in the Brookhaven graphite reactor and the latter (transverse fin) type was used in the British Calder Hall power reactors. However, fins of much more elaborate design, with larger heat-transfer areas, than those shown in Fig. 6.23 have been developed for both gas-cooled and organic-cooled reactors.

6.211. A number of variables must be considered in the design of a finned arrangement.   The temperature distribution in a fin, for example, is determined both by the geometry and the material properties of the fin itself as well as by the hydrodynamic behavior of the coolant in the specific configuration used. The photograph in Fig. 6.24 is of interest in this connection.   It shows the forma-

FIG. 6.24.   Photograph of fluid flow past finned fuel element.
(Courtesy U.K. Atomic Energy Authority)

tion of gas vortices between adjacent fins which produced a marked improvement in the heat-transfer properties.   For the present purpose, it should be emphasized that fin performance is sensitive to geometrical and other conditions, and a careful analysis is necessary for any specific case of interest [39].

6.212. Because of its availability, air is the most obvious choice as a gaseous coolant.*   Unfortunately, as may be seen from Table 6.6, air is not a good heat-transfer material, since a large proportion of the power produced is consumed in pumping the air through the cooling system.   Furthermore, at high temperatures the oxygen (and in some cases nitrogen also) is likely to attack graphite or beryllium, which may be used as moderator, and the structural and fuel-canning materials.   Air is used as the coolant in several natural uranium-graphite research reactors, but these are all large and operate at relatively low power densities.   The heat flux is thus fairly small and excessive volumes of air are not needed to keep the fuel temperatures at reasonable levels.   For research reactors such as these, the inefficiency of air as a coolant is offset by the simplicity

* Nitrogen gas, obtained by the distillation of liquid air, is being studied for use as coolant in reactors for remote locations (§ 13.74); its thermal properties are similar to those of air.

of design.   Air was also used as coolant in the British plutonium-production reactors at Windscale, which are no longer in operation.   After leaving such reactors, the air is always somewhat radioactive, mainly due to the presence of argon-41, but since it is unnecessary to use a closed system, the air is discharged through a high stack.

6.213. Although hydrogen would be the best gaseous coolant, as far as both heat transfer and pumping power are concerned, it would constitute a serious explosion hazard if it escaped into the air.   In addition, its containment, especially under pressure, would require the use of steels not subject to attack or embrittlement by hydrogen.   Consequently, no proposal has been made for the use of hydrogen as coolant except in special cases, such as nuclear rocket reactors.

6.214. From several points of view, helium has much to recommend it as a gaseous coolant.   Although its thermal properties are not quite as good as those of hydrogen, it has a negligible cross section for neutron capture and is not hazardous.   It is also stable to heat and to the action of radiation.   In addition it is inert chemically and has no harmful effects on the moderator or on structural and containing materials.   For this reason, helium is the preferred coolant for graphite-moderated reactors operating at temperatures above 600°C (1100°F). Thus helium gas is the coolant in the Experimental Gas Cooled Reactor (EGCR) at Oak Ridge, Tenn., the High Temperature Gas Cooled Reactor (HTGR) at Peach Bottom, Pa., the United Kingdom DRAGON reactor experiment at Winfrith, England, and the Ultra-High Temperature Reactor Experiment (UHTREX) at Los Alamos, N. M.   High gas pressures are used to minimize pumping power and to decrease the core size.   Because of the high cost of helium, a closed coolant system is used with special precautions taken to reduce losses.   It is of interest to mention that when the first Hanford plutonium production reactors were designed helium was the preferred coolant; it was not adopted, however, mainly because of the difficulties anticipated in the prevention of leakage of helium gas under pressure from such large structures and in the loading and unloading of reactor fuel [40].

6.215. Carbon dioxide is a relatively inexpensive gas which has some advantages as a coolant and it has been used extensively in Great Britain in power reactors of the Calder Hall type.   The carbon and oxygen nuclei in carbon dioxide have small cross sections for neutron capture, and the gas itself is free from the dangers of toxicity or explosion.   Carbon dioxide does not readily attack metals, although it does react with graphite at fairly high temperatures. It is less efficient than helium as regards heat transfer, pumping power, and neutron absorption, but it is, of course, much more readily available.   An uncertain factor in connection with the use of carbon dioxide as a coolant is the effect of the intense radiation field on its interaction with graphite and other materials.   The reaction rate appears to be insignificant at temperatures below 500°C (930°F) and carbon dioxide is probably the most practical gaseous reactor coolant for operation up to 550° to 600°C (1000° to 1100°F), but it is unsatisfactory at higher temperatures.

6.216. Steam has been suggested as a reactor coolant for both central station and marine propulsion plants. At ordinary pressures, its volumetric heat capacity, heat conductivity, and viscosity are not very different from those of carbon dioxide (cf. Table 6.6). Saturated steam, produced in a boiling-water reactor, for example, can be employed as the coolant in another reactor system to yield superheated steam. The principal advantage of nuclear superheating is expected to be the increased thermal efficiency which would result. The superheating reactor may be an integral part of the boiling-water reactor or it may be a separate unit (Chapter 13). In addition, proposals have been made for utilizing the Loeffler cycle, in which part of the superheated steam leaving the reactor serves to heat turbine condensate to form saturated steam in a direct contact heat exchanger [41]. There has also been some interest in the possibility of producing superheated steam in a reactor employing wet steam (fog or spray) as the coolant; the heat-transfer characteristics of the mixture should be superior to those of steam alone [42].

SYMBOLS USED IN CHAPTER 6

A	area
A_f	coolant flow area (cross-sectional area of coolant channels)
A_h	heat-transfer area
a	internal radius of hollow tube or cylinder
a	half-thickness of fuel region
b	external radius of hollow tube or cylinder
b	half-thickness of fuel element
C_1, C_2	arbitrary constants
c	constant in burnout flux equation
c	factor (in boiling heat transfer)
c	cladding thickness
c_p	specific heat (at constant pressure)
D	pipe diameter
D_e	equivalent (or hydraulic) diameter
E	energy per fission
E_c	energy liberated in capture processes
f	Fanning friction factor
G	mass velocity (or flow rate) of liquid ($= \rho v$)
Gr	Grashof number ($= L^3\rho^2 g\beta\Delta t/\mu^2$)
g_c	force to mass ratio constant
H	velocity head
H	height of cylindrical reactor core

H'	effective height of cylindrical reactor
h	heat-transfer coefficient
h_b	boiling heat-transfer coefficient
J_0	Bessel function of zero order
J_1	Bessel function of first order
K_c	pressure loss coefficient in contraction
K_e	pressure loss coefficient in expansion
K_l	pressure loss coefficient in fittings, etc.
k	thermal conductivity
L	length of heat-flow path
L	slab thickness
N	number of nuclei per cm^3
n	number of coolant channels
Nu	Nusselt number $(= hD/k)$
P	power density
P_0	power density at center of reactor
P_{av}	average power density
P_{max}	maximum power density
p	passage perimeter
p	pressure
Δp_f	pressure drop due to friction
Pe	Peclet number $(= \text{Re} \times \text{Pr} = Dv\rho c_p/k)$
Pr	Prandtl number $(= c_p\mu/k)$
Q	volumetric heat source strength (rate of heat release per unit volume)
q	rate of heat flow
q/A	heat flux
R	friction factor multiplier in two-phase flow
R	thermal resistance
R	radius of cylindrical reactor
R'	effective radius of reactor
r	radial distance (coordinate)
r_0	pipe radius for circulating fuel
r_i	inner radius of annular channel
r_o	outer radius of annular channel
Re	Reynolds number $(= Dv\rho/\mu)$
t	temperature
t_m	mixed-mean (bulk) fluid temperature
t_s	temperature at solid surface
t_w	temperature at wall of pipe
Δt	temperature difference
Δt_b	surface temperature difference in boiling system
Δt_c	increase in mixed-mean (bulk) temperature of coolant
Δt_{sub}	amount of subcooling

U	overall heat-transfer coefficient
V_h	volume of heat-transfer channel
v	mean velocity of fluid
w	mass (or weight) rate of fluid flow $(= \rho v A_f)$
x, y, z	coordinates
α	function equal to $\pi x/L$
ϵ	thermal emissivity
ϵ	roughness factor
μ	attenuation coefficient for gamma rays
μ	viscosity
μ_w	viscosity at wall
ρ	density
ρc_p	volumetric heat capacity
ρv	mass flow rate $(= G)$
Σ_f	macroscopic fission cross section
σ_f	fission cross section
σ	Stefan-Boltzmann constant
ϕ	neutron flux
ϕ_{av}	average neutron flux
ϕ_{max}	maximum neutron flux
ϕ_{long}	longitudinal neutron flux
$(\phi_{max})_{long}$	maximum of longitudinal neutron flux
∇^2	Laplacian operator

REFERENCES FOR CHAPTER 6

1. J. J. Newgard and M. M. Levoy, *Nuc. Sci. and Eng.*, **7**, 737 (1960).
2. For further information, standard textbooks should be consulted, e.g., W. H. McAdams, "Heat Transmission," 3rd Ed., McGraw-Hill Book Co., Inc., New York, 1954; M. Jakob, "Heat Transfer," John Wiley and Sons, Inc., New York, Vol. I, 1950; E. R. G. Eckert and R. M. Drake, "Heat and Mass Transfer," McGraw-Hill Book Co., Inc., New York, 1959; J. G. Knudsen and D. L. Katz, "Fluid Dynamics and Heat Transfer," McGraw-Hill Book Co., Inc., New York, 1958; M. W. Rohsenow and H. Y. Choi, "Heat, Mass and Momentum Transfer," Prentice-Hall Inc., Englewood Cliffs, N. J., 1961; etc. See also, C. F. Bonilla (Ed.), "Nuclear Engineering," McGraw-Hill Book Co., Inc., New York, 1957, Ch. 9; P. A. Lottes, "Nuclear Reactor Heat Transfer," U. S. AEC Report ANL-6469 (1961).
3. H. Fenech and M. W. Rohsenow, U. S. AEC Report NYO-2136 (1959).
4. P. J. Schneider, "Conduction Heat Transfer," Addison-Wesley Publishing Co., Inc., Reading, Mass., 1955.
5. P. J. Schneider, Ref. 4; G. M. Dusinberre, "Numerical Analysis of Heat Flow," McGraw-Hill Book Co., Inc., New York, 1949; T. B. Fowler and E. R. Volk, U. S. AEC Report ORNL-2734 (1959).
6. E. R. G. Eckert and T. F. Irvine, Jr., *J. Heat Transfer*, **82**, 125 (1960).
7. F. W. Dittus and L. M. K. Boelter, *Univ. of Calif. Pub. Eng.*, **2**, 443 (1930).
8. A. P. Colburn, *Trans. Am. Inst. Chem. Engrs.*, **29**, 174 (1933).

9. E. N. Sieder and G. E. Tate, *Ind. Eng. Chem.*, **28**, 1429 (1926).
10. W. R. Gambill and R. D. Bundy, U. S. AEC Report ORNL-3079 (1961).
11. R. P. Bringer and J. M. Smith, *Am. Inst. Chem. Engrs. J.*, **3**, 49 (1957).
12. J. A. Lane, Unpublished.
13. R. C. Martinelli, *Trans. Am. Soc. Mech. Engrs.*, **69**, 947 (1949).
14. R. N. Lyon, *Chem. Eng. Progress*, **47**, 75 (1951); *Proc. (First) U. N. Conf. Geneva*, **9**, 290 (1956).
15. B. Lubarsky and S. J. Kaufman, NACA Report 1270 (1956).
16. M. W. Rohsenow and L. S. Cohen, see Ref. 2, p. 193.
17. R. A. Seban, *Trans. Am. Soc. Mech. Engrs.*, **72**, 789 (1950).
18. J. P. Hartnett and T. F. Irvine, Jr., *Am. Inst. Chem. Engrs. J.*, **3**, 313 (1957).
19. R. A. Baker and A. Sesonske, *Trans. Am. Nuc. Soc.*, **3**, 468 (1960).
20. W. H. McAdams, Ref. 2, Ch. 7; M. Jakob, Ref. 2, Ch. 25.
21. J. G. Knudsen and D. L. Katz, Ref. 2, p. 400.
22. W. H. McAdams, Ref. 2, Ch. 14; M. Jakob, Ref. 2, Ch. 29; C. F. Bonilla, Ref. 2, pp. 399 *et seq.;* P. A. Lottes, Ref. 2, Ch. 9; Y. P. Chang, U. S. AEC Report ANL-6304 (1961); V. E. Holt, U. S. AEC Report ANL-6400 (1961); T. B. Drew and W. J. Hoopes (Eds.), "Advances in Chemical Engineering," Academic Press, Inc., New York, Vol. 1, p. 1 (1956); Vol. 2, p. 1 (1958).
23. W. H. Jens and P. A. Lottes, U. S. AEC Report ANL-4915 (1952); see also, *Power Reactor Technology*, **2**, No. 4, 36 (1959); H. Susskind, U. S. AEC Report BNL-636 (1960).
24. D. W. Bell, *Nuc. Sci. and Eng.*, **1**, 245 (1956); P. A. Lottes, M. Petrick, and J. F. Marchaterre, U. S. AEC Report ANL-6063 (1959).
25. For general reviews of burnout, see W. M. Jacobs, *et al.*, *Nucleonics*, **16**, No. 3, 130 (1958); J. G. Collier, *Nuclear Power*, **6**, No. 62, 61 (1961); **6**, No. 63, 64 (1961); also *Power Reactor Technology*, **4**, No. 4, 25 (1961); **5**, No. 1, 17 (1961); *Nucleonics*, **20**, No. 2, 70 (1962).
26. M. T. Cichelli and C. F. Bonilla, *Trans. Am. Inst. Chem. Engrs.*, **41**, 755 (1945).
27. W. H. Jens and P. A. Lottes, U. S. AEC Report ANL-4627 (1951); see also, Ref. 23.
28. H. F. Poppendieck and L. D. Palmer, *Nuc. Sci. and Eng.*, **3**, 85 (1958).
29. C. F. Bonilla, Ref. 2, Ch. 8; V. L. Streeter, "Fluid Mechanics," McGraw-Hill Book Co., Inc., New York, 1951; H. Rouse and J. W. Howe, "Basic Mechanics of Fluids," John Wiley and Sons, Inc., New York, 1953; V. L. Streeter (Ed.), "Handbook of Fluid Mechanics," McGraw-Hill Book Co., Inc., New York, 1961; see also, J. P. Waggener, *Nucleonics*, **19**, No. 11, 145 (1961).
30. L. F. Moody, *Trans. Am. Soc. Mech. Engrs.*, **66**, 671 (1944).
31. R. C. Martinelli and D. B. Nelson, *Trans. Am. Soc. Mech. Engrs.*, **70**, 695 (1948); B. M. Hoglund, *et al.*, U. S. AEC Report ANL-5760 (1961); see also, Ref. 23 and P. A. Lottes, Ref. 2.
32. A. A. Bishop and R. Berringer, U. S. AEC Report YAEC-76 (1958).
33. H. Etherington (Ed.), "Nuclear Engineering Handbook," McGraw-Hill Book Co., Inc., New York, 1958, pp. **12**–56 *et seq.;* L. Green, *Nucleonics*, **19**, No. 11, 140 (1961).
34. H. Etherington, Ref. 33, p. **13**–80; L. Green, Ref. 33; R. N. Lyon (Ed.), "Liquid-Metals Handbook," U. S. Government Printing Office, Washington, D. C., 1952; C. B. Jackson (Ed.), "Liquid-Metals Handbook, Sodium-NAK Supplement, U. S. Government Printing Office, Washington, D. C., 1955; C. R. Tipton, Jr. (Ed.), "Reactor Handbook," 2nd Ed., Vol. I, Interscience Publishers, Inc., New York, 1960, Ch. 49.
35. C. Williams and R. T. Schomer, *Proc. Second U. N. Conf. Geneva*, **10**, 487 (1958).

36. R. A. Jaross, *et al.*, *Proc. Second U. N. Conf. Geneva*, **7**, 82, 88 (1958); H. Etherington, Ref. 34; see also, *Power Reactor Technology*, **4**, No. 2, 59 (1961).
37. For reviews, see *Power Reactor Technology*, **1**, No. 4, 45 (1958); **2**, No. 4, 42 (1959).
38. H. Etherington, Ref. 33, p. **13**–121.
39. W. H. McAdams, Ref. 2, pp. 268 *et seq.*; M. Jakob, Ref. 2, pp. 217 *et seq.*; E. R. G. Eckert and R. M. Drake, pp. 43 *et seq.*; see also, *Power Reactor Technology*, **1**, No. 1, 7 (1957).
40. H. D. Smyth, "Atomic Energy for Military Purposes," U. S. Government Printing Office, Washington, D. C., 1945, § 7.16.
41. H. L. Falkenberry, *et al.*, U. S. AEC Report ORNL-2759 (1959); G. Sofer, *et al.*, U. S. AEC Reports NDA 2148-4 (1961) and 2148-5 (1961); see also, *Power Reactor Technology*, **3**, No. 2, 12 (1960).
42. H. R. C. Pratt and J. Thornton, *Proc. Second U. N. Conf. Geneva*, **7**, 813 (1958); J. G. Collier and P. M. C. Lacey, *Nuclear Power*, **5**, No. 52, 68 (1960); M. Silvestri, *et al.*, *Nucleonics*, **19**, No. 1, 86 (1961); see also, *Power Reactor Technology*, **4**, No. 1, 74 (1960); U. S. AEC Report NDA 2132-6 (1961).

PROBLEMS*

1. In a heat exchanger consisting of steel tubes, with $k = 25$ Btu/(hr)(ft²)(°F/ft), having an internal diameter of 1.1 in. and external diameter 1.3 in., heat is transmitted from a primary coolant, inside the tubes, to a secondary coolant, outside. The respective heat-transfer coefficients are 8400 and 6300 Btu/(hr)(ft²)(°F). Determine the temperature difference between the primary and secondary coolants and the temperature drop across the tube wall when heat is being transferred at a rate of 5000 Btu/hr per ft of tube.

2. A cylindrical, 1-in. diameter, fuel element of natural uranium is clad with aluminum 0.05 in. thick and is cooled with water. The temperature of the water at the surface of the element is 155°F, the heat-transfer coefficient is 6600 Btu/(hr)(ft²)(°F), and the thermal flux in the reactor is 10^{14} neutrons/(cm²)(sec). Assuming the thermal source in the fuel to be uniform, determine the temperature distribution through the fuel and cladding.

3. If the cooling water in the preceding problem flows through an annular channel surrounding the fuel element at a rate of 20 ft/sec, estimate the outside diameter of the annulus which will make the heat-transfer coefficient equal to the value given.

4. A plate-type fuel element, consisting of a uranium-zirconium alloy 3 mm thick clad on each side with a 0.5 mm thick layer or zirconium, is cooled by water (under pressure) at 400°F, the heat-transfer coefficient being 7500 Btu/(hr)(ft²)(°F). If the temperature at the center of the fuel must not exceed 1050°F, determine the maximum specific power attainable, expressed in kw/kg of uranium-235, given that the fuel region contains 10 per cent by weight of the latter. Determine the temperature distribution through the fuel and cladding when the reactor is operating at the maximum power. (The density of the zirconium may be taken as 6.5 g/cm³ and its thermal conductivity as 12 Btu/(hr)(ft²)(°F/ft); the latter value may be taken to apply also to the fuel region.)

5. In the preceding problem the resistance at the fuel-cladding interface is neglected. Investigate the effect of a possible surface conductance of 5000 Btu/(hr)(ft²)(°F).

6. Show that for a sphere of radius a and thermal conductivity k having a uniform volumetric heat source Q, the temperature, t, at a distance r from the center, is given by

---

* If not supplied, data required for the solution of the problems will be found in the text or in the tables in the Appendix.

$$t - t_a = \frac{Q}{6k}(a^2 - r^2),$$

where t_a is the temperature at the exterior.

7. A beryllium moderator ($k = 75$ Btu/(hr)(ft²)(°F/ft)) is cooled by water at 100°F flowing through tubular channels of 1-in. diameter at the rate of 15 ft/sec. If the maximum temperature in the moderator is 180°F, what is the rate, expressed in kw/ft³, at which heat is removed from the beryllium? The moderator may be treated as equivalent to an insulated cylinder of 1 ft diameter with a cooling channel through the center.

8. A thermal reactor employs a ceramic fuel ($k = 10$ Btu/(hr)(ft²)(°F/ft)) in the form of cylindrical rods 1 in. in diameter and 1 ft in length. The thermal-neutron flux varies directly as the radial distance, r, from the central axis of the rod, i.e., $\phi = Cr$. If the maximum temperature which can be tolerated at the center of the fuel is 3000°F and the coolant temperature is 300°F, with a heat-transfer coefficient of 2000 Btu/(hr)(ft²)(°F), calculate the rate of heat production per fuel rod.

9. In the Yankee Atomic Electric Company pressurized-water reactor, the fuel consists of rods (made up of pellets) of uranium dioxide ($k = 1.0$ Btu/(hr)(ft²)(°F/ft)) having a diameter of 0.294 in.; these are jacketed with 0.021 in. of stainless steel with a clearance of 0.002 in. between the pellets and the steel. The fuel rods are arranged in the form of a square lattice with a spacing of 0.422 in. The average temperature of the coolant water is 516°F and its flow rate is 15 ft/sec. The average permissible heat flux is 86,300 Btu/(hr)(ft²). Calculate the corresponding heat generation rate per unit volume in the fuel, and sketch the radial temperature distribution in the fuel element.

10. A hollow cylindrical fuel rod, which is cooled both internally and externally, has an internal diameter r_1 and an external diameter r_2; the mixed-mean coolant temperature and heat-transfer coefficient are, respectively, t_1 and h_1 inside and, respectively, t_2 and h_2 outside. Assuming a uniform heat generation rate, derive expressions for the radial distance at which there is a temperature maximum in the fuel and for the value of this temperature.

11. A reactor cooled with liquid sodium is to produce 50,000 kw of heat. The cylindrical fuel elements are 8 ft long and 0.75 in. in diameter, and the coolant flows through an annular channel of 0.625-in. outer radius at a rate of 25 ft/sec. The liquid sodium enters the reactor at 700°F and the maximum permissible surface temperature of the fuel rods is set at 1050°F. Calculate the minimum number of cooling channels that would be required and the average exit temperature of the coolant. (For the purpose of this calculation the reactor core may be treated as a bare cylinder.)

12. Suppose that in the preceding problem the liquid sodium flows into the channels from a large pipe (or header) and flows into a similar pipe when it leaves the reactor. Calculate the pumping power (a) in kw, (b) in theoretical hp, required to pump the coolant through the reactor.

13. In comparing the fraction of the heat released in a reactor which is used for (friction) pumping power of various coolant gases, it may be assumed that the gas temperature and pressure, and the temperature difference between the fuel elements and the coolant are the same in all cases. Show that the effectiveness of a gas in this connection is determined by $c_p^2 k M^2/\mu$, where M is the molecular weight of the gas.

14. Calculate the values of the heat-transfer coefficient for (a) water at 450°F, (b) sodium at 800°F, at flow velocities of 10, 20, 30, and 40 ft/sec through a circular tube having an internal diamter of 1 in. What conclusions may be drawn concerning the advantage of increasing the flow velocity, taking into account the required pumping power, for the two cases?

15. In a water-cooled reactor with plate-type fuel elements, the plates are 0.25 in. apart and the total flow area for the water is 0.80 ft². The temperature of the water at

the bottom of the fuel plates is 85°F. What would be the temperature at the top of the plates for a rate of heat removal due to convection alone of 100 kw?

16. Calculate the heat-transfer coefficient of helium at 600°F and 300 psia flowing at a rate of 150 ft/sec through a channel of rectangular cross section, 0.2 × 3.0 in. Consider, taking into account the effect on the pumping power, what changes might be made either in the design of the coolant channels or in the conditions of the coolant that would result in a beneficial change in the heat-transfer coefficient.

17. Cooling water, containing 5 ppm of sodium by weight, is circulated through a reactor so that in each cycle the water spends 5 sec in the reactor and 30 min in the external heat exchanger, hold-up tanks, etc. If the average thermal flux of the reactor is 10^{14} neutrons/(cm²)(sec), calculate the equilibrium value, attained after extended operation, of the activity of the water leaving the reactor, expressed in curie/cm³, due to sodium-24 (half-life 15 hr).

Chapter 7

REACTOR STRUCTURAL AND MODERATOR MATERIALS

INTRODUCTION

MATERIALS PROBLEMS

7.1. Although neutron behavior and heat-removal considerations, discussed in earlier chapters, are certainly of major importance in nuclear reactor design, the choice of satisfactory materials of construction is essential for a practical reactor system [1]. It is obvious that a central station power reactor will be useless if it cannot operate for a long time without component failures. Thus, the availability of materials which will withstand the frequently severe environmental conditions that would exist in a given reactor concept, e.g., high temperature, large temperature gradients, and intense radiation field, usually determines whether the concept is technically feasible. Even if structural materials with suitable physical and mechanical properties are available, they must frequently satisfy, in addition, the requirement of having low neutron capture cross sections.

7.2. The importance of the behavior of the reactor materials cannot be overemphasized. In the development of many existing reactors, problems of material selection, which at first appeared to be minor, ultimately proved to be serious. In several cases these difficulties have resulted in long delays, and much research effort and expense were required before solutions were developed. There have, in fact, been a number of otherwise very promising reactor concepts which, for the present at least, have not proved to be practical because satisfactory constructional materials could not be found.

7.3. A wide selection of materials is required to fulfill the requirements of a complex reactor system. However, many of the service conditions, although severe, are not very different from those occurring in other industrial applications. Such structural materials as are not exposed to radiation or to the primary coolant system fall into this category. Although the selection of these materials may require considerable engineering attention, such considerations are not within the scope of this book. Emphasis will therefore be placed on materials and material problems which are unique to fission reactor applications.

418

7.4. Material requirements of the reactor system can vary quite widely, depending on the particular design.   In all cases, however, fuel materials, cladding, moderator (if any), and structural materials must be selected to meet the appropriate requirements.   In addition, materials must be chosen for control, cooling, and shielding the reactor system.   Control materials were described in Chapter 5 and coolants in Chapter 6; fuels will be treated in Chapter 8 and shielding in Chapter 10.   This leaves constructional and moderator materials to be discussed in this chapter, a combination which may be partly justified on the grounds that some moderators, e.g., graphite, are important as structural components of a reactor system.   Fuel-element cladding will be considered in the next chapter, but since some materials, such as stainless steel and zirconium alloys, serve both as structural materials and as cladding, reference to the latter will be made here.

RADIATION EFFECTS ON MATERIALS [2]

GENERAL PRINCIPLES

7.5. The unique characteristic of the reactor environment is the presence of intense radiation.   Before considering specific materials and applications, therefore, the principles underlying the effects of radiation will be described. A large amount of work has been done on the behavior of a variety of materials under the influence of radiation, but the present discussion will be restricted mainly to materials of direct interest in reactor design [3].

7.6. The radiations in a nuclear reactor consist of alpha and beta particles, gamma rays, neutrons, fission fragments, and, possibly, protons.   Although the fission fragments are not strictly nuclear radiations, their behavior is generally similar to that of alpha particles and protons, both of which are charged particles. There are, however, differences in degree arising from much greater mass of the fission fragments.   In addition, the gaseous fission products, e.g., krypton and xenon, introduce many problems of a different kind in connection with solid fuel materials.   An important effect on fuels exposed to high burnup doses is the swelling due to the volume enlargement caused by the accumulation of gaseous fission products (§ 8.51).

7.7. The effect of radiation in crystalline solids depends on the structural type and the nature of the radiation.   Ionization and electronic excitation, for example, produced by beta particles and gamma rays, cause very little permanent change in metals.   The reason is that in a metal the electrons in the conduction band are able to accept very small amounts (quanta) of excitation energy, and so a considerable proportion of the energy of the radiation is absorbed in causing electronic excitation.   Because of the large number of empty energy states available, the excited electrons rapidly lose their excess energy; this is taken up by the atoms of the metal and appears in the form of heat, i.e., as vibrational energy of the nuclei.

7.8. Heavier particles, however, such as protons, neutrons, alpha particles, and fission fragments, do produce significant changes in the properties of metals. As a result of elastic collisions, these particles may transfer appreciable amounts of energy to the nuclei of a solid. If the amount of energy transferred is sufficient to cause the nuclei to be displaced from their normal (or equilibrium) positions in the space lattice, physical changes of an essentially permanent character will be observed in the metal. This effect of nuclear radiation is sometimes referred to as "radiation damage."

7.9. In ionic and covalent (organic) compounds, both solid and liquid, the effect of radiation is generally greater than in metals. With organic materials and water, for example, even gamma rays and beta particles can bring about chemical changes. The ionizing radiations cause damage by breaking chemical (covalent) bonds with, in some cases, the formation of free radicals which result in various molecular rearrangements, e.g., polymerization. The effect of neutrons arises mainly from collisions with hydrogen atoms; the damage is then caused by secondary excitation produced by the recoil protons.

7.10. As a guide to consideration of the specific effects of nuclear radiations on particular materials, to be described later in this chapter, the interaction mechanisms will be reviewed. For metallic and ionic solids the treatment is concerned with the behavior of heavy particles only. But when organic compounds are discussed, the effects of gamma radiations will be included.

ATOMIC DISPLACEMENTS

7.11. If, as the result of an elastic collision between the relatively heavy nuclear particle and an atomic nucleus, the energy transferred to the latter exceeds a certain minimum value (usually assumed to be about 25 ev for a metal), the struck ("knocked-on") atom may be displaced from its normal or equilibrium position in the solid lattice.* If several atoms in close proximity are displaced in this manner, some will be merely transferred from one equilibrium position to another in the lattice. However, if a knocked-on atom is unable to find a vacant equilibrium site, its final location will be a nonequilibrium position. An atom of this kind is said to occupy an interstitial site and is known as an interstitial atom, or in brief, as an "interstitial." For each interstitial atom produced by the action of radiation, there must be a corresponding vacant site in the lattice, possibly at some distance away. The net result is thus a more or less permanent defect in the solid, which, if sufficiently common, may be accompanied by a change in physical properties.†

---

* The displacement of atoms from their positions in the crystal lattice as a result of fast-neutron impact was predicted by E. P. Wigner in 1942. It has consequently been called the "Wigner effect."

† A lattice vacancy and an interstitial atom when considered as a unit are often called a "Frenkel defect."

7.12. In many cases the kinetic energy transferred by the nuclear particle to the knocked-on atom, generally referred to as a "knock-on," is so large that the latter may produce secondary knock-ons by elastic collision with other atoms in the solid. If the kinetic energy of the knock-on is greater than the somewhat indefinite limiting ionization energy, which is on the order of A kev, where A is the mass number of the atom, the probability is that the knock-on will lose much of its energy in causing ionization. However, when the kinetic energy is less than the limiting ionization energy, the knock-on will lose energy by elastic collisions with lattice atoms. In this manner, a cascade of knock-ons may develop until the energy is reduced to the point where displacement cannot occur. If $\bar{E}$ is the average energy transferred by the nuclear particle to the struck atom, then theoretical considerations show that, provided $\bar{E}$ is less than the limiting ionization energy E_i, the average number of displaced atoms, $\bar{\nu}$, per initial knock-on in moderately heavy elements is given approximately by

$$\bar{\nu} \approx \frac{\bar{E}}{2E_d}, \tag{7.1}$$

where E_d is the energy required to displace an atom from the lattice, i.e., about 25 ev. When $\bar{E}$ is greater than E_i, the latter should replace $\bar{E}$ in equation (7.1).

7.13. On the basis of the foregoing considerations and certain simplifying assumptions, the results given in Table 7.1 have been calculated [4]. They show

TABLE 7.1. CALCULATED AVERAGE NUMBER OF DISPLACED ATOMS
PER PRIMARY KNOCK-ON FROM 1-MEV NEUTRONS [4]

Material	Number ($\bar{\nu}$)
Beryllium.............	440
Carbon (graphite).......	900
Iron..................	390

the average numbers of displaced atoms per primary knock-on produced by 1-Mev neutrons in a few elements of reactor interest. It should be noted, however, that not all the displacements will lead to lattice defects since many of the displaced atoms will undoubtedly move to vacant equilibrium positions.

THERMAL AND DISPLACEMENT SPIKES

7.14. The foregoing discussion has been concerned with what may be regarded as the microscopic point of view. Another approach takes into consideration the situation resulting from the transfer of energy by the charged particle to a group of atoms in close proximity. Suppose a struck atom, produced directly by the charged particle or by an energetic knock-on, has sufficient energy to cause it to undergo vibrations of large amplitude without leaving its lattice position entirely. Some of this excess vibrational energy will be rapidly transferred to its immediate neighbors which will, in turn, transfer some energy to their neighbors. The result will be the formation of a limited region in which the atoms have vibrational energies in excess of the normal value corresponding to the bulk temperature of the solid. The situation is then similar to that which would exist if the small volume were suddenly heated to a high temperature. Consequently, the region in which the atoms are in vibrational excited states is referred to as a *temperature spike*.

7.15. If the excitation is relatively small, so that few (if any) of the vibrational atoms leave their equilibrium sites, the disturbance is called a *thermal spike*. It has been estimated that in such a situation a region containing on the order of a thousand atoms is heated to a temperature of around 1000°C (1800°F) for a period of about 10^{-10} sec. At the high temperature of the thermal spike, distortion of the lattice, due to expansion, is to be expected. In the very rapid cooling, resulting from the conduction of heat to the surrounding atoms, a certain amount of the lattice distortion will be "frozen," since there will not be sufficient time for all the atoms to be relocated in equilibrium positions. Consequently, stresses may develop in the material.

7.16. On the other hand, the vibrational excitation may be sufficient to permit a large number of atoms to leave their lattice sites and move about. As a result of collisions with other atoms a *displacement spike* will be produced. It has been suggested that when the kinetic energy of a knock-on falls to a certain value, e.g., some tens of kilo-electron volts in a heavy element, the moving atom is rapidly brought to rest, dissipating its energy in the short distance on the order of 100 A. Several thousands of atoms in a small volume would thus acquire sufficient vibrational energy to permit them to move through the lattice colliding with, and displacing, other atoms. As a consequence, there would be formed a spike containing a large number of interstitials and vacancies and other misoriented regions.

EFFECTS DUE TO NEUTRON CAPTURE

7.17. The capture of a neutron by the nuclei in a solid can produce two effects under suitable conditions; these are (a) the introduction of impurity atoms and (b) atomic displacement due to recoil. When exposed to thermal neutrons, most elements undergo (n, γ) reactions to some extent. The immediate product of such a reaction is isotopic with the absorbing element, and so there should be no damage to the irradiated substance; but if the product is radioactive, as is frequently the case, it will emit beta particles and be thereby converted into a different element. In addition, fast neutrons can produce new elements directly by (n, p) and $(n, 2n)$ reactions. As a result of these nuclear reactions, impurity atoms are introduced into the original crystal lattice. After long exposure to the fast and thermal neutrons in a reactor, a sufficient number of such atoms may accumulate to affect the physical properties of the material.

It should be noted, however, that since only one impurity atom can result from each neutron reaction, the radiation damage is likely to be less than that due to the formation of thermal and displacement spikes by fast neutrons.

7.18. In addition to the production of impurity atoms there is another consequence of (n, γ) reactions which may be more significant [5]. This is the displacement produced by the recoil following the emission of a gamma-ray photon. The magnitude of the recoil energy can be calculated by considering the conservation of momentum in the following manner. The momentum of a photon is equal to E_γ/c, where E_γ is the gamma-ray (photon) energy and c is the velocity of light. If the recoil atom has a mass A, it will recoil with a velocity v such that

$$\frac{E_\gamma}{c} = Av,$$

where all quantities are expressed in cgs units. The recoil energy, E_r, is equal to $\frac{1}{2}Av^2$, so that

$$E_r = \frac{E_\gamma^2}{2Ac^2}.$$

Upon converting the energies into Mev and A into atomic mass (or weight) units, the result is

$$E_r = 5.4 \times 10^{-4}\frac{E_\gamma^2}{A}.$$

7.19. The total energy of the gamma ray accompanying an (n, γ) reaction is roughly 6 to 8 Mev, and so for an element of low atomic weight, e.g., about 10, the recoil energy would be 2 or 3 kev. This is probably an overestimate because it does not take into account the effect of cascade emission of the gamma-ray photons, which occurs frequently. Nevertheless, since it requires only 25 ev to displace an atom, the recoil energy should be capable of causing an appreciable number of atomic displacements. In a thermal reactor, in which the thermal-neutron flux may greatly exceed that of fast neutrons, the radiation damage caused by recoil from (n, γ) reactions may be of the same order as (or greater than) that due to the fast neutrons in a material having an appreciable radiative capture cross section for thermal neutrons. It should be noted, however, that if the cross section were large, the effect of (n, γ) reactions would be mainly near the surface of the material where most of the captures would occur. The damage produced by fast neutrons would extend more deeply into the interior.

PHYSICAL EFFECTS OF RADIATION

7.20. The changes produced by radiation on metals and related substances depend on the nature of the materials and the temperature. Some general results will be described here and more specific cases will be considered later. As a rule, the effects are smaller at elevated temperatures and in some cases the damage can be removed ("annealed") by raising the temperature. At elevated temperatures atoms can diffuse more rapidly from one site to another through the lattice of a solid; consequently, displaced (interstitial) atoms can move into vacancies and thus restore the normal structure existing prior to irradiation.

7.21. Under appropriate conditions, many metals exhibit an increase in hardness and tensile strength and a decrease in ductility as a result of exposure to radiation. The effect of reactor neutrons on a low-carbon steel (§7.51), such as is used for pressure vessels, is shown in Fig. 7.1 [6]. The variation of certain

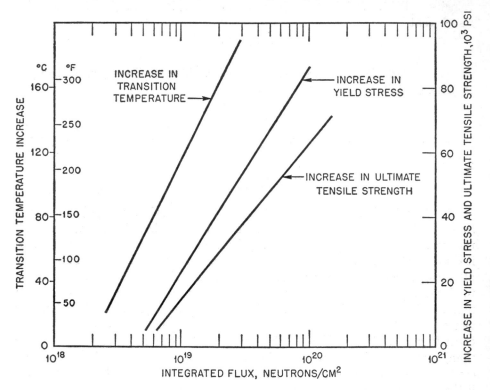

FIG. 7.1. Changes in properties of low-carbon steel [Data from L. F. Porter, ASTM-STP-276, p. 161 (1959) and W. S. Pellini, *et al.*, "Radiation Damage in Solids," IAEA, Vol. II, p. 138 (1962)]

properties is presented as a function of the *integrated flux* of neutrons with energies in excess of 1 Mev. The integrated flux, represented by *nvt*, and expressed in terms of neutrons per cm², is equal to the product of the (average) fast flux and the exposure time. The transition temperature in the figure refers to that at which the fracture changes from brittle to ductile. The indicated increase in this temperature as a result of radiation exposure is accompanied by a decrease in impact strength that may be serious in connection with pressure vessel design (§7.52). However, the exposures to which Fig. 7.1 applies were made at temperatures below 232°C (450°F) and there is evidence that the change is less marked if the material bombarded is at a higher temperature.

7.22. Another interesting example of the effects of reactor irradiation is provided by graphite. The changes in certain properties are indicated qualitatively in Fig. 7.2 [7]. The thermal resistance, lattice constant, and bulk volume

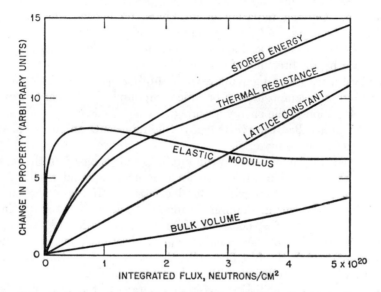

FIG. 7.2.   Variation of properties of graphite with fast-neutron exposure [7]

increase with increasing integrated neutron flux. The elastic modulus is seen to increase rapidly at first, pass through a maximum, and then decrease slowly apparently toward an asymptotic value which is appreciably greater than that prior to irradiation.

RADIATION EFFECTS IN ORGANIC COMPOUNDS [8]

7.23. When a fast neutron interacts with an organic (covalent) material, most of its energy is first transferred to recoil protons; the latter then either cause complete removal of orbital electrons from other atoms, i.e., ionization, or raise these electrons to higher energy states, i.e., electronic excitation. In each case the result is a tendency for carbon-hydrogen and carbon-carbon bonds in the organic compound to break. As a rough average, it has been found that it requires about 25 ev of energy to split a covalent bond. The subsequent recombinations of hydrogen atoms and free radicals leads to the formation of compounds both of lower molecular weight, e.g., hydrogen and simple hydrocarbons, and of higher molecular weight, e.g., various polymers. An organic liquid exposed to radiation will thus release gaseous products as well as accumulate tars and sludge. The increase in viscosity resulting from the formation of substances of high molecular weight and boiling point, known as "high boilers," is often taken as a semiquantitative indication of the extent of radiation damage.

7.24. The effect of thermal neutrons on an organic compound arises mainly from the 2.2-Mev gamma ray which accompanies the (n, γ) capture of neutrons by hydrogen. Gamma radiation of all kinds, no matter what their origin, interact with the organic medium in the manner described in § 2.48 *et seq.* The high-energy electrons produced are able to cause ionization or electronic excitation, followed by the breaking of covalent bonds.

7.25. Since neutrons and gamma rays (and other nuclear radiations) produce the same kind of decomposition of organic compounds, it is usual to express the effects as a function of the energy absorbed. One way is to state the energy in terms of the unit called the *rad* (§ 9.45); this represents an energy absorption of 100 ergs per gram of material.

7.26. The conversion of fast-neutron flux into rads is based on the reasonable assumption that radiation damage in an organic compound results from the interaction of recoil protons with orbital electrons. If N_H is the number of hydrogen atoms per cm^3 in the organic material and σ_s is the scattering cross section of hydrogen for fast neutrons, the rate of formation of recoils for first collisions of the incident neutrons is $N_H \sigma_s \phi_n$ protons/(cm^3)(sec), where ϕ_n neutrons/(cm^2)(sec) is the fast flux. On the average, a neutron loses roughly half its energy in a collision with a hydrogen nucleus, and so the recoil protons from the first collisions dissipate energy at the rate of $\frac{1}{2} N_H \sigma_s \phi_n E_n$ Mev/(cm^3)(sec), where E_n is the energy of the fast neutrons in Mev. Allowance should be made for proton recoils produced in second, third, etc., neutron collisions, but this may be neglected in the present approximation. Since 1 rad is equivalent to 100 ergs, i.e., 6.2×10^7 Mev, per gram of material, it represents an absorption of $6.2 \times 10^7 \rho$ Mev/cm^3, where ρ is the density; consequently,

$$1 \text{ fast neutron}/(cm^2)(sec) \equiv \frac{N_H \sigma_s E_n}{1.24\rho} \times 10^{-8} \text{ rad/sec.} \qquad (7.2)$$

Multiplication of the fast flux by the exposure time gives the integrated flux (§ 7.21) or its equivalent in rads.

7.27. When thermal neutrons are involved, the energy absorption depends on the rate of dissipation of gamma-ray energy. It is convenient, therefore, to consider the absorption of gamma rays in general and then to treat the behavior of thermal neutrons as a special case. If there is a flux ϕ_γ photons/(cm^2)(sec) and E_γ is the photon energy in Mev, the gamma-energy flux is $\phi_\gamma E_\gamma$ Mev/(cm^2)(sec). If μ_e cm^{-1} is the *energy absorption coefficient* of the organic compound, for the given gamma rays,* the rate of energy absorption in the material is $\phi_\gamma E_\gamma \mu_e$ Mev/(cm³)(sec) or $\phi_\gamma E_\gamma \mu_e/\rho$ Mev/(gram)(sec); hence,

$$1 \text{ photon/(cm}^2)(\text{sec}) \equiv \frac{E_\gamma \mu_e}{6.2\rho} \times 10^{-7} \text{ rad/sec.} \tag{7.3}$$

7.28. If ϕ_{tn} is the thermal neutron flux, the rate of formation of capture gamma rays is $N_{\text{H}}\sigma_c\phi_{tn}$ photons/(cm³)(sec) and the equivalent gamma-ray flux is $N_{\text{H}}\sigma_c\phi_{tn}/\mu_e$ photons/(cm²)(sec), where σ_c is the thermal-neutron capture cross section of hydrogen. This value may be substituted for ϕ_γ in the expression obtained above; it follows, therefore, that

$$1 \text{ thermal neutron/(cm}^2)(\text{sec}) \equiv \frac{N_{\text{H}}\sigma_c E_\gamma}{6.2\rho} \times 10^{-7} \text{ rad/sec,} \tag{7.4}$$

where E_γ is the energy (2.2 Mev) of the capture gamma rays. In this calculation capture by nuclei other than hydrogen is neglected since it is very small in materials of present interest.

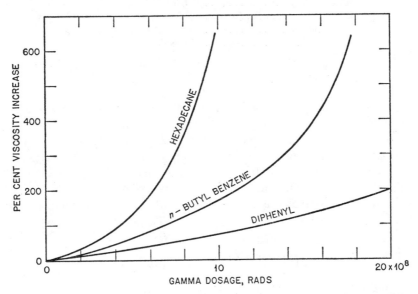

Fig. 7.3.   Effect of gamma radiation on different types of hydrocarbon [9]

7.29. The curves in Fig. 7.3 show the increase in viscosity with radiation exposure (in rads) of three organic compounds which might be considered for use as reactor moderators and coolants [9]. The ordinates represent the viscosity increase relative to that of the material before irradiation (mostly at 100°F), so

---

* The energy absorption coefficient is equal to the total (linear) attenuation coefficient μ, as defined in § 2.61, minus the Compton scattering contribution and it is often represented in the literature by $\mu - \sigma_s$, although $\mu - \Sigma_s$ is strictly correct.

that they give a general indication of the extent of decomposition due to radiation exposure. It is seen that aromatic hydrocarbons are more resistant to radiation damage than are aliphatic compounds. The most resistant of all are the polyphenyls, of which diphenyl is the simplest example.

7.30. The stability of organic (and other covalent) compounds to radiation is frequently expressed by means of the so-called "G" value; this is equal to the number of molecules decomposed or of product formed per 100 ev of energy dissipated in the material. The manner in which the amount of energy absorbed can be calculated from the neutron or gamma-ray fluxes will be obvious from the discussion given above.

7.31. As an example of the use of G values, the data in Table 7.2 [10] are for

TABLE 7.2. RADIOLYTIC DECOMPOSITION OF POLYPHENYLS AT 350°C

Material	$G(gas)$	$G(polymer)$
Diphenyl	0.159	1.13
Ortho-terphenyl	0.108	0.70
Meta-terphenyl	0.081	0.64
Para-terphenyl	0.073	0.54
Santowax-R*	0.080	0.59

* A mixture of the three terphenyls plus a small amount of diphenyl (§ 6.204).

a number of polyphenyls exposed to the radiation in a thermal reactor. The G(gas) value represents the total number of molecules of gas produced and G(polymer) is the number of molecules of polyphenyl destroyed in polymer formation per 100 ev absorbed in the material. It is of interest to note that the terphenyls are even more resistant to radiation than diphenyl and, since they have higher boiling points, a mixture of terphenyls with a relatively low melting temperature was chosen as the moderator-coolant in organic-moderated reactors (§ 13.42).

STRUCTURAL MATERIALS [1]

GENERAL CHARACTERISTICS

7.32. Every reactor system must inevitably include a certain amount of structural material which serves as a mechanical framework for the reactor components. In addition, various types of containers are required for solid or liquid fuels, for coolants, for control rods, and for measuring instruments. Because uranium is readily attacked by air, water, and other fluids which might be used as coolants, it is necessary that fuel elements be protected with suitable cladding. Such cladding is needed, in any event, to prevent escape of fission products from the fuel element.

7.33. The requirements of a structural material will vary to some extent with the type of reactor and with its specific purpose in the reactor. There are, however, some general characteristics which may be enumerated. Mechanical properties, such as tensile strength, impact strength, and rupture stress, must be adequate for the operating conditions. The material must be capable of being fabricated or joined into the required shapes. Thermal conductivity should usually be high and the coefficient of thermal expansion low or well matched with that of other materials.

7.34. In many reactor components there is very considerable internal heating, due either to fission, to the slowing down of fast neutrons, or to the absorption of various radiations. The removal of the heat from the exterior results in high temperature gradients within the material. Such materials must, therefore, be able to maintain stability under severe thermal stress.

7.35. In addition to the physical and mechanical properties referred to above, the nuclear properties must be satisfactory. If the material is to be used in or near the core or reflector of the reactor, it must obviously have a small cross section for neutron capture. It should be noted in this connection that a material that cannot be used in a thermal reactor because of a large cross section may perhaps be employed in a fast reactor, since the cross section for the capture of fast neutrons may then be tolerable.

7.36. As a result of neutron capture, many materials become radioactive and consequently are dangerous to handle. Thus the maintenance or repair of equipment exposed to the neutron flux of a reactor may present a difficult problem. In the choice of materials, preference should be given, if possible, to substances which either do not become radioactive as a result of neutron capture

TABLE 7.3.   INDUCED ACTIVITIES IN METALS
PRESENT IN STRUCTURAL MATERIALS

Element	Isotope Mass No.	Natural Abundance (per cent)	σ_a (barns)	Active Species	Half-life	γ-Ray Energy (Mev)
Chromium.......	50	4.3	17	Cr^{51}	28d	0.32
Manganese.......	55	100	13.2	Mn^{56}	2.6h	2.1
Iron............	58	0.3	2.5	Fe^{59}	45d	1.3
Cobalt..........	59	100	37	Co^{60}	5.3y	1.3
Nickel..........	64	1.2	1.6	Ni^{65}	2.6h	1.5
Copper..........	63	69	4.5	Cu^{64}	12.8h	1.35
Zinc............	64	48.9	0.5	Zn^{65}	250d	1.11
	68	18.6	0.1	Zn^{69}	13.8h	0.44
Zirconium.......	94	17.4	0.1	Zr^{95}	63d	0.75
Molybdenum.....	98	23.8	0.45	Mo^{99}	67h	0.78
Tantalum........	181	100	21	Ta^{182}	112d	1.1
Tungsten........	186	28.4	35	W^{187}	24h	0.77

or for which the induced activity is weak (with no gamma radiation) or of short half-life.   In this connection, the data in Table 7.3 will be of some value.   First there are listed the elements which might have some interest for structural purposes or which might be present as alloying elements or as impurities.   In each case the mass number is given of the isotope (or isotopes) which yields a radioactive species as a result of neutron absorption.   The abundance in the element as it occurs in nature and its thermal (2200-m/sec) absorption cross section then follow.   The next column gives the identity of the active species, and this is followed by its half-life and the energy of the most penetrating (highest energy) gamma rays.   It should be emphasized that the information in the table refers only to radioactive isotopes of appreciable half-life ($>$ 1 hr) emitting gamma radiations of moderate or high energy ($>$ 0.3 Mev), in addition to beta particles. There are other species which are beta-active only, but these do not usually present a serious maintenance problem.

7.37. The elements which may be regarded as making the major contribution to persistent induced activity are chromium, manganese, cobalt, copper, zinc, tantalum, and tungsten.   It will be seen shortly that none of these is likely to be used as a thermal reactor constituent in a moderately pure form.   However, such elements may well be present as impurities or as alloying metals in otherwise acceptable structural materials, and consideration should be given to the activity that will result from exposure to neutrons [11].

MATERIALS FOR THERMAL REACTORS

7.38. The crucial requirement for a material to be used in a thermal reactor is that it shall have a reasonably low capture cross section for neutrons.   If the cross section is large, the material must be rejected no matter what mechanical or other advantages it may possess.   If there are compensating factors, an element with a moderate cross section may be tolerated, but its presence will require the use of additional fissile material in the reactor core.   In a thermal-neutron breeder reactor, where every neutron is significant, parasitic capture would have to be avoided if at all possible.   In a fast reactor, the choice of materials is much less critical for the reason given later (§ 7.45).

7.39. In order to obtain an overall picture of the possibilities with regard to materials for thermal reactors, metallic elements can be divided into three classes: (1) elements with low absorption cross sections for thermal neutrons, i.e., below 1 barn; (2) elements with cross sections of intermediate value, i.e., from 1 to 10 barns; and (3) elements with cross sections exceeding 10 barns.   In general, those in the last category, which include tungsten, tantalum, and cobalt, need not be considered further, except perhaps for intermediate and fast reactors. Of the substances having small or moderate cross sections, soft metals and those with melting points below 500°C (930°F) may be eliminated as not likely to be very useful for structural purposes.   The thermal (2200-m/sec) absorption cross sections and melting points of the remaining elements, which are more or less readily available, are given in Table 7.4.

TABLE 7.4.   PROPERTIES OF POSSIBLE STRUCTURAL MATERIALS

Low Thermal Cross Sections				Intermediate Thermal Cross Sections			
Metal	σ_a (barns)	M.P. °C	M.P. °F	Metal	σ_a (barns)	M.P. °C	M.P. °F
Beryllium.....	0.010	1280	2340	Niobium.......	1.2	2415	4380
Magnesium...	0.069	651	1200	Yttrium........	1.3	1475	2690
Zirconium....	0.185	1845	3350	Iron..........	2.6	1539	2804
Aluminum....	0.24	660	1215	Molybdenum...	2.7	2625	4760
				Chromium......	3.1	1850	3360
				Copper........	3.8	1083	1981
				Nickel.........	4.6	1455	2651
				Vanadium......	5.0	1900	3450
				Titanium.......	5.8	1670	3040

7.40. The number of metals with small thermal-neutron cross sections is seen to be very limited.   Aluminum, usually in the relatively pure (> 99.0 per cent) 2S (or 1100) form, has been extensively used as a reactor structural material, for cladding fuel elements, and for other purposes not involving exposure to very high temperatures.   Aluminum alloys of high creep strength, known as APM (aluminum powder metallurgy) products, have found extensive application in organic-moderated reactors.   Magnesium, in the form of the alloy Magnox,* is employed as cladding in the British Calder Hall (and similar) reactors in which carbon dioxide is the coolant [12].   The high cost of beryllium is a drawback to the use of this element, although it has a low neutron capture cross section and a high melting point.   However, this metal might be used as a structural material in applications where advantage could also be taken of its value as a neutron moderator and reflector.   It might also be employed as a fuel-element cladding in thermal reactors.

7.41. The remaining element zirconium, chiefly in the form of the alloy Zircaloy-2 (§ 7.56), is extensively used as a structural and cladding material for nuclear reactors employing water at high pressure (and temperature) as the coolant.   Another zirconium alloy is being developed for use in sodium-cooled reactors (§ 7.56).   In addition to its low capture cross section for thermal neutrons, zirconium has good mechanical and fabrication properties and is resistant to corrosion in the appropriate environment.   The mechanical strength of pure zirconium falls off to some extent, as also does its resistance to attack, at temperatures above 400°C (750°F), but these drawbacks can be overcome to some extent by suitable alloying.

* Magnox A12 contains 0.8 per cent of aluminum and 0.01 per cent of beryllium.

7.42. Turning to the elements of intermediate cross section, it will be seen that here also the choice is limited. Some of the metals mentioned, especially in the form of alloys, e.g., stainless steel, are being used as structural and cladding materials. Others form components of fuel elements (cf. § 8.53) or are used for cladding purposes.

7.43. The relatively common, but expensive, element titanium and the more scarce yttrium may prove useful in future development of reactor systems. The special interest in titanium has resulted largely from its high strength-to-weight ratio in the temperature range from 100° to 450°C (212° to 842°F), which makes it attractive for uses where weight-saving is important. The fabrication of parts from titanium by forging and welding presents some problems, but continuous progress is being made in overcoming them. The relatively high cross section will limit its use in thermal reactor cores, but it could possibly be utilized to some extent.

7.44. Yttrium metal has potential interest as a thermal reactor material, mainly because of its moderately small thermal-neutron absorption cross section and high melting point. Unfortunately it has poor mechanical properties and reacts with oxygen and hydrogen. However, efforts are being directed toward the development of useful alloys. Incidentally, yttrium is an example of a class of substances, not in general use, which might provide the basis for materials with more favorable properties than those presently available.

MATERIALS FOR FAST REACTORS

7.45. The quantity of fissile material relative to structural components in a fast reactor is much greater than in a thermal reactor. Consequently, the ratio of the macroscopic fission cross section to that for parasitic capture, which is important for the neutron economy, tends also to be larger. As a result, structural materials with fairly large microscopic cross sections, which could not be tolerated in a thermal system, can be used in fast reactors, because their effect on the overall neutron economy is not very significant. The essential requirements for materials are now high melting point, retention of satisfactory physical and mechanical properties at high temperatures, and good corrosion resistance, especially to molten sodium. Metals and alloys of particular interest are stainless steels, niobium, molybdenum, tantalum, and tungsten. The latter two have very high melting points, viz., tantalum 3000°C (5430°F) and tungsten 3400°C (6150°F), but their thermal-neutron cross sections are much too high for use in thermal reactors. Both molybdenum and tungsten are resistant to attack, but they are somewhat brittle and difficult to fabricate. Tantalum, on the other hand, can be fabricated without undue difficulty and it resists the action of sodium up to 1200°C (2200°F). It is the only metal that has been found suitable for containing the highly corrosive molten plutonium-iron fuel in the LAMPRE fast reactor (§ 13.117).

STAINLESS STEELS

7.46. Among the more familiar materials, the austenitic stainless steels are corrosion resistant and have good mechanical properties. Some of the factors governing their selection for reactors and the problems connected with their use are similar to those encountered in other chemical and high-temperature applications. For reactor purposes, however, radiation exposure must also be taken into consideration [13]. In spite of the generally satisfactory performance in a wide variety of circumstances, the choice of an austenitic stainless steel can be made only after a thorough investigation of the potential difficulties which may develop in its use. Among the problems to be considered are the following: (1) susceptibility in certain environments to intergranular corrosion, after sensitizing heat treatments; (2) stress-corrosion cracking; (3) formation of a brittle sigma phase, particularly in the high-ferrite weld deposits which are frequently used to avoid weld cracking; (4) knife-line corrosion attack; (5) high thermal stresses due to poor heat conductivity; (6) accelerated local attack by highly corrosive, rapidly flowing solutions due to removal of protective films; and (7) susceptibility to attack by aqueous solutions containing chloride ions.

7.47. The compositions of a number of austenitic stainless steels which may be useful in reactor work are given in Table 7.5. AISI Type 304, which is a

TABLE 7.5.   AUSTENITIC STAINLESS STEELS FOR REACTOR APPLICATIONS

AISI Type	Carbon, % (maximum)	Chromium, %	Nickel, %	Other Elements
304.........	0.08	18.0 to 20.0	8.0 to 11.0	—
304L........	0.03	18.0 to 20.0	8.0 to 11.0	—
309S Nb.....	0.08	22.0 to 26.0	12.0 to 15.0	Nb (min. $8 \times C$)
316.........	0.10	16.0 to 18.0	10.0 to 14.0	Mo (2.0 to 3.0%)
316L........	0.03	16.0 to 18.0	10.0 to 14.0	Mo (1.75 to 2.5%)
347.........	0.08	17.0 to 19.0	9.0 to 12.0	Nb (min. $10 \times C$)

familiar variety, can be considered for aqueous solutions if the environment does not promote intergranular attack of sensitized stainless steel, if the structure is not to be welded, or if the welded structure can be given a full anneal at about 1000°C (1830°F), followed by fast cooling to avoid sensitization. On the other hand, where intergranular attack of the sensitized material is to be expected or if field welding precludes annealing, Types 304L (low carbon) and 347 may prove useful.

7.48. It should be pointed out that commercial nickel used in the manufacture of stainless steels may contain up to 0.2 per cent of cobalt. Not only does the latter have a high thermal neutron capture cross section (37 barns), but the product, cobalt-60, is radioactive, emitting gamma rays as well as beta particles. Nickel with a lower cobalt content, down to about 0.0012 per cent, can be obtained commercially, but its use greatly increases the price of the resulting stainless steel.

7.49. In addition to the austenitic steels mentioned above, ferritic stainless steels, such as Types 405, 430, and 460, which contain no nickel but about 1 per cent by weight of manganese and of silicon, are useful. They find application where only moderate resistance to chemical attack is required, but where a lower coefficient of expansion and higher thermal conductivity, as compared with the austenitic steels, are desirable.

7.50. Neutron and fission fragments produce lattice disturbances in stainless steels which are similar to those caused by cold working. There is a decrease in ductility and an increase in hardness, probably accompanied by a higher yield strength. Some of the effects of an integrated fast-neutron flux of 2.4×10^{21} *nvt* on Type 347 stainless steel are shown in Table 7.6 [14].

TABLE 7.6. CHANGES IN PROPERTIES OF TYPE 347
STAINLESS STEEL UPON IRRADIATION

Property	Before Irradiation	Irradiated $(2.4 \times 10^{21}$ *nvt*$)$
Hardness (Rockwell C)	16.4	20.5
Tensile strength, 10^3 psi	102	109
Yield strength (0.2% offset), 10^3 psi	66	98
Reduction of area, per cent	74	78
Elongation in 1 in., per cent	57	52

CARBON STEELS

7.51. Carbon steel is an important reactor material for use in pressure vessels and miscellaneous components where the corrosion resistance of stainless steels is not required but an ability to withstand thermal stress is desirable. For pressurized-water reactors, the inside of the vessel is usually clad with a comparatively thin layer of stainless steel to provide corrosion resistance. Steels for thick-walled pressure vessels are specified in the ASME Code (§ 11.64), and are more accurately described as high-strength, low-alloy steels, rather than by the all-inclusive term of carbon steels. These steels contain small amounts of alloying materials to improve mechanical properties. The compositions and some properties of two typical pressure-vessel steels, ASTM A 212-B and A 302-B, are given in Table 7.7.

7.52. It is apparent from Fig. 7.1 that integrated fluxes on the order of 10^{19} to 10^{20} fast neutrons per cm^2 can raise the tensile strength of steel appreciably, while decreasing the ductility and increasing the brittle-to-ductile transition

TABLE 7.7.   COMPOSITION AND MECHANICAL PROPERTIES
OF PRESSURE VESSEL STEELS

Element	Composition (Wt. %, Maximum)	
	A 212-B	A 302-B
Carbon...................	0.35	0.25
Manganese................	0.90	1.50
Phosphorus...............	0.04	0.04
Sulfur...................	0.05	0.05
Silicon..................	0.30	0.32
Molybdenum...............	—	0.60

	Mechanical Properties	
Tensile strength, psi.............	70,000 to 80,000	80,000 to 100,000
Yield strength, psi..............	38,000	50,000
Per cent elongation, in 2 in.......	21	17

temperature.   Although loss of ductility reduces the ability of the steel to accommodate thermal stresses, the increase in the transition temperature is much more serious from the standpoint of the use of the material for pressure vessels.   After being in operation for some time, with accompanying exposure to neutron irradiation, a drop in temperature might cause the vessel to suffer brittle fracture, because of the increase in the transition temperature.   The effects of radiation may be less at the elevated operating temperatures than at ordinary temperatures, but too little is known about this or about the possibility of annealing out radiation damage for definite conclusions to be drawn.   It is important, therefore, to bear in mind that the problem exists and that it may be a serious one for pressure vessels made from ferritic steels in particular [15].

ZIRCONIUM [16]

7.53.   Zirconium, preferably in the form of alloys, is a promising metal for use as a constructional and fuel-cladding material.   Satisfactory alloys are available for water-cooled reactors operating at temperatures up to 400°C (750°F) and others are being developed for sodium-cooled reactors at even higher temperatures, as described below.   The addition of zirconium to uranium and thorium leads to improvements in fuel elements containing these materials.

7.54.   An important fact from the reactor standpoint is that zirconium ores invariably contain from 0.5 to 3.0 per cent of hafnium, and this element has a high cross section (105 barns) for the capture of thermal neutrons.   It is desirable, therefore, that most of the hafnium be removed from zirconium for use in thermal reactors.   Unfortunately, this is not a simple matter because the chemical properties of hafnium and zirconium are very similar; nevertheless, successful methods of separation have been developed, based on liquid-liquid extraction (cf. § 8.135).

Properties

7.55. Between room temperature and its melting point (1850°C or 3360°F), zirconium exists in two allotropic modifications: a close-packed hexagonal (alpha) phase, stable up to 862°C (1584°F), and a body-centered cubic (beta) phase which is stable above this temperature. The theoretical density at ordinary temperatures is 6.49 g per cm^3. In spite of its hexagonal (anisotropic) structure, the alpha phase of zirconium can be prepared in ductile form having mechanical and machining properties similar to those of plain carbon steel. The coefficient of thermal expansion, which is dependent on the direction in single crystals and in rolled material, has a mean value of about 5.9×10^{-6} per °C. The thermal conductivity decreases from 0.050 at 25°C (78°F) to 0.045 cal/(sec)(cm^2)(°C/cm) at 300°C (572°F).*

7.56. The mechanical properties of zirconium and its alloys vary widely because they are markedly affected by the previous treatment and by the presence of impurities. Although the data in Table 7.8 are not definitive, they

TABLE 7.8.   MECHANICAL PROPERTIES AT ROOM TEMPERATURE

Property	Zirconium	Zircaloy-2	3Z1
Tensile strength, 10^3 psi...............	29	70	110
Yield strength (0.2% offset), 10^3 psi....	12	42	85
Elongation, per cent..................	41	16	13
Reduction in area, per cent............	42	45	—

provide a general indication of some of the mechanical properties at room temperature of zirconium and of the alloys Zircaloy-2 [17] and 3Z1 [18]. Zircaloy-2, with about 1.5 per cent tin, 0.12 per cent iron, 0.09 per cent chromium, and 0.05 per cent nickel, has been extensively employed in water-cooled reactors, and 3Z1, containing small amounts of aluminum, tin, and molybdenum, is being developed for use in sodium reactors at temperatures up to 1100°F. The presence of aluminum in 3Z1 makes this alloy unsuitable for water reactor service, but does not affect its behavior in sodium systems. The tensile strength and yield strength of zirconium and its alloys fall off with increasing temperature. However, as seen in Table 7.8, the alloys are stronger than zirconium itself and can be employed at higher temperatures. It should be mentioned that the thermal conductivity of Zircaloy-2 is about 30 per cent less than that of the pure metal.

* For conversion to engineering units, see Table A.1 in the Appendix.

Fabrication

7.57. Conventional fabrication methods, such as machining, hot and cold rolling, forging, extrusion, and drawing can be used with zirconium and its alloys, but oxygen, nitrogen, and hydrogen must be kept out. Provided there has been no access by oxygen, nitrogen, or carbon after annealing, the cold-worked metal is very ductile and can be machined readily. Powder-metallurgical methods can also be used to fabricate zirconium parts. The products are ductile and can be rolled, extruded, and drawn without difficulty.

7.58. Zirconium metal has excellent welding characteristics, but the operation must be performed in an atmosphere of inert gas of high purity, e.g., helium or argon, to prevent contamination. The ability of molten zirconium to dissolve surface films of oxide and nitride is an important factor in the production of bonds having a high degree of integrity. Soldering and brazing by conventional methods are not satisfactory, but special brazing procedures have been devised.

Corrosion

7.59. An outstanding property of zirconium from the standpoint of reactor design is its resistance to corrosion in water and steam at high temperatures. The behavior is, however, sensitive to the state of the surface, the presence of various trace elements, the purity of the water, and the presence of oxygen and nitrogen. Both of the latter elements increase the corrosion rate in high-temperature water, as also do carbon, titanium, and aluminum in the metal. The corrosion resistance of zirconium is indicated by the observation that a pure sample suffered less attack in a 500-hour dynamic test in water at 205°C (401°F) than stainless steel, niobium, beryllium, carbon steel, and aluminum; the superiority over the other metals increased in the order given. By appropriate alloying, the resistance of zirconium to attack by water can be increased still further.

7.60. The corrosion of good quality zirconium (and its alloys) in water is accompanied by the formation of an adherent oxide film which grows slowly in thickness as the metal interacts with the water. After a lapse of a certain time, zirconium ("crystal bar") metal containing detrimental impurities, e.g., nitrogen, aluminum, and titanium, was observed to exhibit a sudden increase in corrosion rate. This phenomenon, which is accompanied by a flaking off of the oxide film, was called "breakaway." With Zircaloy-2, however, the corrosion rate increases gradually and the film remains intact. A so-called breakaway time is estimated from the intersection of two straight lines on a log-log plot representing the corrosion rates at early and late times. This breakaway time provides a rough indication of the corrosion resistance of the alloy.

7.61. Zircaloy-2 has proved to be a valuable corrosion-resistant alloy, its breakaway time and corrosion resistance being insensitive to a variety of impurities which influence the behavior of zirconium itself. Moreover, the resistance of Zircaloy-2 to attack by hot water or steam is generally superior to that of zirconium. In water at 315°C (600°F), no breakaway was observed with Zircaloy-2 after an exposure of 3 years; the gain in weight due to film formation during this period was very small. At a temperature of 360°C (680°F), break-away occurred after 120 days, the total weight gain during this period being about 0.4 mg per cm². Although Zircaloy-2 has good corrosion resistance, even better alloys are being developed. One of the problems with Zircaloy-2, for example, is the retention of hydrogen produced by reaction with water and accompanying embrittlement. The situation can be improved by elimination of the nickel; the resulting alloy, which may contain slightly more iron, is called Zircaloy-4.

7.62. Not only are zirconium and its alloys little affected by water, they are also resistant to attack by liquid metals. Corrosion rates of both unalloyed zirconium and of the alloy 3Z1 have been found to be relatively low in dynamic sodium at high temperatures. After a 5000-hour exposure to sodium ($<$ 10 parts per million of oxygen) at 650°C (1200°F), these materials showed a similar increase in weight of approximately 0.9 mg per cm².

MODERATOR AND REFLECTOR MATERIALS

INTRODUCTION

7.63. The essential requirements for moderator and reflector materials for thermal reactors are low mass number, small capture (or absorption) cross section, and fairly large scattering cross section. The only substances which might be used as moderator or reflector are thus ordinary water, heavy water (deuterium oxide), hydrocarbons, beryllium (either as metal, oxide, or carbide), and carbon (graphite). The main purpose of the present discussion is to consider moderators and reflectors as structural materials. However, the selection is based ultimately on nuclear considerations, and the essential neutronic properties of a number of moderating materials are given in the Appendix (Table A.7). Beryllium carbide has not been included because its chemical reactivity with air and water seems to limit its usefulness. It should be mentioned that the densities of graphite and beryllium oxide vary appreciably with the method of production; the values given in the table are less than the theoretical densities, but they may be taken to be typical of reactor materials.

GRAPHITE [19]

7.64. Graphite has been used extensively as moderator and reflector for thermal reactors. Its nuclear properties are not quite as good as those of heavy water or beryllium, but graphite of satisfactory purity is readily available at reasonable cost. It has good mechanical properties and thermal stability. However, at high temperatures it reacts with oxygen (in air) and with water vapor, and with some metals and metal oxides it forms carbides.   Although it is a nonmetal, graphite is a good conductor of heat, a desirable property for a moderator.   Its chief drawbacks under normal operating conditions are the possibility of oxidation in the presence of air, relatively low impact strength, and the porous nature of the material used in reactors.   In addition, its dimensions and some of its properties are subject to change under the influence of radiation.

7.65. Graphite occurs in considerable quantities in nature, but, since in this state it is relatively impure, the reactor-grade product is made artificially by "graphitization" of petroleum coke.   The following procedure may be regarded as being fairly typical.   The coke (called the "filler") is first heated to drive off volatile matter and is then ground and mixed with a coal-tar pitch "binder."   The mixture is extruded into bars and baked to temperatures up to 1500°C (2730°F) in gas-fired ovens in order to carbonize the pitch and set the binder.   To increase the bulk density of the product, it is impregnated with pitch under vacuum and again baked.   The resultant gas-baked carbon is finally graphitized by electrical-resistance heating.   The temperature is raised to about 2700° to 3000°C (4890° to 5430°F) during the first few days and then about three or four weeks are allowed for the material to cool.   The physical characteristics of the final product vary with the fineness of the ground coke, the type and amount of pitch used for impregnation, and the temperature and duration of the graphitization treatment.

Properties

7.66. Graphite is a black, moderately soft, allotropic form of carbon; it is the stable phase at all reasonable temperatures.   It does not melt upon heating at atmospheric pressure but sublimes at about 3650°C (6600°F).   The crystals have hexagonal symmetry and consist of flat layers or planes of carbon atoms, stacked parallel to one another.*   The planes are relatively far apart, namely, 3.41 A, so that shear occurs easily by a slipping motion of the planes relative to each other.   This type of shear, perpendicular to the hexagonal axis, accounts for the softness and lubricity of flake graphite.   As is to be expected, the thermal, electrical, and mechanical properties of graphite crystals are anisotropic, and this behavior is exhibited to some extent by the polycrystalline material because of preferred orientation of the individual crystallites.

7.67. The physical and mechanical properties of graphite vary widely among different products.   The theoretical density, calculated from the lattice constants, is 2.26 g per cm^3, and this value is approached by natural graphite.   However, the material manufactured for reactors has a density of roughly 1.6 to 1.7 g per cm^3; this low density is due to the porosity of the artificial product.

7.68. The values of a few important properties of graphite at ordinary temperatures for a typical reactor-grade material are summarized in Table 7.9.

---

* Because of small discrepancies in the bond angles, graphite is sometimes regarded as orthorhombic; however, the departure from hexagonal symmetry is very slight.

TABLE 7.9. PROPERTIES OF REACTOR-GRADE GRAPHITE

Property	Longitudinal	Transverse
Thermal expansion coefficient, per °C............	1.4×10^{-6}	2.7×10^{-6}
Thermal conductivity, cal/(sec)(cm²)(°C/cm).....	0.41	0.31
Tensile strength, psi..........................	2000	700
Flexural strength, psi.........................	2100	2100
Compressive strength, psi......................	6000	6000
Modulus of elasticity, psi.....................	1.5×10^6	1.1×10^6

Because the results often vary with respect to the extrusion axis, data are given for directions parallel (longitudinal) and perpendicular (transverse) to this axis. The tensile strength of graphite increases with temperature, attaining a maximum at about 2500°C (4530°F), when the value is roughly twice that at ordinary temperatures. Above 2500°C, the tensile strength drops very rapidly. The creep rate is very small below 1500°C (2730°F), but may become appreciable at higher temperatures. Graphite has excellent resistance to thermal shock, and it is difficult to produce fracture by thermal stress alone.

7.69. Graphite is oxidized in air, but the reaction is usually negligible at temperatures below about 400°C (750°F) or so. The rate of attack by liquid sodium appears to be dependent on the nature of both the graphite and the containing vessel. Even high-density graphite and the highly oriented material known as pyrolytic graphite (§ 7.75) dissolve in sodium in times which are short compared with the reactor life. Consequently, unclad graphite cannot be used in sodium-cooled reactors.

Radiation Effects [20]

7.70. Although radiation exposure increases the strength, hardness, and elasticity of graphite, especially at normal temperatures, the important considerations for reactor design are (a) dimensional changes, (b) decrease in thermal conductivity, and (c) stored energy. The dimensional instability, apparent as distortion and expansion, is probably related, in part at least, to the highly anistropic nature of graphite. Thus, irradiation is accompanied by an increase in length in the longitudinal direction parallel to the extrusion axis but there is a (smaller) decrease in the perpendicular direction. After long irradiation, a longitudinal contraction has been observed to follow the initial expansion; this contraction appears to continue as the exposure increases. With increasing temperature, the dimensional instability of graphite becomes less marked, and at temperatures above about 320°C (608°F), some specimens exhibit an overall contraction even in the early stages of irradiation. The effect of radiation on the thermal conductivity is very marked; thus integrated neutron fluxes of 10^{19} *nvt* reduced the conductivity of graphite by a factor of 40 or more. The increase in stored energy, which is presumably a result of lattice imperfections produced by the radiation, is manifest as an increase in enthalpy.   The heat of combustion of irradiated graphite, for example, is greater than that of the original material.

7.71. At higher irradiation temperatures, the effects are less marked.   More-over, when irradiated graphite is heated, the physical properties tend to return to their initial values.   The rate of recovery (annealing) and the extent to which it occurs depend upon the exposure and the annealing temperature.   Graphite which has suffered considerable irradiation at lower temperatures must be heated to high temperatures, e.g., 1200°C (2200°F) or more, before recovery is anything like complete.

7.72. The effects of radiation on graphite, especially the stored energy, have important implications in connection with the operation of graphite-moderated reactors.   If the graphite temperature is sufficiently high, e.g., above about 350°C (662°F), the radiation damage is small, presumably because the mobility of the carbon atoms in the crystal lattice permits recovery to keep pace with the displacements caused by the radiation.   At lower temperatures, however, the radiation effects increase with time.   The accumulation of the stored energy, sometimes called the Wigner energy, in the graphite could then be disastrous because a point will be reached when the metastable (high energy) material is suddenly transformed to the stable form accompanied by the release of a large amount of energy and a large increase in temperature.   For example, an exposure of 10^{19} *nvt* at ordinary temperature can result in a total stored energy of about 400 cal per g of graphite, which is sufficient to raise the temperature to approximately 1000°C (1800°F).

7.73. It is the practice in graphite-moderated reactors operating at fairly low temperatures to permit a controlled increase of temperature, at predetermined intervals, so that annealing of the radiation damage occurs and much of the stored energy is released.   The temperature is observed to rise, but it is pre-vented from reaching dangerous levels by adjusting the coolant flow.   Local overheating, which took place during the course of a normal periodic release of stored energy, resulted in serious damage to one of the British plutonium produc-tion reactors at Windscale.   It was partly for this reason that operation of the reactor was not resumed after the accident.

Impermeable Graphite [21]

7.74. The so-called impermeable graphite, actually graphite of low permea-bility, is attracting interest as a special-purpose reactor material.   Pure com-mercial graphite is first impregnated under pressure with a carbonaceous liquid such as acrylonitrile, sugar solution, or furfuryl alcohol; of these the last named appears to be the most effective.   After impregnation, the graphite is heated to about 1000°C (1830°F) to carbonize the furfuryl alcohol, and then frequently to a higher temperature to produce graphitization.   The resulting product is not only very much less permeable to gases than the original graphite, but it also has greater flexural and tensile strengths.   Applications which have been proposed for impervious graphite include cladding for fuel elements in high-temperature, gas-cooled reactors and as a means of providing thermal insulation between the coolant gas and ordinary graphite moderator.

Pyrolytic Carbon [22]

7.75. Although pyrolytic carbon (or graphite) has been known for a long time, it is only in recent years that its potential use in reactors has become significant. Pyrolytic carbon is produced as a deposit on a heated surface by the thermal decomposition of a hydrocarbon gas, e.g., methane, propane, benzene, etc. The decomposition temperature may be roughly in the range from 1500° to 2500°C (2730° to 4530°F). The structure and density of the product depend upon the deposition temperature and other factors, and they can be changed to some extent by further heating to higher temperatures. The characteristic of pyrolytic carbon is that it has a highly oriented crystal structure. As a result, the properties of the material in the plane parallel to the surface on which it is deposited are those of a metal, e.g., high thermal conductivity and high tensile strength. In the perpendicular direction, i.e., across the deposition plane, on the other hand, it is a typical ceramic, with low thermal conductivity and tensile strength. Pyrolytic carbon is essentially impermeable to gases, even in thin layers, and so it has the same possible uses as the impervious graphite mentioned above. In addition, it is being considered as a coating for uranium carbide and similar fuel in the form of particles, to prevent the escape of fission product gases.

BERYLLIUM [23]

7.76. Because the slowing down power of beryllium for neutrons is greater than that of graphite, the former material lends itself to the design of smaller reactors; for the same reason it is also a better neutron reflector. The use of beryllium in reactor construction has been limited, however, by the high price of the metal and by its brittleness. After long exposure to fast neutrons, at integrated fluxes of about 10^{21} *nvt*, hot-pressed powder products have shown an increase in hardness and tensile strength and some decrease in ductility. The helium and tritium gases formed in the (n, α) and $(n, 2n)$ reactions between fast neutrons and beryllium may collect as bubbles and cause local swellings.

Properties

7.77. Only one crystalline form of beryllium, with a hexagonal close-packed structure, has been observed between ordinary temperatures and the melting point of 1280°C (2340°F). The lattice constants are $a_0 = 2.281$ A and $c_0 = 3.577$ A, and the theoretical density is 1.848 g per cm³ at 25°C (78°F). The observed densities of different specimens of beryllium range from 1.81 to 1.86 g per cm³, the higher values being due to the presence of beryllium oxide, which has a theoretical density of 3.0 g per cm³. As a result of the hexagonal structure, many of the physical and mechanical properties of the metal are anisotropic in nature.   However, material made by hot-pressing beryllium powder is isotropic, presumably because of the random orientation of the small grains.   At ordinary temperature the coefficient of thermal expansion is about 10×10^{-6} per °C.   The thermal conductivity is 0.35 cal/(sec)(cm²)(°C/cm) at 25°C (78°F), decreasing to 0.25 at 600°C (1110°F).

7.78. The mechanical properties of specimens of beryllium metal vary with the methods of fabrication and treatment.   Since hot-pressing of the powder has proved a very successful fabrication procedure, as will be seen below, the properties recorded in Table 7.10 refer to material obtained by hot-pressing

TABLE 7.10.   MECHANICAL PROPERTIES OF HOT-PRESSED BERYLLIUM

Tensile strength, 10^3 psi	45
Yield strength (0.2% offset), 10^3 psi	25
Elongation in 2 in., per cent	2.3
Modulus of elasticity, 10^6 psi	44
Poisson's ratio	0.024
Unnotched Charpy impact, ft-lb	0.8
Hardness (Rockwell B)	75–85

in vacuum at a temperature of about 1000°C (1830°F).   The small elongation at tensile fracture and the low impact resistance even for unnotched bars indicate that the ductility of the material is relatively low.   The ductility of cast beryllium is, in general, even lower.   By warm-extruding the hot-pressed metal and then annealing at about 700°C (1290°F), the mechanical properties, especially in the direction of extrusion, are greatly improved.   The brittle nature of beryllium has been attributed partly to the presence of impurities, but this is not certain.

Fabrication

7.79. Castings of beryllium have been made by melting the metal in a beryllium oxide crucible and pouring into graphite molds.   However, cracks may develop during or immediately after solidification unless the rate of cooling is carefully controlled.   The cast material does not fabricate easily; because of the large grain size it is difficult to machine without causing surface damage, and cracking may occur below the surface.   Although cold-working is not practical, cast beryllium can be hot-worked to some extent by forging and rolling, provided it is encased in a more malleable metal such as steel.   Cast billets can also be extruded at high temperatures inside a steel jacket.   Pieces of beryllium can be joined by welding or brazing, but the technique is complicated by the tendency for beryllium to oxidize in air.

7.80. The development of powder-metallurgical processes has represented an important advance in the fabrication of beryllium.   The ingots of vacuum-cast beryllium are first chipped and then reduced to fine powder in an attritioning mill in the absence of air. The powder is compacted into the desired shape under vacuum at temperatures in the vicinity of 1000°C (1830°F) and pressures of about 500 psi. Pressings with good mechanical properties (see Table 7.10), weighing up to several hundred pounds, have been produced in this manner. The hot-pressed material can be machined fairly easily and long holes of small diameter can be bored in it.

7.81. Although the hot-pressed beryllium is frequently used without any further treatment, other than machining to size, it can be hot-extruded at 800° to 1100°C (1470° to 2010°F) or warm-extruded at between 300° to 500°C (570° to 930°F). As stated earlier, the resulting products exhibit an increase of ductility in the direction of extrusion.

7.82. The extreme toxicity of beryllium and its compounds complicates production procedures. Although exposure is not usually fatal within a short time, acute irritation of the respiratory tract and lung disease (berylliosis) may result. Beryllium metal is much less of a hazard in this connection than are many of its compounds which form dusts.

Corrosion

7.83. Fresh surfaces of beryllium metal soon tarnish upon exposure to air, indicating the formation of a thin oxide film. This film, however, retards further oxidation. Consequently, in massive form, beryllium does not suffer serious attack in air below 600°C (1110°F), although the powdered metal is oxidized at lower temperatures. Above 600°C (1110°F) the oxide is first formed, and then the nitride (Be_3N_2) at about 1000°C (1830°F). The use of beryllium at temperatures in excess of 600°C (1110°F) in the presence of air is thus not recommended. It is of interest to mention that at high temperatures beryllium is much less readily attacked by oxygen than is titanium or zirconium.

7.84. Highly variable results have been obtained in tests of the corrosion of beryllium in water. The high rates of attack observed in earlier work were undoubtedly due to the presence of impurities in the metal. With the availability of purer material of high density, it has been found that the corrosion rate of beryllium in deaerated, deionized water at 320°C (600°F) is relatively small and actually tends to decrease with time. This effect is probably due to the formation of an adherent film of oxide on the surface of the metal.

7.85. The resistance of beryllium to molten sodium is good in the absence of oxygen. As is the case with water, the results have varied with the nature of the material being tested. It has been reported that beryllium is not seriously attacked by flowing sodium in an Inconel (nickel alloy) system at temperatures up to 650°C (1200°F).

BERYLLIUM OXIDE [24]

7.86. Beryllium oxide is a ceramic moderator material which has found some use in reactors operating at very high temperatures. One reason why it has not been more widely employed is that when hot it vaporizes in moist air due to formation of the hydroxide. The advantages of beryllium oxide, apart from its good moderating and nuclear properties, are high melting point (2550°C or 4620°F), low vapor pressure in a dry atmosphere, and excellent thermal-shock resistance for a ceramic material. Its resistance to large integrated neutron fluxes has not been established.* The theoretical density is 3.025 g per cm³ at ordinary temperatures, but fabricated pieces generally have densities in the range of 2.7 to 2.8 g per cm³, although higher values have been reported. Beryllium oxide pieces are fabricated either by a cold process, involving cold-pressing, extrusion, or slip-casting into shape, followed by sintering, or by a hot process in which forming and sintering are combined in one step.

7.87. Some of the physical and mechanical properties of beryllium oxide pieces are quoted in Table 7.11. As with nearly all ceramics, the tensile strength decreases with increasing temperature (10,000 psi at 1000°C or 1830°F), as also does the compressive strength (about 40,000 psi at 1000°C). The thermal conductivity is exceptionally high for a ceramic, but at 1000°C (1830°F) it has decreased to about one-tenth of the value in Table 7.11.

TABLE 7.11.   PROPERTIES OF SINTERED BERYLLIUM OXIDE

Thermal expansion coefficient, per °C	1×10^{-5}
Thermal conductivity, cal/(sec)(cm²)(°C/cm)	0.60
Tensile strength, psi	15,000
Compressive strength, psi	114,000
Modulus of elasticity, 10⁶ psi	39
Poisson's ratio	0.34

ORDINARY WATER [25]

7.88. Ordinary water is attractive as a moderator because of its low cost, its excellent slowing down power, and its small migration length for thermal neutrons. On the other hand, its absorption cross section for neutron capture is relatively high, so that a critical system can be attained with water as moderator only if uranium enriched in the fissile isotope is used as fuel. However, if enriched material is available, the small migration length of thermal neutrons in water makes it possible to design a reactor of relatively small size. The fact that water can be used as both moderator and coolant, as it is in several reactors, is an added advantage. The water must be free from impurities since these not only capture neutrons but they may become radioactive as a result of (n, γ) reactions, thus constituting a hazard in the cooling system. Water of a high degree of purity is also desirable in order to minimize corrosion and scale formation. The decomposition of water by nuclear radiations (§ 7.93), with the lib-

---

* It is possible that the $(n, 2n)$ reaction with beryllium may cause damage, partially as a result of the formation of helium gas.

eration of gaseous hydrogen and oxygen, is also often facilitated by the presence of ionic impurities.

7.89. One of the main drawbacks to the use of water as a moderator is its relatively low boiling point. This means that, if it is to be utilized in high-temperature reactors, the pressures must be high, thus introducing fabrication and construction problems. Pressures in the vicinity of 2000 psi are used in pressurized-water power reactors, although the values are somewhat lower in boiling-water systems.

HEAVY WATER [25]

7.90. Because of its satisfactory slowing down power and exceptionally high moderating ratio (Table 3.3), heavy water is an excellent moderator and reflector. Under equivalent conditions, both the resonance escape probability and the thermal utilization are larger than when graphite is the moderator. Consequently, as seen in Chapter 4, a heterogeneous natural-uranium reactor is smaller and requires considerably less fuel if heavy water is the moderator rather than graphite. In fact, a homogeneous system of natural uranium and heavy water can become critical, whereas such a mixture of natural uranium and graphite cannot (§ 4.72). Ample supplies of heavy water are now available, but the cost is still high.*

7.91. Pure deuterium oxide melts at 3.82°C (38.87°F) and boils at 101.42°C (214.56°F), the values being a few degrees above the corresponding temperatures for ordinary water. High-temperature systems involving heavy water, like those using ordinary water, would therefore require high pressures. The density of heavy water is about 1.10 g per cm^3 at room temperature.

7.92. The economic incentive for the use of heavy water as moderator has not yet been established. The improvement in fuel-cycle cost, e.g., because natural uranium can be used as fuel, as compared with reactors in which ordinary water is the moderator, appears to be approximately the same as the additional cost of the heavy water. Some possible advantage might be achieved by the increased burnup that may be expected in the spectral shift system in which heavy water is gradually replaced by ordinary water during the operational period of the reactor (§ 13.39).

RADIATION DECOMPOSITION OF WATER [26]

7.93. Because water, either in the ordinary form or as heavy water, is used as moderator, solvent, or coolant in a number of reactors, the action of nuclear radiations is of particular interest. The discussion given below usually applies equally to all isotopic forms of water, i.e., H_2O, HDO, or D_2O, or to mixtures;

* The cost of heavy water (99.75 per cent D_2O) was $28 per lb in 1962.

wherever reference is made to "water" or to "hydrogen" any of the isotopic forms may be implied according to circumstances.

7.94. Upon exposure of water to nuclear radiations, hydrogen and oxygen gases are liberated, and hydrogen peroxide can be detected in the aqueous phase. If the water is contained in a closed system, the gas pressure increases until a steady-state value is attained. The magnitude of this steady-state pressure depends on the nature of the radiation, its density (or flux), the temperature, and the presence of various dissolved substances in the water. In general, the steady-state gas pressure increases with the radiation flux but decreases with increasing temperature. Many dissolved impurities also bring about an increase in the steady-state pressure. The concentration of hydrogen peroxide in the solution is decreased by raising the temperature and by the presence of impurities.

7.95. Slow neutrons probably do not cause direct decomposition of water. However, as a result of radiative capture by (ordinary) hydrogen atoms, gamma rays of 2.21-Mev energy are produced, and these are capable of decomposing the water.* The effect of fast neutrons is probably to be attributed to the high-energy recoil protons produced by the collision of the neutrons with hydrogen nuclei.

7.96. Ionizing radiation apparently causes water to break up into hydrogen atoms and free hydroxyl radicals; thus,

$$H_2O = H + OH$$

This decomposition probably does not occur directly, but the intermediate stages are of no significance here and so will be omitted from consideration. Subsequently the H and OH may either recombine to form water, i.e.,

$$H + OH = H_2O,$$

in which case there is no net decomposition, or they may combine in pairs, i.e.,

$$H + H = H_2$$

and

$$OH + OH = H_2O_2,$$

to yield the observed products, hydrogen and hydrogen peroxide, respectively. If the specific ionization of the radiation is low, as it is with gamma rays, the H and OH formed by the decomposition of a single water molecule will be much closer together than the H and H (or OH and OH) from two different water molecules. Consequently the tendency for recombination to occur will be greater than for radiation having a high specific ionization.

7.97. If part (at least) of the hydrogen peroxide remains undecomposed, the gas produced by radiation decomposition (radiolysis) of water will contain more hydrogen than the normal atomic ratio of two atoms of hydrogen to one of oxy-

---

* This secondary decomposition of water due to slow neutrons would be much less marked in heavy water because of the low neutron capture cross section of deuterium.

gen. As the temperature is raised, the peroxide decomposition into water and oxygen increases and the ratio of hydrogen to oxygen approaches 2 to 1. Impurities which catalyze the decomposition of hydrogen peroxide have the same general effect.

7.98. The steady-state gas pressure mentioned above arises from a balance between the rate of gas formation by the action of radiation and its removal by a chain reaction involving hydrogen atoms and hydroxyl radicals. Increase in the radiation density will be expected to increase the steady-state pressure. On the other hand, impurities present in solution may react with hydrogen atoms and hydroxyl radicals and may destroy them. As a result the reverse (chain) reaction is inhibited and the steady-state pressure is increased.

7.99. It is inevitable that in any reactor using either ordinary or heavy water as moderator, reflector, or coolant there should be some decomposition into hydrogen (or deuterium) and oxygen due to the action of various radiations. Provision must, therefore, be made for the removal or recombination of these gases. The gas evolution is particularly extensive in aqueous homogeneous reactors, because the fission products are released within the solution, and there is danger of an explosion. In pressurized- and boiling-water reactors there is little risk of such an explosion, but the corrosive action of oxygen in the water presents a serious problem. It is of interest to mention that, when operating at a thermal power of 20 Mw, radiolysis in the Experimental Boiling Water Reactor (EBWR) resulted in the loss of 62 lb of water per day, most of which was subsequently recovered in the recombiner.

7.100. In homogeneous reactors, a recombining system is generally used to dispose of the radiolytic gases. This contains a catalyst consisting of alumina pellets coated with finely divided platinum. A recombiner of this type is also employed in the EBWR. In other boiling water reactors, however, the radiolytic gases are removed with the "noncondensibles" through the condenser vacuum jet. In the Pressurized Water Reactor, the oxygen in the cooling water is kept to a low value (less than 0.14 ppm) by injecting hydrogen gas. At the existing high temperature and pressure, the presence of excess hydrogen leads to reaction with the oxygen under the influence of radiation.

ORGANIC MATERIALS

7.101. Organic polyphenyls which have high-temperature stability are used as moderators (and coolants) in the organic-moderated reactors (§ 13.41). The material generally employed for this purpose is a commercial product consisting mainly of a mixture of the isomeric polyphenyls (§ 6.204). The moderating characteristics of the polyphenyls are quite similar to those of ordinary water; the numbers of hydrogen atoms per unit volume are not very different and carbon, like oxygen, has a small capture cross section. As seen earlier (§ 7.29), the polyphenyls are relatively stable under irradiation; nevertheless, radiolytic (and pyrolytic) damage does occur. For this reason, some reactor concepts, while retaining the organic material as coolant, utilize a different moderator.

ZIRCONIUM HYDRIDE [27]

7.102. An unusual, but interesting, moderator is zirconium hydride; its nominal formula is ZrH_2, but the hydrogen content is frequently less than this. It is prepared by heating zirconium metal in hydrogen gas at about 350°C (660°F). When fully hydrided, the hydrogen atom density approaches that in water. Consequently, since the neutron capture cross section of zirconium is small, the hydride has good moderating properties. In powder form, the material has poor thermal conductivity and hence requires special provision for cooling in reactors operating at moderate or high power levels. Zirconium hydride, which is quite stable at temperatures below about 540°C (1000°F), has been considered as a possible moderator for organic cooled reactors.

7.103. Since it has been used as the moderator in the HTRE-3 (Heat Transfer Reactor Experiment) of the Aircraft Nuclear Propulsion program and is employed for the same purpose in the TRIGA (Training, Research, Isotope, General Atomics) reactors (§ 13.162), it may be presumed that the technology of zirconium hydride is reasonably well developed. In the TRIGA reactors, the hydride is incorporated with the uranium in the fuel elements. The thermal conductivity of the material is said to approach that of massive zirconium metal. In the fuel element, the hydrogen (moderator) forms an intimate mixture with the uranium (fuel) and so the moderator temperature follows the fuel temperature without delay. The resulting prompt (negative) temperature coefficient confers an unusual degree of stability upon the TRIGA reactor system.

COOLANT CIRCUIT

CORROSION AND EROSION [28]

7.104. The choice of coolants from the standpoint of their heat-transfer characteristics was discussed in Chapter 6. The technology of handling the coolant, however, is closely related to the selection of materials for the reactor system, and some aspects of the problems encountered will be considered here.

7.105. Corrosion and erosion problems are often unusually serious in the coolant circuit, because of the high temperatures (and sometimes pressures) involved. Furthermore, corrosion rates which would be negligible under other circumstances may become significant in a reactor system. The reason is that the coolant circuit is generally a closed one, so that corrosion products, which may have become radioactive as a result of neutron capture, accumulate. The intense radiation field may also have some influence on the choice of materials for use within the reactor itself. In many instances irradiation appears to have little effect on the corrosion rates at the existing conditions. However, since irradiation often produces surface defects, there is a possibility that these could affect the rate of interaction with the coolant. Studies of corrosion mechanisms may help to explain the role of irradiation and surface defects.

7.106. Although mention has been made earlier in this chapter of the effects of coolants on fuel and cladding materials, the general problem of corrosion resistance has not been treated in detail since it is one that is common in chemical process industry and is not restricted to nuclear engineering applications. It should be noted that corrosion may arise from the galvanic action between two different metals or a metal and a nonmetallic conductor, e.g., graphite, when in contact in an aqueous medium or a fused salt. Even in a single metal, stresses, strains, and other defects produced during fabrication, or the presence of oxide or other inclusions can facilitate corrosion. When there are temperature differences in a circulating system, e.g., of liquid-metal coolant, especially at high temperatures, corrosion of the mass-transfer type can occur. Metal can be dissolved by the circulating liquid in regions where the temperature (and metal solubility) is high and deposited in the cooler parts of the system, where the solubility of metal in the coolant is low. With gaseous coolants, other than helium, chemical reactions may occur and, in addition, there may be erosion as a result of the high flow rates which are necessary for adequate heat removal.

7.107. In pressurized-water reactors corrosion is controlled by appropriate water chemistry, i.e., by maintaining a low oxygen content (0.05 ppm) and a high pH. Such corrosion product ("crud") that does form tends to deposit on the fuel-element surfaces. In the EBWR, where the dissolved oxygen concentration is higher (0.24 ppm), because of greater radiolytic decomposition of the water, a 5-mil deposit of crud accumulated on the fuel plates in two years of operation. This deposit was found to be rich in aluminum and nickel, so that part, at least, originated from the dummy fuel elements, made of an aluminum-1 per cent nickel alloy, which were later removed. Nevertheless, the problem of crud deposition on the fuel elements must not be overlooked.

CHOICE OF MATERIALS

7.108. Containment materials for high-temperature water have been the subject of considerable development in the power industry before the advent of nuclear reactors. Stainless steels are normally employed and corrosion is controlled by maintaining purity of the water, as well as low dissolved oxygen content and moderate alkalinity, as indicated above. Although corrosion rates are small, some corrosion, indicated by crud formation, does occur. The possibility of stress corrosion cracking exists in stainless steels and must be considered in reactor design.

7.109. Stainless steel is usually also specified for sodium-coolant systems. Mass transfer and deterioration in the properties of the steel, mentioned in § 7.50, may be significant at temperatures above about 650°C (1200°F).   Decarburization of ferritic steels and carburization of austenitic steels can occur in sodium at temperatures as low as 540°C (1000°F).   Consequently, systems containing steels of these two types must be carefully designed to avoid transfer of carbon from the ferritic to the austenitic steel.   For operation at higher temperatures, the metals niobium, titanium, and vanadium may be considered as alternatives to stainless steel.   Since the presence of oxygen (as oxide) in the sodium invariably increases the corrosion rate, steps are taken to keep the oxygen content low.   In stainless-steel systems the method of "cold-trapping" is generally adequate.   The cold trap is a region, usually in a bypass of the main coolant circuit, which is maintained at a lower temperature, e.g., about 150°C (300°F).   Because the solubility of sodium oxide in sodium is very low in the cold region, it deposits as a solid and is removed from time to time.   In some situations, e.g., for niobium, titanium, or vanadium circuits, the cold trap does not reduce the oxygen content of the sodium to a sufficient extent to inhibit corrosion.   The use of a "hot trap" is then advisable.   This contains a metal, such as zirconium, which reacts chemically with the sodium oxide at a temperature of 600° to 700°C (1100° to 1300°F) and reduces it to sodium [29].

7.110. In the case of high-temperature, gas-cooled reactors, the possibility of chemical reactions occurring, in the high radiation field, with the materials selected for moderator, fuel, and cladding constitutes a major design problem. For reactors operating at very high temperatures, such as might be used for nuclear rocket applications, this factor becomes a principal design limitation on performance [30].

CHEMICAL REACTIONS [31]

7.111. From the safety point of view, considerable attention in reactor design has been devoted to the hazard arising from the possibility of a chemical reaction between the metallic components of the reactor core and the cooling water. Materials such as uranium, aluminum, and zirconium, in particular, especially when molten, react vigorously with water with the release of large amounts of heat.   This might occur in the event of local overheating followed by melting of fuel or cladding material.   Since the result of such a reaction would be the release of a large amount of fission products, it could be catastrophic.   It is important, therefore, to consider carefully possibilities of this kind in the selection of reactor materials.

7.112. In sodium-cooled power reactors, a significant chemical hazard might be the reaction between sodium and high-temperature water in the event of a failure in the steam generator.   In general, the latter operates in a secondary (nonradioactive) coolant circuit (§ 11.10) and there is little danger of the spread of radioactivity should an accident occur.   In some designs an intermediate fluid such as mercury, which is inert to water, has been used in the steam generator to minimize the danger of interaction between the water and sodium.

SYMBOLS USED IN CHAPTER 7

A mass number

c velocity of light

$\bar{E}$ average energy transferred per struck atom

E_γ gamma-ray (photon) energy

E_d energy of displaced atom

E_i limiting ionization energy

E_n fast-neutron energy

E_r recoil energy

G number of molecules reacting per 100 ev of radiation

N_H number of hydrogen atoms per cm^3

v recoil velocity

μ gamma-ray attenuation coefficient

μ_e gamma-ray energy absorption coefficient

σ_a absorption cross section

σ_c capture cross section

σ_s scattering cross section

ϕ_n fast-neutron flux

ϕ_{tn} thermal-neutron flux

REFERENCES FOR CHAPTER 7

1. See, for example, W. D. Wilkinson and W. F. Murphy, "Nuclear Reactor Metallurgy," D. Van Nostrand Co., Inc., Princeton, N. J., 1958; B. Kopelman (Ed.), "Materials for Nuclear Reactors," McGraw-Hill Book Co., Inc., New York, 1959; A. P. McIntosh and T. J. Heal (Eds.), "Materials for Nuclear Engineers," Interscience Publishers, Inc., New York, 1960; C. R. Tipton (Ed.), "Reactor Handbook," Vol. I, Materials, Interscience Publishers, Inc., New York, 1960; H. Etherington (Ed.), "Nuclear Engineering Handbook," McGraw-Hill Book Co., Inc., New York, 1958, Section 10; "Symposium on Reactor Core Materials," *Nuc. Sci. and Eng.*, **4**, 357 *et seq.* (1959); "Materials in Nuclear Applications," A. S. T. M. Special Publication No. 276 (1960); S. J. Paprocki and R. F. Dickerson, *Nucleonics*, **18**, No. 11, 154 (1960). The *Proc. (First) U. N. Conf. Geneva*, Vol. **8** (1956) and *Proc. Second U. N. Conf. Geneva*, Vols. **5** and **7** (1958) contain many reports on reactor materials. The quarterly U. S. AEC publication *Reactor Core Materials*, the name of which was changed to *Reactor Materials* in 1962, provides regular technical progress reviews in the field.
2. "Radiation Effects Review Meeting," U. S. AEC Report TID-7515 (1956); "Status of Radiation Effects Research, etc.," U. S. AEC Report TID-7588 (1958); "Symposium on Radiation Effects on Materials," Vols. 1, 2, and 3, A. S. T. M. Special Publications No. 208 (1956), No. 220 (1957), No. 233 (1958); G. J. Dienes and G. H. Vineyard, "Radiation Effects in Solids," Interscience Publishers, Inc., New York, 1957; J. J. Harwood, *et al.* (Eds.), "The Effects of Radiation on Materials," Reinhold Publishing Corp., New York, 1958; H. Etherington, Ref. 1, pp. **10**-83 *et seq.*, pp. **10**-126 *et seq.*; C. R. Tipton, Ref. 1, Chs. 2 and 52; D. S. Billington and J. H. Crawford, Jr., "Radia-

tion Damage in Solids," Princeton University Press, Princeton, N. J., 1961; D. S. Billington, *Nucleonics*, **18**, No. 9, 64 (1960); D. O. Leeser, *ibid.*, **18**, No. 9, 68 (1960); see also, *Reactor Core Materials*, Ref. 1; and *Proc. (First) U. N. Conf. Geneva*, Vol. **7** (1956); *Proc. Second U. N. Conf. Geneva*, Vols. **5**, **7**, and **29** (1958).

3. F. Seitz and J. S. Koehler, *Proc. (First) U. N. Conf. Geneva*, **7**, 615 (1956); G. J. Dienes, *ibid.*, **7**, 634 (1956); see also Ref. 2.

4. G. J. Dienes and G. H. Vineyard, Ref. 2, p. 28.

5. R. M. Walker, *J. Nuc. Materials*, **2**, 147 (1960).

6. Personal communication from M. S. Wechsler, ORNL.

7. G. R. Hennig and J. E. Hove, *Proc. (First) U. N. Conf. Geneva*, **7**, 666 (1956).

8. A. J. Swallow, "Radiation Chemistry of Organic Compounds," Pergamon Press, Inc., New York, 1961; C. G. Collins and V. P. Calkins, U. S. AEC Report APEX-261 (1956); J. G. Carroll and R. O. Bolt, *Nucleonics*, **18**, No. 9, 78 (1960); J. G. Burr, *ibid.*, **19**, No. 10, 49 (1961).

9. C. G. Collins and V. P. Calkins, Ref. 8.

10. W. G. Burns, *et al.*, *Proc. Second U. N. Conf. Geneva*, **29**, 266 (1958).

11. C. D. Bopp and O. Sisman, *Nucleonics*, **14**, No. 1, 46 (1956).

12. *Power Reactor Technology*, **1**, No. 1, 32 (1957); A. B. McIntosh and T. J. Heal, Ref. 1, Ch. 7.

13. For requirements of reactor steels, see S. H. Fistedis, U. S. AEC Report ANL-6389 (1961); H. Beeghly, *Nuc. Sci. and Eng.*, **7**, 21 (1960); also "Materials in Nuclear Applications," Ref. 1.

14. M. H. Bartz, *Proc. Second U. N. Conf. Geneva*, **5**, 466 (1958); see also, D. R. Harries, *Nuclear Power*, **5**, No. 47, 97 (1960); **5**, No. 48, 142 (1960).

15. For review, see *Power Reactor Technology*, **3**, No. 4, 38 (1960).

16. B. Lustman and F. Kerze (Eds.), "The Metallurgy of Zirconium," McGraw-Hill Book Co., Inc., New York, 1955; C. A. Hampel, "Rare Metals Handbook," 2nd Ed., Reinhold Publishing Corp., New York, 1961, Ch. 34; C. R. Tipton, Ref. 1, Ch. 32; A. B. McIntosh and T. J. Heal, Ref. 1, Ch. 9.

17. L. S. Rubenstein, *Nucleonics*, **17**, No. 3, 72 (1959); S. Kass and W. W. Kirk, *Bettis Tech. Review, Reactor Technology* (Dec. 1961), p. 27.

18. R. K. Wagner and H. E. Kline, U. S. AEC Reports NAA-SR-3841 (1959) and NAA-SR-6431 (1961).

19. "Nuclear Graphite: Dragon Project Symposium, 1959," European Nuclear Energy Agency, 1961; R. E. Nightingale (Ed.), "Nuclear Graphite," Academic Press, Inc., New York, 1962; A. B. McIntosh and T. J. Heal, Ref. 1, Ch. 6; C. R. Tipton, Ref. 1, Ch. 43; "Conference on Industrial Carbon and Graphite," Soc. Chem. Ind., London, 1958; *Power Reactor Technology*, **2**, No. 2, 36 (1959).

20. See *Proc. (First) U. N. Conf. Geneva*, Vol. **7** (1956); *Proc. Second U. N. Conf. Geneva*, Vol. **7** (1958).

21. W. Watt, *et al.*, *Nuclear Power*, **4**, 86 (1959); D. A. Boyland, *G. E. C. Atomic Review*, **2**, 44 (1959); J. M. Conway-Jones, *Brit. Power Eng.*, **1**, 64 (June 1960).

22. A. R. G. Brown and W. Watt, "Industrial Carbon and Graphite," Soc. Chem. Ind., London, 1958, p. 86; see also, *Materials in Design Engineering*, **51**, No. 2, 16 (1960).

23. D. W. White, Jr., and J. E. Burke (Eds.), "The Metal Beryllium," American Society for Metals, 1955; C. A. Hampel, Ref. 16, Ch. 3; C. R. Tipton, Ref. 1, Ch. 44; A. B. McIntosh and T. J. Heal, Ref. 1, Ch. 8; G. E. Darwin and J. H. Buddery, "Beryllium," Academic Press, Inc., New York, 1960.

24. C. R. Tipton, Ref. 1, pp. 934 *et seq.*; "Proceedings of the Beryllium Oxide Meeting at ORNL, Dec. 1960," U. S. AEC Report TID-7602 (1961); R. E. Taylor, U. S. AEC Report NAA-SR-4905 (1960); J. D. McClelland, *et al.*, U. S. AEC Report

NAA-SR-6454 (1961); J. Lillie, U. S. AEC Report UCRL-6457 (1961); R. P. Shields, *et al.*, U. S. AEC Report ORNL-3164 (1962).
25. C. R. Tipton, Ref. 1, Ch. 42.
26. A. O. Allen, *et al.*, *J. Phys. Chem.*, **56**, 575 (1952); *Chem. Eng. Progress Symposium Series*, No. 12, **50**, 238 (1954); E. J. Hart, *et al.*, *Proc. (First) U. N. Conf. Geneva*, **7**, 593 (1956); C. R. Tipton, Ref. 1, pp. 878 *et seq.*; E. J. Hart, *Nucleonics*, **19**, No. 10, 45 (1961); A. O. Allen, "The Radiation Chemistry of Water and Aqueous Solutions," D. Van Nostrand Co., Inc., Princeton, N. J., 1961.
27. H. M. McCullough and B. Kopelman, *Nucleonics*, **14**, No. 11, 146 (1956); J. D. Gylfe, U. S. AEC Report NAA-SR-5943 (1962).
28. "Conference on Aqueous Corrosion of Reactor Materials," U. S. AEC Report TID-7587 (1960); "Liquid-Metals Corrosion Meeting," NASA Report NASA TN D-769 (1961); D. J. De Paul (Ed.), "Corrosion and Wear Handbook for Water Cooled Reactors," McGraw-Hill Book Co., Inc., New York, 1957; R. N. Lyon, "Liquid-Metals Handbook," U. S. Government Printing Office, Washington, D. C., 1952; C. R. Tipton, Ref. 1.
29. C. Starr and R. W. Dickinson, "Sodium Graphite Reactors," Addison-Wesley Publishing Co., Inc., Reading, Mass., 1958, pp. 166 *et seq.*
30. J. J. Newgard and M. M. Levoy, *Nuc. Sci. and Eng.*, **7**, 736 (1960).
31. See various issues of *Reactor (Core) Materials*.

PROBLEMS

1. A reactor component of stainless steel (density 7.80 g/cm^3) has the following composition in weight percentages: chromium 17.5, nickel 11.0, manganese 1.0, columbium (niobium) 0.75, carbon 0.065, the remainder being iron. What is the total rate of neutron absorption per sec per cm^3 of the steel when located in a thermal flux of 10^{12} neutrons/(cm^2)(sec)? What volume of a zirconium-tin alloy (density 6.50 g/cm^3), containing 2.5 per cent by weight of tin, would capture neutrons at the same rate?

2. Determine the rate of gamma-ray energy emission (in Mev per sec per cm^3) by the steel component referred to in the preceding problem (a) 1 hr, (b) 1 day, after removal from the reactor following a long exposure. (Only those isotopic species and radiations mentioned in Table 7.3 need be considered.)

Chapter 8

REACTOR FUELS

INTRODUCTION

THE FUEL CYCLE

8.1. In the production of useful energy from nuclear fuels, the reactor plant itself is only one part of a complex operation which involves the production of the fuel material, fabrication of the fuel elements, reprocessing of the spent fuel for recovery of fissile and fertile materials, and disposal of the radioactive wastes. The cost of the fuel, including fabrication, reprocessing, depletion, lease (or inventory), and transportation charges, ranges from 35 to 50 per cent of the cost of generating power and is, in general, comparable with the plant capital charges. Hence, the economics of nuclear power is markedly dependent upon operations involving the nuclear fuel, most of which are external to the reactor itself.

8.2. The path followed by the reactor fuel in its various stages is called the "fuel cycle" (Fig. 8.1). The original source of the uranium is from uranium ores which, after mining, are concentrated and shipped to a materials plant. The uranium is extracted and purified either as the nitrate or the hexafluoride, which is converted into various products for fabrication of fuel elements of various types. Since some enrichment of the fuel in the fissile isotope uranium-235 is required for most power reactors, a portion of the natural-uranium feed material, as the hexafluoride, is diverted through an isotope separation plant (§ 8.23).

FUEL MATERIALS

8.3. For the production of fuel elements, the uranium may be in the form of the metal or an alloy or as the dioxide or carbide. This is usually clad with stainless steel or with aluminum, magnesium, or zirconium alloys to protect the fuel from attack by the coolant, to prevent the escape of radioactive fission

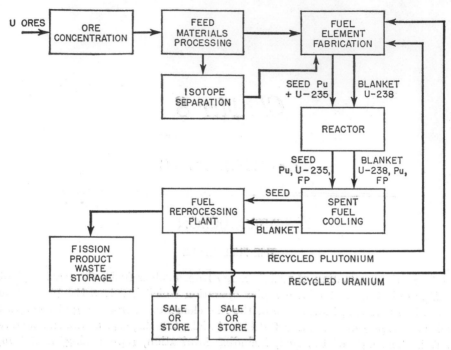

Fig. 8.1.   Reactor fuel cycle

products, and to provide geometrical integrity.   After discharge from the reactor, the spent fuel is highly radioactive ("hot") and so it is frequently allowed to "cool" before reprocessing, especially if solution methods are to be used. During the cooling period the activity decays to a level at which the radiation decomposition of the process chemicals is tolerable.   One of several possible separation methods is then utilized to remove fission products and to recover uranium, on the one hand, and plutonium-239 that has been formed from uranium-238 by neutron capture, on the other hand.   Where desirable, the uranium and plutonium can be recycled to the fuel-element fabrication plant so that they may be used to recharge the reactor.   Uranium which has been depleted in uranium-235 as a result of fission may be converted to the hexafluoride and returned to the isotope separation plant for re-enrichment.

8.4.   In this chapter consideration will be given to the basic principles involved in the methods employed for the production of reactor fuel materials and the reprocessing of the spent (irradiated) fuels.   The treatment is limited to such matters as are necessary for an understanding of the reactor system complex and which have a direct bearing on the cost of nuclear power.   For additional details the literature on the subject should be consulted, the appropriate references being given below.

8.5. In a reactor fuel there must be present at least one fissile species, e.g., uranium-233, uranium-235, or plutonium-239. These are derived from two naturally occurring source materials, namely, uranium and thorium. The former normally contains 0.72 per cent of the fissile uranium-235 and nearly all the remainder is uranium-238 from which plutonium-239 can be obtained by appropriate nuclear reactions (§ 1.134). The fissile uranium-233 is produced from thorium in a similar manner. Consequently, the consideration of reactor fuels will begin with an account of the sources and production of the elements uranium and thorium and some of their compounds.

PRODUCTION OF REACTOR FUELS [1]

SOURCES OF URANIUM

8.6. Uranium is by no means a rare substance; it has been estimated that it is present to the extent of about 4 ppm of the earth's crust. In fact, uranium is more abundant than relatively familiar elements such as silver, mercury, bismuth, and cadmium. The total weight of uranium in the earth's crust, to a depth of 12 miles, is probably of the order of 10^{14} tons. But this figure is apt to be misleading because most deposits are of such low grade, containing 0.001 per cent or less of uranium, that extraction of the metal would appear to be uneconomic.

8.7. Of the uranium sources available to the Western countries, only two consist of relatively high-grade ores, namely, the deposits in the Katanga province of the Congo and in Canada. The ore as mined contains from 1 to 4 per cent of uranium, chiefly in the form of one of the primary minerals *pitchblende* or *uraninite*. These are both oxides of variable composition ranging between UO_2 and U_3O_8, so that they are represented by the general formula $xUO_2 \cdot yUO_3$, the ratio y/x varying from zero to two.

8.8. Uranium ores of medium grade occur in many parts of the world, including the United States, Canada, and Australia. The principal deposits within the United States are those of the Colorado Plateau, an area which includes portions of Arizona, Colorado, New Mexico, and Utah. Although there are considerable variations in grade and composition, the majority of the ores mined in this region contain from 0.1 to 0.5 per cent of uranium. Many uranium minerals are found in the Colorado Plateau, but most of the production has been of secondary minerals, typically of the *carnotite* $(K_2O \cdot 2UO_3 \cdot V_2O_5 \cdot xH_2O)$ and *autunite* $(CaO \cdot 2UO_3 \cdot P_2O_5 \cdot xH_2O)$ families. Some deposits of primary uranium minerals also have been located and developed.

8.9. The carnotite ores of the Colorado Plateau often contain other vanadium minerals. These were formerly mined for vanadium, and the residues, sometimes referred to as "domestic ore concentrate," have been used as a source of uranium. At the present time both the uranium and vanadium are extracted.

If the demand is sufficient the value of the vanadium may compensate to some extent for the cost of recovering the uranium. Similarly, uraniferous copper ores, occurring in the White Canyon district near Hite, Utah, have been processed for both uranium and copper. The gangues in which uranium minerals are found vary widely; they include sandstones, limestones, clays, and asphaltites. The nature of the gangue and the character and grade of the uranium ore determine the type of processing that is applicable.

8.10. As the ores of higher uranium content are depleted, low-grade sources of the element will become of increasing significance. Among these the most immediately important are the South African (Rand) gold-ore residues, containing about 0.02 per cent of uranium, of which large amounts of fairly constant composition are available. The mining costs have already been assigned to the gold and, in addition, the pyrite contained in the residues can be converted into sulfuric acid for use in processing; the gangue consists mainly of inert quartz. Uranium production from the gold-ore residues has already been started and it is expected to provide important amounts of the element.

8.11. Low-grade potential sources within the United States, containing 0.01 per cent or less of uranium, include the extensive phosphate fields of Florida and Idaho, the Tennessee oil shales, and the uraniferous lignites of Wyoming and the Dakotas. From the phosphates, a small amount of uranium is recovered in the production of phosphoric acid. The extraction of uranium from lignite and shales does not appear to be economic at present. Nevertheless, from the long-range standpoint, the large amounts of uranium in these low-grade sources cannot be overlooked, and possible methods of uranium recovery are being studied.

8.12. In conclusion, it should be noted that very large quantities of uranium are present in sea water, but at the extremely low concentration of 1 to 2 parts per billion (10^9). Practicable recovery of materials at this concentration is beyond the reach of any presently known processing method.

8.13. The higher grade ores, containing 2 to 5 per cent uranium (as U_3O_8), are concentrated to about 50 per cent U_3O_8 by standard metallurgical procedures, including crushing, screening, washing, flotation, and gravity separation. The concentration step is usually carried out near the source of the ore and this makes possible economical shipment to central facilities for subsequent purification and conversion to oxide, fluoride, or metal.

8.14. Most uranium ores contain about 1.5 to 10 lb of U_3O_8 per ton, i.e., roughly 0.1 to 0.5 per cent of uranium, and these are concentrated by a chemical process. Several feed material processes have been developed, the method used being determined by the character of the ore. In general, the ore is first leached with acid or with alkali (sodium carbonate) depending on its nature. The uranium can be precipitated from the leach solution in one way or another but it has generally been found more satisfactory to remove the uranium by solvent extraction or ion-exchange techniques (§ 8.150). In one process, the clear leach liquor is passed through a fixed bed of anion-exchange resin. In the alternative ("resin-in-pulp") process the resin is suspended in baskets and passed directly through the leach pulp. The final product generally contains 70 to 80 per cent of U_3O_8 or its equivalent after precipitation and drying.

PRODUCTION OF URANIUM AND ITS COMPOUNDS [2]

8.15. The uranium to be used as a reactor fuel must be especially free from elements having appreciable cross sections for neutron capture. Consequently, the products of the ore concentration stages must be subject to further purification. For this purpose, two different processes are used in the United States: one involves solvent extraction of uranyl nitrate and the other the volatilization of uranium hexafluoride. In the solvent extraction process the finely ground uranium concentrate is digested with nitric acid, so that the uranium passes into aqueous solution in the form of the nitrate. The resulting slurry, without filtration, is fed into the top of an extraction column through which an organic solvent, tributyl phosphate diluted with an inert hydrocarbon, e.g., kerosene, flows upward. The uranyl nitrate is extracted into the organic medium and the solution is scrubbed with dilute nitric acid or water to remove small amounts of impurities. A stripping column follows in which water is used to back-extract the uranium into aqueous solution. This solution is evaporated until its composition corresponds approximately to that of uranyl nitrate hexahydrate, $UO_2(NO_3)_2 \cdot 6H_2O$, known as UNH. The purification process is similar to the "Purex" method for extracting uranium from irradiated fuels, described in § 8.144 *et seq.*

8.16. In the fluoride volatility process, the U_3O_8 concentrate is first ground and sized, so as to make the feed material suitable for treatment in fluidized beds. The sized feed enters a fluidized-bed reactor where it is maintained at a temperature of about 540° to 650°C (1000° to 1200°F) and reduced by hydrogen gas, formed by thermal dissociation ("cracking") of ammonia. The product, consisting mainly of uranium dioxide ("brown oxide"), UO_2, passes on to two successive hydrofluorination fluidized-bed reactors where interaction occurs with anhydrous hydrogen fluoride at temperatures of 480° to 540°C (900° to 1000°F) and 540° to 650°C (1000° to 1200°F), respectively. The reaction which takes place is

$$UO_2 + 4HF \rightarrow 2H_2O + UF_4,$$

so that uranium tetrafluoride ("green salt"), which is a nonvolatile solid of high melting point (about 960°C or 1760°F), is obtained.

8.17. In the next stage the tetrafluoride is treated with fluorine gas at about 340° to 480°C (650° to 900°F) to form uranium hexafluoride ("hex"); thus,

$$UF_4 + F_2 \rightarrow UF_6.$$

Although uranium hexafluoride is solid at ordinary temperatures, it sublimes above 56.4°C (133.5°F) at atmospheric pressure, and so it passes off from the fluorination reactor in the form of vapor. This vapor is condensed in cold traps in which the crude hexafluoride collects as a solid. In the final stage, the uranium hexafluoride is subjected to purification by fractional distillation at a pressure of 50 to 100 psig. The uranium hexafluoride obtained in this manner can be employed as the feed material in the gaseous-diffusion plants for the separation of the isotopes of uranium.

8.18. For the production of normal (unenriched) uranium dioxide or uranium metal, the starting material is generally the concentrated aqueous solution of uranyl nitrate from the solvent extraction process. This solution is heated at about 540°C (1000°F) in a denitrator in which the excess water is removed and the nitrate is decomposed to form uranium trioxide, UO_3 ("orange oxide"). The latter is then reduced in a fluidized-bed reactor at about 600°C (1100°F) by means of hydrogen obtained by cracking ammonia gas. The product is the pure dioxide, which can be used for the fabrication of fuel elements where unenriched material is required.

8.19. To obtain metallic uranium, the uranium dioxide is first hydrofluorinated by means of anhydrous hydrogen fluoride to form the tetrafluoride. The finely powdered tetrafluoride is then heated in a steel mold or bomb with either calcium or magnesium of high purity;* with magnesium the (exothermic) reaction is

$$UF_4 + 2Mg \rightarrow U + 2MgF_2.$$

The calcium or magnesium fluoride forms a slag on the surface of the uranium, from which it can be separated. In order to prevent entry of impurities into the uranium metal, the reaction vessel is lined with a refractory material, such as calcium fluoride (fluorite) or magnesium fluoride. The solid metal is then removed, melted (at 1300°C or 2370°F) in a graphite crucible under vacuum to expel volatile impurities, and poured into molds. The ingots so produced are suitable for various types of fabrication.

8.20. Uranium hexafluoride is an alternative starting material for the production of uranium metal or uranium dioxide. If the hexafluoride is unenriched or has been slightly enriched in uranium-235, it is converted to the tetrafluoride by heating in a tower with hydrogen, derived by cracking ammonia, at about 375°C (700°F); thus,

$$UF_6 + H_2 \rightarrow 2HF + UF_4.$$

The tetrafluoride is then reduced to uranium metal with magnesium, as described above. When the hexafluoride is a highly enriched product from the isotope separation plant, a modified procedure for conversion to the tetrafluoride is employed. Instead of heating the reaction vessel as a whole, the heat required to bring about the reaction between the hexafluoride with hydrogen is produced

---

* In the United States, the magnesium reduction process is now used almost exclusively, but calcium is frequently employed in Europe for uranium production.

by the highly exothermic reaction between hydrogen and fluorine. Hydrogen gas is first introduced into the vessel, then the fluorine gas, and finally the hexafluoride vapor. Solid uranium tetrafluoride is produced and is removed continuously to avoid the possibility of the formation of a critical mass.

8.21. In order to convert uranium hexafluoride, which may be unenriched or enriched, into the dioxide for use in fuel elements, the hexafluoride is hydrolyzed with a dilute solution of ammonia to form a precipitate of ammonium diuranate. This is filtered off, dried, and then heated in a mixture of steam and hydrogen, the latter being produced by the cracking of ammonia, to form uranium dioxide. The product is pulverized and then sintered at 1700°C (3100°F) to increase its density.

8.22. In the reprocessing of spent fuel elements, the uranium is recovered either as the nitrate, in the method most commonly used, or as the hexafluoride. The subsequent treatment will then be the same as described in the preceding paragraphs, according to the nature of the starting material and the desired product.

SEPARATION OF URANIUM ISOTOPES [3]

8.23. Partially enriched uranium, containing from 1 to 5 per cent uranium-235, is used as fuel in many power reactors, and even more highly enriched material is required for special designs. Several procedures for concentrating the uranium-235 were tested during World War II; of these, two are of special interest, namely, the *gaseous-diffusion process*, which was the one adopted, and the *centrifugal method*. The former process will be described more fully below, but the latter is worthy of brief mention because of its potential interest. The basis of the centrifugal method for separating isotopes is that if a gas or vapor containing molecular species having different masses is centrifuged, the gravitational force will bring about a partial separation, the heavier molecules moving toward the periphery and the lighter ones tending to remain near the center. The special feature of the centrifugal process is that the degree of separation (or separation factor) depends on the *difference* in the masses of the isotopic molecules, whereas in the gaseous-diffusion method the square root of the *ratio* of the masses is the determining factor. For a heavy element, such as uranium, the ratio is so close to unity (§ 8.26) that many stages must be used to obtain an appreciable separation of the isotopes. The fact that the masses of the two isotopes of uranium differ by 3 units should make the centrifugal method an effective one. The difficulty has been in the production of high-speed centrifuges suitable for a large-scale operation. Progress is being made in this connection, and if satisfactory machines are developed the whole problem of the separation of the isotopes of uranium may be greatly simplified.

8.24. The gaseous-diffusion method for the separation of the isotopes makes use of the different rates at which gases (or vapors) of different molecular weights diffuse through a porous barrier.† The barriers used in the uranium separation process contain hundreds of millions of pores per square inch, the average pore diameter being about two millionths of an inch. A lighter isotopic molecule will diffuse (or effuse) through such a barrier faster than a heavier one, so that a partial separation of the isotopes occurs. The gas which first passes through the barrier will be relatively richer in the lighter isotope, whereas that remaining will contain a correspondingly higher proportion of the heavier isotopic form.

8.25. For the separation of the isotopes of uranium by diffusion, the gaseous compound employed is the hexafluoride, UF_6. As stated in § 8.17, this substance is a solid at ordinary temperatures, but it has a relatively low sublimation temperature, so that it vaporizes readily. From the standpoint of gaseous diffusion, the use of uranium hexafluoride has two main advantages. First, the atomic weight of fluorine (19.0) is relatively low, so that the molecular weight of the diffusing gas, which is inevitably high because of the atomic weight of uranium, will not be too high. As will be evident shortly, the lower the molecular weight, the better the separation efficiency under given conditions. Second, since fluorine consists of a single species, the only isotopes separated in the diffusion process are those of uranium. The main disadvantages of uranium hexafluoride are its marked corrosive action on many metals and its ready interaction with moisture to form solid uranyl fluoride (UO_2F_2).

8.26. The extent of separation of isotopes in any process can be expressed by means of the *separation factor;* in the present case this is defined as the ratio of uranium-235 to uranium-238 atoms in the enriched state to that in the residual state. The rate of diffusion (or effusion) of a gas is inversely proportional to the square root of its molecular weight, and hence it can be shown that the theoretical separation factor, α^*, is given by

$$\alpha^* = \sqrt{\frac{M(\text{heavy})}{M(\text{light})}},$$

where M(heavy) and M(light) are the molecular weights of the heavier and lighter isotopic diffusion species, respectively. Natural uranium consists essentially of uranium-238 and uranium-235; there is a small proportion of uranium-234 (§ 1.14), but this may be ignored for the present purpose. Hence, for uranium hexafluoride as the diffusing gas, M(heavy) is $(238) + (6 \times 19) = 352$, whereas M(light) is $(235) + (6 \times 19) = 349$. The theoretical separation factor is consequently

$$\alpha^* = \sqrt{\frac{352}{349}} = 1.0043.$$

The theoretical *enrichment factor*, defined as $\alpha^* - 1$, is therefore 0.0043 for uranium-235 in uranium hexafluoride.

---

† The passage of gas through fine pores, small in comparison with the mean free path of the molecules, is properly called *effusion* rather than diffusion. The description "gaseous diffusion" for the isotope separation process is thus not strictly correct.

8.27. Since the separation factor is so close to unity, the degree of enrichment in any diffusion stage is very small, but the effect can be multiplied by making use of a "cascade" consisting of a number of stages, a few of which are indicated in Fig. 8.2. The feed gas, i.e., normal uranium hexafluoride in this case, enters at about the middle of the cascade. The diffusion stages above the feed are often referred to as the "enriching section," and those below as the "stripping section."* These terms imply that the gas moving into the upper stages becomes enriched in uranium-235, whereas that moving into the lower stages is depleted in (or stripped of) this isotope.

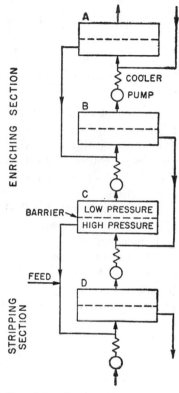

8.28. Consider the stage marked B; the gas enters from the bottom and about half is allowed to pass through the porous barrier. The diffusate, which is richer in the lighter isotope, then flows up into stage A. Here it undergoes further diffusion, so that the diffusate contains a still higher proportion of this isotope. Turning attention once again to stage B, it is seen that the gas which has not diffused, and which is consequently somewhat impoverished in the lighter species, is pumped to stage C where it joins the diffusate from stage D. The gas then undergoes diffusion in stage C, the diffusate going on to stage B, while the residual gas returns to stage D to be recycled. In this way the gas moving upward through the cascade becomes increasingly richer in the lighter isotope, whereas

Fig. 8.2.   Stages in   diffusion cascade

that moving downward contains increasing proportions of the heavier isotope.

8.29. By utilizing theoretical treatments and calculational methods similar to those developed for the study of fractional distillation, the minimum number of diffusion stages in the cascade required to convert natural-uranium hexafluoride into products containing various proportions of uranium-235 can be determined. Some results obtained in this manner are given in Table 8.1; x_p is the atomic fraction of uranium-235 in the product and x_w is that in the waste stream. The percentage recovery gives the proportion of the total uranium-235 present in the

---

* The terms "above" and "below," or "up" and "down," as used here and subsequently, are intended to have a purely figurative significance. The actual gas flow may be from right to left and vice versa. However, the use of "up" and "down" as directions permits a comparison to be made between a diffusion cascade and a fractionation column.

TABLE 8.1. CALCULATED MINIMUM NUMBERS OF STAGES IN DIFFUSION CASCADE FOR URANIUM ISOTOPE SEPARATION

x_p	$x_w = 0.005$	$x_w = 0.002$	$x_w = 0.001$
0.50	1240	1450	1610
0.90	1750	1950	2120
0.95	1920	2130	2290
0.99	2300	2520	2670
Recovery, per cent	30	72	86

feed material which appears in the product. It should be mentioned that the minimum numbers of stages quoted in the table are calculated by postulating that the quantity of product withdrawn from the cascade is negligible. In practice, of course, appreciable amounts would be drawn off and this would have the effect of increasing the number of stages required for any specified product composition.

8.30. It is evident that the production of material highly enriched in uranium-235 from natural uranium by gaseous diffusion of the hexafluoride involves several thousand stages. The largest numbers of stages are, in general, required for the higher degrees of enrichment; thus approximately the same number of stages is needed to increase the fraction of uranium-235 in the product from 0.50 to 0.90 as is necessary to raise it from 0.90 to 0.99. In the design of a diffusion plant, the increased enrichment and percentage recovery resulting from an increase in the number of stages and adjustment of the interstage flow rates must be weighed against the cost of constructing and operating the plant.

THORIUM [4]

8.31. The only ore mineral of thorium available in useful amounts is *monazite* and most commercial supplies are obtained from the monazite-bearing sands. The largest deposits are found in the state of Travancore, India, and there are also considerable quantities of monazite in the Blind River area of Ontario, Canada, and in Brazil; lesser amounts occur in Australia, Madagascar, South Africa, and the United States.

8.32. Monazite is essentially a mixture of rare-earth phosphates, containing 1 to 5 per cent of thorium dioxide, and a smaller proportion of uranium. The alluvial deposits, resulting from the weathering of granites, include a great deal of silica (sand), and these are treated by standard ore-dressing methods to concentrate the monazite. The resulting product may contain from 5 to 8 per cent of thorium, 0.15 to 0.25 per cent of uranium, and some 50 per cent of rare-earth elements, mostly in the form of phosphates. There are also variable amounts of silica as well as of iron, titanium, and other metals.

8.33. The extraction and purification of thorium as the nitrate is carried out in a manner analogous to that used for uranium. After various concentration steps, the thorium in nitric acid solution can be separated from the associated rare earths by multistage solvent extraction with tributyl phosphate. The nitrate is back-extracted with water and from the solution the thorium is precipitated as the oxalate. Upon heating, the dioxide remains and hydrofluorination converts this into thorium tetrafluoride. The latter is reduced to thorium metal by means of calcium. Since thorium has a high melting point (1750°C or 3180°F), zinc is introduced into the reduction system to form an alloy with thorium of relatively low melting point. When reduction is complete, the zinc is removed by heating the alloy in a vacuum.

PLUTONIUM [5]

8.34. Plutonium does not occur in nature to any significant extent, but is obtained by exposure of uranium-238 to neutrons in a reactor. In the processing of spent fuel (or breeder) elements, the plutonium is obtained as a solution of the nitrate in nitric acid. From this solution either plutonium peroxide (Pu_2O_7) is precipitated by means of hydrogen peroxide or plutonium (IV) oxalate by oxalic acid. Upon heating, the peroxide or oxalate yields the dioxide, PuO_2, which may be used as a reactor fuel material. To produce the metal, the peroxide, oxalate, or dioxide is heated with a mixture of hydrogen fluoride and oxygen when the tetrafluoride is obtained. The latter is then reduced with excess calcium in the presence of iodine; the "booster reaction" between calcium and iodine not only supplies additional heating but also introduces calcium iodide into the calcium fluoride slag and thus lowers its melting point.

PROPERTIES OF FUEL-ELEMENT MATERIALS [6]

INTRODUCTION

8.35. The materials used in the fabrication of fuel elements fall into two general classes, namely, the fuel material itself and the cladding; these will be described in turn. A satisfactory fuel material should have the following characteristics:

1. Its heat-removal properties should permit the attainment of a high power density (kilowatts per liter of core volume) and high specific power (kilowatts per kilogram of fuel) in the reactor. A high thermal conductivity is therefore desirable in order to reduce fuel temperature gradients.

2. The material should have a high resistance to irradiation damage. It will be seen in Chapter 14 that the fuel cost, and hence the cost of electricity, in a power reactor is greatly dependent upon the burnup, i.e., upon the amount of fissile material consumed before the fuel element has to be replaced for one reason or another. To obtain optimum burnup, the life of the fuel element in

the reactor should be determined by criticality considerations rather than by radiation effects.

3. The fuel material should be chemically stable, particularly with respect to the coolant, so that there would be little (or no) interaction as the result of a cladding failure.

4. The physical and mechanical properties of the material should permit economical fabrication.

URANIUM METAL [7]

8.36. Uranium, either natural or enriched, is the most common fuel, both as regards availability and use. It can be employed in the form of pure metal, as a constituent of an alloy, as the oxide, carbide, or other suitable compound. Metallic uranium was used as the fuel in most of the earlier reactors, largely because it provided the maximum number of uranium atoms per unit volume. However, because of its poor mechanical properties and great susceptibility to radiation damage, the use of pure uranium metal fuel elements in the United States is largely restricted to reactors for the production of plutonium. In Great Britain, however, the metal is employed as the fuel in the Calder Hall and similar gas-cooled power reactors.

Physical and Mechanical Properties

8.37. One of the difficulties associated with the application of uranium metal arises from its occurrence in three allotropic forms. These are called the alpha, beta, and gamma phases, respectively, and their characteristic properties are recorded in Table 8.2. The cell dimensions and the density, calculated from

TABLE 8.2. PROPERTIES OF THE CRYSTALLINE FORMS OF URANIUM

Property	Alpha Phase	Beta Phase	Gamma Phase
Stability range, °C.......	Below 665	665 to 770	770 to 1130 (m.p.)
°F.......	Below 1230	1230 to 1420	1420 to 2070
Crystalline form.........	Orthorhombic	Tetragonal	Body-centered cubic
Cell dimensions, A	(25°C)	(720°C)*	(805°C)
a_0...........	2.854	10.76	3.524
b_0...........	5.869	—	—
c_0...........	4.95	5.656	—
Density (calc.), g/cm³....	19.04	18.11	18.06
General characteristics...	Soft and ductile	Hard and brittle	Very soft

* The estimated values at ordinary temperatures are $a_0 = 10.59$ A, $c_0 = 5.634$ A; density $= 18.56$ g/cm³.

x-ray measurements, are for the indicated temperatures. The structures of the alpha and gamma phases are relatively simple. The alpha phase may be regarded as having a distorted hexagonal close-packed structure, whereas the gamma phase is body-centered cubic. The beta phase is somewhat more complex, the tetragonal unit cell containing 30 atoms.

8.38. The low symmetry of the orthorhombic system leads to considerable anisotropy in the alpha phase of uranium, as is indicated by the cell dimensions in Table 8.2. One consequence is that the thermal expansion is anisotropic, as may be seen from the approximate average coefficients of thermal expansion in the principal crystallographic directions recorded in Table 8.3; the temperature range is from 0° to 650°C (32° to 1200°F). Thus, while the dimensions parallel to the [100] and [001] directions increase, although to different extents, with increasing temperature, there is a decrease parallel to the [010] plane direction. Upon heating or cooling, therefore, crystals of the alpha phase of uranium undergo considerable distortion. Dimensional changes also result from irradiation effects, as will be seen later (§ 8.48).

TABLE 8.3.　COEFFICIENTS OF THERMAL EXPANSION OF ALPHA PHASE
OF URANIUM FROM 0° TO 650°C (32° TO 1200°F)

Direction	Average linear coefficient, per °C
a [100]...............	36.5×10^{-6}
b [010]...............	-8.6×10^{-6}
c [001]...............	32.9×10^{-6}

8.39. The structure of the alpha phase is not at all typical of a metal but resembles more that of the pseudo-metals, such as tellurium. This is manifested in the relatively low thermal and electrical conductivity of alpha uranium. The thermal conductivity is about 0.060 cal/(sec)(cm²)(°C/cm),* and the electrical resistivity is 25 to 30 $\times 10^{-6}$ ohm-cm at 25°C (78°F).† The values are undoubtedly dependent on the crystallographic direction, and the results quoted are to be regarded as approximate averages. The thermal conductivity increases somewhat with temperature, and is 0.10 cgs unit at 665°C (1230°F), the temperature at which transition to the beta phase occurs. The close-packed cubic structure of the gamma phase is characteristic of many metals, and it is of interest to record that the transition from beta to gamma (and also from alpha to beta) is accompanied by a sharp decrease in electrical resistivity. Curiously enough, there is no corresponding sudden increase in the thermal conductivity, the value increasing steadily through the transition temperatures.

8.40. Although the phase changes take place fairly sharply at the transition temperatures, i.e., 665°C (1230°F) for $\alpha \rightleftharpoons \beta$ and 770°C (1420°F) for $\beta \rightleftharpoons \gamma$, it is possible by suitable alloying to retain the beta and gamma structures at

---

* For conversion to engineering units, see Table A.1 in the Appendix.
† For purposes of comparison, it may be noted that for copper at ordinary temperatures the thermal conductivity is about 1 cal/(sec)(cm²)(°C/cm) and the electrical resistivity is 1.72 $\times 10^{-6}$ ohm-cm.

room temperatures. For example, the presence of a relatively small amount of chromium results in a stabilization of the beta phase, whereas certain solid solutions of uranium with molybdenum, niobium, titanium, or zirconium can retain the gamma structure at ordinary temperatures. These facts have been utilized in efforts to improve the properties of uranium fuel elements (§ 8.53).

8.41. The mechanical properties of metallic uranium are highly dependent on the condition of the specimen and its previous treatment. The values summarized in Table 8.4 may be regarded as roughly applicable to annealed (alpha

TABLE 8.4.   MECHANICAL PROPERTIES OF ANNEALED URANIUM

Hardness (Rockwell B)	90 to 115
Tensile strength, 10^3 psi	50 to 200
Yield strength (0.2% offset), 10^3 psi	25 to 130
Modulus of elasticity, 10^6 psi	15 to 25
Proportional limit, 10^3 psi	10 to 15
Poisson's ratio	0.2 to 0.25

phase) uranium at ordinary temperatures. The upper values for the hardness, tensile strength, and yield strength apply to the cold-worked metal. The Poisson ratio of uranium is seen to be low, in accordance with its "semiplastic" nature.

Fabrication

8.42. Uranium can be fabricated by conventional means, including casting, rolling, extrusion, forging, swaging, drawing, and machining. Hot-rolling of the alpha phase is a useful method for forming the metal. Because of the ease with which uranium oxidizes, especially at higher temperatures, it must be protected from air during fabrication, either by means of a fused salt or by an atmosphere of inert gas. The metal can be machined moderately easily, but precautions are required to prevent oxidation.

8.43. Satisfactory procedures for the production of uranium fuel elements have been developed using powder metallurgical methods, starting with uranium hydride which is converted into uranium by heating. A number of techniques are possible for forming the massive metal, e.g., cold pressing followed by sintering or hot pressing. In this manner high-density shapes, with a random orientation of grain structure, have been produced; these exhibit greater radiation stability than do fuel elements fabricated in other ways.

8.44. Uranium parts can be joined by welding or brazing. Fusion welding is achieved by using a Heliarc torch in an inert atmosphere. Brazing is rendered somewhat difficult by the formation of brittle alloys between the uranium and most of the common brazing alloys. Satisfactory results have been obtained by first plating the uranium with nickel or silver and then brazing the plated metal by conventional methods.

8.45. As an aid to fabrication, cold-worked uranium is frequently heat treated to prepare it for low-temperature forming operations. Annealing in the alpha phase, for example, can provide increased ductility. Some improvement in dimensional stability during irradiation can be achieved by heat treating wrought uranium fuel elements in order to transform them into the beta phase.

Corrosion

8.46. Uranium reacts with air, water, and hydrogen even at ordinary temperatures. The attack of bare uranium metal by water, with the evolution of hydrogen, is quite severe since, in the absence of air, no protective film is present to inhibit the reaction. Even when an oxide film is formed, it soon suffers local breakdown and then the bare areas corrode more rapidly than before. The corrosion rate varies greatly with the conditions, some of which appear to be quite trivial in nature. For this reason, a discussion of the observations on the corrosion of metallic uranium by water is meaningless without going into considerable detail. As an example, however, it may be mentioned that weight losses approaching 1 g per cm² of surface area of uranium have been found to occur in water at 100°C for 20 days.

8.47. Corrosion rates of uranium metal in oxygen-free sodium are quite small. Nevertheless, the chemical reactivity of uranium is such that a protective metal jacket (or cladding) is used to enclose the metallic fuel element when sodium is the coolant. If a leak should develop in the jacket, the interaction between the uranium and sodium would not be significant.

Irradiation Damage [8]

8.48. The dimensional changes which occur upon irradiation represent a drawback to the use of uranium metal fuel elements. The effects are of two types: (a) dimensional instability without appreciable change in density observed at temperatures below about 450°C (840°F), and (b) swelling, accompanied by a decrease in density, which becomes important above 450°C.

8.49. Dimensional instability is a distortion with little or no volume change largely due to the anisotropy of alpha-phase uranium. Studies made with a single crystal have shown that when irradiated there is considerable contraction in the direction paralled to the [100] plane, and almost equal expansion in the [010] plane, with little change along the [001] plane. Since uranium fuel elements fabricated from metal which has solidified from the liquid show preferential orientation of the alpha-phase crystals, it is not surprising that they exhibit nonuniform dimensional changes in a reactor.

8.50. One way in which the distortion under irradiation can be reduced is to use a material in which the crystal grains are small and randomly oriented. As a result, a partial compensation of the dimensional variations in the three crystallographic directions is achieved. Fuel elements prepared by powder metallurgical techniques have this desirable property, and so also do materials obtained by rapid cooling of the beta phase. In the latter case it appears that small amounts of specific impurities must be present. Another approach, as will be seen below, is based on the use of certain alloys.

8.51. At high temperatures, uranium fuel elements in a reactor may swell as the result of an indirect, rather than a direct, consequence of irradiation. The swelling is caused by the accumulation of minute bubbles of fission product gases within the metal. Appropriate alloying or metallurgical treatment may possibly reduce the degree of swelling by providing voids in which the gases can collect.

8.52. There is another reason why uranium fuel elements increase in volume during irradiation in a reactor. Not only do the solid fission product elements have lower densities than the uranium which has undergone fission, but two atoms of fission product are formed for every atom of uranium consumed. Some increase in dimensions is thus inevitable.

Uranium Alloys [7]

8.53. It is possible that the main drawbacks to the use of metallic uranium as fuel, namely, corrosion by water and dimensional instability upon irradiation, may be partly overcome by means of certain alloys. Two general concepts are employed: one is to dissolve relatively small amounts of a suitable metal, such as chromium, molybdenum, niobium, or zirconium, in the uranium so that there is a tendency to stabilize either the beta or gamma phase. When transformation to the alpha phase occurs, the product consists of small, randomly oriented crystals; it has better dimensional stability and is more corrosion resistant than ordinary, cast alpha-phase uranium. Although these alloys are attacked by water to some extent, the reaction would not be serious in the event of failure of the fuel-element cladding. Because the alloying elements, other than zirconium, have appreciable capture cross sections for thermal neutrons, the additions must be kept small.

8.54. Another method for improving the properties of uranium, which is more applicable to fuel elements for thermal reactors, is to add sufficient of the alloying metal to permit retention of the gamma phase. Since this has a cubic structure (see Table 8.2), it does not exhibit the anisotropic behavior which is largely responsible for the distortion of alpha-phase crystals upon irradiation. The presence of 10 per cent by weight of molybdenum, together with suitable heat treatment, gives a fuel material that has good stability when irradiated. This alloy was selected for use in the fast reactor of the Enrico Fermi Atomic Power Plant (§ 13.91).

8.55. In the pyrometallurgical process developed for the reprocessing of fast-reactor fuel, which is described in § 8.166 *et seq.*, the elements molybdenum, zirconium, ruthenium, rhodium, and palladium remain behind. It is of interest to note that this mixture, referred to collectively as "fissium," tends to favor retention of the gamma phase of uranium. Alloys containing artificial fissium have been found to exhibit dimensional stability when irradiated.

8.56. Alloys of aluminum with uranium have been used in the construction of fuel elements for a number of research reactors. The stability of such alloys to radiation is presumably due to the fact that the uranium is not present as the element, but in the form of a compound, e.g., UAl_2, UAl_3, or UAl_4. These alloys have a high thermal conductivity and are readily bonded to aluminum cladding to form a "sandwich" with the fuel layer in the center.

URANIUM DIOXIDE [9]

8.57. Uranium dioxide is an example of a ceramic that is finding extensive application as fuel material, especially in water-moderated power reactors. Ceramics have the advantages of high-temperature stability and general resistance to radiation. In addition, uranium dioxide is chemically inert to attack by hot water. It is this property which makes it attractive for use in water-cooled (including boiling-water) reactors, where the consequences of a cladding failure could be catastrophic if the fuel material reacted readily with the water at the existing high temperature. Another beneficial property of uranium dioxide is its ability to retain a large proportion of the fission gases even at relatively high degrees of burnup. The major disadvantage of uranium dioxide as a fuel material is its low thermal conductivity, although this is partially offset by the fact that very high temperatures are permissible in the center of the fuel element.

Production

8.58. The starting point for the production of massive uranium dioxide bodies is a powder which may be prepared from uranyl nitrate solution. The size and shape of the particles in the powder, which are somewhat dependent on the procedures used in its preparation, have a considerable influence on the properties of the final material. The latter are thus somewhat variable.

8.59. Conventional ceramic procedures, e.g., cold pressing, extrusion, slip casting, etc., have been used to form the uranium dioxide powder into pellets or other desired shapes. These are then sintered in a neutral or reducing (hydrogen) atmosphere at temperatures ranging from about 1300° to 2000°C (2400° to 3600°F). In addition, more unconventional methods, such as hot and cold swaging techniques and vibratory compaction of the powder, show promise. The main procedure used hitherto for the large-scale production of high-density compacts of uranium dioxide is cold pressing followed by sintering. Bodies having 95 per cent of the theoretical density (10.96 g/cm³) have been obtained in this manner. The production of material of maximum density is an important objective for several reasons. A high density means a large number of uranium atoms per unit volume, which is advantageous from the nuclear standpoint, and also a high thermal conductivity. Furthermore, retention of fission product gases is better at higher densities.

Properties

8.60. Some of the important properties of uranium dioxide are listed in Table 8.5. The low thermal conductivity, which decreases further as the temperature

TABLE 8.5.   PROPERTIES OF URANIUM DIOXIDE

Melting point.	2800°C (5100°F)
Crystal structure.	Face-centered cubic
Lattice parameter, A.	5.468
Theoretical density, g/cm³.	10.96
Thermal conductivity, cal/(sec)(cm²)(°C/cm).	0.02 (at 20°C)
Thermal expansion coefficient, per °C.	$\sim$1 $\times$ 10^{-5} (0 to 1000°C)
Fracture strength, psi.	$\sim$10,000
Modulus of elasticity, 10⁶ psi.	25

is raised, at least up to 1100°C (2010°F), is significant in reactor design, since it establishes a limitation in the power obtainable for a fuel rod of given size. In fact, the quantity $\int k(T) \, dT$, where k is the thermal conductivity and T is the absolute temperature, called the *thermal rating parameter* or the *rod power per unit length* and expressed in watts per cm, is often used to indicate the effectiveness of a given fuel-element material [10].   The lower integration limit is the surface temperature of the fuel and the upper limit is the maximum permissible temperature in the interior.

8.61. The variation with temperature, up to about 650°C (1200°F), of the thermal conductivity of sintered uranium dioxide prepared in two different ways is shown in Fig. 8.3; the results have been corrected to the theoretical density in each case.   An extruded specimen, which had been exposed to an integrated flux of 10¹⁹ neutrons per cm² in a reactor at a temperature below 150°C (300°F) had a somewhat lower thermal conductivity than before irradiation, but the curve in Fig. 8.3 indicates that there was some recovery upon heating.   There is a possibility that the thermal conductivity of uranium dioxide may increase at temperatures in excess of 1100°C (2010°F) because of the contribution made by radiation heat transfer at these high temperatures (cf. § 6.62).

8.62. Uranium dioxide must be maintained out of contact with air because it combines readily with oxygen to form a higher oxide, the composition of which depends upon the temperature.   As already mentioned, it does not react with water, and exposure to radiation has no influence on the interaction.   The inertness is not affected by the presence of small amounts of impurities such as carbon and silica.

8.63. Grinding procedures have been used to machine uranium dioxide to close tolerances.   However, inasmuch as this procedure is expensive, an alternative is to form the material into pellets and to provide helium gas bonding (§ 8.104) to improve the heat-transfer characteristics resulting from the presence of gaps between the fuel and the cladding in the fuel element.   Vibratory compaction

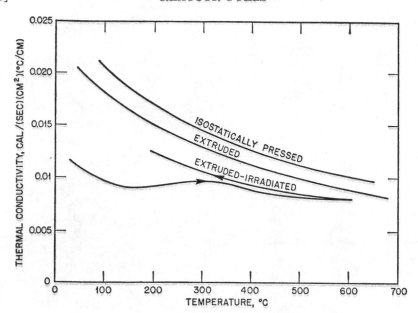

Fig. 8.3.   Thermal conductivity of uranium dioxide

of the uranium dioxide powder may provide good thermal contact with the cladding, so that the helium will not be necessary.

8.64. As a consequence of extended reactor irradiation, uranium dioxide fuel
elements undergo changes in some physical properties and strength characteristics [11].   The high temperatures and extreme thermal gradients, and accompanying stresses, within the fuel material cause grain growth, cracking, and
possibly melting in the interior (Fig. 8.4).   There are, however, no appreciable
dimensional changes.   The effects of irradiation have no significant direct influence on the functioning of the fuel element, provided it has adequate structural
support in the form of suitable cladding or canning.   Uranium dioxide is compatible with the materials commonly used for this purpose, e.g., zirconium alloys
and stainless steel.

8.65. Although cracking, melting, etc., have no direct adverse effect on the
operation of uranium dioxide fuel elements, they have an important indirect
consequence in facilitating the escape of fission product gases.   The resulting
pressures can cause disruption of the cladding.   Moreover, the gases decrease
the thermal conductivity of the helium bonding mentioned above, thus leading
to an increase in the temperature in the interior of the fuel.   This will cause
the release of more fission gases, so that the adverse effects are increased still
further.   The retention of fission product gases by uranium dioxide is favored
by high density and by a composition as close as possible to the stoichiometric

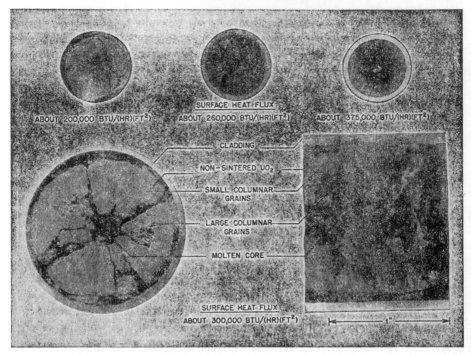

Fig. 8.4.   Cross sections of irradiated uranium dioxide fuel elements

value of two oxygen atoms to one uranium atom.   Products containing even a small excess of oxygen (§ 8.62) tend to release the gases more readily.

URANIUM CARBIDE [12]

8.66. Another ceramic of great potential interest, especially for reactors designed for operation at very high temperatures, is uranium monocarbide (UC). Although this material has not been developed to the extent of the dioxide, it would appear to have some advantages over the latter.   In addition to being stable at high temperatures, there is evidence that it has excellent dimensional stability upon irradiation.   The thermal conductivity of uranium carbide is appreciably greater than that of the dioxide and is, in fact, comparable with that of uranium metal at ordinary temperatures.   Unfortunately, there are indications that the thermal conductivity of the carbide, like that of the dioxide, decreases with increasing temperature.   The higher theoretical density (13.6 g/cm^3) of uranium monocarbide and the larger content of uranium, as compared with the dioxide, are also important in the neutronic aspects of reactor design. Uranium carbide seems to be less capable of containing fission product gases than does the dioxide, but its main drawback is its reactivity with water.   Although this may be overcome, as explained below, it is for this reason that the main interest of uranium carbide fuel elements at the present time is in connection with reactors using a gas (helium) or sodium as coolant. They may also find application in organic-moderated (and cooled) systems.

Production

8.67. Uranium carbide has been prepared in several ways, the simplest being to heat a mixture of uranium dioxide and graphite powders to a temperature of 1600° to 1900°C (2900° to 3450°F) in a vacuum. The product is then ground into powder by mechanical means. Another procedure, which yields powder directly, involves the preliminary formation of uranium metal powder by conversion to the hydride and subsequent decomposition by heat. Reaction of the uranium powder with a hydrocarbon, e.g., methane, at about 600°C (1100°F) then yields uranium monocarbide in powder form. Increase of density, to 85 or 90 per cent of theoretical, can be achieved by sintering or by hot or cold compacting. Products with almost the full theoretical density have been prepared by arc melting and casting.

Properties

8.68. The physical properties of uranium monocarbide are dependent, to a certain extent, on the method of production, and so some of the values summarized in Table 8.6 are approximate. Nevertheless, they do indicate the gen-

TABLE 8.6. PROPERTIES OF URANIUM MONOCARBIDE

Melting point....................................	2380°C (4300°F)
Crystal structure................................	Face-centered cubic
Lattice parameter, A.............................	4.961
Theoretical density, g/cm³......................	13.63
Thermal conductivity, cal/(sec)(cm²)(°C/cm)........	0.078 (at 44°C)
Thermal expansion coefficient, per °C...............	$\sim 1 \times 10^{-5}$ (20 to 1000°C)
Fracture strength, psi............................	9000
Modulus of elasticity, 10^6 psi.....................	31

eral characteristics of the material which may be compared with those of uranium dioxide in Table 8.5. In addition to its higher (theoretical) density and thermal conductivity, the mechanical properties of the carbide are superior to those of the dioxide. The bond strength of sintered uranium monocarbide at room temperature, for example, is about three times that of the sintered dioxide.

8.69. The higher thermal conductivity of the carbide means that, for fuel elements of a given size, higher specific powers are attainable in the reactor than are possible with uranium dioxide. Alternatively, thicker rods or plates of carbide could be used, at the same specific power, thereby achieving economy in a decreased requirement of cladding material per unit mass of fuel. The advantages gained in these respects from the increased thermal conductivity may, however, be partially offset by the somewhat lower melting point of the monocarbide. The indications are that uranium carbide fuel elements have good irradiation properties. There is little cracking and the release of fission-product gases is not serious. The presence of excess carbon in the stoichiometric composition (UC) does not appear to have any deleterious effects.

8.70. Uranium carbide has shown resistance to attack by sodium at temperatures up to 500°C (930°F) and in sodium-potassium alloy at 700°C (1290°F). Organic coolants affect the carbide to a small extent, but it is not certain whether this is significant or not. In powder form the carbide reacts with atmospheric oxygen, but the bulk material is much less reactive and can be stored and handled without difficulty. Of the usual reactor coolants, water interacts with uranium carbide most readily; the material may, therefore, have limited usefulness in pressurized- or boiling-water reactors. However, mixtures (or solid solutions) of uranium monocarbide with other carbides, e.g., molybdenum carbide (Mo_2C) and zirconium carbide (ZrC), appear to be much more resistant to attack by water. Cladding of uranium carbide fuel elements is not likely to be a problem for fuel temperatures up to 600°C (1100°F), but there are indications that the carbide reacts with many metals, including stainless steel, at 1000°C (1800°F).

DISPERSION FUELS [13]

8.71. Dispersion-type fuel elements consist of fuel material, generally a ceramic, such as uranium dioxide or monocarbide, dispersed in a matrix of a metal or a ceramic with desirable physical properties. Among the advantages of dispersion fuels are good integrity under irradiation, with the possibility of high burnups, increased thermal conductivity, and the metallurgical bonding of the cladding material, if the matrix is a metal, thereby improving heat-transfer characteristics. Dispersions of uranium dioxide in such metals as beryllium, aluminum, zirconium, molybdenum, and stainless steel have either been produced or are under development. Plate-type fuel elements consisting of the dioxide in stainless steel are used in some forms of the compact Army Package Power Reactors, designed for power production at remote military bases. Dispersion in nonmetallic materials will be considered in § 8.75 *et seq.*

8.72. In addition to providing good thermal conductivity, excellent resistance to radiation damage, and internal structural support, metal-dispersion fuel elements may be designed to whatever chemical resistance is necessary, since no alloying is required. Many combinations of materials are thus possible, the choice being determined by the particular reactor concept.

8.73. The most useful method for preparing the dispersion fuels is by powder metallurgy, since this provides the best control over the properties of the final product. Dispersions containing up to two-thirds by volume of the dispersed fuel material phase have been prepared. Because of the fuel dilution and neutron absorption by the matrix, dispersion-type fuel elements are limited to situations in which a reasonably high enrichment of uranium-235 is justified.

8.74. An unconventional type of metallic dispersion material which is receiving some attention is a "paste" fuel consisting of a dispersion of uranium metal or dioxide powder in molten sodium at settled density [14]. As in a solid dispersion, the sodium contributes a high thermal conductivity. By removing excess sodium, pastes having densities of approximately 50 per cent of the massive fuel are attainable. A modification of the paste fuel is the so-called "sponge uranium." It consists of metallic uranium powder packed to a density of 70 per cent by volume and infiltrated with liquid sodium [15]. It is hoped that such fuels will permit high burnup combined with low fabrication costs. An element of uncertainty, however, lies in the behavior of the fission product gases which may form thermally insulating bubbles or voids adjacent to the fuel.

8.75. Among possible nonmetallic dispersion media for fuel elements, graphite possesses many desirable properties [16]. Dispersions of uranium monocarbide in graphite can be prepared from the mixed powders or, more simply, by impregnating the graphite with uranyl nitrate solution and heating. The nitrate probably first decomposes to an oxide which then reacts with the graphite to form the carbide. Graphite loaded with uranium carbide is attractive as a high-temperature fuel element, since the physical and mechanical properties are very close to those of graphite, a material which has been used in high-temperature applications for many years (cf. § 7.64 *et seq.*). The graphite can also serve as the moderator in thermal reactors.

8.76. Although there are designs for high-temperature, gas-cooled reactors using unclad graphite-uranium carbide fuel elements, it is probable that in many cases precautions will be necessary to prevent the escape of fission product gases. Some form of cladding will thus be required. An interesting possibility in this connection is the use of the so-called impermeable graphite (§ 7.74) or pyrolytic graphite (§ 7.75).

8.77. Several other ceramic materials, mainly oxides and carbides, have been considered more or less seriously as matrixes for fuel elements intended for use at very high temperatures. Although ceramics have the drawback of being brittle under normal conditions, they tend to become more plastic as the temperature is raised. Their main disadvantage lies in the low thermal conductivities which generally decrease with increasing temperature. Nevertheless, ceramic materials have some interest for fuel-element construction where ability to withstand both high temperature and chemical attack is required. There are indications that, at ordinary temperatures, some ceramic oxides are more susceptible to radiation damage than are metals; however, it is to be expected that annealing will occur at high temperatures.

8.78. Apart from graphite, most of the work on ceramic-based fuel elements has been done with aluminum oxide, beryllium oxide, zirconium dioxide, silicon carbide (Carborundum), and zirconium carbide. Of these, alumina has the lowest thermal conductivity, but mixtures with uranium dioxide have excellent oxidation resistance, which the latter alone does not have, and ability to retain fission product gases. A corrosion- and irradiation-resistant fuel consisting of uranium dioxide (9 per cent), zirconium dioxide (82 per cent), and calcium oxide has been developed and tested in the Experimental Boiling Water Reactor (§ 13.24). Another ceramic of interest is silicon carbide which has been used to fabricate $SiC-UO_2$ fuel elements. They are resistant to oxidation and, in addition, have fair thermal conductivity and thermal-shock resistance.

8.79. Beryllium oxide is a good moderator and, in addition, has other properties which are as good as (or better than) those of other ceramic materials (§ 7.86). Consequently, it was one of the first ceramics to be considered as the matrix for dispersion-type fuel elements for operation at high temperatures. Early studies indicated that the thermal-shock resistance of the available material was not satisfactory, but the situation has now changed, so that serious consideration is again being given to beryllium oxide, both as a moderator and as a fuel-element matrix. A dispersion of uranium dioxide in beryllium oxide has been proposed for the fuel elements in the Maritime Gas Cooled Reactor (§ 13.73).

8.80. Uranium dioxide has been incorporated into silica and the higher oxide, U_3O_8, into various silicate glasses. An interesting form of the latter is a glass fiber, about 1 micron in diameter, containing up to 10 per cent by weight of uranium. Not only do these fibers have a large heat-transfer area, but the fission products escaping into the gaseous coolants can induce useful chemical reactions.

8.81. A special type of dispersion fuel that has been developed is one in which small particles of fuel material, e.g., uranium or uranium oxide, are coated individually with a thin, impervious layer of a metal, e.g., niobium, chromium, nickel, vanadium, molybdenum, etc., or a ceramic, e.g., graphite or alumina [17]. These small coated particles may be used directly in fluid-bed or similar reactors, or they may be consolidated by compacting, pressing, extrusion, or other techniques, or by incorporation into a metallic matrix. When consolidated, the particles constitute small islands of fuel in a continuous, interconnected matrix which provides (a) multiple barriers for corrosion resistance, (b) containment for fission products including gases, (c) restriction of expansion due to radiation damage, and (d) increased thermal conductivity resulting from the continuous metallic network.

PLUTONIUM FUEL MATERIALS [18]

8.82. Although fuel elements containing an appreciable proportion of plutonium 239 have been used hitherto in only a few reactors (or reactor concepts), this isotope must undoubtedly play an important role in some reactors of the future. Small percentages of plutonium-239 would be present, for example, in "recycle fuel," i.e., spent fuel material, originally natural or slightly enriched uranium, which has been treated only for the removal of fission products for re-use in a reactor. There are, however, two other more important ways in which plutonium-239 may be utilized. One is by insertion of a few plutonium fuel elements to provide increased reactivity in a thermal reactor core containing enriched (or slightly enriched) uranium as the main fuel. The other is to use the plutonium as the chief fuel material in a fast (breeder) reactor. The nuclear properties of plutonium-239 are somewhat inferior to those of uranium-235 in a thermal reactor, because of the higher capture-to-fission ratio in the former, but they are very much superior in a fast-neutron spectrum.

Properties

8.83. The physical metallurgy of plutonium is complicated by the fact that six allotropic forms, designated by the Greek letters α, β, γ, δ, δ', and ϵ, respectively, have been reported from room temperature to the melting point, which is only 640°C (1184°F). The stability ranges and some of the properties of the six solid phases are given in Table 8.7. The alpha phase, stable at ordinary tempera-

TABLE 8.7. PROPERTIES OF SOLID PHASES OF PLUTONIUM

Phase	Stability Range	Structure	Density (g/cm³)	Thermal Expansion Coefficient (per °C × 10⁶)
α	Below 120°C (248°F)	Simple monoclinic	19.7 (25°C)	48
β	120° to 206°C (248° to 403°F)	Centered monoclinic	17.8 (150°C)	38
γ	206° to 319°C (403° to 606°F)	Face-centered orthorhombic	17.2 (210°C)	35
δ	319° to 451°C (606° to 844°F)	Face-centered cubic	15.9 (320°C)	−9
δ'	451° to 476°C (844° to 889°F)	Body-centered tetragonal	16.0 (465°C)	−120
ϵ	476° to 640°C (m.p.) (889° to 1184°F)	Body-centered cubic	16.5 (510°C)	36

tures, was at one time thought to be isomorphous with alpha-uranium, but this is not the case. The delta and delta-prime phases of plutonium are remarkable in having negative temperature coefficients of expansion.

8.84. From the data in Table 8.7, it is evident that solid plutonium metal would be very unsatisfactory as a fuel material. The several phase changes, with the accompanying large variations in the thermal expansion coefficients, introduce serious problems of dimensional stability. Moreover, the melting point is relatively low, and the thermal conductivity is exceptionally small, about

0.01 cal/(sec)(cm²)(°C/cm) at ordinary temperatures (alpha phase). The metal is active chemically and reacts with oxygen, hydrogen, and water. Plutonium is a very serious biological hazard and so requires special fabrication facilities (§ 9.63).

Alloys

8.85. The general properties of plutonium metal are so poor that it must be diluted with an alloying material for reactor applications. In selecting suitable alloying elements, the nuclear properties of the latter must be kept in mind. As a guide, it has been suggested that the macroscopic neutron capture cross section of the diluent (other than uranium-238) should be less than 10 per cent of the macroscopic fission cross section of plutonium-239. The cross sections involved are those applicable to the neutron spectrum in the reactor, i.e., thermal or fast, according to circumstances [19].

8.86. Phase equilibrium studies have been made of a number of systems involving plutonium and another metal, but in relatively few cases has it been established that the alloy has possibilities for use as a reactor fuel. Binary plutonium-uranium alloys, which are, at first sight, an obvious choice have proved to be unsatisfactory; in particular they suffer extensive damage upon irradiation. However, the addition of suitable amounts of either molybdenum or the fissium mixture (§ 8.55) effects a remarkable improvement in radiation stability.

8.87. Aluminum alloys containing less than 13 weight per cent of plutonium, in which the latter is present as the compound $PuAl_4$, show considerable promise for use as fuel elements in both thermal and fast reactors. They have good thermal conductivity, a relatively low corrosion rate in water, and exhibit a uniform but small expansion when exposed to reactor radiation. Among other plutonium alloys which might prove useful as reactor fuels, especially for fast reactors, mention may be made of those containing iron, copper, or zirconium.

8.88. For fast reactors, certain alloys of plutonium with iron, cobalt, or nickel may be of special interest. These form eutectic systems, containing roughly 90 atomic per cent of plutonium, with melting points ranging from 410° to 465°C (770° to 870°F), so that reactor designs based on liquid fuel, as in the Los Alamos Molten Plutonium Reactor Experiment (LAMPRE), are possible. Because of their high plutonium content, the eutectic alloys have low thermal conductivities, so that large specific powers are not easily attainable. The addition of cerium, however, to the cobalt-plutonium eutectic results in a considerable improvement in thermal conductivity without too large an increase in the melting point.

Ceramics and Cermets [20]

8.89. Although the use of plutonium dioxide as a fast-reactor fuel material has been proposed, there is little information available concerning its value. It is known, however, that it is dissolved with great difficulty in nitric acid and so presents a special problem in the reprocessing procedure. Of greater interest, because of its applicability to fast breeder reactors, are the PuO_2-UO_2 solid solutions. In the ratio of five moles of uranium dioxide to one of plutonium dioxide, the solid solution has better sintering characteristics than uranium dioxide alone, is readily soluble in nitric acid, and has good radiation stability. Solid solutions of PuC-UC are also being considered as possible fuel materials for fast reactors.

8.90. Cermets (or dispersions) of plutonium dioxide in various metals and in graphite have been studied. The results reported are, however, too preliminary to permit definite conclusions to be drawn. Because of the similarity between the dioxides and carbides of uranium and plutonium, it is possible that useful plutonium fuel elements of the dispersion type may be developed.

THORIUM FUEL MATERIALS [21]

8.91. Thorium, either as metal or oxide, is finding increasing application as the fertile material in thermal reactors, in combination with a uranium fuel. The possibility of the conversion of thorium-232 to uranium-233, with an efficiency exceeding 100 per cent, can provide a marked improvement in the fuel-cycle economics.

Properties

8.92. Metallic thorium melts at about 1750°C (3180°F), and up to about 1400°C (2550°F) the solid has a face-centered cubic structure. The lattice dimension is 5.086 A and the theoretical density is 11.72 g per cm³ at ordinary temperature. Above 1400°C, a body-centered cubic form is the stable phase. The fact that thorium belongs to the cubic system means that its properties are isotropic. As is to be expected, therefore, its radiation stability is greatly superior to that of uranium. The thermal conductivity of thorium metal is 0.090 at 100°C (212°F) increasing to 0.108 cal/(sec)(cm²)(°C/cm) at 650°C (1200°F); these values are not very different from those of metallic uranium, especially at the higher temperature.

8.93. The mechanical properties of thorium are dependent on the method of production and previous treatment. The values given in Table 8.8 are averages applicable to the annealed metal at room temperature. By cold rolling, the tensile strength can be increased to 60×10^3 psi, and the hardness is appreciably increased at the same time. Increase of temperature causes a marked decrease

TABLE 8.8. MECHANICAL PROPERTIES OF ANNEALED THORIUM

Tensile strength, psi	35,000
Yield strength (0.2 per cent offset), psi	28,000
Proportional limit, psi	22,000
Modulus of elasticity, 10⁶ psi	10
Shear modulus, 10⁶ psi	4

in both tensile and yield strengths, and a smaller decrease in the modulus of elasticity.

8.94. The chemical reactivity of thorium is similar to that of uranium metal. It is attacked both by air and water, but appears to be resistant to sodium, if free from oxygen, at temperatures up to 600°C (1100°F), at least.

Fabrication

8.95. Like uranium, metallic thorium can be fabricated by a variety of standard methods. The material is very ductile and considerable reduction is possible before annealing is necessary. It can be forged, extruded, hot- and cold-rolled, swaged, and drawn. Although the metal is semibrittle under impact at ordinary temperatures, it becomes quite tough at about 250°C (480°F). The machining characteristics of thorium are similar to those of mild steel. In order to prevent oxidation of the metal during fabrication, appropriate steps must be taken to prevent access of air.

Alloys

8.96. Since thorium exhibits dimensional stability upon irradiation, the main purpose of alloys would be to improve its corrosion resistance. Attack by water at high temperatures is greatly reduced by the addition of 5 to 30 weight per cent of titanium or zirconium to the thorium.

8.97. The good radiation properties of thorium have been used, however, to confer stability upon uranium. The solubility of uranium in thorium is very small, and alloys containing 5 to 10 weight per cent of uranium, which are of interest as reactor fuels, have a two-phase structure. The uranium-rich phase is present as very small crystals dispersed in a thorium-rich matrix. Such alloys have been found to suffer only little radiation damage, at both moderate and high temperatures.

Ceramics

8.98. Thorium dioxide has the same advantages (and drawbacks) as uranium dioxide as a fuel material. Solid solutions of the two oxides, which have melting points ranging from about 2800° to 3300°C (5100° to 6000°F), can be prepared but only those containing about 5 to 10 weight per cent of the uranium oxide are of immediate interest. The material can be fabricated by powder metallurgical techniques, similar to those used for uranium dioxide, although the reducing atmosphere necessary for sintering the latter is not required for the solid solution in thorium dioxide. The product has excellent resistance to radiation and is as chemically inert as uranium dioxide.

FUEL-ELEMENT CLADDING [22]

8.99. Apart from their mechanical and, in thermal reactors, appropriate nuclear properties, fuel-element cladding (or canning) materials must be able to withstand attack by both the fuel and the coolant at the temperature and radiation field existing in the reactor (see Chapter 7). For pressurized- and boiling-water reactors, the preferred cladding materials have been aluminum for moderate temperatures and stainless steel or Zircaloy-2 for higher temperatures. The latter is more expensive than stainless steel, but the difference in price is offset by the decreased requirement for fissile material because of the smaller neutron capture cross section. When uranium dioxide is the fuel, the high thermal gradients, which cause large thermal stresses, and the intense radiation levels make it desirable to use a material with considerable high-temperature strength. Stainless steel has the advantage in this respect. The deterioration in mechanical properties of zirconium and its alloys and the embrittlement due to hydriding in water make cladding with these materials less attractive where severe conditions may be encountered. The presence of traces of fluoride in uranium dioxide fuel has also been found to have a detrimental effect on zirconium. The hydriding of zirconium and its alloys which occurs in organic moderator-coolants makes these metals unsuitable for use in reactors employing such compounds. Consequently, aluminum, in the form of an APM alloy (§ 7.40), or stainless steel is generally used as fuel-element cladding in organic-moderated reactors. If the fuel is metallic, e.g., uranium-molybdenum alloy, it is coated with a thin layer of nickel or, possibly, niobium for high operating temperatures, to prevent diffusion of uranium into the aluminum.

8.100. The fuel-element cladding in reactors cooled with sodium has generally been stainless steel, but suitable zirconium alloys (§ 7.56) may provide a satisfactory substitute. If it should be desired to go to temperatures above about 600°C (1100°F), it is probable that neither stainless steel nor zirconium alloys will have adequate strength. Furthermore, if metallic uranium or plutonium is used as fuel, a low-melting eutectic can form with iron (and nickel), thus destroying the effectiveness of stainless-steel cladding. Possible materials for use in sodium at high temperatures are tantalum and other refractory metals of high melting point mentioned in § 7.45.

8.101. For gas-cooled reactors operating at low or moderately high temperatures, either aluminum or magnesium (Magnox) has been employed. Beryllium is an alternative (but more expensive) possibility which is being considered. None of these materials can be utilized in the very high-temperature, gas-cooled reactors which may play an important role in the development of economic nuclear power. In such systems, use will probably have to be made of one of the refractory metals, if canning is necessary.

8.102. If the canning is not required to provide structural support or to prevent the escape of fission products, as may be the case in fuel elements having a graphite matrix and perhaps a graphite cladding in addition, but only to prevent corrosion and erosion, a coating of a carbide may be effective. Among the carbides proposed for this application are those of niobium, silicon, tantalum, titanium, and zirconium.

8.103. Cladding materials are, of course, subject to changes in properties as a result of irradiation.   There is no evidence, however, of any significant increase in corrosion rate or decrease in strength and there have been no failures of fuel elements which can be attributed to radiation damage suffered by the cladding. Where failures have been observed, they have been due either to an accident, e.g., overheating due to the inadvertent loss of coolant, or to internal pressures, produced by the accumulation of fission gases and dimensional changes in the uranium metal fuel, generally combined with a defect in fabrication of the cladding.   When radiation damage to the fuel is no longer the limiting factor, it is possible that attention may have to be paid to the effects of radiation on the cladding.

BONDING MATERIALS

8.104. In some fuel-element designs, heat transfer between the fuel itself and the cladding is achieved by simple mechanical contact; this requires close tolerances in production or the use of special techniques, such as vibratory compaction of uranium dioxide powder (§ 8.63) to eliminate gaps.   Thermal contact is frequently improved, however, by bonding the fuel and cladding.   Where possible, metallurgical bonding is used, since this is the most efficient.   Plate-type fuel elements can be clad by rolling or pressure bonding, and cylindrical elements by co-extrusion, provided the materials lend themselves to these fabrication procedures.   Where metallurgical bonding is not feasible, a metal of low melting point, such as sodium, sodium-potassium alloy, or even lead, is introduced in liquid form before the cladding is sealed.   With ceramic fuel materials, such as uranium dioxide, metallurgical bonds are difficult to produce.   In these cases, helium gas, which has moderately good heat-transfer properties, is used as the bonding material.

8.105. There are various ways whereby the bond between the fuel material and cladding may be adversely affected by irradiation.   For example, the mechanical and thermal stress properties of a metallurgical bond may be changed to such an extent that local failure can occur.   Where the bond consists of a material that is liquid at high temperatures, bubbles of fission product gases may be trapped between the fuel and the cladding.   In either case, the heat-transfer capability would be impaired in certain locations, with the result that regions of exceptionally high temperature would develop.   Such "hot spots" could lead to breakdown of the cladding and failure of the fuel element.   If the bonding material is a gas, the change in composition caused by the release of fission gases will alter the thermal properties so that the surface temperatures may differ from the design values.

REPROCESSING OF IRRADIATED FUEL [23]

INTRODUCTION

8.106. There are several reasons why nuclear reactor fuel elements must be discharged for reprocessing long before the fissile and fertile materials are consumed. In the first place, accumulation of fission products and of the isotopes of heavy elements, which act as neutron poisons, and depletion of the fissile species, e.g., uranium-235 and plutonium-239, can decrease the reactivity to such an extent that the operational requirements of the reactor will no longer be satisfied. Furthermore, the changes in dimension and shape, which occur with continued exposure and the accumulation of fission products, may set a limit upon the time a fuel element can remain in the reactor. Finally, accidental circumstances, such as rupture or weakening of the cladding due to thermal shock, weld failures, burnout, corrosion, or embrittlement, may make it necessary to replace fuel elements. Whatever the cause, at the time the element is discharged from the reactor, the spent fuel still contains a considerable quantity of fissile isotopes, and often of fertile species. The objective of the reprocessing operation is to recover in the most economical manner as much as possible of these valuable materials in a form in which they can be utilized for reactor (or other) purposes.

8.107. In a reactor using natural (or slightly enriched) uranium as fuel, so that appreciable amounts of uranium-238 are present, there will inevitably be some formation of fissile plutonium-239. If the latter is allowed to remain in the reactor it will not only contribute to the fission chain, but it will also capture neutrons to produce the nonfissile plutonium-240. This will, in turn, undergo the (n, γ) reaction to yield fissile plutonium-241, and so on. The period during which the fuel elements can remain in the reactor will depend upon whether the primary purpose of the reactor is power or plutonium production. In a power reactor, it may be possible to continue operation for a longer time, since the fissile plutonium-239 and plutonium-241 contribute to the overall neutronic reactivity. On the other hand, if plutonium-239 production is the objective, account must be taken both of the losses and the presence of other plutonium isotopes, i.e., changes in the isotopic purity, which will occur with long exposure times.

8.108. The reprocessing of irradiated fuel elements generally involves three more-or-less distinct phases; they are (a) cooling, (b) head-end treatment, and (c) separation or extraction. The particular procedure which is preferable, although not always possible, in each phase depends on the nature of the fuel element and the treatment to which the recovered material is to be subjected. Before considering the aforementioned aspects of fuel-element reprocessing, there are two general points of great significance to which attention must be called.

First, in all stages of fuel-element treatment, it is necessary to make absolutely certain that a critical mass of fissile material is not attained, either in the solid state or in solution [24]. Second, because of the intense radioactivity of the spent fuel, even after cooling, special techniques and plant designs are required for carrying out the operations and for maintaining the equipment [25].

COOLING IRRADIATED FUEL ELEMENTS

8.109. The purpose of the so-called "cooling" phase is to permit the decay of various radioactive nuclides present in the spent fuel. These nuclides include fission products and isotopes of elements of high mass number, e.g., 92 (uranium) and beyond (§ 8.113). The more significant consequences of cooling are the following: (1) the beta and gamma activity is decreased to a level at which radiolytic decomposition of the processing reagents is not significant, (2) fission products of short half-life decay almost completely and the reprocessing operation is simplified by the decrease in the number of impurity elements which must be removed from the spent fuel, and (3) certain heavy isotopes, which cannot be separated chemically from the final products, decay to the point where their radioactivity is no longer a problem.

8.110. If the extraction (or separation) phase of the reprocessing operation involves the use of solutions, the first two of these objectives are important and, as a general rule, so also is the third. In these circumstances an extended cooling period, on the order of 100 days, is required. For certain separation processes in which solutions are not employed and in which remote handling is feasible, quite short cooling periods may be possible. From the economic standpoint, any decrease in cooling time is highly desirable. Cooling costs, which include charges for storage facilities, for use (or lease) of the fuel, and for the capital invested in the fuel inventory (§ 14.21 *et seq.*), can add appreciably to the overall cost of nuclear power.

8.111. According to equations (2.53) and (2.54), the gross beta and gamma activities of the fission products are roughly proportional to $t^{-1.2}$, where t is the time after removal from the reactor. Consequently, the gross activity after 100 days of cooling is about 10^{-4} of the activity at 1 hr. Most of the decay occurs in the early stages, but the 100-day cooling period is often required for other purposes.

8.112. The major contributions to the total activity made by individual fission products during the cooling period are shown by the curves in Fig. 8.5. The results are based on the supposition that the fuel elements have been in a reactor for an extended period, so that at the time of their removal an approximate condition of equilibrium has been attained. It is seen that after 100 days of cooling only about a dozen elements are responsible for nearly the whole of the radioactivity. These are the elements from which uranium and plutonium must be separated in spent fuel reprocessing. In addition, however, there are a

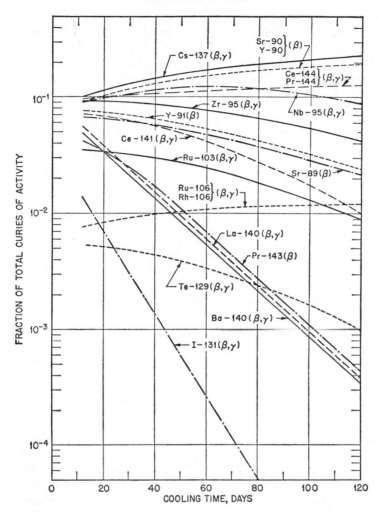

Fig. 8.5.　Activity of fission products during cooling period

few elements, such as the rare earths samarium and neodymium, that must be removed because they have large capture cross sections for thermal neutrons and not on account of their radioactivity.

8.113. The manner in which buildup of isotopes of the heavy elements occurs in a fuel consisting of uranium-235 and -238 is illustrated in Fig. 8.6.　Horizontal arrows pointing to the right represent (n, γ) reactions and those pointing to the left are for $(n, 2n)$ reactions with fast neutrons.　Vertical arrows indicate beta decay.　Of immediate interest is uranium-237 (half-life 6.75 days) which is formed by two (n, γ) stages starting with uranium-235 or by the $(n, 2n)$ reaction with uranium-238.

8.114. Since the isotopes of a given element behave identically in the reprocessing operations, the highly radioactive uranium-237 will be present in the recovered uranium. If the latter is to be fabricated into fuel elements for recycling to a reactor, or is to be converted into hexafluoride for isotopic re-enrichment in a gaseous-diffusion plant, it is generally required that the activity of the uranium-237 should be roughly the same as that of natural uranium in

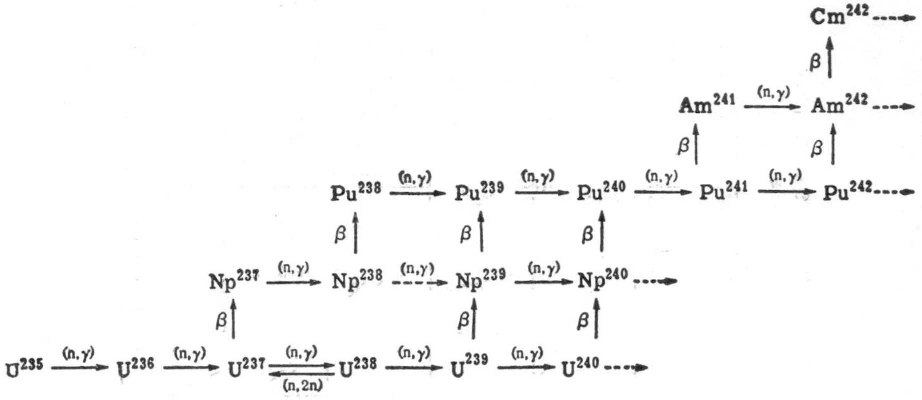

FIG. 8.6. Heavy isotope buildup in uranium

equilibrium with its short-lived decay products. The corresponding beta and gamma activities are as follows:

$$\text{Beta activity} = 0.67 \text{ microcurie per gram.}$$
$$\text{Gamma activity} = 0.16 \text{ microcurie per gram.}$$

The time required for the uranium-237 to decay to these levels varies to some extent with the nature of the fuel, the reactor flux, and the irradiation time, but as a general rule it is 100 to 120 days. During this period most of the uranium-237 decays to neptunium-237 which, being a different element, is removed in reprocessing. It should be noted, too, that neptunium-239 (half-life 2.33 days) decays almost completely into plutonium-239 within 100 days, so that maximum recovery of the latter is possible.

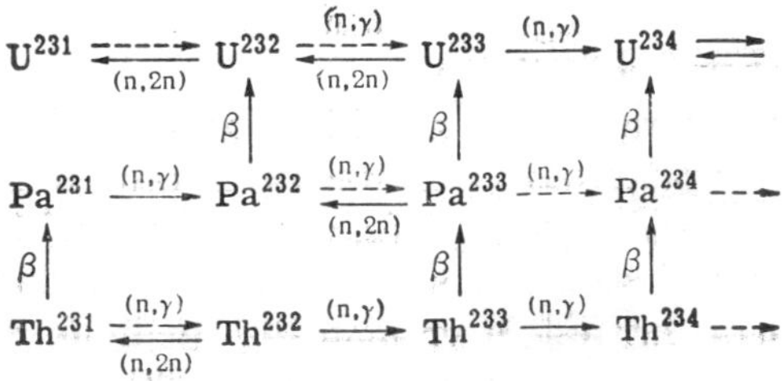

FIG. 8.7. Heavy isotope buildup in thorium

8.115. With thorium-232 as fertile material, a long cooling time would be required to permit decay of the intermediate protactinium-233 (half-life 27.4 days) to the fissile uranium-233. During this period, the activity of thorium-234 (half-life 24.1 days) would also decrease to a permissible level. However, at the same time, thorium-228 (half-life 1.9 years) would accumulate as a result of the alpha-decay of uranium-232 (Fig. 8.7). A possible solution to this dilemma is to separate the thorium and protactinium-233 after a short cooling time, and then permit the latter to decay to uranium-233.

HEAD-END PROCESSES [26]

8.116. The purpose of the head-end process is to prepare the many different types of fuel elements, described earlier in this chapter, so that they can all be treated in the same extraction plant. The recovery procedure in common use requires an aqueous nitrate solution of the fuel materials, and so the following discussion will be concerned mainly with head-end processes appropriate to this method of recovery. Assuming that end boxes, frames, and similar extraneous hardware which hold the fuel elements together have been removed, the head-end processes fall into four categories, namely: (a) removal of cladding (or dejacketing) by mechanical means, (b) dejacketing by chemical or electrochemical methods, (c) dissolution of the fuel but not the cladding, and (d) complete dissolution of the fuel element.

8.117. Mechanical dejacketing is used extensively in Great Britain and France for removal of the Magnox or aluminum cladding from metallic uranium fuel elements. The latter are then dissolved in nitric acid to form the feed solution for the recovery process. In the United States, however, mechanical methods have been developed only for fuel elements in which NaK is employed as the bonding agent, since treatment of such elements with aqueous solutions involves some hazard. Although mechanical dejacketing has the great advantage of simplifying the disposal problem, it does not appear to be practical with uranium dioxide fuel elements because of the expansion of the fuel which occurs upon irradiation. A reasonably complete separation of cladding from fuel would appear to be very difficult.

8.118. Several methods have been developed for chemical dejacketing, depending on the nature of the cladding and fuel materials; these are summarized in Table 8.9. The Darex process cannot be used for removal of stainless steel cladding if the fuel is uranium dioxide, since the latter is soluble in the nitric-hydrochloric acid mixture. The thorium dioxide-uranium dioxide fuel, however, is insoluble.

8.119. In the electrolytic dejacketing process, which is in the experimental stage, the fuel element is attached to a niobium anode in an electrolytic cell containing nitric acid [27]. A current is then passed between the anode and a niobium cathode. Dissolution of several types of stainless steel has been

TABLE 8.9.   CHEMICAL DEJACKETING METHODS

Cladding	Fuel	Reagent
Stainless Steel	U Metal U-Mo Alloy UO_2 UO_2-ThO_2	$6M$-H_2SO_4 (Sulfex process)
Stainless Steel	Th-U Alloy UO_2-ThO_2	$5M$-HNO_3 + $2M$-HCl (Darex process)
Zirconium or Zircaloy	UO_2 U-Zr Alloy U-Zr-Nb	$6M$-NH_4F + $1M$-NH_4NO_3 (Zirflex process) $5.4M$-NH_4F + $0.33M$-HNO_3 + $0.13M$-H_2O_2 (Modified Zirflex process)
Aluminum	U Metal U-Mo Alloy	$5M$-NaOH + $2.5M$-$NaNO_3$

achieved in this manner.   With zirconium cladding, a large fraction of this element is precipitated as the oxide which may trap appreciable quantities of uranium.   If the method is to be satisfactory, complete dissolution of the zirconium is desirable.

8.120. A difficulty encountered in the development of the chemical method of decladding, which will probably also apply to the electrochemical procedure, is the uncertainty regarding the amounts of uranium and plutonium, from the fuel-element core, that either dissolve or remain suspended as oxides.   These would represent losses and if they have to be recovered the advantages of separate dejacketing would be lost.   A possible solution which is being investigated is to chop up the fuel element into pieces and leach out the core material, leaving the cladding behind.   If a small quantity of the latter dissolves, the subsequent treatment will not be affected.

FUEL DISSOLUTION [28]

8.121. Where decladding is undesirable, difficult, or impossible, e.g., where the fuel element consists of a matrix or alloy in aluminum or stainless steel, the element and cladding are dissolved in one operation.   The treatment is then similar in certain respects to dissolution of the fuel core after removal of the cladding.   Some of the procedures which have been developed for various unclad and clad fuels are summarized in Table 8.10.   Attention may be called to the following points: (1) Although uranium-zirconium systems will dissolve in nitric acid, the reaction may become explosive unless fluoride is present.   (2) In the dissolution of uranium-molybdenum alloys, part of the molybdenum may precipitate as the trioxide (MoO_3) which carries down some uranium with it.   The precipitate is therefore digested with sodium hydroxide solution; this dissolves the MoO_3 but leaves nearly all the uranium behind.

TABLE 8.10.   DISSOLUTION OF UNCLAD AND CLAD FUEL ELEMENTS

Fuel Material	Reagent
Unclad	
U, UO_2, U-Mo..................	Nitric acid
U-Zr, U-Zr-Nb, UO_2-ThO_2.......	Nitric and hydrofluoric acids
Aluminum Clad	
U, U-Al, U-Mo................	Nitric acid + mercuric nitrate
Stainless Steel Clad	
U, U-Mo, UO_2, UO_2 in S.S........	Nitric and hydrochloric acids (Darex solution)
Zirconium Clad	
U-Mo, UO_2, U-Zr, U-Zr-Nb......	Nitric and hydrofluoric acids + aluminum nitrate

8.122. Methods are under investigation for the treatment of the newer fuels, e.g., uranium dioxide dispersed in beryllium oxide and uranium carbide either alone or dispersed in graphite.   Graphite-based fuels can be disintegrated and leached by means of 90 per cent nitric acid; alternatively, the carbon may be eliminated by combustion and the residue (uranium dioxide) dissolved in nitric acid, possibly with fluoride as catalyst.   Another approach, which can be applied to various ceramic fuels, is a grind-leach or chop-leach procedure similar to that described in § 8.120.   The uranium in uranium carbide can apparently be extracted with nitric acid.

8.123. It will be observed that all the methods described for dissolving reactor fuels involve the use of nitric acid.   The resulting solutions therefore contain the uranium, plutonium, and fission-product elements—other than iodine and the inert gases—as nitrates.   This is the form required for the extraction process which has been widely used.   There are, however, other procedures which do not employ aqueous solutions, so that the head-end treatment is different, as will be seen later (§ 8.160 et seq.).

SEPARATIONS PROCESSES

INTRODUCTION

8.124. The main purpose of the separation stage of the fuel reprocessing operation is to extract uranium and plutonium in forms which are suitable for further use either as recycle fuels, as feed material in an isotope enrichment plant, or for weapons. In addition, the fission product residues may be processed for the recovery of specific isotopes of scientific or commercial value.

8.125. The most promising separation methods fall into three general categories: (a) aqueous processes with the principal separation being accomplished by solvent extraction techniques; (b) volatility methods depending on the distillation of uranium hexafluoride; and (c) pyrometallurgical (or melt refining) processes, such as removal of impurities as oxides. Other procedures, based upon liquid-metal or fused-salt extraction, metal distillation, zone melting, and electrorefining, have been the subject of experimental study but their potentialities have not yet been established.

8.126. The processes involving extraction of uranium and plutonium from aqueous solution by means of an organic solvent are by far the most advanced in development since they have been in large-scale use since about 1951. The head-end treatments described earlier are, in fact, those employed to prepare the feed solution for this method of extraction. The decontamination factors, i.e., ratios of fission product concentration before treatment to that after treatment, currently attained are on the order of 10^7 to 10^8. The gamma activity due to fission products in the extracted uranium is then less than in the natural element. It can thus be fabricated by direct handling or converted to fluoride for isotope separation, assuming that the uranium-237 has decayed to a sufficient extent (§ 8.114).

8.127. The volatility process of main interest at the present time is based on the fact that uranium hexafluoride is volatile, whereas most of the fission products form fluorides which do not volatilize at all readily. Decontamination factors similar to those attained by solvent extraction can be realized. The volatility method has advantages over aqueous processing in the smaller number of operations and considerable reduction in volume of highly radioactive wastes. There are certain limitations, however, to be discussed later, which restrict the applicability of the volatility separation method.

8.128. In most pyrometallurgical processes the spent fuel is treated in its existing form, so that dissolution or other chemical conversion is eliminated. Since the decontamination factors are on the order of 100 only, the product is highly radioactive and requires remote handling. In view of this situation, the prolonged cooling period for the spent fuel is not required. Thus, about 10 days are probably adequate instead of the 100 or so days for other processes. Pyrometallurgical techniques lend themselves particularly to "closed cycle" procedures which are closely integrated with reactor operation (§ 8.169).

CHEMISTRY OF THE HEAVY ELEMENTS [29]

8.129. The separation processes, the one based on solvent extraction in particular, depend on the somewhat unusual chemical behavior of the heavy elements, i.e., the elements of high mass number. Starting with actinium (atomic number 89), there is a series of elements, called the *actinide series*, in the sixth period of the periodic system which resemble the rare-earth (or *lanthanide*) elements in the fifth period. The lanthanide elements all have similar chemical properties, based on a positive valence of 3, resulting from the presence of three relatively loosely bound outer electrons in each atom. In the analogous actinide series, there are also marked resemblances among the elements, especially in the formation of a tripositive (III) valence state. However, because some of the actinide elements have inner electrons which are not very tightly bound, it is possible to realize tetrapositive (IV), pentapositive (V), and hexapositive (VI) states. It is the existence of these higher oxidation states that facilitates extraction and separation of the heavy elements.

8.130. The known positive valence (or oxidation) states of the first nine members of the actinide series are indicated in Table 8.11. An attempt is made to express the relative stabilities of the different states of each element. The most stable state is represented by four stars, and smaller numbers of stars are used for states of decreasing stability. For the actinide elements beyond berkelium (Bk), atomic number 97, only the (III) state is known.

TABLE 8.11. RELATIVE STABILITIES OF OXIDATION
STATES OF THE ACTINIDE ELEMENTS

Atomic No...... Element...	89 Ac	90 Th	91 Pa	92 U	93 Np	94 Pu	95 Am	96 Cm	97 Bk
III........	****	*	*?	**	**	***	****	****	****
IV........		****	*	***	****	****	*		**
V........			****	*	***	**	*		
VI........				****	***	***	*		

8.131. Although in a given valence state the various actinide elements have similar chemical properties, these properties often are very different in the different oxidation states. For example, the (III) and (IV) states can be precipitated from aqueous solution as the fluorides, but not the (V) and (VI) states.

Furthermore, the fluorides of the (III) and (IV) states are not volatile, but in the (VI) state the fluorides, e.g., uranium hexafluoride, volatilize at fairly low temperatures. Another important difference is in the behavior of the nitrates. Those of the (IV) and (VI) states are appreciably soluble in certain organic liquids but the nitrates of the (III) state are virtually insoluble in these liquids. Facts of this kind have been utilized in the development of separation processes for the actinide elements.

8.132. The relative stabilities of the various oxidation states have been determined quantitatively from measurements of oxidation-reduction electrode potentials. Among the conclusions drawn, some are of special interest in the present connection. It has been found that with increasing atomic number the (IV) state becomes increasingly less stable relative to the (III) state. In uranium, for example, the (IV) state is so stable that even water will oxidize the (III) to the (IV) state; correspondingly, reduction of uranium to the (III) state is very difficult and even if it is formed in aqueous solution it will reoxidize immediately. Neptunium (III) is somewhat more stable than uranium (III), but it is, nevertheless, easily oxidized in solution, e.g., by air, to neptunium (IV). With the next element, plutonium, a somewhat stronger oxidizing agent, e.g., a nitrite, is required to convert the (III) to the (IV) state. The (III) state is fairly stable in solution even in air and is obtained by reducing the (IV) state with a ferrous salt.

8.133. A similar decrease in ease of oxidation with increasing atomic number occurs in connection with the (IV) and (VI) states. Thus, uranium (IV) is very easily oxidized to uranium (VI) and the latter is very difficult to reduce. In fact, as seen in Table 8.11, the uranium (VI) state is the most stable of the oxidation states of this element. On the other hand, oxidation of neptunium and plutonium to the (VI) state requires a moderate oxidizing agent, such as hot bromate or chromate in acid solution. Fairly mild reducing agents, such as iodide or nitrite ions, can then reduce the neptunium (VI) to the (IV) state and plutonium (V) to the (III) state. Uranium (VI), however, would be unaffected by these reagents.

8.134. The relative ease of oxidation and reduction is influenced in some cases by the acidity of the solution and in all instances by the formation of complex ions. If any particular state is present as a complex, then that state acquires additional stability relative to an uncomplexed state. In comparing the (III) and (IV) states of the actinide elements, it has been found that the latter have a greater tendency to form anionic complexes, so that a method of stabilizing the (IV) relative to the (III) state is possible. For example, the addition of complex-forming sulfate ions to a plutonium solution makes it more difficult to reduce the (IV) to the (III) state. Hence, in the reduction of plutonium (VI) solution in the presence of sulfate, the process can be stopped at the (IV) state, whereas in the absence of sulfate it would continue to the (III) state.

GENERAL PRINCIPLES OF SOLVENT EXTRACTION

8.135. The solvent-extraction method for separating the constituents of an aqueous solution can be used when one or more of these constituents are appreciably soluble, whereas the others are much less soluble, in an organic solvent which is essentially immiscible with water. When the organic liquid is brought into intimate contact with the aqueous solution, the substances present will distribute themselves between the organic and aqueous phases. The constituent (or constituents) with the greatest solubility in the organic medium will tend to pass into that phase whereas the others will tend to remain in the aqueous solution. Thus, a partial separation of the constituents of the solution will have been achieved. As will be seen shortly, various procedures can often be used to improve the degree of separation.

8.136. As an engineering problem, solvent extraction is to be regarded as a diffusion operation involving the transfer of a solute from one liquid to another, essentially immiscible, liquid phase. The process is carried out either in extraction columns or in mixer-settler devices operating on the counter-current principle; the organic phase flows in one direction and the aqueous phase in the opposite direction. Liquid-liquid extraction is perhaps one of the simplest unit processes that lends itself to remote control. This is an important consideration in connection with the treatment of the highly radioactive liquids produced by dissolution of spent fuel elements.

8.137. In choosing the organic liquid to be used for a particular solvent-extraction procedure, various factors must be taken into consideration. These usually represent independent properties of the solvent, and the actual substance chosen may involve a compromise giving the best combination of desirable characteristics. The most important property is the selectivity or ability of the organic liquid to extract a particular component (or components) of a solution in preference to all others that are present. The selectivity is expressed by the separation factor, i.e., the ratio of the distribution coefficients of the wanted and unwanted species when equilibrium is attained between the two phases. The distribution coefficient or distribution ratio, D, is defined as

$$D = \frac{\text{Conc. of component in organic phase}}{\text{Conc. of component in aqueous phase}}$$

at equilibrium, and the separation factor, α, is given by

$$\alpha = \frac{D(\text{product})}{D(\text{impurity})}.$$

A good solvent for extraction is one for which the distribution coefficient for one component is large and the separation factor is either large or small. In other words, it is desirable that D(product) shall be large, whereas D(impurity) should be small, or vice versa.

8.138. In addition, a satisfactory organic solvent should be almost immiscible with the solution to be extracted. It should have sufficient chemical stability to withstand the action of fairly high concentrations of oxidizing and reducing agents. Resistance to intense beta and gamma radiation is, of course, desirable, since it may make possible a decrease in the cooling time of the spent fuel. To permit operation of the separation equipment, the density of the solvent should be different from that of the aqueous solution.

8.139. The extraction of an inorganic compound, such as a nitrate, from an aqueous solution by means of an organic solvent is influenced by a number of circumstances. Of particular importance are the presence of (a) salting agents, (b) oxidizing or reducing agents, and (c) complex-forming anions.

8.140. A *salting agent* is either a salt or an acid, having the same anion as the inorganic compound to be extracted, the presence of which in the aqueous solution increases the distribution ratio defined above. In the extraction of uranyl nitrate, for example, either nitric acid or one if its salts, such as sodium, potassium, calcium, or aluminum nitrate, can serve as a salting agent. These substances are soluble in the aqueous phase but not in the organic solvent.

8.141. The distribution coefficient of the nitrate of a particular element is markedly affected by the oxidation (or valence) state of that element. In general, the nitrates of the actinide elements in the hexavalent (VI) and tetravalent (IV) states can be readily extracted by certain organic liquids, but they are much less extractable, i.e., have smaller distribution ratios, in the lower oxidation states. One of the consequences of this situation is that, after an element has been extracted into an organic medium, it can often be back-extracted into aqueous solution by adding a reducing agent to the latter. If two actinide elements, e.g., uranium and plutonium, have been extracted in the (IV) or (VI) state into an organic solvent, this fact can be used as a basis for separating these elements. The separation can be achieved by back-extracting the organic phase with an aqueous solution which will reduce one of the actinide elements but not the other. As seen earlier in this chapter, plutonium can be reduced much more readily than uranium to the (III) state, which is essentially insoluble (or inextractable) in the organic liquid. Hence, the plutonium could be back-extracted with a suitable aqueous reducing agent leaving the uranium (VI) in the organic phase.

8.142. The extraction of a specified element from aqueous solution by an organic medium is dependent upon the particular form in which the element is present in the solution. For example, uranyl nitrate hexahydrate can be extracted by diethyl ether, but the corresponding sulfate is not extractable by this solvent. The addition of a sulfate or other salt of a complex-forming anion to an aqueous solution of uranyl nitrate will thus decrease the extractability of the uranium, since a proportion of the element will be in some form other than the nitrate. The complex-forming anions decrease the distribution coefficient between the organic and aqueous phases, and so their effect is opposite to that of the common-ion salting agents.

8.143. Another aspect of complex formation is its effect on the oxidation-reduction potential, mentioned earlier. Accordingly, salts of complex-forming anions may be able to influence the extraction (or back-extraction) of a given element in the presence of a specified reducing or oxidizing agent. It is apparent, therefore, that the extractability of an element can be affected by variations in the amount and nature of the salting agent, by the nature of an oxidizing or reducing agent, and by the presence of complex-forming ions. The possibility of changes in these three factors thus permits flexibility in the application of the solvent-extraction method for the separation of various elements.

THE PUREX PROCESS [30]

8.144. The "Purex" process, using n-tributyl phosphate (TBP) as the organic solvent, is typical of solvent-extraction procedures employed in the treatment of spent fuel. In the form of the nitrates, uranium (VI) and plutonium (IV) can be readily extracted from aqueous solution by TBP, whereas the fission products are taken up to a much smaller extent. Since TBP is relatively stable in the presence of fairly high concentrations of nitric acid, the latter is used as the salting agent. In the older, but somewhat similar, "Redox" process, the solvent, methyl isobutyl ketone (or hexone), is attacked by nitric acid and so aluminum nitrate was employed as the salting agent. However, nitric acid is preferred because the bulk of solids sent to waste is reduced and the acid is readily recoverable by distillation. Furthermore, TBP is a safer solvent as a result of its lower volatility and higher flash point. In the United Kingdom, $\beta\beta'$-dibutoxydiethyl ether ("Butex") has been used as the solvent, with nitric acid as the salting agent, for the primary extraction of uranium and plutonium and for purification of the uranium [31].

8.145. In the first cycle of the Purex process, of which an outline flow sheet is shown in Fig. 8.8, the feed consists of an aqueous solution containing uranium (VI), plutonium (IV), and fission product (F.P.) nitrates plus an excess of nitric acid. Sodium nitrite is added to make sure that the plutonium is entirely in the (IV) state, since it is in this form that it is best extracted by TBP. The feed solution enters at the middle of the first (extraction) column, while the less dense organic solvent (TBP in a kerosene diluent) entering from the bottom flows upward. The uranium (VI) and plutonium (IV) nitrates are thereby extracted from the aqueous solution and pass into the organic medium. In the upper part of the column the organic phase is scrubbed with nitric acid which acts as a salting agent. Most of the fission products which may have entered the organic solvent are now back-extracted into the aqueous phase, but the salting agent largely prevents back-extraction of the uranium and plutonium. The aqueous

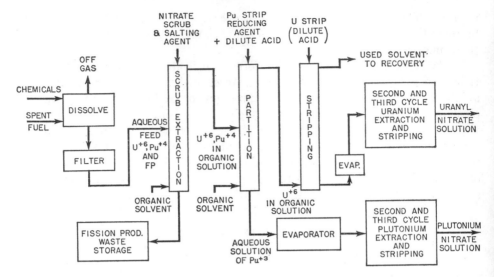

FIG. 8.8.   Purex process flow sheet

effluent (raffinate) from the extraction column contains essentially all the fission products with little or no uranium or plutonium.

8.146. The organic phase containing the uranium and plutonium next passes into the second (partitioning) column where it flows upward and meets the downflowing aqueous strip solution containing a suitable reducing material, such as ferrous sulfamate, together with some salting agent. (The ferrous ion is the reducing agent and the sulfamate removes the nitrite which would otherwise prevent reduction from taking place.) The plutonium is reduced to the (III) state and so is back-extracted into the aqueous phase. As this flows downward it is scrubbed with fresh TBP moving upward from the bottom of the column. Any uranium (VI) that has passed into the aqueous solution is thereby returned to the organic phase. The aqueous medium, containing plutonium (III) nitrate, leaves at the bottom of the partition column.

8.147. The organic solution of uranium (VI) nitrate, from which the plutonium and fission products have been almost completely separated, is now transferred to the bottom of the third (stripping) column where it flows upward and is stripped by dilute nitric acid flowing downward. In the absence of a salting agent the uranium is back-extracted into the aqueous phase and flows out of the bottom of the column. The spent solvent, leaving at the top, is sent to a recovery plant for purification and subsequent re-use in extraction.

8.148. For further purification, both the aqueous uranium (VI) and plutonium (III) nitrate solutions are submitted to a second cycle. The uranium purification cycle is essentially identical with the last two stages of the first cycle shown in Fig. 8.8. The aqueous uranium solution is first extracted into the TBP phase and scrubbed with a reducing solution; the organic phase is then stripped by dilute nitric acid in a second column. The plutonium (III) solution is converted into the (IV) state by means of sodium nitrate and nitric acid, extracted into the TBP medium, and scrubbed with nitric acid in the same column. The organic solution then passes to a stripping column where the plutonium is back-extracted into the aqueous phase by dilute nitric acid. If the plutonium is desired in the (III) state, an aqueous solution of hydroxylamine sulfate is used as the reducing and stripping medium.

8.149. Instead of the second cycle of TBP extraction for the purification of plutonium, two other procedures are possible: one involves ion exchange, which will be described below, and the other utilizes a long-chain tertiary amine, trilaurylamine, as the solvent. In the latter method, which has not yet been fully developed, the plutonium product from the first cycle extraction with TBP is treated with sodium nitrite to convert the plutonium to the (IV) state and is extracted with the amine solvent diluted with diethylbenzene. The organic phase is scrubbed with dilute nitric acid and stripped with acetic acid containing a small concentration of nitric acid.

ION EXCHANGE [32]

8.150. The process of ion exchange has been used both to purify and to concentrate plutonium solutions. Since the ion-exchange technique is used in other parts of the fuel cycle, the general principles will be outlined here. Certain solids, called *ion exchangers*, although themselves insoluble in water, are able to exchange either positive or negative ions with ions of the same charge present in an aqueous solution. Ion-exchange materials of various types are known, but those in most general use are based on synthetic resins, consisting of insoluble organic polymers of high molecular weight. These are called ion-exchange resins.

8.151. An ion-exchange resin consists, in general, of a loosely crossed-linked, polymerized organic structure to which is attached a number of active (or functional) groups. In cation-exchange resins the functional group is acidic in character, capable of splitting off a hydrogen ion; the "weak acid" resins usually contain carboxylic ($-COOH$) or phenolic ($-OH$) groups, and the "strong acid" resins usually contain the sulfonic acid ($-SO_3H$) group. In anion-exchange resins, on the other hand, the active group is basic. In the "weak base" resins this is an amino ($-NX_2$) or an imino ($=NX$) group, where X may be hydrogen or an organic radical, but in those of the "strong base" type it is a quaternary ammonium ($-NX_3^+$) group, where X is an organic radical.

8.152. The ion-exchange separation technique is particularly applicable to dilute solutions. It can be used for such purposes as (1) removal of desired or undesired substances present in solution, (2) concentration (or volume reduction) of solutions, (3) metathesis, i.e., replacement of one ion by another, and

(4) separation of ions from one another. From the engineering standpoint, ion exchange is a diffusion operation involving the transfer of ions from one phase to another in a liquid-solid system. The process is generally carried out with a fixed resin bed through which the solution percolates (see, however, § 8.14). The conditions are such that unsteady-state diffusion theory is applicable, the mathematical treatment being similar to that used in connection with batch distillation.

8.153. When a solution is brought in contact with an ion-exchange resin, an interchange will generally occur between the ions in solution and the mobile ions of the same sign from the resin. For example, a cationic resin, represented by H—R, where R is the insoluble resin and H is the ionizable hydrogen which is part of the functional (acidic) group, will exchange with positive ions, M^{+n}. Thus, the equilibrium

$$n \text{ H—R (resin)} + M^{+n} \text{ (soln.)} \leftrightarrows M\text{—}R_n \text{ (resin)} + n \text{ H}^+ \text{ (soln.)}$$

is established, so that part of the resin is converted into $M\text{—}R_n$ while hydrogen ions are set free in the solution. The amount of hydrogen ions liberated is equivalent to the quantity of M^{+n} ions retained by the resin, so that electrical neutrality is maintained. Similarly, for an anionic resin in the chloride form, i.e., R—Cl, the exchange with anions, A^{-n}, can be represented by

$$n \text{ R—Cl (resin)} + A^{-n} \text{ (soln.)} \leftrightarrows R_n\text{—A (resin)} + n \text{ Cl}^- \text{ (soln.)},$$

so that a portion of the A^{-n} ions are retained by the resin while an equivalent amount of chloride ions is liberated in solution.

8.154. The amount of any particular ion remaining on a given quantity of an ion-exchanger resin increases with the concentration of the ions in solution. However, there is a limit to what is called the "capacity" of the resin; this varies with the nature of the material, but is generally on the order of 2 to 5 gram-equivalents per kilogram of resin.

8.155. The ability of a resin to retain (or its affinity for) different ions, under specified conditions, is determined by two factors; these are (1) the charge on the ion and (2) the size of the hydrated ion as it exists in solution. The greater the charge carried by the ion, either positive or negative, according to circumstances, and the smaller the radius of the ion in the hydrated form the more strongly will the ion be retained by a given resin. For the ions Th^{+4}, La^{+3}, Ba^{+2}, and Na^+, for example, the order of decreasing retention by a given resin from equivalent solutions, i.e., the order of decreasing affinity, is usually

$$Th^{+4} > La^{+3} > Ba^{+2} > Na^+.$$

The order of retention for the alkali metal series of univalent ions is

$$Cs^+ > Rb^+ > K^+ > Na^+ > Li^+,$$

and for the alkaline-earth (bivalent) metals it is

$$Ba^{+2} > Sr^{+2} > Ca^{+2} > Mg^{+2},$$

at least in resins of the strong-acid type.

8.156. The principle of separation by ion exchange is illustrated in Fig. 8.9 for the simple case of a solution containing Na^+ and Mg^{+2} ions, the latter having a greater affinity for the cationic resin. The solution is poured onto the top of the resin, giving the sorption step, shown in Fig. 8.9A. The actual separation occurs in the elution stage. A dilute solution of a simple acid, e.g., hydrochloric

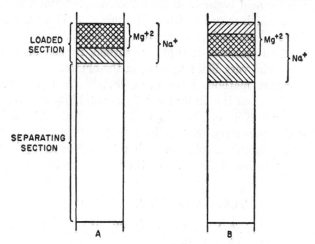

FIG. 8.9.   A. Ion-exchange column before elution
B. Ion-exchange column after elution

acid, is allowed to percolate slowly through the ion exchange column containing the sorbed ions. The less strongly held Na^+ ions will now move steadily ahead of the Mg^{+2} ions (Fig. 8.9B). The first portion of the eluate to leave the column consists almost entirely of dilute acid; this is followed by an eluate containing Na^+ ions, and finally the Mg^{+2} ions are eluted. The same general behavior is observed for solutions containing three or more ionic species, provided they have appreciably different affinities for the resin.

8.157. In addition to cation exchange resins, anion exchangers are available and behave in a similar manner, except that they tend to sorb negative rather than positive ions from solution. For the purification and separation of metallic elements by anion-exchange resins, the element to be extracted must be converted into a negative (anionic) complex ion. In this form it will be taken up by the resin. It is then removed by an eluting solution which converts the complex ion into the simple cationic form which is not sorbed by the resin.

8.158. Both cation and anion exchange have been used to purify plutonium solutions, especially from zirconium and ruthenium, and their respective daughters, niobium and rhodium, which are difficult to remove by solvent extraction

[33]. In the cation-exchange process, the plutonium, which is maintained in the (III) state by means of hydroxylamine, is first sorbed on the resin from a dilute nitric acid solution. It is then eluted with more concentrated nitric acid containing sulfamic acid to prevent oxidation of the plutonium to the (IV) state. For purification with anion-exchange resins, the plutonium must be in the form of an anionic complex of the (IV) state with nitrate ions; this is obtained by means of a concentrated solution of nitric acid containing sodium nitrite. For elution, dilute nitric acid is used, so that the complex anion is converted into the cationic form, which may be either plutonium (III) or (IV), since neither is sorbed by the anion-exchange resin. Zirconium, ruthenium, and their daughters, as well as uranium, if they are sorbed, remain on the resin during the elution process and very little appears in the eluate.

8.159. Ion-exchange methods have been devised for the separation of plutonium from uranium, but the limited irradiation stability of the resins prohibits their use in the first cycle of the separations process. At the present time, therefore, ion exchangers are used only in the later stages, either for purification or concentration of plutonium solutions, when the radioactivity has been greatly reduced by the separation of the bulk of the fission products.

VOLATILITY SEPARATIONS [34]

8.160. The separation of elements by volatilization, making use of differences in vapor pressure of the elements themselves or of suitable compounds, has been employed for many years, especially in the study of radioactive materials. The possibility of extracting plutonium from spent reactor fuel by volatilization was considered in the early stages of the atomic energy program, but this separations method was not pursued in detail because precipitation and, later, solvent extraction processes seemed to offer greater prospects of immediate success. Although the use of solvent extraction has resulted in a considerable saving in cost over the original bismuth phosphate precipitation process, the procedure is still complicated. Interest has therefore been renewed in a number of non-aqueous separations based on the volatility of uranium hexafluoride. Among the inherent advantages of such processes are the following: (1) The uranium is recovered in a form (hexafluoride) that is used as the feed material for the gaseous-diffusion plant and is also closer to the end product (metal or dioxide) for fuel-element fabrication than is the case with aqueous processes. (2) Two chemical engineering unit operations, namely, fractional distillation and absorption, which have achieved a high degree of technological development, can be employed. For reasons which will be apparent in due course, the fluoride volatility process is of special interest for fuels which are either very highly enriched, contain considerable amounts of zirconium, or consist of mixed fluorides (§ 8.165).

8.161. The great majority of the fission products form fluorides which volatilize only at high temperatures. Among those which contribute appreciable activity to irradiated reactor fuel, only the fluorides of tellurium, molybdenum, antimony, niobium, and ruthenium boil or sublime at low or moderate temperatures, as shown by the data in Table 8.12. Of the fluorides which volatilize

TABLE 8.12. BOILING (OR SUBLIMATION) POINTS OF
FLUORIDES AT ATMOSPHERIC PRESSURE

Fluoride	Temperature (°C)
TeF_6	−38.3
MoF_6	35
UF_6	54.6
PuF_6	62.3
SbF_5	150
NbF_5	229
RuF_5	313

more readily than uranium hexafluoride, tellurium hexafluoride is a gas at ordinary temperatures so that it can be readily separated. Any molybdenum hexafluoride that may be volatilized and condensed with uranium hexafluoride is not serious, since the most important radioisotope, molybdenum-99, has a half-life of 67 hours and will decay almost completely in the course of a few weeks. Apart from plutonium hexafluoride, all the other fluorides are so much less volatile than uranium hexafluoride that separation by distillation presents no problem.

8.162. In the methods for converting irradiated fuels into the fluorides, as described below, the plutonium is in the form of the nonvolatile tetrafluoride. Consequently, by controlled volatilization, followed by condensation, the uranium can be obtained as the hexafluoride, free from plutonium and essentially all the fission products. The condensate may be purified by fractional distillation (§ 8.17) or by absorption on granular sodium fluoride. At temperatures below 250°C (480°F), the latter forms a complex with uranium hexafluoride, which is retained while the impurities pass on. Upon raising the temperature to 400°C (750°F) in the presence of a stream of fluorine gas the uranium hexafluoride is released and can be condensed.

8.163. In principle, plutonium can be recovered from the fluoride residues by converting the tetrafluoride into the hexafluoride with fluorine and then volatilizing it. However, the chemical reactivity of plutonium hexafluoride makes it a very difficult material to handle and the problems involved in the plutonium separation step have not yet been solved. For this reason the fluoride volatility process is of special interest for highly enriched fuels where the formation of plutonium is negligible. If sufficient plutonium is present in the residues to make extraction desirable, an aqueous processing method must be used.

8.164. The procedure for conversion of the irradiated fuel material into the fluorides depends on the nature of the fuel. If the latter consists of uranium metal with little or no alloying agent, the fuel element is first declad and the core is treated with a liquid interhalogen compound, such as bromine trifluoride (BrF_3). This reagent converts uranium directly to the hexafluoride.

8.165. Uranium-zirconium alloy fuels are soluble in a fused mixture of sodium fluoride and zirconium tetrafluoride at a temperature of about 600°C (1100°F), and so lend themselves to treatment by the volatility process. The fuel is heated with the fused fluoride mixture in the presence of hydrogen fluoride, when the uranium reacts to form the tetrafluoride. The latter is then converted into the hexafluoride by means of fluorine gas at the existing temperature. Spent fused fluoride fuel mixtures (§ 13.111) can presumably be processed by direct fluorination to produce uranium hexafluoride. Although considerable pilot plant development work has been done on the fluoride volatility separations method, it has not been applied to the treatment of spent fuel on a large scale.

PYROMETALLURGICAL PROCESSING [35]

8.166. In several of the high-temperature (or pyrometallurgical) processes, the fuel may remain in the metallic state throughout. The decontamination factors that have been attained so far are relatively low, on the order of 100, so that remote fabrication is necessary for the recovered material. On the other hand, since chemical conversions in dilute solutions are eliminated, waste volumes are much smaller than for aqueous procedures. Furthermore, irradiated fuel can be treated after a short cooling period with a consequent reduction in inventory requirements for fissile material.

8.167. Several different pyrometallurgical processes, including molten metal extraction, distillation, and crystallization, are under development, but the first pilot-plant scale method is one based on melt refining or oxide drossing. It is being developed to demonstrate the reprocessing of the fuel elements for the fast Experimental Breeder Reactor (EBR)-II [36]. The fuel is melted in a refractory crucible of zirconium oxide and held at 1400°C (2550°F) for several hours. The inert gases (krypton and xenon) and alkali metals (principally cesium) are volatilized, whereas the strontium, barium, yttrium, and rare earth metals, which are oxidized more readily than uranium and plutonium, are converted into oxides; these are removed mechanically from the crucible walls. A block diagram of the process is shown in Fig. 8.10. Among the fission products that are not removed but are permitted to recycle are zirconium, niobium, and molybdenum; these elements are sometimes added deliberately to uranium fuels in order to improve their metallurgical properties and to increase their resistance to damage by radiation (§ 8.53).

8.168. One problem associated with the melt refining process is the appreciable

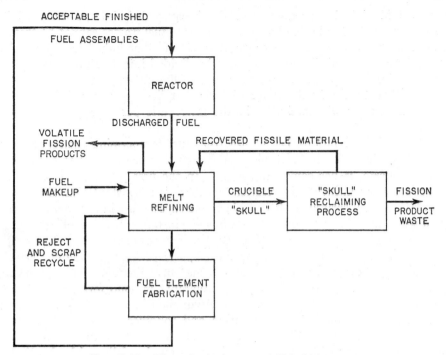

Fig. 8.10.   Flow sheet of pyrometallurgical process

loss of fissile material carried over together with the impurities to the crucible "skull." A method for the recovery of fissile (and fertile) species from the skulls is being developed.

8.169. Because of the short cooling period and the relatively small decontamination factor, the recovered fuel has a considerable radioactivity. It is desirable, therefore, to keep transportation to a minimum. Hence, the melt-refining plant for the EBR-II is closely integrated with the reactor itself in a closed fuel cycle. An interesting (and necessary) feature of the process is the step for refabricating the fuel elements by a remotely operated injection casting technique. The procedure consists in driving the liquid metal into a bundle of Vycor tubes lowered into the melt. The metal solidifies in the molds and when the tubing is broken away the desired pin-type fuel elements (0.144 in. diameter and 14.2 in. long) remain. The elements are then clad and assembled by remote techniques.

PROCESS WASTE DISPOSAL [37]

LIQUID WASTES

8.170. For processes utilizing aqueous solutions, liquid wastes constitute the major aspect of the disposal problem. Because of the possible value of mixed or of individual fission products and also because their intense radioactivity constitutes a potential health hazard, the reactor fuel wastes cannot be treated in the same manner as are the waste products of other industries. From the standpoint of disposal, liquid wastes can be divided into two broad categories which require entirely different methods of handling; these are (a) "high-level" wastes with activities in the range of curies per liter (or per gallon), and (b) "low-level" wastes with activities in the microcurie per liter (or per gallon) range.

8.171. High-level radioactive wastes from fuel reprocessing plants are generally stored in underground tanks. Although this procedure has been satisfactory hitherto, it has some drawbacks. In the first place, the demand for storage space is continually increasing, and there is a constant risk of leakage. In addition, the absorption of beta and gamma rays results in considerable heating of the radioactive solutions, so that in some instances water cooling must be used. With the continued growth of the nuclear power industry, storage will no longer be a suitable method for the disposal of high-level liquid wastes. Consequently, other procedures are being investigated. One promising approach is to concentrate the wastes, preferably in solid form, and then to store or bury the resulting relatively small volume of residues in selected locations. Among methods of concentration and solidification which have been studied are evaporation of the water, addition of the solutions to cement to form concrete, adsorption of the radioactive isotopes on clay followed by heating to "fix" them, fusion into ceramics or glass, and evaporation and calcination in a rotary-ball kiln or fluidized bed. Successful pilot-plant studies have been made of a fluidized-bed calciner for the treatment of high-level wastes resulting from the reprocessing of aluminum-uranium alloy fuels. The solutions contain aluminum nitrate and this is converted in the air-heated fluidized bed into alumina powder containing oxides of the fission product elements [38].

8.172. An alternative possibility, which may be less costly, is direct disposal of the high-level wastes in locations where they will not represent a hazard, e.g., asphalt-lined pits or selected deep underground formations. The hazard associated with disposal of this kind would be greatly reduced if the long-lived fission products strontium-90 and cesium-137 could be removed and stored separately. Although storage of radioactive waste in impervious pits in the ground has temporary value, it is regarded as unsatisfactory for long-term use. Ocean disposal has been the subject of much research, principally because the vast volume of such a repository might be expected to dilute the radioactivity to harmless proportions [39]. It has been concluded, however, that large-scale disposal of radioactive waste solutions to the ocean would not be satisfactory for several reasons. For example, too little is known concerning ocean currents to be able to predict the fate of the wastes. Another serious drawback is the complete lack of control once disposal has taken place. Ocean disposal in selected locations has been used on a small scale for low- and intermediate-level wastes in containers, but this is not practical or desirable where large volumes of high-level wastes are concerned.   A promising method for direct disposal of liquid wastes appears to be deep underground, either in natural salt domes or in cavities left after removal of salt by dissolution, in deep basins (5000 to 10,000 feet below the surface) containing brines and having no connection with potable waters, and in special excavations in shale formations.

8.173. In addition to the wastes of high activity, large volumes of low-level liquid wastes are produced in aqueous reprocessing plants.   Such solutions have been disposed of by discharge into dry wells or in open pits in the ground [40]. The liquids seep out slowly into the surrounding soil where the radioactive ions are adsorbed and fixed by ion exchange.   By this means, permanent disposal of large volumes of wastes of low activity can be achieved.   There is little danger of the spread of radioactivity unless the adsorption capacity of the soil is exceeded and the waste solution enters a water table in general use.   Such a possibility may be avoided by careful evaluation of the site geology before utilization of this method of waste disposal.   In Great Britain low-level process wastes have been disposed of by carefully controlled injection into the sea at a distance of about two miles from shore.

GASEOUS WASTES

8.174. There are three important radioactive waste gases, namely, iodine, and the chemically inert gases krypton and xenon.   These are liberated when the fuel element is dissolved.   If the fuel is in the metallic form, oxides of nitrogen will also be present as a result of the reaction with nitric acid.   In the Hanford process, which may be regarded as fairly typical, the gases from the dissolver vessel pass through a condenser tower where the oxides of nitrogen combine with water to form nitric acid.   The residual gases are heated and then enter the so-called silver reactor, consisting of a packed bed of "saddles" coated with silver nitrate.   Here the iodine is removed in the form of silver iodide.   Finally, the gases, consisting almost entirely of krypton and xenon, are filtered through Fiberglas, or other material, and discharged to the atmosphere through high stacks.   In some processing plants the iodine remaining in the spent fuel after cooling is so small that the silver reactor stage can be dispensed with altogether. If, for meteorological or other reasons, discharge of the radioactive krypton and xenon into the atmosphere is undesirable, these gases can be adsorbed on carbon or silica gel at low temperature.   The resulting radioactive solid can then be disposed of in a suitable manner.

8.175. Under the heading of hazardous gaseous wastes may be included air from processing plants carrying small solid or liquid radioactive particles.   Since this may constitute a serious danger, the air is discharged through filters of sand or, better, of Fiberglas to remove the particulate matter.   These filters have a limited operating life and must be replaced periodically.   The spent filters represent a problem in the disposal of solid waste, to which reference is made below.

SOLID WASTES

8.176. Solid wastes, such as filter and adsorption materials, ion-exchange resins, etc., are usually buried.   Incineration can sometimes be carried out, but careful filtration of the evolved gases is necessary.   Discarded process equipment is decontaminated if possible; if not, it is buried.   The possibility of converting the more radioactive liquid process wastes into solid form was mentioned previously.   Some solid residues, from which the fission products are not easily extracted, could be disposed of in the ocean, whereas others could be stored in natural cavities.

SYMBOLS USED IN CHAPTER 8

D distribution ratio
k thermal conductivity
M molecular weight
T absolute temperature
x_p atomic fraction of uranium-235 in product
x_w atomic fraction of uranium-235 in waste stream
α separation factor for solvent extraction
α^* theoretical separation factor for isotopes

REFERENCES FOR CHAPTER 8

1. D. H. Gurinsky and G. J. Dienes (Eds.), "Nuclear Fuels," D. Van Nostrand Co., Inc., Princeton, N. J., 1956; L. Grainger, "Uranium and Thorium," Pitman Publishing Corp., New York, 1958; W. D. Wilkinson and W. F. Murphy, "Nuclear Reactor Metallurgy," D. Van Nostrand Co., Inc., Princeton, N. J., 1958; C. D. Harrington and A. E. Ruehle, "Uranium Production Technology," D. Van Nostrand Co., Inc., Princeton, N. J., 1959; C. R. Tipton (Ed.), "Reactor Handbook," Vol. I, Materials, Interscience Publishers, Inc., New York, 1960, pp. 87 *et seq.*   Reviews of progress in reactor fuel technology appear regularly in *Reactor Core Materials* (or *Reactor Materials*) and *Reactor Fuel Processing*, published quarterly by the U. S. AEC. See also, *Proc. Second U. N. Conf. Geneva*, Vols. **3** and **4** (1958).
2. Ref. 1; S. M. Stoller and R. B. Richards (Eds.), "Reactor Handbook," Vol. II, Fuel Processing, Interscience Publishers, Inc., New York, 1961, Chs. 8 and 9.
3. S. M. Stoller and R. B. Richards, Ref. 2, Ch. 12; *Proc. Second U. N. Conf. Geneva*, Vol. **4** (1958).
4. Ref. 1; S. M. Stoller and R. B. Richards, Ref. 2, Ch. 10.
5. S. M. Stoller and R. B. Richards, Ref. 2, Ch. 11; W. D. Wilkinson (Ed.), "Extractive and Physical Metallurgy of Plutonium and Its Alloys," Interscience Publishers, Inc., New York, 1960; E. D. Grison, W. B. H. Lord and R. D. Fowler (Eds.), "Plutonium 1960: Proceedings of the Second International Conference on Plutonium

Metallurgy, Grenoble," Cleaver-Hume Press, Ltd., London, 1961; A. S. Coffinberry and W. N. Miner (Eds.), "The Metal Plutonium," University of Chicago Press, 1961.

6. H. H. Hausner and J. F. Schumar (Eds.), "Nuclear Fuel Elements," Reinhold Publishing Corp., New York, 1959; "Fuel Element Conference," U. S. AEC Report TID-7559 (1958); "Fuel Element Fabrication," Proceedings of IAEA Symposium, Vienna, 1960, Academic Press, Inc., New York, 1961; B. Kopelman (Ed.), "Materials for Nuclear Reactors," McGraw-Hill Book Co., Inc., New York, 1959; A. R. Kaufmann (Ed.), "Nuclear Reactor Fuel Elements," Interscience Publishers, Inc., New York, 1962; "Nuclear Metallurgy," The Metallurgical Society, American Inst. of Mining, Metallurgical and Petroleum Engineers, Inst. of Metals Division, Vol. 4, Special Report Series No. 4, 1957; Vol. 5, Special Report Series No. 7, 1958; Vol. 5, Special Report Series No. 9, 1959; "Symposium on Reactor Core Materials," *Nuc. Sci. and Eng.*, **4**, 357 *et seq.* (1959). See also regular issues of *Reactor (Core) Materials* and *Power Reactor Technology*.

7. C. R. Tipton, Ref. 1, Ch. 7; A. D. McIntosh and T. J. Heal, "Materials for Nuclear Engineers," Interscience Publishers, Inc., New York, 1960, Ch. 2; C. A. Hampel (Ed.), "Rare Metals Handbook," 2nd Ed., Reinhold Publishing Corp., 1961, Ch. 31; S. J. Paprocki and R. F. Dickerson, *Nucleonics*, **18**, No. 11, 154 (1960); A. N. Holden, "Physical Metallurgy of Uranium," Addison-Wesley Publishing Co., Inc., Reading, Mass., 1958.

8. L. Grainger, "The Behavior of Reactor Components under Irradiation," IAEA Review Series, No. 6, International Atomic Energy Agency, 1960; see also several reports in *Proc. Second U. N. Conf. Geneva*, Vols. **5** and **6** (1958), and reviews in *Reactor (Core) Materials*.

9. J. Belle (Ed.), "Uranium Dioxide: Properties and Nuclear Applications," U. S. Government Printing Office, Washington, D. C., 1961; A. D. McIntosh and T. J. Heal, Ref. 7, Ch. 5; O. J. C. Runnals, *Nucleonics*, **17**, No. 5, 104 (1959); A. J. Mooradian and J. A. L. Robertson, *ibid.*, **18**, No. 10, 60 (1960). A complete review is given in *Power Reactor Technology*, **5**, No. 1, 54 (1961).

10. J. A. L. Robertson, Atomic Energy of Canada, Ltd., Report AECL-807 (1959); see also Ref. 9.

11. L. Grainger, Ref. 8; J. Belle, Ref. 9, Ch. 9; see also, *Power Reactor Technology*, **5**, No. 1, 54 (1961).

12. A. D. McIntosh and T. J. Heal, Ref. 7, Ch. 5; F. A. Rough, *et al.*, *Nuc. Sci. and Eng.*, **6**, 391 (1959); **7**, 111 (1960); **10**, 24 (1961); *Nucleonics*, **18**, No. 3, 74 (1960); "Proceedings of the Uranium Carbide Meeting," U. S. AEC Report TID-7603 (1960); "Proceedings of Symposium on Uranium Carbides as Reactor Fuel Materials," U. S. AEC Report TID-7614 (1961); *Power Reactor Technology*, **4**, No. 1, 45 (1960).

13. C. R. Tipton, Ref. 1, Ch. 13; D. L. Keller, *Nucleonics*, **19**, No. 6, 45 (1961); see also various issues of *Reactor (Core) Materials*.

14. A. S. Coffinberry and W. N. Miner, Ref. 5, p. 410.

15. U. S. AEC Report NDA-2116-4 (1960).

16. C. R. Tipton, Ref. 1, p. 399; T. M. Benziger and R. K. Rohwer, *Nucleonics*, **19**, No. 5, 80 (1961); see also U. S. AEC Report ORO-240 (1959) and issues of *Reactor (Core) Materials*.

17. U. S. AEC Report TID-11295 and Supplement (1961); U. S. AEC Report BMI-1468 (1960).

18. Ref. 5; C. A. Hampel, Ref. 7, Ch. 18; A. D. McIntosh and T. J. Heal, Ref. 7, Ch. 3; C. R. Tipton, Ref. 1, Ch. 11.

19. R. M. Kiehn, in A. S. Coffinberry and W. N. Miner, Ref. 5, p. 333.

20. Ref. 17; A. S. Coffinberry and W. N. Miner, Ref. 5, p. 416; U. S. AEC Report NDA-2140-2 (1959).

21. C. R. Tipton, Ref. 1, Chs. 9 and 10; A. D. McIntosh and T. J. Heal, Ref. 7, Ch. 4; C. A. Hampel, Ref. 7, Ch. 28; B. Manowitz, *Nucleonics,* **16**, No. 8, 91 (1958).

22. C. R. Tipton, Ref. 1, pp. 477 *et seq.;* see also, *Reactor (Core) Materials* for reviews.

23. M. Benedict and T. H. Pigford, "Nuclear Chemical Engineering," McGraw-Hill Book Co., Inc., New York, 1957; F. S. Martin and G. L. Miles, "Chemical Processing of Nuclear Fuels," Academic Press, Inc., New York, 1958; J. F. Flagg (Ed.), "Chemical Processing of Reactor Fuels," Academic Press, Inc., New York, 1961; S. M. Stoller and R. B. Richards, Ref. 2, Chs. 3 to 7; R. E. Blanco, *Trans. Am. Nuc. Soc.,* **4,** 187 (1961); see also *Proc. Second U. N. Conf. Geneva,* Vol. **17** (1958) and quarterly issues of *Reactor Fuel Processing.*

24. S. M. Stoller and R. B. Richards, Ref. 2, p. 50; J. E. McLaughlin, U. S. AEC Report HASL-66 (1959); U. S. AEC Report TID-7016, Rev. 1 (1961).

25. S. M. Stoller and R. B. Richards, Ref. 2, Chs. 16 and 17.

26. S. M. Stoller and R. B. Richards, Ref. 2, Ch. 3; see also, *Reactor Fuel Processing.*

27. C. M. Slansky, M. W. Roberts, and K. L. Rohde, *Nuc. Sci. and Eng.,* **12,** 33 (1962).

28. Ref. 26; F. L. Culler and R. E. Blanco, *Proc. Second U. N. Conf. Geneva,* **17,** 259 (1958); R. E. Blanco, Ref. 23.

29. J. J. Katz and G. T. Seaborg, "The Chemistry of the Actinide Elements," John Wiley and Sons, Inc., New York, 1957; G. T. Seaborg, "The Transuranium Elements," Yale University Press, New Haven, Conn., 1958.

30. S. M. Stoller and R. B. Richards, Ref. 2, pp. 146 *et seq.;* V. R. Cooper and M. T. Walling, *Proc. Second U. N. Conf. Geneva,* **17,** 291 (1958).

31. G. R. Howells, *et al., Proc. Second U. N. Conf. Geneva,* **17,** 3 (1958).

32. F. C. Nachod (Ed.), "Ion Exchange: Theory and Application," Academic Press, Inc., New York, 1949; R. Kunin and R. J. Myers, "Ion Exchange Resins," John Wiley and Sons, Inc., New York, 1950.

33. S. M. Stoller and R. B. Richards, Ref. 2, p. 445; "Plutonium Ion Exchange Processes," TID-7607 (1961).

34. S. M. Stoller and R. B. Richards, Ref. 2, Ch. 6; G. I. Cathers, *et al., Proc. Second U. N. Conf. Geneva,* **17,** 473 (1958).

35. S. M. Stoller and R. B. Richards, Ref. 2, Ch. 7; L. Burris, *et al., Proc. Second U. N. Conf. Geneva,* **17,** 401 (1958); J. H. Schraidt and M. Levenson, *ibid.,* **17,** 361 (1958).

36. "Melt Refining of EBR-II Fuel," *Nuc. Sci. and Eng.,* **6,** 493 *et seq.* (1959); **9,** 55 *et seq.* (1961); A. S. Shuck, *Nucleonics,* **15,** No. 12, 50 (1957).

37. S. M. Stoller and R. B. Richards, Ref. 2, Chs. 13, 14, and 15; J. C. Collins (Ed.), "Radioactive Wastes: Their Treatment and Disposal," John Wiley and Sons, Inc., New York, 1960; E. Glueckauf (Ed.), "Atomic Energy Waste: Its Nature, Use, and Disposal," Interscience Publishers, Inc., New York, 1961; C. B. Amphlett, "Treatment and Disposal of Radioactive Wastes," Pergamon Press, Inc., New York, 1961; C. A. Mawson, "Processing of Radioactive Wastes," IAEA Review Series No. 18, International Atomic Energy Agency, 1961; see also *Proc. Second U. N. Conf. Geneva,* Vol. **18** (1958); *Power Reactor Technology,* **3,** No. 4, 1 (1960); and issues of *Reactor Core Processing.*

38. "Report of Meeting on Fixation of Radioactivity in Solid Stable Media," U. S. AEC Report TID-7613 (1960); R. F. Domish, *et al., Nucleonics,* **17,** No. 12, 76 (1959).

39. *Nuclear Safety,* **2,** No. 1, 78 (1960); "Radioactive Waste Disposal into the Sea," IAEA Safety Series No. 5, International Atomic Energy Agency, 1961.

40. "Proceedings of Conference on Ground Disposal of Radioactive Wastes," U. S. AEC Report TID-621 (1961).

PROBLEMS

1. A reactor containing 50,000 kg. of 1.5% enriched uranium as fuel operates at a power of 500 Mw (thermal) for 1,000 days. The fuel is then removed and allowed to cool for 120 days. Estimate (a) the activity in curies per pound of fuel and (b) the activity in curies per pound of fission products.

2. In the preceding problem, suppose that at the end of the 120-day cooling period the barium is rapidly separated from the fission products. What would be the approximate activity in curies per pound, assuming it to be due entirely to barium-140?

3. Account for the fact that in Fig. 8.5 the fractional activity for some isotopes increases with time after removal from the reactor, whereas for others it decreases steadily. Would this difference in behavior be expected to continue indefinitely?

Chapter 9

RADIATION PROTECTION AND
REACTOR SAFEGUARDS [1]

RADIATION HAZARDS AND HEALTH PHYSICS

RADIATION PROBLEMS IN REACTOR DESIGN AND OPERATION

9.1. Shortly after the discovery of x-rays and of radioactivity, toward the end of the nineteenth century, it was recognized that nuclear radiations could produce harmful effects on the living organism. Partly as a result of ignorance and partly due to accidental circumstances, a number of cases of injury, ranging from minor skin lesions to bone sarcoma and untimely deaths, were reported among radiologists and others who were exposed to excessive amounts of nuclear radiations. It was not until the early 1920s, however, that organized efforts were made to recommend safety measures to be used in the manipulation of x-rays and radium. About ten years later, maximum permissible levels of exposure to radiation were proposed, and their general acceptance led to a marked decrease in the incidence of radiation injury.

9.2. When plans were made to build the first nuclear reactors, it was realized that radiation hazards of a hitherto unimagined order of magnitude would be involved and that special precautions would have to be taken to protect the operating personnel. Consequently, there was a great expansion of the field of *health physics*, the basic purpose being to devise methods for protection from unwarranted exposure to nuclear radiations. The efforts of health physicists have been so successful that injuries due to radiation have been rare, in spite of the very high levels of activity associated with reactor operation, the treatment of spent fuel, and the handling of fission products.

9.3. Radiation exposure may arise from sources external to the body or from radioactive material that has entered the body, either by inhalation, by ingestion, or by absorption through the skin. There is no bodily sensation to indicate the presence of nuclear radiation except at very high intensities when a burning sensation is experienced. Consequently, instruments must be used to determine

512

where the hazard exists.   If the source of radiation is outside the body, the exposure may be reduced by shielding the source, e.g., a reactor, by limiting the time of exposure, and by using various types of remote-control devices to permit an increase in the distance between the source and the individual.   Internal exposure may be controlled by rigorous cleanliness, avoidance of direct handling of active material, and proper ventilation to prevent inhalation of radioactive gases, aerosols, dust, etc.

9.4. It is impossible, in spite of adequate shielding, to reduce the intensity of neutrons and gamma rays to zero outside the shield during the operation of a reactor.   Furthermore, maintenance and repair of the reactor and its equipment will involve the possibility of exposure to radiation.   The coolant will usually have some induced activity after passage through the reactor, and, if this is recycled, the piping, pumps, tanks, etc., represent sources of radiation exposure for operating personnel.   Even if the coolant is discharged after a single pass, e.g., air or water, its discharge must be controlled in order to avoid the harmful effects of radioactivity.

9.5. The spent fuel elements discharged from a reactor are, of course, highly radioactive, and their treatment requires great care.   Although they are usually allowed to "cool" for a period of about 100 days after their removal from the reactor (§ 8.110), the fuel elements are still very radioactive when they are dissolved in the chemical processing plants.   Thus, the process solutions in such plants represent a serious radiation hazard.   In addition, as indicated in Chapter 8, the various waste products must be disposed of in such a manner as not to jeopardize the health of the surrounding population.

9.6. Apart from the radiation hazards associated with the operation of a reactor and the chemical processing plant, it must be remembered that both uranium and plutonium (and their compounds), which are used as reactor fuels, are radioactive.   In addition, precautions must be taken in handling material that has been placed in the reactor for irradiation.   Dust in a reactor atmosphere is likely to have induced activity, as is also the air pushed out by control rods. All the foregoing considerations apply to normal reactor plant operation and maintenance.   There is, however, always the possibility of a major or minor accident.   Adequate plans for the protection of personnel from radiation injury must take such factors into consideration.

9.7. It can be seen that the necessity for controlling the radiation hazard, in order to protect both operating personnel and those living in surrounding areas, places certain restrictions upon the design and operation of a reactor and its associated plant.   It is in this respect, in particular, that reactor engineering differs from other branches of engineering.

HEALTH PHYSICS ACTIVITIES

9.8. The prime function of the health physics organization is, of course, to safeguard the health of individuals whose work is likely to involve some exposure to nuclear radiations by taking all steps that are considered necessary to minimize such exposure. In addition there is the responsibility of making sure that nothing escaping from the nuclear plant, even in the event of an accident, would represent a radiological hazard to the population of the surrounding area. The achievement of these objectives involves a number of different activities, which may be considered as falling into two broad categories. The first includes advisory and research functions as they affect the overall nuclear energy program, and the second is concerned with the activities of the health physics group at a particular plant. These will be considered in turn.

9.9. In order to establish and maintain quantitative standards of radiation protection, health physicists, together with radiologists, physicians, and others, are members of the International Commission on Radiological Protection, the National (U. S.) Committee on Radiation Protection and Measurement, and the U. S. Federal Radiation Council. These bodies recommend what are now called Radiation Guides for persons whose occupation causes them to be exposed to nuclear radiations.* The Guides represent, in the light of present knowledge, values of radiation exposures and of concentrations of radioactive substances in air and water which are believed to be acceptable by such individuals in view of the nature of their work. The Guides are subject to change periodically as further knowledge and experience are gained of the effects of nuclear radiations on human beings.

9.10. Another of the overall advisory activities of health physicists is in connection with the choice of suitable locations for reactors and chemical processing plants. Due consideration is paid to such factors as population density, the suitability of the geological formation for underground and surface waste disposal, and the effect of local meteorological conditions on the dispersal of airborne radioactivity. Advice is also given concerning the adequacy of shielding, remote-control handling equipment, filters in stacks, and other devices built into the reactor system for the protection of personnel. The design of so-called "hot laboratories," for experimental work with highly radioactive materials, is carried out with the advice of health physicists, so as to ensure incorporation of all features necessary to prevent overexposure of individuals working in such laboratories.

9.11. Turning next to the functions of the health physics organization at a particular reactor or processing plant, one of the main responsibilities is concerned with what is called *radiation monitoring;* this involves, broadly speaking, the determination and recording of radiation dosages and dose rates at numerous locations. It will be seen later that the radiation dosage is measured in terms of the energy absorbed from the radiation, and the dose rate is the time-rate at

---

* The Guides were formerly called "maximum permissible exposures" to nuclear radiations and "maximum permissible concentrations" of radioactive substances in air and water, but in 1960, the newly appointed Federal Radiation Council recommended that these terms be no longer used because they were misleading [2].

which such energy is absorbed.   In general, the total dose (or dosage) received is the product of the dose rate and the exposure time.

9.12. For the purpose of determining radiation exposures, detection and measuring instruments of various kinds must be calibrated regularly and maintained in good operating condition.   All persons likely to be exposed to radiation are provided with meters whereby the dose received can be measured, and any individual who has received an appreciable dose on a given day is notified so that appropriate steps may be taken to avoid overexposure.   In special cases it may be necessary to measure the radioactivity present in the thyroid gland, due to accidental absorption of radioiodine, and analysis of urine and feces may be required to find out if certain radioactive species are accumulating in the body.

9.13. All accessible areas around a reactor and its associated equipment and laboratories are surveyed at frequent intervals to determine the radioactive contamination of surfaces and the amount of activity in the air.   As a result of such measurements the health physicist may recommend the adoption of suitable protective actions.   For example, protective clothing may be deemed necessary when the contamination is excessive.   When appreciable amounts of radioactive dust are present in the air, the use of respiratory protection may be recommended.   In special cases additional shielding or the employment of remote-handling devices may be required.   Advice will also be given concerning the permissible working time in an area of unusually high activity.   A regular survey must be made of the radioactivity of the solid, liquid, or gaseous materials discharged from the plant so as to make certain they will not constitute an environmental hazard.

BIOLOGICAL EFFECTS OF RADIATION [3]

ACUTE AND CHRONIC EXPOSURE

9.14. The nature and extent of the injuries caused by radiation depend on various circumstances.   Considering, first, sources external to the body, the exposure may be either acute or chronic.   *Acute exposure,* which refers to the receipt of a relatively large dose of radiation within a short interval of time, is not very probable under normal operating conditions.   However, it could arise as a result of negligence or an accident.   Since incidents involving acute exposures are rare, it is not necessary to consider the matter in detail here.   Some reference to the probable effects of various acute radiation dosages will be made later (§ 9.64).

9.15. The term *chronic exposure* is used to describe the frequent, e.g., daily, exposure to relatively low radiation intensities, such as might be experienced in various operations connected with nuclear reactors.   In general, for the same total radiation dosage, chronic exposure has less serious consequences than does acute exposure.   Thus a certain dose of radiation received by an individual in a few seconds may cause harm, whereas the same dose spread over a period of years may produce no noticeable effect. In other words, both the total radiation dose and the dose rate, i.e., the amount received in a given time, are important in determining the extent of biological damage, if any. The smaller the total dose and the dose rate, the less will be the injury. For some types of radiation injury, the body appears to have partial recovery powers, so that the damage is not cumulative over a period of time. However, there are certain important exceptions, notably genetic effects, where the injury is probably cumulative.

9.16. Many radiobiologists are of the opinion that there is no lower limit to the amount of radiation which will produce some biological injury. If this is the case, then it is probable that the human body has adapted itself to small, but continuous, doses of nuclear radiations. This is indicated by the fact that living organisms have always been exposed to such radiations. Mention may be made in this connection, of cosmic rays, of radioactive material that is always found in the soil and also in air and water,* and of the radioisotopes carbon-14 and potassium-40 that are present in the body. If radiation injury occurs, the damage may be reversible or irreversible; in the former case complete recovery of function is possible, whereas in the latter case the injury seems to be permanent, e.g., shortening of life span, genetic effects, or scar formation. Skin and bone-marrow damage due to low chronic exposure to radiation, for example, are reversible to some extent, but damage to brain and kidney appears to be largely irreversible.

PARTIAL AND WHOLE-BODY EXPOSURE

9.17. The injury caused by radiation depends on the area (or volume) of the body which is exposed; in this connection a quantity called the *integral dose*, defined as the product of the dose and the mass of the body receiving that dose, is often used. When x-rays or radioisotopes are employed for diagnosis or treatment, limited portions of the body may receive large doses of radiation, so that the integral dose is not excessive. This may result in local injury to the irradiated area, but the overall health of the individual is apparently not seriously affected. The exposure of the whole body or a large portion of the body to the same radiation might well be fatal. In the latter case the radiation would, of course, be absorbed by all parts of the body, and many organs might be affected.

9.18. Different portions of the body show different sensitivities to radiation. Although there are undoubtedly variations of degree among individuals, the lymphoid tissue, spleen, bone marrow, organs of reproduction, and the gastro-

---

* The drinking water in New York City is reported to contain 5×10^{-10} microcuries (μc) of radium per cm^3; in some mineral springs the amount is as high as 10^{-7} $\mu c/cm^3$.

intestinal tract are among the most radiosensitive tissues. Of intermediate sensitivity are the skin, lungs, and liver, whereas the muscles and full-grown bones are the least sensitive.

EFFECTS OF DIFFERENT TYPES OF RADIATION

9.19. Ultimately, the biological effects of nuclear radiation can be largely attributed to ionization and electronic excitation which cause the destruction of various molecules, e.g., proteins, that play an important part in the functioning of living cells. It might therefore appear at first sight that the extent of injury would be determined by the amount of ionization produced and, hence, by the energy absorbed from the radiation. However, this is only partially true since the specific ionization, i.e., the number of ion pairs per centimeter of path (§ 2.28), has a considerable influence. In general, the greater the specific ionization, the greater the damage for a given energy absorption.* For this reason, if an alpha emitter is taken into the body, it produces considerable injury. Not only are the alpha particles released inside the body, perhaps within sensitive tissue, but their energy is dissipated and absorbed within a short distance. Some indication of the relative effectiveness of different types of nuclear radiation will be given in § 9.48.

9.20. Although neutrons do not produce ionization directly, they are able to cause considerable biological damage as a result of secondary effects. With slow neutrons these effects are due almost entirely to the capture of the neutrons by hydrogen and nitrogen nuclei. The (n, γ) capture of neutrons by hydrogen is accompanied by the emission of 2.2-Mev photons, some of which irradiate the surrounding tissue and others escape from the body. The (n, p) reactions with nitrogen result in the production of protons which, like alpha particles, dissipate their energy in short paths. In addition, in each of these reactions a nitrogen atom is replaced by one of carbon, so that the identity of biologically important molecules, e.g., nucleic acids and enzymes, may be destroyed. Slow neutrons also undergo reaction with various other nuclei present in living tissue, but these are believed to be of minor significance.

9.21. When the body is exposed to fast neutrons, the particles lose their energy mainly as a result of elastic collisions with atoms of hydrogen, oxygen, carbon, and nitrogen present in living tissue, the first of these being the most important. In the collision the struck atom acquires kinetic energy which is dissipated by ionization, excitation, and elastic collisions with other atoms. Fast neutrons can undergo the (n, p) reaction with nitrogen-14, mentioned above, and if the energy is in excess of about 1.5 Mev, the (n, α) process also occurs, so that

---

* In biological work, a related quantity called the "linear energy transfer" (or LET) is generally used. It is defined as the rate of energy loss by the particle per unit distance of its path in a given medium.

protons and alpha particles, respectively, are produced. In addition, for energies less than 10^4 ev, the (n, γ) reaction with hydrogen becomes noticeable. The recoil atoms and the protons and alpha particles lose their energy within a very short distance of the point at which a fast neutron interacts with a nucleus. However, in considering the volume in which injury due to fast neutrons may occur, it should be remembered that such neutrons are able to penetrate considerable distances into the body before reacting.

9.22. In referring above to the biological damage caused by alpha particles, it was indicated that the alpha emitter must be in the body. Normally, the range of these nuclear radiations in tissue is so short that alpha particles of energy less than about 7.5 Mev are unable to penetrate the outer protective layer of the skin (epidermis). This means that, with a few rare exceptions, radioactive species which emit only alpha particles do not represent an external hazard. Although beta particles have a longer range, only the most energetic are able to travel more than a few millimeters in tissue. Nevertheless, if close to (or on) the skin, beta emitters are capable of producing serious burns. It must be remembered, in any event, that both alpha and beta activity are frequently accompanied by the emission of gamma rays which are able to penetrate considerable distances into the body. The slowing down of beta particles is also accompanied by bremsstrahlung (x-rays).

EXPOSURE TO RADIATION ORIGINATING WITHIN THE BODY

9.23. There is no reason to suppose that the effects of exposure to radiation from radioactive materials within the body will differ from those resulting from the same kind of radiation originating outside the body, except for the consequences of differences in distribution of the radiation dosage. These differences are the inevitable result of two factors: (1) the tendency of some chemical elements to concentrate in certain types of cells or tissue, e.g., plutonium and strontium tend to concentrate in the bone, whereas iodine is preferentially held in the thyroid gland; and (2) the attenuation of radiation from outside the body with depth of penetration into the body tissue.

9.24. When radioactive material enters the body the emitted radiation is in a position to produce greater injury to the tissues than if the radiation arises from an external source. In the first place, exposure to the radiation is continuous, subject only to depletion of the quantity of radioactive material in the body by physical (decay) and biological (elimination) processes. Furthermore, the tissues in which the injury occurs are nearer to the source of radiation and are not shielded from it by intervening materials. This is of particular importance with alpha particles and low-energy beta particles which cannot reach sensitive tissues from outside the body, but which may dissipate their entire energy within such tissues if the sources are present in the body.

GENERAL CHARACTERISTICS OF RADIATION INJURY

9.25. The observable effects of overexposure to radiation vary widely, depending upon the circumstances involved and, to some extent, upon individual susceptibility.   Among the factors which influence the results are magnitude and time-distribution of the radiation dose, the geometrical distribution within the body, type of radiation, and the age of the individual or the age range through which the overexposure occurs.   From the experience gained by radiologists and others, since the discovery of x-rays and radioactivity toward the end of the nineteenth century, it appears that frequent or chronic exposure of adults to radiation at average rates more than a hundred times as great as that due to cosmic rays and radioactive materials in nature (§ 9.16) may be incurred for many years without observable subsequent effects on health.   However, chronic exposures to radiation at rates several thousand times that from natural sources may result in serious injury if continued over sufficiently long periods of time.   Possible consequences of chronic overexposure are (1) leukopenia, (2) anemia, (3) detrimental changes in tissue structure, (4) leukemia, (5) malignant tumors, (6) cataracts, and (7) increase in the average rate of genetic mutation.   Observations on small animals also suggest the possibility of some reduction in life expectancy without the occurrence of any of these conditions.

9.26. Acute overexposure of the whole body to penetrating radiation can result in latent injury contributing to the probability of the subsequent development of one or more of the effects enumerated above.   If the dose received is sufficiently large there will be, in addition, early nausea and vomiting, decrease in the formed-element content of the blood, epilation, loss of appetite and general malaise, diarrhea, and emaciation.   The most reliable indication of the degree of acute overexposure of the body to radiation is the decrease in the platelet count in the blood.   Unfortunately, this does not become significant until some time after the overexposure has occurred.   The first clinical symptons are generally nausea and vomiting, which may develop within 30 minutes (or less) for very large radiation doses and up to 2 or 3 hours for smaller overexposures.

9.27. In order to correlate the extent and nature of radiation injury with the magnitude of the dose received, it is necessary to express the dose in terms of an acceptable and reproducible unit.   The next section of this chapter will, therefore, be concerned with the matter of radiation dose units.

RADIATION DOSE UNITS [4]

THE ROENTGEN

9.28. In order to express radiation dosage in a quantitative manner, it is necessary to have a suitable unit.   Such a unit should be easily reproducible and should be measurable in terms of relatively simple physical quantities. Since the purpose of knowing the radiation dose is to provide some indication of the possible injury, it would be desirable to have the unit of dosage proportional to the biological damage produced. However, the factors involved in radiation injury are so complex that it is not possible to devise a single unit which will satisfy both physical and biological requirements. Actually three different units are in common use, as will be explained below.

9.29. The first generally accepted unit of radiation is the *roentgen*, represented by the symbol r, which uses the ionization in air as its basis. It is defined as "that quantity of x- or gamma radiation such that the associated corpuscular emission per 0.001293 g of air produces, in air, ions carrying 1 electrostatic unit (1 esu) of quantity of electricity of either sign." The mass of air referred to in the definition is that of 1 cm^3 of dry air at 0°C and 760 mm Hg (1 atm) pressure, i.e., at standard temperature and pressure (STP). It should be noted that the roentgen applies only to x- and gamma radiation in air. The extension of the dose unit to other kinds of nuclear radiation will be considered later.

9.30. In order to appreciate the physical significance of the roentgen, consider 1 cm^3 of dry air at STP exposed to x- or gamma rays. As a result of the interaction of the radiation with the oxygen and nitrogen in the air, there will be produced a number of Compton recoil electrons, photoelectrons, and positron-electron pairs, in various proportions depending on the energy of the radiation (see Chapter 2). These secondary particles, i.e., positive and negative electrons, will produce ion-pairs as they travel through the air. When 1 r of x- or gamma radiation has been absorbed by the 1 cm^3 of dry air at STP, the total charge on all the ions of either sign produced would amount to 1 esu.

9.31. It is important to understand that the ion-pairs under consideration are not all formed in the 1 cm^3 of air. The secondary particles originate in this quantity of air, but, since they have a considerable range, the ion-pairs that they produce, and which are presumed to be collected, will extend over a much larger volume. This fact is of importance in connection with the methods which have been devised for measurement of x- and gamma-radiation dosage in roentgens.

9.32. In principle, the determination of the roentgen requires the use of a large ionization chamber containing air, called a "free-air" chamber. For practical purposes, however, the air can be replaced by a much smaller volume of solid material having an equivalent absorption of x- or gamma radiation. The absorption is largely determined by the atomic number, and so the material should consist of elements with atomic numbers near 7 or 8, the values for nitrogen and oxygen, respectively. Secondary standards for measuring radiation dosage in roentgens are therefore made of approximately air-equivalent material, such as Bakelite, Lucite, or other plastic, consisting largely of carbon and oxygen. Ionization chambers of this type are called *air-wall chambers;* they are also sometimes referred to as *thimble chambers* because of their shape and small size. The response of an actual chamber is, to some extent, dependent on the energy of the x- or gamma radiation. An instrument of this kind should therefore be calibrated with a free-air chamber if it is to be used over a considerable range of energies.

9.33. Most radiation dose measuring instruments, as described later, are not of the air-wall type nor do they measure the charge collected, i.e., the ion current, without amplification. Consequently, if such instruments are to be employed for the determination of radiation doses in roentgens (or dosage rates in roentgens per hour), they must first be calibrated. The obvious procedure would be to use a free-air or air-wall chamber for the purpose, but this is not convenient. In practice, radiation monitoring instruments are calibrated by means of a relatively constant source of gamma radiation. Radium in equilibrium with its short-lived decay products is the primary standard used for this purpose, with the cheaper cobalt-60 being commonly employed as a secondary standard.

9.34. It may be pointed out that the radiation dose expressed in roentgens does not depend on the time during which it is received. The rate of absorption (or dosage rate) is stated in terms of roentgens per unit time, e.g., roentgens per hour or milliroentgens per hour (r/hr or mr/hr). The integrated product of the dosage rate and the exposure time gives the total dose received in roentgens.

ENERGY ABSORPTION IN AIR

9.35. The dose rate is frequently used as a measure of the *radiation intensity* or flux in a certain region. However, this is justifiable only for gamma rays of a specified energy. The radiation intensity is the rate at which the energy flows past a unit area at a given location, but the dosage rate in roentgens per unit time is a measure of the rate at which energy is *absorbed in air* at that point.

9.36. The amount of energy deposited in air by a roentgen of x- or gamma rays can be calculated in the following manner. The unit electrical charge, i.e., the electronic charge, is 4.80×10^{-10} esu, and this is the quantity of electricity carried by each member (positive or negative) of an ion-pair. Consequently, $1/(4.80 \times 10^{-10})$, i.e., 2.08×10^9, ion-pairs are required to give a total charge of 1 esu of either sign. Hence, from the definition of the roentgen (§ 9.29), the absorption of 1 r of gamma radiation in 0.001293 g of air results in the formation of 2.08×10^9 ion-pairs in air. The energy required to produce one ion-pair in air is known to be about 34.0 ev (§ 2.30) and so the energy required for the formation of 2.08×10^9 ion-pairs is $(34.0)(2.08 \times 10^9) = 7.07 \times 10^4$ Mev or 0.113 erg. Hence, this must be the energy deposited in 0.001293 g of air by 1 r of radiation. The energy absorbed per gram of air per roentgen is thus $0.113/0.00129 = 88$ ergs. The energy equivalents of the roentgen in a number of energy units are summarized in Table 9.1.

9.37. It was seen in § 7.27 that, if at a certain location there is a flux of ϕ_γ photons/(cm²)(sec) of energy E_γ Mev per photon, the rate of energy absorption

TABLE 9.1. EQUIVALENTS OF ONE ROENTGEN
OF X- OR GAMMA RADIATION

1 esu of ion pairs produced per cm³ of air
2.08×10^9 ion-pairs produced per cm³ of air
1.61×10^{12} ion-pairs produced per gram of air
7.1×10^4 Mev absorbed per cm³ of air
5.5×10^7 Mev absorbed per gram of air
88 ergs absorbed per gram of air.

is $\phi_\gamma E_\gamma \mu_e$ Mev/(cm³)(sec), where μ_e is the energy absorption coefficient (§ 7.27) of the medium for the specified gamma (or x-) radiation.   Since 1 r leads to the absorption of 7.1×10^4 Mev of energy per cm³ of air, it follows that

$$\phi_\gamma \text{ photons}/(\text{cm}^2)(\text{sec}) \equiv \frac{\phi_\gamma E_\gamma \mu_e}{7.1 \times 10^4} \text{ r/sec}, \qquad (9.1)$$

where μ_e is the energy absorption coefficient of air for the given photons; values

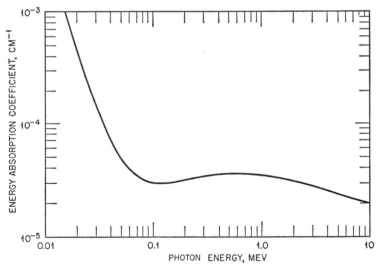

FIG. 9.1. Variation of energy absorption coefficient of air with gamma-ray energy

of μ_e as a function of photon energy are plotted in Fig. 9.1.   Upon rearrangement of equation (9.1), it is seen that

$$1 \text{ r/sec} \equiv \frac{7.1 \times 10^4}{E_\gamma \mu_e} \text{ photons}/(\text{cm}^2)(\text{sec})$$

or, in terms of milliroentgens (10^{-3} r), abbreviated to mr, per hour, which is a more convenient unit for radiation dose rates,

$$1 \text{ mr/hr} \equiv \frac{0.020}{E_\gamma \mu_e} \text{ photons}/(\text{cm}^2)(\text{sec}). \qquad (9.2)$$

The curves in Fig. 9.2 show the gamma-ray equivalents of a dose rate of 1 mr per hr as a function of the photon energy; the values in photons/(cm²)(sec) are derived directly from equation (9.2) and those in Mev/(cm²)(sec) are obtained upon multiplying by E_γ, so that they are equal to $0.020/\mu_e$.

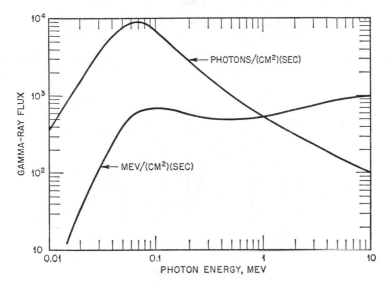

FIG. 9.2.   Gamma-ray photon and energy flux equivalents of 1 mr/hr

DOSE RATE AND RADIOACTIVE SOURCE STRENGTH

9.38. It is sometimes required to know the dose rate to be expected from a given source of gamma rays.   Consider a radioactive *point source* having a strength of C curies (see § 2.15 *et seq.*), so that it decays at the rate of $3.7 \times 10^{10}C$ dis per sec.   It will be postulated that one gamma-ray photon is produced in each act of nuclear disintegration, so that the source yields $3.7 \times 10^{10}C$ photons per sec.   (If not all disintegrations are accompanied by gamma rays, due allowance must, of course, be made.)   Suppose that the radiation from the point source is emitted uniformly in all directions; then at a distance R cm from the source, the photons will be distributed uniformly over the surface of a sphere of area $4\pi R^2$ cm² (Fig. 9.3).   Neglecting the attenuation of radiation by the air from the source to the point of observation, the gamma-ray flux at the distance R from the unshielded source is $3.7 \times 10^{10}C/4\pi R^2$ photons/(cm²)(sec).   Upon combining this result with equation (9.2), it is found that

$$\frac{\text{Dose rate at distance } R \text{ cm}}{\text{from } C \text{ curie source}} = \frac{3.7 \times 10^{10}CE_\gamma\mu_e}{(4\pi R^2)(0.020)} \text{ mr/hr,} \qquad (9.3)$$

the energy E_γ being in Mev.

9.39. An examination of Fig. 9.1 shows that over a considerable range of photon energies, e.g., from 0.07 to 2 Mev, the energy absorption coefficient for air varies only between 3.0×10^{-5} and 3.7×10^{-5} cm^{-1}. For an approximate treatment, μ_e may therefore be taken to have a constant value of 3.5×10^{-5} cm^{-1}; in these circumstances equation (9.3) becomes

$$\text{Dose rate at distance } R \text{ cm from } C \text{ curie source} = 5.2 \times 10^6 \frac{CE_\gamma}{R^2} \text{ mr/hr.} \qquad (9.4)$$

It should be pointed out that the use of a constant value for μ_e makes the dose

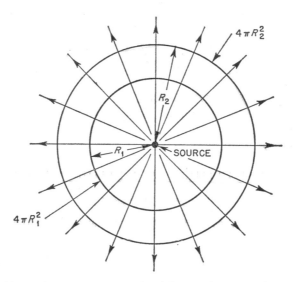

FIG. 9.3.   Inverse square law of spreading of radiation

rate proportional to the radiation intensity, as defined in § 9.35.   Since μ_e actually varies with the photon energy, this result is not exact; however, it is accurate, in the present circumstances, to within ±10 per cent.

9.40. It will be observed that, according to equations (9.3) and (9.4), the dose rate at a given distance from a point source is inversely proportional to the square of that distance.   Consequently, the dose rate falls off with increasing distance in air, even though attenuation by the medium is negligible.   The reason, as may be seen from Fig. 9.3, is that with increasing distance from the point source the same number of photons is spread over a surface which increases in area as the square of the distance.   This result is sometimes referred to as the inverse square law of the geometrical spreading of radiation from a point source.   For a distributed source, e.g., a plane, the inverse square law relationship is not applicable but there is, nevertheless, a falling off of the dose rate with distance from the source.   It is this general result which makes appreciable separation from a radioactive source an important factor in radiation protection.

9.41. A simple and useful form of equation (9.4) is obtained if the distance from the source to the observation point is expressed in feet; if this distance is f ft, then, since 1 ft = 30.5 cm, it is found that

$$\text{Dose rate at distance } f \text{ ft from } C \text{ curie source} \approx 6 \times 10^3 \frac{CE_\gamma}{f^2} \text{ mr/hr}$$

$$= \frac{6CE_\gamma}{f^2} \text{ r/hr.}$$

It follows from this result that at a distance of 1 ft from a point gamma-ray source of strength C curies the dosage rate is approximately $6CE_\gamma$ r per hour.

9.42. The foregoing expressions apply to an unshielded source, but if an absorbing material of thickness x cm and (total) attenuation coefficient of μ cm^{-1} is placed between the source and the point at which the dose rate is being calculated, the general equation (9.4) becomes

$$\begin{array}{l}\text{Dose rate at distance } R \text{ cm from } C \\ \text{curie source with } x \text{ cm absorber}\end{array} = 5.2 \times 10^6 \frac{CE_\gamma}{R^2} e^{-\mu x} \text{ mr/hr,} \quad (9.5)$$

where the exponential factor on the right allows for attenuation of the gamma radiation by the absorbing material (§ 2.61).

9.43. Strictly speaking, the factor $e^{-\mu x}$ should include an allowance for attenuation of gamma radiation in the R cm of air separating the source from the observation point; however, μ for air is on the order of 10^{-4} cm^{-1}, and so, for distances up to a few meters, $e^{-\mu x}$ is so close to unity that attenuation by the air may be neglected.

Example 9.1. A 1.0-mc source of cobalt-60, available commercially for calibrating radiation instruments, is contained in a small capsule at the end of a holder rod so that it is essentially a point source. (a) What is the gamma-ray dose rate at distances of 10 cm, 50 cm, and 1 meter from the source? (b) When not in use, the cobalt-60 source is kept in a lead "pig" having walls 1.5 in. thick; estimate the dose rate at 1 meter from the source inside the pig. (c) Determine the mass of cobalt-60 present in the 1-mc source.

(a) In each act of radioactive decay, a cobalt-60 nucleus emits two gamma-ray photons with energies of 1.17 and 1.33 Mev, respectively. Hence, in the present case E_γ is 2.50 Mev. It follows, therefore, from equation (9.4) that

$$\text{Dose rate at 10 cm} = \frac{(5.2 \times 10^6)(0.0010)(2.50)}{(10)^2} = 130 \text{ mr/hr.}$$

Similarly,

$$\text{Dose rate at 50 cm} = 5.2 \text{ mr/hr}$$
$$\text{Dose rate at 1 meter} = 1.3 \text{ mr/hr.}$$

(b) The attenuation of the gamma rays of each energy should be calculated separately (see Example 2.4), with allowance for the buildup factors (§ 10.67). However, for the purpose of making a rough estimate, the energy difference in the present case may be ignored and an average value taken for the linear attenuation coefficient (0.70 cm^{-1}); the buildup factor is approximately 1.5. Hence, using the result derived above for the dose rate at 1 meter in air, it follows from equation (9.5), noting that 1 in. = 2.54 cm, that

$$\text{Dose rate} = (1.3)(1.5)e^{-(0.70)(1.5)(2.54)}$$
$$= 0.2 \text{ mr/hr.}$$

(c) The mass of cobalt-60 is determined from equation (2.12). Upon dividing the latter by 3.70×10^{10} the result is the number of curies, C, equivalent to g grams of cobalt-60. The half-life of cobalt-60 is 5.3 yr, i.e., 1.7×10^8 sec; hence,

$$g = \frac{CAt_{1/2}}{1.13 \times 10^{13}} = \frac{(0.001)(60)(1.7 \times 10^8)}{1.13 \times 10^{13}}$$

$$= 9 \times 10^{-7} \text{ gram.}$$

9.44. In deriving the equations relating the dose rate to the source strength, the source has been regarded as a point; in other words, it has been assumed that its size is small in comparison with the distance separating it from the position where the dose rate is to be estimated. In many instances, however, the source may be a relatively large sphere, a parallelepiped, a disk, or other shape that cannot be treated as a point. The equations which permit dose rates from a point source to be converted into the values to be expected from sources of other forms will be given in Chapter 10.

THE RAD AND THE REM

9.45. As stated earlier, the roentgen can be used only to express doses of x-rays and gamma rays. A more general unit, which can be applied to all types of nuclear radiation, is the *rad*. It is defined as the quantity of radiation leading to the absorption of 100 ergs of energy per gram of irradiated material.* The rad is thus a quantity capable of physical measurement (or determination), in principle, for any radiation and any exposed material. In soft tissue, an exposure of 1 r leads to the absorption of approximately 1 rad, but in bone, which has a large energy absorption coefficient, more than 1 rad, i.e., more than 100 ergs per g, is absorbed when exposed to 1 r of gamma (or x-) rays.

9.46. Although all ionizing radiations are capable of producing the same kinds of biological effects, the absorbed dose (measured in rads) which will produce a particular effect may vary considerably from one radiation to another. In general, those particles which cause higher specific ionizations along their paths are those having the greater biological effectiveness. For various reasons it is convenient to compare the biological effectiveness of ionizing radiations with that of x-rays of 200-kev energy. On this basis, a quantity called the *relative biological effectiveness* (or RBE) is defined by

$$\text{RBE} = \frac{\text{Physical dose of 200-kv x-rays to produce effect of interest}}{\text{Physical dose of comparison radiation to produce same effect}}.$$

The value of the RBE for a particular radiation may depend upon several factors, e.g., the kind and degree of the biological effect, the nature of the organism or tissue, and the rate at which the dose is delivered.

---

* The rad replaced an older unit called the *rep* (roentgen equivalent physical) to which it is approximately equivalent for radiation absorption in soft tissue. The rep was defined as the energy absorption in 1 g of tissue from 1 r of gamma (or x-) radiation.

9.47. It is evident that the rad, being simply an indication of the amount of energy absorbed from the radiation, irrespective of the nature of the latter, does not provide an adequate measure of the resulting biological injury, since it does not take into account the relative effectiveness of the given radiation.   In an attempt to devise a unit which would provide a better criterion of biological injury when applied to different radiations, the *rem* (roentgen equivalent man) has been introduced.   The rem is a biological dose unit defined by the relationship

$$\text{Dose in rems} = \text{RBE} \times \text{Dose in rads}.$$

9.48. Since, as stated above, the actual value of the RBE depends upon various circumstances, it is apparent that the equivalence implied by the term rem will apply only to the production of the specific effect and under the conditions for which the particular RBE was determined.   However, for purposes of radiation protection, it has become the practice to use a single value of the RBE for different effects of a particular kind of radiation, based on that effect for which the RBE is believed to be highest.   The data in Table 9.2 have been used for radiations of interest in nuclear reactor operation.   The effects considered to be critical in the determination of the various RBE values are given in some cases.

TABLE 9.2.   RECOMMENDED VALUES OF RELATIVE BIOLOGICAL EFFECTIVENESS

Radiation	RBE	Biological Effect
X-rays, gamma rays, beta particles and electrons of all energies	1	Whole body irradiation (blood-forming organs critical)
Fast neutrons and protons up to 10 Mev	10	Whole body irradiation (cataract formation critical)
Thermal neutrons	2.5	Whole body irradiation
Naturally occurring alpha particles	10	Carcinogenesis outside skeleton*

 * For carcinogenesis of the skeleton, alpha-emitting isotopes are compared with radium-226, on the basis that the maximum permissible amount of radium in the skeleton is 0.1 μc.

9.49. Since 1 rem of one kind of radiation will (conventionally, at least) produce the same biological damage to a given tissue as 1 rem of any other kind, doses of different radiations when measured in rems are additive.   This is not true, of course, from the biological standpoint, for doses expressed in rads. However, by making use of the accepted RBE values the measured doses of different radiations expressed in roentgens or rads can be converted into rems and then added.

FLUX EQUIVALENTS OF BIOLOGICAL DOSE RATES

9.50. Since an exposure of 1 r of gamma or x-rays leads to the absorption of approximately 1 rad in soft tissue, and the RBE of these radiations is unity, it follows that the exposure in roentgens is essentially equivalent to the biological dose rate expressed in rems.   Consequently, the curves in Fig. 9.2 indicate the photon or energy flux, for photons of a given energy, which corresponds to a biological dose rate of 1 millirem (mrem) per hour.

9.51. For neutrons, the determination of the biological equivalent of the flux is more complicated than for gamma rays.   Consideration must be given to the

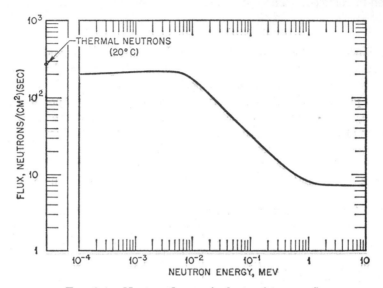

FIG. 9.4.   Neutron flux equivalents of 1 mrem/hr

effects of the scattering by protons and also by carbon, nitrogen, and oxygen nuclei, and also of the (n, p) reaction with nitrogen-14 and the (n, γ) reaction with hydrogen.   The appropriate calculations have been made to determine the flux of neutrons of various energies which would represent a dose rate of 1 mrem per hour, taking into account the RBE values of the neutrons of different energies.   The results for thermal neutrons (0.025 ev) and for faster neutrons, in the energy range from 10^{-4} to 10 Mev, are given in Fig. 9.4 [5].

STANDARDS OF RADIATION PROTECTION

RADIATION PROTECTION GUIDES

9.52. As stated in § 9.4, Radiation Guides have been recommended for the doses which should not normally be exceeded by persons whose work renders them liable to exposure to various nuclear radiations [6].   On the basis of information obtained from radiologists, x-ray technicians, etc., from the biological injury caused by radium, from animal experiments, and from other sources, the following Radiation Protection Guides (RPG) have been suggested. For occupationally exposed individuals, the RPG should not exceed 3 rems over any period of 13 weeks or 12 rems per year. However, in order to minimize the genetic effects of nuclear radiations, the dose accumulated over a period of years should not exceed a total of $5(N - 18)$ rems, where N is the person's age in years.* This means that, effectively, the average dose should not exceed 5 rems per year over a period of years. For limited areas of the body or for specific organs the doses may exceed these values, but from the reactor engineering point of view, specifically the design of radiation shielding (Chapter 10), it is postulated that the radiation dose must not exceed 5 rems in any one year. Assuming a 50-week working year and a 40-hour week, the equivalent maximum dose rate is thus 2.5 mrems per hour for individuals whose work may involve continuous exposure to radiation. This dose rate applies to radiation from external sources; the problem of radioactive material within the body will be considered later.

9.53. Although the materials in a nuclear reactor emit alpha and beta particles, these are so readily absorbed that they can be disregarded in the design of radiation shielding for a reactor installation. In this connection, it is only the fast and thermal neutrons and gamma rays which need to be considered. It is important, however, to note that it is the total dose, due to all these radiations combined, which must not exceed the design value of 2.5 mrems per hour.

9.54. The RPG values given above are for normal occupational exposures. In the event of an emergency, the particular circumstances may make it necessary for certain individuals to accept larger doses. The limit currently recommended for a planned emergency exposure of occupational workers, other than women in the reproductive age, is 12 rems. An accidental exposure of 25 rems or less to the whole body is believed to have no significant effect on the radiation tolerance of the exposed individual provided that it occurs only once in a lifetime. The biological effects of these and larger doses will be described in § 9.64.

9.55. For persons who are not occupationally exposed to radiation, the dose should be limited to 0.5 rem per year to the whole body or to the gonads, and 1.5 rems per year for other organs. The accumulated biological dose up to age 30 should not exceed 1.5 rems to the gonads (or the whole body).

RADIOACTIVITY CONCENTRATION GUIDES

9.56. Since the radiation dose received from a particular radioactive material in the body is proportional to the quantity of that material present, excessive exposure to radiation from internal sources is avoided by limiting the body content of the material. This is achieved by restricting the rate at which the

* In the United States, the age of 18 years is the minimum for employment in any activity in which exposure to radiation is involved.

material enters the system.   In many instances intake into the body results primarily from the occurrence of the radioactive material in the air (or water) supply.   It is then desirable to know the maximum average concentrations of the active species which may be permitted in the air (or water) without eventually resulting in excessive accumulation of the material in the body; these quantities have been called Radioactivity Concentration Guides (RCG) [7].

9.57.   Radioactivity Concentration Guide values for most radioisotopes are not known, since there is insufficient direct experience for establishing levels at which individual isotopes, except possibly radium and radioiodine, become hazardous to human beings.   For most of the radioactive materials which, like radium, accumulate in the skeleton, maximum acceptable levels for the human body are estimated, directly or indirectly, from comparative experimental studies with other species of mammals.   For radioisotopes which concentrate in body organs or tissues other than the skeleton, it is assumed that the body content should not exceed the lowest level which will result in an exposure, in most cases, of 0.1 rem per week.   For less sensitive organs, the dose rate may be larger.

9.58.   If most of a particular radioisotope is fairly uniformly distributed in a single organ or tissue, the maximum acceptable body content is inversely proportional to the mass of that organ or tissue.   For example, since most of the iodine in the body is concentrated in the thyroid gland, the average total activity of iodine-131 permitted is much less than that of sodium-24 which is distributed throughout the body.   This is the case in spite of the fact that the average energy of the radiation from iodine-131 is less than that from sodium-24.

9.59.   The quantity of an active material remaining in the body after its entry is dependent upon its radioactive half-life and also upon its *biological half-life;* the latter is the time taken for the amount of the element in a particular body tissue to decrease to half of its initial value due to elimination by natural processes.   Combination of the radioactive and biological half-lives gives rise to the *effective half-life,* which is the time required for the amount of a specified radioisotope to decrease to half the original value as a result of both radioactive decay and natural elimination.   It is understood, of course, that there is no replenishment of the isotope from any source during the time under consideration.   In the simplest case, the effective half-life in the body as a whole is essentially that in the principal tissue in which the material concentrates.

9.60.   The maximum acceptable average rate of entry into the body of a given radioisotope is determined by (1) initial retention and distribution in the body, (2) maximum acceptable content in the critical organ, and (3) effective half-life. These three factors are not simple ones, since each in turn is dependent upon two or more other factors.   Initial retention, for example, is not only related to the biochemical properties of the chemical element (or, in some cases, the chemical compound), but also to the mode of intake into the body.   Thus, nearly all of the iodine entering the digestive tract may be initially retained in the body, whereas only about 0.01 per cent of ingested plutonium is retained.   On the other hand, it is estimated that as much as 20 per cent of soluble plutonium entering the lungs may reach the circulatory system.

9.61. If other factors are more or less the same, the rate of intake into the body which will eventually result in a specified body content is inversely proportional to the effective half-life of the radioisotope in the critical organ or tissue. If either the radioactive half-life of the given isotope or the biological half-life of the element in the tissue is very short compared to the other, the effective half-life is approximately equal to the shorter of the two, since this essentially determines the rate of loss of the particular active material from the tissue.

9.62. Of the radioactive species involved in reactor operations or in the processing of reactor fuels, plutonium-239, natural uranium (and its isotopes), natural thorium, strontium-90, and iodine-131 have low RCG values, expressed in activity per unit volume, in air breathed by human beings. This is because of the importance of one or more of the three factors mentioned in § 9.60. In the case of plutonium-239, all three factors contribute to limiting the RCG in air to an extremely small value. However, since its uptake from the digestive tract is low, the RCG of plutonium-239 in water is greater than for strontium-90.

9.63. Because of the biological hazard associated with plutonium, which tends to concentrate in the skeleton where it has a long effective half-life, special precautions are always taken in the handling of plutonium-239 and its compounds. Work with these materials is always performed in a closed space, often in an atmosphere of an inert gas since plutonium metal oxidizes so readily. For this purpose, "glove boxes" and remote-operation techniques are employed (§ 9.65).

ACUTE RADIATION DOSES

9.64. On the basis of present knowledge, it is expected that a chronic radiation dose rate of 2.5 mrems per hour (or 20 mrems per working day), referred to in § 9.52, can be accepted for many years without producing any noticeable effects. It has been estimated, in fact, that as much radiation as 500 mrems per day over many months would produce no decrease in efficiency and no significant effect on the life span. Appreciably larger chronic doses continued for a period of time, however, might cause some degree of radiation injury. In the event of a special emergency or an accident, a person might incur, knowingly or otherwise, a fairly large acute (single) dose of radiation. The probable effects of such whole-body doses, up to 1000 rems, are summarized in Table 9.3 [8].

PROTECTION OF PERSONNEL

REMOTE CONTROL

9.65. The protection of personnel from radiation requires the design of suitable shielding for the reactor and for all vessels, piping, pumps, etc., which contain

TABLE 9.3. PROBABLE EFFECTS OF ACUTE WHOLE-BODY RADIATION DOSES

Acute Dose (rems)	Probable Clinical Effect
0 to 25	No observable effects.
25 to 100	Slight blood changes but no other observable effects.
100 to 200	Vomiting in 5 to 50 per cent within 3 hours, with fatigue and loss of appetite. Moderate blood changes. Except for the blood-forming system, recovery will occur in all cases within a few weeks.
200 to 600	For doses of 300 rems and more, all exposed individuals will exhibit vomiting within 2 hours or less. Severe blood changes, accompanied by hemorrhage and infection. Loss of hair after 2 weeks for doses over 300 rems. Recovery in 20 to 100 per cent within 1 month to a year.
600 to 1000	Vomiting within 1 hour, severe blood changes, hemorrhage, infection, and loss of hair. From 80 to 100 per cent of exposed individuals will succumb within 2 months; those who survive will be convalescent over a long period.

or carry radioactive material. In addition, measures must be taken to prevent exposure to the radiations from spent fuel elements after removal from the reactor. Some aspects of the shielding problem will be considered in the next chapter, but, in general, it may be stated that adequate shielding requires the use of considerable thicknesses of material, e.g., several feet of concrete or of water. This means that many phases of operation, instrumentation, maintenance, etc., must be performed by remote control. A great deal of work has been and is being done on the development of remote-control devices for the handling of radioactive material both in the laboratory and in the plant.

9.66. Remote-control operation, in general, involves the use of established mechanical and electrical principles, but the problems are somewhat novel in character, and, consequently, they represent a challenge to the engineer whose interest lies in this direction. Some of the problems are connected with the discharge of spent fuel elements and with the addition of fresh fuel to a reactor. These operations are especially difficult if they have to be performed without shutting the reactor down. Other problems are associated with the chemical plants for the reprocessing of the highly radioactive discharged fuel elements (Chapter 8).

SAFETY MEASURES

9.67. Shielding for a reactor and related equipment is constructed so that operating personnel normally receive considerably less than the RPG dose mentioned earlier. Nevertheless, there is a possibility of leakage, especially of neutrons, through cracks or other openings, particularly if the shield has removable sections. A thorough survey of the entire shield should thus be undertaken with the reactor at low power before it is brought up to full power for the first time. If radiation leaks are found, steps should be taken to repair them or to restrict access to the danger area.

9.68. It is very improbable that personnel connected with reactor and associated operations will receive during normal operation an excessive dose from the radiation that penetrates the shielding. However, there are various maintenance activities, both routine and emergency, which may require the handling of radioactive materials. In such cases some protection may perhaps be attainable by the use of temporary or mobile shielding or, if possible, by performing the operation at a distance with the aid of suitable tools. All work in a hazardous area must be carried out in cooperation with a radiation surveyor (or monitor) who will determine, from the observed dosage rate, the permissible time in that area.

9.69. Access to and egress from an area where there is a possibility of radioactive contamination being present should be by definite routes to prevent spread of the contamination. Personnel entering the area are supplied with expendable clothing, gloves, and shoe covers; these are left at the prescribed exit for decontamination or disposal, according to circumstances. Before being dismissed, each worker is subjected to careful monitoring (§ 9.95).

9.70. Special precautions must be taken to prevent radioactive material from entering the body. This is particularly important where plutonium is involved, as pointed out in § 9.63. Strict personal hygiene must be practiced in areas where there may be radioactive contamination, and there should be no smoking, eating, or drinking in such places. The ventilating system for reactor and other buildings should be designed so that the air always flows from regions of lower radiation level to those of higher level. Dust must be controlled, e.g., by vacuum cleaning if possible; if sweeping is required, there should be enough moisture present to prevent dust being raised. In areas where some dust in the air is unavoidable, operating personnel should wear suitable masks.

MONITORING INSTRUMENTS [9]

9.71. The monitoring operations performed in order to protect personnel from receiving excessive doses of radiation may be divided into three broad classes: (1) general monitoring of air, water, and areas; (2) more detailed survey of working areas and of equipment; and (3) monitoring of personnel, clothing, etc. Although the various functions may be distributed differently in different reactor installations, the foregoing classification is convenient for purposes of description. It should be mentioned, too, that the measurement techniques described below may vary from one place to another and that, in any event, the whole field of radiation instrumentation is always undergoing change.

9.72. In addition to instruments of the ionization chamber and proportional counter type, described in § 5.22 *et seq.*, two other kinds of radiation detection devices are used.   These are the Geiger-Müller (or Geiger) counter, frequently abbreviated to G-M counter, and the scintillation counter.

G-M Counters

9.73. The G-M counter is an ionization device, like the ion chamber and proportional counter, operating in region V of Fig. 5.22, where the applied voltage is high, usually from 800 to 1500 volts.   The actual value is not critical, as long as it is within the Geiger region.   Although the counter (or tube) itself can take various forms, both in size and shape, the electrodes almost invariably consist of a central wire anode, which is the collecting electrode, surrounded by a cylindrical cathode, just as for a proportional counter (§ 5.239).

9.74. The main advantage of the G-M counter lies in the very considerable gas amplification, e.g., as high as 10^8, which is possible.   The output pulses are often of the order of several volts, and little, if any, external amplification is required for the operation of a counter or counting-rate meter.   The size of the pulse is essentially independent of the specific ionization caused by the nuclear radiation, so that the instrument cannot distinguish between different ionizing particles which have entered the counter tube.

9.75. A troublesome feature of the G-M tube is that, when an ionizing particle produces an avalanche of ion-pairs, the resulting discharge may itself initiate a series of further discharges.   One particle may then give rise to several pulses or to a long continuous pulse, so that the observed counting rate may be quite misleading.   Methods have been devised for suppressing (or *quenching*) the discharge so that it stops after each primary pulse.   In tubes of the "self-quenching" type a suitable organic compound or a halogen added to the gas serves to quench the discharge.   Most G-M tubes are of this type; they contain argon, as the main filling gas, and a small proportion of a quenching material, e.g., ethyl alcohol.

9.76. Another drawback to the G-M counter is that a properly quenched tube has a relatively long *dead time;* this is the period between pulses during which the counter is not sensitive to the entry of ionizing particles.   The instrument thus has poor resolving power and is unreliable when the rate of arrival of particles exceeds about 5000 per sec.   For this reason, the G-M counter should not be used, or used only with great caution, when entering an area of high radiation level; the instrument may give a reading that is too low, leading possibly to an overexposure to radiation.

Scintillation Counters

9.77. In solids and liquids, nuclear radiations generally produce electronic excitation, rather than ionization.   The excess energy in certain substances, called *phosphors,* is then emitted, within a very short time, as a flash of light. If the phosphor is transparent, the light can be allowed to fall on the cathode of a photomultiplier tube. The latter produces a pulse of appreciable voltage corresponding to each particle (or photon) of nuclear radiation. The pulses can then be counted in the same manner as with other radiation detecting devices. An advantage of the scintillation counter is that high counting rates are possible. This is one reason why it has replaced the G-M tube in some applications. However, it is more expensive, requires more complicated electronic equipment, and is not so robust as the Geiger counter.

AIR AND WATER MONITORING: FIXED INSTRUMENTS

9.78. An important hazard in air consists of dust particles that have become radioactive, and several different instruments are used for removing the particles and determining their activity. For continuous automatic monitoring of the atmosphere in areas within or surrounding a reactor site, the *constant air monitor* can be used; this is a device for measuring the particulate beta and gamma activity in the ambient air. A vacuum pump draws air continuously through a filter paper wrapped around a G-M tube mounted in a lead shield. The tube is connected with a count-rate meter which is attached to an automatic recorder. A continuous record is thus obtained of the rate of collection of activity and of the total activity collected in a given time. An associated relay-alarm system provides audible warning when a predetermined radiation level is reached. The filter paper upon which the dust particles collect is changed at regular intervals, usually daily.

9.79. Two other instruments are used to collect the dust present in a known volume of air; the activity of the material collected is then determined in a counting laboratory. One is a semiportable instrument consisting of an electric motor driving a blower unit which forces air at the rate of about 5 ft³ per min through a chamber containing a filter paper. The instrument is used to determine beta and gamma activities; it is not recommended for alpha-particle determinations because of spurious counts given by the fiber of the filter. In the second type of instrument the blower forces the air through a corona discharge produced by a high voltage, so that the dust particles acquire electrical charges. They are then precipitated electrostatically onto a thin aluminum foil.

9.80. For checking on the background activity of various zones around a reactor or other source of radiation, gamma-monitoring systems of a permanently installed type are desirable. The dosage rate in various locations can then be recorded at a central point. Thus, not only are personnel entering the area informed in advance of the existing radiation levels, but significant increases can be detected and steps taken to determine the cause and to correct the situation.

9.81. A special monitoring instrument is often used to record the background radiation in working areas. It consists of a large ionization chamber which is coated with carbon, for the detection of gamma radiation only, or with enriched boron (§ 5.211) if it is to be used for thermal neutrons in addition to gamma rays. The chamber is connected with an amplifier circuit and a counting-rate meter which is calibrated to read directly in milliroentgens (or millirads) per hour. A relay system, actuating an alarm, may be set to provide a warning signal at any desired gamma-radiation level. Actual thermal-neutron fluxes are not measured, but relative values are indicated.

9.82. If water is used as either primary or secondary coolant for a reactor, it may be discharged ultimately to a public stream. Monitoring is therefore necessary to avoid the direct hazard that might arise from drinking the water or the indirect hazard resulting from the accumulation of radioactivity in various forms of aquatic life. When water is the primary coolant, it leaves the reactor with a certain amount of induced activity due to the capture of neutrons by traces of dissolved salts. The water is held for several hours in a retention basin in order to allow the short-lived activity to decay.

9.83. The ideal method for monitoring water would be to use an immersion or dipping type of G-M tube, whereby a continuous record of the activity of the effluent could be obtained. Although such instruments exist, the procedure is not too satisfactory because the tube becomes contaminated with continued use. One way of overcoming this difficulty is to take measurements periodically, thus greatly reducing the time of contact of the tube with radioactive contaminants. An alternative is to place a sensitive G-M tube just above the surface of a large volume of water, e.g., a retention basin. In any case, it is desirable that samples of water be taken to the laboratory regularly, where they are evaporated to dryness and the residues counted or subjected to radioanalysis. Consideration must be given to the possible loss of volatile elements during the evaporation.

9.84. In addition to cooling water, it is frequently necessary to discharge dilute aqueous solutions containing traces of radioactive material as an aspect of waste disposal (§ 8.170). These solutions are usually held up in tanks until the activity has largely decayed, and then they are gradually discharged into settling basins from which they find their way into the surrounding soil and possibly into an existing drainage system. As a protective measure, samples of water and soil from various locations are regularly taken to the laboratory for appropriate analysis.

RADIATION SURVEY: PORTABLE INSTRUMENTS [9]

9.85. All exposed surfaces, especially working surfaces, and equipment which has to be approached or handled must be subjected to a detailed radiation survey. In addition, operations involving radioactive material should be monitored so as to make sure that persons performing the operation do not receive an excessive dose of radiation. The instruments used must be portable and reasonably sturdy, and they should respond rapidly, either visually or audibly. The type of instrument chosen for a particular radiation survey depends on (1) the nature of the radiation, e.g., alpha, beta, or gamma radiation or neutrons; (2) the magnitude of the dosage rate; and (3) whether qualitative (detection) or quantitative (measurement) information is required. It may be noted that it is a common practice to make qualitative observations first and to follow them with a quantitative survey if required. Information concerning the type of radiation is generally available from the nature of the operation and of the materials involved. If this is not the case, a preliminary scanning survey may be necessary in order to determine which instrument is most suited to the circumstances.

9.86. Many instruments, made by different manufacturers, are available for radiation surveys, but most of them fall into a few general categories determined by the purpose for which they are intended. The descriptions given here may be taken as fairly typical, although there may be variations due either to personal preferences or to improvements in design which are continually being introduced.

Alpha-Survey Instruments

9.87. Instruments used for making alpha-particle surveys give readings which are semiquantitative in nature. These instruments are primarily intended as detectors and provide an approximate indication only of the quantity of alpha-emitter present. Ionization chamber counter-type meters have been used for alpha surveys because the particles have such a high specific ionization that internal amplification is not necessary. The detector chamber may be included in the instrument box or it may be contained in a separate probe (cf. Fig. 9.5). In either case it must be provided with a thin "window" which can be penetrated by the alpha particles. By adjusting the sensitivity of the pulse counter the instrument will respond to alpha particles only and not to the beta particles or gamma rays which may enter the chamber. Scintillation counters, employing thin layers of an inorganic phosphor, e.g., zinc sulfide activated with silver or sodium iodide activated with thallium, are also used in alpha-survey meters. One such instrument in common use is known as *Poppy*, so called because each pulse causes an audible "pop" in a loud-speaker.

Beta-Gamma Instruments

9.88. The most common type of instrument used for low levels of beta and gamma radiations (less than 5 mrad/hr) is the G-M survey meter (Fig. 9.5). It is employed in decontamination operations and for general survey of personnel, clothing, protective equipment, tools, etc., where a qualitative indication of the activity, often at a relatively low level, is all that is required. The G-M tube is contained in a probe attached to the portable instrument box by means of a cable about 36 in. long, as shown in the figure. The high voltage required to operate the tube is supplied by batteries or by a vibrator power supply contained in the box. With the probe unshielded, the instrument detects beta particles with energies in excess of 0.2 Mev and gamma rays of all energies of interest. On the other hand, with the probe shielded, only gamma rays with energies in

Fig. 9.5.   G-M survey meter (Beckmann Instruments, Inc.)

excess of 0.1 Mev, but no beta particles, can be detected.   It is because of the considerable internal amplification possible in a G-M tube that the device is particularly suited to the detection of low-level activities, but it should not be used for quantitative dose rates.

9.89. For the measurement of very low levels of gamma radiation, a portable scintillation counter has been found useful.   It is designed so as to be insensitive to alpha and beta particles and can measure dose rates in the range of 0.005 to 5 mr (or mrad) per hour.

9.90. For beta and gamma radiation levels in excess of about 5 mrad per hour, ionization chambers are generally used for both detection and measurement. An instrument that is frequently employed is called the *Cutie Pie* (or C.P.) meter Although it is not very accurate (± 10 per cent), it is fairly robust, is light in weight, and can be read very easily (Fig. 9.6).   A thin window permits entry of both beta and gamma radiations, except those of low energy, whereas a thicker window (or shield) eliminates nearly all beta particles.   The C.P. meter can be used for dose rates up to about 20,000 mrad (20 rad) per hr.   For higher radiation fields, various compact instruments of the ionization chamber type are available, with the chamber (or probe) attached to a long pole.

Neutrons

9.91. The detection of slow neutrons can be performed with a proportional-counter survey meter, similar to that used for alpha particles, containing boron

Fig. 9.6.   "Cutie Pie" survey meter (Tracerlab, Inc.)

trifluoride preferably enriched in boron-10. If the walls are not too thin (so that alpha particles cannot penetrate them) and the counting circuit is adjusted so that it is not sensitive to beta and gamma radiation, the instrument will respond only to slow neutrons. Although the meter is calibrated to read thermal-neutron flux, up to about 10^5 neutrons/$(cm^2)(sec)$, the results are semiquantitative only.

9.92. An instrument similar to that described above, without boron trifluoride but lined with paraffin, is used for detecting fast neutrons. The pulses are produced by the recoil protons resulting from elastic collisions of the fast neutrons with the hydrogen atoms in the paraffin. The drawback to this device is that it also counts some of the photons in a strong gamma-ray field.

9.93. For more accurate determination of fast neutrons, a compensated system of two ion chambers is used. One chamber contains methane and records both fast neutrons and gamma rays, whereas the other, containing argon, is sensitive to gamma rays only. The difference in ion currents is then a measure of the fast-neutron flux.

9.94. Scintillation counters, based on the (n, α) reaction with lithium-6, have been developed for the detection of both fast and slow neutrons. The phosphor consists of lithium-6 iodide activated with europium. Crystals of lithium iodide enriched to the extent of 96 per cent in lithium-6 are available commercially.

PERSONNEL MONITORING

9.95. All persons working within areas where nuclear radiations may be encountered must wear individual personnel monitoring devices; two types of such instruments are in general use: (1) pocket (ionization chamber) meters and (2) film badges (or film dosimeters). In addition to these meters which record the total radiation dose received over a period of time, individuals leaving a contaminated area are surveyed by appropriate counters which will be described later.

Pocket Meters [10]

9.96. Pocket meters are actually small ionization chambers similar to a fountain pen in size and shape. The simplest form consists of a cylinder of plastic material lined internally with graphite which acts as the positive electrode. A stout wire, supported at its ends by insulators, runs axially through the cylinder and serves as the negative electrode (Fig. 9.7). The electrodes are charged in a

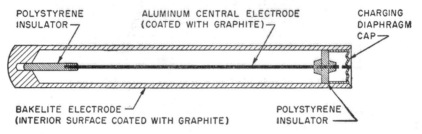

FIG. 9.7.   Pocket ionization chamber

charger-reader (or *minometer*) to a known potential difference, about 150 volts (Fig. 9.8). If there is no exposure to radiation, the charge will remain unchanged. If ionizing radiations enter the chamber, there is a decrease of potential which is proportional to the amount (or dose) of radiation received. The change in potential is measured at the end of the day by means of an electrometer, which is part of the minometer, and the corresponding radiation dose is recorded. Pocket meters measure gamma radiations and such beta particles (energy about 1 Mev or more) as are able to penetrate the walls of the chamber. Their range is usually up to 200 mrad total dosage and the accuracy is within 10 per cent. Similar chambers have been made with boron linings, so that they are also sensitive to slow neutrons.

9.97. Because of the inconvenience involved in charging and reading the pocket meters described above, they are being displaced by the more expensive, but more useful, self-reading pocket dosimeters. These are also ionization chambers resembling a large fountain pen in appearance. The outer electrode is cylindrical and the inner electrode consists of a short wire to which is attached a flexible, metal-coated quartz fiber (Fig. 9.9). When the dosimeter is charged, the quartz fiber diverges from the wire, but when radiation enters the chamber the deflection gradually decreases. The position of the quartz fiber on a cali-

FIG. 9.8.   Charger-reader for pocket chambers (Victoreen Instrument Co.)

brated scale, as observed through a lens system fitted into one end of the instrument, gives the total radiation dose received up to the time of reading.   The self-reading dosimeter usually indicates exposures up to 200 mrad, although instruments with higher ranges are available.

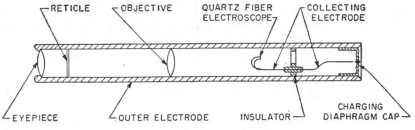

FIG. 9.9.   Self-reading pocket dosimeter

Example 9.2.   A pocket ionization chamber instrument filled with air (assumed to be at standard temperature and pressure) has a volume of 10 cm³ and an electrical capacitance of 50 $\mu\mu$f, i.e., 5.0×10^{-11} farad.   What would be the decrease in voltage resulting from the exposure to a dose of 20 mr of gamma rays?   (This is approximately equivalent to the "guide" dose of 20 mrems per 8-hr day.)

By definition, a dose of 1 r would produce 1 esu of charge in 1 cm³ of air; hence, 20 mr in 10 cm³ of air would yield a charge Q, given by

$$Q = \frac{(20)(10)}{1000} = 0.20 \text{ esu}$$

$$= \frac{0.20}{3 \times 10^9} = 6.7 \times 10^{-11} \text{ coulomb,}$$

since 3×10^9 esu are equivalent to 1 coulomb.

If C is the capacitance of the pocket chamber in farads and V volts is the voltage change, then CV is equal to the charge in coulombs. Hence, in the present case,

$$V = \frac{Q}{C} = \frac{6.7 \times 10^{-11}}{5.0 \times 10^{-11}} = 1.3 \text{ volts.}$$

Film Dosimeters [11]

9.98. The action of the film dosimeter is dependent on the fact that ionizing radiations produce an effect on photographic film similar to that of light; upon development, after exposure to radiation, a general blackening is observed. The extent (or density) of the blackening is a measure of the total amount of radiation to which the film has been exposed, although there is some energy dependence as will be indicated later. Highly ionizing particles, such as alpha particles and protons, form definite linear tracks in their motion through the photographic emulsion. If a special fine-grain (nuclear track) film is employed, the tracks are particularly clear after development and so they can be readily seen (and counted) by means of a microscope. This type of film is also used for the determination of neutron exposure. Fast neutrons produce tracks due to recoil protons, whereas slow neutrons liberate protons as a result of the $N^{14}(n, p)C^{14}$ reaction with nitrogen in the film.

9.99. The film-badge dosimeter consists of a small frame of metal or plastic containing one or two packs of film, about $1\frac{1}{2}$ in. by 2 in. in size; one pack contains beta- and gamma-sensitive film and the other holds nuclear-track film for the determination of neutron exposure, if this is required. For beta- and gamma-radiation measurement two or three films of different sensitivity are generally included in the pack, and these may cover a total dosage range from 50 mrad to 20 rad or to as much as 500 rad,* when exposure is to gamma radiations of energies from 0.3 to 2 Mev. A single film is generally adequate for neutron determination, and this may cover a considerable neutron dose range.

9.100. Film dosimeters of various designs and degrees of complexity are available. The following description covers the general principles of their constructiona and operation. Each of the film packs, wrapped in paper to prevent access of light, is placed in the frame with the neutron-sensitive film at the back. In front, and covering part of the film, are one or more thin metallic shields, one of which consists of cadmium. The latter serves several purposes. In the

---

* Measurement in the high range may prove useful in the event of a major accident. Those in the intermediate range give the dose accumulated over 13 weeks (§ 9.52).

first place, it increases the uniformity of the film response to gamma radiations of all energies.   It happens that the sensitivity of photographic film to gamma radiation increases very markedly for energies less than about 0.3 Mev; thus 1 rad of radiation of 0.1 Mev energy will produce six or more times the blackening caused by the same dose at 1 Mev.   However, passage through a shield of a heavy metal results in the absorption of a considerable proportion of the low-energy radiation (cf. Fig. 2.10), and so the response of the film is more uniform over a range of energies.   The second purpose of the metal shield is to absorb beta particles; hence the beta-gamma film under the shield is affected only by gamma radiations, whereas the unshielded part responds to both gamma rays and to beta particles which pass through the paper enclosing the film, i.e., with energies in excess of 0.1 Mev.   Finally, since cadmium is such a good absorber of slow neutrons, it is penetrated only by fast neutrons.   The shielded portion of the neutron-sensitive film will thus indicate fast neutrons, but the unshielded part will be exposed to both fast and slow neutrons.

9.101.   At certain prescribed intervals, usually at the end of every one or two weeks, the film dosimeters are taken to the laboratory to be developed.   The amount of blackening of the beta-gamma films, as determined by means of an electronic densitometer, gives the total exposure to both beta and gamma radiations and to the latter alone.   These are indicated by the densities of the unshielded and shielded parts of the film, respectively.   The films are calibrated in terms of roentgens by means of a radium or cobalt-60 standard.   The developed films are filed to provide a more or less permanent record of the radiation exposure of the individual.

9.102.   Neutron-sensitive films, after development, are examined with a microscope, and the tracks produced by fast and slow neutrons, under the unshielded portion, and by fast neutrons alone, under the cadmium-shielded portion, are counted.   Weekly inspection of neutron films is necessary only for those persons who are likely to be exposed to such radiations to an appreciable extent.   For others, examination at intervals up to three months or more may be sufficient.

9.103.   It was mentioned in § 9.100 that in some film dosimeters other metallic shields, in addition to cadmium, are used in front of the beta-gamma film; each of these shields then covers a separate part of the film.   The purpose of such shields is to provide information concerning the energy of the gamma radiation. The attenuation coefficients of different metals, e.g., cadmium, copper, and lead, vary differently with energy, and by comparing the densities of the blackening produced under shields of these several metals some conclusions can be drawn concerning the energy of the incident gamma radiation.

PERSONNEL SURVEY INSTRUMENTS

9.104.   In addition to the meters described above, which give the total accumulated radiation dose over a period of time, an aspect of personnel monitoring is the radiation survey of individuals leaving an area where contamination is possible. One method employed is to use a G-M survey meter and to run the probe containing the tube over all parts of the body to detect the possible presence of radioactivity. If the individual or his clothing is found to be contaminated, appropriate decontamination measures must be taken.

9.105. A stationary (nonportable) device for personnel surveying, generally located at a point of egress from an area where contamination may be encountered, is the *hand-and-foot counter.* This consists of an arrangement of five separate G-M tubes and their associated counting-rate meters capable of detecting essentially all gamma radiations and beta particles with energy in excess of 0.2 Mev, but not alpha particles. The individual steps onto a platform and places his hands in two slots provided for the purpose. The counting mechanism is started by pressing down with the hands, and it stops automatically after 24 sec. The five indicators then register the respective amounts of contamination on the palm and back of each hand and on both shoes. If the amounts exceed certain tolerance limits, the individual must report for decontamination.

REACTOR SAFEGUARDS [12]

INTRODUCTION

9.106. The earlier portions of this chapter have been concerned mainly with the radiation protection of individuals during normal operation and maintenance of the reactor plant. Consideration will now be given to what are generally called "Reactor Safeguards." These are the steps which are taken to ensure, as far as is reasonably possible, that the reactor operates safely, and that it will in no circumstances constitute an undue hazard to the health and safety of the inhabitants of the surrounding areas.

9.107. The main aspects of reactor safeguards are as follows: (1) Selection of a site such that, in normal operation, the radioactivity of effluents released will not result in the contamination of air or potable water to levels exceeding the accepted Radioactivity Concentration Guides. (2) Detailed engineering design of all parts of the reactor system, and of the overall system, to minimize the possibility of accidents either as a result of human or mechanical failure, and to deal with the consequences should an accident occur during either normal or abnormal operation. (3) Careful planning of operations and maintenance, with written rules for procedures in both normal and abnormal situations. (4) Precautions taken to prevent the spread of radioactivity to surrounding areas if an accident involving the release of fission products should occur.

SITE SELECTION

9.108. In choosing a site for the reactor installation, it is necessary to make sure that a sufficient area of the environment can be controlled by the operator of the facility and that the general public can be excluded.   The distribution of industry and population in the surroundings is important in assessing the possible consequences of an inadvertent release of radioactivity.   The soil structure and hydrology of the site must be examined to ascertain if there is any danger that leakage or discharge of contaminated liquids from the plant, during normal operation or in the event of an accident, will enter nearby supplies of drinking water or bathing facilities.   The meteorology of the area must be reviewed, with special reference to the possible effects of prevailing winds and the rainfall pattern in carrying airborne radioactivity from the reactor site to populated areas.   If the proposed site is in a region where there is a fairly high probability of hurricanes or tornadoes, it is necessary that this be taken into consideration in the design of the reactor facility and in the plans for its operation.   In areas where earth tremors are relatively frequent, earthquake-resistant construction is mandatory.   In addition, the reactor will have to be designed in such a manner that coolant flow will not be interrupted and that safety (scram) control rods will operate freely should an earth shock occur.

SAFETY FEATURES

9.109. The reactor design will normally include inherent safety features as well as mechanical safeguards of various kinds to deal with specific situations. Among the inherent safety factors are a negative temperature coefficient of reactivity and a negative void coefficient.   In the former case an undesirable increase of temperature will be automatically compensated by a decrease in reactivity which will drop the reactor power and, hence, the temperature.   A negative void coefficient means that there will be a decrease in reactivity should the fuel temperature rise to such an extent as to cause vaporization or excessive boiling of the coolant.   This will generally occur when the coolant is also the moderator, so that the production of a void or an increase in the existing voids, as in a boiling reactor, will tend to make the reactor subcritical because of the decrease in neutron moderation.

9.110. It should be noted, in designing a reactor, that a large negative temperature coefficient can be a source of accidents which must be guarded against. There is, for example, the so-called "cold-water accident" possibility; if the coolant should be returned to the reactor at a lower temperature than normal, there would be a sudden increase in reactivity.   Furthermore, a large negative temperature coefficient means that the cold reactor has considerable excess reactivity.   The possibility of a startup accident is thereby increased.   Precautions must be taken in the design of the reactor to make sure that potentially dangerous situations of this kind do not arise or, if they do, that they can be dealt with in such a manner as to eliminate any hazard to operating personnel and the general public.

9.111. In each aspect of the design of the reactor system, reliability and safety must be prime considerations. For example, the control-rod drive mechanism must be such that an absorber rod can be inserted rapidly but removed only at a slow rate. Several independent scram circuits should be included in the control system, so that different abnormal situations, e.g., a very short period during startup, a dangerously high power level, loss of power, loss of coolant, etc., will automatically cause the reactor to be shut down. In order to prevent burnout of the fuel elements, the maximum permissible heat flux is determined, with an adequate factor of safety (§ 12.25). The characteristics of the coolant circuit are designed to make sure this heat flux is not exceeded in normal operation.

9.112. Safety devices included in the reactor system should be such as to prevent malfunction, as far as possible, or to shut the reactor down should a malfunction occur. In spite of all reasonable precautions, however, potentially hazardous situations may arise. Such foreseeable situations should be examined and steps taken in the overall system design to minimize their consequences. The particular accidental circumstances which might develop in the normal operation of the reactor will, of course, depend on the characteristics of the system, but the following examples will indicate the kind of situations which may have to be considered: startup accident, failure of safety rod or rod drive, power failure, cladding failure, distortion, displacement or fouling of fuel elements, blocking of coolant passage, instrumental errors, pump and valve malfunctions, and pipe or vessel rupture.

9.113. The accidental development of excess reactivity, referred to as a "nuclear excursion," can generally be taken care of by the scram system or by a suitable back-up safety device (§ 5.43). Some consideration has been given to the use of "nuclear safety fuses" which cause a neutron poison to be injected into the core, and thus shut the reactor down, if the temperature exceeds a certain predetermined value [13]. However, in power reactors, in particular, a potentially hazardous situation could arise even if the reactor is shut down very quickly. One of the worst accidents which might occur in a reactor system is the complete loss of coolant; this is more likely to happen in a system operating under considerable pressure than when the coolant pressure is not much above atmospheric. Even if the reactor is shut down immediately, the residual heat may be sufficient to destroy the core. The system design will generally include an auxiliary means for removing heat after shutdown, but if this should fail, the core may melt or collapse. Precautions must be taken in the overall design to make sure that the resulting mass would not be critical.

9.114. In order to minimize the possibility of coolant loss, the specifications for reactor vessels and piping are especially stringent. For nuclear use, the conventional codes have been amended to include special restrictions which must be taken into account in the design of pipes and reactor vessels. Nevertheless, in spite of all reasonable safety measures, there is some possibility that the reactor core may be destroyed. In this event, fission product gases might be released and the overall reactor plant design must include means for preventing these gases from becoming a hazard to the population of the area surrounding the site. The fuel cladding and the reactor vessel will act as barriers against the escape of fission products, but the worst possible situation must be envisaged in which both these barriers are breached.

CONTAINMENT OR CONFINEMENT [14]

9.115. The procedure for containing or confining the radioactive gases released by meltdown or disruption of the reactor core is determined to some extent by the reactor type.   Rupture of the coolant line or reactor vessel of a pressurized-water or boiling-water system would result in a sudden vaporization of the water.   There is also the possibility of the release of hydrogen gas by chemical reaction between uranium metal and aluminum or zirconium cladding with the high-temperature water.   Consequently, for such reactors it has been the practice to enclose the reactor system in a secondary (or vapor) containment structure designed to withstand the maximum pressure which might be attained as the result of an accident with that particular reactor.   The specifications of containment vessels vary from one reactor to another and some examples are given in § 11.38.   The leakage rates of these vessels must be low, usually about 0.1 per cent of the contained volume per day at design pressure.

9.116. The cost of a containment vessel having a volume on the order of a million cubic feet or more, a design pressure of 30 to 35 psig, and a low leakage rate is quite high.   Therefore, proposals have been made for decreasing the pressure of the gas or vapor which has to be contained.   The "pressure suppression" method, involving condensation of the steam from a water-cooled reactor in a pool of water, is described in § 11.39.   Another concept is based on the use of a large containment structure which houses the turbo-generator, condenser, pumps, etc., as well as the reactor and its equipment (§ 11.40).   The maximum pressure which could be attained even if all the coolant water vaporized is no more than a few pounds per square inch; the design pressure is thus decreased accordingly.   A third possibility takes into account the expectation that if there is a loss of coolant, several seconds or even minutes will elapse before the gaseous fission products escape from the core.   A "burp" system is therefore envisaged in which a valve permits venting of the relatively uncontaminated gas and vapor formed in the first few seconds.   Upon sensing radioactivity, the valve would be immediately closed to prevent escape of the contaminated gas.

9.117. Although containment vessels with design pressures of 20 to 30 psig have been used for sodium-cooled reactors, they may not be necessary when the coolant, e.g., sodium or an organic liquid, is at a relatively low pressure.   Such reactors are designed so that the core will remain surrounded by coolant even if there is a pipe rupture or pump failure.   Meltdown, accompanied by the release of fission products, is thus less probable than in a water-cooled system. In any case, in the event of a loss of coolant accident, the pressures developed would not be high. The reactor could then be housed in an ordinary structure which is not gas tight. If an accident should occur, air would be automatically drawn inward from the outside, filtered, scrubbed, and passed through an absorber to remove most of the particulate and gaseous fission products. The largely decontaminated air would be released to the atmosphere in a controlled manner through a high stack, so that there would be no hazard to the surrounding population. This procedure is described as confinement with controlled release. If considered desirable, a double building could be used to contain the reactor, with the space between the structures serving as a holdup volume.

9.118. If moderately high pressures are expected, for example, with a gas-cooled reactor, a burp arrangement could be used to permit release of the initial surge of relatively uncontaminated gas to the atmosphere. Subsequently, the release valve would be closed and a blower started to maintain an inflow of air into the reactor building. An alternative possibility is for the initial pressure loading to be absorbed by an inertial plug. The gas would then be directed to a holdup volume and gradually decontaminated, as described above, and released to the atmosphere.

9.119. There has been much discussion concerning the necessity for building power reactors in isolated areas, distant from centers of population. Although attempts have been made to prescribe specific rules in this connection, it appears that this is not practical. There must be sufficient flexibility to take into account the nature of the reactor and the safeguards included in its design. A reactor of a type with which there has been considerable experience could be constructed closer to a populated area than one involving novel and untried features. Similarly, a large isolation area should not be required for an installation in which adequate precautions are taken to limit the escape of radioactive material in the event of an accident. A criterion which has been proposed is that, in the worst conceivable circumstances, the total radiation dose received by persons in a populated area should be less than 25 rems to the whole body or 300 rems to the thyroid (from radioiodine).

REACTOR LICENSING IN THE UNITED STATES [15]

9.120. In the United States, the Atomic Energy Commission is responsible for the licensing and inspection of reactor plants in order to provide assurance of safety to the public. Licensing of a reactor facility involves two main stages: the first, usually quite early in the project, is concerned with the granting of a Construction Permit, and the second, after construction is completed, is the issuance of an Operating License. The application for a Construction Permit must be accompanied by a Preliminary Hazards Report, the purpose of which is to provide sufficient information for an evaluation to be made of the potential hazards that might arise both from normal operation of the reactor plant and from the consequences of credible accidents. Although there is no prescribed format for this report, it is expected to contain detailed descriptions of the reactor site, of the proposed reactor, with special attention drawn to novel features, and of the auxiliary systems, facilities, and structures. The most important part of the report will consist of a discussion of all conceivable potential hazards together with the steps taken in the design of the system to minimize their consequences.

9.121. A test of the available safeguards is then made by performing a detailed analysis of the effects, especially in the area surrounding the reactor installation, of what is called the "maximum credible accident" [16]. This is a postulated situation which has deleterious consequences not likely to be exceeded by an accident arising from any other credible circumstances. It must be shown that the characteristics of the reactor system and its relationship to the environment are such that the maximum credible accident will not represent a significant hazard to the general public. In this connection, consideration must be given to the different meteorological situations which might develop in the locality of the reactor site. As a general rule, the maximum credible accident is expected to arise from the loss of coolant followed by meltdown of the core and escape of all the volatile fission products and about 30 per cent of the nonvolatile products in particulate form.

9.122. The Preliminary Hazards Report [17] is reviewed by the Hazards Evaluation Staff of the U. S. Atomic Energy Commission and also by a body of appointive experts, called the Advisory Committee on Reactor Safeguards. If these bodies reach a favorable conclusion concerning the proposed reactor facility, a public hearing must be held before a Construction Permit can be granted. The purpose of this hearing is to provide members of the public the opportunity to state their objections to the erection of a reactor plant at the proposed site. If there are no such objections or if they are regarded by the hearing examiner as being invalid, a Construction Permit can be granted. Any design changes made subsequently must be submitted for review and approval in the general manner described above.

9.123. When the facility has been completely designed and construction has progressed to the point that the time for initial operation is being approached, an application is made for an Operating License. This must be accompanied by a Final Hazards Report, which is intended to present information on the reactor system as actually constructed with an outline of anticipated plans and procedures for its operation. Since the plant is now essentially complete, it is possible to make a better assessment of the probability and consequences of potential accidents. A re-examination of the assumptions made in the Preliminary Hazards Report with regard to possible hazards is therefore desirable, including a re-appraisal of what is considered to be the maximum credible accident. The information provided must be such as to permit the responsible evaluating groups to be sure that the facility as designed and constructed, in the particular location where it is situated, will not cause undue hazard to the health and safety of the general public when operated in the manner proposed. The Final Hazards Report is reviewed by the Staff of the Atomic Energy Commission and by the Advisory Committee on Reactor Safeguards, and the facility as constructed is inspected by AEC personnel. If the results of the review and inspection are satisfactory another public hearing is held, and if a favorable conclusion is reached an Operating License is granted. However, should there at any stage be a reasonable doubt concerning the safety of the installation, the license is withheld until the situation is rectified.

SYMBOLS USED IN CHAPTER 9

A	atomic weight or mass number
C	number of curies
E_γ	gamma-ray (photon) energy
f	distance from point source in feet
g	mass of radioactive material in grams
mr	milliroentgen
R	distance from point source in cm
RBE	relative biological effectiveness
RCG	Radioactivity Concentration Guide
RPG	Radiation Protection Guide
$t_{1/2}$	half-life of radioactive species
x	thickness of absorber in cm
μ	linear attenuation coefficient
μ_e	energy absorption coefficient
ϕ_γ	gamma-ray flux, photons/(cm²)(sec)

REFERENCES FOR CHAPTER 9

1. All aspects of radiation protection and reactor safety are reviewed regularly in the U. S. AEC quarterly publication entitled *Nuclear Safety*. General treatments of radiation protection will be found in the following texts: H. Blatz (Ed.), "Radiation Hygiene Handbook," McGraw-Hill Book Co., Inc., New York, 1959; D. E. Barnes and D. Taylor, "Radiation Hazards and Protection," Pitman Publishing Corp., New York, 1958; C. B. Braestrup and H. O. Wyckoff, "Radiation Protection," Charles C. Thomas, Publisher, Springfield, Ill., 1958.
2. Federal Radiation Council Report No. 1, see *Federal Register*, May 18, 1960.
3. S. Glasstone (Ed.), "The Effects of Nuclear Weapons," Rev. Ed., U. S. Government Printing Office, Washington, D. C., 1962, pp. 587 *et seq.*
4. H. Blatz, Ref. 1; D. E. Barnes and D. Taylor, Ref. 1; G. J. Hine and G. L. Brownell (Eds.), "Radiation Dosimetry," Academic Press, Inc., New York, 1956; H. Etherington (Ed.), "Nuclear Engineering Handbook," McGraw-Hill Book Co., Inc., New York, 1958, pp. 7–22, *et seq.*
5. "Protection Against Neutron Radiation up to 30 Mev," National Bureau of Standards Handbook 63 (1957); "Measurement of Absorbed Dose of Neutrons and Mixtures

of Neutrons and Gamma Rays," National Bureau of Standards Handbook 75 (1961), U. S. Government Printing Office, Washington, D. C.

6. Ref. 2; see also, *Reactor Safety*, **2**, No. 2, 6 (1960); *Power Reactor Technology*, **4**, No. 1, 43 (1960); "Radiation Protection: Recommendations of the International Commission on Radiological Protection," Pergamon Press, Inc., New York, 1959. For background information, see "Selected Materials on Radiation Protection Criteria and Standards: Their Base and Use," Joint Committee on Atomic Energy, and "Radiation Protection and Criteria and Standards: Their Basis and Use," Hearings before the Special Subcommittee on Radiation, Joint Committee on Atomic Energy, Congress of the United States, U. S. Government Printing Office, Washington, D. C., 1960.

7. Federal Radiation Council Report No. 2, 1961; see also "Maximum Permissible Body Burdens and Maximum Permissible Concentrations of Radionuclides in Air and Water for Occupational Exposure," National Bureau of Standards Handbook 69 (1959), U. S. Government Printing Office, Washington, D. C.

8. S. Glasstone, Ref. 3, pp. 592 *et seq.*

9. H. Etherington, Ref. 4, p. **7**-45; W. J. Price, "Nuclear Radiation Detection," McGraw-Hill Book Co., New York, 1958; J. S. Handloser, "Health Physics Instrumentation," Pergamon Press, Inc., New York, 1959.

10. H. Etherington, Ref. 4, p. **7**-47; R. H. Dilworth and C. J. Borkowski, U. S. AEC Report ORNL-3058 (1961).

11. H. Etherington, Ref. 4, p. **7**-46; W. T. Thornton, *et al.*, U. S. AEC Report ORNL-3126 (1961).

12. H. Etherington, Ref. 4, pp. **8** 76 *et seq.;* C. R. McCullough (Ed.), "Safety Aspects of Nuclear Reactors," D. Van Nostrand Co., Inc., Princeton, N. J., 1956; "Reactor Safety Conference," TID-7549, Parts I and II (1958); "Theoretical Possibilities and Consequences of Major Accidents in Large Nuclear Power Plants," U. S. AEC Report WASH-740 (1957); see also *Proc. Second U. N. Conf. Geneva*, Vol. **21** (1958) and regular issues of *Nuclear Safety*.

13. *Nuclear Safety*, **1**, No. 1, 17 (1959); **2**, No. 3, 27 (1961).

14. R. W. Bergstrom and W. A. Chittenden, *Nucleonics*, **17**, No. 4, 86 (1959); R. O. Brittan, U. S. AEC Report ANL-5948 (1959); T. H. Smith and B. H. Randolph, *Nuc. Sci. and Eng.*, **4**, 762 (1958); see also, *Nuclear Safety*, **1**, No. 2, 53 (1960); **2**, No. 1, 48, 55 (1960); *Power Reactor Technology*, **5**, No. 1, 33 (1961); and several reports in *Proc. Second U. N. Conf. Geneva*, Vol. **11** (1958).

15. C. K. Beck, *et al.*, *Proc. Second U. N. Conf. Geneva*, **11**, 17 (1958); J. J. DiNunno, Paper presented at the AIEE General Meeting, June, 1960; see also, *Nuclear Safety*, **2**, No. 3, 61 (1961); **2**, No. 4, 63 (1961); **3**, No. 1, 90 (1961).

16. *Nuclear Safety*, **1**, No. 4, 24 (1960).

17. References to Safeguards and Hazards Reports are given regularly in *Nuclear Safety*, together with accounts of actions being taken in connection with all current reactor projects. Earlier references to Hazards Reports are given in a bibliography entitled "Reactor Safety," U. S. AEC Report TID-3073 (1958) and in TID-3525, which is revised periodically.

PROBLEMS

1. If t_b is the biological half-life of an element in a certain organ and $t_{1/2}$ is the radioactive half-life of a particular isotope, derive an expression for the effective half-life of that isotope in the specified organ. Show that if either t_b or $t_{1/2}$ is very short compared to the

other, then the effective half-life is approximately equal to the shorter of the two. Explain the physical significance of the result.

2. The human body consists of 15 per cent by weight of carbon and 1.0 per cent of potassium. The presence of carbon-14 results in the emission of 15.3 beta particles/min per gram of carbon in the living organism. Further, all forms of potassium contain 0.011 per cent of potassium-40, a beta emitter with a half-life of 1.3×10^9 yr. Calculate the number of microcuries of (a) carbon-14 and (b) potassium-40 present in a human being weighing 70 kg.

3. An individual spends a total of 20 hr per week in the vicinity of a reactor where the average fluxes, per cm^2 per sec, are 350 2-Mev photons, 120 thermal neutrons, and 5 fast neutrons. Estimate the total dose received in rems, assuming cataract formation to be the critical injury.

4. A cobalt-60 source containing 5×10^{-6} g of this isotope produced a reading of 7.5 mr/hr when held at a distance of 100 cm from a G-M counter. What is the percentage error of the meter at this reading?

Chapter 10

SHIELDING OF NUCLEAR REACTOR SYSTEMS [1]

REACTOR SHIELDING PRINCIPLES

INTRODUCTION

10.1. In the design of a nuclear reactor and its associated equipment, provision must be made for the attenuation of escaping nuclear radiations by some form of shielding. Not only is such shielding necessary for the protection of personnel, as explained in the preceding chapter, but a relatively high radiation background will interfere with the satisfactory functioning of instruments used in various aspects of reactor operations and control. Furthermore, since the radiations entering the shield from the reactor can produce internal heating, and possibly cause radiation "damage" (Chapter 7) to shield materials, it is necessary to estimate the types and intensities of such radiations throughout the shield. The shielding of nuclear reactor systems is therefore concerned with the study of radiation distribution in the shield, for the dual purpose of reducing the escaping radiation to acceptable levels and of determining certain physical effects on the shield materials.

10.2. The character of the shield suitable for a particular reactor is very largely dependent on the purposes of the reactor, so that shield design is to be regarded as an integral aspect of reactor design. A reactor of very low or zero power, such as might be used for critical experiments, would require little or no shielding for normal operation. In these circumstances, the attenuating effect of distance might be sufficient to reduce the dosage to a safe level. However, some shielding is invariably employed to decrease the consequences of a possible accidental increase in power. For a reactor which normally operates at high power, a shield of considerable thickness, e.g., 6 to 8 ft or more of concrete, will be required. For a stationary system a partial, although expensive, solution may be found in building the reactor underground.

10.3. Even for reactors of similar power, the degree of shielding necessary is influenced by the basic purpose of the reactor. In a power reactor, for example, only occasional access may be required outside the shield for inspection and maintenance. On the other hand, where the reactor is used for research and testing, the apparatus may require continuous attention especially when measurements are being made. The permissible radiation level in the latter case will obviously be less than in the former. In any event, advantage may be taken of the fact that there will generally be some areas around a reactor to which limited access only will be necessary. In these regions the shield may be thinner than in locations which are occupied more frequently by operating personnel. An extreme example is the so-called *shadow shield*, used in connection with mobile reactors, in particular, where essentially all the shielding is required in certain limited directions only, since there is no access in the other directions.

10.4. For mobile reactors, e.g., for propulsion of a submarine or an aircraft, the weight of the shield is of overriding importance; if the shielding is too heavy, the reactor may prove to be unsatisfactory for its intended purpose. In this instance the cost of the shield may be of secondary significance. But for a stationary reactor, such as would be used in a power station, cost becomes the primary factor, and weight is now secondary. The overall weight of the radiation shielding may often be decreased by the use of a *divided shield*, in which part of the shielding is around the reactor itself and part around the operating crew, who may be at some distance from the reactor. The treatment in this chapter will be restricted, however, to what are referred to as *unit shields*, consisting of a single structure. Such shields are commonly used for both research and power reactors.

10.5. The design of a reactor shield involves several phases, of which the initial one is an aspect of the overall design of the reactor. It is necessary, at the outset, to specify the maximum dose rates of radiation that may be permitted to penetrate the shield at various locations. In order to do this the accessibility requirements must be established for different areas and components of the reactor system during operation. Attention must be given, for example, to the facilities for charging and discharging the fuel, and to the locations of coolant loop, control elements, etc. When the basic requirements are known, it is possible to outline the general configuration of the shield. The materials of construction can then be chosen and the optimum arrangement selected to provide most economically, either in weight, dimensions, or cost, the required radiation attenuation in the different locations, as determined by the operational requirements. A subsequent phase of shield design is concerned with many details associated with localized problems, such as the effects of ducts and piping that must pass through the shield. Consideration must also be given to the shielding of pipes and equipment carrying or containing radioactive materials.

SHIELD DESIGN

10.6. The most significant radiations for which shielding is required are the primary neutrons and gamma rays originating within the core itself and the secondary gamma rays produced by neutron interactions with materials external to the core, e.g., reflector, coolant, shield, etc. However, shield design involves more than choosing a suitable shielding material (or materials) and determining the thickness required to decrease the total radiation effectiveness, i.e., the sum of the neutrons and gamma-ray equivalents in rems (§ 9.47 *et seq.*), to an acceptable value. Absorption of the radiations in the shield is accompanied by the liberation of energy which appears as heat. In order to ensure the physical integrity of the shield material, the temperature distribution must be known. Thus, a study of heat generation is an essential aspect of shield design, and this requires a knowledge of the distribution of neutrons and gamma rays throughout the thickness of the shield.

10.7. Since the interactions of neutrons and gamma rays with matter are fairly well understood, it might be possible, in principle, to carry through the design of a shield in a purely analytical manner, provided the cross sections and attenuation coefficients, which are equivalent to cross sections, were available for all interactions at all energies. Because of the complex nature of the radiation source and its distribution in the reactor core and shield, and of the many interactions over a considerable range of energies, this approach is not practical. However, a semianalytic procedure is possible, utilizing relatively simple, but somewhat approximate, mathematical expressions combined with experimental data of a statistical nature. In addition, special consideration is given to the individual behavior of secondary gamma radiations, since these are often the controlling factor in shield design. The general principles of this method will be described later.

10.8. Another approach to shield design, which is considered to be more reliable than the one referred to above, is the comparison method. In this case, the data from a known reactor and shield are used as the basis for predicting the performance of another reactor-shield combination of the same general type. The procedure is obviously limited to cases where the shield, in particular, is of a type for which information is available as a result of previous experience or from experimental measurements.

10.9. Although a great deal is known of the theoretical basis of radiation shielding, the complex nature of the interactions in the reactor-shield system is such that shield design at the present time is partly an "art." It will undoubtedly be possible in due course to use purely calculational methods for shield design; it can, in fact, already be done for some simple, e.g., water, shields. However, the design of many reactor shields cannot currently be achieved by calculations alone, and some use must be made of data obtained from shielding experiments. Such experiments are made to determine the bulk (or macroscopic) properties of various shielding materials, for use in the semiempirical calculations referred to earlier. In addition, experiments are useful for checking mockups of entire systems with full or nearly full shield thicknesses. The information so obtained is not only applicable to the particular system being studied, but it can also be utilized in the comparison procedure for other shields of the same general nature.

10.10. It might appear, at first thought, that the problems of shield design could be simplified by adopting a conservative approach, that is to say, by introducing a factor of safety at each stage. Although this philosophy is satisfactory for preliminary calculations, such as will be described later in this chapter, it is not realistic from the economic standpoint. Because the periphery of the shield increases with increasing thickness, the cost per foot of material increases correspondingly, as also does the cost of erection and of the space occupied. Consequently, a reactor shield must be designed to provide adequate radiation protection, but should not be overdesigned. Satisfying these two requirements necessitates a high degree of precision in an area in which exact calculations are not yet feasible.

10.11. The stages in the design of a reactor shield are performed somewhat along the following lines. First, the maximum permissible radiation dose rates at various locations must be established, as indicated in § 10.5. This implies a knowledge of the physical layout of the reactor and its associated equipment, together with information, if possible, as to which components are fixed in location and which can be changed in position to some extent. The expected power and material distributions in the core are required to determine the radiation source distribution. With a knowledge of the reactor system available, a preliminary choice can be made of reactor materials and shield layout, so that initial attenuation calculations can be carried out. If the system is similar to one for which data are available, the comparison procedure can be used; otherwise, a simple form of the semianalytic method is applied. Both the choice of materials, which may be different in different locations, and the general layout of the shield will be subject to change as the design progresses. The design process is thus iterative in nature, with the calculations becoming more and more refined as the final stages are approached.

10.12. As stated earlier, the attenuation and distribution of both primary and secondary radiations must be considered in the calculations. An accurate treatment of the secondary gamma rays is very difficult, for reasons which will be apparent in due course. However, approximate computations utilizing experimental data can be made to estimate a reasonable upper limit of the extent to which this radiation will penetrate the shield. Up to this point the effects of ducts and voids in the shield have been ignored, since they generally have a minor effect on the overall design. Consideration must now be given to the design of ducts and piping to minimize, as far as possible, the leakage of neutrons, in particular. Additional shielding may then have to be applied at specific locations to decrease the escaping radiation to acceptable levels. In this connection, recourse must generally be had to experiments or to physical models which simulate the actual situation.

10.13. Concurrently with the design of the shield, consideration is given to the heat generated by the deposition of the energy of the absorbed radiation. One way in which excessive heating is avoided is by the use of a thermal shield, as described in § 10.23.

RADIATIONS FROM REACTOR SYSTEMS

10.14. In principle, the radiations which might escape from a reactor system include alpha and beta particles, gamma rays, neutrons of various energies, fission fragments, and even protons resulting from (n, p) reactions. As far as shield design is concerned, however, only gamma rays and neutrons need be considered since these are by far the most penetrating. Any material which attenuates these radiations to a sufficient extent will automatically reduce all the others to negligible proportions.

10.15. Some indication of the complex character of the nuclear radiations arising from a reactor is apparent from Fig. 10.1, which, in accordance with the comment made above, refers mainly to neutrons and various gamma radiations. The fission process itself produces fission fragments, neutrons, and gamma rays. These undergo various reactions both within the reactor itself and in the shield, leading to the formation of a complex variety of other radiations. For the purpose of shield design the neutrons and gamma rays are considered from the standpoint of their place of origin: the *primary radiations* are defined here as those which originate within the reactor core, whereas the *secondary radiations* are produced outside the core as a result of the interaction of the primary radiations, chiefly the neutrons, with nuclei in the reflector, coolant, and shield materials.

10.16. The most significant primary radiations are fast neutrons, the prompt fission gamma rays (§ 2.167), the decay gamma rays from fission products, and capture gamma rays. There are also thermal neutrons, inelastic scattering gamma rays, and decay gamma rays from radioactive products of neutron capture reactions. The gamma rays in this latter group do not significantly affect the shield thickness, but the neutrons captured in the shield produce secondary gamma rays which are very important. Furthermore, the heating effects of the gamma rays—both primary and secondary—must be taken into consideration.

10.17. The energies of the fast neutrons cover a considerable range, but in the semianalytic method of shield design it is convenient to treat them statistically as a single group. As will be seen in § 10.102, the properties of this group are determined largely by neutrons with energy in the range of 6 to 8 Mev. The prompt fission gamma rays have a continuous energy spectrum, from about 0.5 to 10 Mev (Fig. 2.24), but the radiation intensity can probably be neglected for energies in excess of 7 Mev. Except for ordinary hydrogen, which has a single capture gamma ray, most other common elements, at least, exhibit a complex spectrum of capture gamma radiation, the energies ranging up to about 8 Mev

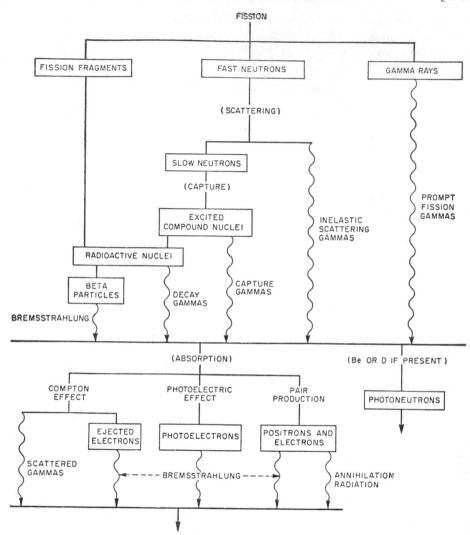

Fig. 10.1. Radiations from nuclear reactor

in many cases. Finally, mention should be made of the decay gamma rays, many of which have energies over 2 Mev, up to 5.4 Mev. It is evident that the primary gamma radiation cannot be treated statistically and each photon energy should be considered individually. Since the attenuation problem would then become extremely complicated, it is the practice to divide the gamma rays into a number, e.g., six or seven, energy groups of appropriate intensity. The attenuation computations are thereby greatly simplified, although they are still lengthy.

10.18. When the fast neutrons which escape from the reactor enter the shield, they are slowed down as a result of scattering collisions. If the collisions are of the inelastic type, the process will be accompanied by inelastic scattering gamma rays. These constitute part of the secondary radiations which must be attenuated in the shield. After suffering a number of collisions, the fast neutrons enter the energy region where they are readily captured. The resulting capture gamma rays make a major contribution to the secondary radiation. Furthermore, some of the capture products will be radioactive and emit beta particles, often accompanied by gamma rays. These and the bremsstrahlung produced by the interaction of the beta particles with nuclei add to the complexity of the secondary radiations in the shield, although the bremsstrahlung do not make any significant contribution to the radiation dose. Apart from the variety of the radiations, both in intensity and energy, the fact that they are produced throughout the shield volume greatly increases the difficulty of making attenuation calculations.

10.19. If beryllium or deuterium is present in the reactor, either as moderator or reflector, the action of gamma rays of energy exceeding 1.62 or 2.21 Mev, respectively, will produce photoneutrons (§ 2.73).* Since many gamma rays in a reactor have energies in excess of these minimum values, e.g., up to about 8 Mev, the photoneutrons can have high energies. They thus behave like fission neutrons and upon escaping from the reactor can give rise to secondary radiations within the shield indentical with those described above.

10.20. There is evidence that photoneutrons can be formed in water shields by interaction of either primary or secondary gamma photons with the deuterium normally present in water. These photoneutrons can have a significant effect on the overall rate of neutron attenuation in the shield.

10.21. The foregoing discussion has referred to the radiations arising from the fission process within the reactor. Reactor coolants which have become radioactive as a result of neutron capture, spent fuel elements, and solutions of fission products emit radiations. In many cases these are limited to beta particles and gamma rays, so that the only shielding which needs to be considered is that against gamma radiation. The absence of any significant neutron emission in these instances greatly simplifies the shielding design calculations, since there are no secondary radiations. In reactors cooled by water, and other oxygen-containing materials, however, neutrons are produced as a consequence of the reaction $O^{17}(n, p)N^{17}$; the product, nitrogen-17, is one of the few unstable nuclides (half-life 4.1 sec) which emits neutrons (§ 6.189). Another special case occurs in connection with circulating fuel systems; after leaving the reactor to enter the heat exchanger, the fuel continues to emit delayed neutrons. In

---

* Actually any nuclear species (except hydrogen) can be made to yield photoneutrons if gamma rays of sufficient energy—equal to the neutron binding energy, i.e., usually more than 6 Mev—are available. Since the cross sections for such (γ, n) reactions are small and photons of 6 Mev (or more) energy are not common in a reactor, photoneutrons emitted by nuclei other than beryllium and deuterium are negligible.

addition to the hazard they themselves represent, they may induce radioactivity, as a result of capture, in the material used as secondary coolant.

THERMAL AND BIOLOGICAL SHIELDS

10.22. Essentially all the energy absorbed in the shield from the fast neutrons and gamma rays is ultimately degraded into heat. This means that, for a reactor of moderate or high power, a considerable amount of heat is generated within the shield. Since the absorption of both neutrons and gamma radiation is, at least approximately, exponential in character, a large proportion of the total heat liberated will be released in those parts of the shield closest to the reactor. For example, suppose that the gamma-ray flux is 10^{13} photons/(cm²) (sec) at the inside face of the shield, i.e., nearest the reactor, and this is reduced to 10^3 photons/(cm²)(sec) at the outer face. If the thickness of the uniform shield is 100 cm, then, assuming an exponential absorption with distance (which will not be strictly the case), it follows that the flux will be decreased from 10^{13} to 10^{12} photons/(cm²)(sec) within the first 10 cm of the shield. This decrease in flux represents a 90 per cent absorption of the gamma-ray energy in the first 10 per cent of the shield.

10.23. To protect the shield from possible damage from the heat liberated upon absorption of radiation, a so-called *thermal shield* is normally introduced close the the reactor [2]. It is made of a substantial thickness of a dense metal of fairly high melting point, e.g., iron, placed between the reactor core and the main shield, referred to as the *biological shield*.

10.24. In the Shippingport Pressurized Water Reactor (PWR), the thermal shield consists of two concentric stainless steel cylinders which surround the reactor core. The inner cylinder, which is 1 in. thick, is separated by an annular space of 0.75 in. from the 3-in. thick outer cylinder. Part of the cooling water entering the reactor flows in the annular spaces between the core and the inner thermal shield and between the outer shield and the containing (pressure) vessel, as well as between the two thermal shields. A similar thermal shield system is used in the Yankee Atomic Electric Co. reactor (§ 12.70). In these reactors, the location of the thermal shields within the pressure vessel greatly decreases the amount of radiation energy absorbed in the walls of the vessel and thereby reduces the thermal stresses in the latter.

10.25. Since the thermal shield is a heavy metal, it is effective for absorbing gamma radiations and for the inelastic scattering of fast neutrons. These two types of radiation carry most of the energy leaking from the reactor core. The absorption of the primary gamma rays and the secondary radiations accompanying the capture of the slowed down neutrons produces a considerable amount of heat in the thermal shield. In power reactors this is removed by the coolant and so contributes to the available energy. In research reactors, on the other hand, the heat is not utilized. For example, in the Materials Testing Reactor the thermal shield consists of two 4-in. thick layers of steel, separated by a space of 4 in. through which air is passed to serve as coolant. The air is then discharged to the atmosphere. When operating at a power of 30,000 kw, the MTR produces 48 kw of heat in the thermal shield and about 2 kw in the biological shield made of special concrete.

10.26. For reactors of low or moderate power, the amount of heat liberated in the shield may not be excessive. Furthermore, if the reactor and the shield are large, the local temperatures may not be sufficiently high for special cooling to be required. In cases of this kind the introduction of a thermal shield may be unnecessary. The natural uranium-graphite reactor at Oak Ridge National Laboratory, for example, does not have a thermal shield, although a special stream of air is directed over the inner faces of the biological shield at the front and back of the reactor. After more than 10 years of operation, most of the time at an average power level of about 3800 kw, there were no indications of damage to the concrete of the shield.

REACTOR SHIELDING REQUIREMENTS

10.27. In principle, the problem of shielding the reactor itself involves three aspects: (1) slowing down of fast neutrons, (2) capture of the slowed down (or initially slow) neutrons, and (3) attenuation of all forms of gamma radiation, including primary radiations from the reactor core and secondary radiations formed as a result of various interactions between neutrons and nuclei in the shield.

10.28. Reactor shielding involves the attenuation of fast neutrons and gamma radiations, so that the problems are essentially the same for both thermal and fast reactors. A significant difference is that the number of neutrons escaping from a fast reactor, for a given operating power, is generally greater than from a thermal reactor. Furthermore, the breeder blanket of a fast reactor can be a major source of radiation, as a result of fast-fission and (n, γ) reactions with uranium-238. The source distribution is therefore more complicated than for a thermal reactor and so the calculations are less amenable to simplified approximations such as are possible for the latter [3].

Slowing Down Fast Neutrons

10.29. Since elements of low mass number are the best moderators, hydrogen, in the form of water, can be used as the shield constituent for slowing down neutrons. The scattering cross section of hydrogen can be represented fairly accurately as a function of neutron energy, in the range from 1 to 12 Mev, by

$$\sigma_s = \frac{10.97}{E + 1.66} \text{ barns,}$$

where E is the energy in Mev. Consequently, for high neutron energies, the scattering cross section is small although still quite appreciable. It is somewhat advantageous, therefore, to introduce an element (or elements) of moderate or high mass number. Such substances are not good moderators, i.e., they do not decrease the neutron energy to an appreciable extent in elastic collisions, but they are able to reduce the energies of very fast neutrons as a result of inelastic-scattering collisions. Elements such as lead, barium, or iron will decrease the neutron energy down to about 1 Mev in a single inelastic collision. At this energy the hydrogen (elastic) scattering cross section is about 4 barns. Hence, a combination of a moderately heavy or heavy element with hydrogen will effectively slow down even neutrons of very high energies.

10.30. It may be pointed out that, although hydrogen alone (as water) could be used as a neutron shield, a shield consisting entirely of heavy elements would be unsatisfactory. A single inelastic scattering collision would decrease the neutron energy to about 1 Mev, but subsequent collisions would be elastic in nature. Many such collisions would then be necessary to reduce the neutron energy to a value at which the capture cross section is significant. In the meantime, the neutrons would have penetrated a considerable distance into the shield. Another drawback to the use of certain heavy elements, especially iron, alone will be mentioned below in connection with neutron capture.

10.31. It is worth noting here that in a thermal reactor the reflector can make an important contribution to fast-neutron shielding. The reflector is invariably a good moderator, e.g., water, heavy water, beryllium, beryllium oxide, graphite or hydrocarbon compound, so that it will slow down an appreciable fraction of the moderately fast neutrons escaping from the reactor core. Because of scattering, many of these slowed down neutrons are returned to the core; if they escape subsequently, they do so as thermal neutrons.

Slow-Neutron Capture

10.32. The next aspect of reactor shielding to consider is that of capturing the neutrons after they have been slowed down. Actually this is a relatively simple matter, even though the shield does not contain an element with a large or fairly large cross section for the capture of slow neutrons. In fact, it is generally accepted that in a shield containing sufficient hydrogen, almost every neutron which has suffered inelastic scattering may be regarded as removed, since the probability of further slowing down and subsequent capture is large.

10.33. The capture cross section (σ_c) of hydrogen for thermal neutrons is only 0.33 barn; nevertheless, the number of these nuclei present in the shield for slowing down purposes is usually sufficient to absorb a large fraction of the neutrons. In addition, iron has an appreciable capture cross section ($\sigma_c = 2.5$ barns) for thermal neutrons. The disadvantage of neutron capture by iron is that the capture gamma-ray spectrum has a strong line at 7.6 Mev and a weaker one at 9.3 Mev energy. Precautions must be taken to attenuate these radiations of high energy. It should be noted in this connection that capture of a neutron by hydrogen is accompanied by a single 2.2-Mev gamma photon which must be taken into consideration in shielding.

10.34. A point of special interest in connection with the use of iron in reactor shields is the fact that it has a minimum (total) cross section (0.15 barn) for neutrons of about 25-kev energy.   Consequently, a shield consisting of iron alone will serve as a kind of filter which allows 25-kev neutrons to pass through readily but removes most of those with higher and lower energies.   This phenomenon is sometimes referred to as *energy streaming* of neutrons, and iron is said to have a "window" for 25-kev neutrons.   Because of this streaming effect, it is not permissible to have long pieces of iron penetrating a reactor shield.   Of course, if the iron is followed by or interspersed with another absorbing material, e.g., water or concrete, the 25-kev neutrons are slowed down and absorbed.   It is of interest in this connection that many stainless steels contain sufficient nickel to prevent energy streaming of neutrons.

10.35. The magnitude of the problem arising from capture gamma rays in a shield can be decreased by the use of boron-10, which has a large cross section for the (n, α) reaction with slow neutrons.   Some gamma radiation accompanies the neutron absorption reaction, but its energy is only about 0.5 Mev and so it is easily attenuated.   Proper location of boron in a reactor shield can often result in a decrease in the overall shield thickness because of the low energy of the secondary gamma rays.   Boron has also been used in fast reactors in conjunction with a moderator, such as carbon, as part of the shield close to the reactor. Many of the neutrons escaping from the reactor are slowed down and captured by the boron, so that a smaller degree of attenuation is required in the remainder of the shield.   Lithium, which contains lithium-6—like boron-10 a good absorber of neutrons by the (n, α) reaction—has also been proposed for use in reactor shields.

10.36. Since cadmium captures slow neutrons so readily, and it also has a fairly high mass number, it would appear that this element could be used as a shielding material.   The difficulty here, as with iron, is that the capture gamma-ray spectrum has lines of high energy.   If, for any reason, a cadmium sheet is used as part of a shield, it should be located near the reactor so that it is backed by adequate material for attenuating the capture gamma rays.   There is another possible drawback associated with the use in a shield of a material, such as cadmium, having a high resonance peak for the capture of neutrons in a limited energy range (below about 1 ev).   Although most of the neutrons within this range are captured, a large proportion of those with higher energies will penetrate the shield.   A hazardous situation might then result from placing too much reliance upon material that is apparently a very good absorber of neutrons.

Attenuation of Gamma Radiation

10.37. The third and final aspect of the shielding problem is the attenuation of the various gamma rays both primary and secondary.   As far as their behavior is concerned, the origin of these radiations is not significant; it is only their energy which determines the attenuation in a given medium (§ 2.65). All substances will attenuate gamma rays to some extent, although the linear attenuation coefficient (or macroscopic cross section) generally increases with the density, for photons of a given energy. This coefficient is very roughly proportional to the density and so, as found in § 2.67, the thicknesses of different materials required to attenuate gamma radiation, of specified energy, to the same extent are inversely proportional to their respective densities. Consequently, where thickness of the shield is an important consideration, a material of high density would be used to attenuate the gamma rays.

SHIELDING MATERIALS [4]

10.38. Materials used in shielding may be divided into three broad categories, according to their function: (1) heavy or moderately heavy elements to attenuate the gamma radiation and to slow down very fast neutrons to about 1 Mev by inelastic collisions; (2) hydrogenous substances to moderate neutrons having energies in the range below about 1 Mev by elastic collisions, and (3) materials, notably those containing boron, which capture neutrons without producing high-energy gamma rays. Although these three classes will be considered separately, it will be seen that the same material may often serve two or all three purposes.

Heavy Elements

10.39. Iron, as carbon steel or stainless steel, has been commonly used as the material for thermal shields; as already seen, such shields can absorb a considerable proportion of the energy of the gamma rays and fast neutrons escaping from the reactor core. In fact, two or three layers of steel with water between them, such as are used in pressurized-water reactors, represent a very effective shield for both neutrons and gamma rays. Apart from the use of massive iron in shield construction, iron turnings or punchings and iron oxides have been incorporated in concrete for shielding purposes. Further reference to special concretes, which incorporate heavy elements such as iron or barium in some form, will be made shortly.

10.40. Because of their high density and ease of fabrication, lead and lead alloys have been used to some extent in nuclear reactor shields. For gamma rays with energies in the region of 2 Mev, roughly the same mass of lead as of iron is required to remove a specified fraction of the radiation. However, at both higher and lower energies the mass-attenuation efficiency of lead is appreciably greater than that of iron. Because of its low melting point, lead can be used only where the temperatures are not too high.

10.41. The elements tantalum and tungsten may be of interest in shield design because they have both high densities and high melting points. However, because of their cost when fabricated they would be of value only in special cases, for example, as thermal shields in mobile reactors operating at high temperatures. Tungsten has a fairly high radiative capture cross section for thermal neutrons and the spectrum of the resulting gamma rays has strong lines in the regions of 6 and 7 Mev.

10.42. From the long-range point of view, carbides, nitrides, and possibly borides of elements of high mass number may be of interest. Ceramic materials of this type are usually refractory, have high melting points, and are relatively dense, although they may prove difficult to fabricate. The presence of a light element, such as carbon, nitrogen, or boron, would contribute to the slowing down of the fast neutrons, and boron would, in addition, capture neutrons.

Hydrogenous Materials

10.43. The value of a hydrogenous material for neutron shielding is determined, first, by its hydrogen content; this is best expressed in terms of the number of hydrogen atoms (or nuclei) per unit volume. In this respect, water, with 6.7×10^{22} atoms per cm^3, ranks high, and it has, in addition, the advantage of low cost. In fact, water is probably the best neutron shield material, although it is a poor absorber of gamma radiations. The low boiling point of water at ordinary pressure and its susceptibility to decomposition by radiation (§ 7.93) are partial drawbacks to the use of liquid water in reactor shields. On the other hand, water provides a ready means for removing the heat generated by radiation absorption. In addition to its employment as the shielding material in research reactors of the pool type, liquid water, as a layer about 2.5 ft thick, is used outside the pressure vessels of the Nuclear Merchant Ship (N.S. Savannah) reactor and the Yankee Atomic Electric Company power reactor (see Chapter 12) to attenuate both fast and slow neutrons.

10.44. Masonite, a compressed wood product, with a density of about 1.3 g per cm^3 and containing about 6 weight per cent of hydrogen, was used as the hydrogenous material in some early reactors. It formed part of laminated shields consisting of alternate layers of Masonite and iron. The number of hydrogen atoms per cubic centimeter of Masonite is as high as about 5×10^{22}, which is not much less than for water. In addition, it contains both carbon and oxygen which can act as moderators. Although it has proved satisfactory in the past, Masonite-iron will probably find little application in the future as the basic reactor shield. However, a number of densified wood laminates, impregnated with polymer resins and sometimes containing boron, are being used for special situations where mobile or light weight neutron shields are required (cf. § 10.52).

10.45. As a general shield material, there is much to recommend concrete; it is strong, is not expensive, and is adaptable to both block and monolithic types of construction. For these reasons it is used more than any other material for the shielding of stationary reactors. Ordinary concrete of density 2.3 g per cm^3 generally contains somewhat less than 10 per cent by weight of water when cured, i.e., about 1.4×10^{22} hydrogen atoms per cm^3. Although this is considerably less than the concentration in water, the larger proportion of oxygen, which acts as additional moderator, compensates for the difference to some extent. The macroscopic cross section for fast-neutron removal by concrete is thus about 0.085 cm^{-1}, compared with approximately 0.1 cm^{-1} for water. However, concrete is very much superior to water for the attenuation of gamma rays; this is mainly a result of the presence of some 50 weight per cent of elements of moderately high mass number, such as calcium and silicon.

10.46. One drawback to the use of concrete as a reactor shield is the considerable variability in composition and water content of the commercially available material. This variation adds a degree of uncertainty to calculations made by the shield designer to predict the radiation distribution and attenuation in the shield. Although the water content of concrete after it cures is important in shielding considerations, the value is not well known. It appears that 7 weight per cent is adequate for neutron attenuation but a higher proportion would be advantageous [5]. However, an increase in the water content has the disadvantage of decreasing both the density and structural strength of ordinary concrete. Another problem arises from the loss of water when the concrete becomes hot, as it does by the absorption of energy from the radiations. The use of the mineral *serpentine*, a hydrous magnesium silicate, as the aggregate is reported to yield a concrete which retains its water up to temperatures of 430° to 480°C (800° to 900°F) [6].

10.47. Various special ("heavy") concretes of higher than normal density, incorporating elements of fairly high mass number, have been developed for reactor shielding in special cases [7].* With these heavy concretes a given amount of attenuation of both neutrons and gamma rays can be achieved by means of a thinner shield than is possible with ordinary concrete. This is an advantage in certain cases. In research reactors, for example, it is desirable to have shield penetrations as short as possible, as an experimental convenience. However, in many reactor applications, where space is not an important consideration, the lower cost of ordinary concrete, compared with that of heavy concrete, makes the former the preferred shield material.

10.48. The heavy concretes which have been mainly used in reactor shielding are *barytes concrete* and *iron concrete*. In addition, some studies have been made with a *ferrophosphorus concrete*. In barytes concrete the mineral *barytes* (or *barite*), consisting mainly of barium sulfate, largely replaces the sand and gravel aggregate of ordinary concrete [8]. The density of barytes concrete is about 3.5 g per cm^3 and the macroscopic cross sections for the removal of fast neutrons and the attenuation of 4-Mev gamma rays are 0.105 and 0.10 cm^{-1}, respectively. The corresponding values for ordinary concrete are 0.085 and 0.066 cm^{-1}. The barytes aggregate may be partly or wholly replaced by *limonite*, an iron ore of composition $2Fe_2O_3 \cdot 3H_2O$; this contributes both iron and bound water to the

---

* It is of interest to record that heavy concretes, containing metallic iron or iron ore, were used in construction work for several years before their application to reactor shielding.

concrete.   If desired, iron punchings can be added to the resulting product. Thus the shield for the Brookhaven uranium-graphite reactor is made of a mixture of iron ore, iron punchings, Portland cement, and water; it has a density of 4.27 g per cm^3, and the cross sections for fast-neutron removal and gamma-ray attenuation are nearly 0.16 and 0.13 cm^{-1}, respectively.   The compositions by weight of two fairly typical heavy concretes for use in reactor shielding are given in Table 10.1.

TABLE 10.1.   COMPOSITION BY WEIGHT OF CONCRETES FOR REACTOR SHIELDS

Barytes Concrete (Density 3.5 g/cm^3)	Weight Per Cent	Iron Concrete (Density 4.5 g/cm^3)	Weight Per Cent
Barytes...............	60	Steel punchings.........	57
Limonite..............	22	Limonite..............	26
Portland cement.......	11	Portland cement.......	13
Water................	7	Water.................	4

10.49. Various other heavy cements and concretes have been devised; one consisting of iron, Portland cement, and water, has a density as high as 6 g per cm^3.   Unless it is kept soaked with water, which is a possibility, the hydrogen content of this material might be too low for it to serve as a neutron shield. Boron compounds, e.g., the mineral *colemanite* ($2CaO \cdot 3B_2O_3 \cdot 5H_2O$), have also been incorporated into concretes in order to increase the probability of neutron capture without high-energy gamma-ray production.   Although colemanite has a deleterious effect on the setting properties of concrete, it can be overcome by the use of proper mixes and certain additives.

10.50. As alternatives to water as the hydrogenous component of a reactor shield, some consideration has been given to metal hydrides, since they possess features of special interest especially for high-temperature shields.   Such materials would combine both inelastic and elastic neutron scattering with gamma-ray absorption properties.   Two hydrides, in particular, have been studied: titanium hydride (TiH_2) and zirconium hydride (ZrH_2), which contain approximately 9.0×10^{22} and 7.5×10^{22} hydrogen atoms per cm^3, respectively. At high temperatures, e.g., above 500°C (930°F), these compounds decompose, giving off hydrogen, but it appears that the dissociation can be greatly diminished by the presence of a small excess of the metal.   Other heavy-metal hydrides, which have been mentioned as worthy of consideration for shielding purposes, are those of lithium (for mobile reactors [9]), chromium, iron and related metals, tungsten, thorium, and uranium.

Boron Compounds

10.51. There are several common boron compounds which are sufficiently soluble to be added to a water shield to increase the rate of neutron capture;

among these, mention may be made of boric acid, borax, and various other borates. The advantage to be gained by the boration of water, however, varies considerably with the character of the shield. In a lead-water shield, with a large proportion of the lead near the core, for example, it would be beneficial to add boron to the water. The fact that there is relatively little gamma-ray absorbing material in the outer regions of the shield would then not be serious since only low-energy gamma rays are produced when boron captures neutrons.

10.52. A boron-containing solid which may have some special applications for neutron shielding is the complex of boron carbide (B_4C) and aluminum known as Boral. As stated in § 5.123, this product is also used as a neutron absorber for reactor control. The composition of Boral is variable but it generally contains from 30 to 50 per cent by weight of boron carbide, and is available in sheets, either $\frac{1}{4}$ in. or $\frac{1}{8}$ in. thick, clad on each side with 0.02 in. of aluminum. The product has a density of 2.5 g per cm^3; it has satisfactory mechanical properties and absorbs neutrons with no accompanying high-energy gamma radiation. Various materials, such as epoxy resins incorporating boron, having good moderating properties and radiation stability, have been developed for shielding purposes. The resin-impregnated wood laminates mentioned in § 10.44, e.g., Hydrobord, Jabroc, Permali, etc., are available in boron-loaded forms [10].

SHIELDING GEOMETRY TRANSFORMATIONS [11]

DISTRIBUTED SOURCES AND THE POINT ATTENUATION KERNEL

10.53. For many aspects of shield design it is useful to have equations which give the intensity of the radiation received at a detector from sources of various shapes, e.g., point, linear, planar, spherical, etc., as a function of the distance between source and detector, i.e., of the shield thickness. Such equations can be utilized to calculate the radiation attenuation in a shielding material. Alternatively, they can be used to determine the flux distribution from a source of one particular shape and size by employing experimental measurements made with a different source. Both types of application of what are frequently called *shielding geometry transformations* will be described in this section. The results obtained are quite general and are equally applicable to neutrons and to gamma radiation.

10.54. The calculation of the radiation flux due to a distributed source, i.e., a linear, surface, or volume source, as a function of distance makes use of the fundamental principle that any distributed source can be treated as a summation of point sources. This principle, sometimes referred to as the theorem of the additivity of radiation effects, states that, at any location, the net effect of a number of radiation sources is the sum of the separate effects of the individual sources. The first step in the derivation of the required equations is then to write a general expression for the radiation flux at a certain distance from a point source of known strength. The resulting expression is then integrated over the whole of the distributed source. In some cases the required integration can be carried out analytically, but in others either approximations must be made or special methods of integration used.

10.55. For the present purpose it is convenient to define a *point attenuation kernel*, $G(R)$; it is the radiation flux or other measurable quantity related to the flux, e.g., dose rate or heating rate, observed at a distance R from a unit point source, i.e., a source emitting one particle of radiation (one neutron or one gamma-ray photon) per second, both source and detector being located within an infinite homogeneous medium. If $G(R)$ is the kernel for radiation flux, then the flux observed at a distance R from a point source of strength S_p particles per second is given by $S_pG(R)$ particles/(cm^2)(sec), i.e., S_p times the appropriate value for a unit point source.*

10.56. In the following treatment it will be assumed that the sources under consideration are isotropic, i.e., they radiate uniformly in all directions, and that the shielding medium is also isotropic, *having the same attenuating properties in all directions.* In these circumstances, for any specified value of the distance R between the radiation source and a detector, i.e., for a specified shield thickness, the attenuation will be the same regardless of position or direction. The point kernel $G(R)$ then depends only on the nature of the attenuating medium, the character, e.g., the type and energy, of the radiation, and the distance R. It will be seen shortly that an expression for the function $G(R)$ which is frequently used is similar to equation (2.18) or (2.32) combined with the inverse-square-law geometrical spreading effect mentioned in § 9.40.

ISOTROPIC CIRCULAR PLANE (DISK) SOURCE

10.57. Although the simplest distributed source is a linear one, it has no direct application in shielding geometry transformations, and so it will be considered later (§ 10.149). The next in order of complexity is an isotropic circular plane (disk) source of radius a, as shown in Fig. 10.2, with the observation point P on the axis. The source strength S_a is the total number of particles emitted (isotropically) per unit area per second. Consider a narrow annulus of radius ρ and width $d\rho$; the area of this annulus is $2\pi\rho\, d\rho$ and it is everywhere distant R from P. The radiation flux at P is equivalent to that from $2\pi\rho\, d\rho$ point sources of total strength S_a, namely, $2\pi\rho\, d\rho\, S_aG(R)$, where $G(R)$ is the point attenuation kernel for flux defined above.

10.58. The value of the radiation flux at P from the whole disk source is then obtained by integrating over all values of ρ from zero to a; the result is

* The symbol S_p particles per second is used for a point source, S_l particles per unit length (1 cm) per second for a linear source, S_a particles per unit area (1 cm^2) per second for a surface (or area) distributed source, and S_v particles per unit volume (1 cm^3) per second for a volume distributed source.

$$\phi = 2\pi S_a \int_0^a G(R)\rho \, d\rho.$$

Since $R^2 = \rho^2 + z^2$, where z is the thickness of the shield, it follows that

$$2R \, dR = 2\rho \, d\rho$$

and hence

$$\phi = 2\pi S_a \int_z^{\sqrt{z^2+a^2}} G(R)R \, dR, \tag{10.1}$$

the limits of integration having been changed to correspond to the new variable.

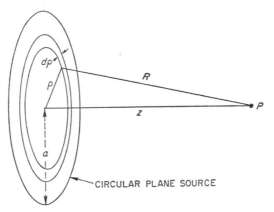

Fig. 10.2. Flux from circular plane source

For a large disk the upper limit may be set equal to infinity, thus simplifying the integration; then

$$\phi = 2\pi S_a \int_z^{\infty} G(R)R \, dR, \tag{10.2}$$

which is strictly applicable to an infinite plane source. Differentiation of this expression with respect to z provides a relatively simple relationship which permits the evaluation of the point kernel from measurements made with a disk source (see § 10.168).

ISOTROPIC SPHERICAL SURFACE SOURCE

10.59. A case of interest, which has a number of important applications, is that of an isotropic source that is spread uniformly over a spherical surface. In order to simplify the treatment, it is postulated that the isotropic absorbing medium extends throughout the inside, as well as the outside, of the sphere. Radiation coming from the surface of the sphere farthest from the observation point P will thus pass through a single medium, and $G(R)$ will be the same function for all values of R.*

* Since only a small proportion of the flux at P is due to emission from the farther surface, differences in the media inside and outside the sphere do not generally have a very significant effect on the result.

10.60. Let r be the radius of the spherical surface and r_0 the distance from the center of the sphere to the point P (Fig. 10.3). Consider a thin annulus, as shown in the figure, all points of which are at a distance R from P. The circumference of this annulus is $2\pi r \sin\theta$, and the width is $r\,d\theta$, so that its area

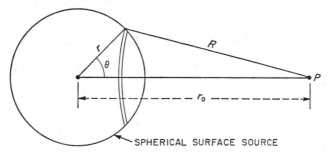

Fig. 10.3. Flux from spherical surface source

is $2\pi r^2 \sin\theta\,d\theta$. If S_a is the source strength, expressed as particles emitted per unit area per second, it is seen that the radiation flux at P from the whole spherical surface is

$$\phi = 2\pi r^2 S_a \int_0^\pi G(R) \sin\theta\,d\theta. \qquad (10.3)$$

The relationship between R, r, and r_0 is given by

$$R^2 = r^2 + r_0^2 - 2rr_0 \cos\theta,$$

so that

$$2R\,dR = 2rr_0 \sin\theta\,d\theta.$$

Hence, upon changing the variables from θ to R, equation (10.3) becomes

$$\phi = 2\pi S_a \frac{r}{r_0} \int_{r_0-r}^{r_0+r} G(R)R\,dR. \qquad (10.4)$$

If $r_0 + r$ is large so that, as an approximation, the upper limit of the integration may be taken as infinity, the integral in equation (10.4) will have the same form as that in equation (10.2), and analytical solution is generally possible.

10.61. In any event, the integral in equation (10.4) may be expressed as the difference between two integrals: one with the limits from $r_0 - r$ to infinity and the other from $r_0 + r$ to infinity. Consequently,

$$\phi = \frac{r}{r_0}\left[2\pi S_a \int_{r_0-r}^\infty G(R)R\,dR - 2\pi S_a \int_{r_0+r}^\infty G(R)R\,dR \right]. \qquad (10.5)$$

Upon comparison with equation (10.2) it is seen that each integral is equivalent to radiation flux from an infinite plane source, the first as observed at a distance $r_0 - r$ and the second at $r_0 + r$. Hence, for the same source strength per unit area, equation (10.5) can be written as

$$\phi(\text{sphere}) = \frac{r}{r_0} \left[\phi(\text{inf. plane at } r_0 - r) - \phi(\text{inf. plane at } r_0 + r) \right], \quad (10.6)$$

so that the flux from the spherical source can be expressed in terms of the values for two infinite planar sources. If the diameter of the sphere is relatively large, the second term on the right of equation (10.6) is negligible, and this equation becomes

$$\phi(\text{sphere}) \approx \frac{r}{r_0} \phi(\text{inf. plane at } r_0 - r). \quad (10.7)$$

Thus the radiation flux to be expected from a spherical source can be calculated from the data for an infinite plane source, or vice versa. It is of interest to note that equation (10.7) implies that the flux received at the detector from the spherical source, after passage through a given shield, is less by a factor of r/r_0 than that due to an infinite plane source (of the same strength per unit area) located at the distance of closest approach, i.e., $r_0 - r$, from the detector.

ISOTROPIC CYLINDRICAL SURFACE SOURCE

10.62. The derivation of the geometrical transformation relationship for an isotropic cylindrical surface source of *infinite length* is given in the appendix to this chapter (§ 10.180). The result is

$$\phi(\text{cylinder}) = \sqrt{\frac{r}{r_0}} \, \phi(\text{infinite plane at } r_0 - r), \quad (10.8)$$

where r is the radius of the cylinder and r_0 is the distance from the axis to the observation point P (Fig. 10.4). Since r/r_0 is here less than unity, $\sqrt{r/r_0}$ is

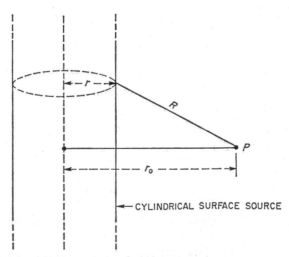

Fig. 10.4.   Flux from cylindrical surface source

greater than r/r_0; the radiation flux from an infinite cylindrical source thus lies between that due to an infinite plane source, on the one hand, and that due to a large spherical source having the same radius as the cylinder, on the other hand. The source strengths per unit area are assumed to be the same in all cases. Although equation (10.8) applies to an infinite cylinder, it is a fair approximation for any cylinder that is not very short in comparison with the distance from the detector to the nearest point on the cylinder. The physical reason for this is that points on the cylinder that are more distant contribute relatively little to the radiation flux at P. Consequently, the radiation received from an infinite cylinder is not much greater than that from a long finite cylinder. However, if the cylinder is short compared with the distance to the observation point, the method for determining the flux described in § 10.151 is to be preferred.

ATTENUATION OF RADIATIONS

BUILDUP FACTORS

10.63. The expressions derived above are completely general in character. They are independent of the nature of the radiation and also of the nature of the point attenuation kernel, $G(R)$, which gives the flux (or related measurable quantity) from a point source as a function of R, its distance from the observation point. In order to make use of the results for semianalytic shielding calculations, it is necessary to specify $G(R)$ explicitly, although this is not required for the comparison method. However, before deriving an expression for the point kernel, it is convenient to consider some general aspects of radiation attenuation.

10.64. It was seen in Chapter 2 that the attenuation of a narrow (or collimated) beam of either neutrons or gamma rays of a given energy can be represented by an exponential expression of the form

$$\phi(t) = \phi(0)e^{-\Sigma t} \quad \text{or} \quad \phi(t) = \phi(0)e^{-\mu t}, \tag{10.9}$$

where $\phi(0)$ and $\phi(t)$ are, respectively, the radiation flux (or current) before and after passage through a thickness t of shield, and Σ is the total macroscopic cross section for the radiation. For gamma rays, Σ is equivalent to μ, the linear attenuation coefficient, as pointed out in § 2.115, footnote.

10.65. The simple exponential attenuation equation (10.9) is based on the tacit assumption that scattered particles are completely removed from the radiation beam. The quantity $\phi(t)$ then gives what is called the *uncollided flux*, i.e., the radiation flux which has not been involved in any collisions in its passage through a thickness t of the shield. In fact, with this definition of $\phi(t)$, equation (10.9) is applicable to a broad collimated beam of radiation.

10.66. For a relatively thin layer of attenuating material, i.e., for a thin shield, equation (10.9) is a good approximation for the measured (or total) flux, even for a broad collimated beam, especially for photons of high energy. This is because the probability that a scattered radiation particle will reach the observation point (or detector) after a single collision is small (Fig. 10.5). The total flux measured is then essentially the same as the uncollided flux. On the other hand, if the shield is relatively thick, some particles which have suffered two or more scattering collisions within the absorber may reach the detector (Fig. 10.6). In this case the scattered particles are not removed, and the flux at the observation point exceeds the uncollided flux; the simple exponential equation (10.9) will then give broad beam values for $\phi(t)$ that are too low.

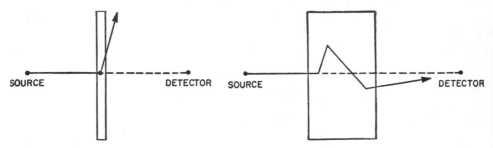

FIG. 10.5.   Scattering of radiation in   FIG. 10.6.   Scattering of radiation in thick
                   thin shield   shield

10.67. The effect of the scattered radiation is allowed for by means of what is called a *buildup factor*, which is a function of the shield material and thickness and the energy of the radiation, and also of the particular quantity being observed. Thus, for a given shield and radiation, the value of the buildup factor would be different for the number flux, i.e., particles/(cm²)(sec), for the energy flux, i.e., Mev/(cm²)(sec), for the dose rate, i.e., roentgens/sec, etc. Assuming the number flux is being observed, the attenuating equation (10.9) would then be modified to take the form

$$\phi(t) = B(\mu t)\phi(0)e^{-\mu t}, \tag{10.10}$$

where $B(\mu t)$ is the appropriate buildup factor; it is generally expressed as a function of μt, since μ depends on the material and energy of the radiation and t is the shield thickness. Although the buildup factor concept can, in principle, be applied to both gamma-ray photons and to neutrons, it is generally used for the former only, and it is for this reason that equation (10.10) was written in the conventional form for gamma-ray attenuation.

10.68. Gamma radiation buildup factors for the number flux have not been determined. Values have, however, been calculated from theoretical considerations, for both point and plane sources, for dose rates and energy absorption. The buildup factors for various substances are tabulated for energies ranging from 0.5 to 10 Mev as a function of μt, ranging from 1 to 20. Some of the data for the dose (or dose rate) buildup factor, for a point isotropic source of gamma rays, for water, ordinary concrete, and iron, are given in Table 10.2.* Since the energy flux, in Mev/(cm²)(sec), is roughly proportional to the dose rate, in roentgens/sec, it is a good approximation to take the values in Table 10.2 as also representing energy-flux buildup factors.

TABLE 10.2. DOSE BUILDUP FACTORS FOR A POINT ISOTROPIC SOURCE

Material	μt	Gamma-Ray Energy, Mev						
		1.0	2.0	3.0	4.0	6.0	8.0	10.0
Water	1	2.13	1.83	1.69	1.58	1.46	1.38	1.33
	2	3.17	2.77	2.42	2.17	1.91	1.74	1.63
	4	7.68	4.88	3.91	3.34	2.76	2.40	2.19
	7	16.2	8.46	6.23	5.13	3.99	3.34	2.97
	10	27.1	12.4	8.63	6.94	5.18	4.25	3.72
	15	50.4	19.5	12.8	9.97	7.09	5.66	4.90
	20	82.2	27.7	17.0	12.9	8.85	6.95	5.98
Concrete	1	2.2	1.7	1.65	1.6	1.55	1.4	1.35
(Density	2	3.6	2.8	2.4	2.25	1.95	1.75	1.65
2.35 g/cm³)	4	7.8	4.9	3.8	3.3	2.75	2.4	2.2
	7	15	8.4	6.2	5.0	4.0	3.3	3.0
	10	24	12.3	8.6	6.8	5.2	4.3	3.8
	15	43	19	12.6	9.9	7.1	5.7	5.1
	20	70	27	17	13	9.1	7.3	6.3
Iron	1	1.87	1.76	1.55	1.45	1.34	1.27	1.20
	2	2.89	2.43	2.15	1.94	1.72	1.56	1.42
	4	5.39	4.13	3.51	3.03	2.58	2.23	1.95
	7	10.2	7.25	5.85	4.91	4.14	3.49	2.99
	10	16.2	10.9	8.51	7.11	6.02	5.07	4.35
	15	28.3	17.6	13.5	11.2	9.89	8.50	7.54
	20	42.7	25.1	19.1	16.0	14.7	13.0	12.4

10.69. In calculations involving gamma-ray attenuation, it is useful to express the buildup factor in analytical form. One convenient expression is

$$B(\mu t) = 1 + b\mu t, \tag{10.11}$$

where b is obtained from the values in Table 10.2; for example, when μt is 10, and the gamma-ray energy is 4.0 Mev, $B(\mu t)$ for water is 6.94, so that b is 0.594. In some calculations b has been taken as unity and in others $B(\mu t)$ has been set equal to μt for simplicity. It is clear from Table 10.2, however, that the results obtained in this manner are very approximate and are suitable only for preliminary estimates of attenuation. The fact that these simple forms of $B(\mu t)$

* For other values, see Ref. 12.

generally overestimate the flux escaping from the shield is an advantage in this connection.

10.70. An alternative, useful expression for the dose buildup factor is as the sum of two exponential terms, namely,

$$B(\mu t) = Ae^{-\alpha\mu t} + (1 - A)e^{-\beta\mu t}, \tag{10.12}$$

which has been found to give results to an accuracy of about 5 per cent or better [13]. Values of the coefficients A, α and β for water, ordinary concrete, and iron are quoted in Table 10.3. Linear attenuation coefficients for water, iron, and other materials for gamma rays of various energies will be found in Table 2.4.

TABLE 10.3. COEFFICIENTS FOR EXPONENTIAL (DOSE) BUILDUP EQUATION

Material	Coeff.	Gamma-Ray Energy, Mev						
		1.0	2.0	3.0	4.0	6.0	8.0	10.0
Water	A	11	6.4	5.2	4.5	3.55	3.05	2.7
	$-\alpha$	0.104	0.076	0.062	0.055	0.050	0.045	0.042
	β	0.030	0.092	0.110	0.117	0.124	0.128	0.13
Concrete	A	10.0	6.3	4.7	3.9	3.1	2.7	2.6
(Density	$-\alpha$	0.088	0.069	0.062	0.059	0.059	0.056	0.050
2.35 g/cm³)	β	0.029	0.058	0.073	0.079	0.083	0.086	0.084
Iron	A	8.0	5.5	4.4	3.75	2.9	2.35	2.0
	$-\alpha$	0.089	0.079	0.077	0.075	0.082	0.083	0.095
	β	0.04	0.07	0.075	0.082	0.075	0.055	0.012

10.71. In the preceding treatment the buildup factors have referred to a single (or uniform) shield material. Some shields, however, involve alternate layers of two different materials, e.g., iron and water. The situation is then complicated because there is uncertainty concerning the value of the buildup factor to be used, especially when the atomic numbers of the components of the multilayer shield are appreciably different. The difficulty arises from the fact that the photon energy distribution (or spectrum) changes with penetration of any medium; this is true whether the incident photons are monoenergetic or not. The change in the spectrum is caused by scattering and absorption of both scattered and unscattered photons. The energy distribution in the emergent gamma rays depends in a complex manner on the initial photon energy, the nature of the medium, and its thickness. One result of this situation is that, for a composite shield consisting of layers of light and heavy materials, the gamma-ray energy spectrum and the attenuation will depend upon whether the light material precedes the heavy one or vice versa.

10.72. A number of "rules of thumb" have been proposed to deal with this situation [14]. If the mass (or atomic) numbers of the materials are not too greatly different, a good approximation is to add the μt values for the separate layers. The buildup factor is then determined from tables for this total μt value for each constituent; the higher value of the buildup factor is used for calculations. Where one material has an appreciably higher mass number than the other, e.g., lead and water, the buildup factor depends on the order of the layers. If the photons pass through the water first and then through the lead, the combined buildup factor is best approximated by taking the value for lead corresponding to the total μt. If the order of the layers is reversed, the buildup factor is generally large, but the actual value depends on the photon energy. For low energies, the overall buildup factor is roughly the product of the values for the two separate materials. When the initial photon energy is high, above the minimum in the attenuation coefficient curve of the heavier material (Fig. 2.10), the photons entering the water may be assumed to have the energy of the minimum and the buildup factor determined accordingly.

THE RELAXATION LENGTH

10.73. Because of the buildup effect due to scattered radiation, the radiation flux of a given energy in a thick shield does not fall off in a simple exponential manner. It is nevertheless convenient for some purposes to retain the exponential character of the attenuation equation and to write

$$\phi(t) = \phi(0)e^{-t/\lambda}, \tag{10.13}$$

where $\phi(t)$ and $\phi(0)$ have the same significance as before and λ is called the *relaxation length* of the shield material for the given radiations. It can be seen from equation (10.13) that, physically, it is the distance (or thickness) of shield in which some property of the radiation, e.g., the current or flux, is decreased by a factor of $1/e$, i.e., $1/2.72$. Comparison of equations (10.9) and (10.13) shows that λ is formally equivalent to $1/\Sigma$ (or $1/\mu$) and hence to the mean free path (§ 2.123) for the photons or neutrons of given energy. However, there is an important difference, since λ in equation (10.13) includes an approximate (empirical) correction for multiple scattering in thick shields for the particular radiation which is being attenuated. In other words, in λ there is an allowance for the buildup factor, so that the latter can be taken to be unity in rough calculations using the relaxation length.

10.74. Since the relaxation length is frequently used in shielding work whether the radiation attenuation is exponential or not, it may be defined more generally by

$$\frac{1}{\lambda} = -\frac{d \ln \phi(t)}{dt} = -\frac{1}{\phi(t)} \cdot \frac{d\phi(t)}{dt} \tag{10.14}$$

at a point, described by t, in a given medium. Then, if $\phi(t)$ is an exponential function of t, as in equation (10.9), the value of λ is constant over the exponential range. But, if the attenuation is not exponential, the relaxation length will vary from point to point in the medium. Experience, based on measurements in various shielding materials, has shown that the deviation from exponential behavior is surprisingly small for thick shields. Consequently, constant relaxation lengths, such as the values obtained from measurements on thick shields given in Table 10.4, can be used for preliminary, approximate attenuation

TABLE 10.4. APPROXIMATE RELAXATION LENGTHS FOR FAST NEUTRONS AND FOR 4-MEV AND 8-MEV GAMMA RAYS

Material	Density (g/cm³)	Relaxation Length (cm)		
		Fast Neutrons	4-Mev Gamma Rays	8-Mev Gamma Rays
Water................	1.00	10	30	40
Graphite..............	1.62	9	19	25
Beryllium.............	1.85	9	20	30
Beryllium oxide........	2.3	9	18	25
Concrete..............	2.3	12	14	18
Aluminum.............	2.7	10	13	17
Barytes concrete.......	3.5	9.5	10	13
Iron concrete..........	4.3	6.3	8	10
Iron..................	7.8	6	3.7	4.4
Lead.................	11.3	9	2.4	1.9

calculations. The results quoted for fast neutrons are for slowing down and do not include absorption unless a hydrogenous material is also present.

10.75. Attention may be drawn to the relatively short relaxation length of fast neutrons in water and the much larger values for gamma rays. This is the basis of the statement made earlier (§ 10.43) that water is a good shielding material for fast neutrons but a poor one for gamma radiations. Concrete is a better overall shield material because the relaxation lengths do not differ greatly for the two important types of radiation which have to be attenuated. The heavy elements, at the other extreme, are very effective for gamma rays, but less satisfactory for neutron shielding.

10.76. It is of interest to note that the quantity t/λ which appears in the exponent of equation (10.13) represents the number of relaxation lengths in the shield thickness. Since λ is formally equivalent to $1/\mu$ (or $1/\Sigma$), the parameter μt used in connection with the buildup factor is also (approximately) equivalent to the number of relaxation lengths in the attenuating medium. In other words, it is usual to express the buildup factor for a given material in terms of the thickness expressed in relaxation lengths.

THE EXPONENTIAL POINT KERNEL

10.77. A simple form of the point kernel is based on the exponential attenuation of radiation as represented by equation (10.9) or (10.10). These expressions apply to collimated beams in which there is no spreading of the radiation. For a point source, however, allowance must be made for the attenuation resulting from geometrical spreading with increasing distance from the source, as described in § 9.40. If a point source emits S_p particles (photons or neutrons) per second, then, assuming isotropic distribution, the flux at a distance R cm, disregarding attenuation by the medium, will be $S_p/4\pi R^2$ particles/(cm²)(sec), since $4\pi R^2$ is the area of a spherical surface of radius R. By combining this result with the equivalent of equation (10.10), to allow for attenuation and buildup of the radiation by a shielding material of thickness t between the point source and the detector, it is found that the flux at the detector is given by

$$\phi = S_p B(\mu t)\, \frac{e^{-\mu t}}{4\pi R^2}.$$ (10.15)

The attenuation due to the thickness $R - t$ of air is here neglected.

10.78. Although equation (10.15) is completely general for the flux received at a distance R from a point source after passage of the radiation through a shield of thickness t, care must be taken in using it in connection with the geometrical transformation equations described earlier. These equations are based on the postulate (§ 10.56) that the shielding medium attenuates equally in all directions. Hence, application of equation (10.15) to derive a point kernel requires either that a uniform shield extend continuously from the source to the point of observation, i.e., t and R in equation (10.15) are the same, or that the shield should be spherically symmetrical about the point source. If the shield is a planar slab of thickness t, which is less than R, neither of these conditions is satisfied. The treatment to be used for shielding calculations in these circumstances is given in § 10.148 *et seq.*

10.79. For the present, it will be assumed that the shield is uniform and continuous between the source and the observation point. This is the situation commonly encountered in reactor shielding. Hence, R and t in equation (10.15) are identical, so that

$$\phi = S_p B(\mu R)\, \frac{e^{-\mu R}}{4\pi R^2}.$$ (10.16)

The right side of equation (10.16) gives the flux in an isotropic medium at a distance R from an isotropic point source. It is consequently equal to $S_p G(R)$ as defined in § 10.55. The expression for the point kernel in terms of flux is thus

$$G(R) = B(\mu R)\, \frac{e^{-\mu R}}{4\pi R^2}.$$ (10.17)

In this form, with or without the buildup factor, equation (10.17) is commonly used for the calculation of gamma-ray attenuation. For fast neutrons, however, it is the general practice to take the buildup factor as unity and to include an allowance for it in the macroscopic (removal) cross section, Σ_r, which replaces μ in the exponent in equation (10.17).

DISTRIBUTED SOURCES

RADIATION FROM AN INFINITE PLANE SOURCE

10.80. With an expression available for the point kernel, it is possible to carry out some of the integrations required for evaluating the flux in an isotropic medium as a function of distance from various distributed isotropic sources. In the first place, it will be assumed that $B(\mu R)$ is unity, and the modifications resulting from the use of equations (10.11) and (10.12) will be given later. For an *infinite plane source*, equation (10.2) now becomes

$$\phi = 2\pi S_a \int_z^\infty \frac{e^{-\mu R}}{4\pi R^2} R \, dR$$

$$= \frac{S_a}{2} \int_z^\infty \frac{e^{-\mu R}}{R} \, dR, \tag{10.18}$$

where S_a is expressed as particles/(cm²)(sec); as seen in Fig. 10.2, z is here the perpendicular distance from the source to the detector and hence is the shield thickness. As is generally done in the shielding literature, the parameter μR, which is strictly applicable to gamma rays, is used. However, for neutrons μ would be replaced by the appropriate Σ_r or $1/\lambda$, where λ is the corresponding relaxation length.

10.81. To evaluate the integral in equation (10.18), a new variable, q, defined by $q = \mu R$, is introduced, so that

$$\phi = \frac{S_a}{2} \int_{\mu z}^\infty \frac{e^{-q}}{q} \, dq. \tag{10.19}$$

This integral is a form of what is known as the *exponential integral function* defined by

$$E_n(x) = x^{n-1} \int_x^\infty \frac{e^{-p}}{p^n} \, dp, \tag{10.20}$$

values of which may be found in tables and handbooks [15].* In the present case, n is unity and x is equal to μz, i.e., the number of relaxation lengths in the perpendicular distance between the infinite plane source and the detector, so that equation (10.19) becomes

---

* If tables are not readily available, a good approximation [16] for $x > 1$ is given by $E_n(x) \approx e^{-x}(1 + x + n)/[x + (x + n)^2]$; for $x > 10$, $E_1(x) \approx e^{-x}/x$.

$$\phi(z) = \tfrac{1}{2}S_a E_1(\mu z), \tag{10.21}$$

where $E_1(\mu z)$ is the first-order exponential integral function for the argument μz. A plot of $E_1(x)$ for various values of x is shown in Fig. 10.7. By ascertaining the appropriate magnitude of $E_1(\mu z)$, either from the figure or from equivalent

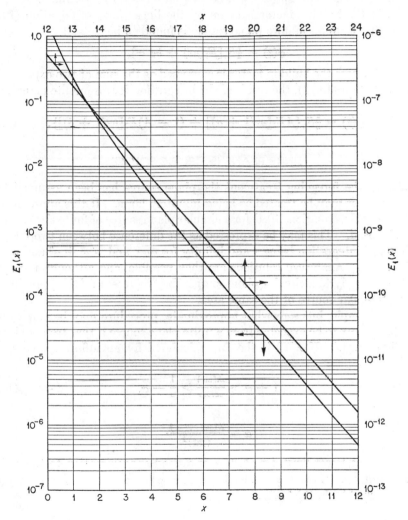

FIG. 10.7. Values of first-order exponential integral function

tables, for a specified shield thickness z, it is possible to determine the flux of a given radiation received from an infinite plane source of known strength.

10.82. A modification of equation (10.21) is often used in connection with the attenuation of fast neutrons by a water shield. The buildup factor is taken as effectively unity, but the point attenuation kernel is the sum of two exponential terms instead of the single exponential employed in deriving equation (10.21). The distribution of the neutron flux in a water shield is then given by an expression involving the sum of two terms of the same form as that on the right of equation (10.21). This subject will be treated more fully in § 10.98.

10.83. If the buildup factor is given by the linear form of equation (10.11), then the flux from an infinite plane source is found to be

$$\phi(z) = \tfrac{1}{2}S_a[E_1(\mu z) + be^{-\mu z}]. \tag{10.22}$$

In the case where the buildup factor is expressed as the sum of two exponentials, i.e., by equation (10.12), the result of the integration for an infinite plane source is

$$\phi(z) = \tfrac{1}{2}S_a\{AE_1[\mu z(1 + \alpha)] + (1 - A)E_1[\mu z(1 + \beta)]\}. \tag{10.23}$$

FINITE PLANE ISOTROPIC SOURCE

10.84. In the foregoing treatment the plane source has been assumed to be large, so that it may be treated as being infinitely large. If this approximation is not justifiable, equation (10.1) for a circular plane source of finite size (radius a) should be used. This may be expressed as the difference between two integrals, the limits of integration being z and infinity for one, and $\sqrt{z^2 + a^2}$ and infinity for the other. Analytical solutions are then possible, and although the results so obtained apply to a circular plane source, they represent a fair approximation for a rectangular source of the same area.

10.85. A simple, but somewhat approximate, procedure may be employed to estimate the flux to be expected from a finite plane source if that from an infinite source is known.* If $\phi(\infty)$ is the flux calculated for an infinite plane source, then

$$1 + \alpha > \frac{\phi(\infty)}{\phi(a)} \geq \frac{1}{2} + \alpha, \tag{10.24}$$

where

$$\alpha \equiv \frac{2(\mu z + 1)}{(\mu a)^2}.$$

In general, the lower limit, i.e., $\tfrac{1}{2} + \alpha$, is a good approximation to the ratio of the radiation dose rates received from infinite and finite source after penetration through a shield z in thickness.

SPHERICAL SURFACE SOURCE

10.86. The calculation of the radiation received from a spherical surface source after passage through a homogeneous (isotropic) shield may be performed by

---

* A more exact treatment, of which the result given here is an approximation, will be found in the appendix to this chapter (§ 10.172).

combining equation (10.4) with the specific expression for the point kernel. In general, however, it is satisfactory to use the relationship, expressed by equation (10.7), between the radiation flux to be expected from spherical and (infinite) plane surface sources. In the simple case where the buildup factor is unity, so that equation (10.21) applies to an infinite plane source, the flux for a spherical surface source is given by

$$\phi(z) = \frac{S_a}{2} \cdot \frac{r}{r_0} E_1(\mu z),\tag{10.25}$$

where r is the radius of the sphere, r_0 is the distance from the center of the sphere to the detector, and z, equal to $r_0 - r$, is the closest distance from the sphere to the detector; S_a is the strength, in particles/(cm²)(sec), of the spherical surface source.

VOLUME-DISTRIBUTED SOURCE WITH SELF-ABSORPTION

10.87. The discussion so far has referred to surface-distributed sources, although in practice most radiation sources are distributed throughout a volume, e.g., a reactor or a vessel containing radioactive material. However, it will now be shown that the radiation to be expected from a volume-distributed source can be expressed in terms of the flux at the surface. Since there is some self-absorption of the radiation within the volume itself, appropriate allowance must be made in the manner indicated below.

10.88. Suppose the volume source is a slab of infinite extent having a finite thickness t, and consider a thin slab element of thickness dx parallel to the plane AB (Fig. 10.8). If S_v particles (neutrons or photons)/(cm³)/(sec) constitute the *uniform* source strength within the slab, the rate of emission per unit area from the surface of the element dx is $S_v\, dx$ particles/(cm²)(sec). The flux from this element at a point on the plane AB, at a distance x, is then given by equation (10.21) as

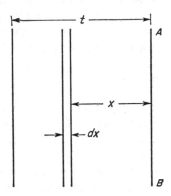

FIG. 10.8. Volume distributed source with self-absorption

Flux at surface from element $dx = \frac{1}{2}S_v E_1(\mu_v x)\, dx$,

where μ_v is the linear attenuation coefficient (or macroscopic removal cross section) of the volume source material for the given particles. The flux at the surface due to the whole thickness t is obtained by integrating from 0 to t; thus,

$$\text{Total flux at surface } AB = \frac{S_v}{2} \int_0^t E_1(\mu_v x)\, dx$$

$$= \frac{S_v}{2\mu_v} [E_2(0) - E_2(\mu_v t)],\tag{10.26}$$

where $E_2(\mu_v t)$ is the second-order integral function for the argument $\mu_v t$. Since $E_2(0)$ is unity, equation (10.26) reduces to

$$\text{Total flux at surface } AB = \frac{S_v}{2\mu_v}\left[1 - E_2(\mu_v t)\right].$$

For a thick slab, i.e., for $\mu_v t$ large, $E_2(\mu_v t)$ is small, so that

$$\text{Total flux at surface } AB \approx \frac{S_v}{2\mu_v}. \tag{10.27}$$

Provided $\mu_v t$ is greater than 3, the error involved in using equation (10.27) instead of (10.26) is less than 1 per cent. Consequently, the former is commonly employed in shielding calculations.

10.89. In order to make use of this result, in conjunction with the equations derived previously, such as equation (10.21), to calculate the flux in a shield at a given distance from the surface of the volume-distributed source, it must be recalled that the equations assume an isotropic source, i.e., with particles emitted uniformly in all directions from every point on the surface. In order to convert the result in equation (10.27) into the value for the equivalent isotropic surface source, it is necessary to multiply by two. Consequently, the strength, S_a, of an isotropic surface source which is equivalent to the volume source is given by

$$S_a = \frac{S_v}{\mu_v}. \tag{10.28}$$

It may be noted that since S_v is expressed as particles/(cm^3)(sec) and μ_v is usually given in cm^{-1}, S_a will have the correct dimensions of a surface source strength, namely, particles/(cm^2)(sec).

10.90. For a *large slab* volume source surrounded by a shield, the radiation flux at a distance z is obtained by inserting equation (10.28) into the appropriate equation for an infinite plane source. Thus, if the buildup factor in the shield is unity, equation (10.21) yields

$$\phi(z) = \frac{S_v}{2\mu_v} E_1(\mu z), \tag{10.29}$$

where μ refers to the shield material and μ_v to the volume source material. When the buildup factor is greater than unity, the appropriate expressions are obtained by substituting equation (10.28) into either equation (10.22) or (10.23).

10.91. If the volume source is *spherical*, then equation (10.25) is used so that, for a buildup factor of unity in the shield,

$$\phi(z) = \frac{S_v}{2\mu_v} \cdot \frac{r}{r_0} E_1(\mu z). \tag{10.30}$$

The corrections required when the buildup factor is not unity can be applied as described above.

10.92. It will be recalled that the derivation of equation (10.28) is based on the postulate that the volume-distributed source is uniform. A reactor, as a general rule, is, however, not a uniform source, so that the result obtained is not strictly applicable. If the source strength (or power), S_v, can be represented by a simple expression, e.g., a cosine function, as is often the case, it may be possible to perform the required integration analytically and obtain a result for the equivalent surface source strength. A more approximate approach is to use an average value for the source strength and assume it to be constant throughout the reactor core. This method tends to overestimate the strength of the equivalent surface source, since it gives equal weight to all parts of the volume-distributed source. In actual fact, most of the radiation reaching the surface arises from the outer regions of the volume where the source strength is generally less than in the interior.

REACTOR SHIELDING

FAST-NEUTRON ATTENUATION [17]

10.93. Much of the work on fast-neutron attenuation in the past has been based on the treatment of these neutrons as a single group. Multigroup and other approaches have been developed and used for attenuation calculations in shields with the aid of computers. Such calculations also provide information to be used in the investigation of special phenomena, such as radiation effects on materials. For the present purpose, however, the single-group method is adequate and the discussion will be restricted to this procedure for studying fast-neutron attenuation in shields.

10.94. Although the reactor core is not a uniform volume source of fast neutrons and primary gamma rays, it is nevertheless satisfactory to treat it as if it were uniform. The neutron source strength, S_v, can then be readily determined. It was seen in § 1.45 that 3.1×10^{10} fissions per sec correspond to 1 watt of power; hence, in a reactor operating at an average power of P watts, there will be $3.1 \times 10^{10} P$ fissions per sec. Since close to 2.5 neutrons are produced per fission, at least in uranium-235, it follows that $2.5 \times 3.1 \times 10^{10} P$ neutrons per sec are produced in the reactor. If V cm^3 is the volume of the core, the volumetric source strength, assumed to be uniform, is then

$$S_v = 7.8 \times 10^{10} \frac{P}{V} \text{ neutrons/(cm}^3\text{)(sec)}. \tag{10.31}$$

10.95. The attenuation coefficient (or macroscopic cross section) of the reactor core for fast neutrons, which is required to convert the volume source into the equivalent surface source using equation (10.28), is not well known. For water-moderated reactors, a rough estimate can be made by using the "removal" cross sections, the precise significance of which will be explained shortly. For the present, the values may be regarded as giving the cross sections for fast-neutron attenuation in a medium containing, at least, a moderate proportion of water (or similar hydrogenous material). For example, for the pool-type reactor known as the Bulk Shielding Reactor (BSR), the constituents are water (volume fraction 0.583), aluminum in the fuel elements (0.415), and uranium (0.0018). The necessary data are then as follows:

Material	Macroscopic Removal Cross Section (cm^{-1})	Volume Fraction	Σ (cm^{-1})
Water	0.0978	0.583	0.0570
Aluminum	0.0789	0.415	0.0327
Uranium	0.17	0.0018	0.0003
		Total	0.0900

The macroscopic cross sections quoted are the values for the individual materials (Table 10.5 given subsequently); these are multiplied by the volume fractions to give the respective contributions to the system as a whole in the last column (cf. § 4.12). The total macroscopic cross section is thus 0.090 cm^{-1} and this is used for μ_v in equation (10.28).

10.96. Where the data are not available or the conditions do not justify the use of removal cross sections, it is a good approximation, at least for thermal reactors, to assume the value 0.10 cm^{-1} for μ_v. The reason is that for the common moderators the relaxation length for fast neutrons is close to 10 cm (Table 10.4). Since the moderator constitutes a large proportion of the reactor core, this length may be taken to apply to the whole core. The corresponding macroscopic cross section is thus 0.10 cm^{-1}, as stated above.

10.97. The procedure used in preliminary calculations of the neutron distribution in a shield depends on the physical nature of the latter. Three general types will be considered here: (a) water shields, (b) shields consisting of a heavy (or fairly heavy) material and liquid water (or similar hydrogenous material), and (c) concrete shields. The water shield is the only one for which satisfactory computations are possible by means of a fairly simple treatment.

WATER SHIELD [18]

10.98. For neutron attenuation in water, experimental data show that the point kernel can be represented by the sum of two exponential terms, which allow for the buildup factor, so that the kernel is of the form

$$G(R) = \frac{1}{4\pi R^2} [Ae^{-aR} + (1 - A)e^{-bR}]. \tag{10.32}$$

The expression for the flux distribution corresponding to equation (10.29) for a large planar slab source is then

$$\phi(z) = \frac{S_v}{2\mu_v} [AE_1(az) + (1 - A)E_1(bz)], \tag{10.33}$$

and for a spherical volume source this would be multiplied by r/r_0, where r is the radius of the sphere and r_0 is $r + z$. The values of the constants A, a, and b, derived from measurements in water at ordinary temperatures, are as follows:

$$A = 0.892, \qquad a = 0.129, \qquad b = 0.091.$$

10.99. The BSR core is a relatively small parallelepiped and it is not satisfactory, except at close distances, to regard the surfaces as infinite planes; a better approximation is to treat the core as a sphere of the same volume. The radius is then found to be 29.4 cm, and the volume is 1.06×10^5 cm^3. Hence, from equation (10.31), S_v per watt of power is

$$S_v = \frac{7.8 \times 10^{10}}{1.06 \times 10^5} = 7.4 \times 10^5 \text{ neutrons/(cm}^3)(\text{sec})(\text{watt})$$

and

$$\phi(z) = \left(\frac{7.4 \times 10^5}{2 \times 0.090}\right)\left(\frac{29.4}{29.4 + z}\right)[0.892\, E_1(0.129z) + 0.108\, E_1(0.091z)].$$

For $z = 100$ cm, for example, this gives

$$\phi(100 \text{ cm}) = 1.2 \text{ neutrons/(cm}^2)(\text{sec})(\text{watt}),$$

compared with the measured value of 1.4 neutrons/(cm^2)(sec)(watt). The agreement is as good as can be expected. By taking a series of values of z, the spatial distribution of fast neutrons from the BSR (or any other reactor) in a water shield can be readily determined in the manner described.

10.100. To estimate the thickness of a shield which will attenuate the fast-neutron flux to an acceptable biological dose rate, it is necessary to convert the flux into the appropriate units, e.g., millirems per hour. Since the energy of the escaping neutrons is not known, it appears to be safe to assume an average energy between 1 and 10 Mev. The fast flux in neutrons/(cm^2)(sec) is then divided by 7 to convert it into a dose rate expressed in millirems per hour (see Fig. 9.4).

REMOVAL CROSS SECTIONS [19]

10.101. The concept of *removal cross sections* for fast neutrons is strictly applicable to shields in which a relatively thin slab of a solid material is placed between a fission source and a moderately thick layer of water. The thickness of the water layer—about 45 cm at least—should be sufficient to ensure that a fast neutron undergoing inelastic scattering or a large-angle elastic scattering collision in the solid slab will be further slowed down and captured in the water. The removal cross section is a measure of the probability of these collisions. It should be noted that the collision in the slab does not remove the neutron but reduces its energy to such an extent that it will be readily thermalized and captured if it subsequently enters the water layer.

10.102. Theoretical considerations show that the removal cross section of a material should be approximately equal to the sum of the inelastic scattering cross section and a fraction of the elastic cross section for the dominant fission neutrons.   The observed values of the removal cross sections are, in fact, roughly equal to two-thirds of the total (inelastic and elastic) scattering and capture cross sections in the given material of neutrons having energies in the range of 6 to 8 Mev.   Because of the opposing effects, with increasing neutron energy, of increasing penetrability and decreasing proportion in the fission spectrum, the major contribution to the neutrons that penetrate farthest into a water shield is made by those which originally had energies of about 6 to 8 Mev. These are the so-called *dominant fission neutrons* that determine the removal cross section.

10.103. The definition and measurement of removal cross sections are based on the following considerations.   Suppose that a fission source emitting one neutron per second is located in an infinite water medium and that the fast-neutron flux, $G(x)$, the point kernel for water alone, is observed at a distance x from the source; then a general expression for $g(x)$ is

$$G(x) = \frac{g(x)}{4\pi x^2}.$$

By measuring $G(x)$ and x, the value of $g(x)$ can be calculated.   Next, the unit point source is placed in the center of a sphere, radius z, of the material whose removal cross section is being determined.   The fast-neutron flux measured in the surrounding water of thickness x is the overall point kernel, given by

$$G(z, x) = \frac{g(x)g(z)}{4\pi(x + z)^2}, \tag{10.34}$$

so that $g(z)$ can be determined, since $g(x)$ is known.   The macroscopic removal cross section, Σ_r, is then defined by

$$g(z) = e^{-\Sigma_r z}, \tag{10.35}$$

so that $g(z)$ has the conventional exponential form.   The corresponding relaxation length is then equal to $1/\Sigma_r$.

10.104. In actual practice, removal cross sections are measured with a small but finite plane source rather than a point source, but the results are adjusted to satisfy the definition given above.   The macroscopic (Σ_r) and microscopic (σ_r) removal cross sections for a number of substances are given in Table 10.5 [20].   Because of uncertainties in the measurements, the probable errors are ±5 to 10 per cent.   For elements with mass numbers greater than 10, the value of σ_r is represented fairly well by the expression

$$\sigma_r \approx 0.35\, A^{0.42} \text{ barns}, \tag{10.36}$$

where A is the mass number.   The removal cross section has been found to be almost independent of the thickness of the sample up to about five relaxation lengths.

TABLE 10.5.   REMOVAL CROSS SECTIONS FOR FISSION NEUTRONS

Material (element)	σ_r (barns/atom)	Σ_r (cm^{-1})	Material (compound)	σ_r (barns/molecule)	Σ_r (cm^{-1})
Aluminum....	1.31	0.079	Boron carbide	5.1	0.12
Beryllium....	1.07	0.128	Deuterium oxide	2.8	0.094
Graphite.....	0.72	0.058	(heavy water)		
Hydrogen....	1.0	—	Hydrocarbon oil	2.8	—
Iron.........	2.0	0.17	(per CH$_2$ group)		
Lead.........	3.5	0.12			
Oxygen.......	1.0	—			

10.105. In applying the removal cross-section concept to determine neutron attenuation from a volume source or its equivalent surface source, the correct treatment is to integrate the point source kernel, equation (10.34), over the whole source. A simple application is to the case of an infinite plane source adjacent to a slab shield of thickness z of a given material which is followed by a layer of x thickness of water. The point kernel may be obtained by combining equations (10.34) and (10.35) to give

$$G(z, x) = \frac{g(x)e^{-\Sigma_r z}}{4\pi R^2},$$

where R is equal to $x + z$. If this is substituted for $G(R)$ in equation (10.2) and the integration performed, the final result, to a good approximation, is

$$\phi(z, x) \approx \phi_{H_2O}(x)E_1(\Sigma_r z), \tag{10.37}$$

where $\phi_{H_2O}(x)$ is the flux which would be received at a distance x from the same source in water alone; this may be calculated by the method described in § 10.98. If there are two or more slabs of different shielding materials, the argument of the exponential integral function is taken as the sum of the appropriate $\Sigma_r z$ terms for all the slabs. The geometrical transformations given earlier in the chapter may be applied to equation (10.37) to obtain approximate values of $\phi(z, x)$ to be expected for various sources of finite size. An illustration of the use of this procedure is given in § 12.85 *et seq.*

10.106. In neutron shielding studies, it is a common practice to measure the dose rate, e.g., in millirads per hour, instead of the flux. The quantities $D(z, x)$ and $D_{H_2O}(x)$, where the D's represent dose rates, are then substituted for the respective fluxes in equation (10.37). Consequently, if the dose rate after penetration of an x-cm layer of water is known, the dose rate for z cm of material followed by x cm of water can be readily evaluated. For measuring fast-neutron dose rates in shielding experiments, a gas-filled proportional counter with polyethylene walls is generally employed; the filling gas is usually ethylene. The response of the gas and walls is then almost the same as that of tissue. The ion current from the chamber thus provides a good indication of the energy that would be deposited by the neutrons in body tissue. If the $D_{H_2O}(x)$ dose rate for the given source is not available, the flux in neutrons/(cm²)(sec) may be calculated, as indicated earlier, and then the conversion factor given in § 10.100 may be employed.

10.107. The removal cross-section values in Table 10.5 are applicable when other hydrogenous materials, such as hydrocarbon oils or polyethylene, replace all or part of the water. The appropriate value of $\phi(x)$ or $D(x)$ in the hydrogenous material to be used in equation (10.37) is taken to be the same as for a distance in water containing the same number of hydrogen atoms. A similar density adjustment is made if water is used at high temperature.

CONCRETE SHIELDS

10.108. Although the data in Table 10.5 refer strictly to layer shields, they have been used to obtain approximate results which can be employed to determine fast-neutron attenuation in homogeneous media, such as concrete, containing an adequate proportion of hydrogen. The basic assumption made in the computations is that the macroscopic removal cross section of the homogeneous mixture is the sum of the values for all the elements present, including hydrogen. It has been found that the removal cross sections estimated in this manner are generally within about 10 per cent of the experimental values.

Example 10.1. Calculate the macroscopic removal cross section and relaxation length of fission neutrons in concrete of the following (approximate) composition in weight per cent: H 1.0%; O 50%; Si 35%; Ca 14% (small amounts of other elements are neglected in order to simplify the calculations). The density of the concrete is 2.3 g/cm³.

If w is the weight per cent of any element of atomic weight A present in a mixture of density ρ in g/cm³, the number N of atoms per cm³ is given by

$$N = \frac{w\rho}{100A} N_a,$$

where N_a is the Avogadro number (0.602×10^{24}). Hence, in the present case,

$$N = \frac{w}{A} \times 0.0138 \times 10^{24} \text{ atoms/cm}^3.$$

For each element present, the contribution to the total macroscopic removal cross section is obtained by multiplying the appropriate N by the microscopic cross section. For H and O the values are obtained from Table 10.5; for Si and Ca, equation (10.36) gives 1.4 and 1.6 barns, respectively. The contributions of the various elements are then as follows,

$$\text{H} \quad \frac{1.0}{1.0} \times 0.0138 \times 1.0 = 0.014 \text{ cm}^{-1}$$

$$\text{O} \quad \frac{50}{16} \times 0.0138 \times 1.0 = 0.042$$

$$\text{Si} \quad \frac{35}{28} \times 0.0138 \times 1.4 = 0.024$$

$$\text{Ca} \quad \frac{14}{40} \times 0.0138 \times 1.6 = \underline{0.008}$$
$$\phantom{\text{Ca} \quad \frac{14}{40} \times 0.0138 \times 1.6 = } 0.088 \text{ cm}^{-1}$$

The estimated macroscopic removal cross section is thus 0.088 cm^{-1}, which is in good agreement with experimental values of 0.083 and 0.086 cm^{-1} for ordinary concretes of different compositions. The calculated relaxation length is $1/0.088 \approx 11$ cm.

10.109. The minimum water content that is desirable in concrete to make it a good neutron shield has not been definitely established. It is known that in ordinary concrete containing 7 weight per cent of water (0.8 weight per cent of hydrogen) there is sufficient hydrogen to slow down and capture the neutrons (§ 10.46). A larger proportion of water is probably beneficial for neutron attenuation, but no information is available for ordinary concrete containing less water. In iron concretes it is necessary that the water content should be high in order to prevent the streaming of 25-kev neutrons described in § 10.34. Provided the temperature does not rise excessively as a result of radiation absorption, concretes would normally contain adequate amounts of water. However, if there is appreciable heating, water may be lost in the course of time.

ATTENUATION OF PRIMARY GAMMA RAYS

10.110. Determination of the distribution of the primary gamma rays (§ 10.15) through a reactor shield is simple in principle but somewhat involved in practice because of the complex nature of these radiations. They include the prompt fission, fission-product decay, inelastic scattering, and capture gamma rays produced in the core, covering a considerable range of energies. In order to simplify the computations, the energy spectrum is first divided into a number of groups, e.g., 0 to 1 Mev, 1 to 2 Mev, etc., and a weighted average value, which may or may not be integral, is assigned to each group. Thus, the continuous gamma energy spectrum is approximated by a number of "lines" corresponding to photons of specified energies. The number of lines chosen is a compromise between the increased accuracy resulting from the use of many lines and the decrease in effort required for the calculations involving a few lines. The choice is often determined by the information that is available. For preliminary calculations the following selection of integral photon energies is satisfactory: 1, 2, 4, 6, and 8 Mev. The appropriate distributions of prompt fission gamma rays and of the equilibrium fission product decay gamma rays in the fission of uranium-235 are given in Table 10.6, in terms of Mev per fission.

10.111. The capture gamma-ray spectra of many elements have been measured. Table 10.7 contains a selection of data for materials which might be found in a reactor core [21]. Here, the spectra are divided up somewhat

TABLE 10.6.   PROMPT FISSION AND FISSION PRODUCT GAMMA-RAY SPECTRA
FROM URANIUM-235

Gamma-Ray Source	Photon Energy (Mev)				
	1	2	4	6	8
Prompt fission, Mev per fission.....	3.45	3.09	1.04	0.26	—
Fission products, Mev per fission ...	5.16	1.74	0.32	—	—

arbitrarily into a set of six discrete energy values to represent, as well as possible, the actual spectra.   For water, the capture gamma-ray spectrum may be taken to be the same as that of hydrogen.

TABLE 10.7.   EQUIVALENT CAPTURE GAMMA-RAY SPECTRA

Material	Photon Energy (Mev)					
	1	2	4	6	8	10
	Mev per Neutron Capture					
Aluminum.......	—	—	2.6	1.0	1.9	—
Beryllium.......	—	—	1.0	5.1	—	—
Carbon.........	—	—	4.6	—	—	—
Hydrogen.......	—	2.2	—	—	—	—
Iron............	0.3	0.3	0.8	1.5	2.9	0.2
Sodium.........	0.8	1.8	1.1	1.2	—	—
Zirconium.......	—	—	4.2	2.2	0.2	0.1

10.112. In calculating the volume source strength of the core, which is assumed to be uniform, it is seen from the arguments in § 10.94 that

$$S_v = 3.1 \times 10^{10} \frac{P}{V} \text{ fissions/(cm}^3)(\text{sec}), \qquad (10.38)$$

where P is the average reactor power in watts and V is the core volume.   Since Table 10.6 gives the spectra in Mev per fission, the source strengths for the various photon energies are readily evaluated.   For the capture gamma rays, however, it is necessary to determine first the number of captures per fission in each material, since the spectral data are given in terms of Mev per capture.   If the composition of the core is known, the required information is readily obtained, since the number of captures per fission is equal to Σ_c/Σ_f in the core, assuming the thermal flux to be the same in each material.   Thus S_v can be determined for each photon energy in units of Mev/(cm³)(sec).   With this information, the gamma-flux energy distribution can be calculated by means of equation (10.29) or (10.30), if the buildup factors are unity, or by the modifications given in equation (10.22) or (10.23) if the buildup factor is greater than unity.

10.113. To use these equations, it is necessary to know μ_v, the linear attenuation coefficient of the core, as well as the μ value for the shielding material, for each photon energy. The latter information is readily available, e.g., Table 2.4, as also are the buildup factors, but μ_v must be calculated from the composition of the core. This is simply done by multiplying the volume fraction of each constituent by its linear attenuation coefficient and adding the results, as in § 10.95. It is now possible to determine the energy flux distribution of each photon group throughout the shield and to convert the results into the corresponding dose rate, e.g., in roentgens per hour, if desired. The total dose rate is then the sum of the values for each photon group.

Example 10.2. Determine the dose rate in roentgens/hr, at 100 cm from the core in a water shield, due to the 2-Mev primary gamma-ray flux from the BSR thermal reactor when operating at an average power of 100 kw. Information concerning the volume, composition, and size of the core has been given earlier. The fuel may be taken to be uranium-235.

The BSR is conveniently treated as a spherical volume source (§ 10.99); hence, from equations (10.23) and (10.28), after applying the transformation from an infinite plane to a spherical source,

$$\phi(z) = \frac{S_v}{2\mu_v} \cdot \frac{r}{r_0} \{AE_1[\mu z(1 + \alpha)] + (1 - A)E_1[\mu z(1 + \beta)]\}.$$

By equation (10.38), with $P = 10^5$ watts and V for the BSR equal to 1.06×10^5 cm^3 (§ 10.99),

$$S_v = 3.1 \times 10^{10} \frac{P}{V} = \frac{(3.1 \times 10^{10})(10^5)}{1.06 \times 10^5}$$

$$= 2.9 \times 10^{10} \text{ fissions/(cm}^3)(\text{sec}).$$

From Table 10.6, the total energy of the 2-Mev photons is $3.09 + 1.74 = 4.83$ Mev per fission; hence,

$$S_v = 2.9 \times 10^{10} \times 4.83 = 1.4 \times 10^{11} \text{ Mev/(cm}^3)(\text{sec}).$$

The value of μ_v is determined by the same procedure as in § 10.95, utilizing linear attenuation coefficients for 2-Mev photons in place of macroscopic cross sections; thus,

Material	μ (cm^{-1})	Vol. fraction	$\bar{\mu}$ (cm^{-1})
Water..............	0.0493	0.583	0.0287
Aluminum...........	0.1166	0.415	0.0484
Uranium............	0.905	0.0018	0.0016

The sum of the values in the last column gives μ_v as 0.0787.

From § 10.99, r is 29.4 cm and in this case $r_0 = r + z = 129.4$ cm, since z is given as 100 cm.

For 2-Mev photons, μ is 0.0493 (as given in the tabulation above) and from Table 10.3, $A = 6.4$, $\alpha = -0.076$, and $\beta = 0.092$; hence,

$$\phi(z) = \frac{1.4 \times 10^{11}}{(2)(0.0787)} \cdot \frac{29.4}{129.4} \{6.4E_1[(4.93)(0.924)] - 5.4E_1[(4.93)(1.092)]\}$$

$$= 1.8 \times 10^9 \text{ Mev/(cm}^2)(\text{sec})$$

From Fig. 9.2, for 2-Mev photons, 1 mr/hr is equivalent to 6.2×10^2 Mev/(cm²)(sec); consequently,

$$\text{Dose at 100 cm from BSR} = \frac{1.8 \times 10^9}{6.2 \times 10^2} = 2.9 \times 10^6 \text{ mr/hr}$$

$$= 2.9 \times 10^3 \text{ r/hr.}$$

DISTRIBUTION OF SECONDARY GAMMA RAYS

10.114. Determination of the attenuation of the secondary gamma rays, resulting from capture of thermal neutrons by the nuclei in the shield, involves a knowledge of the source distribution. This distribution is not uniform, since the thermal flux falls off steadily with increasing penetration of the shield. Before proceeding with the attenuation calculations it is necessary, therefore, to determine the thermal-neutron flux distribution. The thermal flux arises partly from the slowing down of fast neutrons and partly from thermal neutrons originating in the core and reflector.

10.115. Ideally, the neutron flux distribution in the shield should be determined by solving the Boltzmann neutron transport equation (§ 3.8) applicable to the particular problem. Various useful multigroup approaches are being developed for this purpose, including Monte Carlo (random sample) and moments methods [17], as well as others similar to those used in reactor theory calculations (§ 4.64 *et seq.*). All these, however, require the use of high-speed computing facilities. Among the simpler but less accurate treatments, mention may be made of a three-group method, which gives relatively good results for iron-water shields [22]. However, it depends largely on the use of experimentally determined parameters rather than on fundamental considerations. A further simplification, which lends itself to hand calculations, is a modified two-group approach that will be described below. It is a convenient approximation for shields consisting of water, with or without fairly thin layers of a heavy metal, so that neutrons in the intermediate energy group may be neglected [23].

10.116. The first step is to write down the steady-state diffusion equation (§ 3.20) for thermal neutrons,

$$D_s \nabla^2 \phi_s(z) - \Sigma_{as} \phi_s(z) + S = 0, \tag{10.39}$$

where the subscript s refers to the slow (thermal) neutron group, and S is the applicable thermal-neutron source term. Strictly speaking, S is equal to the negative of the divergence of the fast-neutron current, but a reasonable approximation, which simplifies the calculations, is to represent the thermal-neutron source by $\Sigma \phi_f(z)$, where Σ is an effective macroscopic slowing down cross section—actually the reciprocal of a relaxation length—which is determined in the manner given in § 10.118, and ϕ_f is the fast-neutron flux. The latter may be assumed to have an exponential distribution throughout the whole thickness of the shield, so that

$$\phi_f(z) = \phi_f(0)e^{-\Sigma z}, \tag{10.40}$$

where $\phi_f(0)$ is the fast flux at the inner face of the shield.

10.117. Upon making the appropriate substitutions, equation (10.39) thus becomes

$$D_s\nabla^2\phi_s(z) - \Sigma_{as}\phi_s(z) + \Sigma\phi_f(0)e^{-\Sigma z} = 0$$

and the solution for an infinite slab shield is

$$\phi_s(z) = Ae^{\kappa z} + Be^{-\kappa z} + Ce^{-\Sigma z}, \tag{10.41}$$

where κ has the same significance as in § 3.20, namely, $\sqrt{\Sigma_{as}/D_s}$, i.e., the reciprocal of the diffusion length for thermal neutrons in the shield; A and B are arbitrary constants to be determined by the boundary conditions and

$$C = \frac{\Sigma\phi_f(0)}{D_s(\kappa^2 - \Sigma^2)}, \tag{10.42}$$

so that it can be evaluated if $\phi_f(0)$ and the required properties of the shield material are known. Since the thermal flux must go to zero when z is infinite, it is obvious from equation (10.35) that A must be zero; furthermore, when z is zero,

$$\phi_s(0) = B + C.$$

Hence, if C has been calculated, B can be determined if $\phi_s(0)$ is known. It should be noted that these values of A, B, and C are strictly applicable to an infinitely thick shield and can be used as an approximation for shields with a thickness of two or more diffusion lengths (§ 3.45). For thinner shields, the infinite-slab treatment is employed as a first step and the results are then refined by applying the proper boundary conditions to give better values for the thermal flux near the boundaries of the medium.

10.118. The data needed to express $\phi_s(z)$ specifically for a given reactor and shield are $\phi_f(0)$ and $\phi_s(0)$, and D_s, κ, and Σ for the shield material. Values for these three constants for water and for different types of concrete are given in Table 10.8. It should be repeated that the composition and properties of concretes are variable, so that the values for any given concrete may differ from those tabulated here. The removal cross sections in Table 10.8 are thus intended only

TABLE 10.8. PROPERTIES OF SHIELD MATERIALS

Material	Density (g/cm³)	Σ (cm⁻¹)	D_s (cm)	κ (cm⁻¹)
Water...............	1.0	0.14	0.17	0.36
Ordinary concrete.......	2.35	0.085	0.65	0.12
Barytes concrete........	3.50	0.125	0.44	0.21
Iron concrete..........	4.30	0.16	0.29	0.53

as a rough guide. In any particular situation, the appropriate value for Σ should be determined from equation (10.40) in the form

$$\Sigma = \frac{1}{z_2 - z_1} \ln \left[\frac{\phi_f(z_2)}{\phi_f(z_1)} \right], \qquad (10.43)$$

where $\phi_f(z_1)$ and $\phi_f(z_2)$ are values of the fast flux at distances z_1 and z_2, respectively, in the shield which have either been determined experimentally or calculated by the procedure described in § 10.98 et seq.

10.119. The fast-neutron flux at the inner surface of the shield may be obtained by extrapolating the calculated results to $z = 0$. For the BSR, for example, it is found to be 3×10^6 neutrons/$(cm^2)(sec)(watt)$, compared with the experimental value of 3.2×10^6. An alternative procedure is to use equation (10.27) for an infinite slab and to apply a correction factor for a reactor of finite size. Based on equation (10.24), the factor would be about 0.75 for a small reactor, such as the BSR. For this reactor $S_v/2\mu_v$, or the equivalent $S_v/2\Sigma_v$, is $(7.4 \times 10^5)/(2 \times 0.090) = 4.1 \times 10^6$ neutrons/$(cm^2)(sec)(watt)$, from the data in § 10.99. If this is corrected as indicated above, the result is 3.1×10^6, in good agreement with the extrapolated value.

10.120. The determination of $\phi_s(0)$ is based on the continuity of the thermal flux at the boundary between the core (or reflector) and the shield. In other words, $\phi_s(0)$ is equal to the thermal flux at the outer edge of the reactor. The thermal flux distribution in the core and reflector may be obtained by the standard multigroup methods described in Chapter 4. If this information is not available, a preliminary approach is to estimate the average thermal flux in the core, e.g., by means of equation (2.50), and then to follow the distribution through the reflector by diffusion theory, e.g., equation (3.32). In the BSR, the water reflector and shield are the same, so that $\phi_s(0)$ refers to the boundary between the core and shield. From the known amount of uranium-235 in this reactor (3.3 kg), the average thermal flux is found to be close to 10^7 neutrons/$(cm^2)(sec)(watt)$, and this may be taken as equal to $\phi_s(0)$; the experimental value is 1.1×10^7.

Example 10.3. Using data for the BSR already given, and taking Σ which is obtained from equation (10.43) as 0.14 cm^{-1}, estimate the thermal-neutron flux in water at a distance of 60 cm from the core.

The experimental values given in § 10.119 and § 10.120, respectively, are

$$\phi_f(0) = 3.2 \times 10^6 \text{ neutrons}/(cm^2)(sec)(watt)$$

$$\phi_s(0) = 1.1 \times 10^7 \text{ neutrons}/(cm^2)(sec)(watt).$$

For water, Σ is 0.14 cm^{-1}, κ is 0.36 cm^{-1}, and D_s is 0.17 cm (Table 10.8); hence, by equation (10.42), assuming $A = 0$,

$$C = \frac{(0.14)(3.2 \times 10^6)}{(0.17)[(0.36)^2 - (0.14)^2]} = 2.4 \times 10^7$$

$$B = \phi_s(0) - C = 1.1 \times 10^7 - 2.4 \times 10^7 = -1.3 \times 10^7.$$

Hence, by equation (10.41),

$$\phi_s(60) = -1.3 \times 10^7 e^{-(0.36)(60)} + 2.4 \times 10^7 e^{-(0.14)(60)}$$

$$= 5.4 \times 10^3 \text{ neutrons}/(cm^2)(sec)(watt).$$

10.121. If the fast-flux distribution in the shield is not well represented by a simple exponential expression, as in equation (10.40), then an improvement can be obtained by using a sum of two exponentials, as for water (§ 10.98); thus,

$$\phi_f(z) = \phi_f(0)e^{-\Sigma_1 z} + \phi_f'(0)e^{-\Sigma_2 z},$$

where $\phi_f(0)$ has the same significance as before, and $\phi_f'(0)$ and Σ_1 and Σ_2 are adjusted to provide the best fit to the experimental or calculated fast-flux distribution. Recalling that A must be zero, the solution of equation (10.39), corresponding to equation (10.41), is now

$$\phi_s(z) = Be^{-\kappa z} + C_1 e^{-\Sigma_1 z} + C_2 e^{-\Sigma_2 z},$$

where C_1 and C_2 are similar in form to equation (10.42), with Σ_1 and Σ_2, respectively, replacing Σ, and

$$\phi_s(0) = B + C_1 + C_2.$$

The treatment is then similar to that described above for a single-exponential distribution of the fast-neutron flux.

10.122. The simple approach described above can, in principle, be extended to a shield consisting of several layers of different materials. However, it is only for water or other hydrogenous substance that the results are at all useful except for the most preliminary of calculations. The discussion has been given here not so much to present a method to be used in shield design, but to illustrate the basic principles which are employed in some of the more detailed multigroup treatments.

10.123. A thermal-flux distribution estimate, that is sometimes useful for setting rough limits of shield thickness, is based on the application of age theory to the fission neutrons which have suffered a "removal" collision in the shield [24]. Theoretical considerations lead to the conclusion that the thermal-flux distribution is approximately proportional to the fast-flux distribution displaced a distance $\tau\Sigma$ into the shield; τ is the Fermi age (or its experimental equivalent) in the shield and Σ has the same significance as in equations (10.40) and (10.43). The quantity $\tau\Sigma$, generally written as τ/λ, where $\lambda(=1/\Sigma)$ is the relaxation length of the fast neutrons, has the dimensions of length and is called the *age displacement*. The values for concretes, like their compositions, are variable. For ordinary concrete the age displacement is about 25 cm and for heavy concretes about 15 cm. On the basis of the age-displacement theory, the thermal-flux distribution is given by

$$\phi_s(z) \approx \frac{\Sigma}{\Sigma_c} \phi_f(0)e^{-\Sigma(z-\tau/\lambda)},$$

where Σ_c is the thermal-neutron capture cross section in the shield (see Table 10.9 and § 10.110). The very approximate nature of this result should be borne in mind.

10.124. With the thermal-flux distribution in the shield known, the distribution of capture gamma rays can be determined.* If the thermal flux can be expressed in the form of a simple exponential, i.e.,

$$\phi_s(z) = \phi_s(0)e^{-\kappa z}, \tag{10.44}$$

then the distribution of capture gamma-ray photons of a *specific energy*, with a buildup factor of unity, is given by [25]†

$$\phi_\gamma(z) = \frac{S}{2\kappa} e^{-\kappa z} \left\{ e^{\kappa z}E_1(\mu z) + Ei[\mu z(\nu - 1)] + \ln\left(\frac{\nu + 1}{\nu - 1}\right) \right\}, \tag{10.45}$$

where μ is the linear attenuation coefficient of the given photons in the shield, $Ei(x)$ is the exponential integral of the argument x, where

$$Ei(x) = -E_1(-x)$$

and

$$\nu \equiv \frac{\kappa}{\mu}.$$

10.125. The source factor S in equation (10.45) is defined by

$$S = \Sigma_c N_\gamma \phi_s(0),$$

where Σ_c is the macroscopic capture cross section of the shield material and N_γ is the number or energy of the photons of the specified energy produced per neutron capture. If N_γ is the number of photons, then equation (10.45) gives the flux distribution in photons/(cm²)(sec), but if N_γ is the energy of the photons in Mev per capture, as in Table 10.7, the result will be the energy flux distribution in Mev/(cm²)(sec). Either flux can then be converted into the dose rate, e.g., in milliroentgens per hour, by utilizing Fig. 9.4. In deriving equation (10.45), the buildup factor was taken to be unity. Because of the continuous variation in the actual number of relaxation lengths traveled by the photons, according to their point of origin, the actual buildup factors are uncertain, but suitable values can be included if desired.

10.126. As in § 10.110, the calculation of the total gamma flux is simplified (and approximated) by using a selected number of lines or photon energies. Data for the three typical concretes (see Table 10.1) are recorded in Table 10.9 [26]; the values for iron and hydrogen (water) were given in Table 10.7. The macroscopic (thermal average) capture cross sections for the concretes are also included in Table 10.9; for water it is 0.020 cm⁻¹ and for iron 0.22 cm⁻¹. The distribution of each photon energy is calculated by means of equation (10.45) and the results added to give the total capture gamma energy flux. If the capture of neutrons in the shield results in the formation of radioactive

---

* The assumption is made here that all the gamma rays are produced by the capture of thermal neutrons, the neutrons of other energies being neglected. This can be justified for thermal reactors, but in the shielding of fast reactors allowance must be made for gamma rays resulting from the absorption of neutrons of all energies.

† This form is applicable provided κ/μ is greater than unity, as it generally is in shield materials. Different expressions are used when κ/μ is equal to or less than unity.

TABLE 10.9.   CAPTURE GAMMA-RAY SPECTRA IN SHIELDING CONCRETES

Type of Concrete	Σ_c (cm^{-1})	Photon Energy (Mev)			
		2	4	6	8
		Mev per Neutron Capture			
Ordinary.........	0.0085	1.5	3.0	1.8	1.0
Barytes..........	0.020	0.3	1.1	1.4	3.5
Iron.............	0.082	0.4	1.0	1.6	3.6

nuclides which emit decay gamma rays, allowance for the latter should be included. As a general rule, these radiations are not significant for shielding calculations.

Example 10.4. The thermal-neutron flux in the first 100 cm of a shield of ordinary concrete can be represented by $\phi_s(z) = 10^8 e^{-0.13z}$, for a given reactor. Determine the energy flux of 4-Mev gamma rays at a depth of 90 cm.

If the thermal flux is represented by equation (10.44), then $\phi_s(0)$ is 10^8 and κ is 0.13. From Table 10.9, for ordinary concrete Σ_c is 0.0085, and for 4-Mev gamma rays, N_γ is 3.0 Mev per neutron capture; hence,

$$S = (0.0085)(3.0)(10^8) = 2.55 \times 10^6 \text{ Mev/(cm}^3\text{)(sec)}.$$

For 4-Mev photons in (average) ordinary concrete, μ is about 0.074 cm^{-1} (cf. Table 2.4); hence,

$$\nu = \frac{\kappa}{\mu} = \frac{0.13}{0.074} = 1.8.$$

Upon substituting into equation (10.45), it is found that for $z = 90$ cm,

$$\phi_\gamma(90) =$$

$$\frac{2.55 \times 10^6}{(2)(0.13)} e^{-(0.13)(90)} \left\{ e^{(0.13)(90)} E_1[(0.074)(90)] + Ei[(0.074)(90)(0.8)] + \ln \frac{2.8}{0.8} \right\}$$

$$= 5500 \text{ Mev/(cm}^3\text{)(sec)}.$$

SHIELDING EFFECT OF REFLECTOR

10.127. In the BSR, which has been used for illustrative purposes because of its simplicity, the shield serves as the reflector. A similar situation exists in all pool-type reactors and also, to a great extent, in other water-moderated systems, e.g., pressurized-water reactors. When the moderator is a solid, however, it is the general practice to employ a separate reflector. Although the reflector is not part of the shield, it has, nevertheless, a shielding effect since it decreases the rate at which fast neutrons, in particular, and gamma rays escape from the reactor. The attenuation necessary in the shield itself is thereby reduced.

10.128. The quantities which are required for the calculations described above are (1) the surface equivalent of the volume source, i.e., of the reactor as a whole, including core and reflector, of fast neutrons and core gamma rays, and (2) the thermal-neutron flux at the reflector-shield interface. For fast neutrons and core gamma rays, a simple approach is to treat the reflector as a shield in the same manner as described previously. The flux calculated at the outer surface is then multiplied by two to provide the equivalent isotropic surface source for determining the fast-neutron and gamma-ray flux distributions in the actual shield. The procedure is illustrated in the following example.

Example 10.5. A large natural uranium-graphite reactor operating at a core power density (P/V) of 0.06 watt/cm³ has a graphite reflector 60 cm thick. Estimate the isotropic surface source strength at the outside of the reflector of (a) fast neutrons, and (b) 2-Mev gamma rays. The capture gamma rays in the core and reflector may be disregarded.

(a) Since P/V is 0.06, the volume source strength of the fast neutrons in the core is $(0.06)(7.8 \times 10^{10}) = 4.7 \times 10^9$ neutrons/(cm³)(sec). The value of Σ_v (or μ_v) for fast neutrons in the core is approximately the same as for the graphite moderator alone. According to Table 10.4, the relaxation length of fast neutrons in graphite is about 9 cm, so that μ_v is 0.11 cm⁻¹. Similarly, μ for the graphite reflector is 0.11 cm⁻¹. Since the reactor is large, the surface may be treated as an infinite plane and equation (10.29) used to give the fast flux at the outer surface of the reflector, i.e., for $z = 60$ cm. Hence,

$$\phi_f = \frac{4.7 \times 10^9}{(2)(0.11)} E_1[(0.11)(60)]$$

$$= 3.6 \times 10^6 \text{ neutrons/(cm}^2)(\text{sec}).$$

The surface isotropic source for determining the attenuation of fast neutrons in the shield is then roughly 7.2×10^6 neutrons/(cm²)(sec).

(b) If the capture gamma rays are disregarded, it is seen from Table 10.6 that 4.83 Mev of 2-Mev photons are produced per fission; hence, using equation (10.32), the volume source strength of 2-Mev gamma rays is $(3.1 \times 10^{10})(0.06)(4.83) = 9.1 \times 10^9$ Mev/(cm³)(sec). For 2-Mev photons in graphite (density 1.65 g/cm³) μ_v is 0.074 cm⁻¹, so that the equivalent isotropic surface source strength is $(9.1 \times 10^9)/(0.074) = 1.2 \times 10^{11}$ Mev/(cm²)(sec).

For the reflector μ for 2-Mev photons is also 0.074 cm⁻¹ and μz is $(0.074)(60) = 4.4$, for which the buildup factor is about 4.7; hence, by equation (10.11), the value of b is about 0.85. It follows, therefore, from equation (10.22), that

$$\phi_\gamma = (\tfrac{1}{2})(1.2 \times 10^{11})[E_1(4.4) + 0.85e^{-4.4}]$$

$$= 7.6 \times 10^8 \text{ Mev/(cm}^2)(\text{sec}).$$

The required surface isotropic source strength at the outer surface of the reflector is then approximately 1.5×10^9 Mev/(cm²)(sec).

10.129. If the reactor physics calculations have proceeded to the extent of determining the thermal-neutron distribution throughout the core and reflector, the flux at the outside of the reflector, required as the boundary condition in § 10.117, will be known. If this information is not available, an approximate estimate may be made by starting with the average thermal flux in the reactor, which will be known from the average power and the amount of fissile material present, and applying simple diffusion theory to determine the attenuation through the reflector. For this purpose, use may be made of equation (3.32) or the approximate form

$$\phi(x) \approx S_a \frac{e^{-\kappa x}}{2\kappa D},\qquad(10.46)$$

which applies to diffusion from a large plane source into an infinitely thick medium; κ and D refer to the reflector material and have their usual significance (§ 10.117). The surface source strength S_a may be taken as twice the thermal-neutron flux in the core. From this equation, the thermal flux, $\phi(x)$, after penetration through the reflector x cm in thickness, can be determined.

Example 10.6. The average thermal-neutron flux in the core of the reactor referred to in Example 10.5 is 5×10^{12} neutrons/(cm²)(sec). Determine the flux at the outer surface of the 60-cm thick graphite reflector, i.e., $\phi_s(0)$ required for shielding calculations.

For reactor-grade graphite, κ is 0.185 cm^{-1} and D is 0.94 cm (Table 3.1). In the present case, S_a is $2 \times 5 \times 10^{12} = 10^{13}$ neutrons/(cm²)(sec) and x is 60 cm; hence, by equation (10.46),

$$\phi_s(0) = \frac{10^{13} \times e^{-(0.185)(60)}}{(2)(0.185)(0.94)}$$

$$= 4.6 \times 10^8 \text{ neutrons/(cm²)(sec)}.$$

10.130. Determination of the attenuation of the primary gamma rays in the reflector presents no special problems. The procedure is exactly as described in § 10.110 *et seq.*, with the required gamma-ray spectra in Tables 10.6 and 10.7. The calculated flux at the outer face of the reflector is multiplied by two to provide the surface source for the shield. The secondary gamma rays produced in the reflector as a result of neutron capture can generally be neglected.

RESULTS FOR IRON-WATER SHIELD [27]

10.131. As an illustration of the type of results obtained in shielding calculations, Figs. 10.9 and 10.10 are presented for a pressurized-water (N.S. Savannah) reactor, operating at a thermal power of about 70 Mw. One shows the estimated fast- and thermal-neutron flux distributions through a series of radial water and steel layers which represent the thermal shield and part of the biological shield. The second shows the primary and secondary gamma dose-rate distributions. It will be observed in Fig. 10.9 that the fast flux decreases regularly, although somewhat more rapidly in the steel layers because the macroscopic removal cross section is slightly larger than for water. The thermal flux increases at first upon entering a water layer, as a result of moderation, and then decreases due to neutron capture. In the steel layers, however, the thermal flux first decreases, because capture is significant, and then increases as the fast neutrons are slowed down.

10.132. The primary-gamma dose rate is seen in Fig. 10.10 to decrease steadily, the attenuation, as expected, being more rapid in the steel than in the water

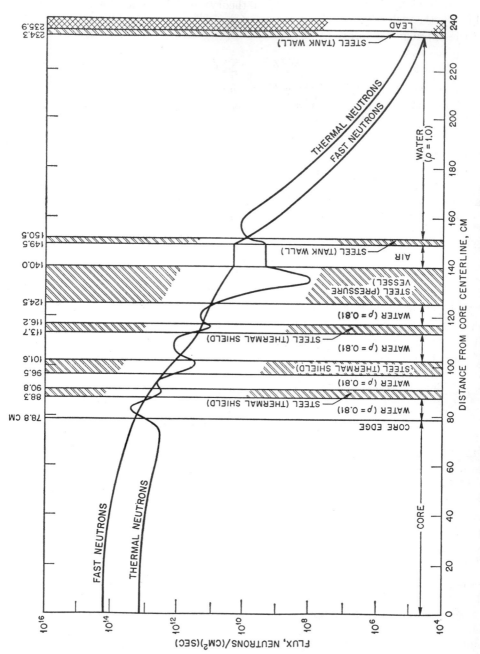

Fig. 10.9.   Fast and thermal flux distribution in shield from a 70-Mw reactor

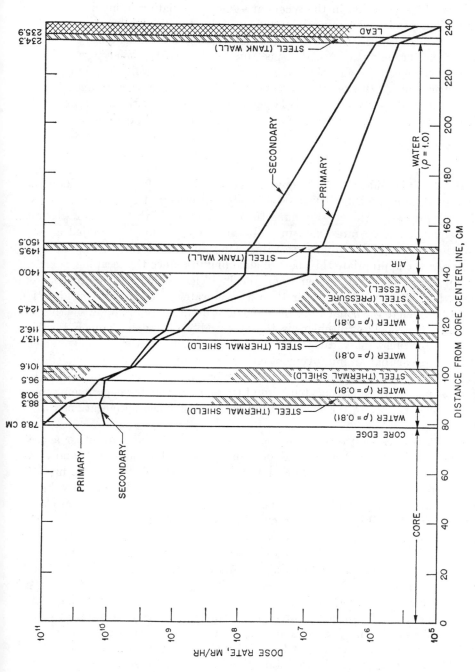

Fig. 10.10. Primary and secondary gamma-ray dose rates in shield from a 70-Mw reactor

layers.   The variation in the secondary-gamma radiation is much less simple since it depends upon several factors, namely, the thermal-neutron flux distribution, the capture cross sections in the water and steel, the energy spectrum of the capture gamma rays, the spectrum of the decay gamma rays resulting from neutron capture, and the attenuation of the various radiations in the water and steel layers.   It is of interest to note that in this reactor, as in many others, the secondary gamma-ray flux is larger than that of the primary gamma rays. It is consequently a very important consideration in shield design.

GENERAL CONCLUSIONS

10.133. It is evident from the preceding discussion and from the results shown in Figs. 10.9 and 10.10 that the semianalytical design of a reactor shield is a lengthy process.   After determining the distributions of the various radiations, the equivalent dose rates, e.g., in millirems per hour, must be added at each location to obtain the total dose rate.   Only in this manner is it possible to arrive at the appropriate thickness of shielding to reduce the escaping radiation to an acceptable level.   Even when the biological shield consists of concrete only, and is apparently simpler than a steel-water layer shield, consideration must be given to the behavior of the radiations in the reflector and in the thermal shield.   The shielding problem is thus inevitably an involved one.   Furthermore, after preliminary calculations have been made with a particular shield, it may appear that an improvement either in cost or in the space occupied may be achieved by changes in either materials or configuration, or in both.   The process may have to be repeated several times, and probably experiments made with mockups, before the final decision concerning shield specifications can be made.

10.134. In addition to the attenuation of radiation, there are other aspects of shield design.   The "streaming" of radiation through various ducts and voids, such as coolant channels and other penetrations through the shield, must be considered [28].   Then the possible effects of scattered radiation must be examined [29].   For example, suppose access to the top of a reactor is completely prohibited, it might appear, at first, that no shielding was necessary in this region.   However, this is not the case because neutrons and gamma rays escaping from the top of the reactor would be scattered downward by the nuclei (or atoms) present in the air and by the roof and walls of the building.   As a result, operating personnel might be exposed to radiation which reached them in an indirect manner.   Some shielding is thus necessary for inaccessible faces of a reactor and, in fact, for any source of radiations, whether they are neutrons or gamma rays or both.   Methods for calculating the effects of ducts and of scattering have been developed, but they lie outside the scope of this book.

10.135. Even after all the radiation phenomena associated with a shield have been treated as completely as possible, there still remains the question of heating due to the absorption of radiations by the shield material. A particular design may be quite adequate from the standpoint of radiation protection, but the temperatures developed may lead to unacceptable thermal stresses or to the loss of water from concrete. Steps must therefore be taken to ensure that there is no excessive increase of temperature in the shield. This will require an appropriate distribution of radiation attenuation between the thermal shield, which is cooled, and the biological shield, which generally does not have provisions for cooling. Some aspects of heat generation in shields will be considered later in this chapter.

COMPARISON METHOD OF SHIELD DESIGN

INTRODUCTION

10.136. A simple and reliable method of shield design is based on the comparison in which the experimental results for a given reactor and shield are used for predicting the performance of another reactor-shield configuration, provided the latter shield is of a similar type to the experimental one [30]. If a convenient radiation source, such as the one to be described below, is available, a mockup of the proposed shield design may be constructed and this can be used to obtain data which can be applied in the comparison method. The transformation from experimental results to the values to be expected for an actual design is carried out by means of the geometry equations developed earlier. An illustration of the calculational procedure is given later.

10.137. Although the comparison method is one of the most reliable at present available for shield design, it suffers from the drawback that every fresh proposal requires the construction of a new mockup and a repetition of the measurements. However, as information and experience are accumulated, the problem of shield design becomes considerably simplified and fewer tests are necessary.

EXPERIMENTAL FACILITIES

10.138. Several facilities for experimental studies on reactor shields exist in the United States and elsewhere [31]. Some of these consist of reactors of special types, such as the BSR already mentioned, whereas others are associated with reactors constructed for general research activities, such as the large uranium-graphite reactors at the Oak Ridge and Brookhaven National Laboratories. The experimental system to be described here is the Lid Tank Shielding Facility (LTSF) at Oak Ridge which has been used extensively in shield design and for the determination of removal cross sections. It consists of a large tank of water, 11 ft wide, 7 ft broad, and 7 ft 6 in. deep, attached to the side of the reactor where a hole extends through the entire thickness of the concrete shield (Fig. 10.11). At its outer end the dimensions of the hole are 28.5 in. by 32.5 in.

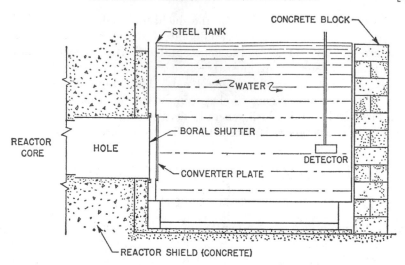

FIG. 10.11. Lid Tank Shielding Facility at Oak Ridge National Laboratory

10.139. A converter plate, 28 in. in diameter (35.6 cm radius) and 0.06 in. thick, of uranium enriched to 20.8 per cent in uranium-235, covers the hole. Thermal neutrons escaping from the reactor cause the uranium-235 to undergo fission, and a large proportion of the neutrons and gamma radiation produced enter a tank of water in which the shield to be tested is inserted. The instruments used for determining the intensities of various radiations are placed in the water behind the shield.

10.140. The LTSF has a number of advantages. Many shields of interest incorporate water as an important component; such shields are very easily mocked-up in the tank. The radiation background due to leakage from the reactor itself is considerably reduced by the water, so that it is possible to measure the low radiation fluxes observed at appreciable distances in the shield. The fact that the source is essentially a flat disk simplifies the transformation calculations, as shown in § 10.145. Its main disadvantage is its relatively low source strength.

10.141. The total power output of the source has been determined in various ways and found to be 5.2 watts. However, not all the radiations produced enter the water tank, because about 10 per cent is absorbed in the walls of the box containing the source, the walls of the tank, etc.; hence, the effective (isotropic) power of the source is 4.7 watts. The area is 3970 cm², and so the surface source strength may be represented as 4.7/3970, i.e., 1.2×10^{-3} watt per cm². This is too low to permit measurement of the fast-neutron fluxes that have penetrated moderately thick shields. In using an experimental facility for shield design by the comparison method, a mockup of the proposed shield is constructed, and the radiation fluxes or dose rates at various distances in the shield are determined.

10.142. Fast-flux measurements are generally made with a tissue-equivalent proportional counter, as described in § 10.106, so that the data are expressed as dose rates, e.g., millirads per hour. Gamma-ray intensities are similarly determined with a carbon-lined ionization chamber filled with carbon dioxide gas or by means of a sensitive anthracene scintillator for low dose rates. The results are also given in terms of millirads per hour or an equivalent quantity. For measuring the thermal-neutron flux, large proportional counters containing boron trifluoride are generally employed. Near the source, however, where the interfering gamma-ray flux is large, parallel-plate fission chambers are preferred. The data are expressed in terms of thermal neutrons/$(cm^2)(sec)$.

APPLICATION OF LTSF MEASUREMENTS

10.143. Data obtained with the finite disk source of the LTSF can be readily transformed to dose rates or fluxes for other sources. As an illustration of the procedure employed, suppose that Fig. 10.12 represents the fast-neutron dose rates, as measured in the LTSF, for various thicknesses (z cm) of a particular type of shield. The values are given in millirads per hour (mrad/hr), and so the subsequent discussion will be in terms of dose rates, although the treatment is exactly the same if neutron fluxes are employed.

10.144. Suppose it is required to transform the results in Fig. 10.12 so as to give the fast-neutron dose rates to be expected for the same shield when used in connection with a spherical reactor of 50-cm radius, having a power density of 1 watt per cm^3. If the Σ_v (or μ_v) of fast neutrons in the reactor core is 0.10 cm^{-1}, the equivalent surface source strength, by equation (10.28), is 10 watts per cm^2. The problem is now to transform the dose rates obtained with a circular disk of radius 35.6 cm and source strength 1.2×10^{-3} watt per cm^2 to the case of a spherical surface source of radius 50 cm and strength 10 watts per cm^2.

10.145. Let $D_{LT}(z)$ be the dose rate observed in the LTSF at a distance z from the source plate in a specified shield. The value of Σ (or μ) for fast neutrons in the shield at this point is estimated from the slope of the line in Fig. 10.12. The result so obtained is used to calculate the correction factor $\frac{1}{2} + \alpha$, given in § 10.85, for converting $D_{LT}(z)$ into the dose rate to be expected at z from an infinite plane reactor having the same source strength per unit area as the disk in the LTSF. To obtain the dose rate at the distance z when the shield is used in connection with a spherical reactor of the same source strength, then, in accordance with equation (10.7), this result is multiplied by r/r_0, which in this case is $50/(50 + z)$.

10.146. Finally, allowance is made for the difference in the source strengths

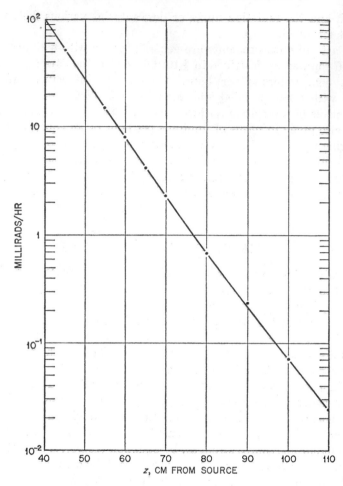

FIG. 10.12. Fast-neutron dose rates measured in a shield

of the LTSF and the specified spherical reactor by multiplying by the ratio of
the source strengths, i.e., $10/(1.2 \times 10^{-3})$. Thus,

$$D_R(z) = [D_{\mathrm{LT}}(z)](\tfrac{1}{2} + \alpha) \left(\frac{r}{r_0}\right) \left(\frac{10}{1.2 \times 10^{-3}}\right), \tag{10.47}$$

where $D_R(z)$ is the dose rate which would be expected from the spherical reactor
with a shield z cm thick, and $D_{\mathrm{LT}}(z)$ is the observed LTSF dose rate, in the same
units, for the same shield thickness. By performing this calculation for a
number of values of z, it is possible to plot a curve, equivalent to the one in Fig.
10.12, giving the fast-neutron dose-rate distribution throughout the shield for
the specified spherical reactor.

10.147. In the determination of gamma-ray dose rates, the instruments do not discriminate between primary and secondary radiations, nor between photons of different energies. It is the *total* gamma-ray energy flux, or the equivalent dose rate, which is measured. The source strength factor should be taken as the ratio of the actual reactor power densities, rather than of the surface source densities. However, since the values are usually not greatly different, the latter may be employed, as in the fast-neutron calculations. The value of μ to be used in determining the correction factor $\frac{1}{2} + \alpha$ is somewhat indefinite, because of the complex nature of the gamma radiations and the uncertainty as to whether μ should refer to the attenuation coefficient of the various gamma rays or to the neutrons which generate the secondary radiations. For calculations involving large distances in the shield, μ may be taken as an average value, e.g., for about 4-Mev energy, of gamma rays in the shield material. It should be noted that this value is used only in a correction factor and has a minor effect on the final results.

Example 10.7. Calculate the fast-neutron dose rate to be expected from the spherical reactor referred to above (radius 50 cm, equivalent surface source strength 10 watts/cm²) for a thickness of 100 cm of shield for which Fig. 10.12 gives the dose rates in the LTSF.

For $z = 100$ cm the LTSF dose rate in Fig. 10.12 is close to 7×10^{-2} mrad/hr, and the relaxation length for fast neutrons in the shield is approximately 9 cm. (This is the distance in which the dose rate is decreased by a factor of $1/2.72$ in the vicinity of $z = 100$ cm.) Then, as given in § 10.85,

$$\alpha = \frac{(2)(9)^2}{(35.6)^2}\left(\frac{100}{9} + 1\right) = 1.52,$$

since a, the radius of the disk source, is 35.6 cm. Consequently $\frac{1}{2} + \alpha$ is 2.02, and this is the factor for converting the LTSF dose rates to those for an infinite plane source. For a spherical source of radius 50 cm, the conversion factor for a shield thickness of 100 cm is now $50/(50 + 100)$, and the source strength ratio is $10/(1.2 \times 10^{-3})$. The required dose rate is then given by

$$\text{Dose rate at 100 cm from spherical reactor} = (7 \times 10^{-2})(2.02)\left(\frac{50}{150}\right)\left(\frac{10}{1.2 \times 10^{-3}}\right)$$

$$= 400 \text{ mrad/hr.}$$

SHIELDING BY ANISOTROPIC MEDIUM

LINE SOURCE

10.148. In the discussion of reactor shielding, in the preceding sections of this chapter, it has been assumed that the shield material was continuous between the reactor and the observation point. The shielding medium could thus be treated as isotropic, and the point kernel expressed by equation (10.17). In some cases of interest, however, the shield is in the form of a slab with a considerable air space between it and the object to be shielded. In these situations the shield is anisotropic and the geometrical transformation relationships derived earlier are not strictly applicable. However, appropriate expressions can be derived for different geometries by using equation (10.15) for the flux from a point source and integrating over the actual source [11].

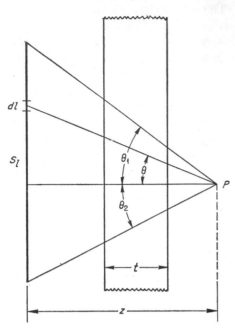

FIG. 10.13. Calculation of flux (or dose rate) from a line source

10.149. The first case to be examined is that of a line source, which represents a good approximation to a cylindrical fuel element or a pipe. Such a source, emitting S_l particles per cm per sec, is shown in Fig. 10.13. A slab shield of thickness t is interposed between the source and the observation point P, which is at a perpendicular distance z from the line source. Consider a small element dl of the line source, so that the line joining dl to P makes an angle θ with the perpendicular. Assuming simple exponential attenuation and inverse-square-law spreading, as in equation (10.15), with a buildup factor of unity, the flux, $d\phi$, at P due to emission from dl will be

$$d\phi = \frac{S_l\,dl}{4\pi(z\sec\theta)^2}\,e^{-\mu t\sec\theta}, \quad (10.48)$$

where μ is the attenuation coefficient of the shield. In this expression the attenuation by the air is neglected. It will be noted that, although the shield thickness is t, the length of the radiation path is $t\sec\theta$. Since

$$dl = z\sec^2\theta\,d\theta,$$

equation (10.48) becomes

$$d\phi = \frac{S_l}{4\pi z}\,e^{-\mu t\sec\theta}.$$

Integration over the whole line source then gives

$$\phi = \frac{S_l}{4\pi z}\left(\int_0^{\theta_1} e^{-\mu t\sec\theta}\,d\theta + \int_0^{\theta_2} e^{-\mu t\sec\theta}\,d\theta\right). \quad (10.49)$$

10.150. The function $F(\theta, x)$, sometimes called the secant integral,* is defined by

$$F(\theta, x) \equiv \int_0^\theta e^{-x\sec\theta}\,d\theta,$$

and so equation (10.42) can be written as

* The name Sievert's integral [32] has been proposed, after R. M. Sievert who first used it in shielding calculations [33].

$$\phi = \frac{S_l}{4\pi z} \left[F(\theta_1, \mu t) + F(\theta_2, \mu t) \right]. \tag{10.50}$$

Plots of the function $F(\theta, x)$ for various values of θ and x are available in the literature [15].

CYLINDRICAL SOURCE

10.151. A cylindrical volume source with a slab shield at the side can be treated, with fairly good accuracy, as an equivalent line source located at a "self-absorption distance," d, within the cylinder. The flux at a point P similar to that in Fig. 10.14 is given by

$$\phi = \frac{S_v r^2}{4(z+d)} \left[F(\theta_2, \mu t + \mu_v d) + F(\theta_1, \mu t + \mu_v d) \right], \tag{10.51}$$

where r is the radius of the cylinder, z is the perpendicular distance from P to the cylinder, and the other symbols have the same significance as before. The

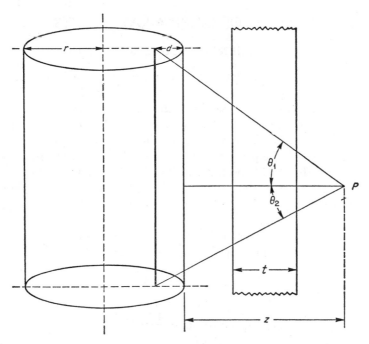

FIG. 10.14.   Calculation of flux (or dose rate) from a cylindrical source

appropriate value of d, which depends on the ratio z/r, can be found in handbooks [34].   An example of the application of equation (10.51) will be given shortly.

DISK SOURCE

10.152. For a circular disk source and a slab shield, integration over the whole area as in § 10.58, using a point kernel based on equation (10.15) with unit buildup factor, gives for the flux at a point on the axis at a distance z (Fig. 10.2),

$$\phi = \tfrac{1}{2}S_a\{E_1(\mu t) - E_1[\mu t(1 + a^2/z^2)^{1/2}]\}, \tag{10.52}$$

where a is the radius of the disk. The same expression may be used for a cylinder with a slab shield at the end; in this case, S_a in equation (10.52) is replaced by S_v/μ_v, in the usual manner. For a disk of large radius, i.e., when a is large, equation (10.52) reduced to simple form, analogous to equation (10.21),

$$\phi = \tfrac{1}{2}S_a E_1(\mu t).$$

10.153. In the equations given above the buildup factor is assumed to be unity. For the attenuation of gamma rays by a thick shield, however, allowance for the buildup factor must be included. The appropriate correction can be made along the lines indicated by equation (10.22) or (10.23).

SHIELDING OF GAMMA-RAY SOURCES

CALCULATION OF SOURCE STRENGTH

10.154. In nuclear reactor plants it is often necessary to shield pure gamma-ray sources, such as spent fuel elements and tanks and pipes containing radioactive substances. The accompanying beta particles can be neglected, since they present a very minor hazard compared with that of the gamma rays.* The principles of shielding gamma-ray sources of this kind are the same as those used in connection with the primary gamma rays from a reactor core. However, as a general rule, the shields used are slabs which are not in contact with the system to be shielded, and the situations encountered are of the type in which the shielding medium is anisotropic. The equations derived in the preceding section are then applicable.

10.155. If a pipe, cylinder, or other vessel contains g grams of a radioactive species of mass number A and half-life $t_{1/2}$ sec, the total rate of decay is given by equation (2.12) as $4.17 \times 10^{23} g/At_{1/2}$ dis/sec. If γ is the number of gamma-ray photons of a given energy E_γ Mev produced in each act of radioactive decay, then

$$\text{Rate of gamma energy emission} = \frac{4.17 \times 10^{23}\gamma E_\gamma g}{At_{1/2}} \text{ Mev/sec.} \tag{10.53}$$

If photons of two or more different energies accompany the radioactive decay,

---

* When using water as the shield, especially for freshly discharged spent fuel elements, consideration must be given to the possibility of neutron formation by the (γ, n) reaction of gamma rays with energies in excess of 2.2 Mev with the deuterium in the water.

an equation of this type will be applicable to each energy value. In the event that several radionuclides are present in the vessel, one or more expressions similar to equation (10.53) would be used for each such nuclide.

10.156. For a pipe, which can be considered as a line source located at the center of the pipe, division of equation (10.53) by the length of the pipe gives the source strength in Mev/(cm)(sec). If the gamma-ray emitter is contained in a vessel, e.g., a cylinder, the volume-distributed source strength in Mev/(cm³)(sec) units is obtained upon dividing equation (10.53) by the volume. From this, the corresponding surface source strength can be determined if required.

10.157. In the cases just cited, it is assumed that the weights of the various radioelements in the vessel are known. In many cases, however, the activity in curies is known, rather than the weight. The source strength calculations will then be based on the definition (§ 2.15) that the curie represents a decay rate of 3.7×10^{10} dis per sec. The expression for the rate of gamma-ray energy emission is then

$$\text{Rate of gamma energy emission} = 3.7 \times 10^{10} \gamma E_\gamma C \text{ Mev/sec}, \quad (10.54)$$

where C is the number of curies of a given radioactive species present; the product γC is referred to as the activity in gamma-curies (§ 2.188). An expression of this same general form applies to each photon energy for each gamma emitter.

10.158. Although these two examples cover many situations encountered in pure gamma-ray shielding, other circumstances arise from time to time. One of these is concerned with the activity induced in a coolant, as a result of neutron capture, during its passage through a reactor.

ATTENUATION CALCULATIONS

10.159. By combining the gamma-ray energy source strength, as determined above, with the appropriate equation for the flux distribution, it is possible to calculate the dose rate to be expected at a prescribed distance from the source, with a shield of known material and thickness. It will be observed that equations (10.50), (10.51), and (10.52) all contain the parameter z in such a way that ϕ decreases as z increases. In other words, for a given shield thickness, the dose rate received decreases with distance z from the source. This is known as the attenuating effect of distance and arises from the geometrical spreading of the radiation. For a point source, this spreading is related to $1/z^2$, but for distributed sources the effect is less marked but nevertheless significant.

Example 10.8. A cylindrical tank, 4 ft long and 2 ft in diameter, contains 1 gamma-curie per cm³ of a radioactive material which emits 2.8-Mev gamma rays. A 6-ft thick shield of ordinary concrete surrounds the tank; determine the dose rate in mr/hr at a distance of 10 ft from the middle of the side of the tank. The linear attenuation coefficient of the material in the tank for 2.8-Mev gamma rays is 0.036 cm⁻¹.

The first step is to determine the self-absorption distance, *d*, referred to in § 10.151;

since $r = 1$ ft (30.5 cm) and $z = 10$ ft (305 cm), z/r is 10.   Since μ_v is 0.036 cm^{-1}, $\mu_v r$ is (0.036)(30.5) = 1.1; for this value of $\mu_v r$ and $z/r = 10$, the handbooks show that $\mu_v d$ is 0.58, so that d is 0.58/0.036 = 16 cm.

The observation point is opposite the middle of the tank, so that θ_1 and θ_2 are the same and equal to tan^{-1} [61/(305 + 16)], i.e., tan^{-1} 0.19, so that $\theta_1 = \theta_2 = 10.8°$.

For concrete μ for 2.8-Mev photons is 0.099 cm^{-1}, and the shield thickness, t, is $6 \times 30.5 = 183$ cm; hence,

$$F(\theta, \mu t + \mu_v d) = F[10.8°, (0.099)(183) + (0.036)(16)]$$

$$= F(10.8°, 18.68) = 1.3 \times 10^{-9}.$$

Since the activity of the material in the tank is 1 gamma-curie/cm^3, it emits 3.7×10^{10} photons/(cm^3)(sec), so that S_v is $(3.7 \times 10^{10})(2.8) = 1.0 \times 10^{11}$ Mev/(cm^3)(sec); hence, by equation (10.51),

$$\phi = \frac{(10^{11})(30.5)^2}{4(305 + 16)} [(2)(1.3 \times 10^{-9})]$$

$$= 1.9 \times 10^2 \text{ Mev/(cm}^2)\text{(sec)}.$$

From Fig. 9.2, for 2.8-Mev photons, 1 mr/hr is equivalent to a flux of 7.0×10^2 Mev/(cm^2)(sec), and so the dose rate in the present case is 1.9/7.0 = 0.27 mr/hr.

(It is of interest to note that if the self-absorption distance is taken as equal to the radius of the cylinder, θ_1 and θ_2 would be 10.3°, and the final result would be within 10 per cent of that calculated above.   This is true, in general, provided z/r is moderately large, e.g., 10 or more.)

RADIATION HEATING IN SHIELDS [35]

INTRODUCTION

10.160. Nearly all the energy of the fast neutrons and of the gamma rays entering the shield from a reactor is deposited within it as heat.   In addition, essentially all the neutrons are slowed down and captured, and the energy of the capture gamma rays (2.2 Mev per capture by hydrogen and about 7 to 8 Mev per capture by other elements) is also liberated as heat.   The total amount of heat generated in the shielding material is thus quite considerable.   The effect can be significant, especially as most of the heat is produced in the layers of the shield nearest to the reactor (§ 10.22).   The determination of the heat distribution is an important aspect of shield design which is carried out simultaneously with that of the radiation distribution.   The calculations are very lengthy, and it is not possible to do more here than to indicate the general lines along which they are made.

10.161. Much of the heat is deposited in the thermal shield which can be cooled.   In a pressurized-water reactor, which is contained within a thick-walled, steel pressure vessel, the design of the thermal shield is determined by the thermal stress developed as a result of the temperature gradient across the wall.   Roughly, the total energy flux, from both neutrons and gamma rays, impinging on the interior of the pressure vessel must not exceed about 10^{11} Mev/(cm^2)(sec).   To achieve this reduction, two or more thermal shields are placed between the reactor core and the wall of the vessel. The water coolant circulating between the separate thermal shields removes the heat generated within them. In order to maintain the thermal stresses in these shields within tolerable limits, the thickness of the innermost thermal shield, which is exposed to the highest radiation flux, is restricted to about an inch. An increasing thickness is permissible for each subsequent thermal shield.

HEATING BY NEUTRONS

10.162. Neutrons lose energy to the shield as a result of both inelastic and elastic collisions. In a material consisting of heavy elements, such as the iron of a thermal shield, the former predominate and the latter are insignificant; in water, the reverse is true. For elastic scattering, the volumetric heat generation rate at any location z is given approximately by

$$Q(z) \approx \Sigma_s \phi_f(z) \xi E_n \text{ Mev/(cm}^3)(\text{sec}),$$

where Σ_s is the macroscopic elastic scattering cross section of the fast neutrons of flux $\phi_f(z)$ and energy E_n Mev; and ξ is the average logarithmic energy decrement per collision (§ 3.61). In making the calculations, an appropriate energy spectrum for fission neutrons should be used. Neutrons with energies less than about 1 Mev can be neglected, since their contribution to the heating is small in comparison with that of neutrons with higher energies.

10.163. In an elastic collision, the kinetic energy of the neutron is transferred directly to the struck atom, so that the heat is produced essentially at the site of the collision. The energy lost by the neutron in an inelastic collision appears as a gamma ray; the heat is then liberated over a distance as the radiation is absorbed. However, the inelastic scattering heating is generally not large and the gamma rays have moderate energies, so that they are absorbed in a short distance. As an approximation, it may be assumed that, as with elastic scattering, the heat produced in an inelastic scattering is generated at the collision point. The corresponding heat generation rate can then be represented by

$$Q(z) = \Sigma_i \phi_f(z) f E_n \text{ Mev/(cm}^3)(\text{sec}),$$

where Σ_i is the inelastic scattering cross section and f is the fraction of the neutron energy lost in the collision. Again, the fission-neutron energy spectrum must be taken into consideration. If the rate of heat production turns out to be significant, the assumption that the inelastic scattering gamma rays are absorbed at the point of creation will not be justifiable.

HEATING BY GAMMA RAYS

10.164. Considerable heat is released by the absorption of both primary and secondary gamma rays in the shield. The treatment of the primary radiations is comparatively simple because they all enter the shield, from the reactor, at the same location. If $\phi_\gamma(z)$ is the uncollided flux of primary gamma rays of a given energy E_γ Mev, the volumetric heat generation rate at the location z is

$$Q(z) = B_e \mu_e E_\gamma \phi_\gamma(z) \text{ Mev}/(\text{cm}^3)(\text{sec}), \qquad (10.55)$$

where B_e is the *energy absorption* buildup factor for photons of energy E_γ [12] and μ_e is the energy absorption coefficient, defined in § 7.27 as the (total) linear attenuation coefficient minus the contribution of Compton scattering. Since the values of $\phi_\gamma(z)$ represent the uncollided flux, which is approximately a broad collimated beam, it is possible to write

$$\phi_\gamma(z) = \phi_\gamma(0)e^{-\mu z}, \qquad (10.56)$$

where $\phi_\gamma(0)$ is the gamma-ray flux of energy E_γ entering the shield and μ is the (total) linear attenuation coefficient of these photons in the shield material. The total heat generation due to primary gamma rays is obtained by adding the contributions over the energy spectrum of prompt fission gammas, fission product decay gammas, and capture gamma rays from the core. If the shield is thin, e.g., for a thermal shield, the buildup factor in equation (10.55) may be taken as unity; the heat generation rate, for gamma rays of a specified energy, can then be represented simply by

$$Q(z) = Q_0 e^{-\mu z}. \qquad (10.57)$$

This is the basis for the relationship used in § 6.82 for determining the temperature distribution.

10.165. Finally, an estimate must be made of what is generally the most significant contribution to the heat generated in a reactor shield, namely that due to the absorption of the secondary gamma rays resulting from neutron capture. Equation (10.55) is applicable to this case also, but the uncollided flux distribution, for each energy value, is given by equation (10.45). If the energy absorption buildup factor in equation (10.55) is expressed in exponential form, it can be factored directly into this equation, thereby simplifying the computations. In the definition of the source factor, S, the quantity N_γ is the *number* of photons of the specified energy produced per neutron capture, since the photon energy, E_γ, is included in equation (10.55). The total heat produced is the sum of the values over the capture gamma-ray energy spectrum.

10.166. The heating in a shield subjected to a pure gamma-ray source from the outside is given by equation (10.55), with $\phi_\gamma(z)$ defined by equation (10.56). For a thin shield, equation (10.57) can be used to determine the volumetric heat generation rate, and the temperature distribution can be computed by the procedure described in Chapter 6.

APPENDIX [11]

10.167. This appendix contains the detailed derivations of certain results used or mentioned in earlier parts of this chapter. Since many of the equations to be derived are used to make transformations of actual measurements, the quantity treated will be the dose rate rather than the flux. However, the results are equally applicable to the latter, since it is proportional to the dose rate for a specified radiation (§ 10.106).

GEOMETRICAL TRANSFORMATIONS: PLANAR SOURCES

10.168. Much of the experimental work in shielding has been carried out, in the LTSF, with a nearly isotropic circular (disk) source of finite radius (§ 10.139). Hence, equations permitting transformation from such a source are very useful in shielding calculations. Consider, first, an infinite plane source; the dose rate at a distance z will be represented here by the symbol $D_{pl}(z, \infty)$, and this is given by equation (10.2) as

$$D_{pl}(z, \infty) = 2\pi S \int_z^\infty G(R)R \, dR.$$

Upon differentiating with respect to z, it is found that

$$\frac{d}{dz} D_{pl}(z, \infty) = -2\pi SzG(z)$$

or

$$G(z) = -\frac{1}{2\pi Sz} \cdot \frac{d}{dz} D_{pl}(z, \infty). \tag{10.58}$$

The point kernel, $G(z)$, can thus be determined, provided $D_{pl}(z, \infty)$ is known for several values of z, so that the derivative can be evaluated. As seen above, the point kernel is fundamental to radiation dose-rate calculations; consequently the data obtained for an infinite plane source can be utilized to treat sources of other shapes.

10.169. For a disk source of finite radius a, the dose rate at a distance z, represented by $D_{pl}(z, a)$, is given by equation (10.1); thus,

$$D_{pl}(z, a) = 2\pi S \int_z^{\sqrt{z^2+a^2}} G(R)R \, dR.$$

If the integral is now expressed as the difference of two integrals, in which the lower limits are z and $\sqrt{z^2 + a^2}$, respectively, and the upper limits are infinity in each case, it is readily seen with the aid of equation (10.58) that, upon differentiation with respect to z,

$$\frac{1}{2\pi Sz} \cdot \frac{d}{dz} D_{pl}(z, a) = G(\sqrt{z^2 + a^2}) - G(z), \tag{10.59}$$

where $G(\sqrt{z^2 + a^2})$ and $G(z)$ are the values of the point kernels for source-to-detector separations of $\sqrt{z^2 + a^2}$ and z, respectively.

10.170. If a function $A(z)$ is defined by

$$A(z) \equiv -\frac{1}{2\pi Sz} \cdot \frac{d}{dz} D_{pl}(z, a), \tag{10.60}$$

then it is found from the recursion formula of equation (10.59) that

$$G(z) = A(z) + A(\sqrt{z^2 + a^2}) + A(\sqrt{z^2 + 2a^2}) + \cdots$$

$$= \sum_{\nu=0}^{\infty} A(\sqrt{z^2 + \nu a^2}), \tag{10.61}$$

where ν is zero or integral. This means that, if the function $A(z)$ is determined for a series of z values, i.e., distances from disk source to detector, equal to z, $\sqrt{z^2 + a^2}$, $\sqrt{z^2 + 2a^2}$, $\sqrt{z^2 + 3a^2}$, etc., the sum will give $G(z)$, the point kernel at z. For this procedure to be useful, the series must converge rapidly; such is the case when the disk radius a is large compared with the relaxation length of the radiation. If a is small, then the disk is essentially equivalent to a point source, and $D_{\mathrm{pl}}(z, a)$ may be taken as equal to $D_{\mathrm{pt}}(z)$, from which $G(z)$ is obtained directly by definition (§ 10.55).

10.171. The use of equation (10.61) may be illustrated by means of Fig. 10.15. The top curve represents experimentally determined dose rates at distances, z, ranging from 40 to 120 cm along the axis of a circular disk source of radius $a = 50$ cm. The next lower curve gives the negative slope of the $D_{\mathrm{pl}}(z, 50)$ curve as a function of z, whereas the third curve represents the corresponding values of $A(z)$ for unit source strength, i.e., for $S = 1$. Taking z as 55 cm, the arrows point to the values of $A(\sqrt{z^2 + \nu a^2})$ for $\nu = 0, 1, 2, 3$, and 4, respectively. It is seen that the convergence is quite rapid, so that it is not necessary to go beyond $\nu = 4$. The sum of the five A terms is about 790, and this gives the point kernel $G(z)$ for $z = 55$ cm.

10.172. An analogous transformation to that given above may be used to relate the observed dose rates for a finite disk source to those to be expected from an infinite plane source. By combining equations (10.58), (10.60), and (10.61) and integrating, it is found that

$$D_{\mathrm{pl}}(z, \infty) = D_{\mathrm{pl}}(z, a) + D_{\mathrm{pl}}(\sqrt{z^2 + a^2}, a) + D_{\mathrm{pl}}(\sqrt{z^2 + 2a^2}, a) + \cdots$$

$$= \sum_{\nu=0}^{\infty} D_{\mathrm{pl}}(\sqrt{z^2 + \nu a^2}, a). \tag{10.62}$$

Consequently, if the dose rates recorded from a disk source of radius a at distances z, $\sqrt{z^2 + a^2}$, $\sqrt{z^2 + 2a^2}$, $\sqrt{z^2 + 3a^2}$, etc., from the source are added, the sum will give the dose rate at z from an infinite plane source having the same strength per unit area.* The series converges rapidly when a is large in comparison with the relaxation length of the radiation.

10.173. In some circumstances it is not possible to obtain a sufficient number of reliable measurements to perform the summation required by equations (10.61) and (10.62). An approximate relationship, which is derived from equation (10.62), can then be used, provided the series converges rapidly and the dose rate in the region of interest can be expressed in the form of a simple

---

* This procedure is known as the *Hurwitz transformation*.

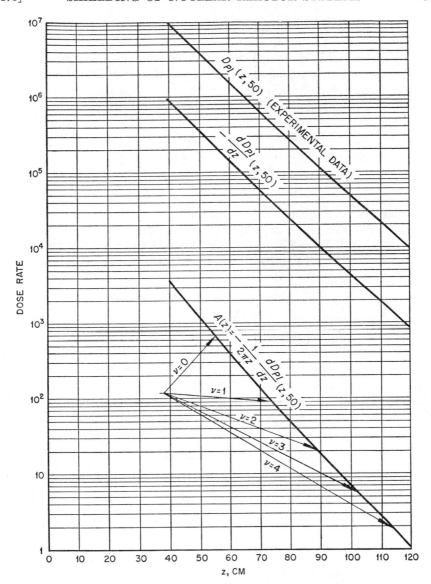

FIG. 10.15.   Dose rates from circular disc source

exponential equation in terms of an apparent relaxation length λ.   It is then found that

$$1 + \alpha > \frac{D_{\mathrm{pl}}(z, \infty)}{D_{p^{\backprime}}(z, a)} \geq \frac{1}{2} + \alpha, \tag{10.63}$$

where

$$\alpha \equiv \frac{2\lambda^2}{a^2}\left(\frac{z}{\lambda} + 1\right).\tag{10.64}$$

This is the relationship given, without proof, in § 10.85. In general, the lower limit, i.e., $\frac{1}{2} + \alpha$, is a very good approximation to the ratio of the dose rates from infinite and finite plane (disk) sources.

10.174. The use of the transformation of equation (10.63) requires a knowledge of the apparent relaxation length of the radiation in the absorbing medium. This can be obtained from the data in Fig. 10.15. From the general definition of relaxation length [equation (10.14)],

$$\frac{1}{D_{p1}(z, a)} \cdot \frac{d}{dz} D_{p1}(z, a) = -\frac{1}{\lambda},$$

and so λ at any point z is readily obtained upon dividing the value of $D_{p1}(z, a)$, from the top curve, by the corresponding $-dD_{p1}(z, a)/dz$, from the second curve. Thus it is found that, for $z = 55$ cm, λ is about 11 cm. Since a, the radius of the disk source to which Fig. 10.15 applies, is 50 cm, the value of α, defined by equation (10.64), is 0.58. Consequently from equation (10.63)

$$\frac{D_{p1}(55, \infty)}{D_{p1}(55, 50)} \approx 1.08.$$

From Fig. 10.15, $D_{p1}(55, 50)$ is about 2.2×10^6, so that $D_{p1}(55, \infty)$ is approximately 2.4×10^6. This result is somewhat low, as may be seen by applying equation (10.62) to Fig. 10.15. Upon adding the values of $D_{p1}(z, 50)$, from the top curve, corresponding to the points for $\nu = 0, 1, 2$, etc., it is found that $D_{p1}(55, \infty)$ is somewhat greater than 2.7×10^6. For many rough calculations the difference between 2.4×10^6 and 2.7×10^6 is not very significant.

GENERALIZED CURVED SURFACE SOURCE

10.175. In the following section there will be derived an expression for the dose rate from a generalized curved surface source. In two special cases, namely, when the surface is that of (1) a sphere or (2) a right circular cylinder, the results obtained reduce to the expressions given earlier.

10.176. Consider an isotropic radiation source, spread uniformly over a surface; let the point O on the surface nearest to the detector be the origin. The xy plane at the origin is tangent to the surface (Fig. 10.16). For a point, such as Q, not far from the origin, let the distance z_1 from the xy plane be given approximately by

$$z_1 \approx \frac{1}{2}\left(\frac{x^2}{a} + \frac{y^2}{b}\right),\tag{10.65}$$

where a and b are the normal curvatures at the surface. Although this approximation does not apply to points at an appreciable distance from O, very little error is introduced by assuming the expression to hold for the whole surface.

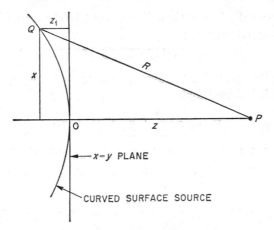

FIG. 10.16.   Generalized curved surface source

This is because most of the radiation received at the detector comes from sources near the origin, whereas the contribution of the more distant sources is much less important.

10.177. The dose rate received at a distance z from the origin is represented by

$$D(z) = S \int_{\text{surface}} G(R) \, ds, \tag{10.66}$$

where ds is a small element on the surface, S is the uniform source strength per unit area, and $G(R)$ is the point kernel; the integration is carried over the whole of the surface.   The value of ds is given by

$$ds = dx \, dy \sqrt{\left(\frac{\partial z_1}{\partial x}\right)^2 + \left(\frac{\partial z_1}{\partial y}\right)^2 + 1}$$

$$\approx dx \, dy \left(\frac{x^2}{2a^2} + \frac{y^2}{2b^2} + 1\right), \tag{10.67}$$

using equation (10.65) for z_1.

10.178. It is now postulated that $G(R)$ can be expressed as the product of an unspecified, and presumably exact, kernel $G(z)$ for the distance z, and an exponential term for the additional distance $R - z$; thus

$$G(R) = G(z)e^{-(R-z)/\lambda}, \tag{10.68}$$

where λ is the apparent relaxation length which makes this expression correct. Actually, λ will vary slowly with distance, but a satisfactory value for the region of interest can be obtained from experimental data.   Upon combining equations (10.66), (10.67), and (10.68), the result for an infinite surface is

$$D(z) = SG(z) \int_{-\infty}^{\infty} \int_{-\infty}^{\infty} e^{-(R-z)/\lambda} \left(\frac{x^2}{2a^2} + \frac{y^2}{2b^2} + 1\right) dx \, dy. \tag{10.69}$$

In order to perform the integration, $R - z$ must be expressed in terms of x, y, and z. This is done by writing

$$R = \sqrt{(z_1 + z)^2 + x^2 + y^2}$$

and substituting the approximate value $R - z$ for z_1; upon expanding and using equation (10.65) it is found that

$$R - z \approx \frac{x^2}{2z} + \frac{y^2}{2z} + \frac{x^2}{2a} + \frac{y^2}{2b}.$$

Upon inserting this into the exponential term in equation (10.69), the integration can be performed analytically. If a and b are large compared with λ, the result reduces to

$$D(z) \approx 2\pi SG(z) \frac{\lambda z \sqrt{ab}}{\sqrt{(a + z)(b + z)}} \quad \text{for} \quad a, b \gg \lambda. \tag{10.70}$$

10.179. For the case of a *spherical surface source*, $a = b = r$, the radius of the sphere, and z is equivalent to $r_0 - r$, as may be seen by comparing Fig. 10.3 and Fig. 10.16; consequently equation (10.70) now becomes

$$D_{sp}(r_0, r) \approx 2\pi SG(z)\lambda z \frac{r}{r_0} \tag{10.71}$$

or, in view of equation (10.58),

$$D_{sp}(r_0, r) \approx -\frac{d}{dz} D_{pl}(z, \infty)\lambda \frac{r}{r_0}. \tag{10.72}$$

If the relaxation length λ employed above for the spherical surface applies also to the infinite plane source, then, since $d \ln D_{pl}(z, \infty)/dz$ will be equal to $-1/\lambda$, it follows that

$$-\lambda \frac{d}{dz} D_{pl}(z, \infty) = D_{pl}(z, \infty)$$

and, consequently, from equation (10.72),

$$D_{sp}(r_0, r) \approx \frac{r}{r_0} D_{pl}(z, \infty)$$

$$\approx \frac{r}{r_0} D_{pl}(r_0 - r, \infty),$$

which is exactly equivalent to equation (10.7).

10.180. The transformation for a *cylindrical surface source* can be derived in an analogous manner; this is achieved by setting $a = \infty$, $b = r$, and $z = r_0 - r$ in equation (10.72), where r is the radius of an *infinitely long* cylinder and r_0 is the distance from its axis to the detector (cf. Fig. 10.4 and Fig. 10.16). The result is

$$D_{cy}(r_0, r) \approx 2\pi SG(z)\lambda z \sqrt{\frac{r}{r_0}},$$

and by using the same procedure as with equation (10.71), it is seen that

$$D_{cy}(r_0, r) \approx \sqrt{\frac{r}{r_0}}\, D_{pl}(r_0 - r, \infty),$$

which was given as equation (10.8).

SYMBOLS USED IN CHAPTER 10

A	atomic weight, mass number
A	coefficient in exponential buildup equation
$A(z)$	function defined by equation (10.60)
a	radius of disk source
A, B, C	constants
a, b	coefficients in exponential kernel
a, b	normal curvatures of surface
B_e	energy-absorption buildup factor
$B(\mu t)$	dose buildup factor for μt relaxation lengths
b	coefficient in linear buildup equation
C	number of curies
D	radiation dose rate
D	thermal diffusion coefficient (in reflector)
D_s	diffusion coefficient (slow neutrons)
d	self-absorption distance
E	energy
E_γ	energy of gamma-ray photons
E_n	neutron energy
$Ei(x)$	exponential integral of the argument x
$E_n(x)$	exponential integral function of order n and argument x
e	base of natural logarithms
f	fraction of neutron energy lost per inelastic collision
$G(R)$	point kernel for distance R
$G(x)$	fast-neutron flux point kernel for thickness x of water
$G(z, x)$	fast-neutron point kernel for thickness z of material and x of water
g	mass of radioactive species in grams
N	atoms per cm^3
N_a	Avogadro number
N_γ	number (or energy) of photons of specified energy per neutron capture
P	reactor power level (watts)
Q	volumetric heat generation rate
q	variable $(= \mu R)$
R	distance from radiation source to observation point
r	radius of cylinder
r	radius of sphere

r_0	distance from center of sphere to observation point
r_0	distance from cylinder axis to observation point
S	source strength, source term, or source factor
S_a	equivalent surface source strength (particles per cm^2 per sec)
S_l	linear source strength (particles per cm per sec)
S_p	point source strength (particles per sec)
S_v	volume source strength (particles per cm^3 per sec)
t	shield thickness
$t_{1/2}$	half-life of radioactive species
V	volume of reactor core
x	distance
x, y, z	coordinates
z	distance
α	function defined by equation (10.64)
α, β	coefficients in exponential buildup equation
γ	number of gamma-ray photons per decay
θ	angle
κ	reciprocal of diffusion length
λ	relaxation length (approximately equivalent to $1/\mu$ or $1/\Sigma$)
μ	linear attenuation coefficient ($= \Sigma_a$)
μ_e	energy absorption coefficient
μ_v	linear attenuation coefficient of volume source ($= \Sigma_v$)
ν	zero or integer
ξ	average logarithmic energy decrease per collision
ρ	radius of annulus
ρ	density
Σ	macroscopic cross section
Σ_{as}	macroscopic absorption cross section for slow neutrons
Σ_c	macroscopic capture cross section
Σ_f	macroscopic fission cross section
Σ_r	macroscopic removal cross section
Σ_s	macroscopic elastic scattering cross section of fast neutrons
Σ_v	macroscopic absorption cross section of volume source
σ_r	removal cross section
σ_s	scattering cross section
τ	Fermi age of thermal neutrons
ϕ	flux (neutrons or gamma-ray photons)
ϕ_f	flux of fast neutrons
ϕ_s	flux of slow neutrons
ϕ_γ	gamma-ray flux

REFERENCES FOR CHAPTER 10

1. For general reviews, see T. Rockwell, "Reactor Shielding Design Manual," U. S. AEC Report TID-7004 (1956) and D. Van Nostrand Co., Inc., Princeton, N. J.; B. T. Price, C. C. Horton and K. T. Spinney, "Radiation Shielding," Pergamon Press, Inc., New York, 1957; J. R. Harrison, "Nuclear Reactor Shielding," Simmons-Boardman Books, New York, 1958; H. Goldstein, "The Attenuation of Gamma Rays and Neutrons in Reactor Shields," U. S. Government Printing Office, Washington, D. C., 1957; H. Goldstein, "Fundamental Aspects of Reactor Shielding," Addison-Wesley Publishing Co., Inc., Reading, Mass., 1959; M. Grotenhuis, "Lecture Notes on Reactor Shielding," U. S. AEC Report ANL-6000 (1962); A. F. Avery, et al., "Methods of Calculation for Use in Design of Shields for Power Reactors," U. K. AEA Report AERE-2-3216 (1960); E. P. Blizard (Ed.), "Reactor Handbook," Vol. III, Part B, Shielding, Interscience Publishers, Inc., New York, 1962.
2. N. F. Lansing, *Nucleonics*, **13**, No. 6, 58 (1955).
3. For description of a specific fast-reactor shield design, see H. E. Hungerford and R. F. Mantey, *Nucleonics*, **16**, No. 11, 120 (1958).
4. C. R. Tipton (Ed.), "Reactor Handbook," Vol. I, Materials, Interscience Publishers, Inc., New York, 1960, Chs. 50, 51, and 52; H. S. Davis, *Nucleonics*, **13**, No. 6, 60 (1955); *J. Am. Concrete Inst.*, **29**, 965 (1958); N. Hodge and R. G. Sowden, *Nucleonics*, **19**, No. 11, 158 (1961); A. N. Komarovskii, "Shielding Materials for Nuclear Reactors" (trans. by G. V. M. Newton), Pergamon Press, Inc., New York, 1961; "Concrete for Radiation Shielding," 2nd Ed., American Concrete Institute, 1962.
5. E. P. Blizard and J. M. Miller, U. S. AEC Report ORNL-2195 (1958); see also, J. A. Dyson and J. R. Harrison, U. K. AEA Report AERE-RP/R-1942 (1956); A. F. Avery and J. E. W. Simmons, U. K. AEA Report AERE-R/R-2782 (1959).
6. H. E. Hungerford, et al., *Nuc. Sci. and Eng.*, **6**, 396 (1959).
7. Ref. 4; for compositions and radiation attenuating properties of various concretes, see R. L. Walker and M. Grotenhuis, U. S. AEC Report ANL-6443 (1961).
8. W. J. Grantham, U. S. AEC Report ORNL-3130 (1961).
9. C. W. Hamill, et al., U. S. AEC Report Y-1366 (1961).
10. See N. Hodge and R. L. Sowden, Ref. 4.
11. E. P. Blizard, Ref. 1, Ch. 11, also in H. Etherington (Ed.), "Nuclear Engineering Handbook," McGraw-Hill Book Co., Inc., New York, 1958, Section **7–3** (E. P. Blizard); T. Rockwell, Ref. 1, Ch. 9.
12. See Ref. 1; R. L. Walker and M. Grotenhuis, Ref. 7; H. Etherington, Ref. 11; H. Goldstein and J. E. Wilkins, U. S. AEC Report 3075 (1954).
13. J. J. Taylor, U. S. AEC Report WAPD-RM-217 (1954); see also T. Rockwell, Ref. 1, Chs. 9 and 10.
14. H. Goldstein, Ref. 1; see also T. Rockwell, Ref. 1, p. 7.
15. T. Rockwell, Ref. 1; E. P. Blizard, Ref. 1; M. Grotenhuis, Ref. 1; H. Etherington, Ref. 11.
16. F. Clark, unpublished, quoted from E. P. Blizard, Ref. 1, Ch. 11.
17. E. P. Blizard, Ref. 1, Ch. 9; see also, E. P. Blizard, *Proc. Second U. N. Conf. Geneva*, **13**, 3 (1958).
18. M. Grotenhuis, Ref. 1, see also D. S. Duncan and H. O. Whittum, U. S. AEC Report NAA-SR-2380 (1958), p. 15.
19. R. D. Albert and T. A. Welton, U. S. AEC Report WAPD-15 (1950); see also T. Rockwell, Ref. 1, p. 48; E. P. Blizard, U. S. AEC Report ORNL CF-54-6-164 (1954); E. P. Blizard, Ref. 1, Ch. 9.

20. G. T. Chapman and C. L. Storrs, U. S. AEC Report AECD-3978 (1958) or ORNL-1843 (1958).
21. Adapted from M. Grotenhuis, Ref. 1; see also E. Troubetzkoy and H. Goldstein, *Nucleonics*, **18**, No. 11, 171 (1960).
22. C. Cooper, J. E. Jones, and C. C. Horton, *Proc. Second U. N. Conf. Geneva*, **13**, 21 (1958); see also T. Rockwell, Ref. 1, p. 62; U. S. AEC Report GNEC-187 (1961).
23. M. Grotenhuis and J. W. Butler, U. S. AEC Report ANL-5544 (1956); M. Grotenhuis, Ref. 1; D. S. Duncan and H. O. Whittum, Ref. 18.
24. H. Etherington, Ref. 11, p. 7–90 (E. P. Blizard); E. P. Blizard, Ref. 1, Ch. 9.
25. H. Etherington, Ref. 11, pp. 7–87 *et seq.*
26. Adapted from R. L. Walker and M. Grotenhuis, Ref. 7.
27. W. R. Smith and M. A. Turner, U. S. AEC Report BAW-1144-1 (1959).
28. T. Rockwell, Ref. 1, Ch. 8; E. P. Blizard, Ref. 1, Ch. 12.
29. E. P. Blizard, Ref. 1, Ch. 15.
30. E. P. Blizard, Ref. 1, Chs. 7 and 11.
31. E. P. Blizard, *Proc. Second U. N. Conf. Geneva*, **13**, 3 (1958); E. P. Blizard, Ref. 1, Ch. 7; W. J. McCool and D. R. Otis, *Nucleonics*, **18**, No. 4, 98 (1960).
32. E. P. Blizard, Ref. 1, Ch. 11 (Appendix 11A).
33. R. M. Sievert, *Acta Radiol.*, **1**, 89 (1921).
34. E. P. Blizard, Ref. 1, Ch. 11; T. Rockwell, Ref. 1, Ch. 9; H. Etherington, Ref. 11.
35. Ref. 25; M. Grotenhuis, Ref. 1; E. P. Blizard, Ref. 1, Ch. 13.

PROBLEMS

1. After traversing an isotropic shield of 6 ft thick, in which the relaxation length of a certain radiation is 12 cm, the dose rate received from a spherical surface source, radius 2 ft, was equivalent to 1 mrem/hr. For the same shield what would be the dose rate from a plane circular isotropic source, radius 2 ft, of the same strength per unit area, located at the central plane of the sphere?

2. Estimate the effective removal cross section of magnesium for fast neutrons. Compare the masses and thicknesses of magnesium and iron required to attenuate fast neutrons to the same extent in a water shield.

3. For approximate attenuation calculations, it is sometimes possible to take the buildup factor as unity and to include the effect of scattering in an adjusted exponential attenuation coefficient; the reciprocal would then be the equivalent relaxation length. Estimate such a coefficient for 4-Mev gamma rays in thick shields of iron. Over what range of thickness would this coefficient be reasonably satisfactory?

4. Compare the relative dose rates derived from equations (10.22) and (10.23) for the passage of 2-Mev gamma rays through a concrete shield 6 ft thick.

5. A pool-type reactor core has a volume of 100 liters and the following composition in volume per cent: water 61.0, aluminum 38.8, uranium 0.20; it operates at a power of 10 kw. What thickness of water would be required to attenuate the fast flux to 10 neutrons/$(cm^2)(sec)$?

6. For the reactor in Problem 5, what would be the total primary gamma dose rate at the calculated thickness of water? (Use the data in Table 10.6.)

7. The fast-neutron flux at the surface of the pool reactor in Problem 5 is 2×10^{10} neutrons/$(cm^2)(sec)$ and the thermal flux is 6×10^{10} neutrons/$(cm^2)(sec)$. Estimate the thermal-neutron dose rate received through the thickness of water determined in Problem 5.

8. Suppose that, instead of water, the shield in Problem 7 were 4 ft of barytes concrete. What would be the thermal-neutron dose rate in this case?

9. Evaluate the relaxation length for fast neutrons, by utilizing the conventional removal theory, in a heavy concrete of the following (approximate) composition: H 0.023, O 0.640, Al 0.027, Si 0.19, Ca 0.26, and Fe 3.39 g/cm^3.

10. A reactor with a cylindrical core having a radius of 50 cm is surrounded by a shield consisting of 25 cm of iron and 150 cm of water. The equivalent isotropic surface source strength of the core is 5×10^{13} fast neutrons/(cm^3)(sec). Estimate the fast-neutron flux which penetrates the shield.

11. A cylindrical tank, 6 ft in height and 4 ft in diameter, contains a gamma emitter, the linear attenuation coefficient (μ_v) within the source volume being 0.040 cm^{-1}. A vertical shield 4 ft thick, in which the attenuation coefficient for the gamma rays is 0.10 cm^{-1}, is to be located 4 ft from the outside of the tank. Compare the expected dose rates 2 ft above the ground when (a) the tank is standing upright and (b) the tank is on its side with the shield 4 ft from the end.

12. A cylindrical reactor 5 ft in diameter (including the reflector) is to be surrounded by a radial shield of concrete, with an air space in between. A thicker shield may be constructed closer to the reactor or a thinner shield would be equally effective at a further distance; the radial dimensions, however, would be greater in the latter case. Make a semiquantitative study of the different situations and indicate the conclusions to be drawn.

13. Calculate the fast-neutron dose rate to be expected through a 3-ft thick shield, of the same material as that to which Fig. 10.12 applies for the LTSF, from a spherical reactor of 100-cm radius having an equivalent isotropic surface source strength of 5 watts/cm^2.

14. From Fig. 10.15 determine the values of the point kernel $G(z)$ for $z = 45$ and 65 cm. By combining these values with the result in the text for $z = 55$ cm, derive an analytical expression for $G(z)$ as a function of z.

15. A vertical pipe 10 ft long contains a material which emits gamma rays with an average energy of 1 Mev. Compare the dose rates received (a) opposite an end of the pipe and (b) opposite its midpoint, through a concrete shield 6 ft thick which is located 4 ft from the pipe.

Chapter 11

MECHANICAL AND STRUCTURAL COMPONENTS

INTRODUCTION

SPECIAL REQUIREMENTS FOR REACTOR SYSTEMS

11.1. A nuclear reactor consists of a large number of mechanical and structural components [1]. Although many of the mechanical components are not unusual for a power producing facility, some are unique to nuclear power plants. For example, specialized mechanisms have been designed for insertion and removal of fuel elements and also for moving control and safety rods. Since each individual reactor will require numerous mechanical devices, the exact nature of which will depend on design details, no attempt will be made here to give a detailed description of the appropriate "hardware." However, a few examples of some general types of reactor mechanisms will be considered in order to indicate the role played by these components in the overall design of a reactor system.

11.2. As far as structural components are concerned, here again many of the items are similar to those used in a conventional power plant. The description of such components will therefore be omitted. Several aspects of reactor pressure vessels and of containment structures (§ 9.115), however, are somewhat specialized in nature and these will be considered. Among the unusual situations encountered are the following: necessity for leak-tightness where radioactive liquids are present, thermal stresses in metals arising from the absorption of gamma radiation, and problems of radiation damage.

MECHANICAL COMPONENTS

POWER-CYCLE COMPONENTS

11.3. In power reactors, some form of thermodynamic cycle is necessary to transform the thermal energy, generated by fission, into electricity.* In many

---

* It is possible that an efficient method of direct conversion of fission heat into electricity may eventually be developed (§ 13.125).

respects the process is the same as that employed in fossil fuel plants and the power components are not unique to nuclear systems.   There are, however, some special features which should be mentioned.

Water-Cooled Reactors

11.4. Coolant pumps for nuclear reactors are generally of special design.   In pressurized (nonboiling) water reactors, for example, the heat acquired by the coolant is in the form of sensible heat, whereas in a conventional boiler it is mainly latent heat.   This means that the rate of coolant circulation is relatively high; pumps with capacities of 10,000 to 20,000 gal per min of water are in common use, and some of 40,000-gal per min capacity have been constructed for nuclear power plants.   The pumping problem is further complicated by the need to maintain a sealed system, since the coolant may become contaminated by neutron-induced activation of impurities or as the result of failure of fuel-element cladding.   If the liquid being pumped is fairly expensive, e.g., heavy water, there is an additional necessity for minimizing losses.

11.5. The requirement of zero leakage has led to the development of sealed (or "canned") motor designs, especially for pumping water at high pressures (Fig. 11.1).   These pumps represent a considerable capital investment, and so alternatives are being investigated.   One possible way of reducing costs is to make use of very large pumps of more conventional design having a controlled leakage seal.   In such a device there is a small flow of a buffer fluid, e.g., water if this is the liquid being pumped, at a pressure slightly in excess of the system pressure.   Some of this buffer liquid is allowed to leak into the coolant system while the remainder discharges to a drain or to a recirculation system [2].

11.6. Even in a water-to-water heat exchanger (steam generator), special problems are encountered.   In a conventional heat exchanger (or reboiler), the maximum pressures are generally from 200 to 300 psi, and there is little pressure difference between the primary and secondary sides.   For a pressurized-water reactor, however, the pressure on the primary side may be 2000 psi or more, whereas that on the secondary (or steam) side is generally about 600 psi.   Furthermore, the steam generator must be able to withstand the rapid temperature changes which are characteristic of a nuclear system.   Finally, there is the requirement that the primary system, at least, shall be leak tight.   Heat exchangers for pressurized-water reactors have thus been made of stainless steel instead of carbon steel, and with the tubes welded to the tube sheet to prevent mixing of primary and secondary fluids.   Assembly under exacting conditions has also added to the overall cost.   Effort is being devoted to the development of steam generators for nuclear power plants which are reliable but less expensive than those employed hitherto [3].

11.7. A schematic representation of a typical reactor coolant system is shown in Fig. 11.2.   Each coolant loop consists of a pump, heat exchanger, and the appropriate piping.   In a large power facility, several such loops in parallel

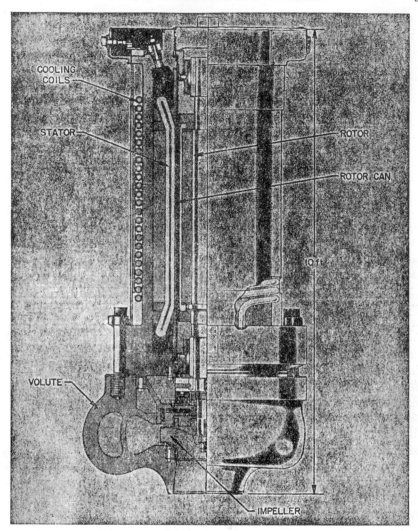

Fig. 11.1. Sectional diagram of canned motor pump used in Shippingport PWR

would be included in the design.   The Shippingport and Yankee pressurized-water reactors, for example, have four independent coolant loops.   The pressurized coolant water is circulated around the loops from the reactor pressure vessel through the steam generators and back to the reactor vessel.

11.8. The steam produced in a simple water-cooled (including boiling water) reactor is essentially unsaturated, whereas supersaturated steam is obtained from conventional steam plants.   It is desirable in the former situation, therefore, to employ special turbines with interstage moisture separators in order to utilize the wet steam effectively.   Because of the advantages associated with

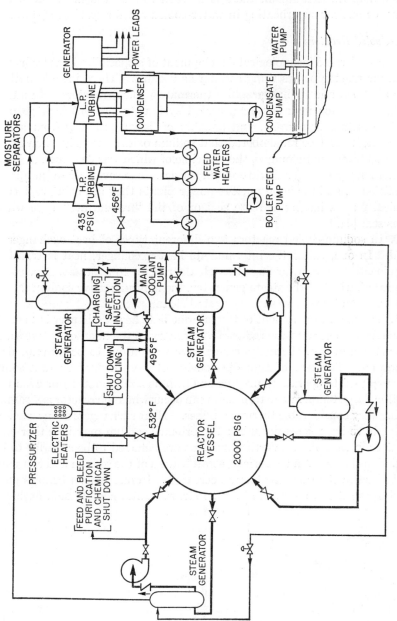

Fig. 11.2. Flow sheet of pressurized water reactor system

the use of superheated steam, there is a trend toward the inclusion of either fossil-fuel or nuclear superheating in water-cooled reactor systems (§ 13.29).

Sodium-Cooled Reactors

11.9. Considerable technological development of a specialized nature has been required for reactor plants using sodium as coolant. These include both fast breeder (§ 13.78) and sodium-graphite thermal (§ 13.50) reactors. In addition to the necessity for maintaining a sealed system to prevent leakage of the radioactive isotope sodium-24, and possibly of fission products from a faulty fuel element, it must be remembered that sodium oxidizes readily when exposed to air. Electromagnetic pumps, the principle of which was described in § 6.201, have been largely used for liquid-sodium service. However, these pumps have a low efficiency, and for large nuclear power plants they are being replaced by mechanical pumps having a freeze seal or of the "free-surface" type with an inert-gas seal [4].

11.10. In sodium-cooled reactors two different types of heat exchangers are required. In one, called the intermediate heat exchanger, heat is transferred from radioactive sodium flowing through the reactor to nonradioactive sodium. In the second, which is the steam generator, the nonradioactive sodium transfers heat to water and produces supersaturated steam.

11.11. The main requirement of the intermediate heat exchanger is that there should be no leakage from the radioactive primary to the nonradioactive secondary circuit. However, thermal stresses have turned out to be a major design consideration. The temperature differences between inlet and outlet sodium, during steady-state operation, may be as much as 250°C (450°F) or even more. In addition, large temperature transients can occur when the reactor is scrammed, producing very considerable thermal stresses. Heat exchangers of the shell-and-tube type have been generally used; the primary sodium may be either on the shell side or in the tubes. Vertical systems are believed to be superior to horizontal designs because, in the former, stratification of the sodium and consequent thermal stress in the tube sheets are decreased. In one type, a bellows joint is included in the shell to alleviate thermal stress. Intermediate heat exchangers of other types are under development [5].

11.12. In the sodium-to-water steam generator, the main design problems arise from thermal stresses and the possibility of chemical interaction between the sodium and water. In order to minimize this hazard and to provide an instant detection of leakage, heat exchangers of the tube-and-shell type with double-wall tubes have been employed. The annular space between the two concentric tubes may contain mercury or helium which facilitates heat transfer and also provides a means for leak detection. In one sodium-water steam generator design, the pressure of the helium gas is intermediate between that of the sodium (tube) side and of the water and steam (shell) side. If a sodium leak were to develop, the helium pressure would drop, whereas a water leak would result in an increase of pressure [63].

11.13. In the steam generator of the Enrico Fermi Atomic Power Plant (§ 13.91), single-wall, rather than double-wall, tubes are used, in the expectation that in the former case the equipment can be constructed with greater assurance that leaks will not occur. A small amount of leakage of water into the sodium, i.e., from higher to lower pressure, can be tolerated, from the corrosion standpoint, and soon detected and rectified. Relief vent ports and a rupture diaphragm provide protection in the event of a major sodium-water reaction.

11.14. The Fermi plant steam generator differs from others in the respect that the evaporator and superheater are in a single, once-through system. The unit is vertical with a central sodium well and an annular region containing the water and steam tubes. The secondary sodium enters the annular shell near the top and flows out of a nozzle at the bottom. The feed-water flows down 1200 tubes, arranged in the form of a ring around the outside of the central well, and then upward, in coils, through the annulus. It is in this region that the water first forms steam and then the later, as it ascends and encounters sodium at higher temperature, is superheated. An inert-gas space is provided above the surface of the sodium. Both tube sheets and tube-sheet joints are in the inert gas, so that the danger of leakage is minimized [7].

FUEL HANDLING MECHANISM

11.15. An example of the many unique auxiliary mechanisms required for the operation of a reactor system is the fuel handling equipment. The nature of this mechanism varies greatly in complexity with the reactor design. Its characteristics also depend upon whether refueling is performed when the reactor is shut down or while it is still operating. Water-cooled reactors, which operate under high pressure, are usually refueled while the system is shut down. An ideal arrangement in such reactors is to provide a means for transferring spent fuel-element assemblies, while under water, to a storage pit located in the reactor area. To accomplish this, cranes with attachments which are manipulated in an unusual manner and other specialized devices are required. In some cases, underwater transfer is not possible. It is then necessary to use a shielded "coffin" to move the highly radioactive spent fuel from the reactor to appropriate storage.

11.16. In sodium-cooled reactors, insertion and removal of fuel elements is generally carried out remotely under liquid sodium which, in turn, is under an inert atmosphere. Such a mechanism may be a major part of the installation, as seen in Fig. 11.3 which is a representation of the reactor vessel of the Enrico Fermi Atomic Power Plant. A fuel transfer device of this kind requires consider-

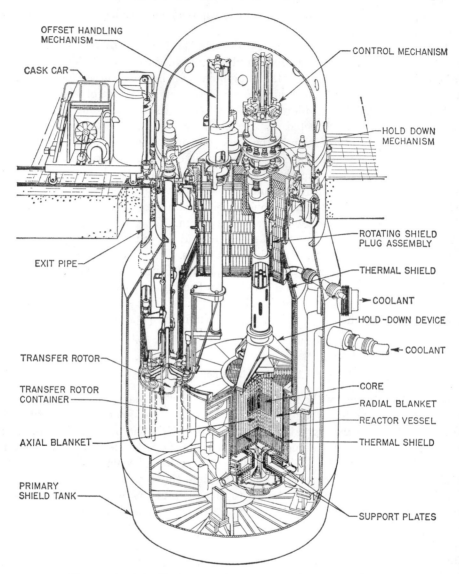

FIG. 11.3.   The Enrico Fermi reactor vessel and components

able design and development effort, including "mockups" to assure operational reliability [8].

11.17. In many reactors, such as those referred to above, the discharge and recharging of fuel are performed with the reactor shut down.   But shutdown time is expensive and contributes significantly to the cost of nuclear power (§ 14.51).   Attention has therefore been devoted to means for refueling reactors while they are operating at their normal power.   Gas-cooled reactors, in particular, lend themselves more readily to this approach, and all the more recently constructed power reactors in the United Kingdom have facilities for discharging spent fuel and recharging during normal operation.   The Ultra-High Temperature Reactor Experiment (UHTREX) reactor (§ 13.71) was specifically designed with a rotating core to permit refueling without shutdown.   A special fuel handling device has been developed for the Canadian NPD-2 heavy-water reactor (§ 13.37) in which the pressure is about 1000 psi.   The recharging machine is operated hydraulically with heavy water as the working fluid.

MAINTENANCE AND DISASSEMBLY MECHANISM

11.18. In designing a reactor, consideration must be given to the methods of maintenance.   Ordinary procedures are not always possible, since many of the components, including the reactor vessel, become highly radioactive during operation as a result of activation by neutron capture and the accumulation of radioactive deposits.   Suitable devices for remote maintenance, and perhaps also for assembly and disassembly, must be incorporated in the design of the reactor facility.   These may include special cranes, manipulators, and other mechanisms.   Since they contribute to the capital cost, they must be specified with care.   Major maintenance may be required only a few times during the life of the reactor, and so it may be possible to delay installation of the devices until they are needed.   Provision for them must, nevertheless, be made in the design of the reactor components.

CONTROL ROD MECHANISMS

11.19. The equipment required to move the control rods is another example of a special mechanism developed for nuclear application [9].   In some reactors, the same rods are used for shim control and safety (scram) purposes (§ 5.142).   When functioning in the former capacity they must be withdrawn slowly, but for the latter purpose they have to be inserted rapidly.   One way of achieving this objective is to provide a magnetic coupling between the rod and drive mechanisms, as described in § 5.142.   Such a system takes advantage of the force of gravity, sometimes enhanced by means of a spring, to scram the reactor by rapid insertion of the rods.

11.20. A dilemma sometimes facing the reactor designer is the need to provide space for the control rod drive mechanisms without interfering with the loading and removal of fuel elements.   To do this with the control rod drives at the top of the reactor is generally difficult, although it is often done.   In some designs, however, the control rods are moved in and out from the bottom of the reactor, e.g., in the Experimental Boiling Water Reactor (§ 13.25).   In order to take advantage of gravity for rapid shutdown, the rods consist of an upper poison section with a "follower," consisting of a weak neutron absorber, below. When the rods are withdrawn, to increase the reactivity, they are pushed upward, so that the poison is above the reactor core. To achieve a scram, the rods are allowed to fall under the influence of gravity, so that the poison section drops into the core. In the boiling-water reactor of the Dresden Nuclear Power Station, the control rods are inserted at the bottom, but for rapid shutdown the safety rods are forced upward hydraulically by means of a high-pressure water accumulator. When the control rods are at the bottom of the reactor, consideration must be given in the design to the possibility of sediment falling into and interfering with the drive mechanism.

STRUCTURAL COMPONENTS

INTRODUCTION

11.21. Structural components of particular interest are the reactor vessel and the containment (or confinement) structure [10]. The reactor vessel serves to contain the core, including the coolant, and sometimes part of the reflector, whereas the containment structure is intended to prevent the escape of radioactive material in the event of an accident. The reactor plant includes many other structural items, of course, but these are either the same as are found in nonnuclear power plants or they are too specialized to be treated here.

11.22. Thermal and biological shields may be regarded as structural components; they were discussed in Chapter 10, but certain aspects of stresses in thermal shields will be mentioned here. Thermal stresses, which are important in the design of pressure vessels, are also significant in connection with fuel elements, as well as in the design of heat exchangers. The problem of thermal stress in pressure vessels will be considered later in this chapter.

REACTOR VESSELS

11.23. For water-cooled reactors, with or without boiling, the pressures are high and the reactor vessel, then often called a *pressure vessel*, is designed for a substantial operating pressure. It must also be able to withstand large thermal stresses. For sodium- and organic-cooled reactors, the operating pressures are fairly low and so also are the design pressures of the reactor vessels. Special problems in the construction of reactor vessels arise from the necessity for including fuel handling facilities and control rod drives. Among the general features which must be considered in the design of reactor vessels are the need for extremely high structural integrity to prevent escape of radioactive materials, the effect of irradiation on mechanical properties, and the effect of vessel design limitations on the economics of the power system. These and other features will be considered in connection with the vessel requirements for various reactor types.

Water-Cooled Reactors

11.24. Because the boiling point of water at atmospheric pressure is relatively low, power reactors moderated and cooled by water operate at high pressures. This makes it possible to attain a higher steam temperature and, hence, a reasonable efficiency of the thermodynamic cycle. Design pressures for pressurized-water reactors are generally higher than those for boiling-water reactors, although the criteria for these two reactor types are tending to approach one another (§ 13.10). The vessel for the Shippingport (PWR) reactor is designed for a pressure of 2500 psi; it has an internal diameter of 109 in. and a wall thickness of $8\frac{3}{8}$ in. For the Dresden (Boiling Water) reactor, on the other hand, the design pressure is 1250 psi; the vessel has an internal diameter of 146 in. and a wall thickness of $5\frac{5}{8}$ in. High-strength, low-alloy steels have been specified for pressure vessels, but the effect of neutrons in raising the brittle-to-ductile transition temperature of these steels may be a problem (§ 7.52). For water-cooled reactors, the interior of the pressure vessel is lined with about 0.1 to 0.25 in. of stainless steel to prevent corrosion.

11.25. An important problem associated with pressure vessels is the requirement, in certain cases, that the head be removable to permit refueling. When the operating pressure is very high, e.g., 2000 psi or more, a bolted joint closure is sometimes used in conjunction with a seal weld and an omega-type seal. However, the process of disassembly and assembly is troublesome and time-consuming; as already seen, down time has an adverse effect on the plant economics. The breaching of the seal also introduces a possible hazard which needs to be taken into consideration. As an alternative to seal-welding, double O-ring gaskets have been used in the Yankee Atomic Electric Company (pressurized water) reactor, shown in Fig. 11.4. The region between the inner and outer O-rings is monitored periodically for water leakage. It is of interest to mention that 52 studs of $5\frac{1}{4}$-in. diameter are used to bolt down the head of the pressure vessel [11].

11.26. Large pressure vessels, such as those being considered here, must be fabricated under controlled shop conditions. This places a restriction upon the maximum size of vessel that can be moved and installed. The vessel size is particularly limited if rail transportation is necessary, although much larger vessels can be shipped by water. In any case, it is difficult to fabricate vessels with wall thicknesses greater than 9 to 10 in. and maintain satisfactory physical properties through the welds. It is generally accepted that 350 tons is the maximum weight for a pressure vessel, and this tends to restrict the reactor output to about 1000 thermal (or 300 electrical) megawatts, because of power-density, i.e., heat-removal, considerations. The limitations upon size and weight of the pressure vessel thus affect the economic potential of water-cooled reactors. However, the use of high-tensile-strength materials in the future may make

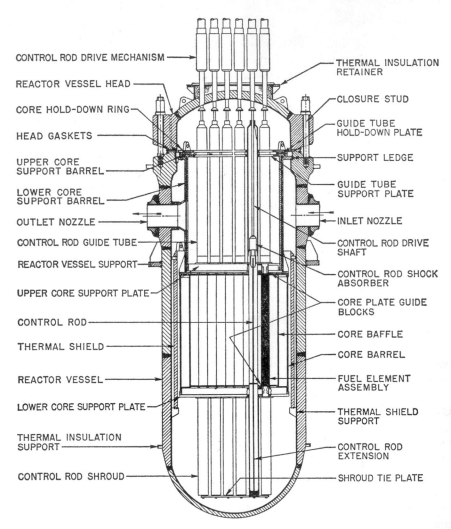

CONTROL ROD DRIVE MECHANISM

REACTOR VESSEL HEAD

CORE HOLD-DOWN RING

HEAD GASKETS

UPPER CORE SUPPORT BARREL

LOWER CORE SUPPORT BARREL

OUTLET NOZZLE

CONTROL ROD GUIDE TUBE

REACTOR VESSEL SUPPORT

UPPER CORE SUPPORT PLATE

CONTROL ROD

THERMAL SHIELD

REACTOR VESSEL

LOWER CORE SUPPORT PLATE

THERMAL INSULATION SUPPORT

CONTROL ROD SHROUD

THERMAL INSULATION RETAINER

CLOSURE STUD

GUIDE TUBE HOLD-DOWN PLATE

SUPPORT LEDGE

GUIDE TUBE SUPPORT PLATE

INLET NOZZLE

CONTROL ROD DRIVE SHAFT

CONTROL ROD SHOCK ABSORBER

CORE PLATE GUIDE BLOCKS

CORE BAFFLE

CORE BARREL

FUEL ELEMENT ASSEMBLY

THERMAL SHIELD SUPPORT

CONTROL ROD EXTENSION

SHROUD TIE PLATE

FIG. 11.4. Sectional diagram of pressure vessel of Yankee Atomic Electric Co. reactor

possible the construction of vessels which are larger or have higher design pressures, or both.

Gas-Cooled Reactors

11.27. The specifications of the pressure vessels for gas-cooled reactors are greatly dependent on the reactor characteristics, particularly the dimensions of the core.   In the United Kingdom natural-uranium power reactors (§ 13.60), for example, the operating temperature is limited to relatively modest values by the integrity of the fuel elements, e.g., attack of the carbon dioxide coolant gas on the Magnox cladding (§ 7.40).   A relatively high power output, therefore, requires a large number of fuel elements, i.e., a large core, and a high pressure of the coolant gas in order to facilitate heat removal.   These two requirements are not economically compatible, since the construction cost of large vessels capable of withstanding high pressures increases rapidly with increasing size and operating pressure.   The pressure vessel specifications are thus a compromise between size and design pressure, and hence represent a limitation on the power output.

11.28. In general, the most economical dimensions of the spherical, steel pressure vessels, for an operating pressure of roughly 150 psig, were found to be a diameter of some 67 ft and a wall thickness of about 3 in.   Vessels of this type are used in the Berkeley, Bradwell, Hunterston, and Hinkley Point power stations in the United Kingdom [12].   For the somewhat more advanced Trawsfynydd reactor, the pressure vessel has a diameter of 61 ft and a 3.5-in. plate thickness; it was designed for a gas pressure of 250 psig.   Vessels of this large size must be fabricated in the field and this tends to increase their cost.*

11.29. Advances in the design of gas-cooled reactors have been in the direction of more compact cores, made possible by the use of enriched uranium fuel, and higher operating temperatures.   The pressure vessels can thus be smaller than those described above; this permits an increase in the design pressure at a reasonable cost.   In the High Temperature Gas Cooled Reactor (§ 13.66), to operate at a thermal power of 115 Mw, with the helium coolant gas at 350 psia, the cylindrical pressure vessel made of A-212-B steel has a length of about 35 ft and a diameter of 14 ft.   Except where additional strength is required, the wall thickness is $2\frac{1}{2}$ in.   The vessel for the Experimental Gas Cooled Reactor (§ 13.65) is somewhat larger and the minimum wall thickness is $2\frac{3}{4}$ in.   For the UHTREX system, the pressure vessel is roughly a sphere about 12 ft in diameter with a general wall thickness of $1\frac{3}{4}$ in.   The helium gas pressure is 500 psia, but the thermal power of 3 Mw is much less than for the aforementioned reactors.

11.30. Problems still exist in the design of pressure vessels for gas-cooled reactors of advanced types, but they are mainly concerned with such matters as thermal stresses and seals, and not necessarily with their size.   When helium is the coolant, as it is in several designs for high-temperature operation, the cost of the gas makes a low leakage rate mandatory.   This requirement must be included in the specification of the reactor vessel.   However, it appears that in the advanced gas-cooled reactors, the rate of heat removal from the fuel elements, rather than the pressure vessel, represents the limiting factor in the operating power level of the reactor.

---

* In France, prestressed concrete vessels, erected *in situ*, have been used for gas-cooled reactors of the same general type.

Organic- and Sodium-Cooled Reactors

11.31. When the coolant is an organic liquid or sodium, the pressure requirements of the reactor vessel are fairly modest. In the Organic Moderated Reactor Experiment (§ 13.45) the coolant pressure is about 200 psig, and in the Piqua (organic) reactor it is 135 psig. A design study for a still larger reactor of this type, with an electrical capacity of 300 Mw, specifies a vessel 17 ft in diameter and an operating pressure of 80 psig. The combination of low design pressure and relatively minor chemical reactivity makes it possible to fabricate vessels of carbon steel rather than of the more expensive high-alloy steels.

11.32. In the EBR-II and the Fermi reactor the maximum sodium pressures are about 50 to 110 psig, with the pressure of the inert cover gas close to atmospheric. However, the requirements for sodium-cooled reactor vessels are generally based on integrity, that is, on the ability to retain hot liquid sodium and the radioactive cover gas, rather than on the pressure they can withstand. The vessel for the Fermi reactor has some interesting features. It is made of Type 304 stainless steel, with a maximum wall thickness of 2 in., and is divided into three parts: a lower reactor vessel, an upper vessel, and a container at one side for the rotor used in transferring fuel elements (see Fig. 11.3). The upper vessel, containing the fuel-handling equipment and control rod drives, is 14 ft in diameter. In the $9\frac{1}{2}$ ft diameter lower reactor vessel, in which the core is located, a section is provided for holding the fuel in a subcritical geometry in the event of a fuel meltdown (§ 9.113). For the Hallam (sodium-graphite) reactor, however, a much simpler vessel is used, although it is also made of Type 304 stainless steel. It is cylindrical in form, 19 ft in diameter and 33 ft high. The bottom of the vessel and the regions around the coolant inlet and outlet nozzles, where piping reactions must be resisted, are 2 in. thick; most other parts are less than 1 in. in thickness.

CONTAINMENT STRUCTURES

11.33. It was seen in § 9.115 *et seq.*, that all reactors are required to have facilities for retaining fission products and gases that might be produced as a result of the maximum credible accident [13]. In some plants, this takes the form of a secondary containment structure, capable of withstanding the maximum expected pressure and having a low leak rate. Although such containment structures are neither required nor used in all reactor plants, the fact that they are fairly common justifies some description of them here.

11.34. In determining the design load for the containment vessel, consideration must be given to the internal pressures produced by vaporization of the coolant, e.g., water, by chemical reactions of various kinds, e.g., uranium, zirconium, or aluminum with water or sodium with oxygen, and by the fission product gases. In addition, allowance must be made for concentrated loads caused by various components acting as missiles upon disruption of the reactor core, and from the displacement of equipment, e.g., structures and cranes, in actual contact with the secondary containment vessel.   Very low leak rate is, of course, an important requirement.

11.35. In the past, containment vessels for power reactors in the United States have usually been constructed in accordance with the general regulations of the Boiler and Pressure Vessel Code of the American Society of Mechanical Engineers.   For reactors that do not have coolants under high pressure, the American Petroleum Institute standards for low-pressure storage vessels have been used [14].   However, a new code is being prepared specifically for reactor (secondary) containment structures which takes cognizance of both the ASME and API codes.

11.36. When containment vessels are used for nuclear power reactors, the most common types employed hitherto are either a sphere or a vertical cylinder with rounded caps at top and bottom [15].   For equal volumes, a sphere uses less steel; on the other hand, the space in a cylinder can generally be used to better advantage.   The cost of erecting a sphere may also be greater than for a cylindrical vessel.   At this stage of development, it is not possible to say if either type of containment structure is definitely superior to the other for high design pressures.   For lower pressures, dome-shaped containment structures may have significant advantages (§ 11.40).

11.37. For the Shippingport PWR a different type of containment system is used.   The heat exchangers (or steam generators) are enclosed in three horizontal cylinders but the reactor is contained in a spherical chamber.   All four vessels are underground, the tops of the cylinders being approximately at grade level.   It may be noted in this connection that a few reactors have been located underground.   However, as the result of a study, it has been concluded that, for reactors of present-day sizes, the cost of excavation is greater than that of a containment structure.   For large nuclear power plants, on the other hand, there is a possibility that underground installations may prove to be economical.

11.38. The steel used for containment vessels should not be susceptible to brittle failure.   Most such vessels have thus been fabricated from A-201 Grade B steels conforming to ASTM Specification A-300.   In some cases A-212 steel has been used.   In general, it is desirable to keep the vessel wall thickness small enough to make stress relief unnecessary for the welds.   This requirement sets the maximum thickness at $1\frac{1}{2}$ in.   The specifications of some typical containment vessels for various power reactors are given in Table 11.1 [16].   The specified leakage rates at the operating pressure are generally about 0.1 per cent of the contained volume per day, although they are somewhat higher in a few instances.   Photographs of spherical and cylindrical containment structures are shown in Figs. 11.5 and 11.6, respectively.

11.39. Various procedures have been proposed which will permit considerable relaxation of the design pressure of the containment structure for water-cooled

TABLE 11.1.   SPECIFICATIONS OF CONTAINMENT VESSELS FOR POWER REACTORS

Reactor	Coolant	Power (Thermal Mw)	Diam. (ft)	Height (ft)	Volume (10^6 ft³)	Design Pressure (psig)	Plate Thickness (in.)
Dresden*.......	Boiling water	626	190	—	3.6	30	1.25, 1.4
Yankee*........	Press. water	392	125	—	1.02	35	0.875, 1.25
Indian Point*...	Press. water	585	160	—	2.14	25	0.89, 1.03
Fermi†	Sodium	300	72	120	0.42	32	1.03, 1.25
EBR-II†	Sodium	62.5	80	139	0.58	24	0.5, 1.0
Elk River†	Boiling water	73	74	115	0.40	21	0.5, 0.7

* Spherical vessels.   † Cylindrical vessels.

reactors.   One of these is the pressure suppression concept referred to in § 9.116. It was first employed in the SM-1A pressurized-water reactor built for the U.S. Army in Alaska [17] and is utilized for the Humboldt Bay, California, boiling-water reactor [18].   A simplified schematic diagram of the containment system is shown in Fig. 11.7.   The reactor, inside its pressure vessel, is supported

FIG. 11.5.   Yankee Atomic Electric Co. nuclear power plant showing spherical containment.   (Courtesy Yankee Atomic Electric Co.)

FIG. 11.6. Enrico Fermi nuclear power plant showing cylindrical containment. (Courtesy Detroit Edison Co.)

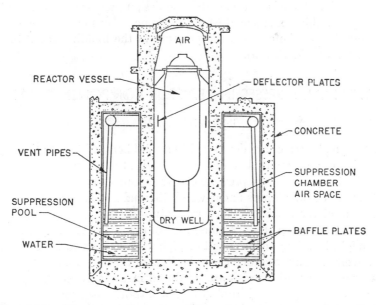

FIG. 11.7. Diagrammatic representation of pressure suppression system

in an air-filled cylindrical concrete tank, called the "dry well," from which vent pipes lead to a surrounding pool of water. The pool is covered by means of a vapor-tight metal container. Should a leak or rupture occur in the reactor pressure vessel, a steam-water mixture would flow out into the dry well and then, after displacing the air, into the pool of water. The steam will condense in the pool and if this occurs rapidly enough, the pressure buildup within the vapor-tight enclosure will be small. A number of tests indicate that the system should operate as expected; furthermore, the water pool has proved to be effective in retaining fission products. With this scheme for vapor condensation, the pressure specification for the containment vessel would be much less than for a similar reactor system without the pool, thereby reducing capital costs.

11.40. Another approach to the containment problem is being made for the Boiling Nuclear Superheater (BONUS) Power Station, Puerto Rico (§ 13.31). As a general rule, the reactor and the primary coolant system are housed in the gas-tight containment vessel, but the electrical generating equipment is in a separate building. In the BONUS facility, the reactor, fuel-storage pool, turbo-generator, condenser, pumps, etc., are all enclosed in a hemispherical geodesic dome, made of steel, 190 ft in diameter. The dome is anchored and sealed to a circular, reinforced concrete wall, about 26 ft high, in which a $\frac{1}{4}$-in. thick steel plate is embedded. A gas-tight, concrete floor slab, upon which all components are mounted, completes the containment system. The basic idea is that by having a building of such large volume (2.5×10^6 ft^3) for the thermal power of 50 Mw, the pressure arising from the maximum credible accident will not exceed 3 psig. This is consequently the design pressure of the structure, so that steel $\frac{3}{16}$-in. thick is adequate for the dome. By means of a cold-water spray operated from the outside, the pressure in the building could be rapidly reduced after an accident [19].

THERMAL STRESSES IN REACTOR COMPONENTS [20]

INTRODUCTION

11.41. Thermal stresses are developed in a solid body whenever the expansion or contraction of a differential volume element, that would normally result from a change in temperature, is prevented. It is convenient to distinguish two different sets of circumstances under which thermal stresses occur. In the first case, the form of the body and the temperature conditions are such that there would be no stress except for the constraint of external forces. An example of this type might be a long fuel element or pipe with little or no freedom of movement at the ends. In the second case, the form of the body and the temperature conditions are such that stresses are produced in the absence of external constraints because of the incompatibility of the natural expansions or contractions of the different parts of the body. Both of these circumstances frequently arise in reactor systems as a result of nonuniform temperature distributions, e.g., in fuel elements, thermal shields, pressure vessels, etc., and of mechanical constraints in assembled components.

11.42. The conditions leading to thermal stresses may be clarified by imagining the body in question to be composed of a number of unit volumes. If the temperature of all the unit volumes is increased to the same extent and the outer boundaries of the body are free, i.e., not restrained, each unit volume will expand an equal amount in all directions. The units will still fit together as before, and no stresses will be present. However, should the temperature increase not be the same in all the unit volumes, the expansions will also be different. Since each volume is attached to its neighbors, there will be a tendency for one unit to oppose the expansion of those surrounding it. Stresses will therefore arise, unless the temperature gradient throughout the body is uniform and the restraints are such that regular expansion is possible.

11.43. A simple form of the general equation which relates the thermal stress to certain physical properties of the solid body and the temperature changes may be derived in the following manner. The basic procedure is to determine first the dimensions (or shape) the body would assume if unconstrained, and then to calculate the stress, i.e., the force per unit area, required to restore it to its original shape. Consider a rectangular element of an isotropic, elastic material, and suppose, in the first place, that it is free to move in the y and z directions, but is constrained in the x direction. The element is then heated uniformly, so that expansion tends to occur. If there were no restraint in the x direction, the dimensional increase per unit length in this direction would be $\alpha\Delta t$, where α is the thermal coefficient of linear expansion and Δt is the temperature increase. The unit strain in a given direction, represented by ϵ, is defined as the change in length per unit length, and so, in the present case, the strain in the x direction is given by

$$\epsilon_x = \alpha\Delta t.$$

Consequently, if the material is elastic, the stress, σ_x, required to restore the original x dimension is obtained from Hooke's law as

$$\sigma_x = -E\epsilon_x = -\alpha E\Delta t,$$

where E is Young's modulus for the material. The stresses in the y and z directions are, of course, zero.

11.44. When there are restraints upon more than one coordinate, it is necessary to employ the generalized Hooke's law in order to obtain the required relationships. One of the factors which must now be taken into consideration is the contraction that tends to occur in the directions normal to an elongation. For example, if there is a strain in the x direction, then the accompanying strains in the other two directions are

$$\epsilon_y = -\nu\frac{\sigma_x}{E} \quad \text{and} \quad \epsilon_z = -\nu\frac{\sigma_x}{E},$$

where ν is Poisson's ratio of the material. For most metals, Poisson's ratio is in the range from 0.25 to 0.35.

11.45. The strain in any direction actually consists of two parts. First, there is the strain arising from external loads and the need to maintain continuity of the body. This includes a contribution from the normal strains mentioned above. Second, there is a strain proportional to the temperature increase, i.e., $\alpha \Delta t$. Consequently, in Cartesian coordinates,

$$\epsilon_x = \frac{1}{E} \left[\sigma_x - \nu(\sigma_y + \sigma_z) \right] + \alpha \Delta t$$

$$\epsilon_y = \frac{1}{E} \left[\sigma_y - \nu(\sigma_x + \sigma_z) \right] + \alpha \Delta t$$

$$\epsilon_z = \frac{1}{E} \left[\sigma_z - \nu(\sigma_x + \sigma_y) \right] + \alpha \Delta t.$$

11.46. The general Hooke's law relationships also include three expressions for shearing strain, namely,*

$$\epsilon_{xy} = \frac{\sigma_{xy}}{2G}, \qquad \epsilon_{yz} = \frac{\sigma_{yz}}{2G}, \qquad \epsilon_{zx} = \frac{\sigma_{zx}}{2G},$$

where G is the modulus of elasticity in shear or modulus of rigidity and is related to Young's modulus and Poisson's ratio by

$$G = \frac{E}{2(1 + \nu)}.$$

The shearing strain relationships are not affected by temperature since angular distortion is not produced in an isotropic material by free thermal expansion.

11.47. In addition to the six equations given above, there are six "compatibility conditions." These are obtained by writing six strain relationships for $\epsilon_x, \epsilon_y, \ldots, \epsilon_{xy}, \ldots$, etc., in the form

$$\epsilon_x = \frac{du_x}{dx}, \qquad \epsilon_{xy} = \frac{1}{2} \left(\frac{du_x}{dy} + \frac{du_y}{dx} \right), \text{ etc.,}$$

where u_x, u_y, etc., are the displacement components, and differentiating them twice with respect to two of the three coordinates, x, y, z. Since the stresses are expressed in terms of displacements in these three directions, the compatibility conditions are necessary to make the problem determinate. There are, finally, three "equilibrium conditions" which must be satisfied at every differential volume of the body. As the stresses vary within the body, they must be such as to maintain equilibrium with the external forces.

11.48. It can be shown that the combination of the 15 thermoelasticity relations described above with the postulated temperature distribution and bound-

---

* Some writers use a different method of representation in which the 2 is omitted from these expressions.

ary conditions leads to a unique solution which includes one set of stress and displacement components. It is therefore possible to derive stress equations for any combination of geometry and boundary conditions of interest. Although the outline given here has been in terms of Cartesian coordinates, the results can, of course, be expressed in other coordinate systems, i.e., cylindrical or spherical, as may be convenient.

11.49. An interesting application of the results is to a body subjected to a uniform temperature increase with complete restraint, that is to say, the displacements are zero in all directions. In these circumstances,

$$\epsilon_x = \epsilon_y = \epsilon_z = \epsilon_{xy} = \epsilon_{yz} = \epsilon_{zx} = 0$$

and

$$\sigma_x = \sigma_y = \sigma_z = -\frac{\alpha E}{1 - 2\nu} \Delta t.$$

For constraints in the x and y directions only,

$$\epsilon_x = \epsilon_y = \epsilon_{xy} = 0$$

and

$$\sigma_x = \sigma_y = -\frac{\alpha E}{1 - \nu} \Delta t \quad \text{and} \quad \sigma_z = 0.$$

11.50. The foregoing analysis may be extended to include restraints in **any of** three directions, and the results may be summarized in the form

$$\sigma = -\frac{\alpha E}{1 - c\nu} \Delta t,$$

where c is 0, 1, or 2, depending upon whether restraints exist in one, two, or three directions, respectively. More generally, the stress is given for a finite system by an expression of the form

$$\sigma = \frac{\alpha E}{1 - c\nu} F(r, t), \tag{11.1}$$

in which $F(r, t)$ is a function of the geometry of the body and the temperature distribution within it; examples of some cases of interest are presented below.

11.51. In the design of reactor components it is frequently desirable to make a preliminary estimate of the order of magnitude of a thermal stress to determine if it is large enough to warrant more precise calculation. For this purpose a relationship similar to equation (11.1) may be used; this is

$$\sigma \approx \frac{\alpha E}{1 - c\nu} (\Delta t)_{max}, \tag{11.2}$$

where $(\Delta t)_{max}$ is the maximum (numerical) difference between the average temperature and the temperature of any other part of the system. It can be seen from equation (11.2) that thermal stresses may be kept small by using, as far as possible, materials having low values of the coefficient of thermal expansion,

Young's modulus, and Poisson's ratio. Furthermore, it would appear to be advantageous for the material to have a high thermal conductivity since this would tend to decrease temperature differences within the body.

11.52. One objective of this section is to present some thermal stress relations in a form useful to the reactor designer. Although these relations may all be developed from the equations of thermoelasticity, the procedures are complicated and will not be given here. For the details the literature of the subject should be consulted. Tabulations of stress and temperature distributions for various geometries are also available [21].

THERMAL STRESS IN SPHERE WITH UNIFORM SOURCE

11.53. A fairly simple case which is of interest in certain reactor designs is that of a sphere, e.g., of moderator or reflector, in which the heat source is essentially uniform. If a is the radius of the sphere, the temperature t at any point at a radial distance r from the center is given by*

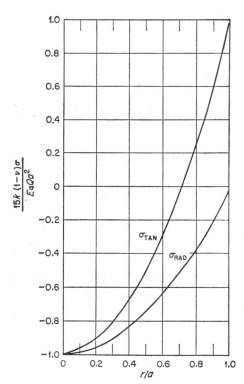

$$t - t_a = \frac{Q}{6k}(a^2 - r^2),$$

where Q is the source strength, i.e., the rate of heat generation per unit volume, k is the thermal conductivity, and t_a is the temperature at the surface of the sphere. If this equation is inserted into the appropriate form of equation (11.1) for a solid sphere and the required integrations performed, it is found that

$$\sigma_{\text{rad}} = \frac{\alpha EQ}{15k(1 - \nu)}(r^2 - a^2) \qquad (11.3)$$

and

$$\sigma_{\text{tan}} = \frac{\alpha EQ}{15k(1 - \nu)}(2r^2 - a^2), \qquad (11.4)$$

FIG. 11.8. Calculation of thermal stress in solid sphere

where σ_{rad} and σ_{tan} refer to the thermal stresses in radial and tangential directions, respectively. It will be observed that the radial stress becomes zero at the outer surface of the sphere, i.e., where $r = a$, as is to be expected. The tangential stress is zero at a point

* See Chapter 6, Problem 6.

within the sphere where $2r^2 = a^2$, i.e., where $r = 0.707a$. As an aid to the use of the foregoing equations, the dimensionless quantity $15k(1 - \nu)\sigma/\alpha EQa^2$ has been plotted in Fig. 11.8 as a function of r/a for both σ_{rad} and σ_{tan}.

Example 11.1. Calculate the tangential stresses (*a*) at the surface and (*b*) at a radial distance of a quarter of the radius, in a 3-in. diameter sphere of a material in which heat is generated uniformly at the rate of 10 watts/cm³ due to gamma-ray absorption. The following values of the physical constants of the material may be taken as applicable at the existing temperatures:

$$\alpha = 5.33 \times 10^{-6}/°F \qquad\qquad E = 39 \times 10^6 \text{ psi}$$
$$k = 21.5 \text{ Btu/(hr)(ft}^2)(°F/ft) \qquad \nu = 0.34.$$

(*a*) At the surface of the sphere, $r = a$, so that

$$\sigma_{\text{tan}} = \frac{\alpha EQa^2}{15k(1 - \nu)},$$

and since a is 1.5 in. = 0.125 ft, and Q is 10 watts/cm³ = 9.66×10^5 Btu/(hr)(ft³), it follows that

$$\sigma_{\text{tan}} = \frac{(5.33 \times 10^{-6})(39 \times 10^6)(9.66 \times 10^5)(0.125)^2}{(15)(21.5)(1 - 0.34)}$$

$$= 14{,}700 \text{ psi}.$$

(*b*) For $r = 0.25a$, i.e., when $r/a = 0.25$, the ordinate for the σ_{tan} curve is found from Fig. 11.8 to be close to -0.86; that is,

$$\frac{15k(1 - \nu)\sigma_{\text{tan}}}{\alpha EQa^2} = -0.86$$

so that

$$\sigma_{\text{tan}} = \frac{-0.86\alpha EQa^2}{15k(1 - \nu)}$$

$$= -12{,}600 \text{ psi}.$$

THERMAL STRESS IN HOLLOW CYLINDER WITH ZERO HEAT SOURCE

11.54. The cladding of a cylindrical fuel element represents the important case of a hollow cylinder with no internal heat generation. The temperature difference Δt is the difference between the higher inner temperature at $r = a$ and the lower outer temperature at $r = b$; its value is given by $t_1 - t_2$ of Example 6.3. The tangential stresses are compressive near the inner surface and tensile near the outer surface.

11.55. For most design purposes the following approximate treatment is adequate for cylinders having small wall thicknesses, as is the case for fuel element cladding. A quantity m is defined by

$$\frac{b}{a} = 1 + m,$$

and if m is small, i.e., the wall thickness is small, the stress value at the inner surface at which $r = a$ is given by

$$\sigma_{\text{tan}} = -\frac{E\alpha\Delta t}{2(1 - \nu)}\left(1 + \frac{m}{3}\right)$$

and at the outer surface, where $r = b$,

$$\sigma_{\text{tan}} = \frac{E\alpha\Delta t}{2(1 - \nu)}\left(1 - \frac{m}{3}\right).$$

For most fuel element rods, m lies between 0.1 and 0.2, and then it is a good approximation to neglect $m/3$ in comparison with unity, so that the foregoing equations reduce to

$$\sigma_{\text{tan}}\,(r = a) = -\frac{E\alpha\Delta t}{2(1 - \nu)} \tag{11.5}$$

and

$$\sigma_{\text{tan}}\,(r = b) = \frac{E\alpha\Delta t}{2(1 - \nu)}. \tag{11.6}$$

Example 11.2. A fuel rod, in which heat is being generated uniformly at a rate of 1.035×10^7 Btu/(hr)(ft^3), is clad with stainless steel, the inner radius of the cladding being 0.0328 ft and the outer radius 0.0338 ft. At the operating temperature of about 1100°F, the properties of the stainless steel are as follows:

$$\alpha = 10 \times 10^{-6}/°F \qquad\qquad E = 22 \times 10^6 \text{ psi}$$
$$k_c = 13 \text{ Btu}/(\text{hr})(\text{ft}^2)(°F/\text{ft}) \qquad \nu = 0.32.$$

Calculate the thermal stress in the cladding during operation.

The temperature drop across the cladding, i.e., Δt, is equal to $t_1 - t_2$ in Example 6.3; thus,

$$\Delta t = \frac{Qa^2}{2k_c}\ln\frac{b}{a}$$

$$= \frac{(1.035 \times 10^7)(0.0328)^2}{(2)(13)}\ln\frac{0.0338}{0.0328}$$

$$= 12.8°F.$$

The numerical value of the thermal stress at the cladding surface is given by equation (11.5) or (11.6) as

$$\sigma = \frac{E\alpha\Delta t}{2(1 - \nu)} = \frac{(22 \times 10^6)(10 \times 10^{-6})(12.8)}{(2)(0.68)}$$

$$= 2070 \text{ psi.}$$

THERMAL STRESS IN HOLLOW CYLINDER WITH UNIFORM HEAT SOURCE

11.56. An internally cooled element or a moderator piece with regularly spaced cooling holes may often be treated as an unrestrained hollow cylinder, insulated at the outside, with heat being withdrawn from the inner surface. The results of inserting the radial temperature distribution for this geometry, assuming a uniform source, into appropriate forms of equation (11.1) are shown in Figs. 11.9, 11.10, and 11.11, for a hollow cylinder for which a and b are the

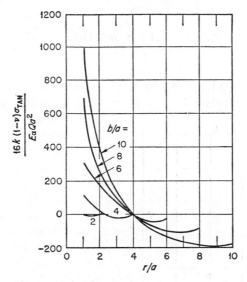

FIG. 11.9. Calculation of tangential thermal stress in hollow cylinder

internal and external radii, respectively. In these figures the dimensionless quantity $16k(1 - \nu)\sigma/\alpha EQa^2$, for tangential, radial, and longitudinal stresses, is plotted as a function of r/a for various values of the parameter b/a. General inspection of the curves shows that for a given value of r/a, the stresses all increase as b/a, the ratio of the external to the internal radius, increases.

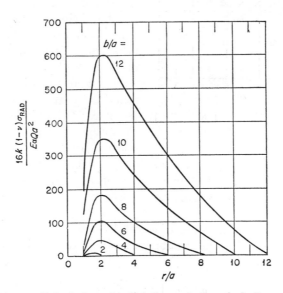

FIG. 11.10. Calculation of radial thermal stress in hollow cylinder

THERMAL STRESS WITH EXPONENTIAL HEAT SOURCE

Hollow Cylinder

11.57. For exponential internal heat generation (cf. § 6.81) in long thick-walled cylinders, such as would arise from gamma-ray absorption in a thermal shield and in a pressure vessel, suitable design curves are given in Fig. 11.12 for the case of an outside adiabatic (insulated) wall [22]. The nondimensional quantity σ_T is plotted as a function of the radius ratio b/a, with the dimensionless product μa, where a is the inner radius of the cylinder, as a variable parameter. In this relation, μ is an exponential coefficient which is used to describe the decrease in heat generation rate with distance through the material. In practice, this decrease, like the gamma-ray attenuation, is rarely exactly exponential, and so some approximation is necessary. For simple gamma-ray energy absorption, μ may be taken as equal to the reciprocal of the relaxation length, defined in § 10.73, if the latter is known. If not, an appropriate value of μ may be estimated from the narrow-beam (collimated) total attenuation coefficient and the buildup factor for the specific case under consideration. When the situation is complicated by the production of gamma rays within the material, e.g., as a result of neutron capture, an approximate approach is to take the calculated values of the gamma-ray flux at inner and outer surfaces of the cylinder and to assume that the decrease within the wall is exponential.

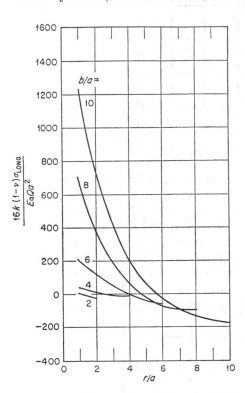

FIG. 11.11. Calculation of longitudinal thermal stress in hollow cylinder

11.58. With μ known, the value of σ_T can be obtained from Fig. 11.12 for the given radii, a and b. The maximum tangential stress, σ, is then obtained by means of the relationship

$$\sigma = \sigma_T \left[\frac{\alpha E Q_0}{k\mu^2(1 - \nu)} \right], \qquad (11.7)$$

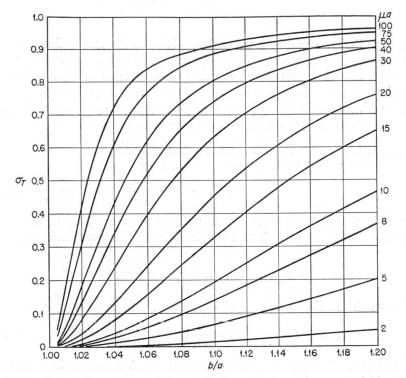

Fig. 11.12.   Calculation of thermal stress in hollow cylinder with exponential heat source

where Q_0 is the heat generation rate, per unit volume, at the inner wall of the cylinder.

Example 11.3.   The steel, cylindrical pressure vessel of a pressurized-water reactor has an outer radius of 62.4 in. and a wall thickness of 7.9 in.   The volumetric rate of heat generation at the inner surface, due to the absorption of gamma rays, is 8.63×10^3 Btu/(hr)(ft³).   Assuming an outer adiabatic wall, calculate the maximum (tangential) stress at a temperature of about 500°F.   The linear attenuation coefficient of the gamma rays in steel may be taken as 0.27 cm⁻¹ and other physical properties of steel at the existing temperatures are as follows:

$$\alpha = 8.0 \times 10^{-6}/°F \qquad E = 26.4 \times 10^6 \text{ psi}$$
$$k = 10.7 \text{ Btu/(hr)(ft}^2)(°F/ft) \qquad \nu = 0.28.$$

The inner radius a is $62.4 - 7.9 = 54.5$ in., and so the radius ratio is $62.4/54.5 = 1.145$. The value of μa is $0.27 \times 54.5 \times 2.54 = 37.4$, where the factor 2.54 is used to convert inches to centimeters and make μa dimensionless.   From Fig. 11.12, it is seen that with $b/a = 1.145$ and $\mu a = 37.4$, σ_T is about 0.83.   In order that the units in equation (11.7) may be consistent, μ is expressed in ft⁻¹, i.e., $0.27 \times 30.5 = 8.2$ ft⁻¹; hence,

$$\sigma = 0.83 \left[\frac{(8.0 \times 10^{-6})(26.4 \times 10^6)(8.63 \times 10^3)}{(10.7)(8.2)^2(0.72)} \right]$$

$$= 2900 \text{ psi.}$$

Slab

11.59. Cylinders of fairly large radius, with the internal radius more than about ten times the wall thickness, may be treated as slabs, to a good approximation. The following procedure for calculating thermal stresses provides an alternative to that given above. The temperature distribution in a slab of thickness L with an exponential heat source is expressed by equation (6.32), and combination with the appropriate thermal stress equations gives

$$\sigma_y = \sigma_z = \frac{\alpha E}{1 - \nu} \cdot \frac{Q_0}{k\mu^2} \left\{ \frac{2x}{L} \left[\frac{3(e^{-\mu L} + 1)}{\mu L} + \frac{6(e^{-\mu L} - 1)}{(\mu L)^2} \right] \right. $$
$$\left. - \frac{2(e^{-\mu L} + 2)}{\mu L} - \frac{6(e^{-\mu L} - 1)}{(\mu L)^2} + e^{-\mu x} \right\},$$

where x is the distance from the inner face of the slab. If the rate of heat removal from the outer face of the slab is small, the maximum temperature is

$$t_{max} = \frac{Q_0}{k\mu^2} \left[1 - (\mu L + 1)e^{-\mu L} \right]$$

and the maximum thermal stress, which occurs at the inner face, is

$$\sigma = \frac{\alpha E Q_0}{2k\mu^2(1 - \nu)} \left[2 - \frac{2}{\mu L} + \left(\frac{2}{\mu L} + \mu L \right) e^{-\mu L} \right]. \tag{11.8}$$

Example 11.4. Calculate the maximum thermal stress in a steel slab 7.9 in. thick using the data in Example 11.3.

Since the dimensionless quantity μL appears frequently in equation (11.8), this is first evaluated; it is equal to $0.27 \times 7.9 \times 2.54 = 5.4$. Hence,

$$\sigma = \frac{(8.0 \times 10^{-6})(26.4 \times 10^6)(8.63 \times 10^3)}{(2)(10.7)(8.2)^2(0.72)} \left[2 - \frac{2}{5.4} + \left(\frac{2}{5.4} + 5.4 \right) e^{-5.4} \right]$$

$$= 2900 \text{ psi.}$$

The exact agreement with the maximum stress calculated in Example 11.3 for a cylinder of the same thickness is partly fortuitous.

THERMAL STRESS IN UNRESTRICTED CLAD FLAT-PLATE FUEL ELEMENT [23]

11.60. For flat-plate (sandwich type) fuel elements clad on each side (cf. § 6.73), in which the thermal and mechanical properties of the fuel and cladding do not differ greatly, the thermal stress in the y (and z) direction (Fig. 11.13), due to a uniform heat generation in the fuel, is given at the *central plane* of the fuel region by

$$\sigma_y = \sigma_z \atop (\text{at } x = 0) = -\frac{\alpha E Q a^2}{2k(1 - \nu)} \left[\frac{c}{a} + \frac{a}{3(a + c)} \right] \tag{11.9}$$

and at the *outer surface* of the cladding by

$$\sigma_y = \sigma_z = \frac{\alpha E Q a^2}{2k(1-\nu)}\left[1 + \frac{c}{a} - \frac{a}{3(a+c)}\right], \qquad (11.10)$$
(at $x = a + c$)

where a is the half-thickness of the fuel region and c is the cladding thickness. (It may be noted that c is equivalent to $b - a$ of Fig. 6.5.)

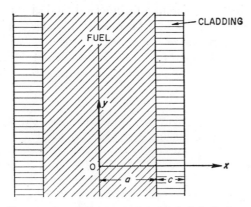

FIG. 11.13. Representation of clad fuel plate

Example 11.5. Calculate the thermal stress (*a*) at the central plane and (*b*) at the outer surface of a sandwich-type fuel element consisting of a 0.06-in. thick fuel region clad on both sides with a 0.03-in. layer of a metal having essentially the same mechanical and thermal properties as the fuel region. These properties are as follows:

$$\alpha = 9 \times 10^{-6}/°F \qquad\qquad E = 10 \times 10^6 \text{ psi}$$
$$k = 10 \text{ Btu/(hr)(ft}^2)(°F/ft) \qquad \nu = 0.30.$$

Assume that the heat is generated uniformly so as to give a heat flux of 0.5×10^6 Btu/(hr)(ft^2) at the surface of the plate, which is cooled equally on both sides.

It follows from the statement in § 6.74 that the heat flux, i.e., q/A, is equal to aQ, so that in the present case, since $2a$ is 0.06 in., i.e., $a = 2.5 \times 10^{-3}$ ft,

$$Q = \frac{0.5 \times 10^6}{2.5 \times 10^{-3}} = 2 \times 10^8 \text{ Btu/(hr)(ft}^3)$$

(a) *Central plane:* From equation (11.9),

$$\sigma_y = \sigma_z = -\frac{(9 \times 10^{-6})(10 \times 10^6)(2 \times 10^8)(2.5 \times 10^{-3})^2}{(2)(10)(0.70)}\left[\frac{0.03}{0.03} + \frac{0.03}{(3)(0.06)}\right]$$

$$= -9400 \text{ psi.}$$

(b) *Outer surface:* From equation (11.10),

$$\sigma_y = \sigma_z = \frac{(9 \times 10^{-6})(10 \times 10^6)(2 \times 10^8)(2.5 \times 10^{-3})^2}{(2)(10)(0.70)}\left[1 + \frac{0.03}{0.03} - \frac{0.03}{(3)(0.06)}\right]$$

$$= 14,700 \text{ psi.}$$

COMPARISON OF MATERIALS

11.61. It is of interest to compare the expected behavior of different materials from the thermal stress viewpoint. An examination of the various equations given above shows that in all cases the factor $\alpha E/k(1 - \nu)$ appears, and the value of this quantity for different substances should give a relative indication of the thermal stresses that would be produced under equivalent conditions. Some data for a temperature of about 600°F, together with the approximate tensile strengths at this temperature, are given in Table 11.2. A high value of the thermal stress can be compensated, to some extent, by a large tensile strength.

TABLE 11.2. RELATIVE THERMAL STRESS FACTORS
AND TENSILE STRENGTHS AT 600°F

Material	$\alpha E/k(1 - \nu)$	Tensile Strength (psi)
Aluminum (2S)	1.4	3,500
Beryllium (extruded)*	6.0	41,000
Beryllium oxide*	14.6	15,000
Carbon steel (A212-B)	12.9	70,000
Graphite*	0.1	3,200
Stainless steel (Type 347)	33.8	69,000
Uranium	24.8	29,000
22% U-Al alloy	2.0	8,100
Zirconium	4.6	15,600

* These materials are generally considered as semibrittle, so that direct comparison of their tensile strengths with those of ductile materials is not valid.

11.62. It is seen from the table that, for a given temperature difference in the material, the corresponding thermal stess will be lowest for graphite, aluminum, and the uranium-aluminum alloy; zirconium and beryllium are in the medium range, whereas beryllium oxide, uranium, and stainless steel (Type 347) will exhibit the largest stresses. It should be noted, however, that the thermal stresses are calculated on the assumption that the material is stretched below the yield point. The ductility of the metal, which permits it to yield without cracking, is thus an important consideration, e.g., for stainless steel. Beryllium, on the other hand, is relatively brittle. It is seen, therefore, that the calculated thermal stresses alone do not provide a complete guide to the expected behavior. Although the values are important in establishing temperature conditions under which cracking may occur, they do not indicate whether this actually happens, since other factors are equally significant. Temperature cycling, for example, even below the elastic limit may cause thermal fatigue failure, although the thermal stresses may not be excessive for static conditions.

CREEP

11.63. The preceding treatment has been based on the assumption of elastic stress-strain behavior. If the elastic limit is slightly exceeded, plastic flow may occur and thermal stresses partially relieved, so that the actual values are lower than expected. This flow phenomenon is referred to as "creep." High-temperature creep behavior can also result in the relief of thermal stress. For example, if the temperature of a bar under a constant tensile force load is raised to a higher level, the elongation of the bar will increase continuously with time. Similarly, consider the situation when the bar is subjected to a constant elongation. In elastic behavior at low temperature, the force required to produce this elongation is constant. At high temperature, however, the force will decrease with time. This increasing tendency toward plastic flow with time has the effect of relaxing the thermal stress. In fact, in some instances, creep rather than stress may be the limiting design consideration [24].

11.64. Various theories of inelastic or viscoelastic behavior have been proposed and have been applied to thermal stress analysis. With creep present, the summation of permissible and mechanical stresses may indeed be more than the usually allowable design stress, and several different methods have been suggested for taking this into account. One is to assume that the mechanical stresses alone may be up to the allowable stress, with the sum of mechanical and thermal stresses being less than the yield point. A better procedure, however, is to require that the combination of mechanical and thermal stresses at any level of power generation, including heat generation, shall not exceed 1.5 times the allowable design stress value as required by the ASME Boiler and Pressure Vessel Code [25]. Another approach to the problem is indicated below, where the importance of other considerations in pressure vessel design are pointed out.

PRESSURE VESSEL DESIGN [26]

INTRODUCTION

11.65. The design of a thick-walled pressure vessel, such as is used for a pressurized, water-cooled reactor, normally proceeds through several phases. First, certain basic requirements are established: these include the height, internal diameter, location of inlet and outlet pipes, methods of support, head configuration, closure type, and pattern for attachments and penetrations. The structural material, its mechanical properties, and allowable stress values are assumed to be known. A preliminary design is then prepared in which the main metal thicknesses are established. In this step primary consideration is given to mechanical or nonthermal loadings, with thermal and cyclic loadings being taken into account in an approximate manner based on the designer's experience with similar vessels. Next, a detailed stress analysis is carried out of all major structural components, with changes made in the design as found desirable. Examples of stresses to be considered are those at the inside and outside of the vessel wall from internal pressure and thermal effects; at discontinuities such as head-to-shell junctures, changes in section, and openings; in flanges, seals, bolts, and closure assemblies; stresses due to external loads, such as pipe and foundation eactions; and the effect of stress concentrations, if any.

11.66. The analysis must take into account the effects, mostly thermal, of various startup and shutdown requirements, as well as those of combinations of steady-state and transient conditions resulting from normal plant operating procedures. In this connection, for example, fatigue effects may prove a design limitation. It is evident that the design of a reactor pressure vessel is a highly complex and specialized problem. Consequently, only the more elementary considerations in the preliminary design will be described here.

STRESS INTENSITY

11.67. The mathematical determination of the stresses in a particular system is of no value to the designer unless the calculated values can be related to some criteria for "failure" of actual materials. A number of theories establishing criteria for elastic failure have been proposed, but the subject is still somewhat controversial. For the design of pressure vessels, the *maximum shear stress theory* has been recommended. It is the basis of a code prepared for application in the U. S. Naval Reactor Program and has been generally accepted by reactor designers [27]. According to this theory, the algebraic difference between the largest and the (algebraic) smallest of the principal stresses at a given location is the determining factor. The term "stress intensity" is used as an abbreviation for the expression "equivalent intensity of combined stress" to designate this algebraic difference. Values of maximum allowable stress intensities for some steels of interest for pressure vessel construction are given in Table 11.3. The quantities S_m, the membrane stress intensity, and S_p, the primary plus secondary stress intensity, are described below; S_y and S_u are the minimum values of the yield strength and the ultimate tensile strength of the material.

11.68. For a thick-walled cylindrical vessel in the range of elastic behavior, the maximum stress is at the inner surface, and this value of the circumferential (or hoop) tensile stress is

$$\sigma_{max} = \frac{P[(b/a)^2 + 1]}{(b/a)^2 - 1}, \tag{11.11}$$

where P is the internal pressure and b/a is the radius ratio. The minimum stress is the radial (compressive) component at the inner surface of the cylinder, i.e.,

$$\sigma_{min} = -P.$$

TABLE 11.3. MAXIMUM ALLOWABLE STRESS INTENSITIES

Material	Temperature (°F)	S_m*	S_p*	S_y*	S_u*
ASTM A 302-B.....	70	27	45	50	80
	400	27	43	47	80
	700	27	38	43	80
ASTM A 212-B.....	70	23	34	38	70
	400	20	29	32	70
	700	16	23	25	66
AISI Type 304......	70	19	27	30	75
	400	13	19	21	62
	700	10	15	16	59
AISI Type 347......	70	19	27	30	75
	400	15	22	25	62
	700	14	20	22	59

* In thousands of pounds per square inch.

The primary stress intensity, S, is the algebraic difference of these quantities, so that

$$S = \frac{2P(b/a)^2}{(b/a)^2 - 1}.$$ (11.12)

11.69. The membrane stress intensity, the maximum allowable value of which is given in Table 11.3, is based on the average primary stress across the thickness of the vessel wall. The average maximum (hoop) stress may be taken as the "thin-wall" value to which equation (11.11) reduces when the wall thickness, i.e., $b - a = t$, is small; thus,

$$\sigma_{max}(\text{av.}) = \frac{Pa}{t}.$$ (11.13)

The average minimum (radial) stress is half the value at the inner radius, i.e., $-\frac{1}{2}P$. Hence, in this case, the membrane stress intensity, S_m, is given by

$$S_m = \frac{P(2a + t)}{2t}.$$ (11.14)

Example 11.6. Calculate the principal stress intensity and the membrane stress intensity for the pressure vessel referred to in Example 11.3; the internal design pressure is 2500 psi.

The inner radius is 54.5 in. and the wall thickness is 7.9 in., so that the outer radius is 62.4 in., and the radius ratio, b/a, is 1.145. Hence, from equation (11.12), with P equal to 2500,

$$S = \frac{(2)(2500)(1.145)^2}{(1.145)^2 - 1} = 21,200 \text{ psi.}$$

The membrane stress intensity is obtained from equation (11.14) as

$$S_m = \frac{(2500)[(2)(54.5) + 7.9]}{(2)(7.9)} = 18,500 \text{ psi.}$$

It is of interest to note that the circumferential stress at the inner surface, as derived from equation (11.11), is 18,300 psi. This is very close to the membrane stress intensity, as is generally the case for cylinders of the pressure vessel type.

The data in this example refer to the Yankee Atomic Electric Company reactor pressure vessel which is fabricated from A 302-B (low carbon) alloy steel. It is seen that the calculated value of S_m is less than the maximum allowable given in Table 11.3.

11.70. The quantity S_p is the stress intensity due to both primary and secondary stresses; it results from mechanical forces, structural discontinuities, and loads and moments induced by supports and piping, in addition to the internal pressure. The primary stress has been calculated above, but the determination of the secondary stress is complicated. The procedure is described in the Naval Reactors Code [27], but for preliminary design purposes the evaluation of S_p may be omitted.

11.71. In addition to the maximum allowable values of S_m and S_p, the Code gives data for the permissible stress intensities arising from a combination of steady-state and various types of transient loading conditions. In this connection, thermal stresses are treated with other transients. An adequate design or satisfactory operating condition is determined by the use of a so-called "fatigue diagram" derived from the steady and variable components of the fluctuating load. For preliminary design, however, the criterion for thermal stress based on the ASME Code mentioned in § 11.64 is suggested.

11.72. It may be mentioned, in conclusion, that in some cases there are many penetrations of the pressure-vessel head, in particular, to permit the entrance of control mechanism housings and fuel-element assemblies. Reliable computation of the stresses to be expected is then very difficult, if not impossible. In these circumstances a scaled-down model of the vessel is constructed, making use of preliminary design calculations. The stresses and deflections in this mockup are then measured by means of strain gages and other instruments under simulated operating conditions. The results of these measurements are used as the basis for the final design [28].

SYMBOLS USED IN CHAPTER 11

a	radius of sphere
a	inner radius of hollow cylinder
a	half-thickness of fuel portion of clad plate fuel element
b	outer radius of hollow cylinder
c	thickness of cladding of clad plate fuel element

c	factor ($=0$, 1, or 2)
E	Young's modulus
G	modulus of elasticity in shear
k	thermal conductivity
k_c	thermal conductivity of cladding
L	slab thickness
m	quantity ($=1 - b/a$)
P	internal pressure
Q	rate of heat generation per unit volume
Q_0	rate of heat generation per unit volume at surface
r	radial distance
S	primary stress intensity
S_m	membrane stress intensity
S_u	minimum ultimate tensile strength
S_y	minimum yield strength
t	thickness of "thin" wall
t	temperature
Δt	temperature difference
$u_x, u_y, \ldots$	displacement in $x, y, \ldots$ direction
x, y, z	coordinates
α	thermal coefficient of linear expansion
ϵ	unit strain (change in length per unit length)
$\epsilon_x, \epsilon_y, \ldots$	strain in $x, y, \ldots$ direction
$\epsilon_{xy}, \epsilon_{yz}, \ldots$	strain in $xy, yz, \ldots$ plane
μ	linear attenuation coefficient of gamma rays
ν	Poisson's ratio
$\sigma_x, \sigma_y, \ldots$	stress in $x, y, \ldots$ direction
$\sigma_{xy}, \sigma_{yz}, \ldots$	stress in $xy, yz, \ldots$ plane
σ_{max}	maximum stress
σ_{min}	minimum stress
σ_{rad}	radial stress
σ_{tan}	tangential stress
σ_T	factor for calculating stress in hollow cylinder

REFERENCES FOR CHAPTER 11

1. For reviews, see *Nucleonics*, **13**, No. 6, 41–72 (1955); H. Etherington (Ed.), "Nuclear Engineering Handbook," McGraw-Hill Book Co., Inc., New York, 1958, Section **13**.
2. H. Etherington, Ref. 1, p. **13**-24; for reviews, see *Nucleonics*, **17**, No. 5, 138 (1959); **19**, No. 7, 55 (1961).
3. H. Etherington, Ref. 1, p. **13**-27; G. T. Lewis, *et al.*, *Nucleonics*, **19**, No. 7, 70 (1961).
4. H. Etherington, Ref. 1, p. **13**-84; *Power Reactor Technology*, **4**, No. 2, 59 (1961); R. A. Jaross and A. H. Barnes, *Proc. Second U. N. Conf. Geneva*, **7**, 82 (1958); O. S. Stein and R. A. Jaross, *ibid.*, **7**, 88 (1958); R. W. Dickinson and J. B. Williams, *Nucleonics*, **19**, No. 1, 65 (1961).

5. R. W. Dickinson and J. B. Williams, Ref. 4.
6. R. W. Dickinson and J. B. Williams, Ref. 4; F. Boni and P. S. Otten, *Nucleonics*, **19**, No. 6, 58 (1961).
7. F. Boni and P. S. Otten, Ref. 6.
8. J. R. Dietrich and W. H. Zinn (Eds.), "Solid Fuel Reactors," Addison-Wesley Publishing Co., Inc., Reading, Mass., 1958, pp. 321 *et seq.;* A. Amorosi and J. G. Yevick, *Proc. Second U. N. Conf. Geneva*, **9**, 358 (1958).
9. W. J. Kann and J. M. Harrer, *Nucleonics*, **16**, No. 5, 84 (1958); H. Etherington, Ref. 1, pp. **8**-67 *et seq.; Nuclear Safety*, **2**, No. 4, 17 (1961).
10. "Nuclear Reactor Containment Buildings and Pressure Vessels," Proc. of Int. Symposium at Royal Coll. of Sci. and Technology, Glasgow, 1960, Butterworths, London, 1960.
11. *Power Reactor Technology*, **4**, No. 3, 47 (1961); *Nucleonics*, **19**, No. 3, 53 (1961).
12. *Power Reactor Technology*, **4**, No. 4, 51 (1961).
13. R. A. Johnson and I. Nelson, U. S. AEC Report SL-1868 (1961), Supplement 1 (1962); *Nuclear Safety*, **2**, No. 1, 55 (1960); *Power Reactor Technology*, **5**, No. 1, 33 (1961).
14. R. O. Brittan, U. S. AEC Report ANL-5851 (1958); R. O. Brittan and J. C. Heap, *Proc. Second U. N. Conf. Geneva*, **11**, 66 (1958).
15. R. O. Brittan and J. C. Heap, Ref. 14; R. N. Bergstrom and W. A. Chittenden, *Nucleonics*, **17**, No. 4, 86 (1959).
16. R. N. Bergstrom and W. A. Chittenden, Ref. 15.
17. H. Roberts, *Nucleonics*, **20**, No. 4, 96 (1962); see also, G. I. Staber, *et al., Power*, **99**, 75 (September, 1955)
18. *Nuclear Safety*, **2**, No. 1, 49 (1960); *Power Reactor Technology*, **4**, No. 2, 32 (1961).
19. U. S. AEC Report GNEC-168 (1961).
20. For detailed treatments of thermal stress theory, see S. Timoshenko and J. N. Goodier, "Theory of Elasticity," McGraw-Hill Book Co., Inc., New York, 1951; B. E. Gatewood, "Thermal Stresses," McGraw-Hill Book Co., Inc., New York, 1957; B. A. Boley and J. H. Weiner, "Theory of Thermal Stresses," John Wiley and Sons, Inc., New York, 1960; C. F. Bonilla (Ed.), "Nuclear Engineering," McGraw-Hill Book Co., Inc., New York, 1957, Ch. 11 ("Thermal Stress Analysis and Mechanical Design," by A. M. Freudenthal).
21. R. Hankel, *Nucleonics*, **18**, No. 11, 168 (1960).
22. U. S. AEC Report WAPD-BT-1 (1957).
23. P. C. Zmola, unpublished; see also, C. O. Smith, *Nuc. Sci. and Eng.*, **2**, 213 (1957); **3**, 540 (1958).
24. N. H. Triner, *Nucleonics*, **17**, No. 10, 58 (1959); see also, U. S. AEC Report PWAC-354 (1961).
25. *ASME Boiler and Pressure Vessel Code*, Case Interpretation No. 1273N (1959).
26. For reviews of Nuclear Code Cases, see *Power Reactor Technology*, **4**, No. 2, 44 (1961); **4**, No. 4, 50 (1961); *Nuclear Safety*, **1**, No. 3, 9 (1960).
27. "Tentative Structural Design Basis for Reactor Pressure Vessels and Directly Associated Components," U. S. Bureau of Ships, PB 151987, December 1, 1958, revision with addenda, February 27, 1959.
28. See, for example, R. E. Vining, U. S. AEC Report NYO-2856 (1961); B. L. Greenstreet, *et al.*, U. S. AEC Report ORNL-3157 (1961).

PROBLEMS

1. A fuel rod, 90 in. long, for a pressurized-water reactor contains uranium dioxide pellets 0.294 in. in diameter clad with 0.021-in. thick stainless steel, with a small gap

between, which is initially filled with helium gas at 1 atm pressure. Determine the stresses in the cladding from both the thermal gradient and the internal pressure of the fission product gases. The maximum heat output of a rod is 11.65 kw/ft. Assume that 0.25 atom of stable fission gases are produced per fission and that 28 per cent of the gas formed is released from the fuel during 10,000 hr of operation to a "hot void" over the entire length of the rod having a volume of 0.20 in.^{3}.

2. A pressure vessel wall, consisting of 6-in. thick steel, is exposed on its inside surface to a gamma-ray flux of 5×10^{13} Mev/(cm^2)(sec). Use data given in the text to estimate the thermal stress developed; the energy absorption buildup factor may be taken as unity, assuming that the coefficient used has been adjusted appropriately.

3. Calculate the thermal stress in a steel flange 0.63 ft thick subject to a gamma-ray heat generation rate at the inside wall of 1.8×10^4 Btu/(ft^3)(hr), which decreases exponentially. The required physical properties of the steel will be found in the text.

4. Hydraulic loading must frequently be taken into consideration in stress analysis. In a given reactor, the coolant water flows upward through holes in a support plate at the bottom of the core at a velocity of about 25 ft/sec. The pressure loading may be taken as the difference between the impact pressure on the upstream surface and the flow-hole static pressure which may be considered to act on the downstream surface. Calculate the loading for a pressure drop of 2.0 ft of water with a support plate having a solid area of 1.26 ft^2.

Chapter 12

PRELIMINARY REACTOR DESIGN

GENERAL DESIGN PROCEDURES

INTRODUCTION

12.1. The design of a nuclear reactor system is a complex project involving a number of engineering disciplines. Although a detailed description of a design effort is too involved for inclusion in this book, an insight into the procedures will be given. A design example, in somewhat simplified form, will be used to illustrate the application of the basic concepts developed in this text. The particular approach used is consistent with the general treatment presented earlier and is not necessarily that which would be employed in actual practice. A highly experienced design organization might approach the task somewhat differently, but the fundamental principles used would be essentially the same [1].

12.2. A nuclear power plant includes a number of components similar to those found in nonnuclear power facilities or in other industrial operations, e.g., the chemical industry. Such items as heat exchangers, turbo-generators, switchgear, and water purification systems do not involve any different engineering principles, although they require special attention to certain aspects for nuclear applications. The present discussion will therefore be restricted to those features which are unique to reactor power plants, particularly the reactor core itself and components closely associated with it.

DESIGN PHASES

12.3. The design of a nuclear reactor system usually proceeds through several distinct phases. First, after objectives are defined and criteria established, a preliminary design may be prepared which outlines the principal features of the proposed reactor. At this stage it may be necessary to consider preliminary designs for several different concepts in order to develop sufficient information for a choice of design which can best meet the established criteria. If novel features are involved, some research and development may be justified in this early phase. The next stage will generally involve a preliminary design effort during which the principal features of the reactor and its associated plant are defined in sufficient detail to permit a fairly accurate cost estimate to be made. It is customary for the preliminary design to be detailed enough to permit the main problems to be recognized so that an indication can be obtained of the necessary development work.

12.4. The detailed design activities required for construction follow the usual engineering procedures. A consulting firm, experienced in nuclear reactor design, frequently works in close association with the architect-engineer responsible for detailed engineering and construction supervision.

12.5. This chapter will be largely concerned with the preliminary design aspects of the reactor core, its pressure vessel, and shielding. Where it is desirable to illustrate certain important principles, some calculations will be given based on an existing reactor design. Under normal design procedures certain of the results quoted might evolve at a later stage.

12.6. Before design criteria for the reactor can be established, even in a preliminary manner, decisions must be reached as to the main objectives which the reactor is expected to meet. The purpose for which the reactor system is required will, of course, be apparent from the outset. For example, it may be intended for naval propulsion, for a nonremote central station power plant, for a remote power plant, or for process heat. There is the possibility that the prime use of the reactor is not for power but for plutonium production or research. In every case, different design criteria and even different reactor concepts may be involved. The present chapter is based on the assumption that the reactor is required for a standard central-station power plant.

REACTOR DESIGN CRITERIA

12.7. Decisions concerning design objectives of a power reactor are primarily a function of management, whereas the establishment of design criteria is an engineering responsibility. The design objectives might well include the desired power output of the plant, although this may be subject to modification as the reactor criteria are established. First, a choice of reactor concept must be made, e.g., nature of fuel, moderator, coolant, and general configuration. The number of variations is large and, as will be seen in Chapter 13, it is unfortunately not possible at the present time to state that any one concept is definitely better than the others. In the United States, the choice has been narrowed down to some half dozen or so types and these are undergoing practical tests in large-scale plants. Until a definite conclusion has been reached, the choice is largely a matter of personal taste of the design group.

12.8. Suppose the concept selected is a pressurized-water reactor; that is to say, it is a heterogeneous (lattice) reactor in which ordinary water under pressure is to be used as moderator and coolant. For such a reactor, the fuel must be slightly enriched in uranium-235, and a decision has to be made concerning the degree of enrichment and the chemical form, such as metal alloy, oxide, carbide, etc., of the fuel. Then a choice must be made among different fuel-element geometries, the two most common being cylindrical rods and flat plates.

12.9. As is the case in other industries, extensive "parametric studies" are necessary before definite conclusions are reached even with regard to the preliminary design [2]. In the parametric calculations, an attempt is made to estimate the effect on the total power cost of changing various parameters, e.g., fuel enrichment, shape, size, and number of fuel elements, nature and thickness of cladding, ratio of fuel to moderator (in a thermal reactor), coolant temperatures and flow rate, pressure drop, etc. The selection of an optimum design is often rendered complicated by the interrelationships among the variables. Thus, the consequences of changing one parameter must be considered in the light of the effects on others (see Chapter 14). Furthermore, there are often practical problems which make it difficult, if not impossible, to realize the optimum design. Apart from economic aspects, there are two other matters which must be taken into account: these are concerned with stability and safety. Detailed analyses of a proposed design must be made to be sure that the system is stable (§ 5.200 *et seq.*) and that there are adequate safeguards against accidents that might arise during both normal and abnormal operation (§ 9.120 *et seq.*). The final choice of the preliminary design is thus usually a compromise based upon many different considerations.

12.10. With the basic criteria of the core determined, its dimensions and neutron flux distribution must be obtained quite early, as most of the preliminary (and conceptual) design is dependent upon these values. Since considerable iteration may be required before the core specifications are finally fixed, it is customary to establish some preliminary values to serve as a guide to persons responsible for other aspects of the design.

12.11. In order to simplify the present discussion, it will be postulated that the fuel elements in the pressurized-water reactor are to be rods composed of uranium dioxide pellets enclosed in stainless steel tubes. By means of calculations along the lines indicated below, the rod diameter, moderator-to-fuel ratio, etc., can be chosen and then parametric calculations made to establish the optimum core design. A preliminary estimate of the uranium-235 enrichment of the fuel can now be made, although it may be necessary to adjust it later to satisfy reactivity requirements (§ 5.149 *et seq.*). For large central-station power plants, intended to produce electricity as economically as possible, parametric calculations indicate that the enrichment should be between 2 and 4 per cent of uranium-235. If the reactor being constructed is not particularly novel, the designer will have recourse to a substantial body of literature describing theoretical studies, experimental results, and probably some parametric data which will greatly facilitate the establishment of the characteristics of the core. With these important constants prescribed, preliminary calculations can be made of the buckling and reflector savings so as to determine the approximate core dimensions.

12.12. If the reactor is similar to others with which the design group has had experience, it may be possible to omit the very preliminary design efforts. However, if the reactor core has novel features, a detailed study may be necessary to establish even preliminary design values. For the final neutron flux and lattice parameter calculations, it will probably be necessary to make use of a digital computer (§ 4.64 *et seq.*). The availability of codes, which require merely the insertion of certain variables, should be recognized in this connection. Since the purpose of the present treatment is to emphasize and illustrate basic principles, and to show how preliminary calculations can be made in a relatively simple manner, secondary attention only will be devoted to computing machine procedures.

PRESSURIZED-WATER REACTOR

FUEL-ELEMENT CHARACTERISTICS

12.13. In order to make the subsequent discussion fairly specific, a hypothetical pressurized-water reactor will be considered, since this is one of the basic types of power reactor (see Chapter 13) and has certain unique characteristics of interest to the design engineer. Where numerical examples are required, they will be based on the design values for the 110-Mw electrical (initial core) Yankee Atomic Electric Company reactor (abbreviated to "Yankee reactor") at Rowe, Mass. [3]. It should be mentioned that, although the fission energy in a pressurized-water reactor is mainly removed as sensible heat of the coolant water, a certain amount of localized (but not bulk) boiling is regarded as acceptable and has only a minor effect on the basic design considerations.

12.14. As seen in Chapter 4, a reactor moderated by ordinary water must use enriched fuel. The optimum degree of enrichment, from the overall economic standpoint, depends upon several factors; these include the size of the core, as determined by pressure vessel (§ 11.26) or other considerations, the operating thermal power, the moderator-to-fuel ratio, the diameter of the fuel elements, the nature of the cladding, the expected burnup (or life) of the elements, and fuel-cycle costs. The conversion of uranium-238 into fissile plutonium-239 is involved in some of these factors and must be taken into consideration. Both low enrichment and high enrichment have their advantages and disadvantages, and the final choice will be determined by a parametric study of all the variables. For a reactor of the type under discussion, the proportion of uranium-235 is likely to be below 4 per cent, possibly down to 2 per cent. With this degree of enrichment, the characteristic close-packed lattice with a water-to-uranium (element) volume ratio (§ 4.123) in the vicinity of 3 to 1 is usually employed, as explained in § 12.22 *et seq.*

12.15. To reduce the effects of radiation damage and the consequences of jacket (canning) failure in a water-cooled reactor, the preferred choice of fuel material at the present time is some form of high-density uranium dioxide clad with either stainless steel or a zirconium alloy. Uranium dioxide has the disadvantage of a relatively low heat conductivity, and investigations are proceeding with the object of finding better fuel materials (see Chapter 8).

12.16. The selection of the diameter of the fuel rods again involves a number of variables. If the rod diameter is small, the heat-transfer area per unit mass of fuel will be large and a high power density will be possible in the core. On the other hand, the increase in the number of fuel elements will mean an increase in fabrication costs. Another factor to be borne in mind is the effect of the rod diameter on the resonance escape probability and the associated change in the conversion of uranium-238 to plutonium-239.

12.17. From the viewpoint of heat transfer there are certain limitations to be considered. First, the central temperature in a cylindrical fuel element should not exceed a value at which undesirable physical changes may occur. Although the specification of a maximum temperature requires further research, a temperature of 4800°F, which is less than 200°F below the melting point of uranium dioxide, has been chosen for present purposes. Second, the heat flux at the interface between the fuel jacket and the water coolant should be sufficiently low so that the danger of burnout due to film boiling is avoided (§ 6.141). Furthermore, the temperature of the cladding must not be so high as to jeopardize its physical integrity [4].

12.18. The curves in Fig. 12.1 show the dependence of the maximum values of specific power, power density, and heat flux at the jacket-water interface for uranium dioxide fuel rods of various external diameters [5]. In each case there is an annular gap of 0.0015 in. between the fuel pellet and the zirconium jacket which has a wall thickness of 0.020 in.; the annular space is filled with helium to facilitate heat transfer. The maximum temperature at the center of the fuel is taken to be 4800°F, as stated above, and the ambient water temperature is 550°F. The volume ratio of water to uranium dioxide in the lattice is 2.5 to 1. As expected, the specific power (power per unit mass of fuel) and power density (power per unit core volume) increase as the diameter of the fuel rod, sometimes called a fuel "pin," is made smaller. However, the surface heat flux increases at the same time; and burnout considerations, as well as rapidly increasing fabrication costs, set a lower limit for a given water-to-uranium ratio.

12.19. Another factor involved in the determination of the optimum fuel-rod diameter is the cladding. The choice of cladding material is determined by its chemical reactivity, neutron capture cross section, and cost, and the thickness must be sufficient to provide structural support for the fuel material and to be able to withstand the pressure of the fission product gases, from the inside, and of the coolant, from the outside. In deciding the cladding thickness, allowance must be made for corrosion which will occur during the lifetime of the fuel element. Both the fission product gas pressure and corrosion will increase with time and so the cladding thickness will be related to the dimensions, including diameter, of the fuel rod. In general, it is to be expected—although it is not necessarily always the case—that the quantity of cladding per unit mass of fuel will increase with decreasing diameter of the fuel elements. In this event, the neutron poisoning due to the cladding will increase correspondingly. The advantages of rods of small diameter may thus be offset by the requirement for additional fuel or for higher enrichment.

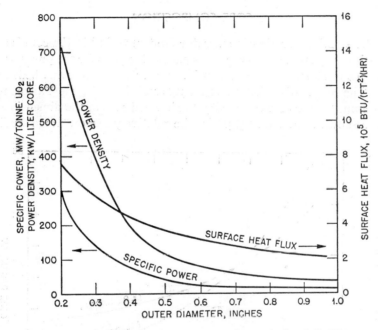

Fig. 12.1.　Dependence of core properties on fuel-rod diameter

12.20. In a pressurized-water reactor, the jacket material for the fuel elements may be a zirconium alloy, which has good neutron economy, i.e., low capture cross section, but is expensive. Stainless steel is less expensive, can be fabricated more easily, and appears to be somewhat more reliable, but it has a fairly large neutron capture cross section. Fabrication costs are generally quite high, even when steel is the jacket material, since the close-packed lattice requires rigid manufacturing tolerances. Precise spacing between fuel elements is necessary in order to avoid neutron flux peaks and regions of abnormally high temperature ("hot spots") where burnout could occur.

12.21. As the result of a detailed study, taking into account possible variations in core length, lattice spacing, hot-channel factors (§ 12.46 *et seq.*), etc., a value of 0.340 in. was finally chosen for the outer diameter of the fuel rods for the

Yankee reactor. Among the limitations which had to be kept in mind were the maximum temperature in the center of the hottest fuel element, the maximum value of $\int k \, d\theta$, i.e., the thermal power per unit length of fuel rod (§ 8.60), and the burnout flux. The jacket material is Type 348 stainless steel, which is similar to Type 347 (§ 7.47), and the wall thickness selected was 0.021 in. This value was based mainly on the need for maintaining dimensional stability of the cladding and the ability to withstand the external coolant pressure and the stresses arising internally from accumulated fission product gases.

CORE COMPOSITION

12.22. The optimum ratio of moderator (water) to fuel in the core is selected on the basis of the interaction of various parameters, but for the present purpose the problems can be illustrated by considering certain nuclear and heat-removal factors. The results of critical experiments carried out for the Yankee reactor, with uranium (2.5 per cent enriched*) dioxide fuel rods of the dimensions given above, are shown in Fig. 12.2 [6]. In the relatively narrow range studied, an

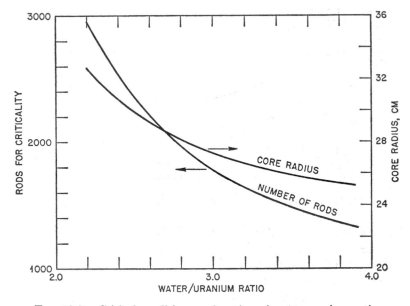

Fig. 12.2.   Critical conditions as function of water-uranium ratio

increase in the ratio of water to uranium is seen to decrease the number of rods of the specified size required for criticality and also the overall core radius. This result can be attributed partly to the improved moderation as the proportion of water is increased, with a corresponding increase in the fast neutron

* The value finally selected was an enrichment of 3.4 per cent uranium-235 (§ 12.28).

nonleakage probability. Beyond a certain value of the water-to-uranium ratio, the increase in neutron capture by the hydrogen, i.e., the decrease in the thermal utilization, will require an increase in the number of fuel rods and in the radius of the critical core, for the given enrichment. It should be noted, however, that minimum critical mass or size is not the only nuclear consideration. An increase in the ratio of water to fuel increases the resonance escape probability and, consequently, the infinite multiplication factor of the system. However, the accompanying decrease in the conversion of uranium-238 to fissile plutonium-239 will decrease the core life for a given initial mass of fuel.

12.23. Consider, next, the situation from the point of view of heat removal. For a fixed thermal flux, determined by burnout conditions, i.e., for a fixed power output per fuel rod, the power density of the reactor will decrease as the ratio of moderator to fuel is increased. A large power density is desirable, however, in order to minimize the reactor dimensions for a specified total power output. An optimization study is necessary, therefore, in order to determine the proper ratio of water to uranium for a given reactor. In the Yankee reactor, the value selected for this ratio was 3.0 to 1, to provide the best compromise between core size and specific power. The proportions of the principal materials in the reactor core, expressed in volume per cent in the cold condition, are given in Table 12.1. The zirconium is present in the "followers" attached to the lower ends of the control rods; they serve two purposes, namely, to reduce peaking of the neutron flux in the regions adjacent to the control-rod channels and to prevent excessive by-pass flow of coolant through these channels [7].

TABLE 12.1. PRINCIPAL MATERIALS IN YANKEE REACTOR COLD CORE

Material	Volume (per cent)
Uranium dioxide	35
Water	50
Stainless steel	12
Zircaloy	3

NUCLEAR DESIGN [8]

12.24. The development of a final core layout is an iterative process, involving a compromise between thermal and nuclear parameters. For simplicity, it will be assumed that the number of fuel rods of a specified size required is determined by the maximum permissible thermal (heat removal) flux. Recognizing that the core must have a reasonable length-to-diameter ratio—1.2 in the case of the Yankee reactor—a core layout can be based solely on heat-removal considerations. The results must, of course, satisfy nuclear reactivity requirements, but these are taken into account in the preliminary design in which the fuel enrichment is a variable. The actual core layout of the Yankee reactor, in which both thermal and hydraulic design aspects are taken into account, is shown in cross section in Fig. 12.3 (see also Fig. 11.4).

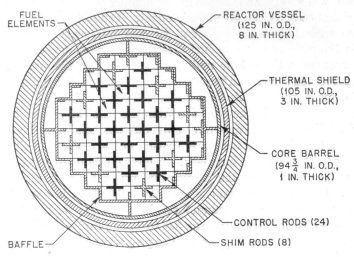

FUEL ELEMENTS

REACTOR VESSEL (125 IN. O.D., 8 IN. THICK)

THERMAL SHIELD (105 IN. O.D., 3 IN. THICK)

CORE BARREL ($94\frac{3}{4}$ IN. O.D., 1 IN. THICK)

CONTROL RODS (24)

BAFFLE

SHIM RODS (8)

FIG. 12.3.   Cross section of Yankee reactor showing fuel element and control rod arrangement

12.25. For the planned operating conditions, the maximum heat flux (see Fig. 12.1) is calculated to be about 4.46×10^5 Btu/(hr)(ft²), which is less than 50 per cent of the estimated burnout value (§ 6.141) of approximately 10^6 Btu/(hr)(ft²). In order to allow for local deviations from average behavior the maximum flux is reduced by the so-called "hot-channel" factor, which will be described later. In the present case, this factor is 5.17, so that the average heat flux is 8.63×10^4 Btu/(hr)(ft²).* The initial thermal power of the Yankee reactor is 392 Mw, i.e., 1.34×10^9 Btu/hr, so that the total surface area of the fuel elements must be

$$\frac{1.34 \times 10^9}{8.63 \times 10^4} = 1.55 \times 10^4 \text{ ft}^2.$$

The length of a fuel rod in the core is 90 in. and, since the outside diameter is 0.340 in., the area per rod is found to be 0.67 ft²; hence,

$$\text{Number of fuel rods required} = \frac{1.55 \times 10^4}{0.67} \approx 23{,}000.$$

12.26. The actual number of fuel rods in the Yankee reactor is 23,142. The rods are held in rigid subassemblies, and nine of these subassemblies are strapped together to form a complete fuel-element assembly. The basic pattern of each subassembly is six by six, but the total number of fuel pins is less than $9 \times 6 \times 6$,

* It may be noted that for preliminary calculations of pressurized-water reactor cores it is customary to take the average heat flux as 10^5 Btu/(hr)(ft²) as a first approximation.

i.e., 324, per complete assembly, because spaces must be left for the cruciform control rods (Fig. 12.3).   Thus, there are 76 fuel assemblies, half containing 304 fuel rods each and the other half 305 fuel rods.

12.27. The pitch (spacing) between the fuel rods, corresponding to the water-to-uranium ratio of 3.0, is 0.422 in.   The unit cell of fuel rod plus moderator thus has a cross section of $(0.422)^2 = 0.178$ in.^{2} or 1.24×10^{-3} ft^2.   The total cross section for each subassembly, allowing for the space for the control rod, is $36 \times 1.24 \times 10^{-3}$ ft^2, and for each fuel element it is nine times this amount, i.e., 0.402 ft^2.   The cross-sectional area of the 76 fuel elements is thus 30.6 ft^2, so that the corresponding diameter of a cylinder would be 6.2 ft.   The actual value for the Yankee core diameter is approximately 6.3 ft.

12.28. The reactor designer must now determine a fuel enrichment which will yield an effective multiplication factor (or excess reactivity) in the "cold, clean" condition, i.e., a freshly charged core at ordinary temperature, sufficiently high to compensate for the decrease in reactivity upon heating the core to the operating temperature, the Doppler effect loss, and the decrease during the life of the core due to depletion of fissile atoms and accumulation of fission product poisons. Some allowance must also be provided for the buildup of xenon and samarium. Taking into account the expected average lifetime of the fuel elements (§ 12.34) and the excess reactivity which can be compensated by control rods and burnable poison (boric acid), the necessary excess reactivity in the Yankee reactor was determined to be 20 per cent, as shown in Table 5.7.   In other words, the required effective multiplication factor is about 1.20 in the cold, clean condition. With the optimized core layout described above, it is then determined that a fuel enrichment of 3.4 per cent uranium-235 is required to satisfy the design conditions.

12.29. The approximate method described in § 4.117 *et seq.* for water lattices may now be used to check that the specifications given above will indeed give the required effective multiplication factor.   Since the details of the procedure were given in Chapter 4, it is sufficient to state here that, using the dimensions given above to calculate the core buckling, the effective multiplication factor for the assembly is found to be 1.18.   This is in satisfactory agreement with the more exact calculations indicating an excess reactivity of 20 per cent.

12.30. The excess reactivity at the operating temperature of the moderator (516°F) can be determined in the same manner as indicated above, utilizing nuclear constants corrected to this temperature.   The value of the effective multiplication factor at 516°F is found to be 1.13, and this leads to the average isothermal temperature coefficient of approximately -1.6×10^{-4} per °F given in § 5.83.   An additional temperature coefficient arises from the power Doppler effect (§ 5.84).   This is estimated to decrease the reactivity by 0.025 in going from the hot (516°F), clean, zero power, where k_{eff} is 1.13, to hot, clean, full power, so that k_{eff} in the latter condition is 1.105.

12.31. Finally, consideration must be given to the effect of xenon and samarium poisoning during operation. The average thermal-neutron flux in the Yankee reactor is 2×10^{13} neutrons/(cm^2)(sec) and hence the allowance for the poisoning is approximately 0.032 (see Table 5.5). The values of the effective multiplication factor for various core conditions are then as indicated in Table 12.2 with all control rods withdrawn. The average fuel burnup of 8000 Mw-days per tonne (cf. § 12.34) represents a total operating time of 10,000 hr at an initial thermal power level of 392 Mw. The excess reactivity necessary to compensate for losses during this period is calculated to be approximately 7 per cent.

TABLE 12.2. EFFECTIVE MULTIPLICATION FACTORS OF YANKEE REACTOR

Condition	k_{eff}
Cold, clean	1.20
Hot, clean, zero power	1.13
Hot, clean, full power	1.105
Hot, beginning of life, full power, equilibrium xenon and samarium	1.073

CONTROL RODS

12.32. The control rods, together with the burnable poison, must be adequate to compensate for the 20 per cent excess reactivity plus a shutdown margin when the reactor is cold and clean. In addition, there is a safety requirement for all reactor systems that the complete withdrawal of a single rod should not make the reactor critical. This means that shutdown should be possible even if one rod sticks completely in the "out" position. In considering the design of the control system for a reactor moderated by ordinary water, it should be recalled that the distance over which a control rod is effective is roughly equal to the diffusion length of the neutrons in the core (§ 5.158). In pressurized-water reactors, L is about 2 cm or less, so that many control rods must be distributed within the core.

12.33. Since it is desirable to minimize variations in the neutron flux throughout the core, as well as to maintain the regular, square lattice for the fuel rods, the cruciform type of control rod absorber element has been specified for many water-moderated reactors. The Yankee reactor has 24 movable control rods, made of silver-indium-cadmium alloy, plus eight "fixed" shim rods, of boron in stainless steel; the arrangement of the rods is shown in Fig. 12.3. The "fixed" rods are used for adjusting reactivity as required, but are not moved while the reactor is operating. This represents a practical compromise between the need for a large number of absorbing elements distributed uniformly throughout the core and the engineering requirement to reduce the number of complicated rod mechanisms to a minimum. The control rods are held by electromagnets during operation, and switching off the energizing currents, either deliberately or as the result of loss of power, will scram the reactor.

12.34. The 24 movable control rods are divided into six independent groups, arranged in a more or less concentric manner.   During operation of the reactor all the rods in any one group are withdrawn together in order to compensate for fuel burnup and the accumulation of fission products.   When one group is completely withdrawn, another group takes up the control function, in a definite, prescribed manner.   The purpose of the control rod programing is to attempt to make the fuel burnup as uniform as possible throughout the core.   However, because the rods are withdrawn upward, it is inevitable that burnup will be greatest in the lower part of the core.   The reactivity during operation is thus less in the lower than in the upper portion.   At the end of the estimated core lifetime (10,000 hr), the fuel consumption or burnup, expressed in terms of the megawatt-days of energy produced per tonne of uranium in the core (cf. Example 1.2 and § 12.99), is expected to average 8000 Mw-days per tonne; the corresponding maximum burnup, which will occur near the center of the core, will then be about 15,000 Mw-days per tonne.

12.35. In designing the control system, it is necessary to make detailed calculations of the reactivity equivalence (or worth) of the rods.   For preliminary design purposes, the approximate poison cell method described in § 5.159 *et seq.* may be used.   The data in Example 5.5 actually refer to the control rods of the Yankee reactor, and the result obtained shows that one control element will reduce the reactivity of the surrounding fuel super-cell by roughly 14 per cent. If the 32 rods in the Yankee reactor were distributed through the core so as to maintain a uniform flux, the total rod worth would be about 14 per cent.   Because of the approximate nature of the calculations, this result is somewhat low. A more accurate treatment shows that the reactivity equivalent of the control rods is actually close to 17 per cent.   This amount, together with 8 per cent of reactivity in the burnable poison, is sufficient to provide adequate safety during the whole of the core lifetime, under both hot and cold conditions.

FUEL MANAGEMENT PROGRAM

12.36. The average burnup attainable from the fuel can be increased by a suitable fuel management program.   As will be described later in this chapter, such a program involves the nonuniform distribution of fuel throughout the core with changes made from time to time to provide an optimum overall degree of burnup.   However, in considering such a program, the beneficial effects of increased burnup must be weighed against the additional cost involved in the required shutdown necessary for rearranging the fuel.   For the Yankee reactor, as is generally the case, a single-region (uniform) core has been specified for the initial startup.   Although this type of fuel loading is not the most efficient for obtaining maximum burnup or flattening of the power distribution, the simplicity of its theoretical analysis makes it a logical first step.   As operating experience is gained and information on the actual burnup is obtained, more efficient arrangements of the fuel will be tested.

FUEL SUBASSEMBLY DESIGN

12.37. Although the core layout has been used as the basis for the nuclear design described earlier, it is important to recognize that the mechanical design of the core is the subject of considerable engineering effort.   These mechanical features are closely related to both the nuclear and the thermal-hydraulic design (§ 12.40 *et seq.*).   For example, the specification of the stainless steel (or other material) for encapsulation of the uranium dioxide fuel pellets must take into consideration the necessity for keeping the amount of poison material in the core to a minimum.   At the same time, the wall thickness must be sufficient to withstand the more than 2000 psi external pressure of the water as well as the internal pressure of the fission product gases which accumulate during reactor operation.

12.38. In the Yankee reactor, stainless steel was chosen as the cladding because both the material and fabrication costs are less than those for Zircaloy-2. It was also considered to have greater reliability, especially in unforeseen circumstances.   Zircaloy-2 has the important advantage of a much lower thermal-neutron absorption cross section than stainless steel and the consequent possibility of lower fuel-cycle costs.   However, the use of zirconium alloys is not entirely free from problems, as indicated in § 7.56, although they can undoubtedly be overcome.

12.39. An important aspect of mechanical design is the method of joining the fuel rods to the subassemblies.   In addition to the poisoning effect of the material, which would necessitate an increase in fuel enrichment, consideration must be given to the need for minimizing the coolant pressure drop through the core.   In the Yankee reactor, boxes or cans containing the fuel assemblies, which normally provide structural support, have been eliminated.   The individual fuel rods are therefore constructed so as to have the necessary mechanical strength.   This open-lattice design, while reducing the amount of neutron poison, has the possible drawback of permitting redistribution of the coolant flow with changing resistances in the flow channels (§ 6.139).

THERMAL AND HYDRAULIC DESIGN

GENERAL CONSIDERATIONS

12.40. Since the function of a power reactor is to convert thermal energy generated in the core into useful power, the thermal design is of great importance. The attainable heat-removal rate from the fuel to the coolant frequently limits the core design and establishes such specifications as fuel rod diameter, number of rods, core diameter, etc., as in the example treated in § 12.24 *et seq.*   Furthermore, the specifications of various components of the reactor system outside the core are largely dependent on the thermal design.

12.41. In designing the heat-removal system for a reactor, it is necessary to bear in mind certain limitations imposed mainly by such factors as temperature, thermal stresses, and corrosion and erosion. For example, there are definite limits set upon the temperature in the interior of the fuel element, the surface temperature of the cladding, as well as upon the temperature of the coolant. In addition, the possibility of a limitation upon the rate of flow of the coolant, e.g., due to pressure drop through the reactor or to erosion, will have an important influence on the problem of heat removal.

12.42. The maximum surface temperature of the fuel-element cladding depends upon the nature of both the cladding material and the coolant.   There is always the danger of corrosion (and erosion) of the cladding by a liquid coolant and this may increase with temperature and irradiation.   The maximum permissible temperature is chosen, therefore, as that at which the attack is tolerable under the conditions, e.g., pressure, water purity, oxygen content, etc., existing in the reactor.   Another requirement, for liquid-cooled systems in general, is that the surface temperature should be below the saturation temperature of the coolant, e.g., water, at the existing pressure, in order to prevent extensive boiling where this is considered undesirable.   However, in most pressurized-water designs allowance is made for a certain amount of local boiling.   The maximum surface temperature may also be related to the temperature in the interior by the allowable thermal stresses in the fuel element, especially where large temperature fluctuations may occur or brittle material is present.

12.43. The surface temperature will, of course, have some influence on the maximum coolant temperature, but, in addition, factors external to the reactor core may have to be considered in connection with the latter.   The coolant temperature must be such that there will be no vapor binding of pumps, and no appreciable corrosion of vessels, piping, or other external equipment.   Furthermore, the overall change in temperature of the coolant between inlet and exit from the reactor may be limited by thermal stresses or distortions which may develop in the interior of the reactor, particularly as a result of sudden changes in power level or abnormal conditions in the coolant loop.

12.44. The flow of coolant, either water or liquid metal, at high velocity (and high temperature) is likely to cause erosion of fuel-element cladding in the reactor core, as well as of the tubes in the heat exchanger.   However, it seems that, as a general rule, the flow rate of the coolant in the reactor is limited by the friction pressure drop, which increases according to the square of the velocity (§ 6.154).   As a rough, order-of-magnitude indication it may be taken that the maximum flow rate for water at 450° to 650°F and for liquid sodium at 900° to 1000°F is about 25 to 35 ft per sec.

12.45. The methods of calculating some of the temperature gradients associated with the aforementioned design limitations have been discussed in

Chapter 6. For instance, a maximum fuel-element temperature may be calculated from the specified coolant temperature, following the procedure used in Examples 6.3 and 6.4. However, in fuel rods containing uranium dioxide, a clearance normally exists between the outside of the fuel pellet and the inner wall of the cladding, which is filled initially with an inert gas, e.g., helium. The dimensions of this clearance space will change with core temperature, while the composition of the gas will change as fission gases are released during burnup of fuel. This gap consequently introduces some complication in the calculation of the temperature gradient between the coolant and the center of the fuel.

HOT-CHANNEL FACTORS [9]

12.46. In a reactor employing a liquid coolant, there is a maximum heat flux above which departure from nucleate boiling will occur at the cladding surface. As mentioned in § 6.132, this condition may result in steam blanketing the surface, with a consequent low heat-transfer coefficient and a very high surface temperature which will eventually lead to burnout. A design criterion frequently used is based on the establishment of a maximum permissible heat flux in the so-called "hot channel" which is a fraction, generally about 50 per cent, of the calculated burnout flux. The relationship of the conditions at the "hot spot" in the hot channel to the average conditions specified for the core is the subject of the design procedure which depends upon the evaluation of the "hot-channel (or hot-spot) factors." This procedure is commonly used in thermal design to determine the expected deviations from the average behavior in a reactor core, regardless of the nature of the coolant.

12.47. The hot-channel concept is based on the assumption that a reactor, having solid fuel elements with coolant passages (or channels) between them, will have one channel in which a combination of dimensions and power density will produce a fuel temperature above that existing at any other place in the reactor core. This temperature represents a significant design limitation and so it defines a useful criterion.

12.48. The most important deviation from uniform thermal behavior arises from the variation of the thermal-neutron flux in both the radial and axial directions in the core. In addition to the cosine distribution expected for a bare, uniformly loaded reactor, the reflector and control elements introduce local perturbations. Since the flux variation can be calculated with reasonable accuracy, many reactor designers maintain that this variation should not be considered as a hot-channel factor. The uncertainty in calculating the flux distribution, however, should be introduced as one of the factors.

12.49. The various temperature differences which contribute to the maximum fuel temperature are as follows:

1. The temperature rise of the coolant.
2. The difference between the average coolant temperature (at a given height

in a specified channel) and that of the outer surface of the fuel element (or rod).

3. The temperature drop through the cladding.

4. The temperature drop within the fuel itself.

Deviations in these temperature distributions can result from (a) uncertainty in neutron flux and a maldistribution of fissile material and moderator, (b) a maldistribution of coolant flow arising from a number of different causes, e.g., deviations from nominal spacing dimensions, fuel-element eccentricities, surface roughness, warping, etc., (c) uncertainties in heat-transfer film coefficients, (d) uncertainties in thermal conductivity of the fuel, (e) uncertainty in fuel-element (or rod) dimensions, and (f) variations in power due to control instrument limitations.

12.50. A number of these factors lend themselves readily to calculation. For example, from manufacturing tolerances on the width of the spacing between fuel elements it is possible to determine the associated flow variation and resulting temperature effects. Where the coolant channels are in parallel, a spacing that is less than average in a particular channel will lead to a lower flow rate in that channel. This lower flow rate will affect both the temperature rise of the coolant and the temperature difference between the coolant and the fuel-element surface in the manner shown below.

12.51. For turbulent flow, with the pressure drop, Δp_f, being principally due to drag friction, it follows from equation (6.67) and § 6.157 that

$$\Delta p_f = 4f \frac{L}{D_e} \cdot \frac{\rho v^2}{2g_c},$$

where f is the (Fanning) friction factor, ρ and v are the density and velocity of the coolant, respectively, L is the length and D_e is the equivalent diameter of the flow channel (§ 6.111), and g_c is the force-to-mass ratio constant. If the channels are in parallel, the pressure drop through each channel must be the same, so that a comparison between a "hot" channel and a "nominal (or average)" channel leads to

$$\frac{v_{\text{hot}}}{v_{\text{av}}} = \left(\frac{f}{D_e}\right)_{\text{av}}^{0.5} \left(\frac{D_e}{f}\right)_{\text{hot}}^{0.5}. \tag{12.1}$$

12.52. For turbulent flow, the friction factor, according to equation (6.68), is approximately proportional to $\text{Re}^{-0.25}$, where Re is the Reynolds number, and hence to $(D_e v)^{-0.25}$, in view of the definition of Re by equation (6.50). Hence, equation (12.1) reduces to

$$\frac{v_{\text{hot}}}{v_{\text{av}}} = \left[\frac{D_{e(\text{hot})}}{D_{e(\text{av})}}\right]^{0.71}. \tag{12.2}$$

For a rectangular coolant flow channel, e.g., between parallel plate elements, $D_e \approx 2d$, where d is the spacing between the elements. In these circumstances, equation (12.2) becomes

$$\frac{v_{\text{hot}}}{v_{\text{av}}} \approx \left(\frac{d_{\text{hot}}}{d_{\text{av}}}\right)^{0.71}. \tag{12.3}$$

12.53. If the energy input per unit channel length is constant, the coolant temperature rise will vary inversely as the mass flow rate, $v\rho A$; hence,

$$\frac{\Delta t_{\text{hot}}}{\Delta t_{\text{av}}} = \frac{(vA)_{\text{av}}}{(vA)_{\text{hot}}}, \tag{12.4}$$

where A is the cross-sectional area of the flow channel, the density being taken as constant. For parallel plate elements, as postulated above, A is equal to the product of a constant channel width and the spacing d. Combination of equations (12.2) and (12.4) then leads to

$$\frac{\Delta t_{\text{hot}}}{\Delta t_{\text{av}}} \approx \left(\frac{d_{\text{av}}}{d_{\text{hot}}}\right)^{1.71}. \tag{12.5}$$

Example 12.1. In plate-type fuel elements, such as are used in the Materials Testing Reactor (MTR) and various other research reactors, the plate spacing is a nominal 0.100 in. with a maximum tolerance of ± 0.008 in. Determine the hot-channel factor for the coolant temperature rise.

The hot channel will be the one in which the spacing has the minumum value of $0.100 - 0.008 = 0.092$ in. Thus, taking d_{av} as 0.100 and d_{hot} as 0.092, it follows from equation (12.5) that

$$\frac{\Delta t_{\text{hot}}}{\Delta t_{\text{av}}} \approx \left(\frac{0.100}{0.092}\right)^{1.71} = 1.16.$$

The hot-channel factor for coolant temperature rise due to fuel plate spacing deviations is thus 1.16.

12.54. If the temperature difference between the coolant and the fuel-element surface, i.e., the film temperature drop, is assumed to be inversely proportional to the heat-transfer coefficient then, from the correlation given by equation (6.52),

$$h = \text{constant} \times \frac{v^{0.8}}{D_e^{0.2}},$$

the quantities c_p, k, ρ, and μ being included in the constant factor; hence,

$$\frac{\Delta t_{\text{hot}}}{\Delta t_{\text{av}}} = \frac{h_{\text{av}}}{h_{\text{hot}}} = \left(\frac{v_{\text{av}}}{v_{\text{hot}}}\right)^{0.8} \left(\frac{d_{\text{hot}}}{d_{\text{av}}}\right)^{0.2}$$

for a rectangular channel of spacing d. Upon introducing equation (12.3) this reduces to

$$\frac{\Delta t_{\text{hot}}}{\Delta t_{\text{av}}} = \left(\frac{d_{\text{av}}}{d_{\text{hot}}}\right)^{0.368}. \tag{12.6}$$

Example 12.2. Use the data in the preceding example to estimate the corresponding hot-channel factor for the temperature difference between the coolant and the fuel element surface.

From equation (12.6)

$$\frac{\Delta t_{\text{hot}}}{\Delta t_{\text{av}}} = \left(\frac{0.100}{0.092}\right)^{0.368} = 1.03.$$

The hot-channel factor for the film temperature drop due to spacing deviations is thus 1.03.

12.55. A number of other hot-channel factors have been employed in the thermal design of a reactor core. Sufficient information has been given here to indicate the general procedure for calculating such factors. In each case it is necessary to make an estimate of the maximum uncertainties or deviations in the particular parameter under consideration. From this, the appropriate hot-channel factor can be determined.

12.56. Since the hot-channel concept takes into consideration uncertainties of various types, including that of understanding the problems, it is probable that the magnitude of many of the factors will be reduced as design experience is gained and improvements are achieved in fabrication techniques. In this event, it will be possible to increase the output from a reactor of given size, thereby reducing the cost of the power produced.

PRODUCT METHOD FOR HOT-CHANNEL FACTORS

12.57. The simplest, but not necessarily the best, procedure for the application of hot-channel factors to the thermal design of a reactor core utilizes the "factor product" method. In this approach, the overall hot-channel factor is taken as being equal to the product of the individual factors. Thus, every deviation is given equal weight.

12.58. For purposes of illustration, the analysis of the Yankee reactor will be used. The effects of various deviations are related to three hot-channel factors

TABLE 12.3.   HOT-CHANNEL FACTORS

Fuel Rod Characteristics	$F_{\Delta t}$	F_q	F_θ
Pellet diameter	1.002	1.003	1.003
Pellet density	1.027	1.05	1.05
Pellet enrichment	1.011	1.022	1.022
Rod diameter, pitch, and bowing	1.097	—	1.134
Coolant Flow Characteristics			
Flow distribution in plenum	1.07	—	1.07
Flow distribution due to local boiling	1.05	—	1.05
Product of engineering factors	1.28	1.08	1.37
Nuclear Factors			
Local power peaking	1.3	1.3	1.3
Overall maximum to average	1.8	2.9	2.9
Product of nuclear factors	2.4	3.8	3.8
Overall hot-channel factor	3.1	4.1	5.2

concerned with coolant enthalpy increase, heat flux, and film temperature drop, respectively. These factors are defined as follows:

$F_{\Delta t}$ = Ratio of maximum to average coolant enthalpy increase
F_q = Ratio of maximum to average heat flux
F_θ = Ratio of maximum to average film temperature drop.

The components of these three factors, as used in the design of the first core of the Yankee reactor, are listed in Table 12.3 [10]. The factor $F_{\Delta t}$ is an integrated effect over the entire channel length, and so it was estimated by using average deviations from the nominal channel dimensions. On the other hand, F_q and F_θ were determined from the total difference between minimum and maximum values in order to take into account the possibility of the condition at any point being outside the specified design tolerance. This was considered to be a conservative, but not unrealistic, approach because it is difficult to predict accurately the dimensional variations of fabricated components. As further experience is gained, it is possible that some of the individual hot-channel factors may be decreased.

12.59. The hot-channel factor used for design purposes is the largest of the three overall factors, namely 5.2. This is the value (actually 5.17) given in § 12.25 which, together with the desired heat output of the reactor, was used in the preliminary calculation of the number of fuel rods required and the core diameter. The other two hot-channel factors in Table 12.3 may be utilized to determine whether or not the degree of local boiling permitted is within tolerable limits and whether the pressure drop characteristics of the coolant parallel flow are stable between the hot channel and other channels. It might appear, at first sight, that the overall hot-channel factors are surprisingly large. It should be noted, however, that the main contribution is made, in each case, by the ratio of maximum to average power along the length of a fuel element. This is a direct consequence of the particular (uniform) fuel distribution in the first core of the Yankee reactor. With a different distribution, the ratio of maximum to average power could be made closer to unity and the overall hot-channel factors would be reduced accordingly.

STATISTICAL METHOD FOR HOT-CHANNEL FACTORS

12.60. The factor-product method described above assumes that there is at least one channel in which all possible adverse deviations occur. The probability that this will be the case is extremely small, and so the design based on the product of the hot-channel factors may be unnecessarily conservative. A more logical procedure would be to make use of statistical considerations to determine the probability of the occurrence of various deviations and to design the heat-removal system accordingly. This particular approach is based on the assumption that the significant hot-channel factors arise from variables which are statistical in nature, as appears to be the case.

12.61. In applying the statistical method, it is first necessary to decide upon the number of standard deviations appropriate to each individual hot-channel factor. Thus, one standard deviation, σ, corresponds to a probability of 16 per cent that the actual deviation will exceed σ; two standard deviations imply a probability of 2.3 per cent, and three standard deviations 0.13 per cent, i.e., 1.3 chances in 1000, that the deviation will be greater than σ. Since a degree of conservatism in the design of a reactor is necessary, it will be postulated that the actual deviations represent two standard variations, i.e., 2σ. Suppose, for example, that the estimated hot-channel factor for the maldistribution of coolant flow is 1.12; the total deviation is 0.12 and so the standard deviation is 0.06. Coolant flow distribution will mainly affect the coolant temperature rise; if this is 300°F, then the standard temperature deviation, σ_t, will be $300 \times 0.06 = 18$°F. The same considerations are applied to all the uncertainties, such as those listed in Table 12.3, and the temperatures affected by such deviations, e.g., film temperature drop, temperature drop in cladding and in the fuel, as well as the change in coolant temperature. The various σ_t's are squared, to give σ_t^2, and then added together; the square root of the sum then yields the overall standard temperature deviation.

12.62. It is now necessary to decide the degree of confidence which can be placed upon this result. As a general rule, again in order to be conservative, the actual temperature deviation at the "hot spot" is taken to be three times the standard deviation. Thus, in a particular set of calculations, the standard temperature deviation was found to be 137°F. If the temperature in the fuel, without allowance for uncertainties, is estimated to be 1000°F, then the probability is only 1.3 in 1000 that the actual temperature will exceed $1000 + (3 \times 137) = 1411$°F. The application of the factor-product method to the same data indicated a possible hot-channel temperature of 2117°F.

12.63. Although the statistical treatment of hot-channel factors is analytically sound, there is still the problem of assessing the appropriate number of standard deviations applicable to each situation. If a very conservative approach is adopted throughout, the final results may not differ greatly from those given by the factor-product method. Some engineers feel that an uncertainty of one in a thousand is still too large in designing a reactor core. Most heat removal systems, to date, have therefore been designed on the assumption that all possible deviations could occur in one channel, but it is probable that the situation will change as greater experience is gained in reactor construction and operation.

12.64. A more rigorous approach to the statistical evaluation of hot-channel factors is possible in connection with the fuel assembly sub-factors. The core performance is affected by variations in the uranium-235 enrichment, the weight of the fuel pellets, the outer diameter of the fuel rod, and the fuel-rod spacing. Sample measurements have shown that these variations have normal distributions and from these the appropriate standard deviations can be determined. By using conventional statistical procedures, it is possible to calculate the probability that particular hot-channel factors will be exceeded [11].

HYDRAULIC DESIGN [12]

12.65. A study of the hydraulic resistance of the reactor system is necessary in order to determine specifications for pumps to provide the desired coolant flow. Furthermore, the interrelationships between the pressure drops, flow distribution, and mixing determines the temperature pattern in the core. Finally, such matters as the possibility of a cold-water accident (§ 9.110) and the potential for heat removal at emergency shutdown are sensitive to the core hydraulic characteristics.

12.66. In the reactor vessel shown in Fig. 11.4, the water flows in through four inlet nozzles, then downward past the core barrel (which serves to confine the coolant flow within the core region) and thermal shield into a lower plenum. It then flows up through the core, past fuel assembly and core support plates, into an upper plenum and leaves by four outlet nozzles. The calculation of the total pressure drop thus involves a total of 13 stages, in some of which allowance has to be made for losses due to changes in flow direction as well as to various obstructions.

12.67. Pressure drop calculations for the hydraulic system are made along the lines indicated in Chapter 6. However, because of the three-dimensional flow in and out of the plenum regions and other complexities of the flow pattern in the reactor vessel, model studies are desirable to confirm the validity of the calculations. These were carried out for the Yankee reactor and the results served as a basis for some of the assumptions made in the final estimates upon which the

TABLE 12.4.   PRESSURE DROPS IN YANKEE REACTOR VESSEL

Region	Pressure Drop (psi)	
Inlet nozzles	0.02	
Inlet annulus	4.31	
Core baffle and thermal shields	4.17	
Losses in lower plenum	3.48	
Lower core support plate	1.74	
Lower fuel assembly end plates	1.90	
Fuel assemblies in core	7.35	Core 15.1
Upper fuel assembly end plates	2.69	
Upper core support plate	1.40	
Upper plenum	4.38	
Outlet nozzle	1.63	
Total	33.1	

hydraulic design was based. The results of the pressure drop calculations are given in Table 12.4. The total pressure drop across the reactor vessel was found to be 33 psi, with 15 psi of this across the core, including the various supports.

THERMAL AND HYDRAULIC DESIGN DATA FOR YANKEE REACTOR

12.68. The thermal and hydraulic design data for the initial core of the Yankee reactor are given in Table 12.5. All the dimensions are for the initial cold core.

TABLE 12.5.　THERMAL AND HYDRAULIC DATA

Total (initial) heat output $\begin{cases}\text{Btu/hr} \\ \text{Mw}\end{cases}$	1.338×10^6	
	392	

Coolant Flow

Core flow rate, lb/hr	34.0×10^6
Flow area in core cross section, ft²	15.4
Velocity along fuel rods, ft/sec	14.0

Pressure

Operating pressure (normal), psig	2000
Design pressure, psig	2500
Pressure drop across reactor vessel, psig	34
Pressure drop across core, psig	16

Heat Transfer

Heat-transfer surface, ft²	15,500
Average flux, Btu/(hr)(ft²)	86,300
Maximum flux, Btu/(hr)(ft²)	446,000
Average heat-transfer (film) coefficient, Btu/(hr)(ft²)(°F)	6050

Temperature

Coolant in core (average) °F	516
Coolant at vessel inlet, °F	499
Coolant rise in core (average), °F	33
Film drop (average), °F	14.3
Fuel-rod jacket surface (maximum), °F	663
Center of fuel (maximum), °F	4330
Outlet of hot channel, °F	603

Steam

Temperature, °F	475
Pressure, psig	525

SHIELD DESIGN

SHIELDING CRITERIA

12.69. As shown in Chapter 10, a biological shielding system around the reactor is necessary to reduce the radiation to tolerable levels for operation and maintenance functions. The shield must be capable of absorbing both fast and thermal neutrons as well as primary and secondary gamma rays. To accomplish the desired degree of attenuation, the shielding system normally consists of two main components, namely, thermal and biological shields, although each of these may have several components. Before attempting the design of the shields, however, it is necessary to establish the criteria which the shielding system is expected to meet. The proposed criteria for the Yankee reactor plant are given in Table 12.6.

TABLE 12.6. SHIELDING CRITERIA FOR YANKEE REACTOR PLANT

Location	Dose Rate (mrem/hr)
Continuous working stations	0.75
Intermittent working stations	2.0
Short-term exposure (e.g., during fuel transfer operations)	16.0

12.70. The thermal shield system for the Yankee reactor consists essentially of three parts, namely, the core barrel, which is of steel 1-in. thick; the main thermal shield, 3 in. in thickness; and finally the reactor pressure vessel itself with a wall thickness of 7.9 in. Outside the pressure vessel is 3 ft of water, contained in a tank, to serve as a neutron shield; it is designed to reduce the neutron flux to 1×10^3 fast neutrons/(cm²)(sec) and 5×10^3 thermal neutrons/(cm²)(sec). The water tank is surrounded by a wall, approximately 5 ft thick, of ordinary reinforced concrete, which attenuates both neutrons and gamma radiation down to the level of the main coolant system. A secondary shield of concrete, immediately inside the outer containment vessel (Table 11.1), also approximately 5 ft thick, surrounds the entire plant. The arrangement

TABLE 12.7. SHIELD DIMENSIONS

Region	Inner Radius (cm)	Thickness (cm)
Reflector (water)	97.1	20.7
Core barrel (steel)	117.8	2.6
Water layer	120.4	5.3
Thermal shield (steel)	125.7	7.6
Water layer	133.3	5.1
Pressure vessel (steel)	138.4	20.0
Water layer	158.4	91.6
Concrete	250	152

of the various materials in a radial direction from the core, with dimensions in centimeters, is shown in Fig. 12.4, and the values summarized in Table 12.7.*

* A 1.2-cm thick baffle between the core and the reflector has been neglected, since it is of little significance in the present connection.

APPROXIMATE ATTENUATION CALCULATIONS

12.71. The calculation of the attenuation of gamma rays and neutrons by a composite laminated shield, such as the one described above, is a very lengthy and specialized procedure. In the present treatment all that will be done is, first, to make an estimate of the total gamma-ray energy incident upon the inner wall of the reactor pressure vessel, since this is required to determine the thermal stress resulting from gamma-ray absorption in the walls, and second, to make an order-of-magnitude estimate of the neutron flux at the inner face of the concrete shield. Both of these calculations give results that are reasonably good in view of the relatively simple methods used.

12.72. The primary gamma flux incident on the pressure vessel is obtained by means of a procedure similar to that described in Chapter 10 but with considerable simplification which is adequate for the present purpose. It should be emphasized, however, that this treatment should not be used in connection with shield design for the determination of gamma-ray dose rates. The same considerations apply to the simplified calculation of the secondary gamma rays which is described below. Instead of using diffusion theory (§ 10.115) to determine the neutron fluxes, the values are obtained from published data on similar water-steel shields (Fig. 10.9). The appropriate secondary gamma flux, resulting from the capture of thermal neutrons in the water and steel, is estimated in the following manner. If the layer (or slab) of material is thin, e.g., less than one relaxation length, all the secondary gamma rays produced in that layer are assumed to originate at the midplane. For thicker layers, an isotropic surface source at the outer boundary, equivalent to the volume source, is calculated by means of equation (10.28).

12.73. The attenuation of the flux by the water and steel, up to the pressure vessel surface, is then calculated for each layer separately in the same manner as is used for the primary gamma radiation. Only photons of 2.2-Mev energy result from the capture of thermal neutrons in water, but the capture gamma rays in steel cover a range of energies. However, the most significant are those with energies from 5.9 to 7.6 Mev, and it is sufficiently accurate for preliminary calculation of the heating in the pressure vessel to assume a single energy value of 7 Mev. Because of the approximations employed, extension of the computations beyond the interior wall of the pressure vessel is not justified.

12.74. The attenuation of fast neutrons in the water and steel layers between the core and the inner face of the concrete shield is estimated by using fast-neutron removal theory, as outlined in § 10.101 *et seq.* Here again, further extension of the calculations does not appear to be justifiable.

Primary Gamma Rays

12.75. To determine the attenuation of the primary gamma rays, i.e., all the gamma photons arising in the core, it is necessary first to calculate the equivalent surface source strength, using equation (10.28). The thermal power of the reactor under consideration is 392 Mw, i.e., 3.92×10^8 watts: hence, since there are 3.1×10^{10} fissions per sec per watt,

$$\text{Fission rate} = 3.92 \times 10^8 \times 3.1 \times 10^{10} = 1.22 \times 10^{19} \text{ fissions/sec.}$$

The cylindrical core has a diameter of 6.3 ft and a length of 7.5 ft, so that

$$\text{Core volume} = \tfrac{1}{4}\pi(6.3)^2(7.5) \text{ ft}^3 = 6.57 \times 10^6 \text{ cm}^3.$$

The total energy from both the prompt (or instantaneous) gamma rays emitted in the fission process and the gamma radiation accompanying the decay of the fission products is about 14 Mev (Table 10.6). In addition, roughly 6 Mev per fission are liberated in the form of capture gamma rays and decay gamma rays of the products resulting from neutron (nonfission) capture within the core. Each fission thus yields a total of about 20 Mev gamma-ray energy in the reactor core, and the volume source strength is consequently

$$S_v = \frac{(1.22 \times 10^{19})(20)}{6.57 \times 10^6} = 3.7 \times 10^{13} \text{ Mev/(cm}^3)\text{(sec).}$$

12.76. The energies of the core gamma rays are spread more or less evenly over the range from 1 to 8 Mev, and, for the present preliminary heating calculations, it may be assumed that all the photons have an energy of 4 Mev. For this energy, an average value of the mass attenuation coefficient for the core materials is about 0.032 cm^2 per g. From the data in Table 12.1, the mean density of the core at 515°F, the approximate average temperature during operation, is estimated to be 5.0 g per cm^3. Consequently, the value of μ_v to be used in equation (10.28) may be taken to be $(5.0)(0.032) = 0.16$ cm^{-1}. The surface source strength equivalent to the volume source calculated above is then

$$S_a = \frac{S_v}{\mu_v} = \frac{3.7 \times 10^{13}}{0.16} = 2.3 \times 10^{14} \text{ Mev/(cm}^2)\text{(sec).}$$

12.77. To determine the primary gamma-ray energy flux incident on the pressure vessel, equation (10.21), which is applicable when the buildup factor is unity, will be modified to take the form

$$\phi = \tfrac{1}{2}BS_aE_1(\mu z), \tag{12.7}$$

where B is the buildup factor appropriate to the situation, and

$$\mu z = \mu_1 z_1 + \mu_2 z_2,$$

where μ_1 and μ_2 are the linear attenuation coefficients for water and iron, respectively, and z_1 and z_2 are the total thicknesses of the water and iron layers between the reactor and the pressure vessel, i.e., 31.1 cm and 10.2 cm (Table 12.7).

12.78. Assuming an average photon energy of 4 Mev, μ_1 is 0.027 cm^{-1} and μ_2 is 0.27 cm^{-1} at the reactor temperature of 515°F; then

$$\mu z = (0.027)(31.1) + (0.27)(10.2) = 3.6.$$

The buildup factors for $\mu z = 3.6$ at a photon energy of 4 Mev are not very different for water and iron, but the larger of the two, namely that for water, will be chosen here since this will yield a slightly higher energy flux. The appropriate value of B is then 3.0, and so equation (12.7) gives

$$\phi = \tfrac{1}{2}(3.0)(2.3 \times 10^{14})E_1(3.6) = 2.1 \times 10^{12} \text{ Mev}/(\text{cm}^2)(\text{sec}),$$

since $E_1(3.6)$ is 6.2×10^{-3}. This flux is based on the assumption of an infinite plane source; to transform the result to a finite cylindrical source, use is made of equation (10.8), so that

$$\phi = 2.1 \times 10^{12} \left(\frac{97.1}{138.4}\right)^{1/2} = 1.8 \times 10^{12} \text{ Mev}/(\text{cm}^2)(\text{sec}).$$

This gives the required primary gamma energy flux incident on the pressure vessel.

Secondary Gamma Rays

12.79. As stated above, the approximation is made of assuming that the secondary photons are all produced, by thermal neutron capture, at the midplane of the appropriate layer of water or iron, provided the thickness is less than one relaxation length. For the present calculation of the secondary gamma-ray dose at the pressure vessel it is only the thermal shield that does not satisfy this requirement. However, in order to calculate the volume source strength, the thermal-neutron flux at the midplane of this layer is required as well as the values for the other layers. The appropriate fluxes are derived by comparison with Fig. 10.9.

12.80. The average thermal-neutron flux in the core of the Yankee reactor is 2×10^{13} neutrons/$(\text{cm}^2)(\text{sec})$ and the fast flux is somewhat higher. Using the procedure to be described shortly (§ 12.85), the attenuation of the latter can be estimated and the thermal-neutron variation adjusted to be in agreement. The approximate values used for the various layers preceding the pressure vessel are as follows:

Midplane of	Thermal Flux (ϕ) (Neutrons/$(\text{cm}^2)(\text{sec})$)
Reflector (water).................	3×10^{13}
Core barrel (steel)...............	3×10^{12}
Water layer......................	1×10^{12}
Thermal shield (steel)............	5×10^{11}
Water layer......................	3×10^{11}

The rate of production of secondary gamma photons (per cm² per sec) in any layer is then equal to $\Sigma_a \phi z$, where Σ_a is the appropriate macroscopic absorption cross section for thermal neutrons at 515°F (0.0117 cm^{-1} for water and 0.186 cm^{-1} for steel), ϕ is the thermal flux at the midplane, and z is the total thickness of the given layer.

12.81. For example, for the reflector layer,

Rate of secondary photon production $= (0.0117)(3 \times 10^{13})(20.7)$
$$= 7.2 \times 10^{12} \text{ photons/(cm}^2\text{)(sec)}.$$

Since these are produced by capture in hydrogen, the photon energy is 2.2 Mev, and so the surface source strength of secondary gamma rays at the midplane of the reflector is

Surface source strength $= 1.6 \times 10^{13}$ Mev/(cm²)(sec).

This is now treated in a manner exactly equivalent to that described above for S_a in calculating the primary gamma rays incident on the pressure vessel.

12.82. At the reactor temperature of 515°F, the linear attenuation coefficients for the 2.2-Mev secondary gamma rays produced in water and for the 7-Mev (average) gamma rays originating in the steel are as follows:

LINEAR ATTENUATION COEFFICIENTS

Photon Energy	Water	Iron
2.2 Mev	0.037 cm⁻¹	0.32 cm⁻¹
7	0.022	0.24

12.83. For the thermal shield, the effective uniform volume source strength of secondary gamma rays is taken to be $\Sigma_a\phi$; the linear attenuation coefficient for 7-Mev gamma radiation in iron is 0.24 cm⁻¹ and this is consequently the value of μ_v. The equivalent isotropic source strength is thus given by equation (10.28) as

$$S_a = \frac{S_v}{\mu_v} = \frac{(0.186)(5 \times 10^{11})}{0.24}.$$

Since it is only the gamma-ray flow in the outward direction which is required for the present calculations, it follows that

Surface source strength $= 2.0 \times 10^{11}$ photons/(cm²)(sec).

12.84. The results of the calculations of the secondary gamma rays incident upon the inner wall of pressure vessel, originating in various layers, are summarized below.

Originating Layer	Buildup Factor	Energy Flux at Pressure Vessel
Reflector	4.2	1.0×10^{11} Mev/(cm²)(sec)
Core barrel	2.2	2.4×10^{11}
Water layer	3.6	3.9×10^{9}
Thermal shield	1.0	1.0×10^{12}
Water layer	1.0	0.3×10^{11}
Total		1.4×10^{12} Mev/(cm²)(sec).

The incident energy flux of secondary gamma rays is thus 1.4×10^{12} Mev/(cm²)(sec).*

Fast Neutrons

12.85. The number of neutrons produced, on the average, per fission in the slightly enriched uranium is close to 2.5; using the data in § 12.75, the volume source of fission neutrons is

$$S_v = \frac{(1.22 \times 10^{19})(2.5)}{6.57 \times 10^6} = 4.6 \times 10^{12} \text{ neutrons/(cm}^3\text{)(sec)}.$$

The macroscopic removal cross section in the core can be calculated from the known core composition and the cross sections in Table 10.5; the result, which is equivalent to μ_v, is found to be 0.103 cm⁻¹. Hence, the equivalent isotropic surface source strength, based on an infinite volume source, is

$$S_a = \frac{S_v}{\mu_v} = \frac{4.6 \times 10^{12}}{0.103} = 4.5 \times 10^{13} \text{ neutrons/(cm}^2\text{)(sec)}.$$

12.86. The attenuation of fast neutrons in the composite shield consisting of alternate layers of iron and water is calculated by using fast-neutron removal theory. An infinite (isotropic) plane source of strength 4.5×10^{13} neutrons/(cm²)(sec) is assumed with attenuation by infinite slabs; the results are then corrected to a cylindrical source in the usual manner. In the following approximate treatment, the fast-neutron flux incident on the concrete shield is estimated.

12.87. To the degree of accuracy which is possible in the following calculation, the iron-water shield shown in Fig. 12.4 may be treated as consisting simply of $2.6 + 7.6 + 20 = 31.2$ cm $(= z)$ of iron and $20.7 + 5.3 + 5.1 + 91.6 = 122.7$ cm $(= x)$ of water, the departure of the density from the normal value being neglected. In order to utilize equation (10.37), it is first necessary to determine $\phi_{H_2O}(x)$ and this may be derived from equation (10.33) with S_v/μ_v replaced by S_a, which was calculated in § 12.85. Hence, for an infinite plane source, the fast-neutron flux is given by

$$\phi(z, x) = \tfrac{1}{2}S_a[AE_1(ax) + (1 - A)E_1(bx)]E_1(\Sigma_r z),$$

where Σ_r for iron is 0.17 cm⁻¹ for iron (Table 10.5); A is quoted in § 10.98 as

---

* A more exact calculation of the flux from an infinite-slab volume source of thickness t and an absorber of thickness z (attenuation coefficient μ) makes use of the relation

$$\phi = \frac{BS_v}{2\mu_v} [E_2(\mu z) - E_2(\mu z + \mu_v t)],$$

where B is an appropriate buildup factor. The result obtained with this equation is within less than 10 per cent of that given above.

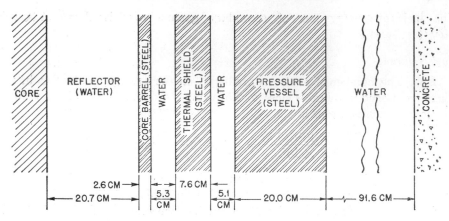

FIG. 12.4.   Dimensions of steel-water shield of the Yankee reactor

0.892, a is 0.129 and b is 0.091. Thus, for the infinite plane source, the fast-neutron flux at the surface of the concrete would be

$$\phi(z, x) = \frac{4.5 \times 10^{13}}{2} [0.892\ E_1(15.8) + 0.108\ E_1(11.2)]E_1[(0.17)(31.2)]$$

$$= 3.0 \times 10^3 \text{ neutrons/(cm}^2)(\text{sec}).$$

The correction factor for a finite cylindrical source is given by equation (10.8) which, in the present case, is $(97/250)^{1/2}$, i.e., roughly 0.6. The actual fast-neutron flux is thus approximately 1.8×10^3. The fact that this is nearly twice the design value of 10^3 fast neutrons/(cm^2)(sec) given in § 12.70 is an indication of the kind of accuracy obtainable by calculations of this type.

MECHANICAL DESIGN

PRESSURE VESSEL

12.88. In addition to the mechanical features of the internal structure of the core, a large amount of additional "hardware" and structure components are present in the reactor pressure vessel which must receive design attention. A partial indication of the complexity can be obtained from Fig. 11.4. It is seen that there are many components which represent a major design effort if the reactor system is to operate efficiently. Although attention here will be confined to some general aspects of pressure vessel design, the magnitude of the overall problem should be appreciated.

12.89. The design of a reactor pressure vessel, the basic requirements of which were outlined in § 11.65 *et seq.*, is complicated by the presence of thermal stresses arising from heat deposited as a result of gamma-ray absorption and the slowing down of fast neutrons. The introduction of one or more thermal shields between the reactor core and the inner wall of the vessel helps to alleviate this situation. Although the thermal shields are subject to thermal stresses, as a result of the temperature gradients arising primarily from the exponential absorption of gamma rays, they are not subject to any pressure. Furthermore, flow of the coolant past the shields serves to minimize the temperature difference between inner and outer walls. Two thermal shields separated by a layer of coolant are thus preferable to one shield of the same total thickness. As mentioned earlier, in the Yankee reactor the so-called core barrel, which is close to the core, also serves as a thermal shield.

12.90. For design purposes, the thermal shields and pressure vessel must be considered together and an iterative approach, guided by engineering practice, is employed to minimize both thermal stresses and overall cost. As a starting point, a pressure vessel thickness is specified to meet the requirements of an accepted code (§ 11.67), with some consideration being given to the presence of thermal stresses. With the temperature of the coolant inside the vessel specified, and the thermal loss properties known for the insulation on the outside, a calculation can be performed to determine the thermal stress which can be tolerated. From this, a permissible heat generation rate in the reactor wall can be determined by reversing calculations of the type given in § 11.57 *et seq.* By evaluation of the radiation emitted from the reactor core, using normal shielding calculation procedures, a thermal shield system can be designed.

12.91. It may be assumed that such computations were made in choosing the thicknesses of the core barrel and the thermal shield given earlier. It is of interest, therefore, to see if these shields, together with the water in the reactor, do indeed attenuate the gamma radiation to such an extent that the thermal stress in the walls of the pressure vessel are well within the tolerance limit. The primary gamma energy flux incident on the inside of the vessel is 1.8×10^{12} and the secondary flux is 1.4×10^{12} Mev/(cm^2)(sec), as estimated in § 12.78 and § 12.84, respectively. The energy absorption coefficient of iron, which is required for these calculations, does not vary greatly in the photon energy range from 1 to 10 Mev, and for the present purpose a constant value of 0.18 cm^{-1} may be used with a buildup factor of unity. In these circumstances, it is permissible to add the primary and secondary gamma-ray energy fluxes, giving a total of 3.2×10^{12} Mev/(cm^2)(sec) incident on the vessel. With a buildup factor of unity, the rate of gamma energy absorption at the surface is $\phi_\gamma \mu_e$ (cf. § 10.164), where ϕ_γ is the gamma energy flux, i.e., $(3.2 \times 10^{12})(0.18) = 5.8 \times 10^{11}$ Mev/(cm^3)(sec). This is equal to the volumetric heat source Q_0 at the inner wall of the pressure vessel. Since 1 Mev is equivalent to 1.6×10^{-6} erg or 1.6×10^{-13} watt-sec, it follows that Q_0 is 0.093 watt per cm^3 or 8.3×10^3 Btu/(hr)(ft^3).

12.92. More accurate calculations indicate that for the Yankee reactor pressure vessel, Q_0 is 8.63×10^3 Btu/(hr)(ft^3), and this was the value postulated in Examples 11.3 and 11.4. The wall thickness of 7.9 in. of steel used in these examples is also that of the Yankee pressure vessel. The thermal stress calculated was about 2900 psi, which is small compared with the allowable value. The absorption of the secondary gamma rays produced within the walls of the vessel as a result of thermal-neutron capture will contribute to some extent to the thermal stress, but the effect in the present case is minor.

12.93. In addition, the mechanical stress due to the design value of 2500 psi for the internal pressure was seen in Example 11.6 to be compatible with the particular steel used in the construction of the vessel. A complete analysis of the reactor vessel requires that consideration be given also to power surges, decay heat, and other factors, as well as to the mechanical stress caused by the coolant pressure and the thermal stress resulting from internal heat generation.

12.94. A significant limitation in the design of reactor pressure vessels may be the integrated fast-neutron flux to which it will be exposed over its lifetime. As seen in § 7.21, neutron irradiation results in an increase in the brittle-to-ductile transition temperature in steels of the type generally used for the fabrication of pressure vessels. The transition temperature may thus increase to the point at which the material will become brittle when the reactor is shut down.

FUEL UTILIZATION

INTRODUCTION

12.95. In a broad sense, the philosophy of nuclear fuel utilization should be considered in relation to reactor design. At the present time there exist systems ranging from "burners" of highly enriched uranium, in which there is essentially no conversion of uranium-238 to plutonium-239, to breeders (§§ 1.58, 1.62, 4.145), which produce more fissile nuclei than are consumed. If full use is to be made of the energy available from natural uranium and thorium, a certain amount of breeding of fissile material will be necessary. At the present time, however, the cost of power generation from breeder reactors is greater than from several highly developed reactors that use nuclear fuel less effectively. The greatest promise for the production of power, which will be economically competitive with that from fossil-fuel plants in the not too distant future, is shown by reactors using slightly enriched uranium as fuel. In these reactors there is some conversion of fertile material, but more fissile material is consumed than is produced during operation.

12.96. From the short-term economic standpoint, therefore, there is a tendency to select an existing, well-proven design that is known to be satisfactory for power production. An example is the type of thermal reactor considered earlier in this chapter, namely, a system using ordinary water under pressure as the coolant and moderator with slightly enriched uranium as the fuel. Other reactors which are rapidly being developed and fall into the same general category are the boiling-water systems and those using either sodium, an organic liquid, or a high-pressure gas as coolant.   Although these reactors are of more-or-less immediate economic interest, they are unable to make the maximum possible use of nuclear fuel.   On the other hand, advanced breeder reactors, probably operating on fast or epithermal neutrons, require research and development, but ultimately such reactors may play an important role if nuclear fission is to make its maximum contribution to the world's energy resources.

12.97. It should be mentioned that some authorities are of the opinion that there is no urgency for the development of breeder reactors.   It is argued that when the cost of power from other sources increases to the point where fast breeders are reasonably competitive, there will still be ample supplies of natural uranium to provide the uranium-235 required to start such reactors (as pseudo-breeders).   In addition, other reactors will have produced plutonium-239 which can be used as fuel in a fast reactor.   Moreover, there is a possibility that methods for producing fast neutrons cheaply, e.g., by thermonuclear reactions with deuterium, may be developed, so that fission energy can be obtained from uranium-238 and thorium-232 in a system that does not require a self-sustaining chain reaction [13].

FUEL BURNUP

12.98. An important item in the cost of fuel per kilowatt-hour of electricity produced by a nuclear power plant is the expense incurred in the fabrication of fuel elements.   This represents a unit charge each time a spent fuel element is removed from the reactor core.   It is obviously desirable, therefore, in the interest of economy, to minimize the frequency with which the reactor fuel is discharged; in other words the degree of fuel *burnup* should be as large as possible.

12.99. The burnup is a measure of the amount of fissile material consumed (or power produced) before the fuel element is removed from the reactor for processing.   Its magnitude has been expressed in various ways, the most common being the number of megawatt-days of heat energy produced per metric ton (or tonne), i.e., 2200 lb, of uranium, abbreviated to Mw-days per tonne or MWD/T.*   Although this manner of stating the degree of burnup has been widely used, it has its limitations when applied to special fuels, e.g., those containing plutonium or thorium.   Alternative proposals which have been made for expressing burnup and which are preferable in many respects are (a) number of fissions per cm³ of fuel and (b) percentage of fissile nuclei consumed, both of which can be related to the Mw-days per tonne unit [14].   The latter will, however, be used here since it is the one generally found in the literature.

12.100. The maximum average fuel burnup for thermal reactors is currently about 10,000 Mw-days per tonne, but values of 20,000 to 30,000 Mw-days per tonne may be expected within a few years.   Several factors determine the

* It is of interest to note that 1 megawatt-day of heat corresponds to the fission of approximately 1 g of material.   The amount consumed is larger than 1 g because of nonfission captures.

possible degree of burnup.   One of these is the effect of irradiation in causing dimensional changes and physical damage to both the fuel material and cladding. Progress is being made in the development of materials for which stability is not the limiting factor, as was seen in Chapter 8.   Where burnup is not restricted by radiation damage, the net loss of reactivity, due to consumption of fissile material, the accumulation of fission products, and the buildup of heavy isotopes, as well as to the capture of neutrons by control materials and other poisons, determines the lifetime of the fuel elements.   When this is the case, it is possible to compute the burnup corresponding to a given amount of excess reactivity or, as is generally the case, to prescribe the expected average burnup and then to determine the excess reactivity required.   The general procedure used in making these calculations is indicated below.

12.101. Consider a reactor in which the fuel is uranium slightly enriched in uranium-235.   The latter isotope will undergo fission and in addition there will be neutron capture by both uranium-235 and uranium-238.   The product in the latter case will lead to the formation of plutonium-239 which exhibits both fission and neutron capture.   As a result of capture, plutonium-240 is formed and this also captures a neutron to produce fissile plutonium-241, and so on.   The first step in determining the relationship between the reactivity and the burnup (or irradiation time) is to write down a series of differential equations describing the changes with time of the concentrations of the various isotopes of interest [15].

12.102. As a general rule, as in § 2.190 *et seq.*, the rates of simple nuclear reactions are written in the form $dN/dt = N\sigma\phi$, where N is the number of nuclei per cm³, σ is the cross section of the reaction, and ϕ is the neutron flux, the latter being treated as a constant.   In a reactor, the flux in a given fuel element will vary with time and with its location.   It is convenient, therefore, to consider the nuclear reaction rate in terms of the integrated flux, θ, sometimes called the "integrated flux-time" in this connection, where θ is defined by

$$\theta = \int_0^t \phi(t) \, dt.$$

The rate equation in the simplest case then takes the convenient form $dN/d\theta = -N\sigma$.   This will be used in the following treatment which refers, in particular, to a thermal reactor.

12.103. The rate of change in the number of uranium-235 nuclei, i.e., N_5,* per unit volume is given by

$$\frac{dN_5}{d\theta} = -N_5\sigma_{a5},$$

where σ_a is the absorption, i.e., fission plus capture, cross section.   For uranium-236, the corresponding differential equation is

---

* To simplify the representation, a nuclide is identified by the final integer in its mass number.

$$\frac{dN_6}{d\theta} = N_5\sigma_{c5} - N_6\sigma_{a6},$$

where σ_{c5} is the capture cross section in uranium-235. A similar expression can be written for uranium-237, but this and its decay products can be neglected because their effect on reactivity is small. The uranium-238 captures neutrons in both thermal and resonance regions and undergoes a certain amount of fission by fast neutrons. However, as a first approximation, the concentration of uranium-238, which changes by a few per cent only during normal reactor operation, may be taken as constant.

12.104. An important consideration in evaluating reactivity changes is the accumulation of fissile plutonium-239; this is produced as a result of the capture in uranium-238 of thermal neutrons and of resonance neutrons arising from the slowing down of fast neutrons produced by the fission of uranium-235 and plutonium-239. For the present purpose, neutrons produced by fission of plutonium-241, etc., may be neglected. If N_f represents the nuclear concentration of any fissile species and σ_{af} is its absorption cross section for thermal neutrons, the rate of production of fast neutrons is $N_f\sigma_{af}\eta_f\epsilon$, where η_f is the number of fission neutrons produced per absorption in a fissile nucleus (§ 4.5) and ϵ is the fast-fission factor. Of these, a fraction P_1, where P_1 is the neutron nonleakage probability in slowing down into the resonance region, reaches resonance energies. Finally, if p is the resonance escape probability (§§ 3.79, 4.8), a fraction $1 - p$ of these neutrons in the resonance region will be captured by uranium-238 to form plutonium-239. Hence, the overall rate equation is

$$\frac{dN_9}{d\theta} = N_8\sigma_{c8} + [\epsilon P_1(1 - p)](N_5\sigma_{a5}\eta_5 + N_9\sigma_{a9}\eta_9) - N_9\sigma_{a9},$$

where the first and second terms are due to plutonium-239 formation by thermal-neutron capture in uranium-238 and by resonance capture, respectively, and the third to neutron absorption (fission plus capture) in plutonium-239.

12.105. The rate equation for plutonium-240 is

$$\frac{dN_0}{d\theta} = N_9\sigma_{c9} - N_0\sigma_{a0},$$

where the first term gives the rate of formation from plutonium-239 and the second is the rate of removal by neutron absorption. Similarly, for plutonium-241,

$$\frac{dN_1}{d\theta} = N_0\sigma_{c0} - N_1\sigma_{a1},$$

where σ_{a0} and σ_{c0} are essentially the same because plutonium-240 is not fissile. Unless recycle plutonium, which may contain an appreciable amount of plutonium-240, is used, it is not necessary to consider any nuclides beyond plutonium-241.

12.106. Finally, for reactivity calculations, it is required to know the rate of formation of fission products. The appropriate rate equation is

$$\frac{dN_F}{d\theta} = 2(N_5\sigma_{f5} + N_9\sigma_{f9} + N_1\sigma_{f1}),$$

where N_F refers to total number of fission product nuclei, two being formed in each act of fission; the σ_f's are the fission cross sections for the three indicated fissile nuclides present.

12.107. The initial amounts of the various species in the fuel may be presumed to be known, as also are the various cross sections and other nuclear data. The values of P_1 and p are dependent on the core design and so are specific to a particular reactor. However, all the information can be obtained which will permit the solution of the differential equations given above. The concentration (nuclei per cm³) of each nuclide of interest can thus be calculated as a function of the integrated flux.

12.108. In order to evaluate the variation of reactivity with time, it is necessary to carry out a thermal-neutron balance. On the one hand, there is the rate of thermal-neutron production by fission of uranium and plutonium isotopes; on the other hand, neutrons are lost by absorption in these isotopes, in fission product nuclei, coolant, structural materials, burnable poison (if any), and control rods, and by leakage. In determining the neutrons captured by fission products the simplification is generally made of treating all nuclides other than xenon-135 and samarium-149 as a single group with a weighted average cross section. The magnitude of this cross section varies with the nature of the nuclide undergoing fission, and for uranium-235 in a thermal reactor it is about 64 barns [16]. The excess neutrons available per neutron absorbed in fuel is then a measure of the reactivity. Thus, it is possible to calculate the variation of the reactivity as a function of the integrated flux. The value of the latter at which the reactivity decreases to zero, so that the effective multiplication factor with all the control rods withdrawn is unity, corresponds to the maximum burnup for the given fuel. To convert the burnup in terms of integrated flux into Mw-days per tonne, the energy produced during the exposure is calculated from the total fission products formed. In each act of fission two nuclei are produced and approximately 200 Mev of energy released; hence each fission product nucleus formed corresponds to 100 Mev of thermal energy.

12.109. The computational procedure outlined above is based on the tacit assumption that the reactor core behaves as a uniform system. This is, of course, not the case. Spatial variations in the neutron flux will result in changes in burnup, as already seen (§ 12.34). Although the treatment is the same in principle as that already described, it is much more complicated [16]. The core is divided into a number of regions and in each region an initial neutron-flux pattern is postulated. This is assumed to remain constant over a period of time and the composition changes occurring during this period are determined. A new flux pattern is then evaluated on the basis of the new composition and the calculations are repeated, and so on until the excess reactivity has fallen to zero. The use of fuel shifting, as prescribed by various fuel management schemes which provide improved fuel utilization (§ 12.114), tends to further complicate burnup calculations. Methods employing digital computers have therefore been developed.

THE CONVERSION RATIO

12.110. If the fuel contains a large proportion of fertile material, such as is the case for the natural or slightly enriched uranium used in most central-station power plant reactor concepts, conversion into fissile material makes an important contribution to the burnup. An important quantity in this connection is the *conversion ratio*, defined as the number of fissile nuclei (plutonium-239) formed to the number (uranium-235) destroyed. The rate of plutonium-239 formation can be readily obtained from the arguments in § 12.104, where the rate at which uranium-235 is removed is $N_5\sigma_{a5}$; hence, soon after startup, when fission in plutonium-239 is negligible,

$$\text{Conversion ratio} = \frac{N_8\sigma_{c8} + \epsilon P_1(1 - p)N_5\sigma_{a5}\eta_5}{N_5\sigma_{u5}}$$

$$= \frac{\sigma_{c8}}{r\sigma_{a5}} + \epsilon P_1(1 - p)\eta_5,$$

where $r = N_5/N_8$ is the ratio of uranium-235 to uranium-238 nuclei in the fuel. If the Fermi age theory of slowing down is assumed to apply, then P_1 is equal to $e^{-B^2\tau}$, where τ is the age of neutrons at uranium-238 resonance energy.

Example 12.3. In a natural uranium-graphite reactor, p is 0.905, ϵ is 1.03, and B^2 is 10^{-4} cm^{-2}; the Fermi age of thermal neutrons in this reactor is about 400 cm^2. Estimate the expected initial conversion ratio.

Since the age of thermal neutrons is 400 cm^2, the age of neutrons at the uranium-238 resonance region is roughly 300 cm^2; hence, $B^2\tau$ is 3×10^{2} and $e^{-B^2\tau}$ is close to 0.97. For uranium-235, η is 2.06 (Table 4.1) and the cross sections σ_{c8} and σ_{a5} are 2.7 and 683, barns, respectively, for 2200-m/sec neutrons (Table 2.8). Since the expression for the initial conversion ratio involves σ_{c8}/σ_{a5}, this may be taken to be 2.7/683 for a thermal reactor. For natural uranium, r is close to 0.0072, so that

$$\text{Initial conversion ratio} = \frac{2.7}{(0.0072)(683)} + (2.06)(1.03)(0.095)(0.97)$$

$$= 0.55 + 0.20 = 0.75.$$

12.111. The expression derived above gives what is called the "initial conversion ratio" for a fuel consisting of uranium-235 and uranium-238, since it applies only to conditions soon after startup. As operation of the reactor proceeds, fissions occur in plutonium-239, as well as in uranium-235, and neutron capture in the resonance region is not restricted to uranium-238 because of the buildup of fission products and higher isotopes including plutonium-239. The conversion ratio thus generally tends to decrease. However, a large value of the initial conversion ratio indicates a high conversion efficiency throughout the operating period of the reactor and is conducive to a high degree of burnup.

12.112. One of the consequences of conversion of uranium-238 into plutonium-239 is that when the fuel is natural or very slightly enriched uranium, the reactivity actually increases in the early stages of operation, although the conversion ratio is less than unity. The reason is that both the fission cross section and the number of fission neutrons produced per neutron capture in fuel are greater for plutonium-239 than for uranium-235. Hence, if the conversion factor is larger than about 0.80, as is frequently the case, there will be an initial increase in the reactivity. However, after a certain time (or integrated flux), the net decrease in the total number of fissile nuclei causes the reactivity to reach a maximum value and then to decrease steadily. For fuels containing roughly 2 per cent or more of uranium-235 the initial conversion ratio is probably too small to permit an initial increase in the reactivity; the latter thus decreases continuously from the beginning of reactor operation.

12.113. The value of the initial conversion ratio is dependent to some extent on the reactor design [18]. For example, increasing the enrichment of the fuel increases the value of r and this decreases the conversion ratio, as indicated above. On the other hand, the conversion ratio can be increased by a decrease in the resonance escape probability. Where the infinite multiplication factor of the system is sufficiently large, the design may be deliberately altered to permit an increase in resonance capture and, hence, in the conversion ratio. This is the situation in the "spectral shift control" concept described in § 5.154.

FUEL MANAGEMENT [19]

12.114. For fuel of a given enrichment, the burnup can be extended by suitable fuel loading schedules, generally referred to as "fuel management." A number of such procedures, with their advantages and disadvantages, will be described. Some of these require changing the location of the fuel elements within the core according to a predetermined schedule. Although these schemes may increase the burnup and reduce the expenditure of fuel, they may add to the cost of reactor operation, especially if shutdown is necessary for rearranging the fuel.

Batch Irradiation

12.115. The reactor is loaded uniformly with a complete core at one time and this is irradiated without moving the fuel elements. When the system ceases to be critical the whole core is discharged and reloaded. This method leads to smaller degrees of burnup than other schemes. If the fuel enrichment is uniform throughout the core, there are large variations in power density, both in space and time, so that heat-removal problems are increased. The batch irradiation method, with uniform fuel distribution, is not favored for large reactors, although it has some merit for small units.

Center to Outside Loading

12.116. From time to time, fresh fuel material is added to the center of the core, where the leakage of neutrons is low and hence their importance is high, and progressively moved to outer radial positions. The spent fuel is discharged from the outer radius of the core. Because the neutrons are used most efficiently, this fuel management scheme gives the highest burnup of any of the methods proposed. However, since both the neutron flux and density of fissile nuclei are greatest at the center of the core and decrease radially outward, the variation in power density is even larger than for batch irradiation. Design of the heat-removal system is thus especially difficult.

Outside to Center Loading

12.117. In this procedure, the fresh fuel is charged near the outer edge and moved progressively toward the center from which it is discharged. Although considerably better burnup is attained than in batch irradiation, the neutron utilization is not as efficient as in the preceding scheme because the highest reactivity, i.e., fresh fuel, is introduced in a region of low neutron importance, whereas the lowest (or possibly negative) reactivity occurs where the flux is highest. The advantage of outside to center loading is that it leads to a fairly uniform power density distribution in the radial direction.

Bidirectional Loading

12.118. In bidirectional loading, charging and discharging of the fuel is achieved by pushing relatively short fuel slugs through the core from one side of the reactor to the other, but from opposite directions in adjacent rows of fuel elements. In one row, for example, fresh fuel enters at the left side and spent fuel is discharged at the right; in an adjacent row the situation is reversed. The result is a burnup and power density gradient somewhere between center to outside and outside to center loading schemes.

Graded Irradiation

12.119. The same general result as that just described can be obtained by dividing the core into a number of regions; each region contains fuel elements of varying degrees of irradiation. When any particular element has received a predetermined exposure, it is discharged and replaced by a fresh one. In this manner, positive reactivity is added and negative reactivity removed as required in regions throughout the core, rather than at specific locations.

Zonal Core Loading

12.120. As a modification of the simple batch irradiation scheme, the core is divided into two or three radial zones, with fuel of different enrichment in each zone. The complete core would remain in place for 2 or 3 years and then be discharged. In determining the degree of enrichment of the fuel in the various zones, the efficiency of burnup must be weighed against the nonuniformity of the power density. One advantage of this loading procedure over those procedures requiring shifting of the fuel elements is that shutdown of the reactor is much less frequent.

Axial Distribution

12.121. Any system in which there is only a radial variation or shifting of the fuel elements will not fully utilize the fuel at the axial ends of the core, i.e., at the top and bottom. This can be overcome by varying the enrichment along the length of the fuel element or by dividing the core into upper and lower halves and exchanging the elements in them at a predetermined stage in the fuel burnup cycle.

Seed-Blanket Loading

12.122. A variation of the zonal core loading is the so-called seed-blanket scheme. The core consists of an annular "seed" of highly enriched fuel, with internal and external blankets of natural uranium; numerous variations in the geometry are possible. The advantage of this method of loading is that it avoids peaking of the neutron flux in the center of the core (§ 13.16). The economic potential of the seed-blanket arrangement appears to be roughly equivalent to that of cyclic three-zone core loading.

BREEDING [20]

12.123. It was stated earlier that breeding, i.e., the production of more fissile nuclei of a given kind than are consumed, is necessary if the maximum use is to be made of nuclear fuel materials. Two main types of breeding cycles are theoretically possible; one is based on a fast reactor using plutonium-239 as the fissile material and uranium-238 as the fertile species, and the other on a thermal (or fast) reactor utilizing uranium-233 and thorium-232 as the corresponding components. The fertile species is employed in the form of a blanket which surrounds the core, and so a large proportion of the neutrons escaping from the latter are captured in the blanket. In most breeder reactor designs, especially for fast reactors, some fertile material, e.g., uranium-238, is included in the core. Fissile nuclei are then formed both in the core and the blanket.

12.124. The fissile material produced in the core is useful as fuel at the place of formation. That generated in the blanket, however, must be returned to the core in some manner if it is to be utilized efficiently. An appropriate blanket processing scheme is therefore an essential part of the breeding system. The amount of fissile material which is allowed to accumulate in the blanket, and hence the reprocessing rate, is subject to economic optimization. For example, if the permissible plutonium-239 buildup is set at a low value, fissile material inventory charges will be reduced; on the other hand, the greater mass of blanket material which will have to be reprocessed per unit mass of fissile species will increase the cost of recovery.

12.125. In reprocessing the core of a breeder reactor, the determining factor may well be the radiation damage suffered by the fuel elements. The effect of fission product poisoning is relatively small in a fast reactor, and so reprocessing for the removal of these impurities will be required at rare intervals only. In order to maintain steady-state conditions in a breeder operating at constant power, the blanket must be processed at such a rate as to provide fissile atoms to compensate for those consumed in the core. In other words, it should not be necessary to obtain fissile material from an outside source after the initial loading. After the requirements of the core have been met, additional fissile atoms will be available as a result of breeding.

12.126. In the theoretical consideration of the breeding process, two quantities are of interest, namely, the doubling time and the breeding ratio. The *breeding ratio*, which is treated more fully below, is defined in a manner similar to that for the conversion ratio (§ 12.110); it is the ratio of the number of fissile atoms produced to the number of the same kind that have been consumed. The excess of the breeding ratio over unity, i.e., the number of fissile atoms gained for each one consumed, is called the *breeding gain* and is represented by G. This is related to the doubling time in the following manner.

12.127. In its simplest form, the *doubling time* is defined as the time required for a breeder reactor to produce a surplus amount of fissile material equal to that required for the initial charge of the reactor. However, this definition does not take into consideration the time the material spends in other parts of the fuel cycle, e.g., processing, storage, etc. It is preferable, therefore, to regard the doubling time as that in which the surplus fissile material produced would equal the total quantity in the fuel cycle. Suppose this amount is M grams and let g grams be the quantity of fissile material consumed in the breeder reactor per day; the doubling time, D, is then given by

$$D \text{ (days)} = \frac{M}{gG},$$

where G is the breeding gain defined above. Of the g grams consumed, $g/(1 + \alpha)$ grams have undergone fission, where α is the ratio of capture to fission cross sections (§ 4.143). It was seen in § 1.46 that the fission of 1 g of fissile material per day produces approximately 1 Mw of power; hence, the power of the reactor under consideration is $g/(1 + \alpha)$ megawatts. If the power is represented by P megawatts, then g is equal to $P(1 + \alpha)$, and upon making this substitution in the expression given above, it is seen that

$$D \text{ (days)} = \frac{M}{GP(1 + \alpha)}.$$

A short doubling time means a rapid increase in the production of surplus fissile material, and this desirable situation can be achieved by a high breeding gain and by a large value of P/M. The latter is favored by a high specific power, i.e., power per unit mass of fissile material in the core, and by having as small a quantity of material as possible in the fuel cycle outside the core. This is one reason why efforts are being made to develop spent fuel reprocessing methods which do not involve long cooling periods (§ 8.166).

12.128. A completely general expression for the breeding ratio is

$$\text{Breeding ratio} = \frac{\int_{\text{core}} \phi \Sigma_c^{\text{fertile}} \, dV + \int_{\text{blanket}} \phi \Sigma_c^{\text{fertile}} \, dV}{\int \phi (\Sigma_f + \Sigma_c)^{\text{fuel}} \, dV},$$

where Σ_c and Σ_f are the macroscopic capture and fission cross sections, respectively; the numerator gives the total number of fissile nuclei formed in core and blanket, and the denominator the number consumed.* Integration over core and blanket volumes are required because, in addition to the normal spatial variation of the flux, there are also spatial variations in the cross sections due to differences in the neutron energies. Fast (fission) neutrons originating in the core are slowed down to some extent, especially by inelastic scattering, before they are absorbed. Moreover, since some of the neutrons leaking into the blanket from the core are reflected back, the core spectrum will be further degraded, especially near the core-blanket boundary.

12.129. A useful simplified approach to calculation of the breeding ratio is to assume that all neutrons leaking from the core will eventually be utilized in breeding. This assumption is fairly reasonable since some neutrons are formed by fission in the blanket to compensate for neutrons lost in parasitic (nonbreeding) processes and by escape. If the total breeding ratio is divided into two parts, namely, external (or blanket) and internal (or core), it follows that

$$\text{External breeding ratio} = \frac{\text{Core leakage}}{\int_{\text{core}} \phi (\Sigma_f + \Sigma_c)^{\text{fuel}} \, dV}. \tag{12.8}$$

From neutron balance considerations in the core, it is found that

$$\text{Core leakage} = \int_{\text{core}} \phi (\nu \Sigma_f - \Sigma_f - \Sigma_c)^{\text{core}} \, dV,$$

where the first term in parentheses determines the rate of neutron production by fission and the other two represent losses by neutron absorption; ν is the number of neutrons produced per fission (§ 2.164). In this expression for core leakage, the macroscopic cross sections include the contributions of all the core materials, not merely the fuel. If now the spatial variations are neglected, equation (12.8) reduces to

---

* As used here, the term "fuel" refers to the fissile species only.

$$\text{External breeding ratio} \approx \frac{(\nu\Sigma_f - \Sigma_f - \Sigma_c)^{\text{core}}}{(\Sigma_f + \Sigma_c)^{\text{fuel}}}$$

$$= \frac{\nu - (1 + \alpha^*)}{1 + \alpha},$$

where

$$\alpha^* = \frac{\Sigma_c}{\Sigma_f} \quad \text{for the core}$$

and

$$\alpha = \frac{\Sigma_c}{\Sigma_f} \quad \text{for the fuel (fissile) species only.}$$

If α^* and α are assumed to be not very different, then

$$\text{External breeding ratio} \approx \frac{\nu}{1 + \alpha} - 1$$

$$\approx \eta - 1,$$

since η is equal to $\nu/1 + \alpha$ (§ 4.146).

12.130. An indication of the effect of neutron energy on the potential for external breeding (or conversion) is thus given by the corresponding value of $\eta - 1$. The data for pure plutonium-239, uranium-235, and uranium-233 are quoted in Table 12.8. It is seen that only in a comparatively fast-neutron

TABLE 12.8. BREEDING (OR CONVERSION) POTENTIAL

Nuclide	Thermal	Neutron Energy			
		1 to 3000 ev	3 to 10 kev	0.1 to 0.4 Mev	0.4 to 1 Mev
Plutonium-239.......	1.09	0.62	0.93	1.61	1.94
Uranium-235.........	1.06	0.68	0.90	1.12	1.33
Uranium-233.........	1.27	0.88	1.29	1.30	1.58

spectrum does plutonium-239 have any potentiality for breeding, whereas with thorium-233 significant breeding should be possible in thermal reactors and also at energies above 3 kev. Conversion of uranium-235 into plutonium-239 to an extent appreciably in excess of 100 per cent, i.e., pseudo-breeding, also requires a fast spectrum.

12.131. The internal (or core) breeding ratio is given by

$$\text{Internal breeding ratio} = \frac{\int_{\text{core}} \phi\Sigma_c^{\text{fertile}} \, dV}{\int_{\text{core}} \phi(\Sigma_f + \Sigma_c)^{\text{fuel}} \, dV}$$

$$\approx \frac{\Sigma_c^{\text{fertile}}}{\Sigma_f^{\text{fuel}}} \cdot \frac{1}{1 + \alpha}, \tag{12.9}$$

where α is the same as defined above. Apart from its contribution to the total breeding ratio, internal breeding has an important bearing on fuel-cycle costs since fissile material produced in the core can be utilized directly without the necessity of going through reprocessing and fabrication stages. It is apparent from equation (12.9) that the internal breeding ratio for a given fertile-fissile nuclide system depends on the core composition, since this determines the macroscopic cross-section ratio. There is a general decrease in breeding ratio as the core is diluted with inert material because of spectral softening which results in a greater parasitic capture of neutrons in inert diluents. Comparison of different cases, however, must take into consideration changes in critical mass and neutron spectrum and it is therefore difficult to express the conclusions in simple form.

12.132. An indication of the expected effects of some parameters is given in Table 12.9. The calculated data apply to a spherical reactor having a volume of

TABLE 12.9. CHARACTERISTICS OF FAST BREEDER REACTORS

Core Material	Atomic Ratio Pu/U^{238}	Critical Mass (kg Fuel)	Internal Breeding Ratio	Total Breeding Ratio
Pu and U metals.....	0.128	431	0.73	1.82
PuC and UC........	0.222	396	0.46	1.62
PuO$_2$ and UO$_2$.......	0.336	372	0.31	1.55

800 liters, of which 25 per cent (by volume) is fuel plus fertile material, 25 per cent steel structural material, and 50 per cent coolant (sodium). The spherical blanket has a thickness of 45 cm, contains 60 per cent (by volume) of uranium-238, 20 per cent steel, and 20 per cent sodium; this is surrounded by a 30-cm thick reflector of 60 (volume) per cent iron and 40 per cent sodium [21].

12.133. The lower breeding ratios, especially the internal values, for the carbide and oxide fuels, can be ascribed to two main causes. First, the presence of elements of low mass number, which act as moderators, causes a softening of the neutron spectrum, so that the capture to fission ratio is increased. Second, in order to obtain a critical mass within the prescribed volume of 800 liters, it is necessary to decrease the amount of uranium-238 in the core. Consequently, there is a decrease in both neutron capture, leading to the production of plutonium-239, and of "bonus" neutrons resulting from fast fission of uranium-238.

12.134. The calculations were extended to estimate the effects of replacing the steel structural material in the core by other metals. It appears that with titanium or vanadium both the critical mass and total breeding ratio are not very different from those given in Table 12.9. Vanadium, niobium, and molybdenum are increasingly harmful, as compared with steel, and complete substitution of the latter by tantalum would decrease the breeding ratio almost to unity as a result of parasitic neutron capture.

CONCLUSION

12.135. In concluding this chapter, it must again be emphasized that the material presented is intended to provide no more than a guide or outline of the preliminary effort involved in the design of a nuclear reactor. Every aspect of the problem treated here is the subject of detailed and refined calculations and frequently of experiments with critical assemblies and mockups. Moreover, much of the design work is concerned with matters which are not at all specific to the nuclear aspects of the system but involve engineering problems which would be required of any complex industrial installation of the same magnitude. Consequently, many interrelationships are encountered between nuclear and traditional engineering parameters. Since the overall objective of the design of a central-station power plant is to produce electricity at a minimum cost, considerable attention will have to be devoted to iterative optimization procedures along the lines of the parametric studies mentioned earlier in the chapter. The design of a nuclear power plant is thus a highly complex procedure, much more complicated than could possibly be treated in this book.

SYMBOLS USED IN CHAPTER 12

A	coefficient in neutron flux equation
A	cross section of flow channel
a	coefficient
B	buildup factor
b	coefficient
D	doubling time in days
D_e	equivalent diameter of coolant flow channel
d	spacing between fuel elements
E_1	exponential integral function of first order
E_2	exponential integral function of second order
F_q	ratio of maximum to average heat flux
$F_{\Delta t}$	ratio of maximum to average coolant enthalpy increase
F_θ	ratio of maximum to average film drop temperature
f	Fanning (friction) factor
G	breeding gain
g	mass of fissile material in grams
g_c	force to mass conversion constant
h	heat-transfer coefficient
L	length of flow channel

M	mass of fissile material in fuel cycle
N	number of nuclei per cm^3
P	reactor power in megawatts
P_1	nonleakage probability for the resonance energy region
p	resonance escape probability
Δp_f	pressure drop due to friction
Q_0	Volumetric heat source at pressure vessel surface
r	ratio of U^{235} to U^{238}
S_a	strength of surface source (particles/cm^2)
S_v	strength of volume source (particles/cm^3)
t	time
Δt	increase in temperature of coolant
v	flow velocity of coolant
x	thickness of water layer in shield
z	distance (or thickness) in shield
α	ratio of capture to fission in fuel
α^*	ratio of capture to fission in core
ϵ	fast-fission factor
η	neutrons produced per neutron absorption in fissile nuclei
θ	integrated flux (or flux-time)
μ	linear attenuation coefficient
μ_e	linear energy absorption coefficient
μ_v	linear attenuation coefficient of volume source
ν	neutrons produced per neutron causing fission
ρ	density
Σ_c	macroscopic capture cross section
Σ_f	macroscopic fission cross section
Σ_r	macroscopic removal cross section
σ	standard deviation
σ	microscopic cross section
σ_a	microscopic absorption (fission + capture) cross section
σ_c	microscopic capture cross section
σ_f	microscopic fission cross section
ϕ	neutron flux
ϕ_γ	gamma-ray (energy) flux

REFERENCES FOR CHAPTER 12

1. N. J. Palladino and H. L. Davis, *Nucleonics*, **18**, No. 6, 85 *et seq.* (1960).
2. See, for example, R. E. Behmer and B. L. Hoffman, *Nuc. Sci. and Eng.*, **2**, 14 (1957); U. S. AEC Report APDA-133 (1960); U. S. AEC Report Y-1396 (1962).
3. W. E. Shoupp, *et al.*, *Proc. Second U.N. Conf. Geneva*, **8**, 492 (1958); *Power Reactor Technology*, **4**, No. 3, 47 (1961); *Nucleonics*, **19**, No. 3, 53 (1961).
4. See, for example, W. A. Sutherland, *Nucleonics*, **19**, No. 1, 80 (1961).

5. *Power Reactor Technology*, **2**, No. 1, 20 (1958).
6. Data obtained from W. H. Arnold, U. S. AEC Report YAEC-152 (1959).
7. R. W. Deutsch, *Nucleonics*, **16**, No. 6, 95 (1958).
8. H. W. Graves, *et al.*, U. S. AEC Report YAEC-136 (1961).
9. "Reference Material on Atomic Energy," Vol. III, Reactor Handbook: Engineering, U. S. Government Printing Office, Washington, D. C., 1955, pp. 133 *et seq.*; B. W. Le Tourneau and R. E. Grimble, *Nuc. Sci. and Eng.*, **1**, 359 (1956); C. F. Bonilla, "Nuclear Engineering," McGraw-Hill Book Co., Inc., New York, 1957, pp. 442 *et seq.*; P. A. Rude and A. C. Nelson, *Nuc. Sci. and Eng.*, **7**, 156 (1960); F. H. Tingey, *ibid.*, **9**, 127 (1961); for reviews, see *Nucleonics*, **17**, No. 8, 92 (1959); *Power Reactor Technology*, **2**, No. 4, 42 (1959); **3**, No. 3, 7 (1960); **4**, No. 1, 30 (1960).
10. E. A. McCabe, U. S. AEC Report YAEC-106 (1960).
11. H. Chelemer and L. S. Tong, *Trans. Am. Nuc. Soc.*, **3**, 517 (1960).
12. R. T. Berringer and A. A. Bishop, U. S. AEC Report YAEC-74 (1959).
13. W. B. Lewis, *Nuclear News*, **4**, No. 3, 10 (1961).
14. R. L. Stover and G. K. Moeller, U. S. AEC Report MIT-OR-06 (1961); P. R. Huebotter, *Nucleonics*, **18**, No. 11, 176 (1960).
15. M. Benedict and T. H. Pigford, "Nuclear Chemical Engineering," McGraw-Hill Book Co., Inc., New York, 1957, Ch. 7; see also, B. Spinrad, *et al.*, *Proc. (First) U.N. Conf. Geneva*, **5**, 125 (1956); T. H. Pigford, *et al.*, *Proc. Second U.N. Conf. Geneva*, **13**, 198 (1958); R. L. Murray, *et al.*, *Nuc. Sci. and Eng.*, **6**, 18, 455 (1959); R. T. Shanstrom and M. Benedict, *ibid.*, **11**, 377, 386 (1961).
16. A. G. Ward, *Nucleonics*, **18**, No. 10, 69 (1960); see also, G. R. Hopkins and G. P. Rutledge, U. S. AEC Report WAPD-R(L)-6 (1959); C. R. Greenhow and E. C. Hansen, U. S. AEC Report KAPL-2172 (1961); *Power Reactor Technology*, **3**, No. 4, 12 (1960).
17. See, for example, B. F. Rider, *et al.*, U. S. AEC Report GEAP-3373 (1960); C. Graves, *et al.*, U. S. AEC Report NDA 2131-24 (1961); M. Robkin, U. S. AEC Report YAEC-164 (1959); J. D. McGaugh, U. S. AEC Report YAEC-183 (1960).
18. J. Bengston, *et al.*, U. S. AEC Reports CEND-137, CEND-145 (1961); *Power Reactor Technology*, **4**, No. 1, 5 (1960).
19. *Power Reactor Technology*, **2**, No. 4, 49 (1959).
20. "Conference on Physics of Breeding," U. S. AEC Report ANL-6122 (1959); *Power Reactor Technology*, **3**, No. 2, 21 (1960); **4**, No. 1, 5 (1960); **4**, No. 4, 84 (1961).
21. S. Yiftah and D. Okrent, U. S. AEC Report ANL-6212 (1960); see also, W. B. Lowenstein and D. Okrent, *Proc. Second U.N. Conf. Geneva*, **12**, 16 (1958); D. B. Hall, *ibid.*, **13**, 300 (1958); W. H. Roach, *Nuc. Sci. and Eng.*, **8**, 621 (1960); for review, see L. J. Koch and H. C. Paxton, *Ann. Rev. Nuc. Sci.*, **9**, 437 (1959).

PROBLEMS

1. A preliminary design for a 200-Mw (thermal), pressurized-water reactor calls for the following specifications: a single-region core consisting of uranium (3.7 per cent U^{235}) dioxide rods 0.313-in. diameter clad with 0.020-in. thick stainless steel, arranged on a square pitch 0.494 in. center to center; the core is 65 in. in diameter and 60 in. high. The exit temperature of the coolant water is 570°F.

(a) Calculate the mass of uranium dioxide (density 10.7 g/cm^3) in the core, and the specific power in kilowatts per kilogram of uranium.

(b) Estimate the average thermal-neutron flux.

(c) Determine the average heat flux at the fuel surface in Btu/(hr)(ft^2).

2. By utilizing estimated hot-channel factors to account for deviations from the aver-

age neutron flux, coolant flow, and other variables, estimate the maximum heat flux for the reactor in Problem 1.

(a) If experiments indicate a burnout heat flux of 850,000 Btu/(hr)(ft²), what is the "burnout ratio," i.e., ratio of actual to burnout flux?

(b) What coolant flow rate would be required if the water enters the reactor at 540°F?

(c) Suppose the increase in the coolant temperature in its passage through the reactor was 60°F. Discuss the problems which would be introduced and the advantages gained.

3. As an alternative to the reactor described in Problem 1, a boiling-water reactor system is being considered to operate at the same turbine conditions. Indicate how the specifications of the reactor will be changed and estimate as many as possible of the new values.

4. A pressurized-water reactor contains 20,000 cylindrical uranium dioxide fuel rods, 6 ft long and 0.350 in. diameter clad with 0.015 in. of stainless steel, arranged in a square lattice with a pitch of 0.450 in. The maximum temperature at the center of the fuel is 4000°F during operation. The average temperature of the coolant water is 510°F and its flow rate is 20 ft/sec. Estimate (a) the thermal power of the reactor and (b) the expected burnout flux assuming a hot-channel factor of 5.0 has been used in the design calculations. (The thermal resistance at the fuel-cladding interface may be neglected and average values may be employed for the thermal conductives of fuel and cladding.)

5. In a preliminary design for a nuclear rocket reactor, the hydrogen propellant (coolant) leaves the reactor at a temperature of 4000°R, the exit flow rate being 300 lb/sec and the pressure 60 atm. The reactor is to be 12 ft in diameter, with an appropriate length, and its nominal (thermal) power is expected to be 3000 Mw; the flow velocity of the coolant in the core channels is on the order of 1000 ft/sec. Starting with a guess that the coolant channel area should be about half the total core cross section, and the fuel might be in the form of uranium-impregnated graphite plates 0.25 in. thick, develop a preliminary core design. Check, as far as possible, such factors as heat transfer to the coolant, temperature gradient in the fuel, pressure drop, and thermal stress.

The following physical properties are to be used:

Graphite at 4000°R: Density 1.70 g/cm³; tensile strength 4000 psi; $k = 15$ Btu/(hr)(ft²)(°F/ft); $E = 1.2 \times 10^6$ psi; $\alpha = 6 \times 10^{-6}/°F$; and ν (Poisson's ratio), 0.25.

Hydrogen: See Appendix, Table A.7.

6. Review the fuel management procedures described in the text and consider in some detail the advantages and disadvantages of (a) the seed-blanket and (b) the bidirectional loading procedures.

7. The relative (atomic) composition of the fuel (initially 1.5 per cent uranium-235) of the Dresden (boiling water) reactor after an irradiation of 10,000 Mw-days/tonne and a decay time of 120 days is estimated to be as follows:

Uranium-235	0.85
Uranium-236	0.11
Uranium-238	99.10
Plutonium-239	0.36
Plutonium-240	0.09

The average thermal flux in the reactor is 3.2×10^{13} neutrons/(cm²)(sec). Calculate the expected composition of the spent fuel, using the approximation that the concentration of uranium-238 remains constant, and compare the results with those given above.

8. If the two cladding layers of a flat plate (sandwich type) fuel element have unequal thicknesses, a larger proportion of the total heat generated in the fuel (central) region will pass out through the thinner clad. A hot-channel effect could arise in a channel where the two (different) clad thicknesses are a minimum for the adjacent plates making

up the channel. Derive a relation for the hot-channel factor in terms of the pertinent dimensions and physical properties.

9. A reactor designed to operate at a thermal power of 1000 Mw provides for vertical rod uranium dioxide fuel elements of 0.436 in. outer diameter and 100 in. length. The maximum central temperature permitted for the oxide fuel is 4500°F and that for the surface is 650°F. How many fuel rods would be required for this reactor? (The value of k for the uranium dioxide may be assumed to be constant at 1 Btu/(hr)(ft²)(°F/ft).)

Chapter 13

NUCLEAR REACTOR SYSTEMS [1]

POWER REACTORS

INTRODUCTION

13.1. Although certain other reactor types will be mentioned later, the main portion of this chapter will be devoted to a consideration of reactors designed for the most economical production of nuclear power. A power reactor consists of three essential components, namely, the fuel, the moderator (if the neutrons are to be thermalized), and the coolant. Even after ruling out materials that are unsatisfactory from the nuclear (or neutronic) standpoint, a fairly large choice is available for each of these components. Fissile material of three nuclear species may be used as fuel over a wide range of concentrations in several different physical and chemical forms. Almost any material containing light atoms, other than lithium-6 and boron-10, will serve as a neutron moderator. Finally, several different liquids, liquid metals, gases, and even fused salts are available as coolants. It is possible, therefore, to write down a large number of "reactor concepts" based on the many combinations of the three basic components. Physical and chemical incompatibilities may make some of these impractical; nevertheless, there will still be a large number that are applicable, in principle, as the basis for reactor designs.

13.2. The task of making a choice among the many possibilities is rendered difficult by the fact that every system that has been proposed has some disadvantages to offset its desirable characteristics. For example, water is a cheap and good moderator and coolant, but it has a fairly high thermal-neutron capture cross section, so that enriched fuel is required; it also has a relatively low boiling point, which means that pressurization is necessary. Heavy water has a very small cross section for the capture of neutrons and can be used with natural uranium fuel, but it is expensive. Liquid sodium is a good heat-transfer medium, but its chemical reactivity and its activation, by the formation of the radioisotope sodium-24, introduce complications. Similarly, opposing considerations arise in connection with all other reactor concepts. It is for this reason that there are such wide differences of opinion concerning the "best" reactor system.

13.3. Economic and other factors, which vary with local conditions, play an important part in determining a choice of reactor type at any particular time. The urgent need for power in the United Kingdom, for example, has led to the use of gas-cooled reactors of the Calder Hall type, operating at relatively low temperatures and pressures (§ 13.60). Because of previous experience, these systems could be readily utilized without extensive research and development. In the United States, however, the abundant supply of fossil fuels makes such reactors of little practical interest. Another consideration may arise in connection with reactors for special purposes. A small, compact power plant, e.g., for a submarine or for erection in a remote location, will require an entirely different reactor than would a large, central-station electrical facility.

13.4. It should be borne in mind that nuclear power reactor technology is relatively new and is still undergoing development. Problems which initially appear to render some concepts impractical may be solved in the course of time, thus bringing about a change in the situation. For many years, for example, organic coolants were considered to be unsuitable for reactor use because of their relatively poor thermal and radiation stability. When it was recognized that moderately stable compounds existed and that a certain amount of decomposition could be tolerated, a practical reactor concept developed.

13.5. An illustration of a future possibility is provided by lithium, which would be a good coolant because of its high specific heat, high thermal conductivity, and low melting point. However, the thermal-neutron absorption cross section of natural lithium (71 barns) is much too large to be acceptable in a reactor coolant; even for fast neutrons the cross section is still fairly large. The absorption of neutrons in lithium is due almost entirely to the lithium-6 isotope. If this could be removed at a reasonable cost, the lithium-7, which has a thermal cross section of only 0.033 barn, would be an interesting reactor coolant.

13.6. In recent years, major attention has been paid to a limited number of power reactor concepts. This has not necessarily been because these are the "best," but for more immediate practical reasons. In some cases the concept has not required any very advanced technology, whereas in others it has appeared to have promising economic potential as well as reasonable technical feasibility. It should be realized that the technology of power reactors is still far from mature, and many possibilities still exist for new, advanced concepts which require development. For the purpose of illustrating power reactor engineering principles and pointing out some of the technical problems, attention will be limited here to those reactor types to which the main effort has been devoted in recent years [2].

13.7. The power reactors to be considered may be divided into the following categories:

1. *Ordinary-Water Moderated Reactors.*   This category includes both pressurized-water and boiling-water reactors, as well as modifications utilizing nuclear superheating.

2. *Heavy-Water Moderated Reactors.*   These reactors are generally also cooled with heavy water, but this is not essential.   Both sodium and gases have been proposed as coolants, but the latter will be considered in connection with gas-cooled systems.

3. *Organic-Cooled Reactors.*   The organic liquid may serve as moderator, as well as coolant, or a different moderator can be employed.

4. *Sodium-Graphite Reactors.*   In these reactors, graphite is the moderator and sodium the coolant.

5. *Gas-Cooled Reactors.*   These may utilize either natural or enriched uranium as fuel.   Although this category overlaps the heavy-water moderated systems to some extent, the gas cooling, rather than the moderating, characteristics are of major interest.

6. *Fast Breeder Reactors.*   The coolant is generally sodium, although sodium-potassium alloy has been used in an experimental system.

7. *Fluid-Fuel Reactors.*   This category includes thermal breeders (aqueous homogeneous reactors), liquid-metal and fused-salt fueled thermal reactors, and even some fast-reactor concepts.

13.8. The essential plant parameters for a number of reactor types are given in Table 13.1; the "current" values are based on conditions in the early 1960's. The "potential" parameters are those that may be expected by about 1970, after more experience has been gained and a number of technological problems have been solved [3].

13.9. It will be apparent from the subsequent descriptions that many reactor concepts are practical, but each has its inherent advantages and disadvantages. So far there has been insufficient development to determine which type of reactor might be capable of producing electrical power at the lowest cost, in the foreseeable future, say toward the end of the 1960's.   This is the justification for the parallel development work that is being done on six or seven concepts which appear to show most promise at the present time.   Fast-reactor systems are perhaps in a special category.   Because of the fairly high current fuel-cycle costs, it may be some years before electricity can be produced from such reactors as cheaply as from several different thermal concepts now under study.   However, the potentiality that fast reactors have for breeding fissile material is likely to make them ultimately the most economical sources of nuclear power.   In any event, breeding in fast reactors may become essential, in due time, if it is necessary to make maximum use of uranium-238 as a source of energy (see, however, § 12.97).

ORDINARY-WATER MODERATED REACTORS [4]

13.10. The pressurized-water reactor and, more recently, the boiling-water reactor are the principal types in this category.   As will be seen below, develop-

TABLE 13.1. PLANT PARAMETERS

Reactor	Primary Coolant Conditions		Steam Conditions		Cycle Efficiency (Per cent)		Core Power Density (kw/liter)		Fuel Material		Average Fuel Exposure (Mw-days/tonne)	
	Current	Potential	Current	Potential	Current	Potential	Current	Potential	Current	Potential	Current	Potential
Pressurized water	2000 psia subcooled	2000 psia bulk boiling	600 psia (sat.)	1000 psia (sat.)	28	30	55	80	UO₂	UO₂	13,000	19,000
Boiling water	1000 psia (sat.) dual cycle	1400 psia (sat.) direct cycle	1000 psia (sat.)	1400 psia (sat.)	29	30	30	50	UO₂	UO₂	11,000	19,000
Superheat		1400 psia 1000°F		1400 psia 1000°F		36		50		UO₂		19,000
Organic cooled	120 psia 575°F	300 psia 725°F	600 psia 550°F	1000 psia 700°F	29	34	20	44	U-3½% Mo	UO₂	4500	19,000
Sodium-graphite	30 psia 900°F	30 psia 950°F	800 psia 850°F	2400 psia 1000°F	34	41	5	8	U-10% Mo	UC	11,000	19,000
Fast breeder	30 psia 900°F	30 psia 900°F	800 psia 850°F	800 psia 850°F	34	34	850	850	U-10% Mo	PuO₂	1½ w/o	50,000
Heavy water	750 psia subcooled	800 psia boiling dir. cycle	150 psia (sat.)	750 psia (sat.)	23	26	26	35	Nat U		3960	7000
Gas cooled (natural uranium)			500 psia 650°F		24		0.75		Nat U		3000	
Gas cooled (enriched fuel)	300 psia 1050°F	400 psia 1200°F	950 psia 950°F	950 psia 950°F	33	33	0.75	1.28	UO₂	UO₂	10,000	18,000

ments in each case have tended to bring the two concepts closer together.   Some local boiling is now tolerated in the pressurized-water systems, and boiling reactors tend to operate at higher pressures.   A useful distinction, however, is to designate those reactors in which steam is produced without a heat exchanger in the primary coolant loop as "direct cycle" systems.   Reactors requiring a heat exchanger to produce steam are then labeled as "indirect cycle."

13.11.   In the original pressurized-water reactors, the coolant was maintained at a sufficiently high pressure so that the bulk temperature of the coolant leaving the reactor was below the saturation temperature.   The operating pressure is usually about 2000 psia, but some improvement is possible if this is increased to about 2500 psia.   Above this pressure, little is to be gained in plant efficiency, as is shown by the upper curve in Fig. 13.1, which gives the efficiency as a

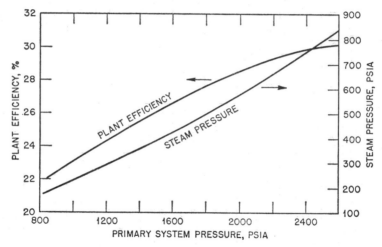

Fig. 13.1.   Plant efficiency and steam pressure as function of system pressure in water-cooled reactors

function of the system pressure.   The inlet and exit temperatures of the reactor coolant were assumed to be 470° and 511°F, respectively.   The lower curve in the figure indicates the variation in steam pressure [5].

13.12.   The basic flow diagram of the reactor and the steam generating system of a pressurized-water reactor power plant were shown in Fig. 11.2.   The cooling water under pressure is pumped through the reactor core and steam is generated in a heat exchanger.   In a large-scale central-station facility, four such steam generators, i.e., four coolant loops, are generally associated with the reactor.

13.13.   The pressurized-water concept, as well as much of the basic technology, evolved from the Naval Submarine Reactor Program which produced the "Nautilus" family of reactors.   The full-scale, land-based prototype of the

Nautilus reactor, called the STR Mark 1, achieved initial criticality in March 1953; the U.S.S. Nautilus commenced its sea trials in January 1955.   Since that date, numerous submarine reactors of the same type have operated very satisfactorily for extended periods.

13.14. The first civilian power application of a pressurized-water reactor was in the Shippingport (Pa.) Atomic Power Station [6].   Criticality was first attained in December 1957 and full power operation, 60 Mw electrical, was realized three weeks later.   This installation was followed by other reactors of the same type including the Yankee Atomic Electric Company reactor [7] which was described in Chapter 12, the Indian Point (N. Y.) Reactor [8], and the reactor for the first merchant ship, the N.S. Savannah, to be propelled by nuclear power [9].   In addition, several compact pressurized-water reactors, capable of being transported by air, have been constructed for use at military posts in remote locations.

13.15. The fuel currently used for pressurized-water reactors is generally enriched uranium dioxide jacketed with stainless steel or Zircaloy.   For the very compact reactors, a high proportion (about 90 per cent*) of uranium-235 is required; otherwise, the enrichment is generally 2.5 to 4 per cent.   The initial-core fuel of the Indian Point reactor contains some thorium dioxide to test the possibility of thermal breeding of uranium-233.

13.16. The core of the Shippingport PWR is of special interest as it was the first to utilize the seed-blanket loading system (§ 12.122).   Two types of fuel elements are used; they are collectively called the "seed" and "blanket," respectively.   There are 32 flat-plate assemblies in the seed, made of a uranium-zirconium alloy clad with Zircaloy-2; the fuel contains more than 90 per cent of uranium-235.   The blanket consists of a total of 113 assemblies of rods of natural uranium, as the dioxide, jacketed in Zircaloy-2.   The seed assemblies are arranged in a square annular array with blanket assemblies both inside and outside.   The seed alone cannot sustain a chain reaction and part of the neutron reflecting character of the blanket is required for criticality.   Thus, some of the neutrons escaping from the seed are reflected back, whereas others are absorbed by the blanket and either cause fission of the uranium-235 or are captured by uranium-238 and lead to the formation of plutonium-239.   An advantage of the seed-blanket system is that it avoids the power "peaking" in the center of a uniformly loaded core.   On the other hand, since most of the fission energy is released in the seed assemblies, orificing of the coolant water is necessary to ensure the proper rate of heat removal.

13.17. The use of such a familiar material as water for both moderator and coolant is attractive, but the necessity for high pressures to attain temperatures which will yield even a moderate thermal efficiency introduces many problems. The vapor pressure of water determines the maximum temperature attainable

---

* Uranium consisting of 90 per cent or more of uranium-235 is frequently referred to as "fully enriched."

in the system and hence the thermal efficiency. At operating pressures of 2000 to 2500 psia, which are close to the practical limit, the gross electrical efficiency of pressurized-water reactors is 28 to 30 per cent. Since pressure vessels of large size are difficult and expensive to build, the maximum power output for these reactors appears to be about 1600 thermal Mw or roughly 500 electrical Mw at present.

13.18. Another problem arises from the fact that at elevated temperatures water is quite corrosive. Similar difficulties are encountered in conventional, high-temperature steam plants and new materials have been developed to overcome them, but most of the problems are more serious in nuclear power plants. This is particularly the case in connection with cladding materials for the fuel. Furthermore, certain corrosive products, in addition to being deposited on heat-transfer surfaces (§ 7.107), may become radioactive; this may necessitate the provision of additional shielding and also complicates maintenance. Some limitation on the use of otherwise desirable materials exists because they have large cross sections for the capture of thermal neutrons. In addition, the possibility of decomposition by radiation or neutron capture makes it difficult to find suitable corrosion inhibitors.

13.19. Despite these drawbacks, the water-cooled reactor in the United States is based on more highly developed technology than any other type. Pressurized-water reactors have a relatively simple design and they have been found to be safe, dependable, and easily controlled. Although nuclear considerations necessitate a close-packed lattice, this does have the advantage of compactness, leading to a high power density. Water-cooled systems have a large negative temperature coefficient, due to the expansion of the water, and hence are very stable. An increase in turbine demand results in a lower temperature in the water returning to the reactor core and this tends to increase the reactivity because of the negative temperature coefficient. The system is thus a "demand follower."

13.20. Although the pressurized-water concept is comparatively highly developed, significant advances are possible which will reduce the cost of power generation. By allowing bulk boiling to occur in some channels, less conservative hot-channel factors may be used in designing the heat-removal system, thus making possible higher power densities. In addition, some improvement in thermodynamic cycle efficiency would be achieved by an increase in bulk temperature of the coolant with reference to the saturation temperature. A reduction in fuel-cycle cost could result from improvement in materials so as to permit increased fuel burnup; present exposures are on the order of 10,000 Mw-days per tonne, but values of 20,000 Mw-days per tonne and even higher are considered to be feasible. Developments in other system components, e.g., pumps and pressure and containment vessels, could also result in some reduction in power costs. By including a device for vapor suppression, e.g., by venting into a pool of water, as described in § 11.39, the cost of the containment vessel could be decreased.　Other possibilities which have been proposed are the use of open, water-filled vessels within the containment space or water sprays which scrub and cool (or condense) the steam and hot water released in an accident, thus reducing the pressure.

BOILING-WATER REACTORS [10]

13.21. Early in the reactor program, the advantages of a boiling coolant system were recognized.　The direct generation of steam within the reactor vessel is not only a very efficient method of removing heat produced in the core but, in addition, the need for a separate heat exchanger (or steam generator) is eliminated.　Moreover, more energy can be removed by the same quantity of water as latent heat than as sensible heat.　Consequently, the coolant circulation requirements for a boiling-water reactor are modest compared with the large pumping rates necessary for a conventional pressurized-water system.

13.22. For many years, the concept of boiling water within the reactor core was considered to be impractical because it was felt that fluctuations in the boiling process would lead to uncontrollable and unstable oscillations in the power output.　It was also recognized, however, that the boiling-water system had some self-controlling characteristics.　Because of its lower density, steam is a poorer moderator than liquid water.　Hence, an increase in power would be accompanied by a loss of reactivity due to an increase in the steam (or void) fraction, i.e., in the volume fraction of water in the core present as steam.　In addition, the factors leading to a negative temperature coefficient in a pressurized water reactor, e.g., expansion of the water, would also be present.

13.23. Whether or not these safeguards would operate rapidly enough to confer stability on a boiling-water reactor could only be demonstrated by experience. A series of tests at Argonne National Laboratory were climaxed in 1953 and 1954 by the Boiling Reactor (BORAX-I) Experiments at the National Reactor Testing Station in Idaho.　The results proved that the formation of steam in the reactor core would reduce the reactivity and cause the system to become self-regulating under reasonable operating conditions.　Thus, a power excursion, due to a sudden relatively large increase in reactivity, would shut down the reactor.　In order to determine the limitations of the self-regulating process, larger and larger amounts of excess reactivity were added to the core by sudden ejection of the control rods.　Ultimately, in 1954, an excess reactivity of 4 per cent resulted in a power excursion which destroyed the BORAX-I reactor.

13.24. Much valuable information regarding the characteristics of boiling-water reactors was subsequently obtained from additional BORAX cores [11]. In 1955, BORAX-II was constructed for operation at a higher pressure than BORAX-I; later, electrical power facilities were added (BORAX-III).　The Experimental Boiling Water Reactor (EBWR), which began operation at Argonne National Laboratory late in 1956, was the first demonstration boiling-water reactor power plant to be completed [12]. It was designed to produce 20 thermal Mw, but its operation was so satisfactory that the thermal power was increased to 60 Mw. As part of the development for the 626 thermal (180 electrical) Mw Dresden Nuclear Power Station near Chicago, which commenced operation in 1960 [13], the Vallecitos Boiling Water Reactor (VBWR) began to supply electrical power in 1957 [14]. There are also a number of later boiling-water power reactors, e.g., Humboldt Bay, Calif., Elk River, Minn., etc. [15].

13.25. Boiling-water reactor designs are similar to those of pressurized-water systems. In the EBWR, the fuel elements were of the plate type and contained the uranium, enriched to 1.44 per cent in uranium-235, in the form of an alloy with zirconium and niobium. The fuel for the Dresden reactor, however, is uranium dioxide (1.5 per cent enrichment) in rod form, as in the later pressurized-water reactors. The fuel element cladding may be made of Zircaloy or stainless steel. Because the need for a separate steam generator can be avoided, boiling-water reactors can operate at lower pressures than equivalent nonboiling systems. Since there are more steam voids in the upper part of the core, there is tendency for the neutron flux (and power) to peak in the lower portion. This flux distortion is partially compensated by operation of the control rods from the bottom in several boiling-water reactors. Provisions are made for rapid insertion of the poison (control) rods in the event of a scram being required (§ 11.20).

13.26. Boiling-water reactors are only slightly less advanced technologically than are pressurized-water systems, since it has been possible to take advantage of fuel and materials development work performed for the latter concept. There are several variations of the boiling-water reactor, but they fall into two main categories, namely, direct-cycle and dual-cycle systems (Fig. 13.2 A and B). In the direct-cycle system the water is boiled in the reactor pressure vessel and steam is fed directly to the turbine. The coolant may either be forced through the core by means of a circulating pump or natural convection may be used. The direct-cycle reactor has the disadvantage of not following turbine demand. As more steam is called for, the reactor pressure is decreased, thus increasing the steam void fraction and decreasing the reactivity, although an increase is actually required. In the dual-cycle system, part of the energy from the reactor forms steam directly which is fed to the turbine, but part of the hot water produced in the reactor is pumped to a (secondary) steam generator. Additional steam is formed in this generator and is used to supply the turbine, the cooled water being returned to the reactor. An increase in turbine demand thus results in a decrease in temperature of the water and, consequently, an increase in reactivity. The demand-following characteristic of the pressurized-water reactor is thus possible in a dual-cycle boiling system [16].

13.27. Although the boiling-water reactor has the advantage, in principle, of eliminating the heat exchanger as steam generator, the introduction of the dual-cycle concept reduces this to some extent and brings the boiling-water and

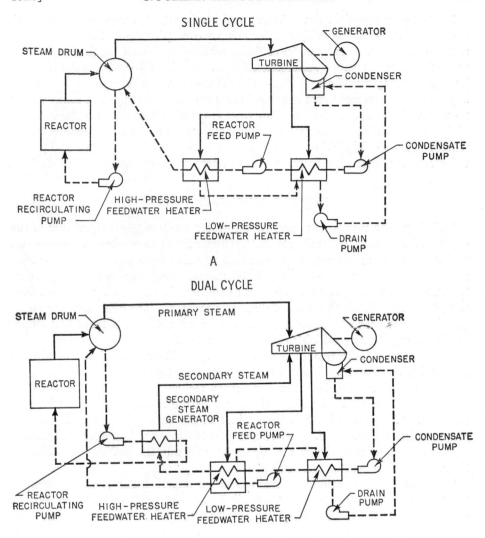

SINGLE CYCLE

A

DUAL CYCLE

B

FIG. 13.2.   Flow diagram of (A) single cycle, (B) dual cycle boiling-water reactor system

pressurized-water reactor types closer together.   The fact that bulk boiling may
be permissible in pressurized water-cooled systems enhances the general similarity.   Consequently, the basic advantages and disadvantages of boiling-water
reactors are essentially identical with those given above for nonboiling systems;
corrosion problems are important and the size of the reactor vessel is limited.
The maximum electrical power attainable from a boiling-water reactor, at
reasonable cost, is believed to be about 300 Mw.

13.28. Some technical development and study are still necessary to exploit the full potential of the boiling-water reactor concept. As in the case of pressurized-water reactors, economies could be achieved by increasing the fuel burnup and by the use of vapor suppression in the containment vessel. An increase in power density would also decrease the net cost of the power produced. Some improvement in components is needed and optimization studies of many of the process variables, e.g., pressure, flow rate, effect of self-cooling on the stability of the system, etc., are desirable. The attainment of higher pressures in the reactor vessel would yield a more efficient steam cycle. The present and potential steam conditions for boiling-water reactors are summarized in Table 13.1.

NUCLEAR SUPERHEATING [17]

13.29. In both pressurized- and boiling-water reactors the temperature of the steam is limited and, hence, so also is the thermodynamic efficiency. Moreover, the steam produced is saturated and this requires the use of more expensive turbine equipment than for superheated steam. One approach to improving the efficiency, which has been adopted in the Indian Point, N. Y. (pressurized water) power plant, is to use an oil-fired superheater. The oil supplies approximately 40 per cent of the total thermal energy input. This fossil fuel superheating is said to be justifiable on economic grounds.

13.30. Studies indicate that if superheating could be carried out within the reactor itself, an appreciable decrease in power cost would be achieved. Such superheating, however, involves a different technology from that utilized in water-cooled reactors. Although the superheating region may be an integral part of a boiling-water reactor, the coolant in this region is now a gas, namely steam, rather than a liquid. Severe service conditions are therefore encountered with a lower power density in the superheater than in the boiling region. Furthermore, the moderator and coolant are now separate.

13.31. Basic information concerning nuclear superheating is to be obtained from the experimental (20-Mw thermal) BORAX-V system, in which boiling and superheating can take place in a single core [18]. The reactor has been designed so as to provide considerable flexibility; the boiling region can be either in the center or at the periphery of the core, with superheating at the periphery or center, respectively. The fuel elements are in the form of rods in the boiling part of the core and plates in the superheating region. In addition to this experimental reactor, nuclear superheating is incorporated in two power prototype systems, namely the Pathfinder Station, near Sioux Falls, S. D. [19], and the BONUS (Boiling-Nuclear Superheat) reactor in Puerto Rico [20]. Although significant reductions in power costs are expected for large-scale power facilities of this type, considerable research and development are required to overcome the problems likely to be encountered.

13.32. For boiling-water reactors, the nuclear superheater may be either of the "integral" type, with the superheating being achieved in the same reactor as produces steam, as described above, or of the "separate" type which utilizes steam from another reactor. The latter system can, of course, be added to any steam-generating system. Although separate or "add-on" superheating is simpler in some ways, it would be necessary, for economic reasons, to build a large superheater reactor and this would require two or three ordinary reactors to supply the steam. The total capital investment would be very considerable and the amount of power produced could be handled by only the largest utility networks. It would appear, therefore, that the integral type of nuclear superheater has a better economic potential but the technological difficulties are greater. Among them is the problem of preventing liquid water from the boiling region from entering the superheater and causing the latter to become supercritical.

13.33. In one design for an integral boiler superheater reactor, the core consists of two regions: an outer (or peripheral) region where the water boils and produces steam; and an inner (or central) region, in which higher neutron fluxes and higher temperatures can be attained, where some water is boiled and the steam is superheated. In the peripheral (or boiler) region, fuel rods of the usual type can be used, but in the center (boiler-superheater region) it is proposed to employ double-annular fuel elements, shown in cross section in Fig. 13.3; water is boiled at the outer surface and steam is superheated at the inner surfaces [21]. A rod of burnable poison may be inserted in the center of the inner fuel annulus. With this arrangement, the saturated steam and superheated steam

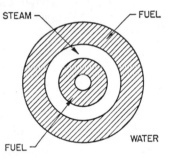

FIG. 13.3. Section of nuclear superheater fuel element

are separated by fuel material, and the introduction of a neutron-absorbing substance, like stainless steel, is avoided. Feedwater entering the reactor vessel is divided into two streams: one flows to the peripheral (boiler) region and the other to the outer surfaces of the boiler-superheater annular fuel elements. The saturated steam thus produced is led to the inner channels of these elements where it is superheated. Other concepts for integral superheaters have been proposed [17].

HEAVY-WATER MODERATED REACTORS [22]

13.34. The principal advantage of moderation by heavy water is that natural uranium can be used as the reactor fuel, and so such reactors are independent of sources of enriched uranium. For long exposure times, however, some enrichment of the uranium or addition of recycled (or recovered) plutonium may prove to be desirable. Heavy-water reactors appear to have an economic potential comparable with other types, especially where natural uranium can be obtained cheaply.

13.35. Although heavy-water moderated power reactors are still a novelty, considerable experience has been gained with reactors of this type designed for research or for plutonium production. The Canadian NRX and NRU experimental reactors, for example, have been in successful operation for several years. In the United States, the production reactors at Savannah River (Aiken, S. C.) have provided many advances in heavy-water technology. One of the disadvantages of heavy-water reactors is the capital investment in the moderator; this may amount to as much as 20 per cent of the total investment in a large power plant. In addition, special precautions must be taken in the design of the plant to minimize loss of the expensive heavy water ($28 per pound) and to prevent absorption of ordinary water. In the Savannah River reactors it has been possible to maintain a purity of about 99.7 per cent D_2O by distillation.

13.36. Since heavy-water lattices need not be as closely packed as those using ordinary water as moderator, it is possible to utilize the "pressure tube" approach and thus avoid the problem of large pressure vessels inherent in systems in which ordinary water is the moderator-coolant. In this scheme, a large tank of heavy water, at ordinary pressure, is pierced with a number of tubes; the coolant flows in the annular channels surrounding the cylindrical fuel elements placed in these tubes. The latter are maintained at a pressure of 500 to 1500 psi to prevent boiling; it is for this reason that the term "pressure tube" is used. The heavy water in the tank acts as the moderator and remains at low temperature. To take effective advantage of the good moderating properties of heavy water, however, it is necessary to minimize neutron absorption in structural components. The selection of pressure-tube and cladding materials may thus present some problems. It appears that the use of zirconium alloys would be mandatory, since the maximum moderated flux must pass through the tubes and cladding to reach the fuel. Where a pressure vessel is used, the choice of materials is less critical, although neutron absorption must be kept low if natural uranium is the fuel.

13.37. A number of prototype reactors are under construction and many others are in the design phase. The essential characteristics of some of these are outlined below.

Heavy-Water Components Test Reactor (HWCTR), at the Savannah River Plant, is to have a thermal power of 16 Mw but will produce no electricity. It uses heavy water as both moderator and coolant in a pressure-tube system, with a natural uranium-zirconium alloy as the fuel material. The basic purpose of the HWCTR is to provide information on the technology of heavy-water moderated power reactors [23].

Plutonium Recycle Test Reactor, located at Hanford, Wash., is intended to further the development of recycled plutonium as a fuel. The reactor is of the pressure-tube type cooled and moderated by heavy water.   It is designed for a thermal output of 70 Mw with no electrical power [24].

Carolinas-Virginia Tube Reactor (CVTR), at Parr Shoals, S. C., is of the pressure-tube type, rated at 60.5 thermal Mw and 17 electrical Mw, using slightly enriched uranium dioxide as fuel.   It is expected to be the first heterogeneous heavy-water reactor to produce electric power in the United States [25].

Nuclear Power Demonstration Reactor (NPD-2) in Canada has a thermal power of 89 Mw and a gross electrical output of 22 Mw; the fuel is natural uranium dioxide pellets jacketed with Zircaloy-2.   A unique feature of the NPD-2 is that the pressure tubes are horizontal and a special machine permits bidirectional charging (and discharging) of fuel elements while the reactor is in operation (§ 11.17).   This procedure permits efficient fuel utilization without the expense of shutdowns.   The purpose of the NPD-2 reactor is to determine the feasibility of a large-scale CANDU (Canadian Deuterium Uranium) power plant (794 thermal Mw, 208 electrical Mw) of essentially the same design [26].

Halden Boiling Heavy Water Reactor, in Norway, is the only known reactor in which the coolant is boiling heavy water.   The operating pressure is about 575 psia and the temperature of the heavy-water steam is approximately 250°C (480°F); the thermal power of the reactor is 10 Mw.   In order to maintain criticality at the higher temperatures, it is necessary to use slightly enriched uranium as the fuel material [27].

13.38. In all the reactors described above, heavy water is both the moderator and coolant; steam is produced from ordinary water in a heat exchanger of the usual type.   Other concepts which have received (or are receiving) serious consideration employ heavy water as the moderator but with another material, such as sodium or an organic liquid, as the coolant [28].   The core is of the pressure-tube type, although the coolant is not under any significant pressure. Heavy-water moderated, gas-cooled reactors have also been proposed and these are discussed in the section on gas-cooled systems.

THE SPECTRAL SHIFT CONTROL CONCEPT

13.39. Mention has been made earlier (§ 5.154) of the Spectral Shift Control Reactor (SSCR) concept in which the core is essentially a tight lattice of slightly enriched fuel, such as is used in a pressurized-water reactor [29].   In order to avoid the wasteful use of control rods and burnable poisons to compensate for the excess reactivity at the beginning of the core life, the initial moderator-coolant consists of heavy water (about 70 mole per cent D_2O).   The under-moderated core has little excess reactivity, since there is an increased absorption of thermal neutrons by uranium-238, or preferably thorium-232 because of breeding possibilities, with the formation of fissile material.   As fuel is consumed and fission product poisons accumulate, the addition of ordinary water to the heavy water improves the moderation and restores the reactivity. This dilution is continued until the D_2O content of the moderator-coolant is about 5 mole per cent. The capture of excess neutrons in fertile material, instead of in poisons, should make possible fuel burnups on the order of 20,000 Mw-days per tonne.

13.40. Another aspect of the SSCR is that, once full power operation has been attained, conventional control rods can be withdrawn and control maintained by adjustment of the D_2O/H_2O ratio in the moderator. The elimination in this manner of flux distortion by control rods together with suitable loading of the fuel should result in a considerable improvement in flattening of the power distribution. In this way, the maximum thermal flux and the specific power of the reactor can be increased. The benefits of increased burnup and specific power are partly offset by the cost of recovery of the heavy water by distillation, and the replacement of losses. It is claimed, however, that the net cost of power can be decreased by means of the SSCR concept.

ORGANIC-COOLED REACTORS [30]

13.41. The advantage of organic fluids as coolants and moderators was recognized as early as 1944. The moderating properties of diphenyl ($C_{12}H_{10}$), for example, are not very different from those of ordinary water, and both require fuel of slight enrichment, at least, to make a self-sustaining fission chain possible. The concept was not pursued for several years, however, because decomposition of the organic liquid in the combined radiation and thermal environment of a reactor core was thought to be intolerable. The unavailability of enriched uranium at the time contributed to the minor interest in this reactor concept.

13.42. As corrosion and pressure-vessel problems became apparent for water reactors, attention was again paid to the possibility of using organic materials of low vapor pressure as moderator-coolants. Since these substances are noncorrosive and have high boiling points, relatively thin reactor vessels of carbon steel should be adequate. Studies were therefore made of the thermal and radiation stability of polyphenyls, including diphenyl and terphenyl isomers and mixtures of these compounds, because it was known that they are more resistant to radiation than any other organic materials which have been tested. Furthermore, bulk temperatures of 700°F were found to be possible without excessive thermal decomposition.

13.43. It was realized that there would be some decomposition of the organic material (§ 7.29) and this was taken into consideration in the design. The effect of irradiation and high temperature is to produce polymers of high molecular weight and high boiling point, called "high boilers." The relationship between the total quantity of high boilers formed, per megawatt-hour of reactor exposure, and the high-boiler content of the irradiated material is shown in Fig. 13.4 for various single polyphenyls and mixtures. It is seen that, when the high-boiler content reaches about 30 per cent by weight, the degree of radiolysis, for a given reactor energy, approaches an asymptotic value [31].   Consequently, a high-boiler content of 30 per cent was accepted for the circulating fluid in the reactor. This composition is maintained by continuous purification, using low-pressure distillation, and make-up with fresh material.

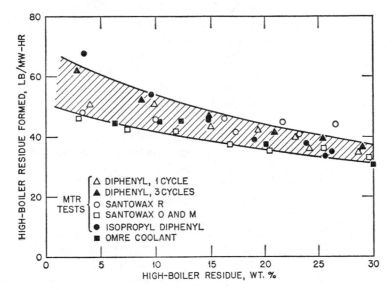

Fig. 13.4.   Rate of decomposition of organic coolant-moderator

13.44. There is a possibility that certain additives may be able to inhibit partially the radiolytic decomposition of polyphenyls.   If these prove to be practical, a decrease in the high-boiler content of the coolant may be feasible.   The consequent decrease in viscosity and increase in specific heat would improve the heat-transfer properties of the coolant and reduce the pumping power.

13.45. The Organic Moderated Reactor Experiment (OMRE) has a maximum design power of 12 thermal Mw, but no provision for electrical power production; it began operation at the National Reactor Testing Station in September 1957 [32].   The coolant (Santowax) was a mixture of diphenyl and the three (ortho, meta, and para) terphenyls, with a boiling point at atmospheric pressure of about 640°F.   Operation was satisfactory and coolant outlet temperatures of over 700°F were attained with a system pressure of 200 psig or less.   The OMRE thus proved the technical feasibility of this reactor concept.   A considerable amount of information was obtained about coolant stability, heat-transfer conditions, and operating characteristics.   The corrosion of carbon steel used in the organic circulating system was negligible at the coolant flow velocity of 15 ft per sec, which was chosen to minimize the pressure drop through the reactor while providing adequate heat-transfer capability.

13.46. The first organic moderated and cooled reactor designed to produce electricity is the one constructed in 1961 for the city of Piqua, Ohio [33]. The thermal power is 45.5 Mw and steam is generated at 550°F and 450 psia; the electrical output is 11.4 (net) Mw. The fuel material is an alloy of metallic uranium (1.94 per cent uranium-235) with 3.5 weight per cent of molybdenum and about 0.1 weight per cent of aluminum, metallurgically bonded to aluminum cladding with a diffusion barrier of nickel about 1 mil thick. Each fuel element is in the form of two concentric tubes, the inner and outer surface of each tube being finned. The coolant, which is a mixture of terphenyls, flows in parallel through the central hole in the inner tube, in the annular space between the tubes, and around the outside of the outer tube. The pressure in the reactor vessel of low-carbon steel, 27 ft high, 7.6 ft diameter, and 2 in. wall thickness, is 135 psig.

13.47. In order to learn more about the potentialities of organic-cooled and moderated systems, the Experimental Organic Cooled Reactor (EOCR) has been constructed at the National Reactor Testing Station [34]. It does not produce electric power and has been designed as an experimental facility for testing new organic coolants and various types of fuel elements. The thermal power of the EOCR is 40 Mw.

13.48. Among the advantages of the organic reactor concept are the comparatively low capital costs, resulting from the use of inexpensive materials and the fairly low pressure requirement for the reactor vessel and the primary coolant loop. Design pressures of about 150 psi, for example, are common. Because of the high temperatures which are possible even at these moderately low pressures, superheated steam can be produced directly in the heat exchanger. As with water, the radioactivity induced in the pure coolant by neutron capture is low. Furthermore, the absence of chemical reaction with the organic material should make possible the use of fuels of various types, e.g., metallic alloys and uranium oxide and carbide. Of these, it is expected that the oxide will provide the highest burnup.

13.49. The low capital cost of an organic-moderated reactor is partially offset by the cost of purification of the coolant and the replacement to make up for radiolytic and thermal decomposition. The coolant is flammable in air at temperatures above about 415°F, and the highest operating temperature, regardless of the pressure, is about 800°F. Since the organic fluid has relatively poor heat-transfer properties, it is necessary to design the fuel elements with extended areas, e.g., by the use of fins. Zirconium and its alloys cannot be used as cladding, because of hydrogen embrittlement, and stainless steel is not too satisfactory because of its low thermal conductivity. Consequently, the APM alloys, described in § 7.40, are specified for fuel-element cladding in organic cooled reactors. Fouling of the fuel elements by organic decomposition products is a problem that arises during operation of the reactor.

SODIUM-GRAPHITE REACTORS [35]

13.50. Another approach to the attainment of high coolant temperatures without pressurization is by the use of liquid metals as coolants. Of these, sodium is the most favorable since it meets the requirement of excellent heat-transfer properties without excessive neutron absorption. Its relatively high melting point (98°C, 208°F) and its chemical reactivity with air and, especially, with water are its more serious disadvantages.

13.51. Studies were commenced in 1949 of a thermal reactor concept, based on sodium as the coolant and graphite as the moderator, which would produce plutonium-239 at low cost. Although such a reactor would operate on natural uranium, it appeared that an improvement in performance would result from the use of a slightly enriched fuel. These early studies indicated that a dual-purpose plant, producing both plutonium and electric power, would be practical but it was decided that further efforts should be concentrated on a reactor system for power alone. The work led to the development of the Sodium Reactor Experiment (SRE) which attained criticality early in 1957 and first produced power in July of that year [36]. In its design, advantage was taken of the experience gained with sodium-cooled systems in the work on the Submarine Intermediate Reactor, which was the prototype of the first power plant installed in the U.S.S. Sea Wolf, and also in connection with the Experimental Breeder Reactor program (§ 13.83).

13.52. The core of the SRE is composed of hexagonal graphite "logs" canned in zirconium; each log has a central channel containing a fuel element. In the first core the fuel consisted of metallic uranium (2.8 per cent enriched) slugs jacketed in stainless steel. In subsequent cores, thorium plus 7.6 weight per cent of highly enriched uranium and slightly enriched uranium carbide will be tested as fuel materials. The sodium coolant flows in a single pass through the annular space between the fuel element and the surrounding graphite log. The cross section of the reactor vessel in Fig. 13.5 shows some of the design features of the SRE.

13.53. An intermediate heat exchanger is used to transfer the reactor heat to a secondary sodium circuit which is connected to the steam generator. The maximum outlet temperature of the sodium from the reactor is 1010°F with a rise of 480°F across the core; the average steam temperature is about 950°F. With sodium as the coolant, it is possible, of course, to produce superheated steam directly. The maximum thermal power attained was 20 Mw, although higher values were possible; the electrical output was 6 electrical Mw, in accordance with the design specifications.

13.54. As expected from an experimental prototype of a new reactor concept, experience has demonstrated both the strong points and the weaknesses of the design. The reactor, with its turbo-generator, has proved to be exceptionally

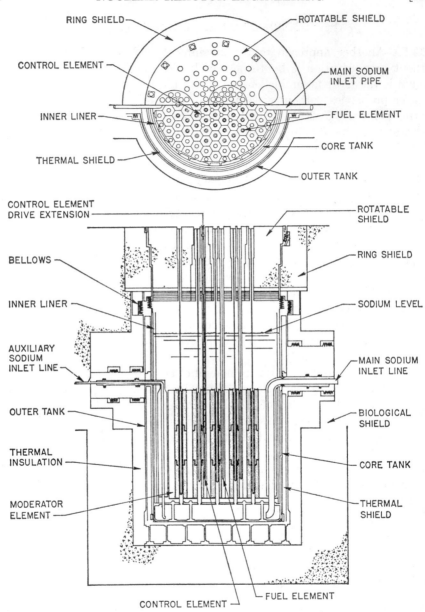

FIG. 13.5.   The Sodium-Graphite reactor (SGR)

stable and easy to control under transient conditions.   It has been possible
to follow easily the changes in load by adjusting the reactor power level.   Numerous techniques have been developed for removal of components which are
immersed in sodium and maintenance has proved to be relatively simple.

13.55. Most of the difficulties encountered in the early stages were mechanical in nature, e.g., failure of valves, improper operation of cold traps in the sodium, and leakage into the primary circuit of tetralin, which was used to cool the seals of the sodium freeze-seal pumps. These difficulties have been overcome by appropriate development work and replacement of the tetralin by other coolants. Most of the problems with the reactor core arose from radiation damage to the metallic uranium fuel, which was not unexpected, and failure of the fuel cladding on one occasion.

13.56. The Hallam Nuclear Power Facility (HNPF), in Nebraska, is essentially a scaled up version of the SRE; it is designed for a thermal power of 254 Mw with a gross electrical output of 80 Mw [37]. The main difference between the HNPF reactor and that of the SRE is that stainless steel is used, instead of zirconium, for the moderator cladding. The fuel channels have been moved to the corners of the hexagonal graphite logs in order to improve the neutron economy. The fuel for the initial loading of the HNPF reactor is an alloy of uranium (3.55 per cent uranium-235) with 10 weight per cent of molybdenum. The swelling of this material with exposure, at a maximum central temperature of 1260°F, limits the anticipated burnup to an expected average of 8000 Mw-days per tonne. In subsequent loadings it is planned to use uranium carbide as the fuel because of its high thermal conductivity and radiation stability (§ 8.66). The burnup should then be increased to about 10,000 Mw-days per tonne with a central temperature of 1600°F.

13.57. From the economic point of view, the potential of the sodium-graphite reactor concept lies in the possibility of constructing large plants. Since the system operates at essentially normal pressures, there are no great problems in connection with the size of the reactor vessel. In addition to reducing the capital costs, a large power station has a higher net plant efficiency than a smaller one because of the inherent advantages of large, reheat-cycle turbo-generators and improvement in the feedwater heating cycle. A 300-electrical Mw sodium-graphite power station, incorporating all conceivable improvements, is under consideration for the late 1960's.

13.58. The most important single item in the economic success of the sodium-graphite reactor concept is the need for the development of a fuel element which will provide a long burnup at the high temperatures and specific powers that are possible with sodium as coolant. It is, of course, these factors which make cooling by sodium of such great interest. The fuel should have a high thermal conductivity and the ability to withstand exposures of 17,000 Mw-days per tonne or more at a surface temperature of 1200°F. Uranium dioxide does not appear to be satisfactory because of its low thermal conductivity, and so attention is being devoted to carbides and similar ceramic materials.

GAS-COOLED REACTORS [38]

13.59. The use of a gas for cooling a large reactor was considered seriously in the United States as far back as 1943. When the Hanford (graphite moderated) reactors for plutonium production were first designed, helium was the preferred coolant; it was not adopted, however, largely because of the difficulties anticipated from the prevention of leakage of the expensive helium gas from the large pressure vessels that would be necessary to contain the reactors. The coolant actually employed in the Hanford reactors was water, an abundant supply of which was available from the Columbia River. In the United Kingdom air was employed as the coolant for the Windscale plutonium production reactors. There were two main reasons for this choice: first, the lack of a suitable water supply, and second, the possibility of an accident, due to an increase in reactivity, that might result from loss of the neutron-absorbing water coolant in a graphite-moderated reactor.

13.60. Because of the successful operation of the Windscale reactors, a natural evolution was the development of the dual-purpose Calder Hall and Chapel Cross stations in the United Kingdom for the production of plutonium and electrical power. These served as prototypes for a series of gas-cooled reactor power plants with an installed electrical capacity that is expected to total 5000 electrical Mw by 1968 [39]. The sectional drawing in Fig. 13.6, which is of the Hinkley Point installation, is typical of these nuclear power systems. In all the reactors, which are graphite moderated, the coolant is carbon dioxide at a pressure of 150 to 250 psia, and the fuel is natural uranium metal clad with a thin layer of

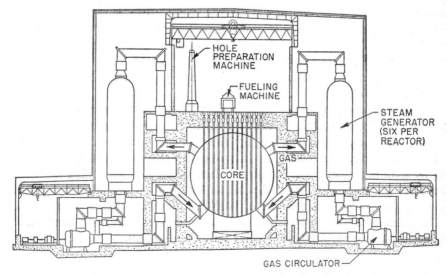

Fig. 13.6. Sectional representation of Hinkley Point reactor system

Magnox (§ 7.40).   The choice of the fuel material was originally determined by the shortage of enriched uranium in the United Kingdom, but it has been retained because it proved to be economically satisfactory.   Special finned designs of fuel elements of various types have been developed in order to facilitate heat removal.   Steam temperatures are limited to about 650°F and fuel burnup, which is determined by metallurgical damage, is about 3000 Mw-days per tonne. It is estimated that this would be almost doubled if reactivity loss, rather than irradiation, were the limiting factor.

13.61. The reason why natural uranium, gas-cooled power reactors have not attracted attention in the United States is because the plants are more expensive to construct than corresponding water-cooled systems.   As a result the fixed charges on capital are greater.   In the United Kingdom the difference is less significant and is compensated by the use of a cheaper fuel material, because financial charges are only 6 to 8 per cent compared with 14 per cent in the United States.   The price of nuclear power in the United Kingdom is thus not greatly different from that derived from fossil fuels.   Incidentally, these conventional fuels are relatively more costly than in the United States.

13.62. Since 1957 there has been a revival of interest in the United States in gas-cooled reactors using somewhat enriched fuel.   Such reactors would be smaller, for a given electrical output, than those fueled by natural uranium, and so would be more economical because of the lower capital charges.   However, the full potentialities can be realized only in plants of large size, partly because the high cost of auxiliary systems is largely independent of the total output. Additional benefits can be gained by substituting uranium dioxide for uranium metal, which would permit both a larger burnup and higher steam temperatures. With metallic uranium, the temperature of the fuel-element surface, and hence of the coolant, is also limited by the possibility of a change from the alpha to the beta phase (§ 8.37) in the center of the element.   Magnox could not be used as cladding for the fuel elements, owing to the increased temperature, but other materials are available.   Because of the chemical reaction between carbon dioxide and graphite, the maximum coolant temperature for this system is about 650°F.   Since higher temperatures are desirable, it would be necessary to use either a different moderator, e.g., beryllium oxide, or a different gas, e.g., hydrogen or helium.   Although hydrogen is much cheaper, the risk of explosion has made helium the choice for advanced, gas-cooled reactors in the United States.

13.63. In countries where helium is either not available or costly, the use of this gas may not be practical, especially as it is virtually impossible to eliminate loss by leakage.   In the Calder Hall reactors, for example, the initial leakage rate was 2 tons of carbon dioxide per day for each reactor; this has now been reduced to 0.5 ton per day.   There is thus considerable incentive in Europe and elsewhere to develop satisfactory methods for using carbon dioxide at higher temperatures than are now feasible.   An experiment in this connection is the Advanced Gas-Cooled Reactor (AGR) in the United Kingdom, designed to produce 100 Mw of heat and 28 Mw of electricity [40]. The fuel rods are made of 1.2 per cent enriched uranium dioxide canned in beryllium, but stainless steel will also be tested as cladding material. The carbon dioxide coolant is at a pressure of 270 psi and its exit temperature is 500°C (930°F), which will later be increased to 575°C (1070°F). In the latter circumstances, the steam conditions will be 455°C (850°F) at a pressure of 650 psi.

13.64. Despite the extensive experience with gas cooling in the United Kingdom and with various fuels in the United States, the technology of economic gas-cooled reactor using partially enriched fuel requires further development. Useful information has been obtained from the so-called Heat Transfer Reactor Experiment (HTRE) series [41], carried out at the Nuclear Reactor Testing Station as part of the program to study the feasibility of using a direct-cycle reactor for the propulsion of aircraft. The series involved compact reactors, moderated with water or zirconium hydride, operating at thermal power levels up to 32.4 Mw with compressed air as the coolant. The HTRE reactors were designed for somewhat extreme conditions, such as gas exit temperatures up to 1640°F. The fuel elements consisted of uranium dioxide, enriched in uranium-235 to 93 per cent, in a matrix of 80 per cent nickel–20 per cent chromium, and clad with a niobium-stabilized form of the same alloy.

13.65. The Experimental Gas Cooled Reactor (EGCR), at Oak Ridge, Tennessee, is intended largely as a facility for testing advanced fuel materials and components and also to provide data on the economics of gas-cooled power reactors [42]. The fuel elements consist of rods made up of uranium dioxide pellets (12.5 per cent uranium-235) jacketed in stainless steel, and graphite is the moderator. The helium coolant, at a pressure of 330 psia, enters the core at 510°F and leaves at 1050°F. Steam at a temperature of 900°F and a pressure of 1300 psia will be produced in a heat exchanger. The design thermal power of the EGCR is 85 Mw and the electrical output 29.5 Mw (gross) or 22.0 Mw (net). The system is expected to be the prototype of a larger reactor, with a net electrical power of 225 Mw, which will be contained in a spherical pressure vessel 50 ft in diameter. Each of the four steam generators would be 20 ft in diameter and 60 ft high.

13.66. A second approach is adopted in the proposed High Temperature Gas Cooled Reactor (HTGR) for Peach Bottom, Pa. [43]. The fuel is to be 13.8 per cent uranium-235, about 1 per cent of uranium-238, and the remainder normal thorium, in the form of the respective carbides dispersed in graphite. The cladding is a dense, impervious graphite and the moderator is also graphite. Helium at a pressure of 350 psia will enter the core at 660°F, leaving at 1380°F, and will provide 1000°F steam at 1450 psia at the turbine. The thermal design power is 115 Mw and the net electrical power 40 Mw, giving an unusually high net thermal cycle efficiency of nearly 35 per cent. The main specifications of the DRAGON reactor experiment at Winfrith, in the United Kingdom, are similar to those of the HTGR; however, the thermal power will be 20 Mw and no electricity production is planned. The reactor is a joint project of twelve European countries and is designed to provide information for a High Temperature Gas Cooled (HTGC) reactor [44].

13.67. There are some important differences between the EGCR and the HTGR concepts. First, the former contains uranium-238 as the fertile species, whereas in the latter it is thorium-232. The objective in the HTGR is thus to attempt thermal breeding of uranium-233 as a contribution to lower power costs. It is expected that advanced gas-cooled reactors may contain an even larger proportion of thorium. Second, the use of a graphite-impregnated fuel is expected to permit a higher specific power, higher power density, and larger burnup than is possible with a clad uranium dioxide fuel. In fact an exposure of 75,000 Mw-days per tonne of uranium plus thorium is expected for the HTGR. If this is achieved, it will mean a significant reduction in power costs.

13.68. The third difference lies in the use of 2.5 per cent enrichment in the EGCR and 13.8 per cent in the HTGR. It is not possible to determine at this time which will be the more economical, and so both approaches are being pursued. An interesting fact to bear in mind is that the cost of enrichment is greater at the lower concentrations. For example, the difference in cost per gram of contained uranium-235 is about the same in going from 0.72 per cent, i.e., natural uranium, to 2.5 per cent as it is from 2.5 to 93 per cent enrichment (see Fig. 14.5). On a fissile atom basis, therefore, it makes only a minor difference whether the fuel used contains 5 or 95 per cent of uranium-235. However, the highly enriched material may be difficult to process because of criticality considerations.

13.69. High enrichment has the advantage of permitting some under-moderation of the lattice, so that the size of the core can be reduced. In addition, since the premium on neutron economy is not as important in a highly enriched as in a slightly enriched system, a graphite of poorer quality and lower cost may be adequate. Other benefits arise from the high specific powers that are possible with enriched fuel. For example, a decrease in the size of the reactor will result in a decrease in capital costs. An increase in the bulk gas outlet temperature, which would be possible, would reduce the heat-exchange requirements and lead to additional savings in capital expenditure. Furthermore, fuel rods of comparatively large diameter capable of large burnup can be used, thereby decreasing fuel-cycle costs.

13.70. The gas-cooled reactors described above do not present any particularly unconventional features. There are, however, some that do, and these will be reviewed. In the Pebble-Bed reactor concept, the fuel is in the form of ceramic spheres 1.5 to 2.5 in. in diameter [45]. The spheres contain fissile and fertile material, e.g., uranium carbide or uranium and thorium carbides, dispersed in graphite. A coating of impervious graphite provides resistance to the escape of fission-product gases. The proposed coolant is generally helium gas. The expected advantages of the pebble-bed concept arise from the use of simple mobile fuel elements; they permit relatively easy charging and discharging, and reprocessing and fabrication. The large heat-transfer surface and the absence of metal cladding should make operation possible at high temperatures (1250° to 1500°F) and at high power densities. Although a number of pebble-bed reactor designs have been proposed, only one such reactor is under construction; this is the Brown Boveri-Krupp (40 thermal Mw) installation in Germany. The coolant is a mixture of helium and neon obtained in the distillation of liquid air.

13.71. An advanced gas-cooled reactor concept of unusual design is the Ultra-High Temperature Reactor Experiment (UHTREX), which was originally called the TURRET system because refueling is to be carried out during operation by rotating the entire core to align it with a refueling machine [46]. This reactor is being designed and constructed at the Los Alamos Scientific Laboratory. The purpose of UHTREX is to test the feasibility of using unclad, graphite fuel elements impregnated with enriched uranium carbide. The coolant is helium at a pressure of 500 psia and the fission product gases released to the gas stream are to be continuously removed by a charcoal purification system. The design thermal power of the reactor is 3 Mw. An interesting application of the plant is as a source of process heat rather than of electricity.

13.72. The very high temperatures possible in gas-cooled reactors opens up the possibility of using the coolant gas in a closed-cycle gas turbine to produce electricity without the necessity for generating steam. In the HTRE series (§ 13.64), for example, the reactor was coupled with a turbojet engine. Although the direct production of electricity in this manner is feasible, it is not clear whether there would be any significant advantage for large-scale, central-station power plants. There are, however, certain special applications for which a closed-cycle gas turbine system may be of special interest, namely, for ship propulsion and for small power plants in remote locations.

13.73. In addition to its use in conjunction with a closed-cycle gas turbine, the Maritime Gas Cooled Reactor (MGCR) is unusual in the respect that it is being designed to use beryllium oxide as moderator [47]. The advantages of the latter over graphite are expected to be better neutron economy in small cores, giving a higher conversion ratio and even the possibility of thermal breeding of uranium-233, greater safety in the event of a loss of coolant accident, and compatibility with various structural materials. In the MGCR the proposed fuel elements are rods of 5 per cent enriched uranium dioxide diluted with beryllium oxide and clad probably with stainless steel. The moderator will consist of large blocks of beryllium oxide contained in a cylindrical core 6.4 ft in diameter and height, surrounded by a 7-in. thick reflector. The helium coolant, at a pressure of about 1100 psia, will enter the reactor at 880°F and leave at 1500°F. The design power is 64 thermal Mw with an output of 30,000 shaft horsepower, which is equivalent to 23 electrical Mw. Before constructing the MGCR, experience will be gained with the fabrication and use of beryllium oxide as moderator in a flexible, gas-cooled experimental facility called the Beryllium Oxide Reactor Experiment (BORE) with a thermal power of 10 Mw [48].

13.74. The purpose of the Gas Cooled Reactor Experiment (GCRE) is to provide the basis for the design of the ML-1, a mobile plant for the production of 300 to 500 kw of electrical power for use by the U. S. Army in remote locations [49]. The ML-1 is to be gas cooled and water moderated, and will be coupled with a closed-cycle gas turbine. The fuel is to be highly enriched (93 per cent) uranium dioxide, and the working fluid is nitrogen gas containing 0.5 per cent by volume of oxygen. Nitrogen gas was chosen because it can be produced in the field and also permits the use of existing technology on gas turbines and compressors. The purpose of the small amount of oxygen is to prevent nitriding and consequent embrittlement of piping and structural materials. The nominal gas temperatures are to be about 800°F at the reactor inlet and 1200°F at the exit. The first core of the GCRE used four concentric cylindrical fuel elements, but as a result of a design change in the ML-1 reactor, they were replaced by pin-type elements consisting of either uranium dioxide alone or mixed with beryllium oxide contained in tubes of Hastelloy-X.

13.75. It is probable that nuclear rocket development will contribute significantly to gas-cooled reactor technology. The first of the Kiwi reactor series, for example, was a graphite-moderated reactor of this type using flat plate fuel elements with hydrogen gas as the coolant. The plates were approximately 0.25 in. thick with 0.05-in. wide coolant passages separating them. Although performance specifications of this and subsequent Kiwi reactors have not been released, it may be assumed that to develop the thrust required for rocket propulsion the exit gas temperature must have been very high. Furthermore, since propulsion systems should be compact, extremely high power densities must be attained [50].

HEAVY-WATER MODERATED, GAS-COOLED REACTORS [51]

13.76. The combination of good moderating power of heavy water and the low neutron capture cross section and high-temperature capability of a gaseous coolant would appear to make possible a practical power reactor design using natural uranium as fuel. One serious problem in such a reactor system is that of insulating the gas channels from the heavy water, in order to minimize the flow of heat into the moderator. Heat leakage of this kind cannot be utilized at high efficiency, and so the overall plant efficiency would be reduced.

13.77. In one design, the heavy water is contained in a tank penetrated by pressure tubes containing the fuel elements, with the coolant flowing in the annular channels in between, as described in § 13.36. The pressure tubes are insulated from the surrounding moderator to minimize heat losses. In order to obtain reasonable power densities, i.e., high rates of heat removal, the gas coolant is usually at high pressure, e.g., 500 psia. Since the moderator is not pressurized, there are no serious problems in the design of a reactor of large size. Both carbon dioxide and steam have been considered as coolants for a system of this type; the specific power expected is about 15 thermal Mw per tonne of uranium, which is about twice that attained in corresponding graphite-moderated reactors. Although the gas-cooled, heavy-water moderated reactor has considerable potential, some improvement in neutron economy of the cladding material may be necessary if natural uranium is to be the fuel material. Beryllium is being considered for this purpose in design studies for a 50 electrical Mw prototype power reactor, which might use un-enriched uranium dioxide as the fuel.

FAST REACTORS [52]

13.78. In a fast reactor no effort is made to slow down the neutrons before they are captured by fissile material. In fact, the principal objective of such a reactor is the breeding of plutonium-239 from uranium-238, which is possible only in a fast-neutron spectrum (§ 12.130); consequently, slowing down must be minimized. Average neutron energies before capture, however, are on the order of several hundred kilovolts, since inelastic collisions with structural and fuel materials lead to some loss of energy. In order to avoid slowing down by elastic scattering and also to take advantage of the high temperatures which are possible in a fast reactor, sodium appears to be the most practical coolant at the present time. Sodium-potassium alloy (NaK) has been used because of its lower melting point, but its heat-transfer properties are inferior to those of sodium. However, other metals, such as lithium which is an excellent coolant and has a small elastic scattering cross section, may prove advantageous if suitable structural and cladding materials can be developed for high-temperature operation.

13.79. Fast reactors have a number of characteristics which will be reviewed briefly. Because of the absence of moderator, the core is generally small and the power density, i.e., power per unit volume, is high. This makes it essential to use a coolant with good thermal properties, e.g., a liquid metal. The specific power, i.e., power per unit mass of fuel, is not necessarily high since a large amount of fissile material is required for criticality in a fast reactor. It is, nevertheless, desirable to design the reactor so as to have a high specific power, since this decreases both fuel charges and the doubling time for breeding (§ 12.127). Because of the large critical mass, the macroscopic capture cross sections of fission products and other poisons is small in comparison with the macroscopic fission cross section of the fuel. Consequently, in a fast reactor there is more flexibility in the choice of constructional materials, and the accumulation of fission products does not represent a limitation on burnup. For this reason, and also because the temperature coefficient of reactivity is low, the excess reactivity requirement of a fast reactor is small (cf. Table 5.7). As a result of the Doppler effect, it is possible for a fast reactor to have a positive temperature coefficient, but in reactors designed for economical power production the proportion of fertile material should be large enough for the fuel coefficient to be negative (§ 5.85).

13.80. Since the reactivity temperature coefficient of a fast reactor is small, it is desirable that the coolant provide a negative reactivity effect. However, the contribution of the coolant should not be large, so that the excess reactivity in the cold core can be minimized in the interest of safety. Increase of temperature causes the coolant to expand and this has two opposing effects: first, there is an increase in neutron leakage, which leads to a decrease in reactivity; and second, the hardening of the energy spectrum favors fission of both uranium-238 and plutonium-239, so that the reactivity tends to increase. Inasmuch as small core size, in particular, favors the first effect, the requirement that the coolant have a negative temperature coefficient of reactivity sets a limit upon the size of the reactor. This limitation can be overcome by increasing the ratio of diameter to height of a cylindrical core and thereby increasing the neutron leakage. There is an accompanying decrease in the internal breeding ratio (§ 12.131), but since most of the neutrons leaking from the core will be captured in the blanket, the overall breeding ratio will not be greatly affected.

13.81. Fast reactors can be controlled satisfactorily on delayed neutrons in a manner similar to thermal reactors. With plutonium-239 (or uranium-233) as fuel, however, the delayed neutron fraction is small (§ 2.169) and the prompt neutron lifetime in a fast reactor is short, e.g., 10^{-7} to 10^{-8} sec compared with 10^{-3} to 10^{-4} sec in most thermal reactors. Special attention must therefore be given to the control system design. The small excess reactivity in a fast reactor provides an element of safety in this connection.

13.82. The first known fast reactor, as well as the first to use plutonium-239 as fuel, was constructed at the Los Alamos Scientific Laboratory for experimental work requiring a good source of fast neutrons [53].* It began operation in November 1946 at a power of 10 kw, which was increased to 25 kw in March 1949. The cylindrical core was approximately 6 in. in diameter and 6 in. high and contained a number of vertical plutonium rods clad with steel. The coolant was mercury which was satisfactory in this case because the temperatures attained were relatively low (175°F). The reactor was dismantled during 1953, partly because of the failure of at least one of the fuel elements and the resulting contamination of the mercury.

13.83. The first prototype fast power reactor was the Experimental Breeder Reactor-I (EBR-I) at the National Reactor Testing Station which became critical in August 1951 [54]. Its purpose was to obtain information on the possibilities of breeding with fast neutrons with simultaneous power production using a liquid metal (NaK) as coolant. The first known generation of electricity

---

*The reactor was nicknamed "Clementine" because of the association of "49," the code symbol for plutonium-239, with the location of the reactor in a canyon near Los Alamos, N. Mex.

from nuclear power was realized from the EBR-I in December, 1951. The core was made up of a number of fuel rods of highly enriched uranium-235 and this was surrounded by a blanket of natural uranium. The thermal power of the EBR-I was 1.4 Mw. Although the reactor was a pseudo-breeder, it achieved its objective in proving that breeding and power production, using liquid metal cooling, were feasible.

13.84. During four years of trouble-free operation, the EBR-I proved to be very stable. The experience with both Clementine and the EBR-I has demonstrated that fast reactors can be controlled satisfactorily on delayed neutrons in essentially the same manner as thermal reactors. However, in the fast reactors control is usually, although not in all cases, achieved by varying the leakage of neutrons, e.g., by moving rods of core or blanket material, rather than by neutron poisons.

13.85. During the course of a series of experiments designed to study the behavior of the EBR-I under very extreme conditions, a partial meltdown of the fuel occurred in the second (Mark II) core. It was the result of a unique combination of nuclear effects caused by bowing of the fuel elements during the course of tests with no coolant flow. A power excursion occurred and the temperature rose sufficiently high to melt the uranium. It is believed, however, that the meltdown would have been avoided if the operator had activated the fast-scram mechanism when a reactor period of 1 sec was reached.

13.86. After this incident a new core (Mark III) was designed for the EBR-I in order to continue the studies of fast reactor stability. This core consists of fuel rods made of an alloy of highly enriched uranium with 2 per cent zirconium which is metallurgically bonded to a zirconium jacket. In the earlier cores, the fuel consisted of slugs held loosely in stainless steel tubes with liquid NaK as the heat-transfer bond, so that they were less rigid. Ribs are used in the Mark-III fuel elements to prevent bowing. With this core design, the EBR-I has operated in a stable manner. The Mark IV core utilizes plutonium-239 as the fuel in the form of an alloy containing 1.25 weight per cent of aluminum.

13.87. The Experimental Breeder Reactor-II (EBR-II), with a thermal power rating of 62.5 Mw and an electrical output of 20 Mw (gross) or 16.5 Mw (net), is planned for operation in 1963 [55]. This reactor plant, located at the National Reactor Testing Station, includes a complete fuel processing and fabrication facility in order to determine the technical feasibility of integrating pyrometallurgical processing and remote fabrication of spent fuel elements with operation of the reactor (§ 8.166 *et seq.*). The reactor itself has been designed to demonstrate high thermal performance, efficient breeding, and the use of prototype components suitable for a central-station power plant. The EBR-II plant is a flexible, experimental reactor facility for investigating various engineering design configurations and fuel materials.

13.88. The fuel in the first EBR-II core contains 49 per cent uranium-235, 46 per cent uranium-238, with the remaining 5 per cent of fissium mixture

(§ 8.55).   Later cores will have fuel containing 20 per cent of plutonium, in addition to uranium and fissium.   Each fuel element has attached to it, above and below the fuel section, uranium depleted in uranium-235 which forms part of the upper and lower blankets.   In addition there are inner and outer blanket sections of natural uranium surrounding the core.   All fuel and blanket elements have stainless steel jackets.   The coolant is sodium and, since a significant fraction of the reactor power is produced in the blanket, adequate cooling must be provided.   The sodium flows at the same rate through the core and the inner blanket, but at a lower rate through the outer blanket.   Control is achieved by the movement of core (fuel) rods rather than of the blanket.

13.89. An interesting feature of the primary sodium system is that all the components, i.e., core, blanket, primary pump, intermediate heat exchanger, and fuel-transfer and storage system, are submerged in an 8000-gal cylindrical tank, 26 ft high and 26 ft in diameter, containing liquid sodium.   Except for the unlikely development of a serious leak in this tank, accidental loss of coolant from the core is thus not possible.   Furthermore, the large volume of sodium prevents temperature changes caused by low power demand or by conditions in the secondary system from quickly affecting the reactor.

13.90. The primary sodium flow rate of 8200 gal per min (26 ft/sec) through the core is supplied by two main pumps operating in parallel.   The coolant enters the reactor at 700°F and leaves at an average temperature of about 900°F.   Steam at a temperature of 850°F and a pressure of 1325 psia is produced in the secondary heat exchanger.

13.91. The Enrico Fermi Atomic Power Plant, near Monroe, Mich., is to be the first full-scale fast-reactor installation for the production of electricity [56]. Basically, the design of the reactor is the same as that of EBR-I and -II, but there are some differences in detail (see Fig. 11.3).   The core of the Fermi reactor is made up of subassemblies of fuel pins, 0.158 in. outer diameter, consisting of an alloy of uranium (28 per cent uranium-235) and 10 weight per cent of molybdenum, metallurgically bonded to a thin cladding of zirconium.   The blanket—above, below, and surrounding the core—is in the form of subassemblies of rods, 0.443 in. outer diameter, of natural uranium alloyed with 2.75 weight per cent of molybdenum clad with stainless steel.   Sodium acts as the heat-transfer bonding in the blanket elements.   The subassemblies of core and blanket are arranged to approximate a cylinder of circular cross section, about 80 in. high and 70 in. in diameter; the core itself is approximately 31 in. by 31 in. A large pool of sodium above the blanket and core serves to reduce the effect of transient temperature changes, as in the EBR-II.

13.92. The control of the Fermi reactor is somewhat unusual for fast reactors in the respect that a neutron poison, namely boron carbide (B_4C), is used for both control and safety rods.   The principal reason for selecting this method, although it results in a loss of neutrons, is that the mechanical design is less complicated than that involving movement of reflector or core material.   With poison control, less mass has to be moved and, in addition, shutdown is achieved more easily by rapid insertion of poison into the core.

13.93. The sodium coolant enters the reactor at 550°F and leaves at 800°F, the flow velocity being somewhat more than 30 ft per sec. It is of interest to mention that the average heat flux in the Fermi fast reactor exceeds 650,000 Btu/(hr)(ft²), which may be compared with less than 100,000 Btu/(hr)(ft²) for a pressurized-water reactor of similar total thermal output. Thus, as mentioned earlier, fast reactors have high power density and specific power. Steam conditions at the turbine are 740°F temperature and 600 psia pressure. The total thermal power of the reactor is 300 Mw, of which 35 Mw is produced in the blanket, with an electrical output of 104 Mw (gross) or 94 Mw (net).

13.94. The reactor vessel for the Fermi reactor has some unusual aspects. The fuel is introduced and removed from the core by means of an offset handling mechanism called the transfer rotor, contained in a compartment at the side of the reactor vessel (Fig. 11.3). Fuel-element subassemblies can be lifted out of the core and deposited in a pot adjacent to the reactor; after cooling, the spent fuel is removed through a vertical exit tube into a cask car for transportation. By this means, the reactor can be discharged and reloaded without removing the head of the vessel. Another interesting feature is the provision of a section at the bottom of the vessel to provide for dispersion and poisoning of fissile material in the unlikely event of a fuel meltdown.

13.95. Experimental fast reactors have been constructed in both the United Kingdom, at Dounreay, Scotland [57], and in the U.S.S.R. [58]. The unique feature of the Dounreay reactor is that the fuel element is in the form of a hollow rod, to provide increased heat-transfer surface, clad on the outside with niobium and on the inside with vanadium. The purpose of the two different cladding materials is that, in the event of accidental overheating, the inner coating will fail first and then the outer tube will act as a guide for the flow of molten fuel and prevent it from agglomerating within the core. The fuel in the first Dounreay reactor core was enriched uranium metal, but the second core is an alloy with 10 weight per cent of molybdenum.

13.96. The chief advantage of the fast-reactor concept lies in the possibility of breeding, thereby greatly reducing the cost of the fuel cycle as well as achieving efficient utilization of nuclear energy resources. The use of sodium (or other liquid metal) as coolant confers the additional benefit of operation at high temperatures without the need for high pressure. To maintain criticality in a fast-neutron spectrum with a reasonable mass of fissile material, the volumetric concentration of the latter in the core must be fairly high. For power reactors, about 20 or 25 per cent by volume of fissile material in the core appears to be satisfactory. Since the core also contains coolant and structural material, the fuel enrichment may be higher, but it is low enough for the temperature coefficient to be negative. Since fast-reactor cores are generally small and the power densities high (§ 13.79), careful design of the heat-removal system is essential.

Special attention must be given to large temperature gradients, thermal stresses, and any factors which might reduce the requisite high heat-transfer coefficients. A high power density also necessitates a large amount of fuel subdivision to provide a large heat-transfer area, with associated fabrication problems and high manufacturing costs.

13.97. The potential economic advantage of the fuel cycle in a breeder reactor is also not without its problems. The fast reactor requires the use of fuel containing about 20 per cent of a fissile species, whereas an equivalent thermal reactor can employ fuel enriched to the extent of 3 per cent or less. The critical mass of uranium-235 in a fast-neutron spectrum is larger than in a similar thermal system, but if plutonium-239 is the fissile species, the core inventory is not very different from that of a thermal reactor of the same power. A high specific power is desirable to minimize fuel charges, and in a fast reactor using plutonium-239 as fuel, for example, a specific power of at least 500 watts per gram of fissile material is a design objective. A high burnup, together with an efficient and inexpensive method for reprocessing fuel and blanket materials, is also necessary to realize the inherent potential of low fuel-cycle costs.

13.98. Because of the high volumetric concentration of fissile material in a fast-reactor core, a compaction from melting might result in a large increase in reactivity, thus creating a dangerous situation. The designer must take extraordinary precautions to prevent loss of cooling ability by the heat-removal system and make suitable arrangements for removal of fission product decay heat when the reactor is shut down.

13.99. Many of the mechanical and "hardware" problems common to all nuclear reactors are complicated by the small core of a fast reactor. Devices for fuel handling, control-rod drive mechanisms, and instrumentation are often elaborate. One reason for such complication is the necessity to design the devices to be operative when immersed in hot liquid sodium.

13.100. The foregoing discussion has referred in particular to fast reactors of moderate power. Because of the long-range economic potential of sodium-cooled breeders which can produce surplus fissile material as well as high-quality, superheated steam, consideration has been given to the design of reactors having thermal powers on the order of 1000 Mw. These may utilize ceramic (oxide or carbide) instead of metallic fuel elements. To overcome the problem of the positive Doppler temperature coefficient of plutonium-239 in such large reactors they would be designed so as to increase core leakage (§ 13.80). However, because of diluents in the fuel, which may be necessary to extend the burnup, and of considerable amounts of structural material, the neutron spectrum in a large reactor would probably be degraded. As a result, the breeding ratio would be decreased. Uranium-233 breeders, using thorium-232 as the fertile material, do not suffer from this drawback, as may be seen from Table 12.8. Furthermore, the temperature coefficient of reactivity is negative provided the ratio of thorium to uranium is greater than 0.6. The prospects of uranium-233 breeding in a fast-neutron spectrum are being studied under the Advanced Epithermal Thorium Reactor program [59].

FLUID-FUEL REACTORS

INTRODUCTION

13.101. Several reactor concepts in which the fuel is a fluid (or mobile) have a number of advantages and problems in common. Although they often have very different nuclear characteristics, it is convenient, therefore, to consider them together. Included in the category of fluid-fuel reactors are the various forms of the aqueous homogeneous reactors, molten salt reactors, and reactors using a liquid metal as fuel.

13.102. A fluid-fuel design offers the possibility of reducing fuel cycle costs by avoiding some of the factors which normally limit burnup, e.g., dimensional changes due to irradiation and the poisoning effect of gaseous fission products. Fabrication costs, which are significant for solid fuel reactors, are completely absent. The possibility of integrating a continuous (or batch) reprocessing facility into the reactor operation is also an important advantage. The question of technical feasibility has been the most serious uncertainty in all proposed fluid fuel reactor concepts.

AQUEOUS HOMOGENEOUS REACTORS [60]

13.103. The aqueous homogeneous reactor, using a solution of an enriched uranium salt in ordinary water as the fuel, is one of the oldest reactor concepts. The first homogeneous reactors were the Los Alamos "Water Boiler" experiments used for research. The LOPO (low power) and HYPO (high power) versions were operated in 1944, and these were followed by SUPO (super power), a 45-kw modification, completed in 1951 [61].

13.104. At Oak Ridge National Laboratory, efforts were initiated in 1950 toward the development of a homogeneous reactor for power production and thermal breeding of uranium-233. These efforts resulted in the Homogeneous Reactor Experiment-I (HRE-I), which operated successfully from 1952 to 1954 at a maximum thermal power level of 1.6 Mw [62]. This reactor produced electricity in February 1953, representing the second known instance of electrical power derived from nuclear fuel. In 1954, the HRE-I was dismantled to provide room for HRE-II [63]. The latter, with a design thermal power of 5 Mw, was completed in 1957. Its operation, terminated in 1961, was interrupted by problems arising from the corrosive nature of the fuel solution.

13.105. Both HRE-I and -II were two-region reactors. In HRE-I the inner region (or core) contained a solution of highly enriched uranyl sulfate in ordinary water, but in HRE-II a dilute solution (less than 6 grams uranium-235 per liter) in heavy water was used, thus providing better neutron economy. In both reactors, the outer (reflector) or blanket region contained heavy water. For breeding purposes, it was the intention to use a blanket containing a concentrated slurry of thorium dioxide in heavy water to absorb neutrons leaking from the core region, but this aspect of the experiment was not attempted. Radiolytic decomposition of the water in the core was a serious problem, but it was minimized in two ways; first, by pressurizing the solution—1000 psi in HRE-I and up to 2000 psi in HRE-II—and second, by the addition of a small amount of copper sulfate, which is able to catalyze recombination of hydrogen and oxygen gases in the solution. Steam was generated by passing the hot fuel solution through a heat exchanger, but since the maximum practical temperature, for various reasons, was 482°F, the thermodynamic efficiency of the system was relatively low.

13.106. A remarkable feature of the HRE reactor is the simplicity of control. Although HRE-I had neutron-absorbing control plates, HRE-II was constructed without any control rod mechanisms. Criticality was achieved by increasing the uranium concentration of the fuel solution and shutdown by diluting it with heavy water. Changes in power level were realized by altering the demand. If the load at the turbo-generator was increased, the fuel solution returned to the reactor at a lower temperature. As a result of the large negative temperature coefficient the reactivity increased to meet the demand. The reverse behavior occurred if the load was decreased. Rapid shutdown could be achieved by dumping the solution to a tank in which it had a noncritical configuration.

13.107. A somewhat different approach to an aqueous homogeneous reactor was the Los Alamos Power Reactor Experiment-I and -II (LAPRE-I and -II) begun in 1953 [64]. The fuel was enriched uranium trioxide dissolved in 53 weight per cent (aqueous) phosphoric acid in LAPRE-I and uranium dioxide in 95 weight per cent phosphoric acid in LAPRE-II. With these concentrated acid solutions the temperature limitation mentioned above, e.g., due to the critical temperature of water being exceeded, did not exist and solution temperatures above 800°F were possible. The design of LAPRE-I featured a forced fuel circulating system and a steam generator (heat exchanger) all within the reactor pressure vessel. The extremely corrosive nature of the fuel solution necessitated the use of gold or platinum cladding for components in contact with it. Criticality was reached in 1956, but as a result of various corrosion difficulties encountered during operation the reactor was soon shut down and dismantled.

13.108. LAPRE-II was similar to LAPRE-I, although the use of a more concentrated phosphoric acid solution permitted operation at lower pressures. It also had a self-contained steam generator, but cooling was by natural convection so as to avoid moving parts. The reactor operated intermittently at power levels up to 0.8 Mw (thermal) during 1959. Tests demonstrated the unusual stability of the reactor system and the chemical stability of the highly acid fuel solution at high temperatures.

13.109. In addition to the advantage arising from the simplicity of reprocessing aqueous fuel solutions, reactors using such solutions have marked inherent safety and self-control characteristics. Experiments with the HRE-I showed that a sharp increase in reactivity up to about 0.8 per cent Δk per sec, caused by introduction of cold fuel solution into the core, would result in the power rising to several times the normal value. It then decreased rapidly to an equilibrium level. It was also found that delayed neutrons exert a powerful damping influence on power oscillations. As mentioned previously, safe operation is possible even without control rods. The fuel circulation feature of the HRE reactors facilitated concentration adjustment, including that required for startup and shutdown, removal of gaseous fission products, and chemical processing. Since additional fuel can be added as required and xenon can be removed continuously, the excess reactivity is low and loss of neutrons by capture in control materials is obviated. In the HRE-II system in particular, the use of heavy water in both core and blanket regions provided efficient utilization of the neutrons, so that the breeding gain or conversion ratio could be high.

13.110. The most important drawbacks to the circulating solution type of reactor is the highly corrosive nature of uranyl sulfate solutions. Stability tends to decrease with increasing temperature, and above a certain temperature the homogeneous solution can break up into two liquid phases one of which is much more dense than the other. Furthermore, thorium oxide slurries, which were to be used in the breeding blanket because of the low solubility of the sulfate and the relatively large neutron capture cross section of the nitrate, possess unusual non-Newtonian flow characteristics which result in settling tendencies and lead to pumping problems. The vapor pressure of the aqueous solutions is high so that pressurization is necessary. Finally, the relatively low maximum practical temperature limits the thermodynamic efficiency. The use of phosphoric acid solutions decreases some of these difficulties but increases others, especially corrosion.

MOLTEN-SALT REACTORS [65]

13.111. The molten-salt reactor concept evolved from work at Oak Ridge National Laboratory on an experimental reactor for aircraft propulsion. Various reference designs have been proposed for a power reactor type in which the fuel is uranium tetrafluoride dissolved in a molten-salt mixture of beryllium fluoride and either lithium-7 fluoride or sodium fluoride. The lithium-7 fluoride has better neutronic properties but would be much more expensive, since the neutron absorbing lithium-6 nust be removed from normal lithium. Similar mixtures with thorium tetrafluoride could be used as the blanket material in a thermal breeder reactor. The melting points of these salt mixtures are between 850° and 950°F.

13.112. In some designs the system is completely homogeneous, with the various light elements in the molten salt acting as moderators. However, since the slowing down power of fluorine, which constitutes a large proportion of the fuel, is only about half that of carbon, an additional moderator, such as graphite, is included in some forms of the molten-fuel reactor concept. In the Molten-Salt Reactor Experiment (MSRE), for example, at the Oak Ridge National Laboratory, the reactor core consists of a number of graphite stringers which are machined to provide when assembled 1064 fuel channels of 1.2 in. by 0.4 in. cross section [66]. The fuel is a mixture of 70 mole per cent lithium-7 fluoride, 23 per cent beryllium fluoride, 5 per cent zirconium tetrafluoride, and 1 per cent each of enriched uranium tetrafluoride and of thorium fluoride. The circulating molten fuel enters the reactor at 1175°F and leaves at 1225°F. It then passes to an intermediate heat exchanger where heat is transferred to a secondary "coolant" consisting of 66 mole per cent normal lithium fluoride and 34 per cent beryllium fluoride. The containing material proposed for the molten salts is an alloy (INOR-8) of nickel (66 to 71 weight per cent) and molybdenum (15 to 18 per cent), with a few per cent each of iron and chromium.

13.113. In addition to the usual advantages of fluid-fuel systems outlined above, the molten salt mixture is stable to radiation and temperature and has a low vapor pressure. The latter property simplifies the design of the reactor vessel and the primary heat exchanger. However, the intense radioactivity of the fuel makes the use of a secondary coolant necessary and an additional heat exchanger would be required to produce steam. The engineering problems associated with the manipulation of a circulating fuel having such a high melting point are considerable. The primary objectives of the MSRE are to investigate the reliability and compatibility of the reactor components and to demonstrate the safety of the system.

LIQUID-METAL FUEL REACTOR

13.114. Although a thermal liquid-metal fuel reactor (LMFR) has not yet been constructed, a considerable amount of development and design work has been done, especially at Brookhaven National Laboratory [67]. The basic LMFR concept utilizes a liquid metal, such as bismuth, as the solvent or suspending medium for both fuel and fertile materials. Bismuth has the advantage of low neutron capture cross section, low vapor pressure, moderately low melting point, fairly good heat transfer properties, and satisfactory uranium solubility. Graphite can be used as the moderator in direct contact with the molten fuel up to about 1050°F. At this temperature, the fuel may be pumped through an intermediate heat exchanger where the heat is transferred to sodium, the latter then being used to produce steam.

13.115. Several variations and alternative designs of the LMFR concept have been proposed, including both one- and two-region breeder versions. In the former, the fertile material contains thorium as the compound Th_3Bi_5 suspended in liquid bismuth, in addition to the uranium. In the two-region concept, the solution of uranium in bismuth is circulated through the graphite-moderated core, whereas the suspension of Th_3Bi_5 flows through a separate blanket region. In another form the fluid fuel is stationary and heat is removed by circulating a coolant, e.g., liquid sodium, through the core in the conventional manner. Such an internally cooled reactor has a lower fuel inventory than the externally cooled version, and so may have some economic advantage.

13.116. The LMFR concept has the merit of providing a high temperature, for high thermal efficiency, and low pressure, so that a pressurized system is not required. An important problem, however, is the uncertainty regarding container materials. No definite solution has yet been found for the excessive mass-transfer corrosion rates that have been observed with the proposed bismuth system, although addition of small amounts of certain metals, e.g., zirconium or magnesium, may be beneficial (§6.196).

FLUID-FUEL FAST REACTORS

13.117. The possibility of combining the advantages of the fluid-fuel approach with the favorable breeding potential of fast-reactor systems, with plutonium-239 as the fuel and uranium-238 as the fertile material, has long been recognized. The Los Alamos Molten Plutonium Reactor Experiment (LAMPRE), which started operation in 1961, is a prototype of such a reactor [68]. The fuel is a eutectic of plutonium with 2.4 weight per cent of iron, having a melting point of 772°F, which is contained in capsules made of high-purity tantalum. These fuel capsules constitute the reactor core which is cooled by circulating sodium. The operating temperature is about 1000°F. As is common to most fluid-fuel concepts, the major limitation is the behavior of the fuel container materials, since the molten plutonium-iron alloy is extremely corrosive. An objective of the LAMPRE system is therefore to permit an evaluation of constructional materials.

13.118. One of the problems associated with fast reactors is the possibility of meltdown of the fuel elements arising from external failure, e.g., in the coolant system. In addition to the destruction of the core and the danger of super-criticality, the release of accumulated fission products, especially gases, can represent a severe hazard. When the fuel is already liquid, as in LAMPRE, the meltdown problem does not exist.

13.119. A number of other modifications of the molten plutonium (alloy), fast-reactor concept are under development at Los Alamos Scientific Laboratory. These include a liquid plutonium alloy fuel which will be in direct contact with the molten sodium coolant in the reactor core, with mixing accomplished by a jet-pump arrangement. The fuel material gives up its heat to the sodium and then, being considerably more dense, is allowed to settle outside the reactor and returned to the core. With this system, the problem of containing the corrosive molten plutonium alloys can be solved, since thick-walled vessels can be used, whereas in the LAMPRE concept described above the capsules must have thin walls for heat-transfer purposes.   On the other hand, extraction by the sodium of the halogens, which are the delayed neutron precursors (§ 2.170), may increase the difficulty of control. For the breeding blanket a "paste" of densified uranium dioxide powder in sodium has been proposed (§ 8.74).   This material has the advantage of mobility combined with the good thermal conductivity of the sodium carrier medium.

SPACE POWER UNITS

INTRODUCTION

13.120. An interesting application of nuclear reactor systems is to provide electric power aboard space vehicles of relatively long life.   Some of the designs proposed involve novel features which may perhaps indicate future trends in conventional power reactors.   For use on a space vehicle, the energy content per pound is very important and this consideration tends to limit the power source of interest to those sources depending on solar radiation, nuclear fission, or radioactive decay.   Chemical energy sources are too heavy and expensive for a life of more than a week.   Consequently, the nuclear reactor system with a possible energy content of about 10^3 kw-hr per pound gross weight is attractive, especially as its operation, unlike that of a solar power unit, is independent of location or orientation of the carrying vehicle.   To meet the anticipated need for power supplies in space, a series of Systems for Nuclear Auxiliary Power (SNAP) reactors are being developed.   Conceptual designs have also been proposed for nuclear power plants to be assembled on the earth and operated on the moon.

GENERAL REQUIREMENTS

13.121. A space power system has the usual components, namely, the reactor heat source, an energy conversion unit, and a means for rejecting heat.   The only practical method of rejecting heat in space is by radiation.   Since the rate of heat loss by radiation is proportional to the fourth power of the absolute temperature (§ 6.62), the rejection temperature should be high.   Consequently, the operating temperature of the reactor must be even higher in order to achieve reasonable thermodynamic efficiency of energy conversion.   For a given rejection temperature, the power output can be increased by increasing the radiating surface.   As the power level is raised, the weight of the radiator, in fact, becomes the dominant weight factor.   At lower power, however, the reactor and shield make the major contributions in this respect.   Various methods have been proposed for converting into electric power the heat produced in the reactor core.   Some of these will be mentioned below.

13.122. An important requirement for space power systems is the need for unattended operation for a year or more. This complicates the already difficult problems of material selection. Small amounts of corrosion and mass transfer, which could be accepted under conditions where periodic inspection and maintenance were possible, could result in failure when attention is not feasible. The need for low weight and high operating temperatures provides further problems. The development of a practical power unit for space application may well depend on the availability of suitable materials.

SNAP REACTORS [70]

13.123. In the SNAP-2 reactor (Fig. 13.7) the core is a hexagonal array of 61 cylindrical elements, 10 in. long and 1 in. diameter, consisting of a uniform mixture of zirconium hydride moderator and highly enriched uranium fuel. This is surrounded by a beryllium reflector. Control is achieved by the rotation of

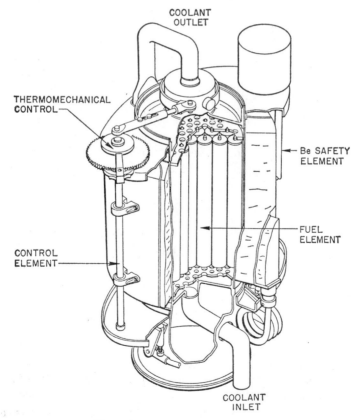

Fig. 13.7. The SNAP-2 reactor

two semicylindrical beryllium drums which are part of the reflector.   The NaK coolant enters the core at 1000°F and leaves at 1200°F, removing 50 kw (thermal) from the fuel.   An electromagnetic pump, which has no moving parts (§ 6.202), circulates the coolant through the core and a heat exchanger where mercury is boiled.   The mercury vapor is used in a Rankine cycle to drive a small turbo-generator operating at normal boiling and condensing temperatures of 900°F and 600°F, respectively.   The power system, weighing 200 lb (or 600 lb with shielding), should be capable of producing 3 kw (electrical) continuously for a year or more.   The SNAP-8 design is a scale-up of SNAP-2 operating at 250 kw thermal; the maximum temperature of the coolant is 1400°F.

13.124. A modification of the foregoing system has been proposed, utilizing thermoelectric conversion, that would have no moving parts.   Heat would be transferred from the reactor to a hot junction, at a temperature of about 850°F, by means of NaK circulated by an electromagnetic pump which derives current from a thermocouple operating between the coolant inlet and outlet pipes of the reactor.   The cold junction would be relatively high, around 580°F, in order to minimize the dimensions and weight of the radiator (140 ft² area per kw).   As a further simplification, the pump could be eliminated and heat conducted from the reactor to the hot junction by means of beryllium plates.

THERMIONIC CONVERSION [71]

13.125. Thermionic conversion has the potential of considerably higher effi-ciencies than conventional thermoelectric systems for producing electricity from heat.   In its simplest form, the thermionic converter consists of two electrodes in a space containing an easily ionized gas (plasma) at low pressure.   One electrode, called the emitter, is at a temperature high enough for the emission of electrons thermionically.   The other electrode, known as the collector, is main-tained at a lower temperature.   Under the influence of the driving force arising from the temperature difference, electrons emitted by the hotter electrode are directed to the colder one.   The system is analogous to a conventional thermo-electric arrangement in which the emitter-plasma interface is the hot junction while the collector-plasma interface serves as the cold junction.   If heat is continuously supplied to the emitter, to maintain its temperature, and removed, e.g., by radiation, from the collector, there will be a continuous flow of electrons, and hence of current, through an external circuit.   For the maximum conversion of heat into electricity in this manner, the thermionic work function of the emitter should be high and that of the collector low.

13.126. In one concept for a direct conversion reactor power source, the thermionic emitter material is heated by adjacent fuel elements.   Cesium vapor forms the plasma between the electrodes.   A simplification is possible by utilizing the fact that uranium carbide (in the form of a solid solution with zirconium carbide) can act as both the hot emitter electrode and the fuel material in a reactor core. In principle, therefore, a compact reactor design, suitable for space use, is conceivable in which the fission heat is converted directly into electricity. The reactor is thus a self-contained source of electric power.

LUNAR POWER PLANT [72]

13.127. One design for a stationary power plant of 1 Mw electrical capacity has a number of interesting features. The source of heat is a fast reactor, which is compact, and boiling mercury, used as the coolant, is incorporated into a Rankine cycle to drive a turbo-generator. For condensation, large radiators extend from the reactor unit. Cooling of the generator is necessary because of the high operating temperatures required for high efficiency, and hydrogen gas, also cooled by the radiators, serves this purpose. Apart from the radiators, all components are installed in a single cylindrical unit about 26 in. in diameter and 30 ft long. The total weight would be nearly 4500 lb.

RESEARCH, TEST, AND TRAINING REACTORS [73]

INTRODUCTION

13.128. Although the production of power represents the major effort in the design and construction of nuclear reactors, there are numerous reactors in use in all parts of the world for research and other experimental purposes. In this connection, the reactor is used basically as a source of neutrons.* The main activities associated with reactors of this type are (a) reactor technology, mostly related to the irradiation behavior of various reactor materials and components, (b) basic research in the physical and biological sciences in which neutrons are used, and (c) experiments primarily useful for teaching and training purposes.

13.129. Since the reactors under consideration are to be regarded as neutron sources, it is convenient to classify them, in the first place, according to the neutron flux, viz., high flux, medium flux, and low flux. These three categories correspond very approximately to the three kinds of application given in the preceding paragraph. However, the correspondence is not exact because reactors of moderate flux are used in teaching institutions and low-flux reactors can be employed in research. Although it is the objective here to examine these reactors from the engineering design viewpoint rather than that of the experiments or research performed with them, it is important nevertheless to give some consideration in the design to the type of work for which the reactor is to be used.

13.130. The important variables in a reactor intended for irradiation experi-

---

* Because of their prime purpose, these reactors are differentiated from those designed for the development of the technology of a particular reactor concept. Such reactors, e.g., EBR, HRE, OMRE, etc., are properly termed "reactor experiments."

ments are the neutron flux and the irradiation volume. A selection of the neutron energy is also important in providing for the desired experimentation. In a well-moderated reactor, most of the neutrons fall into two broad energy groups: one group consists of neutrons in the high-energy range, e.g., 0.1 Mev or more, and the other consists of thermal neutrons, with energies of 1 ev or less. The proportion of neutrons in the intermediate energy range is usually small in comparison with the fast and thermal neutrons. The average thermal neutron flux in the reactor core per unit power is generally inversely proportional to the mass of fissile material, whereas the fast flux per unit power varies inversely as the core volume. Consequently, to attain a high thermal flux per kilowatt, the critical mass should be small and the core volume relatively large. Frequently these requirements are conflicting and a compromise is necessary. Consideration must also be given to heat removal from the reactor; the heat dump facilities are often expensive and so it is desirable to obtain a maximum neutron flux per unit of power produced.

13.131. The specification of irradiation space must take cognizance of the flux available in various holes in the core and reflector where experimental materials or devices are to be placed. Not all experiments need the maximum flux, but it is important that sizes and locations of the spaces be adapted to the expected use. High-flux reactors are sometimes employed for testing fuel element-coolant configurations and may involve a circulating coolant loop passing through or near the reactor core. Holes of relatively large size are then required in locations where the thermal neutron flux is high, e.g., 10^{15} neutrons/(cm²)(sec), which will provide adequate neutron exposure without excessive gamma heating. This represents a difficult engineering design problem. Finally, the effect of spaces and the materials in them on the reactivity of the reactor must not be overlooked.

HIGH-FLUX TEST REACTORS

13.132. High-flux reactors, with neutron fluxes on the order of 10^{15} neutrons/(cm²)(sec), are of interest for a number of experimental purposes [74]. They can be used to study radiation damage in various reactor (and other) materials and to investigate the behavior of fuel-coolant systems under extreme irradiation conditions. Furthermore, the production of the isotopes of transplutonium elements by successive neutron captures and beta decays, starting from plutonium isotopes, requires very high thermal neutron fluxes. Since the intermediate stages are often highly unstable, e.g., due to radioactive decay or spontaneous fission, it is essential that the neutron capture reaction occur rapidly and this can be facilitated by using a high-flux environment. Strong sources are also desirable for physics experiments with neutrons. A summary of the maximum (or desirable) neutron flux requirements for various purposes is given in Table 13.2; however, lower fluxes than those indicated are generally useful.

13.133. Examples of high-flux reactors are the Engineering Test Reactor

TABLE 13.2. MAXIMUM FLUX REQUIREMENTS

Type of Research	Unperturbed Flux neutrons/(cm²)(sec)	Energy Range
Radiation damage in materials.........	2×10^{15}	Fast
Loop (fuel-coolant) experiments........	$1\text{-}2 \times 10^{15}$	Thermal and fast
Isotope production; neutron spectrometer......................	1×10^{15}	Thermal
Neutron velocity selector..............	$1\text{-}2 \times 10^{15}$	Thermal to 10 kev
Transplutonium isotope production.....	5×10^{15}	Thermal

(ETR) and the Materials Testing Reactor (MTR) at the National Reactor Testing Station. These are both heterogeneous, water-cooled and -moderated reactors of the "tank" type. The core and reflector (or part of the reflector) are enclosed in a tank through which cooling water is pumped at a high rate. The high enrichment (more than 90 per cent uranium-235) of the fuel and the water moderator make possible a small core with a low critical mass, so that both fast and thermal neutron fluxes are high. Other similar high-flux reactors are the Oak Ridge Research Reactor (ORR) and the General Electric and NASA (Plumbrook, Ohio) Test Reactors [75].

13.134. The MTR, completed in 1952, was the first reactor intended primarily for the study of the effects of nuclear radiation on materials required for reactor construction [76]. The fuel elements consist of slightly curved plates of an alloy of 20 per cent by weight of highly enriched uranium in aluminum with a thin cladding of aluminum on both sides, leading to a sandwich type of construction. Eighteen such sandwich plates, about 0.116 in. apart, held together in a 2 ft-long, boxlike (3 in. by 3 in.) assembly, form a fuel element (Fig. 13.8). Both ends of the element are open to allow the cooling water to flow downward between the plates. The inner reflector is of beryllium and this is surrounded by a graphite reflector, which is outside the water tank, as shown in the sectional drawing in Fig. 13.9.

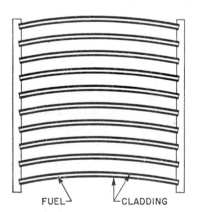

FUEL   CLADDING

FIG. 13.8. Sandwich (MTR) type fuel element

13.135. The MTR operates at a power of 40 thermal Mw, and because of the small size of the core the thermal neutron flux and power density are high. The cooling water is therefore circulated at a rate of 22,000 gal per min. The average thermal flux is 3×10^{14} and the maximum is about 4.5×10^{14} neutrons/(cm²)(sec). Numerous experimental spaces are available in the MTR;

some go into and through the primary beryllium reflector right up to the reactor core whereas others extend only into the graphite zone. One of the faces is penetrated by a "thermal column" of graphite blocks the object of which is to provide a source of low-energy neutrons for experimental purposes.

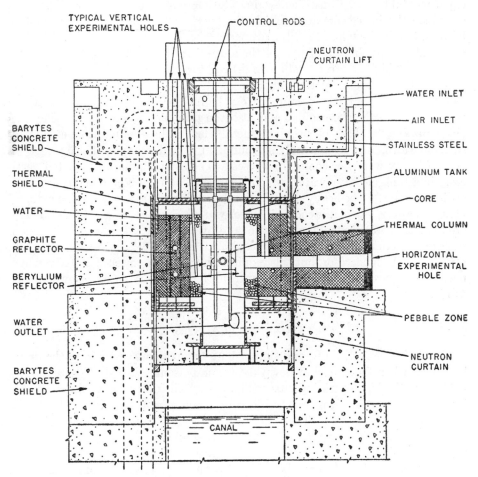

TYPICAL VERTICAL
EXPERIMENTAL HOLES

CONTROL RODS

NEUTRON
CURTAIN LIFT

WATER INLET

AIR INLET

STAINLESS STEEL

BARYTES
CONCRETE
SHIELD

THERMAL
SHIELD

WATER

GRAPHITE
REFLECTOR

BERYLLIUM
REFLECTOR

WATER
OUTLET

BARYTES
CONCRETE
SHIELD

ALUMINUM TANK

CORE

THERMAL COLUMN

HORIZONTAL
EXPERIMENTAL
HOLE

PEBBLE ZONE

NEUTRON
CURTAIN

CANAL

Fig. 13.9. Sectional drawing of the MTR at the National Reactor Testing Station, Arco, Idaho

13.136. The ETR is, in general, similar to the MTR, except that its design rating is 175 thermal Mw; is cooled by water circulating at the rate of 44,000 gal per min [77]. The average thermal flux is 4×10^{14} and the maximum 6×10^{14} neutrons/(cm²)(sec). The main differences between the ETR and MTR designs are as follows: (a) the ETR has flat-plate, rather than curved-plate, sandwich-type fuel elements 3 ft long, (b) the outer reflector of the ETR is of aluminum instead of graphite and there is no thermal column, and (c) the control rods in the ETR enter at the bottom in order to leave the top of the reactor accessible for insertion of materials or equipment to be tested.

13.137. The major objective of the ETR was to provide ample space for experimental loops. Consequently, in addition to numerous smaller holes, there are nine holes in the core and eight in the beryllium reflector ranging from 3 in. by 3 in. to 9 in. by 9 in. Additional spaces where the neutron flux is fairly high are available by removal of part of the aluminum reflector.

13.138. A high-flux test (and research) reactor of a different type to the MTR and ETR, and probably the only one of its kind, is the Canadian National Research Council's NRU reactor (1957); the fuel is natural uranium and heavy water is the moderator-coolant and also the primary reflector [78]. The fuel elements are flat bars of uranium metal clad with aluminum; five bars, of suitable width, form an assembly surrounded by an aluminum tube to direct the flow of heavy water for cooling purposes. The assemblies are supported vertically in a tank of heavy water, which is surrounded by a secondary reflector of ordinary water. The heavy water is circulated through the tank and an external heat exchanger, where it is cooled by river water. Numerous experimental holes are available, some extending into the core by the use of re-entrant thimbles in the core tank. The maximum thermal flux is 3×10^{14} neutrons/(cm²)(sec), attained in the moderator, but this requires an operating power of 200 Mw which may be compared with 4.5×10^{14} neutrons/(cm²)(sec) at 40 Mw power in the MTR.

FLUX-TRAP REACTORS

13.139. The MTR, ETR, and similar high-flux reactors suffer from the drawback that when they are shut down the xenon poisoning increases quite rapidly because of the high operating fluxes (Fig. 5.7). The difficulty can be overcome by sufficient built-in excess reactivity or by loading fresh fuel elements into the core to permit startup. The "flux-trap" reactor design, however, largely obviates both the xenon and heat-removal problems by having the maximum thermal-neutron flux in the reflector rather than in the core. However, the fast flux is high in the core, but this is not involved in xenon poisoning. The flux-trap reactor thus provides the maximum thermal flux with the minimum fission (and heat-removal) rate.

13.140. The fuel in the flux-trap reactor is in an annular region with moderating (and reflecting) material both in the central island and in the space surrounding the annulus. The central region is referred to as the moderator and the outer one as the reflector. Fission neutrons produced in the annular core escape into the moderator, and as the slow neutrons accumulate they diffuse back and are ultimately absorbed by the fuel. The thermal-neutron density in the central island therefore builds up to a steady-state value when the rate at which fast neutrons leak into the central region from the core is equal to the rate at which thermal neutrons diffuse back to the core. The result is a buildup of thermal flux in the island (or flux-trap) by a factor of five to six times greater than would be available in a reactor of the MTR-type operating at the same power.

13.141. Several different forms of the flux-trap concept have been proposed and some are under construction. These designs generally feature heavy water or beryllium as the reflector and water as the moderator in the central island. Thermal fluxes up to 5×10^{15} neutrons/(cm²)(sec) are feasible for a 100-Mw power. Since the degree of flux peaking in the moderator depends on the rate at which neutrons escape from the core, a high-power density is desirable. This means that efficient heat removal is required. Fuel elements of the sandwich form, with either flat or curved plates, are convenient but more efficient types have been designed.

13.142. The Advanced Test Reactor (ATR), planned for completion in 1964 at the National Reactor Testing Station, has a unique flux-trap design to provide a very high power density in the core with provision for the adjustment of the neutron flux levels in the individual test spaces [79]. The core is shown schematically in Fig. 13.10; it is seen to consist of a number of separate cores or

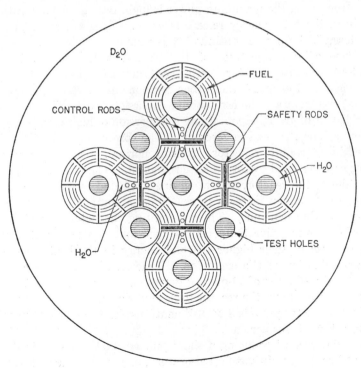

FIG. 13.10. Cross section of the Advanced Test Reactor (ATR)

"lobes" surrounding the test "islands." At a thermal power of 250 Mw, a thermal flux of up to 1.5×10^{15} neutrons/(cm²)(sec) is expected in some of the islands together with an average fast flux of 1.5×10^{15} neutrons/(cm²)(sec).

13.143. The Brookhaven High-Flux Beam Reactor (HFBR) is a compromise between the ordinary water-moderated tank-type reactor and the flux-trap concept. It uses flat-plate, sandwich fuel elements, similar to those in the ETR, with heavy water as the coolant, moderator, and reflector. The general design of the core is similar to the ETR and MTR, there being no central island. With heavy water, rather than ordinary water, as the moderator, the leakage rate of fast and epithermal neutrons into the surrounding reflector is high. Consequently, a peak thermal flux about ten times that in the center of the core is attained in the reflector a few inches from the core face. The maximum thermal flux is 3 to 4×10^{13} per megawatt of thermal power, and at a design power of 40 Mw a flux of over 10^{15} neutrons/(cm²)(sec) should be available.

PULSED TEST REACTORS

13.144. For a number of test purposes, an extremely high neutron flux is required for a very short period of time. In a reactor designed to produce a strong neutron pulse, the maximum power is high, but since the duration of the pulse is short, the total energy release is small. The heat-removal limitation upon reactors which operate continuously is thus not applicable.

13.145. One of the objectives of the Transient Reactor Test (TREAT) Facility at the National Reactor Testing Station is to investigate the possibility of meltdown of fuel elements and other components at the high temperatures that might arise from a sudden reactivity excursion [80]. The fuel-moderator consists of blocks of graphite containing a dispersion of highly enriched uranium dioxide (or carbide), canned in Zircaloy. The core made up of these blocks is surrounded by a graphite reflector. During normal, steady-state operation at 100 kw the core is cooled by forced air circulation. The poison control rods are at the bottom of the reactor, and when one is ejected rapidly by a pneumatic device there is a sudden increase in reactivity accompanied by the enhanced production of fission neutrons. The self-limiting characteristics of the system, due to the normal negative temperature coefficient and the greater leakage of neutrons, because of their higher thermal energy in the heated graphite, results in a termination of the excursion. Hence, the neutrons are emitted in the form of a pulse or burst of 50-millisec duration. Immediately after the burst, a scram mechanism causes the reactor to be shut down. The peak flux attained in the TREAT reactor pulse is 3×10^{16} neutrons/(cm²)(sec) and the integrated flux is 3×10^{15} neutrons per cm².

13.146. A device for producing a short pulse of fast neutrons is the Godiva assembly at Los Alamos Scientific Laboratory, so called because it consists of a bare core of highly enriched uranium [81]. It has two (leakage) control rods and a "burst" rod of uranium which enter the core from the bottom. The control rods are adjusted for steady-state operation with the burst rod in the out position. Rapid insertion of the burst rod, e.g., by means of a pneumatic cylinder, causes the core to become supercritical and there is a burst of fission neutrons. The duration of the burst, with an integrated flux of about 10^{14} neutrons per cm^2, is automatically limited by neutron leakage due to thermal expansion of the core. A fast-acting mechanism then causes a safety block of core material to fall away so that shutdown occurs within about 0.04 sec.

13.147. It may be mentioned here that significant pulses consisting of both thermal and fast neutrons can be obtained with some laboratory-type reactors. These include the Water Boiler reactor (§ 13.159) used in the Kinetic Experiments on Water Boilers (KEWB) and the TRIGA reactor (§ 13.162).

MEDIUM-FLUX RESEARCH REACTORS [82]

13.148. Apart from the fast reactor Clementine (§ 13.82) and the so-called Water Boiler reactor, all the research reactors in use before 1950 used natural uranium as fuel and either graphite or heavy water as the moderator. Of the graphite-moderated reactors, several were of the low-power, low-flux type and are now mainly of historical interest. Typical of the more important research reactors in this category are the graphite (X-10) reactor at Oak Ridge National Laboratory, the original form of the graphite reactor at Brookhaven National Laboratory, and the British Experimental Pile* O (BEPO) at Harwell. These have horizontal channels in the graphite matrix in which are placed cylindrical slugs of natural uranium metal jacketed with aluminum. Cooling is achieved by forcing air through the channels by means of fans. Because of the limited cooling capability and the necessity for keeping the temperatures of the fuel cladding and the graphite below those at which oxidation would occur, the power densities of the natural uranium-graphite, air-cooled reactors are low. The average thermal fluxes are thus on the order of a few times 10^{12} neutrons/(cm^2)(sec), but these reactors are classified here with those of medium flux, i.e., 10^{13} neutrons/(cm^2)(sec) or more, because of their use as research tools.

13.149. One of the advantages of the graphite-moderated natural uranium reactor is that its large size provides ample space for experiments (Fig. 13.11). The dimensions of the Brookhaven reactor, for example, including its high-density concrete shield, is 38 ft by 55 ft by 30 ft high, and the others have similar dimensions. The combination of negative temperature coefficient and mass of material makes these reactors insensitive to transient reactivity changes. Furthermore, because of their small excess reactivity, about 0.7 per cent $\Delta k/k$, a serious power excursion is almost impossible. For the same reason, however, they are easily poisoned. The main drawback to the natural uranium-graphite research reactor is the low ratio of flux to power as a consequence of the small

* For the origin and use of the term "pile," see § 1.68.

FIG. 13.11.   West face of the Brookhaven National Laboratory graphite reactor

power density.   Hence, for a given thermal-neutron flux, they cost more to build than reactors employing enriched fuel.   It is of interest in this connection that, when the fuel in the Brookhaven reactor was changed to highly enriched uranium, in 1957, the operating power was reduced from 28 to 20 Mw, but the

thermal flux was increased about sevenfold to an average of approximately
2.5×10^{13} neutrons/(cm²)(sec).

13.150. Several of the early research reactors in which natural uranium was
the fuel and heavy water the moderator operated at low power and had a low
neutron flux. The best known reactor of this design in the medium-flux range
is the National Research (Council's) Experimental (NRX) Reactor in Canada,
completed in 1947 [83]. The heavy-water moderator is contained in an aluminum tank, called a calandria, 8 ft in diameter and 10 ft high, surrounded by a
graphite reflector. In the tank are supported the vertical fuel rods which are of
aluminum-clad natural uranium, surrounded by two concentric aluminum tubes.
The inner annulus is the coolant channel through which flows ordinary water,
whereas the outer annulus serves as an air gap to prevent contamination of the
moderator by the coolant if a leak should develop. An interesting feature of the
NRX reactor is that coarse (or shim) control is achieved by varying the level of
the heavy-water moderator in the calandria. After reconstruction, following
an accident in 1952, the power was increased from 30 to 40 Mw, with an average
thermal flux of about 5×10^{13} neutrons/(cm²)(sec).

13.151. Because reactors with enriched fuel have a higher ratio of neutron
flux to power, a number of heavy-water moderated research reactors use fuel of
this type. These include the CP-5 reactor (1954) at Argonne National Laboratory, the DIDO (1956) and PLUTO (1957) reactors at Harwell, England, and
several others. All use highly enriched uranium fuel in the form of an alloy
with aluminum, clad on both sides by aluminum, as described in § 13.134. The
fuel elements were originally of the plate type, but those in the CP-5 reactor
were changed to concentric tubes. The fuel assemblies are supported vertically
in a tank of heavy-water moderator. Heavy water within the tank and graphite
surrounding it serve as neutron reflectors. Cooling is achieved by pumping the
heavy water through the channels between the fuel plates (or cylinders) and
then to an external heat exchanger in a closed circuit. The maximum authorized
thermal power of the CP-5 reactor and of the British reactors is 10 Mw. The
average thermal neutron flux ranges from 2×10^{13} to 10^{14} neutrons/(cm²)(sec).
At the higher operating powers, the systems under consideration are virtually
high-flux reactors.

13.152. Since 1950, many research reactors have been built with highly enriched fuel and ordinary water as coolant, moderator, and, in part at least, as
reflector. These reactors are classified in two general categories, called the
"pool" and "tank" types, respectively. Both kinds have compact cores consisting of assemblies of sandwich plates, either flat or curved, similar to those in
the MTR and ETR, described above. In fact, the pool type of reactor developed directly from work done on a mockup, at Oak Ridge National Laboratory, used for mechanical and hydraulic studies related to the MTR [84].
The basic difference between the pool and tank types is that, as the name implies,

the reactor core in the former type is suspended near the bottom of a pool of water, some 20 ft deep (Fig. 13.12). The water serves as moderator, reflector, coolant, and as a radiation shield in the upward direction. Shielding at the sides is provided partly by the water and partly by the concrete walls of the pool. In some forms of the pool reactor, beryllium, beryllium oxide, or graphite, contained in boxes to simulate the fuel element assemblies, are placed in contact with the core.

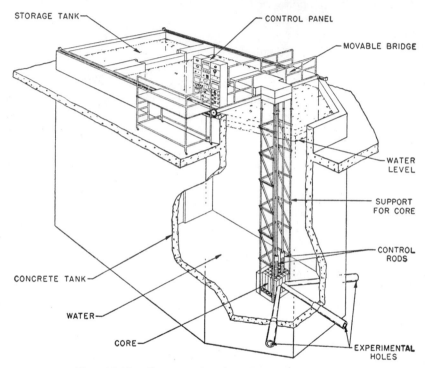

FIG. 13.12. Cutaway drawing of pool type reactor

13.153. For operation at powers up to 100 kw, which yields maximum thermal fluxes of roughly 10^{12} neutrons/(cm²)(sec), natural (or free) convection in the pool of water provides all the cooling that is necessary. An accidental reactivity excursion would cause the water between the fuel plates to boil and the displacement of moderator would then rapidly decrease the reactivity. Cooling by natural convection is feasible for powers up to 1 Mw, at least for intermittent operation, but flow divertors must then be used to prevent the radioactive nitrogen-16 in the rising water, formed by the (n, p) reaction with oxygen, from reaching the surface of the pool. For operation at powers in the range of about 1 Mw to 5 Mw (or above), forced downward circulation of the water, in conjunction with an external heat exchanger, and possibly deeper pools are necessary.

13.154. When the power is in excess of 5 Mw and even at lower values, such a large depth of water would be required for shielding that it becomes more convenient to use a solid, e.g., concrete, shield. The core is then placed in a closed tank through which water is circulated in a manner similar to that in the MTR and ETR. This is the tank type of research reactor, examples of which are the Omega West Reactor (OWR) at Los Alamos Scientific Laboratory and the Risö reactor in Denmark. The tank is surrounded by a concrete shield, as in the MTR and other reactors. One of the advantages of the tank-type reactor is that the closed system permits a more effective handling of the problem arising from the formation of nitrogen-16 in the water.

13.155. In certain respects, tank-type reactors may be regarded as intermediate in design, power, and neutron flux between pool-type reactors and high-flux reactors used for testing purposes. However, the pool reactors have an advantage in the respect that fuel elements can be removed and added from the top, using long-handled tools, since the core is accessible at all times. This not only facilitates charging and discharging, but it makes possible a degree of flexibility not attainable in a tank reactor. In the latter, the cover must be removed in order to reach the core. Moreover, experiments can be inserted from the top of a pool reactor, as well as through beam holes at the sides.

13.156. A comparison of light and heavy water as moderator for a research reactor using enriched fuel is of interest. Both are very stable in operation and the large negative temperature coefficients make them self-limiting to a power excursion of reasonable magnitude. It may be noted, however, that the prompt neutron lifetime in a deuterium-moderated reactor is approximately ten times that in ordinary water; hence, a higher excess reactivity is permissible in the former case for a given (short) reactor period. Because of the high cost of heavy water, it must be enclosed in a tank to prevent loss and to protect it from contamination by ordinary water in the atmosphere, which it absorbs readily. The core is thus not accessible, as it is in a pool reactor. For a given average flux, a heavy-water reactor is more expensive, but the larger core and the peaking of the thermal flux in the reflector provide more space in the primary reflector for experiments requiring a high thermal-neutron flux than is possible with a reactor using ordinary water. Because the proportion of moderator to fuel is usually considerably less in an ordinary-water reactor, the fast flux is generally greater in proportion to the thermal flux than in a heavy-water system.

LOW-FLUX TRAINING REACTORS

13.157. Pool reactors, similar to those described above, but operating at powers in the range of 1 to 100 kw are in use for training and research in a number of educational institutions.   In addition, several other reactor types of low power and a thermal flux from 10^6 to 10^{12} neutrons/(cm²)(sec) have been developed and can be obtained commercially at prices ranging from approximately \$100,000 to \$300,000.

13.158. The Water Boiler reactor (§ 13.103), which uses an aqueous solution of uranyl sulfate (or nitrate) as fuel, was one of the first reactor types to attract interest for teaching purposes [85].   The design is very simple, consisting of 1 kg (or less) of uranium-235, in the form of an enriched uranium salt, dissolved in about 13 liters of water contained in a stainless steel sphere of 1 ft in diameter. In one case, the vessel is cylindrical, rather than spherical.   Cold water is passed through a stainless steel coil in the core vessel to keep the temperature of the fuel solution down to about 175°F or below during operation.   In spite of its name, the water in a Water Boiler reactor does not boil.*   The core is surrounded by a reflector of graphite and the usual biological shield of concrete with layers of lead in certain locations.   Control is achieved by poison rods, which may be in the core or in the reflector or in both.   Water boilers are in operation at powers ranging from 1 watt to 50 kw, with respective average thermal fluxes of 5×10^7 to 10^{12} neutrons/(cm²)(sec).   For the very low powers, the ambient air provides all the cooling that is necessary.

13.159. The advantages of the Water Boiler reactor include simplicity and compactness of design, low inventory of fissile material, relatively low cost, and continuous removal of fission product gases.   The Kinetic Experiments on Water Boilers (KEWB) studies on the dynamic behavior of homogeneous systems have shown that Water Boiler reactors are extremely safe against the accidental release of large amounts of reactivity [86].   This safety is due to the large negative temperature coefficient and to the formation of bubbles of hydrogen and oxygen gases resulting from the radiolytic decomposition of the water by fission products.   These factors provide a rapid response opposing a reactivity excursion.

13.166. On the other hand, the small size of the Water Boiler core means that the space in which the neutron flux is highest is limited.   The evolution of radioactive fission product gases necessitates adequate shielding of ducts and provision for storage of the gases to permit their decay.   The KEWB studies have shown that explosion of the radiolytic gases is not a significant hazard.   Nevertheless, it is the practice in reactors of moderate and high power to employ a catalytic bed to bring about recombination of the hydrogen and oxygen gases; the water so produced returns to the reactor core.   Traces of chloride in the fuel solution have been known to cause serious corrosion of the stainless steel containment vessel.   At one time, difficulty was experienced with loss of reactivity in Water Boilers resulting from the precipitation of uranium peroxide by hydrogen peroxide; the latter was formed by the radiolytic decomposition of water (§ 7.93).

* Evolution of hydrogen and oxygen gas bubbles, resulting from radiolytic decomposition of the water, gives an appearance of gentle boiling.

The addition to the core solution of catalysts for hydrogen peroxide decomposition has overcome this problem [87].

13.161. The first Argonaut (Argonne Nuclear Assembly for University Training) reactor was built at Argonne National Laboratory in 1956 [88]. It was designed to be a safe, flexible, and low-cost facility that might be of interest for teaching purposes. The reactor core consists of assemblies of fuel plates made of 20 per cent enriched U_3O_8 dispersed in an aluminum matrix. These assemblies are arranged in an annular tank, 2 ft internal and 3 ft external diameter and 4 ft high. Both graphite and water serve as moderators; the graphite is in the form of segments between the fuel-plate assemblies, and the water, which is also the coolant, is in the channels between the plates. The central island contains a 2-ft diameter cylinder of graphite to serve as moderator and reflector, as in a flux-trap reactor, and an external graphite reflector surrounds the core vessel. Finally, there is biological shield of high-density concrete. The maximum operating power of the Argonaut reactor is 10 kw with water circulating in the coolant at a rate of 6.5 gal per min. The average thermal flux is then about 2.5×10^{11} neutrons/(cm^2)(sec).

13.162. The TRIGA reactor, which was mentioned in § 13.147, was first introduced in 1958 [89]. The cylindrical fuel rods consist of an intimate mixture of enriched (20 per cent) uranium with solid zirconium hydride moderator clad with aluminum. Burnable poison, in the form of thin wafers of samarium oxide and aluminum, is included in the fuel elements. Some 65 of these elements form a core 17 in. in diameter and 14 in. high which is submerged in a pool beneath 16 ft of water. The water acts as coolant, by natural convection, and also partly as moderator in the spaces between the fuel rods. Graphite enclosed in aluminum is the primary reflector and the water in the tank also functions as a secondary reflector. The maximum power level of the original model of the TRIGA reactor is 10 kw and the corresponding average thermal flux is about 10^{11} neutrons/(cm^2)(sec). However, designs of higher power, up to 1 Mw, have been developed. A special feature of the reactor is its self-limiting property arising from the negative temperature coefficient of the fuel; as seen in § 5.78, this provides a prompt response to reactivity changes. As a result, a form of the TRIGA reactor, called the FLASH reactor, can be used to produce a strong pulse of neutrons which is rapidly terminated as the temperature in the fuel rod increases.

13.163. A compact assembly of low power, and reasonable cost, suitable for teaching laboratories, is the Aerojet-General Nucleonics (AGN) reactor (1956). In the AGN-201 model, the core is made up of a number of disks of polyethylene, the moderator, in which uranium dioxide (20 per cent enrichment) powder is dispersed [90]. The assembled core, 10 in. in diameter and 9.5 in. high, is surrounded by a graphite reflector about 8 in. thick, all being contained in an aluminum tank to prevent leakage of fission gases which might possibly escape from the fuel. Around this tank is a 4-in. layer of lead, enclosed in steel.

Finally, the whole is surrounded by a steel vessel, 6.5 ft in diameter, containing boric acid solution to provide additional shielding. At the top of this tank is a compartment which can serve as a thermal column when filled with graphite or with ordinary or heavy water. Control is by poison rods inserted through the bottom of the tank. A safety fuse (§ 9.113) causes the lower half of the core-reflector assembly to drop a few inches, thus making the system subcritical, if its temperature exceeds 212°F. The normal operating power of the AGN-201 reactor is 0.1 watt and the maximum thermal flux is 4.5×10^6 neutrons/(cm²) (sec).

13.164. The AGN-211, which also uses a dispersion of enriched uranium dioxide in polyethylene as the fuel-moderator system, has a core of a somewhat different design. By placing the reactor in a pool of water, operation at a power of 100 watts is possible; the thermal flux is then 3×10^9 neutrons/(cm²)(sec).

REFERENCES FOR CHAPTER 13

1. H. S. Isbin, "Catalogue of Nuclear Reactors," *Proc. Second U.N. Conf. Geneva,* **8**, 561 (1958); "ASME Nuclear Reactor Plant Data," Vols. I and II, McGraw-Hill Book Co., Inc., New York, 1959; "International Atomic Energy Agency Directory of Nuclear Reactors," Vols. I and II, IAEA, 1959; "Power Reactors and Experiments the World Around," *Nucleonics,* **19**, No. 11, 132 (1961); "Nuclear Reactors Built, Being Built, or Planned in the United States," U. S. AEC Report TID-8200 (Rev.), 1962; "Small Power Reactors," U. S. AEC Report TID-8535 (1960); "Power Reactor Experiments," Vols. I and II, International Atomic Energy Agency, 1962; see also, *Proc. (First) U.N. Conf. Geneva,* Vols. **2** and **3** (1956) and *Proc. Second U.N. Conf. Geneva,* Vols. **8** and **9** (1958) or summaries in J. K. Pickard (Ed.), "Nuclear Power Reactors," D. Van Nostrand Co., Inc., Princeton, N. J., 1956; J. K. Pickard (Ed.), "Power Reactor Technology," D. Van Nostrand Co., Inc., Princeton, N. J., 1958.
2. "Civilian Power Reactor Program," Part III, Books 1 to 8 and Index, U. S. AEC Report TID-8518, U. S. Government Printing Office, Washington, D. C., 1959; addendum to the preceding, entitled "Core-Parameter Studies for Selected Reactor Types," U. S. Government Printing Office, Washington, D. C., 1961; see also U. S. AEC Report TID-8504 (1959); *Nucleonics,* **17**, No. 8, 59 *et seq.* (1959).
3. "Civilian Power Reactor Program," Part II, Economic Potential and Development Programs, U. S. AEC Report TID-8517, U. S. Government Printing Office, Washington, D. C., 1960; *ibid.,* Part IV, Plans for Development as of February, 1960, U. S. AEC Report TID-8519, U. S. Government Printing Office, Washington, D. C., 1960.
4. "Status Report on Pressurized Water Reactors," U. S. AEC Report TID-8518, Book 2, U. S. Government Printing Office, Washington, D. C., 1960; see also, U. S. AEC Report TID-8513 (1959).
5. Ref. 4, U. S. AEC Report TID-8518, Book 2, p. 4.
6. "The Shippingport Pressurized Water Reactor," Addison-Wesley Publishing Co., Inc., Reading, Mass., 1958; *Nucleonics,* **16**, No. 4, 53 *et seq.* (1958); J. W. Simpson and H. G. Rickover, *Proc. Second U.N. Conf. Geneva,* **8**, 40 (1958); P. G. DeHuff, *et al., ibid.,* **8**, 47 (1958); P. A. Fleger, *et al.,* U. S. AEC Report WAPP-T-1429 (1961).
7. W. E. Shoupp, *et al., Proc. Second U.N. Conf. Geneva* **8**, 492 (1958); *Power Reactor Technology,* **4**, No. 3, 47 (1961); *Nucleonics,* **19**, No. ?, 53 (1961).

8. G. R. Milne, *Proc. Second U. N. Conf. Geneva*, **8**, 483 (1958).

9. *Power Reactor Technology*, **2**, No. 2, 3 (1959); W. H. Zinn and R. P. Godwin, *Proc. Second U. N. Conf. Geneva*, **8**, 110 (1958); J. H. McMillan, U. S. AEC Report BAW-1117 Rev. (1958); *Nucleonics*, **20**, No. 7, 37 (1962); A. W. Kramer, "Nuclear Propulsion for Merchant Ships," U. S. Government Printing Office, Washington, D. C., 1962.

10. "Status Report of Boiling Water Reactor Technology," U. S. AEC Report TID-8518, Book 5, U. S. Government Printing Office, Washington, D. C., 1960; A. W. Kramer, "Boiling Water Reactors," Addison-Wesley Publishing Co., Inc., Reading, Mass., 1958; J. M. West and G. M. Roy, *Nucleonics*, **17**, No. 1, 42 (1959); see also, U. S. AEC Report TID-8510 (1959); *Power Reactor Technology*, **3**, No. 3, 64 (1960).

11. A. W. Kramer, Ref. 10, Ch. 3.

12. U. S. AEC Report ANL-5607, U. S. Government Printing Office, Washington, D. C., 1957; J. M. Harrer, *et al.*, *Proc. (First) U. N. Conf. Geneva*, **3**, 250 (1956); A. W. Kramer, Ref. 10, Chs. 5 and 6; *Nucleonics*, **15**, No. 7, 53 *et seq.* (1957).

13. A. W. Kramer, Ref. 10, Ch. 9; *Nucleonics*, **17**, No. 12, 65 (1959); *Power Reactor Technology*, **4**, No. 4, 56 (1961).

14. A. W. Kramer, Ref. 10, Ch. 7; *Nucleonics*, **16**, No. 2 (1958).

15. A. W. Kramer, Ref. 10, Ch. 10.

16. S. Untermyer, *Nucleonics*, **13**, No. 7, 34 (1955).

17. *Power Reactor Technology*, **2**, No. 3, 51 (1959); **3**, No. 4, 68 (1960); **4**, No. 3, 71 (1961); "Report of 4th Nuclear Superheating Meeting," U. S. AEC Report TID-7617 (1961); *Power Reactor Technology*, **5**, No. 3, 71 (1962); see also, U. S. AEC Report TID-8536 (1961).

18. M. Novick and R. E. Rice, *Proc. Conf. on Small and Medium Power Reactors*, p. 111, International Atomic Energy Agency, 1960; R. E. Rice, *et al.*, U. S. AEC Report ANL-6120 (1960); *Power Reactor Technology*, **4**, No. 1, 71 (1960).

19. *Power Reactor Technology*, **2**, No. 3, 51 (1959); *Nucleonics*, **18**, No. 3, 108 (1960); U. S. AEC Report ACNP-5917 (1959).

20. *Power Reactor Technology*, **3**, No. 4, 68 (1960); **4**, No. 3, 76 (1961).

21. U. S. AEC Report ACNP-5917 (1959).

22. "Status Report of Heavy Water Moderated Reactors," U. S. AEC Report TID-8518, Book 4, U. S. Government Printing Office, Washington, D. C., 1960; *Power Reactor Technology*, **1**, No. 4, 59 (1958); **4**, No. 2, 62 (1961); see also, U. S. AEC Reports TID-8529, 8530 (1960); U. S. AEC Reports DP-480, 510, 520 (1960); U. S. AEC Report SL-1873 (1961); J. R. Dietrich and W. H. Zinn (Eds.), "Solid Fuel Reactors," Addison-Wesley Publishing Co., Inc., Reading, Mass., 1958, Ch. 5.

23. J. R. Dietrich and W. H. Zinn, Ref. 22, p. 428; U. S. AEC Report DP-383 (1959); *Power Reactor Technology*, **3**, No. 2, 54 (1960); **4**, No. 2, 74 (1961).

24. J. R. Dietrich and W. H. Zinn, Ref. 22, Ch. 8; R. M. Fryar, *Proc. Second U. N. Conf. Geneva*, **9**, 221 (1958); U. S. AEC Report HW-61236 (1959); *Power Reactor Technology*, **3**, No. 3, 53 (1960).

25. P. G. DeHuff, *Proc. Conf. on Small and Medium Power Reactors*, p. 179, International Atomic Energy Agency, 1960; *Power Reactor Technology*, **3**, No. 2, 54 (1960); **4**, No. 2, 73 (1961).

26. I. N. MacKay, *Proc. Second U. N. Conf. Geneva*, **8**, 313 (1958); H. A. Smith, *et al.*, **9**, 3 (1958); W. B. Lewis, Atomic Energy of Canada, Ltd., Report AECL-785 (1959); *Nucleonics*, **18**, No. 10, 54 *et seq.* (1960); *Power Reactor Technology*, **4**, No. 2, 64 (1961).

27. N. Hidle and O. Dahl, *Proc. Second U. N. Conf. Geneva*, **9**, 255 (1958); see also, *Nucleonics*, **17**, No. 8, 59 *et seq.* (1959); *Power Reactor Technology*, **4**, No. 2, 65 (1961); U. S. AEC Reports TID-8504 (1959), TID-8529 (1960).

28. J. R. Dietrich and W. H. Zinn, Ref. 22, pp. 611 *et seq.;* M. J. McNelly, *Proc. Second U. N. Conf. Geneva,* **9,** 79 (1958); E. Bernsohn, *et al., Nucleonics,* **17,** No. 5, 112 (1959); I. MacKay, *ibid.,* **18,** No. 10, 78 (1960).

29. M. C. Edlund and G. K. Rhode, *Nucleonics,* **16,** No. 5, 80 (1958); M. C. Edlund, *Proc. Conf. on Small and Medium Power Reactors,* p. 165, International Atomic Energy Agency, 1960; U. S. AEC Report BAW-1241 (1961); *Power Reactor Technology,* **1,** No. 2, 62 (1958).

30. "Status Report on Organic Cooled Reactors," U. S. AEC Report TID-8518, Book 7, U. S. Government Printing Office, Washington, D. C., 1960; "Proceedings of the Organic Cooled Reactor Forum," U. S. AEC Report NAA-SR-5688 (1960); C. A. Trilling, *Nucleonics,* **17,** No. 11, 113 (1959); S. Siegel and R. F. Wilson, *ibid.,* **17,** No. 11, 118 (1959); S. Siegel, *et al., Proc. Conf. on Small and Medium Power Reactors,* p. 295, International Atomic Energy Agency, 1960; B. Balent, *et al., ibid.,* p. 317; J. R. Dietrich and W. H. Zinn, Ref. 22, Ch. 7; U. S. AEC Report ANL-6360 (1961); see also, U. S. AEC Reports TID-8511, 8512 (1959).

31. C. A. Trilling, *et al., Proc. Second U. N. Conf. Geneva,* **29,** 292 (1958); R. H. J. Gercke, U. S. AEC Report NAA-SR-5688 (1960), p. 5.

32. J. R. Dietrich and W. H. Zinn, Ref. 22, pp. 696 *et seq.;* C. A. Trilling, *Proc. Second U. N. Conf. Geneva,* **9,** 468 (1958).

33. E. F. Weisner, *et al., Proc. Second U. N. Conf. Geneva,* **9,** 99 (1958); U. S. AEC Report NAA-SR-5688 (1960), p. 211; *Proc. Conf. on Small and Medium Power Reactors,* p. 339, International Atomic Energy Agency, 1960.

34. M. R. Dusbabek, U. S. AEC Report NAA-SR-5688 (1960), p. 189; J. H. Rainwater and W. E. Nyer, U. S. AEC Reports IDO-16570 (1959), IDO-24034 (1960); *Power Reactor Technology,* **5,** No. 1, 88 (1961).

35. "Status Report on Sodium Graphite Reactors," U. S. AEC Report TID-8518, Book 6, U. S. Government Printing Office, Washington, D. C., 1960; C. Starr and R. W. Dickinson, "Sodium Graphite Reactors," Addison-Wesley Publishing Co., Inc., Reading, Mass., 1958; *Power Reactor Technology,* **4,** No. 3, 88 (1961).

36. C. Starr and R. W. Dickinson, Ref. 35, Ch. 2; *Nucleonics,* **15,** No. 12 (1957); *Power Reactor Technology,* **3,** No. 2, 59 (1960).

37. C. Starr and R. W. Dickinson, Ref. 35, Ch. 8; R. L. Olson, *et al., Proc. Second U. N. Conf. Geneva,* **9,** 161 (1958); *Power Reactor Technology,* **5,** No. 3, 39 (1962).

38. "Status Report on Gas Cooled Reactors," U. S. AEC Report TID-8518, Book 8, U. S. Government Printing Office, Washington, D. C., 1960; J. R. Dietrich and W. H. Zinn, Ref. 22, Ch. 6; "Gas-Cooled Reactors," Proceedings of a Symposium, Franklin Institute Monograph No. 7, 1960; W. S. Banks, *Nucleonics,* **17,** No. 9, 96 (1959); *Power Reactor Technology,* **1,** No. 4, 53 (1958); **2,** No. 1, 26 (1958); **2,** No. 3, 55 (1959); **3,** No. 4, 58 (1960).

39. *Nucleonics,* **14,** No. 12, S1–S32 (1956); *Proc. Second U. N. Conf. Geneva,* **8,** 416–478 (1958); K. Jay, "Nuclear Power: Today and Tomorrow," Methuen and Co., Ltd., London, 1961, Ch. 6 and Appendix III.

40. R. V. Moore, *et al., Proc. Second U. N. Conf. Geneva,* **9,** 104 (1958); K. Jay, Ref. 39, pp. 108 *et seq.*

41. G. Thornton and R. Blumberg, *Nucleonics,* **19,** No. 1, 45 (1961); see also, *Nuc. Sci. and Eng.,* **2,** 797 *et seq.* (1957).

42. U. S. AEC Report AECU-4701 (1959); *Power Reactor Technology,* **4,** No. 1, 63 (1960).

43. "High Temperature Gas-Cooled Civilian Power Reactor Conference," U. S. AEC Report TID-7611 (1961); P. Fortescue, *et al., Nucleonics,* **18,** No. 1, 86 (1960); F. de Hoffman and P. Fortescue, *Proc. Conf. on Small and Medium Power Reactors,* p. 441, International Atomic Energy Agency, 1960.

44. L. R. Shepherd, *et al.*, *Proc. Second U. N. Conf. Geneva*, **9**, 289 (1958); *Nuclear Power*, **4**, No. 36, 102 (1959); K. Jay, Ref. 39, pp. 113 *et seq.*

45. A. Boettcher, *et al.*, *Proc. Second U. N. Conf. Geneva*, **7**, 748 (1958); R. Schulten, *ibid.*, **9**, 306 (1958); A. P. Fraas, *et al.*, U. S. AEC Reports ORNL CF-60-10-63, CF-60-12-5 (1960); L. Goldstein and C. W. Monroe, U. S. AEC Report NDA-2159-2 (1961); U. S. AEC Report NYO-8753 (1958); *Power Reactor Technology*, **2**, No. 3, 56 (1959).

46. R. P. Hammond, *et al.*, *Nucleonics*, **17**, No. 12, 106 (1959); U. S. AEC Reports LA-2303 (1959); LA-2689 (1962).

47. U. S. AEC Report GA-1612 (1960); F. de Hoffman, *et al.*, *Proc. Conf. on Small and Medium Power Reactors*, p. 431, International Atomic Energy Agency, 1960.

48. U. S. AEC Reports GA-1738 (1960), GAMD-1565 (1960).

49. *Power Reactor Technology*, **3**, No. 4, 59 (1960), **4**, No. 2, 7 (1961); *Nucleonics*, **19**, No. 2, 82 (1961).

50. R. W. Bussard and R. D. DeLauer, "Nuclear Rocket Propulsion," McGraw-Hill Book Co., Inc., New York, 1958; M. M. Levoy and J. J. Newgard, *Nucleonics*, **16**, No. 7, 60 (1958); F. P. Durham, *J. Am. Rocket Soc.*, **4**, 10 (1959); J. J. Newgard and M. M. Levoy, *Nuc. Sci. and Eng.*, **7**, 337 (1960); H. B. Finger, *et al.*, *Nucleonics*, **19**, No. 4, 58 *et seq.* (1961).

51. J. R. Dietrich and W. H. Zinn, Ref. 22, pp. 611 *et seq.*; *Power Reactor Technology*, **4**, No. 2, 74 (1961); U. S. AEC Report GNEC-141 (1960).

52. "Status Report on Fast Reactors," U. S. AEC Report TID-8518, Book 1, U. S. Government Printing Office, Washington, D. C., 1960; *Nucleonics*, **15**, No. 4, 61 *et seq.* (1957); L. J. Koch, *Nucleonics*, **16**, No. 3, 68 (1958); J. R. Dietrich and W. H. Zinn, Ref. 22, Chs. 2, 3, and 4; *Power Reactor Technology*, **1**, No. 3, 56 (1958); **4**, No. 4, 84 (1961); U. S. AEC Report TID-8523 (1960).

53. E. T. Jurney, *Nucleonics*, **12**, No. 9, 28 (1954); *Chem. Eng. Prog. Symposium Series*, No. 13, **50**, 191 (1954).

54. H. V. Lichtenberger, *et al.*, *Proc. (First) U. N. Conf. Geneva*, **3**, 345 (1956); *Chem. Eng. Prog. Symposium Series*, No. 13, **50**, 139 (1954).

55. L. J. Koch, *et al.*, *Proc. Second U. N. Conf. Geneva*, **9**, 323 (1958); U. S. AEC Reports ANL-5719 (1957), ANL-6383 (1961); J. R. Dietrich and W. H. Zinn, Ref. 22, Ch. 3.

56. *Nucleonics*, **15**, No. 4, 68 (1957); J. R. Dietrich and W. H. Zinn, Ref. 22, Ch. 4; A. Amorosi and J. G. Yevick, *Proc. Second U. N. Conf. Geneva*, **9**, 358 (1958); *Nucleonics*, **19**, No. 2, 64 (1961).

57. J. W. Kendall and T. M. Fry, *Proc. (First) U. N. Conf. Geneva*, **3**, 193 (1956); H. Cartwright, *et al.*, *Proc. Second U. N. Conf. Geneva*, **9**, 316 (1958); K. Jay, Ref. 39, pp. 165 *et seq.*

58. A. I. Leipunsky, *et al.*, *Proc. Second U. N. Conf. Geneva*, **9**, 348 (1958).

59. J. R. Beeley, *Nucleonics*, **16**, No. 11, 162 (1958).

60. "Status Report on Aqueous Homogeneous Reactors," U. S. AEC Report TID-8518, Book 3, U. S. Government Printing Office, Washington, D. C., 1960; J. A. Lane, H. G. MacPherson, and F. Maslan, "Fluid Fuel Reactors," Addison-Wesley Publishing Co., Inc., Reading, Mass., 1958, Chs. 1 to 10; *Power Reactor Technology*, **1**, No. 2, 46 (1958).

61. J. A. Lane, *et al.*, Ref. 60, pp. 341 *et seq.*; L. D. P. King, *Proc. (First) U. N. Conf. Geneva*, **2**, 372 (1956).

62. S. E. Beall and C. E. Winters, *Chem. Eng. Prog.*, **50**, 256 (1954); R. B. Briggs and J. A. Swartout, *Proc. (First) U. N. Conf. Geneva*, **3**, 175 (1956); J. A. Lane, *et al.*, Ref. 60, pp. 348 *et seq.*

63. S. E. Beall and J. A. Swartout, *Proc. (First) U. N. Conf. Geneva*, **3**, 263 (1956);

J. A. Lane, *et al.*, Ref. 60, pp. 359 *et seq.; Proc. Second U. N. Conf. Geneva*, **9**, 509 (1958).

64. J. A. Lane, *et al.*, Ref. 60, pp. 397 *et seq.;* D. Froman, *et al., Proc. (First) U. N. Conf. Geneva*, **3**, 283 (1956); R. J. Thamer, *Proc. Second U. N. Conf. Geneva*, **7**, 54 (1958).

65. J. A. Lane, *et al.*, Ref. 60, Chs. 11 to 17; H. G. MacPherson, *et al., Proc. Second U. N. Conf. Geneva*, **9**, 188 (1958).

66. J. H. DeVan, *Trans. Am. Nuc. Soc.*, **4**, 199 (1961); A. L. Bloch, *et al., ibid.*, **4**, 331 (1961); S. E. Beall, *et al.*, U. S. AEC Report ORNL CF-61-2-46 (1961).

67. C. Williams and R. T. Schomer, *Proc. Second U. N. Conf. Geneva*, **10**, 487 (1958); *Power Reactor Technology*, **1**, No. 3, 42 (1958).

68. R. M. Kiehn, *et al., Proc. Second U. N. Conf. Geneva*, **9**, 411 (1958); U. S. AEC Report LA-2327 (1959); *Power Reactor Technology*, **3**, No. 2, 65 (1960).

69. *Nucleonics*, **18**, No. 1, 104 (1960); W. R. Corliss, *ibid.*, **18**, No. 8, 58 (1960).

70. H. M. Dieckamp, *et al., Nucleonics*, **19**, No. 4, 73 (1961).

71. K. G. Hernqvist, *Nucleonics*, **17**, No. 7, 49 (1959); G. M. Grover, *ibid.*, **17**, No. 7, 54 (1959); *Power Reactor Technology*, **3**, No. 1, 6 (1959); **3**, No. 3, 48 (1960).

72. R. H. Armstrong, U. S. AEC Report ANL-6261 (1960).

73. "Selected Reference Material on Atomic Energy," Vol. 1, Research Reactors, U. S. Government Printing Office, Washington, D. C., 1955; "U. S. Research Reactors," U. S. AEC Technical Information Service, 1957; "Symposium on High-Flux Materials Testing Reactors," U. S. AEC Report TID-7548 (1960); "Proceedings of the University Reactor Conference," U. S. AEC Report TID-7608 (1961).

74. U. S. AEC Report TID-7548 (1960).

75. T. E. Cole and J. A. Cox, *Proc. Second U. N. Conf. Geneva*, **10**, 86 (1958); *Nucleonics*, **16**, No. 8 (1958).

76. J. R. Huffman, *Nucleonics*, **12**, No. 4, 21 (1954); A. M. Weinberg, *et al., Proc. (First) U. N. Conf. Geneva*, **2**, 402 (1956).

77. *Nucleonics*, **15**, No. 3, 41 *et seq.* (1957); R. L. Doan, *ibid.*, **16**, No. 1, 102 (1958); J. Barnard, *et al., Proc. Second U. N. Conf. Geneva*, **10**, 75 (1958); U. S. AEC Report TID-7552 (1958).

78. W. Boyd, *et al., Proc. Second U. N. Conf. Geneva*, **10**, 128 (1958).

79. D. R. de Boisblanc, *et al.*, U. S. AEC Reports IDO-16666 (1960), IDO-16667 (1960); R. S. Marsden, *et al.*, U. S. AEC Report IDO-16668 (1961).

80. G. A. Freund, *et al., Proc. Second U. N. Conf. Geneva*, **10**, 461 (1958); U. S. AEC Report ANL-6034 (1960).

81. T. F. Wimett and J. D. Orndoff, *Proc. Second U. N. Conf. Geneva*, **10**, 449 (1958).

82. See *Proc. (First) U. N. Conf. Geneva*, Vol. **3** (1956).

83. W. B. Lewis, *Physics Today*, **4**, No. 11, 12 (1951); F. W. Gilbert, *Chem. Eng. Progress*, **50**, 267 (1954); D. G. Hurst, *Proc. (First) U. N. Conf. Geneva*, **5**, 111 (1956).

84. W. M. Brezeale, *Nucleonics*, **10**, No. 11, 56 (1952); *Proc. (First) U. N. Conf. Geneva*, **2**, 420 (1956).

85. See Ref. 61.

86. J. W. Flora and R. K. Stitt (Eds.), U. S. AEC Report NAA-SR-5415 (1962).

87. C. K. Beck, *Nucleonics*, **13**, No. 7, 58 (1955).

88. R. H. Armstrong and C. N. Kelber, *Nucleonics*, **15**, No. 3, 62 (1957).

89. S. L. Koutz, *et al., Proc. Second U. N. Conf. Geneva*, **10**, 282 (1958); *Nucleonics*, **16**, No. 8, 116 (1958).

90. A. T. Biehl, *et al., Nucleonics*, **14**, No. 9, 100 (1956).

91. A. T. Biehl, *et al., Proc. Second U. N. Conf. Geneva*, **10**, 368 (1958).

PROBLEMS

1. Examine in detail the advantages and disadvantages of the following reactor concepts mentioned in the text:
   (a) Pebble-bed fuel elements, gas cooled.
   (b) Heavy-water moderator, sodium coolant.
   (c) Heavy-water moderator, carbon dioxide coolant.
   (d) Heavy-water moderator, organic coolant.
Outline a design for each type, including choice of fuel, cladding, reflector, etc., and explain the reasons which determine the selection made. Problems of safety and economy must be kept in mind.

2. The following are reactor systems of a somewhat unusual type; consider their advantages and disadvantages and show how they might be realized in practice. Indicate if there is any specific item, in each case, which is likely to have a major effect in determining the success or failure of the concept if applied to the production of economic electric power.
   (a) A boiling slurry fuel, consisting of UO_2-ThO_2 particles (solid solution) suspended in heavy water.
   (b) Heavy-water moderator and boiling sulfur coolant.
   (c) Organic moderator and sodium coolant.
   (d) Heavy-water moderator and ordinary water "fog" as coolant.
   (e) Organic moderator and coolant with fine particles of uranium carbide in a fluidized bed as fuel.

3. A so-called "coupled breeding superheating" system features a central, steam-cooled inner fast core and an outer thermal region cooled by nonboiling pressurized water, combined with a Loeffler cycle (§ 6.216). Analyze this concept in detail and outline a design for the core, blanket, and coolant system.

4. Discuss in detail the requirements of a nuclear reactor plant for supplying electric power to a station on the moon. Design a possible system for this purpose, bearing in mind that cost would be a secondary consideration.

Chapter 14*

NUCLEAR POWER COSTS [1]

INTRODUCTION

IMPORTANCE OF ECONOMIC CONSIDERATIONS

14.1. From time to time, the economic aspects of the various types of power reactors have been mentioned in the preceding chapters. An attempt will now be made to provide a more quantitative examination of the factors that contribute to the cost of nuclear power. In this way, it should be possible to identify some of the particular directions in which research and development could help to improve the economics of a specific change. Furthermore, economic considerations frequently play a major role in the choice of a reactor concept for a given application. A design which appears best for a government-owned desalting plant may not be suitable for a privately-owned utility since the interest charges for capital funds may be quite different. Similarly, differences in economic factors from one country to another may affect many of the engineering design decisions that must be made in achieving a minimum cost reactor operation.

14.2. The reactor engineer must therefore concern himself with economics as well as with such topics as neutron behavior, energy removal, etc., described earlier in this book. In fact, as has been indicated in Chapter 13, the technological and economic variables are often in conflict in the design of a nuclear reactor system. It is therefore not satisfactory merely to design and build a commercial power reactor that will operate and produce electricity in an acceptable manner. The design parameters and specifications must be chosen so as to minimize the net cost of the power. Since the technological and economic factors are interrelated, the reactor engineer must be aware of their relative importance.

PROBLEMS IN MAKING ECONOMIC ESTIMATES

14.3. Nuclear power economics cannot be treated with the same degree of precision as is possible with the other topics in this book. Although numerous

* Revised September 1966.

cost studies have been published, actual cost and performance data are available from only a limited number of commercial plants, mostly of smaller size than those likely to be competitive with fossil fueled plants (§14.5). Furthermore, the cost information developed by equipment manufacturers is used by them in the preparation of competitive bids, and hence is considered proprietary. A rigorous economic treatment is further complicated by the rapid changes resulting from new developments, due to the very rapid expansion of the nuclear power industry and, finally, by sensitivity to general business conditions and related economic factors. The relative importance of fixed charges for capital funds can vary somewhat, for example, as general interest rates and security market levels change.

14.4. In spite of the uncertainties, by the use of standardized procedures it is possible to make comparative estimates which serve to indicate the current strong and weak aspects of various reactor designs. In addition, standardized procedures are useful for determining the relative importance of individual cost items. For example, although the absolute value of a given item may be uncertain, the relation that it bears to the total cost may be considered fairly reliable. A framework may therefore be established for parametric studies in which the effects of systematic changes in engineering, physics, and economic variables can be explored. It should be realized, however, that because the field of nuclear power reactor engineering is relatively new, changes can occur rapidly which may alter this economic framework. Nevertheless, the discussion given here should serve two main purposes: first, to show, in a general way, how nuclear power cost estimates are made, and second, to point up the areas in which cost changes might be significant.

COMPETITIVE NUCLEAR POWER

14.5. Before proceeding with a detailed analysis of the factors contributing to nuclear power costs, it is of help, for orientation purposes, to consider the economic situation in a conventional power plant using fossil fuel. Power costs are customarily broken down into three major categories: (a) a fixed charge which can be related to the capital cost of the facility, (b) a fuel cost, and (c) a charge for operation and maintenance. A breakdown into these groups of the costs for a very modern coal-fueled plant of 600-Mw (electrical) capacity is given in Table 14.1.* This size is fairly typical of new nuclear and fossil-fueled plants planned for completion about 1970 in the United States. The costs shown represent an average over the estimated 30-year lifetime of the plant based on investor-owned electric utility practice of calculating depreciating fixed charges year by year. Furthermore, it is assumed that the plant factor (§14.17) decreases from an initial

---

* Adapted from Jersey Central Power and Light Company Report on Oyster Creek Station. 1964.

TABLE 14.1   600-MW (ELECTRICAL) FOSSIL FUEL PLANT COSTS

Plant Description

Electrical output, megawatts...................	600	
Heat rate, Btu/kw-hr.......................	9,110	
Capital cost, $/kw........................	114	

Power Costs

Fixed charges, mills/kw-hr...................	1.45	1.45
Fuel cost, mills/kw-hr......................	1.79*	2.33†
Operation and maintenance,		
mills/kw-hr	0.41	0.41
Total mills/kw-hr...........	3.65*	4.19†

* Fuel at 20 cents per million Btu   † Fuel at 26 cents per million Btu

8-year value of 88 per cent to a final value of 40 per cent due to diminishing use of the facility since newer units will probably be more efficient. Costs are given for coal both at 20 cents per million Btu, a "minimum" value representative of an area close to a plentiful supply, and 26 cents per million Btu, a typical price for large-scale users on the Atlantic seaboard.

14.6. The total costs in Table 14.1 are quite close to those estimated for proven designs of the boiling water or pressurized water type of nuclear plant of similar size. Local economic factors specific to the application and the resulting interplay with engineering parameters therefore become important in the detailed economic analysis that is required for the selection decision. In 1966, both the capital cost and fuel costs for fossil plants appeared to be rising whereas the corresponding costs for nuclear plants were remaining at a constant level or, in fact, were declining. As a result, a number of utilities in the United States decided to build nuclear-fueled plants in preference to coal-fired stations.

NUCLEAR POWER COST ANALYSIS

COST NORMALIZATION

14.7. Although exact cost information may be incomplete, it is possible to make comparative estimates of the cost of nuclear power from different reactor concepts or for different situations by utilizing a standard accounting procedure. Many items which do not make a large contribution to the total and which are not dependent upon the variables under study are assumed to be constant from one estimate to another. Comparative data obtained in this manner through "cost normalization" are also useful to the designer who must take into account the interplay between engineering and economic parameters in the development of a preliminary design but need not be concerned with accurate costs until later. In the United States, a useful uniform treatment of cost parameters is described in the "Guide to Nuclear Power Cost Evaluation" [2] to be referred to subsequently as the Guide.

CAPITAL COSTS

Direct Construction Costs

14.8. In the analysis of nuclear power costs, the same three categories are used as for conventional power plants (§ 14.5). The first matter to consider, therefore, is the capital cost of the reactor plant, since this is responsible for an important contribution to the fixed charges. To systematize the listing of the capital costs of various components and to simplify accounting procedures, industry has adopted a "code of accounts" in which the main items are identified by numbers in a standard manner. For purposes of illustration the estimated construction costs of a 1000-Mw (electrical) pressurized water reactor facility are shown in Table 14.2, following the code of accounts system. The subdivisions are given only in connection with Account No. 22, since this is of direct interest to the nuclear reactor engineer, although subdivisions exist under all the other account numbers. It should be mentioned that the table includes only the direct construction costs; indirect costs will be considered shortly. In the following paragraphs is given a description of the major categories and component items of Table 14.2.

TABLE 14.2 CONSTRUCTION COST BREAKDOWN FOR 1000 MW
(ELECTRICAL) PRESSURIZED-WATER REACTOR PLANT [4]*

	Account Number and Description		*Capital Cost*
20	Land and Land Rights..		$ 360,000
21	Structures and Improvements.................................		9,000,000
22	Reactor Plant Equipment.....................................		44,200,000
	221	Reactor equipment...................	$14,000,000	
	222	Heat transfer equipment...............	19,000,000	
	223	Fuel handling and storage facilities......	900,000	
	224	Fuel reprocessing and refabrication......	—	
	225	Waste disposal.......................	300,000	
	226	Instrumentation and control............	2,600,000	
	227	Feedwater supply and treatment........	4,400,000	
	228	Steam condenser and feedwater piping.............................	3,000,000	
23	Turbo-generator Units.......................................		26,700,000
24	Accessory Electrical Equipment..............................		3,800,000
25	Miscellaneous Power Plant Equipment.......................		800,000
	Total Direct Construction Costs..........................		$84,860,000

* Costs adjusted to correspond with 1966 commercial bids for complete systems.

Land and Land Rights: In this category is the cost of the land required for the plant site as well as for the adjacent exclusion area (§ 9.11). For purposes of cost normalization, a "standard" typical site 35 miles from a city and 40 ft above river level is described in the Guide. A cost of $360,000 for the 1200 acre site represents almost an insignificant proportion of the total reactor plant construction cost and hence need not be considered in detail. In actual cases, however, the cost of land and site development may amount to about $2 million [3]. It should be noted that this item does not change with time whereas the other items in the direct construction cost are subject to depreciation. As a result, the land cost is handled separately in the calculation of the annual fixed charges.

Structures and Improvements: This represents the cost of preparing the site and constructing all the buildings required for the reactor plant, including the reactor building, shielding, and the containment vessel. Depending on the details of the reactor system, this item may range from $8 to $12 million for a 1000-Mw (electrical) facility.

Reactor Plant Equipment: In this category is included the reactor proper, reactor vessel (and thermal shields), coolant, moderator,* foundation, and fuel handling system. The costs of the heat-transfer system, comprising heat exchangers, auxiliary piping, tanks, etc., as well as of various support systems, such as instrumentation, water treatment, hot cells, remote maintenance, waste disposal, and special ventilation equipment are in this category. The costs of the plant equipment are not very sensitive to the reactor type, at least for water-cooled, heavy water-cooled, high pressure gas-cooled, and sodium-cooled reactors [5]. A total cost of $50 to $55 million for a 1000-Mw plant is appropriate for this category.

Turbo-generator Plant: The turbo-generator, its foundations, and all accessories are in this accounting category. The differences in cost are dependent on the quality of steam available from the given reactor system, but the variations are not large. The cost for a 1000-Mw (electrical) nuclear plant would be from $22 to $27 million, which is a substantial proportion of the total.

Accessory Electrical Equipment, Miscellaneous Power Plant Equipment, and Main Power Transformer: These three categories, which are taken together, do not involve a large fraction of the capital expenditure, amounting to about $4 million for the 1000-Mw installation. Included are (1) switchgear, wiring, and conduit work, (2) other equipment, such as compressed air and refrigeration systems, and (3) main power transformer, for the three respective accounting items.

Indirect Construction Costs

14.9. In addition to the direct construction costs, there are a number of indirect costs which must be added; these include costs of contracting, design, engineering, contingencies, inspection, startup, and interest during construction. Such indirect construction costs often vary widely, sometimes because of different accounting systems, but also as a result of more or less experience by the contractor in building a specific plant. Design expenses and allowances for contin-

---

* Heavy water is treated as a non-depreciating asset; the procedure is similar to that used for natural uranium (§ 14.23).

gencies, for example, can be markedly reduced for the second, third, etc., plant of a given design. To serve as a guide, however, some typical indirect costs for power reactors for which the direct construction costs are over $75 million are listed in Table 14.3.

TABLE 14.3   INDIRECT CONSTRUCTION COSTS FOR LARGE NUCLEAR PLANTS*

Item	Cost
General and administrative........	6% of direct cost
Miscellaneous construction.........	1% of above
Architect-engineer fees............	5% of above two items
Nuclear-engineer fees..............	2% of above three items
Startup...........................	35% of first year's nonfuel operating and maintenance cost
Contingency.......................	10% of subtotal of all of above, including startup
Interest during construction........	10.8% (based on 48-month construction schedule in all cases)

* Some reduction is possible when more than one plant is to be built.

14.10. Another allowance sometimes made in cost estimates is for the predictable increase in prices during the period of equipment procurement and of construction. Although the amount allowed for this purpose will depend upon price trend forecasts, a reasonable value for "escalation," as it is called, of 10–15 per cent of the direct cost may be included in the preliminary estimates.

14.11. For the 1000-Mw (electrical) pressurized water reactor, to which the data in Table 14.2 apply, the startup cost is estimated to be $700,000, and the total indirect costs $35 million. The total estimated capital cost for the installation is thus close to $120 million, representing $120 per kilowatt.

Effect of Size

14.12. As the electrical capacity of the power plant is increased, the total construction cost when expressed on a dollars per kilowatt basis shows a significant decrease. A ten-fold increase in capacity, for example, results in a decrease in unit construction costs by a factor of about two. Figure 14.1 shows this trend for highly-developed pressurized water and boiling water reactors ordered in the mid 1960s. Costs for several earlier reactors, such as Yankee (§ 13.14) and Dresden (§ 13.24), which may be considered as prototypes, lie well above the curve. Plants ordered prior to 1965 may be considered as intermediate between the prototype stage and the commercially developed stage represented by the curve in Fig. 14.1. Although capital costs for other concepts cannot be estimated with the same degree of reliability, it is possible to compare design features with those of the PWR-BWR systems and thereby predict probable cost variations from the "known" system. Since a large fraction of the construction costs for the entire reactor plant system arises from items which are external to the reactor proper (§ 14.18), however, total costs do not vary markedly from one concept

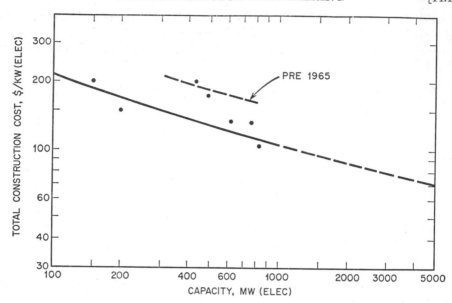

FIG. 14.1. Total construction costs per kilowatt of electricity as a function of plant capacity.

to another at the same respective stage of development. Moreover, the sensitivity to size is likely to be similar.

14.13. Several factors contribute to the decrease in construction costs on a unit power basis. Some features of the design, such as instrumentation, fuel handling, etc., need not be very much more expensive or more complicated for a large plant than for a small one. There is consequently a substantial decrease in the cost of "auxiliary systems" on a unit power basis for the large system. Another factor is the possibility of increasing the design capacity of process-type equipment, e.g., pumps, heat exchange systems, etc., by utilizing additional materials but with only modest additional cost for fabrication and manufacturing overhead. In fact, a useful rule-of-thumb in the chemical and petroleum industries is that the total investment will vary as the 0.6 power of throughput or capacity. On the other hand, the cost of the turbine-generator increases roughly in proportion to the plant capacity. Finally, many of the indirect costs are relatively insensitive to size. About the same engineering effort is required for a small plant, for example, as for a large one. An additional saving is realized when several plants of the same design are ordered. The manufacturer may then distribute the indirect costs over several plants. In this respect, the competitive bidding procedure for securing a reactor order from a utility company may well encourage the manufacturer to minimize his allowance for indirect costs; acceptance of the resulting low bid may secure for him enough business to spread out his costs over a number of orders.

FIXED CHARGES

14.14. An annual fixed charge of 12.5 per cent on depreciating capital costs is representative for investor-owned, conventional power plants in the United States as shown in Table 14.4. This includes interest on borrowed money, return

TABLE 14.4 ANNUAL FIXED CHARGE BASIS

	Per Cent
Return on money invested..............	6.50
30-year depreciation..................	1.25
Interim replacements..................	0.35
Federal income taxes	1.80
Other taxes...........................	2.40
Insurance (excluding liability)	0.20
	12.50

on equity, depreciation, interim replacements, taxes, and insurance. The depreciation allowance is equivalent to setting aside each year a fixed percentage of the total capital cost and investing it in a sinking fund to provide for plant replacement. For the determination of federal income tax, however, it is permissible to use a different method such as the "sum-of-the-years digits" approach [6] which will result in a lower tax than the sinking fund method. It is therefore customary to use one depreciation method for actual plant replacement requirements and another for tax purposes. "Return on money invested" includes both the interest paid on bonds and the dividends paid to stockholders. The percentage will therefore vary from year to year depending on both security market conditions and the actions of state regulatory agencies.

14.15. Fixed charges of another type are associated with requirements for funds necessary to operate the plant. Working capital is required for payments which can include, for example, charges associated with fabrication of the core, etc., before revenue is received from the sale of electricity. Working capital may therefore be considered as a working fund needed to provide for day-to-day operations. Since no depreciation is involved in such a fund, a lower annual fixed charge rate applies than for depreciating capital. Actual rates vary, of course, depending upon market conditions, the ratio of debt to equity capital (bonds to stocks), taxes, etc. For preliminary estimation purposes, an annual rate of 11.5 per cent may be assumed. Land and land rights should also be included in the non-depreciating capital category. Another such item is an allowance for nuclear liability insurance which is entered separately in the tabulation of fixed charges [2]. For estimation purposes, the cost may be assumed to be $260,000 per year for $60 million of coverage from private sources plus $30 per year per megawatt (thermal) for government indemnification.

14.16. In the United States the annual fixed charge rate for depreciating capital for government-owned utilities, municipal power plants, and those constructed with the aid of loans from the Rural Electrification Administration may be from 5.5 to 7.0 per cent because of favorable interest rates and tax exemption. Similarly, non-depreciating capital annual fixed charge rates range from 3.5 to 6.0 per cent. Fixed charge rates also vary from one country to another. The charges for capital in the United Kingdom, for example, are between 6 and 8 per cent, which accounts for the difference in some British and American estimates of power costs from plants of the same type. This difference is significant because the fixed charges for nuclear power represent from 40 to 60 per cent of the final cost of the electricity produced. Since the cost of electricity is determined by the expenses during the period of generation, including the fixed charges, the fixed charge rate is very important to the designer in the selection of the most economical concept for a given service and also in the determination of optimum operating conditions. In cases where operating costs can be reduced, e.g., by spending additional money on construction, additional capital requirements can be more easily tolerated under public ownership, where a low fixed charge rate applies, than under private ownership.

14.17. Since fixed costs are constant per unit of time and are independent of electrical output, the contribution of the fixed costs to the cost of energy, expressed in mills/kw-hr, is dependent upon the *plant factor* (or *use factor*). This is defined as the actual energy production during a given period divided by the maximum possible amount of energy that could have been produced during the same period. Although an 80 per cent plant factor is assumed for estimating purposes, a higher value would be appropriate for a highly developed type of reactor plant, e.g. PWR and BWR, whereas for a new concept it may not be possible to achieve such a high factor for the first few years of operation. System fuel costs will be minimized if the most efficient units are utilized as much as possible with less efficient, probably older, units operated to meet peak load requirements during a fraction of the period in question. As a result, the plant factor will tend to decrease somewhat during the life of the plant as more efficient units are built and assume "base-load" service. It should be noted, however, that this tendency is not considered as strong for nuclear plants as it is likely to be for fossil-fueled plants.

14.18. From the viewpoint of the utility company operation planning, the *plant availability factor* is probably of greater interest than the plant factor [7]. The former may be defined as the ratio of the integrated megawatt-hour output capability for a given period to the total rated megawatt-hours during the period. Loss in plant availability occurs as a result of planned outages for maintenance and refueling in addition to forced outages caused by equipment breakdowns. Since a plant that is available may be held in reserve, depending upon the load demand of the system, the plant factor may be less than the availability factor.

A high availability factor tends to reduce the need for additional installed reserve capacity.

14.19. Although capital costs have been estimated, and corresponding fixed costs calculated, for a number of different reactor types, the uncertainties involved make comparison difficult. Information tends to be much more reliable, however, for a reactor concept that is highly developed than for a design that has yet to be built. Although fixed charges are not likely to vary by more than 20 per cent from concept to concept, some general trends are apparent. Heavy water and graphite moderated reactors using low enrichment fuel tend to have large cores and, hence, a high construction cost for the reactor shielding, etc. Since the steam conditions are poorer for water-cooled reactors than for high temperature sodium- or gas-cooled reactors, the cost of the turbine generator tends to be greater in the former case. The development of new technology or new design approaches may also affect capital costs. The use of pre-stressed concrete for containment, for example, is said to reduce containment costs for a gas-cooled reactor [5]. In presenting an introductory cost component framework such possibilities need not be considered.

FUEL CYCLE COSTS [8]

14.20. Equal in importance to fixed charges in determining the cost of nuclear power are the fuel cycle costs. In fact, since it is possible to reduce costs through both improvements made in fuel cycle technology and fuel design during the lifetime of the plant, the analysis of fuel costs is of continuing interest to the electric utility, whereas fixed charges are not subject to change after the plant is built. The analysis of fuel cycle expenses is complicated, however, since a number of operations are involved (Fig. 8.1) and economic factors are sensitive to irradiation time in the reactor as well as to numerous financial and process variables. Furthermore, with the advent of private fuel ownership, there is some diversity of opinion regarding methods of accounting for some of the charges [9]. For the present purposes it is adequate to identify the principal cost components and how they may be affected by changes in the reactor design. Such an analysis is useful in the early stages where only comparisons may be desired. A more detailed study can follow using economic procedures and information available at the time to meet the requirement for actual cost projections. It is useful to consider fuel cycle costs under six headings, namely (a) fabrication, (b) reprocessing, (c) depletion, (d) use (or lease) charge, (e) plutonium credit, and (f) transportation. For accounting purposes, all fuel charges must be included in one or other of these categories. The choice is in some cases fairly arbitrary but the following discussion indicates the considerations which are involved.

Fuel Fabrication Costs

14.21. The direct and indirect fuel fabrication charges together represent an important item in fuel cycle costs. The direct cost includes charges for conversion of uranium hexafluoride into the oxide, metal, alloy, or other form to be used as fuel, as well as shipping and actual fabrication costs. The latter varies, depending on the choice of materials, including cladding, and the difficulty of fabrication, e.g., complexity of the design. In some cases rigid manufacturing tolerances are required, and this adds to the cost because of the need for inspection and the high percentage of rejections. Recycle fuel containing plutonium may require remote handling which also increases the cost. As is true in most manufacturing activities, efficiencies are realized as the throughput is increased. Unit fabrication costs, normally expressed as dollars per kilogram of heavy metal, therefore tend to decrease as the requirement (kilograms per day) is increased. Estimated fabrication costs for fuel elements in several types of power reactors are given in Table 14.5 [5]. Since the core of a central station reactor may contain some 100,000 kilograms (metal basis), the direct cost of fabricating the fuel material for a complete core may range from $5 million to $15 million.

TABLE 14.5 ESTIMATED FUEL FABRICATION COSTS AND AVERAGE BURNUP

Reactor Type	Fuel	Cladding*	Fabrication Cost† ($/kg U)	Average Burnup (Megawatt-days/tonne)
Pressurized water............	UO_2	Zr-4	75	25,000
Boiling water...............	UO_2	Zr-2	70	22,000
Heavy water-natural U.......	UO_2	Zr-2	50	10,000
Fast breeder				
Fuel.....................	PuO_2–UO_2	SS	120‡	75,000
Blanket	UO_2	SS	65	

* Zr-2 and Zr-4 are Zircaloy-2 and -4, respectively (§ 7.61); SS is stainless steel.
† Based on manufacturing throughput of 1000 kg/day [5].
‡ Based on manufacturing throughput of 500 kg/day [10].

14.22. In addition to the direct costs, allowance must be made for indirect costs on the capital invested in fabricated fuel. Working capital requirements for nuclear power plants are much greater than for those employing fossil fuels largely as a result of the considerable investment in fabricated fuel, both before and during reactor operation. Hence an annual charge on working capital is assigned as an indirect cost on the fuel cycle. The average investment by the utility in fabricated fuel may be about 60 per cent of the fabrication cost for one full reactor loading. The annual fixed charge for working capital is taken to be 11 per cent of the average investment in fuel, consistent with other rates used for non-depreciating capital. Although this rule-of-thumb was useful during the period when the utility could not own the fissile material in the fuel (§ 14.23), with the advent of private ownership, it is probably more appropriate to consider all of fuel cycle capital charges as a separate item. Furthermore, some depreciation may be appropriate at least amounting to the value of one core over the plant lifetime. For the present purpose, however, the foregoing secondary considerations may be disregarded.

Fuel Use (Inventory) Charge

14.23. Since enriched uranium has a certain value, it is necessary that a charge be applied for its rental or use. If the fuel is owned by the U. S. Atomic Energy Commission, a lease charge, currently 4.75 per cent per annum, is applied. When the fuel material is privately owned, fixed charges apply based on the funds invested in the material. According to the procedure in the Guide, the fixed charges are considered as working capital and reported separately. To simplify the desired identification of cost components, however, capital charges for privately owned fuel will be treated in the same manner as the lease charge but using an annual rate of 11 per cent, as appropriate for non-depreciating capital. The annual capital or use charge is applied to the cost of enriched uranium as uranium hexafluoride (UF_6) according to an official schedule of prices [2]. These are based on the cost of isotope separation in the gaseous diffusion plants (§ 8.23), with the value of normal uranium hexafluoride being taken to be \$23.50 per kilogram of uranium. Some typical costs of uranium hexafluoride of various degrees of enrichment, expressed in terms of dollars per kilogram of uranium, are given in Table 14.6. From these it is possible to calculate the corresponding costs

TABLE 14.6.   COST OF URANIUM HEXAFLUORIDE*

Weight Per Cent of U-235	Dollars per kg of Uranium	Weight Per Cent of U-235	Dollars per kg of Uranium
0.75.	26.50	4.0	365.80
0.80.	30.50	5.0	479.40
0.90.	38.90	7.0	710.50
1.0.	47.70	10.0	1062.00
1.5.	95.30	30.0	3456.00
2.0.	146.50	70.0	8329.00
2.5.	200.00	90.0	10,808.00
3.0.	254.30	93.0	11,188.00

* U. S. AEC schedule effective July 1, 1962.

per gram of contained uranium-235; the results are plotted in Fig. 14.2. It is seen that it costs about as much, per gram of uranium-235, to produce material of 2.5 per cent enrichment starting from natural uranium hexafluoride as it does to produce 93 per cent uranium-235 starting from 2.5 per cent material. This fact has a bearing on the choice of fuel enrichment.

14.23a. It should be noted that the use charge is payable not only on the fissile material in the core, but also on the fuel being fabricated, stored, cooled, and reprocessed. Since one fuel cost analysis approach is to calculate separately all of the costs associated with each step, such calculation must take into consideration the "time-value investment product" (§ 14.26) needed for determining

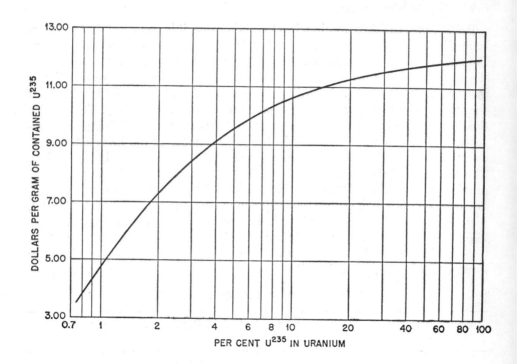

FIG. 14.2.   Cost per gram of uranium-235 as a function of enrichment

the respective inventory charges. Although enriched material can be privately owned (and must be after 1971), the corresponding fixed charges are treated throughout the fuel cycle in a manner similar to that used for leased material but at a different annual rate. Plutonium is treated in the same manner as enriched uranium; natural uranium, thorium, and depleted uranium, on the other hand, are treated in the same manner as ordinary purchased materials with the charge included in the fabrication cost.

Fuel Depletion Cost

14.24. The depletion cost is the value of fissile material that has been consumed during reactor operation. The charge is the difference between the value of the uranium loaded into the reactor and that discharged; it is equal to the product of the weight of uranium and its unit value as determined by the enrichment. Consumption of uranium-238 during reactor operation, by fast fission and radiative capture, will reduce the total weight of uranium discharged and also the ratio of uranium-235 to uranium-238. Both of these affect the value of the uranium, and so the burnup charge involves more than just the consumption of uranium-235.

Other Costs

14.25. Of the other three cost categories, chemical processing and plutonium credit tend to be about equal in importance, at least for non-breeder reactors, whereas transportation is generally an incidental cost. Spent fuel processing costs may be determined from a rate scale published by the only commercial processor (Nuclear Fuel Services, Inc.) in the United States in 1966. On the other hand, such estimates may be overly conservative since the growth of the nuclear industry will require additional processing facilities, probably of larger capacity, by the time a reactor currently being planned is operative. The projected processing rate therefore becomes a variable. For purposes of identifying cost components, however, the published rates are adequate. The plutonium credit depends upon the value of reactor grade plutonium, fixed by the U. S. Atomic Energy Commission at $10 per gram (fissile) until 1971. For estimating purposes, it is assumed that this value will prevail after that time.

14.26. The calculation of fuel cycle costs is a complex undertaking. In fact, some of the accounting methods are not uniform. For the present purposes, however, the procedure described in the Guide, and outlined below with minor modifications, is adequate for preliminary estimates and the analysis of the effects of changes in the cost contributions. In considering the various items listed, it is helpful to keep in mind the pattern. The fuel cycle is divided into three principal operational areas shown in Fig. 14.3, namely, fuel fabrication, the reactor, and chemical processing. Within each area, the cost contributions can, in turn, be divided into three categories: the cost of carrying out the operations in question, capital charges, and losses. In order to determine the capital charges, it is necessary to obtain first the product of the value of the material being held or processed in the particular operational area and the corresponding time, on a yearly basis. This quantity, called the "time-value investment product," is then multiplied by the fixed charge rate per annum to obtain the capital charge. Numerous assumptions of processing times, losses, etc., are listed in the Guide to simplify the procedure but need not necessarily apply in a given situation. In fact, it has been shown that substantial changes may be appropriate, particularly for non-

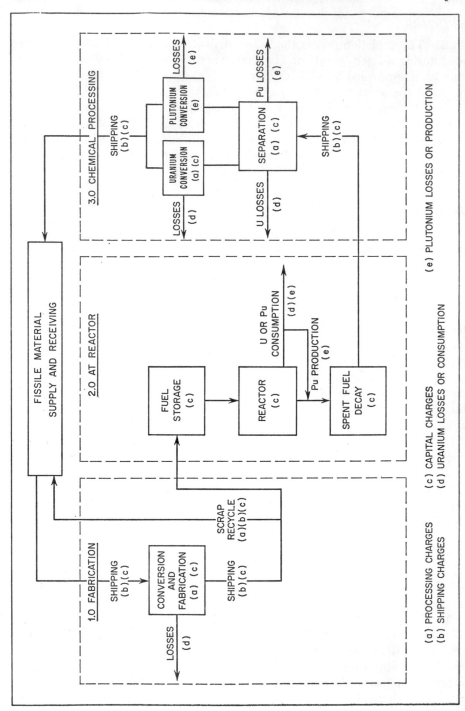

Fig. 14.3.   Fuel cost areas

PWR-BWR concepts [5]. The procedure, including these assumptions, is illustrated in Example 14.1.

Example 14.1. Find the fuel cycle costs in mills/kw-hr for the following PWR:
ENRICHMENT: 2.2% U-235 (initial)
           0.63% U-235 (final); 4.42 g Pu (fissile) discharged/kg U charged.
BURNUP: 21,000 Mw-days/tonne U, i.e., megawatt days (thermal) per tonne (1000 kg) on a U metal basis.
CORE LOADING: 104,000 kg (104 tonnes) U; 0.975 kg of U discharged per kg U charged into reactor. Fuel is oxide but all values are expressed on a U metal basis.
POWER: 3333 Mw(thermal)
EFFICIENCY: 30%
PLANT FACTOR: 80%
FABRICATION PRICE: $80/kg U

Assume private ownership but include fuel capital charges as a constituent of the fuel cycle cost.

$$\text{Average residence time} = \frac{21,000 \text{ Mw-days (thermal)/tonne U}}{0.8 \times 3333 \text{ Mw(thermal)}/104 \text{ tonnes U}} = 819 \text{ days} = 2.25 \text{ yr.}$$

When only a fraction of the core is charged and discharged at one time, the cycle time (or time between these operations) will be the corresponding fraction of the residence time.

$$\text{Annual throughput} = \frac{104 \text{ tonnes U}}{2.25 \text{ yr}} = 46.3 \text{ tonnes U/yr} = 46,300 \text{ kg U/yr.}$$

1.0 FABRICATION OPERATION

(a) (b) *Processing and Shipping*
Base charge (conversion of UF_6 to U at $13.50/kg U + $3.33/kg U for shipping to fabricator and to reactor = $16.83/kg U) + fabrication charge

Base charge ($16.83/kg U) + fabrication charge ($80/kg U) = $96.83/kg U*

(c) *Capital Charge*
Value: Initial cost of enriched material ($/kg U as UF_6) multiplied by assumed inventory factor of 1.12 to allow for losses and scrap recycle
Time: Base time (transit to conversion and fabricator and to reactor = 40 days, i.e., 0.1095 yr) + conversion and fabrication time (= annual throughput $\times 2.06 \times 10^{-5}$ yr, assuming 4000 kg/month capacity)
Capital charge at 11% = value $\times$ time $\times 0.11$

Base charge for UF_6 enriched to 2.2% U^{235} (from Table 14.6) = $167.60/kg U
Value = 167.60×1.12
    = $187.71/kg U
Annual throughput = 46,300 kg
Time = $0.1095 + (46,300 \times 2.06 \times 10^{-5})$
    = 1.06 yr

Capital charge =
  $187.71 \times 1.06 \times 0.11 = $ 21.90/kg U

(d) *Losses*
Base charge ($/kg U) $\times 0.02$

Losses = 167.60×0.02 = $ 3.35/kg U
Total fabrication charge = $122.08/kg U

* Unless otherwise indicated (by $/kg U (dis)), the kg U refers to the material charged into the reactor.

2.0 REACTOR OPERATION

(a) (b) *Processing and Shipping:* none
(c) *Capital Charge*
  Initial value charge basis
(i) Pre-irradiation period charge (assume 60 days, i.e., 0.164 yr) applied to charging batch

Pre-irradiation charge
  = 0.164 yr × base charge ($/kg U)
  = 0.164 × $167.60
  = 27.55 $-yr/kg U

(ii) Spare fuel inventory charge (assume period equal to average residence of fuel in reactor only; applied to 0.025 kg U per kg U charged)

Spare fuel charge
  = 0.025 × base charge × residence time (yr)
  = 0.025 × $167.60 × 2.25
  = 9.41 $-yr/kg U

  Irradiated fuel: average value charge basis
(iii) "Gross" discharge value (value of discharged U as UF_6 + value of Pu at $10.00/g); allow for ratio of kg U discharged to the kg U charged into reactor

U: value of UF_6 in discharged U containing 0.63% U^{235} = ($17.48/kg U(dis)) × (0.975 kg U(dis)/kg U) = $17.04/kg U
Pu: (4.42 g Pu/kg U) × ($10.00/g Pu) = $44.20/kg U
"Gross" discharge value
  = $17.04 + $44.20 = $61.24/kg U

Reprocessing cost (less capital charges): total of Items 3(a), (b), (d), and (e) below
"Net" discharge value = "Gross" value − reprocessing cost
Average value = ½ (Initial value + "net" discharge value)
Irradiation charge = Average value × residence time (yr)

Reprocessing cost = $50.31/kg U

"Net" discharge value
  = $61.24 − $50.31 = $10.93/kg U
Average value
  = ½($167.60 + $10.93) = $89.27/kg U
Irradiation charge = $89.27 × 2.25
  = 200.55 $-yr/kg U

  Net discharge value basis
(iv) Spent fuel decay charge (assume period of 120 days, i.e., 0.328 yr, applied to "net" discharge value)

Decay charge = ($10.93/kg U) × 0.328
  = 3.60 $-yr/kg U

Total reactor operation capital charge
Capital charge in $/kg U (assume fixed rate of 11 per cent on pre-irradiation, spent fuel, and decay charges in $-yr/kg U)

Capital charge = (27.55 + 9.41 + 200.55 + 3.60) × 0.11 = $26.52/kg U

(d) *Uranium Consumption*
Initial value of UF_6 − discharge value (from Item 2(c, iii)) per kg U

Initial value of UF_6 = $167.60/kg U
Discharge value of UF_6 = $17.04/kg U
Cost of U^{235} consumed = $167.60 − $17.04
  = $150.56/kg U

(e) *Plutonium Credit*
Value of Pu (from Item 2(c, iii)) per kg U

Value of Pu = $44.20/kg U
Value of "losses" = $150.56 − $44.20
  = $106.36/kg U

3.0 CHEMICAL PROCESSING

(a) Processing Charges

(i) Separation: Base charge ($23,500 per 1000 kg daily processing "unit," i.e., $23.50/kg day)

Time: Processing time ($= W/P$ days) and turnaround time ($= T$ days), where W = batch size (taken here as annual throughput) in kg; $P = 1000$ kg for $U^{235} \leq 3\%$; $T = 8$ for $W/P \leq 24$ and $T = W/3P$ for $W/P > 24$.

Separation charge per kg U charged
$$= (\text{kg U(dis)/kg U}) \times \$23.50$$
$$\times \left(1 + \frac{T}{W}\right)$$

$W = 46,300$ kg

Since $U^{235} < 3\%$, $P = 1000$ kg

Processing time $= W/P = 46.3$ days

Since $W/P > 24$, $T = W/3P = 15.4$ days

Separation charge $= 0.975 \times 23.50$
$$\times \left(1 + \frac{15.4}{46.3}\right) = \$30.51/\text{kg U}$$

(ii) Uranium conversion: Base charge of $5.60/kg U applied to discharged material with 1% loss assumed in chemical separation

U conversion charge $= (\$5.60/\text{kg U(dis)})$
$$\times (0.975 \text{ kg U(dis)/kg U}) \times 0.99$$
$$= \$5.41/\text{kg U}$$

(iii) Plutonium conversion: Base charge of $1.50/g Pu with 1% loss assumed in chemical separation

Pu conversion charge
$$= (\$1.50/\text{g Pu}) \times (4.42 \text{ g Pu/kg U}) \times 0.99$$
$$= \$6.56/\text{kg U}$$

Total processing charges
$$= \$30.51 + \$5.41 + \$6.56 = \underline{\$42.48/\text{kg U}}$$

(b) Shipping

(i) Reactor to chemical separation site: Base charge $6.00/kg U(dis)*

Shipping to chemical separation site
$$= (\$6.00/\text{kg U(dis)})$$
$$\times (0.975 \text{ kg U(dis)/kg U})$$
$$= \$5.85/\text{kg U}$$

(ii) Separation site to receiving site: Base charge $1.00/kg U(dis), allowing 1% loss in separation and conversion

Shipping to receiving site
$$= (\$1.00/\text{kg(dis)})$$
$$\times (0.975 \text{ kg U(dis)/kg U}) \times 0.99$$
$$= \$0.96/\text{kg U}$$

(c) Capital Charges

Transit to processing site: Assume 20 days, i.e., 0.055 yr; chemical separation time $= W/P$ days, as in 3(a, i) plus 30 days lag; conversion time $= W/P$ days plus 5 days lag. Capital charges at 11% based on net discharge value of fuel; allow 1% for losses in separation.

Transit charge $=$ "Net" discharge value of fuel $\times 0.055 = \$10.93 \times 0.055$
$$= 0.60 \text{ \$-yr/kg U}$$
Separation time $= (W/P) + 30$
$$= 76.3 \text{ days}$$
$$= 0.209 \text{ yr}$$
Separation charge $= \$10.93 \times 0.209$
$$= 2.29 \text{ \$-yr/kg U}$$
Conversion time $= (W/P) + 5$
$$= 51.3 \text{ days}$$
$$= 0.144 \text{ yr}$$

---

* A base charge of $16/kg U is recommended in the 1963 edition of the Guide but subsequent studies [5] yield costs for 1000-Mw (electrical) PWR systems of the order of $3.00/kg U. The value recommended is therefore considered to be realistic and conservative.

$$\text{Conversion charge} = \$10.93 \times 0.144$$
$$\times 0.99$$
$$= 1.52 \; \$\text{-yr/kg U}$$
$$\text{Total processing operation capital charge}$$
$$= (0.60 + 2.29 + 1.52) \times 0.11$$
$$= \underline{\$0.49/ \text{kg U}}$$

(d) *Uranium-235 Loss*

Assume 1% separation loss, followed by 0.3% conversion loss

Separation loss
$$= \$10.93 \times 0.01 = \$0.11/\text{kg U}$$
Conversion loss
$$= \$10.93 \times 0.99 \times 0.03 = \$0.03/\text{kg U}$$

(e) *Plutonium Loss*

Assume 1% separation loss, followed by 1% conversion loss

Separation loss
$$= \$44.2 \times 0.01 = \$0.44/\text{kg U}$$
Conversion loss
$$= \$44.2 \times 0.99 \times 0.01 = \$0.44/\text{kg U}$$
Total loss $= 0.11 + 0.03 + 0.44 + 0.44$
$$= \underline{\$1.02/\text{kg U}}$$

The foregoing results are summarized in Table 14.7. The costs in terms of mills/kw-hr, based on a 46,300 kg U/yr throughput, 1000 Mw (or 10^6 kw) electrical, and a 0.8 plant factor is given by:

$$\frac{\$}{\text{kg U}} \times \frac{1000 \; \text{mills}}{\$} \times \frac{46,300 \; \text{kg U/yr}}{(10^6 \; \text{kw})(365 \times 24 \times 0.8 \; \text{hr/yr})} = \frac{\text{mills}}{\text{kw-hr}}.$$

In addition, there is a Working Capital charge for Fuel Cycle Operations; this is not given in Table 14.7, but is included in Table 14.8, under Nondepreciating Capital, Working Capital, Item (b). It is based on the following:

Complete core fabrication charge = Fabrication charge ($/kg U) from

Item 1 × core loading (kg)
$$= \$122.08 \times 104,000 = \$12,700,000$$

Average investment: 60% of above (§ 14.22)
$$= \$12,700,000 \times 0.6 = \$7,620,000$$
Annual working capital charge: 11% of above
$$= \$7,620,000 \times 0.11 = \$838,000$$

TABLE 14.7. FUEL COST SUMMARY FROM EXAMPLE 14.1
(IN $/KG U CHARGED)

	Processing and Shipping	Capital Charge	Value of Losses	Total	Equivalent Cost mills/kw-hr
1. Fabrication.........	$ 96.83	$21.90	$ 3.35	$122.08	0.806
2. Reactor.............		26.52	106.36	132.88	0.878
3. Chemical Processing...					
Processing........	42.48 ⎱	0.49	1.02	50.80	0.336
Shipping..........	6.81 ⎰				
Totals..............	$146.12	$48.91	$110.73	$305.76	2.020
Equivalent Cost mills/k-whr........	0.965	0.323	0.732		

VALIDITY OF FUEL CYCLE COST ESTIMATES

14.27. The procedure described above and the example given are primarily intended to serve as a "framework" for so-called cost allocation studies in which the effects of various changes in parameters can be evaluated rather than to provide an accurate estimate for the system used. In fact, some published [11] values for PWR total fuel costs are about 15 per cent less than those cited here, providing an indication of the probable range of validity of the estimation procedure. Caution must certainly be exercised when applying the procedure to other reactor concepts, particularly those involving unusual fuels.

14.28. The relative costs listed for the categories in Table 14.7 provide a guide for possible benefits (or lack of them) from attempts at cost reduction. Chemical processing, for example, contributes only about 16 per cent of the cost. Hence, improvements in the processing operation would not have a marked effect on the total fuel cycle cost. A change in fixed cost rate would affect only the capital charge category, again comprising only about 15 per cent of the total. A more complete analysis, however, would take into account the investment required for fuel manufacturing and chemical separation facilities, and, in turn, the effects of corresponding fixed charge parameters on the fuel cost.

14.29. An uncertainty also arises from the market price of raw materials. For example, the value of uranium concentrate, upon which the costs in Table 14.6 are based, is $8.00 per pound of contained U_3O_8. It has been possible, however, to purchase the concentrate on the world market at a price as low as $5 per pound. In these circumstances, it may be more economical, in the operation of a reactor utilizing natural uranium as fuel, to store the irradiated fuel elements, since the reprocessing cost would probably exceed the value of the recovered product. As more power reactors are built and absorb existing supplies of low cost uranium, a rise in the market price to about $7 per pound of U_3O_8 is expected in the 1970s [12]. There is considerable uncertainty as to subsequent costs since the availability of new supplies and the possible influence of the introduction of fast breeder reactors (§ 12.123) become pertinent. But a doubling of the price from $7 to $14 per pound of U_3O_8 appears possible during the 1980s. Since a 30-year operational period is anticipated for reactors starting up in the early 1970s, such projections are very important to the reactor designer concerned with planning a fueling strategy likely to lead to minimum costs.

14.30. The cost of recovered plutonium is another factor which can affect the economic evaluation of proposed reactor concepts. In the United States, the Atomic Energy Commission has guaranteed a buy-back price of $10 per gram (fissile) until 1971. After that time, the price will presumably depend on the free market value, which, in turn, will depend upon the type of reactor in which it is used. Although plutonium has a higher value than uranium-235 in a fast reactor because of its more favorable physics characteristics in a fast spectrum

(§ 4.143), studies have shown that in a thermal reactor it is about equal to that of fully-enriched uranium-235 on a dollar-per-gram basis [13]. The market price of plutonium will therefore be related to that of enriched uranium which is a function of ore and separation costs (§ 14.31). Although the economics of plutonium fuels is complicated, the $10 per gram value still appears to be the most realistic for cost estimation purposes. It is also well to recognize that, until significant quantities of plutonium recovered from power plants become available, presumably in the 1970s in the United States, plutonium for fuels development must be purchased from the Atomic Energy Commission at a cost of $43 per gram of fissile isotopes.

14.31. An analysis of the gaseous diffusion cascade [14] provides a value of enriched uranium product in terms of the cost of raw material, the cost of separative work, and the optimum waste stream composition as follows:

$$D = C_s \left[(2X - 1) \ln \frac{X(1 - X_w)}{X_w(1 - X)} + \frac{(X - X_w)(1 - 2X_w)}{X_w(1 - X_w)} \right]$$

where D is the value in $/kg U of enriched uranium with a weight fraction X of uranium-235, C_s is the unit cost of separative work in $/kg U, and X_w is the weight fraction of uranium-235 in the waste stream. The "optimum" waste stream composition is that obtained from this equation when C_F, the unit cost of natural uranium fuel is substituted for D, and $X_F = 0.00711$, the composition of natural uranium, is substituted for X. This waste stream composition corresponds to material which has zero value. The relation given above serves as the basis for the charge schedule in Table 14.6. The effect of variations in the parameters on the calculated cost of uranium-235 in 93.5 per cent enriched uranium is shown in Fig. 14.4. Variations in waste stream composition have little effect on the curves. "Toll enrichment," by means of which enrichment services are provided at a fuel charge per unit of separative work, is being made available during the 1970s by the U. S. Atomic Energy Commission to operators of enriched fuel reactors. As commercial needs grow, however, a privately-owned enrichment plant may well prove to be economically feasible.

14.32. A detailed cost analysis of chemical processing requirements is desirable when a specific reactor is under study, and the various assumptions of the uniform cost estimation procedure need not necessarily apply. A non-aqueous processing method (§ 8.160), for example, which might require only a short cooling time and permit a rapid processing schedule, might provide some cost savings. The actual planning of satisfactory processing arrangements and appropriate waste disposal should also be considered as part of the reactor system design requirement. Furthermore, special cost analyses are necessary, even at the earliest stages, for reactor systems having unusual fuels. Fast reactor systems, with comparatively highly enriched fuels, and high fission product loading as a result of long burnups, are likely to fall in the non-aqueous processing category.

14.33. In reviewing the situation regarding fuel cycle costs, it can be concluded

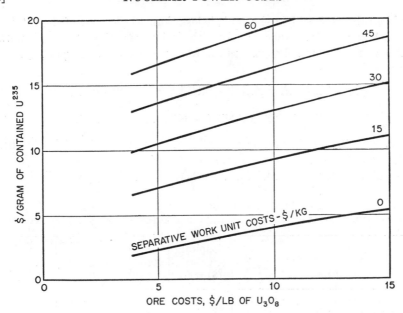

FIG. 14.4.   Calculated costs of uranium-235 in 93.5 per cent enriched uranium
(after M. F. Kanninen)

that the standard method provides a useful guide in most instances for arriving
at a preliminary estimate. There are, however, dangers inherent in most normalization procedures, because of the tendency to mask significant cost differences
when deviations from the standard approach are required.

OPERATION AND MAINTENANCE COSTS

14.34. During the early 1960s nuclear power plants required more operating and maintenance personnel than coal-fired plants of similar size, primarily
because they were somewhat experimental in nature. Furthermore, highly specialized training for reactor operators, radiation monitors, etc., is necessary. In
the case of several large nuclear plants ordered in 1966, however, and planned
for 1971 operation, estimated operation and maintenance costs are less than those
for coal-fired plants of the same capacity [11]. Although a higher level of training
is required for the nuclear plant personnel, the substantial staff associated with
coal handling and maintenance is eliminated. Although there is likely to be some
variation from one project to another, it can be concluded that operation and
maintenance costs for large commercial nuclear power plants of proven design
are about the same as those for fossil-fueled plants.

14.35. Actual operating costs including nuclear insurance amount to about $2
million per year for a 1000 Mw (electrical) plant; these represent a unit energy

cost of less than 0.3 mill/kw-hr. This is less than 10 per cent of the total power cost and tends not to change greatly with design variations; hence, it normally does not warrant detailed consideration at the conceptual design stage.

FACTORS AFFECTING NUCLEAR POWER COSTS

TOTAL COSTS AND ECONOMIC POTENTIAL

14.36. The total estimated cost per kilowatt-hour of electricity is obtained by adding the amounts in the various categories according to the procedure shown in Tables 14.8 and 14.9 for a 1000-Mw (electrical) pressurized water re-

TABLE 14.8. POWER GENERATION COST SUMMARY
1000 MW (ELECTRICAL) PRESSURIZED WATER REACTOR PLANT (80% PLANT FACTOR)

Cost Item	Capital Cost ($1,000)	Rate (%)	Annual Cost ($1,000)	Unit Cost (mills/kw-hr)
FIXED CHARGES				
Depreciating Capital				
Total Capital Cost (less				
Land and Land rights).........	$119,500	12.5	$14,938	2.13
Nondepreciating Capital				
Land and Land Rights				
Working Capital................	360	11.0	40	0.01
(a) Plant Operation and				
Maintenance...............	703	11.0	77	0.01
(b) Fuel Cycle Operations......	7,620	11.0	838	0.12
Nuclear Liability Insurance........	—	—	357	0.05
Subtotal—Annual Fixed Charges...	—	—	$16,250	2.32
OPERATING COSTS				
Operating and Maintenance Cost....	—	—	$ 1,700	0.24
*Fuel Cost**......................	—	—	14,157	2.02
Subtotal—Operating Costs........	—	—	$15,857	2.26
TOTAL POWER GENERATION COSTS....			$32,107	4.58

* Fuel Cost includes capital charges (see Table 14.7).

actor. The listed total power generation cost of 4.58 mills per kw-hr is about 0.5 mill/kw-hr higher than some total costs published for large reactors sold in 1966. About 0.2 mill/kw-hr of the difference can be assigned to a difference in plant thermal efficiency while the remainder is probably a result of limitations of the Guide method used for the fuel cycle cost estimate. Although the costs in Table 14.8 are also higher than the fossil fuel plant costs shown in Table 14.1, the difference is primarily a result of a lower fixed charge rate used for the latter. It should

TABLE 14.9. WORKING CAPITAL FOR 1000-MW (ELECTRICAL) REACTOR

A. *Plant Operation & Maintenance*
  1. Average net cash required (assumed as 2.7% of annual operating
     expense, including fuel)
       0.027 × $15,857,000 . . . . $428,000
  2. Materials and supplies in inventory (25% of annual cost of maintenance
     materials and operating supplies)
       0.25 × $1,100,000   275,000
                                                                                                                    ————————
       Subtotal $703,000

B. *Fuel Cycle Operations*
   Core Fabrication
       (60% of entire core cost) = $7,620,000
   Nuclear Materials
       (included in fuel cycle cost)

be emphasized that, for each of the many nuclear power plants sold in 1966,
careful studies were made comparing costs with a comparable coal-fired plant.
In each case the nuclear plant was found to provide energy at less cost. As shown
above, however, differences in total generation costs can be expected from one
set of estimates to the next from a variety of causes. As more and more nuclear
units are built, a continuation of a downward trend in costs is anticipated.

COST OPTIMIZATION

14.37. In the design of commercial power plants, there is considerable incentive to seek a combination of engineering, physics, and economic specifications
which will yield a minimum cost product under acceptable safety conditions.
Design improvements which reduce the cost of electrical energy by even a fraction of a mill per kilowatt-hour are well worth while since 0.1 mill/kw-hr is
equivalent to about $700,000 annually for a 1000-Mw (electrical) plant with an
80 per cent plant factor. As seen from Table 14.8, such an amount represents
only about 2.5 per cent of the power generation cost, a variation which could
result from changes in secondary design parameters which at first glance might
not appear worthy of consideration.

14.38. Optimization of experimental and commercial prototype reactor plants
has generally been carried out as a result of a series of parametric calculations
(§ 12.9). Such calculations are often highly complex because of the interactions
among the different parameters which may be varied. It has been possible to
obtain practical results through such trial-and-error type procedures because,
often, there are many basic limitations to the parameters for any given reactor
concept. For example, operating temperatures may be limited by the choice of
materials available and coolant conditions, and heat flux may be limited by the
possibility of burnout. In slightly enriched, water-moderated reactors, only a

limited range of moderator-to-fuel ratio is satisfactory from the neutronics point of view. On the other hand, as it becomes necessary to devote more and more attention to economic variables, sometimes in fine detail, the number of parameters that must be studied simultaneously becomes difficult to manage.

14.39. The application of so-called formal optimization techniques to the design of power reactor systems is therefore receiving considerable attention [15]. These involve the establishment of a technical-economic model of the reactor system or part of it to which various methods for seeking optimum conditions may be applied, normally involving digital computer programming. But it is important to emphasize that optimization theory, including such techniques as dynamic programming, the maximum principle, etc., is a complicated discipline in its own right, which has been applied with considerable success to space vehicles and the chemical process industry. Application to complete reactor systems involves substantial difficulties and is still in the course of development. Some success has been achieved, however, in the analysis of fuel management programs [16].

TECHNICAL-ECONOMIC INTERRELATIONSHIPS

14.40. To illustrate some of the many interrelationships existing among both the technical and economic parameters of a reactor system, the problems associated with the reduction of the fuel cycle costs for a given design will be examined. It is important to realize, however, that a change in one cost category may well affect another category. Since reduction in fuel cycle cost might con-

TABLE 14.10. FUEL COST BURNUP EFFECTS [5]
(1000-MW (ELECTRICAL) PRESSURIZED WATER REACTOR)

Burnup			
(Mw-days/tonne U).................	13,000	22,000	31,000
Fuel enrichment			
(wt % U-235)......................	1.70	2.20	3.00
Discharge enrichment			
(wt % U-235)......................	0.65	0.63	0.60
Discharge Pu			
(g/kg U fuel)......................	5.00	6.22	7.30
Fuel cycle costs*			
(mills/kw-hr)			
Fabrication.......................	1.33	0.81	0.61
Depletion (Burnup)			
(including Pu credit).............	0.65	0.70	0.78
Working capital			
(fabrication + inventory).........	0.28	0.30	0.39
Chemical processing................	0.36	0.33	0.29
Total.............................	2.62	2.14	2.07

* Costs for 13,000 and 31,000 Mw-days/tonne U cases were derived from 22,000 Mw-days/tonne reference case following trends given in [5].

ceivably be offset by an increase in capital costs, the design variables which are related to the fuel cycle must be chosen so as to provide the lowest nuclear power cost, not the lowest fuel cycle cost.

14.41. As a basis for discussion, typical contributions to the fuel cycle cost are given in Table 14.10 and plotted in Fig. 14.5 for a 1000-Mw (electrical)

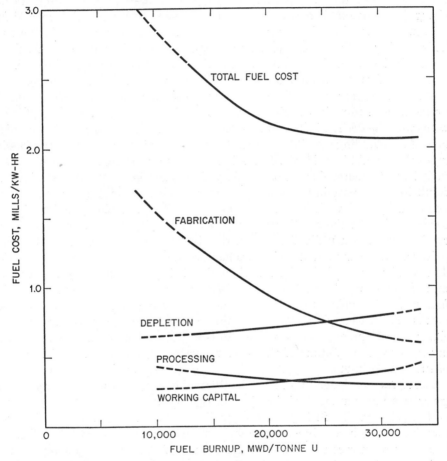

Fig. 14.5.   Typical fuel cycle costs as a function of burnup

pressurized water reactor, following the pattern in Example 14.1, with burnup as a variable. Several competing factors contribute to the shape of the curves in Fig. 14.5. As burnup is increased, the cost of fabricating the core and reprocessing it can be allocated to a larger number of kilowatt-hours produced during the longer exposure time. Hence, such costs, which apply to a batch of fuel elements when expressed on a unit energy produced basis, tend to decrease. On the other hand, in order to maintain reactivity of the reactor during the longer exposure period, it is necessary to start with fuel having a higher enrichment than that required for a shorter exposure. Because of the increased cost of this enrichment, the atoms of uranium-235 that fission in the longer burnup fuel are more expensive. The necessary higher investment per fissile atom of fuel also results in higher interest charges. This somewhat simplified picture is complicated, however, by the influence of the bred plutonium. Although the amount of plutonium discharged, and available for sale, does not increase very much with increased fuel exposure, more plutonium atoms are fissioned while they remain in the core. This sharing of the energy production load by plutonium therefore results in somewhat less depletion of uranium-235 than would otherwise be the case. Despite this effect, the depletion cost rises slightly with increased exposure.

14.42. The foregoing preliminary picture does not take into consideration the probable need to make certain design changes in the fuel element so that it can withstand the desired increased exposure. For example, an increase in the thickness of the cladding may be necessary to contain the increased quantity of fission gases that would be released. In addition to adding to the cost of fabricating a fuel element, such a change, particularly if the cladding is stainless steel, would result in an increase in neutron absorption; this in turn would have to be compensated by fuel with higher enrichment. Similarly, the need to provide for compensation for the reactivity change associated with long burnup fuels could result in inefficient utilization of neutrons by capture in control rods, burnable poisons, and accumulated fission products with consequent economic penalties. The conflicting trends in Fig. 14.5 show that it is possible to design for an optimum fuel burnup. An analysis of all of the economic, neutronic, and engineering factors which contribute to the determination of this optimum point is by no means simple. Changing economic conditions, i.e. interest rates, separative work charges, uranium ore costs as well as shifts in reactor load requirements, all during the projected lifetime of the reactor, further complicate the determination of such an optimum.

14.43. As mentioned in § 14.36, the thermal energy that must be produced to generate a given amount of electrical energy depends upon the thermodynamic cycle efficiency. The fuel cycle cost contribution when expressed as mills/kw-hr therefore varies inversely with the efficiency. Attempts to improve the efficiency in a given reactor concept may involve the use of higher coolant temperatures, and this might require the use of more expensive structural materials. A balance is consequently achieved between the desire to reduce fuel cycle costs by using higher temperatures, on the one hand, with the possible accompanying increase in the construction cost, on the other hand.

14.44. For a given kilowatt rating, a plant with a higher thermal efficiency will reject less waste heat to the environment. The heat rejection process may involve significant costs in the case of a large reactor if a river or other body of water is not available and cooling towers must be used. A rise in temperature acceptable to aquatic life may also limit the thermal absorption capacity of a river, even if available. All of these factors therefore contribute to the selection of a suitable cycle efficiency.

14.45. A number of technical-economic interrelationships exist in the design of the fuel element. In the case of a pressurized water reactor, for example, core properties depend upon the rod diameter in the manner shown in Fig. 12.1. As the diameter is decreased, fabrication costs per mass of fuel increase, the proportion of cladding material will tend to increase, and various mechanical problems are introduced with the design of a fuel assembly containing a very large number of very thin elements. The resulting compact core size, however, would result in a desirably smaller pressure vessel and surrounding shielding, although coolant pumping requirement would increase. A design compromise is thus necessary between these conflicting parameters.

14.46. The location of a large nuclear power plant may also involve a compromise. If the plant is located close to a large center of population, the containment specifications and the requirements for engineered safeguards are stricter than if a more remote site were selected. On the other hand, higher transmission charges, including both operating and fixed costs, must be paid for electricity generated at a greater distance from the load center which is normally where the population is greatest.

14.47. All of the economic parameters that contribute to the type of technical-economic compromise involving construction costs or investments are, in turn, affected by changes in the applied fixed charge rate. For example, if a decrease in fuel cycle charges can be obtained by increasing the construction cost (§ 14.43), the point of compromise would have to be shifted toward a higher fuel cycle cost if interest rates should increase. Such a complication is a result of the dependence of the annual "fixed" charge on the interest rate for a given construction cost. Furthermore, marked differences in design choices might result in the case of plants owned by government groups which not only are able to secure capital at low interest rates but also may not be subject to taxes. In such cases, it is advantageous to select a design that may involve large capital requirements if non-fixed charges are thereby reduced.

EFFECT OF SPECIFIC POWER AND POWER DENSITY

14.48. A high specific power, i.e., power per unit mass of fuel, is a desirable design objective because it leads to a low inventory charge contribution to the fuel cost, when expressed in terms of mills per kilowatt-hour of electricity. Similarly, a high power density, i.e., power per unit core volume, is advantageous since a small reactor will have comparatively low capital charges and fuel inventory. It has been shown, however, that these objectives are frequently attainable only by changing other design variables in a direction which will tend to increase the total cost of the power produced. Determination of the optimum specific power and power density for a given reactor design thus requires consideration of all the factors involved, not merely the more obvious ones.

14.49. An effect of specific power that may be overlooked is the shift in the equilibrium concentration of various isotopes resulting from both radioactive decay and neutron capture. For example, in thorium-fuel reactors, in which conversion into fissile uranium-233 is desired, an increase in the neutron flux accompanying an increase in specific power may be expected to favor the capture of neutrons by the intermediate protactinium-233, to form protactinium-234, instead of decaying to yield uranium-233. The effect is not simple, however, since protactinium-233 is itself formed by neutron capture in thorium-232, and hence at a rate which depends on the specific power. A comprehensive fuel cycle analysis has shown that the overall effect of specific power on the formation of uranium-233 is small, but this result would not have been apparent without careful study [17].

FUEL, MANAGEMENT PROGRAMS

14.50. The effect on burnup, etc., of various schemes for loading the reactor core was considered in § 12.114 *et seq.* Some additional factors pertinent to overall power costs will be mentioned here. Fuel cycle costs are highest for batch loading, where the whole core is discharged at one time and replaced with a complete loading of fresh fuel elements. The neutron flux generally has a cosine spatial variation and this leads to unequal burnup, especially if the fuel is uniformly distributed in the core, as is frequently the case. Another disadvantage is that a large excess of initial reactivity is required to provide a reasonably long life. In addition to the costs of the necessary extra enrichment, compensation by more control rods or burnable poisons increases capital costs and leads to a wastage of neutrons.

14.51. Fuel cycle costs can be reduced by moving the fuel progressively, either from the outside of the reactor to the center or the reverse. The former is probably preferable, as explained in § 12.117, because the advantages of the flattening of the power distribution more than offset the slightly lower burnup. Except for the simplest fuel management schemes, e.g., outside-to-center or bi-directional loading, which could be carried out while the reactor is operating for gas-cooled systems, shutdown would be required for rearranging the fuel. In planning a suitable management scheme, therefore, the effect of needed downtime must be considered. But such planned downtime is only one contribution to the plant availability factor (§ 14.18). Although a two-week shutdown period may be required for fuel changes in a pressurized water reactor, scheduled maintenance can also be accomplished. Since the economic importance of the availability factor depends upon the requirement for reserves, its evaluation is not simple. For an 800-Mw (base load) plant with fixed charges at 12.5 per cent, an increase in the availability from 80 to 90 per cent is equivalent to a decrease in energy cost of the order of 0.27 mill/kw-hr, a saving amounting to about $1.5 million annually [7].

14.52. In the case of large (1000 Mw-electrical) water-moderated reactors having very large cores, there may be little coupling in the reactivity sense between different parts of the core. Since the core is close to critical on a local basis, relatively small local reactivity effects can cause a large change in flux and power distribution [4]. A major objective of fueling scheme studies for such reactors, therefore, is to provide for as uniform a power distribution as possible and yet accommodate high burnup fuels which have a large difference in reactivity between the fresh and discharged states. So-called "roundelay" or "scatter-loading" methods, which provide for the placing of a fresh fuel assembly adjacent to an older one with accompanying averaging out of reactivity, are widely used.

14.53. The power distribution flattening achieved by a movement scheme can be related approximately to the capital cost. Suppose that the ratio of the peak to the average power can be reduced from 2:1 to 1.5:1; it should be possible to increase the power output of the core by roughly 35 per cent. If the investment required for the reactor only is assumed to be $150 per kilowatt of electricity, the effect of the power increase would be to reduce the fixed charges by roughly $2.19 million for a 1000-Mw (electrical) system per year.

14.54. It must be emphasized, in concluding this section, that each reactor project requires individual evaluation. Only a few of the variables have been mentioned as examples of some of the technical and economic interrelationships which must be recognized by the reactor designer. These interrelationships frequently involve several factors that contribute to nuclear power costs. It would not be out of place, therefore, to repeat the statement made earlier that optimization procedures must be based on total costs and not on one category, such as fuel cycle costs, only.

ECONOMIC POTENTIAL FOR NUCLEAR POWER

14.55. Following a trend beginning in 1964, the majority of large new electric generating plants ordered in 1966 by utilities in the United States were nuclear fueled. Since many of these were in areas where coal costs were below average, it appeared likely that the rate of growth of nuclear power in other countries where fossil fuel costs are high would indeed be great. Predictions have been made [18] for a cumulative installed nuclear generating capacity by 1980 of about 130,000 megawatts for the foreign "free world" compared with 100,000 megawatts for the United States. Since a large fraction of the plants will be of the water-moderated type, having relatively low conversion ratios, the requirement for uranium fuel would, when projected to future years, absorb most of the known low-cost ore supplies. It appears, therefore, that such a picture should encourage the development of breeder reactors which can assume a larger and larger fraction of the nuclear load to permit the production of nuclear energy at low cost for many years.

REFERENCES FOR CHAPTER 14

1. Discussions of the economics of nuclear power and related topics with appropriate examples will be found in the following reports: U. S. AEC Report TID-8531 Rev. (1961) "Costs of Nuclear Power"; L. E. Crean, R. A. Laubenstein, and L. S. Mims, "Comments on Some Aspects of Nuclear Power Reactor Economics," Review Series No. 9, International Atomic Energy Agency, 1961; "Introduction to the Methods of Estimating Nuclear Power Generating Costs," Technical Report Series No. 5, International Atomic Energy Agency, 1962; J. A. Lane, "Economics of Nuclear Power," *Ann. Rev. Nuc. Sci.*, **9**, 473 (1959); L. Geller, J. F. Hogerton, and S. M. Stoller, *Nucleonics*, **22**, No. 7, 64 (1964); J. Happel, "Chemical Process Economics," John Wiley and Sons, Inc., 1958; C. W. Bary, "Operational Economics of Electric Utilities", Columbia Univ. Press, 1963; R. L. Loftness, "Nuclear Power Plants," Chap. 12, D. Van Nostrand Co., Inc., 1964.
2. U. S. AEC Report TID-7025 (1962), "Guide to Nuclear Power Cost Evaluation."
3. U. S. AEC Report CONF-650201 (1965), "Proc. National Topical Meeting on Nuclear Power Reactor Siting."
4. U. S. AEC Report WCAP-2385 (1963), "1000-Mw Closed Cycle Water Reactor Study."
5. M. W. Rosenthal, U. S. AEC Report ORNL-3686 (1965).
6. E. L. Grant and W. G. Ireson, "Principles of Engineering Economy," 4th Ed., Ronald Press, 1964.
7. G. S. Vassell and N. Tibberts, *Trans. Am. Nuclear Soc.*, **9**, 217 (1966).
8. U. S. AEC Report NYO-9992 (1963), "Euratom Reactor Fuel Management Program"; Dietrich, J. R., *Power Reactor Technology*, **6**, No. 4, (1963); "Down Come Nuclear Fuel Costs," *Power Engineering*, **70**, No. 1, 30 (1966); J. M. Vallance, "Fuel Cycle Economics of Uranium-fueled Thermal Reactors," *Proc. Third U. N. Conf. Geneva*, 1964, Paper No. P/247; L. E. Link, G. E. Fischer, and E. L. Zebroski, "Fuel Cycle Economics of Fast Reactors," *ibid.*, Paper No. P/248.
9. P. Dragoumis, J. Cademartori, and S. Milioti, *Nucleonics*, **24**, No. 1, 40 (1966).
10. M. J. McNelly, U. S. AEC Report GEAP-4418 (1964).
11. "The TVA Analysis," *Nuclear Industry*, **13**, No. 7, 5 (1966).
12. J. F. Hogerton, L. Geller, and A. Gerber, "The Outlook for Uranium," S. M. Stoller Associates, New York, 1965.
13. E. A. Eschbach and M. F. Kanninen, U. S. AEC Report HW-72219 (1964).
14. M. Benedict and T. H. Pigford, "Nuclear Chemical Engineering," McGraw-Hill Book Co., 1957.
15. N. Kallay, *Nuc. Sci. and Eng.*, **8**, 315 (1960); W. Boettcher, F. Lafontaine, and P. Tausch, *Trans. Am. Nuclear Soc.*, **7**, 520 (1964).
16. T. Wall and H. Fenech, *Nuc. Sci. and Eng.*, **22**, 285 (1965).
17. E. A. Mason and J. R. Larrimore, *Nuc. Sci. and Eng.*, **9**, 332 (1961).
18. U. S. AEC Report TID-22973 (1966), "The Growth of Foreign Nuclear Power Plants," Arthur D. Little, Inc.

PROBLEMS

1. Prepare a construction cost breakdown similar to the one in Table 14.2 for a pressurized-water reactor plant having a capacity of 500-Mw (electrical). Indicate the main differences arising from the smaller capacity.

2. A 500-Mw (electrical) nuclear power plant is being planned for (a) an investor-owned company, for which an annual fixed charge of 12.5 per cent on plant investment is appropriate and (b) a government-owned utility which is able to operate at an annual fixed charge of 7 per cent. In each case, the choice is between a boiling light water reactor and a heavy water-moderated, organic-cooled reactor. What type of plant is likely to prove most attractive economically for each owner? Explain in detail the reasons for the selections made.

3. The management of an investor-owned utility is considering the possibility of constructing either (a) a 300-Mw (electrical) pressurized water reactor plant close to its load center in an urban area or (b) a 1000-Mw (electrical) plant in a more remote area. The smaller plant is just adequate to meet load growth requirements whereas the larger plant will provide excess capacity which can be useful in subsequent years to meet future requirements. Identify and discuss some of the factors which should be considered in reaching a decision.

4. Make a preliminary estimate of fuel cycle costs for a 1000-Mw (electrical) heavy water-moderated, organic-cooled reactor based on the following information:

Specific Power	4.42 Mw(el)/tonne U in reactor (UO_2 fuel)
Feed Enrichment	1.10 wt% U-235
Discharge Exposure	15,900 Mw-days/tonne U
Discharge Enrichment	0.142 wt% U-235; 3.05 g fissile Pu/kg U
Fabrication Cost	$60/kg U

5. Identify the components of the cost per kw-hr in Example 14.1 which are appreciably affected by fuel burnup, and estimate the change in fuel cycle costs resulting from irradiation levels of 15,000, 25,000, and 35,000 Mw-days per tonne. Plot a curve of cost per kw-hr as a function of burnup.

APPENDIX

Multiply	By	To Obtain
Length		
centimeters	0.03281	feet
centimeters	0.3937	inches
feet	30.48	centimeters
inches	2.540	centimeters
Volume		
cubic centimeters	3.532×10^{-5}	cubic feet
cubic centimeters	0.06102	cubic inches
cubic feet	2.832×10^{4}	cubic centimeters
cubic feet	7.481	gallons
cubic feet	28.32	liters
cubic inches	16.39	cubic centimeters
gallons	0.1337	cubic feet
gallons	3.785	liters
liters	0.03532	cubic feet
liters	0.2642	gallons
Mass		
grams	2.205×10^{-3}	pounds
kilograms	2.205	pounds
pounds	453.5	grams
pounds	0.4535	kilograms
kilograms	10^{-3}	tonnes
tonnes	1000	kilograms
tonnes	2205	pounds
atomic mass units	1.66×10^{-24}	grams
Density		
grams/cubic centimeter	62.42	pounds/cubic ft
pounds/cubic ft	0.01601	grams/cubic centimeter
Heat and Energy		
British thermal units	251.8	calories
calories	3.968×10^{-3}	British thermal units
electron volts	1.603×10^{-12}	ergs (or dynes-cm)
electron volts	1.603×10^{-19}	watt-seconds
horsepower-hours	2544	British thermal units
kilowatt-hours	3413	British thermal units
million electron volts	1.603×10^{-6}	ergs
million electron volts	1.603×10^{-13}	watt-seconds
million electron volts	1.520×10^{-16}	British thermal units
million electron volts	4.45×10^{-20}	kilowatt-hours
watt-hours	3.413	British thermal units
Power		
Btu/hour	2.931×10^{-4}	kilowatts

TABLE A.1. CONVERSION FACTORS (continued)

Multiply	By	To Obtain
Btu/hour	0.2931	watts
Btu/hour	3.930×10^{-4}	horsepower
horsepower	2544	Btu/hour
horsepower	0.7457	kilowatts
kilowatts	3413	Btu/hour
kilowatts	1.341	horsepower
watts	3.413	Btu/hour
calories/second	4.187	watts
calories/second	14.29	Btu/hour
Specific Heat		
Btu/(pound)(°F)	1.000	calories/(gram)(°C)
calories/(gram)(°C)	1.000	Btu/(pound)(°F)
Thermal Conductivity		
Btu/(hr)(ft²)(°F/ft)	4.134×10^{-3}	calories/(sec)(cm²)(°C/cm)
calories/(sec)(cm²)(°C/cm)	241.9	Btu/(hr)(ft²)(°F/ft)
watts/(cm²)(°C/cm)	57.78	Btu/(hr)(ft²)(°F/ft)
watts/(cm²)(°C/cm)	0.239	calories/(sec)(cm²)(°C/cm)
Heat Flux		
watts/cm²	4.187	calories/(sec)(cm²)
watts/cm²	3170	Btu/(hr)(ft²)
Btu/(hr)(ft²)	3.155×10^{-4}	watts/cm²
Heat-Transfer Coefficient		
Btu/(hr)(ft²)(°F)	5.68×10^{-4}	watts/(cm²)(°C)
Btu/(hr)(ft²)(°F)	1.356×10^{-4}	calories/(sec)(cm²)(°C)
watts/(cm²)(°C)	1761	Btu/(hr)(ft²)(°F)
Power Density		
watts/cm³	9.662×10^4	Btu/(hr)(ft³)
cal/(sec)(cm³)	4.045×10^5	Btu/(hr)(ft³)
cal/(sec)(cm³)	4.187	watts/cm³
Viscosity		
poise (gram/(sec)(cm))	241.9	lb_m/(hour)(foot)
centipoise	2.419	lb_m/(hour)(foot)
lb_m/(hour)(foot)	4.134×10^{-3}	poise
lb_m/(hour)(foot)	0.4134	centipoise
lb_m/(sec)(foot)	1488	centipoise
lb_m/(hour)(foot)	2.778×10^{-4}	lb_m/(second)(foot)
Universal Constants		
Avogadro number (or constant), N_a		0.6025×10^{24} atoms/mole
Boltzmann constant, k		1.3804×10^{-16} erg/°K
		8.617×10^{-5} ev/°K
Gas constant, Nk		0.08206 liter-atm/(mole)(°C)
		1.986 Btu/(lb_m-mole)(°F)
		1545 ft-lb_m/(lb_m-mole)(°F)
Planck's constant, h		6.625×10^{-27} erg-sec
1 curie		3.700×10^{10} disintegrations/sec
Mass-Energy Equivalence		
1 atomic mass unit = 931.1 Mev		

$E = h\upsilon$ The energy carried by a photon is a function (h) of its frequency (υ) in cm

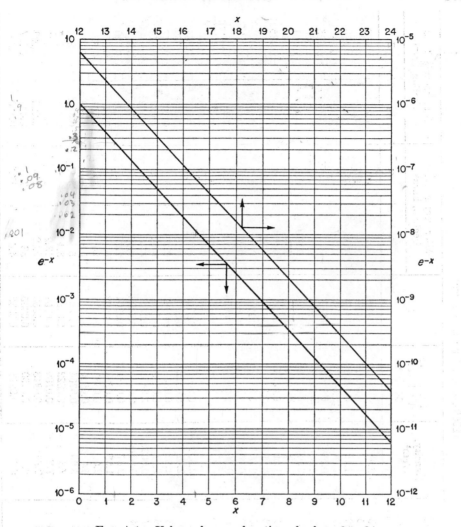

FIG. A.1.   Values of e^{-x} as function of x from 0 to 24

$$e^{-1.355} = 0.258$$

TABLE A.2. CROSS SECTIONS FOR NATURALLY OCCURRING ELEMENTS
(2200-METERS/SEC NEUTRONS)

Atomic No.	Element or Compound	Atomic or Mol. Wt.	Density, g/cm³	Nuclei per Unit Vol. × 10⁻²⁴	ξ	Microscopic Cross Section, barns		Macroscopic Cross Section, cm⁻¹	
						σ_a	σ_s	Σ_a	Σ_s
1	H	1.008	8.9[a]	5.3[a]	1.000	0.33	38	1.7[a]	0.002
	H₂O	18.016	1	0.0335[b]	0.948	0.66	103	0.022	3.45
	D₂O	20.030	1.10	0.0331[b]	0.570	0.001	13.6	3.3[a]	0.449
2	He	4.003	17.8[a]	2.6[a]	0.425	0.007	0.8	0.02[a]	2.1[a]
3	Li	6.940	0.534	0.0463	0.268	71	1.4	3.29	0.065
4	Be	9.013	1.85	0.1236	0.209	0.010	7.0	124[a]	0.865
	BeO	25.02	3.025	0.0728[b]	0.173	0.010	6.8	73[a]	0.501
5	B	10.82	2.45	0.1364	0.171	755	4	103	0.346
6	C	12.011	1.60	0.0803	0.158	0.004	4.8	32[a]	0.385
7	N	14.008	0.0013	5.3[a]	0.136	1.88	10	9.9[a]	50[a]
8	O	16.000	0.0014	5.3[a]	0.120	20[a]	4.2	0.000	21[a]
9	F	19.00	0.0017	5.3[a]	0.102	0.01	3.9	0.05[a]	20[a]
10	Ne	20.183	0.0009	2.6[a]	0.0968	<2.8	2.4	7.3[a]	6.2[a]
11	Na	22.991	0.971	0.0254	0.0845	0.525	4	0.013	0.102
12	Mg	24.32	1.74	0.0431	0.0811	0.069	3.6	0.003	0.155
13	Al	26.98	2.699	0.0602	0.0723	0.241	1.4	0.015	0.084
14	Si	28.09	2.42	0.0522	0.0698	0.16	1.7	0.008	0.089
15	P	30.975	1.82	0.0354	0.0632	0.20	5	0.007	0.177
16	S	32.066	2.07	0.0389	0.0612	0.52	1.1	0.020	0.043
17	Cl	35.457	0.0032	5.3[a]	0.0561	33.8	16	1.7[a]	80[a]
18	Ar	39.944	0.0018	2.6[a]	0.0492	0.66	1.5	0.002	3.9
19	K	39.100	0.87	0.0134	0.0504	2.07	1.5	0.028	0.020
20	Ca	40.08	1.55	0.0233	0.0492	0.44	3.0	0.010	0.070
21	Sc	44.96	2.5	0.0335	0.0438	24	24	0.804	0.804
22	Ti	47.90	4.5	0.0566	0.0411	5.8	4	0.328	0.226
23	V	50.95	5.96	0.0704	0.0387	5	5	0.352	0.352
24	Cr	52.01	7.1	0.0822	0.0385	3.1	3	0.255	0.247
25	Mn	54.94	7.2	0.0789	0.0359	13.2	2.3	1.04	0.181
26	Fe	55.85	7.86	0.0848	0.0353	2.62	11	0.222	0.933
27	Co	58.94	8.9	0.0190	0.0335	38	7	3.46	0.637
28	Ni	58.71	8.90	0.0913	0.0335	4.6	17.5	0.420	1.60

29	Cu	63.54	8.94	0.0848	0.0309	3.85	7.2	0.326	0.611
30	Zn	65.38	7.14	0.0658	0.0304	1.10	3.6	0.072	0.237
31	Ga	69.72	5.91	0.0511	0.0283	2.80	4	0.143	0.204
32	Ge	72.60	5.36	0.0445	0.0271	2.45	3	0.109	0.134
33	As	74.91	5.73	0.0461	0.0264	4.3	6	0.198	0.277
34	Se	78.96	4.8	0.0366	0.0251	12.3	11	0.450	0.403
35	Br	79.916	3.12	0.0235	0.0247	6.7	6	0.157	0.141
36	Kr	83.80	0.0037	2.6ᵃ	0.0236	31	7.2	81ᵃ	19ᵃ
37	Rb	85.48	1.53	0.0108	0.0233	0.73	12	0.008	0.130
38	Sr	87.63	2.54	0.0175	0.0226	1.21	10	0.021	0.175
39	Y	88.92	5.51	0.0373	0.0223	1.31	3	0.049	0.112
40	Zr	91.22	6.4	0.0423	0.0218	0.185	8	0.008	0.338
41	Nb	92.91	8.4	0.0545	0.0214	1.16	5	0.063	0.273
42	Mo	95.95	10.2	0.0640	0.0207	2.70	7	0.173	0.448
43	Tc	98			0.0203	22			
44	Ru	101.1	12.2	0.0727	0.0197	2.56	6	0.186	0.436
45	Rh	102.91	12.5	0.0732	0.0193	149	5	10.9	0.366
46	Pd	106.4	12.16	0.0689	0.0187	8	3.6	0.551	0.248
47	Ag	107.88	10.5	0.0586	0.0184	63	6	3.69	0.352
48	Cd	112.41	8.65	0.0464	0.0178	2450	7	114	0.325
49	In	114.82	7.28	0.0382	0.0173	191	2.2	7.30	0.084
50	Sn	118.70	6.5	0.0330	0.0167	0.625	4	0.021	0.132
51	Sb	121.76	6.69	0.0331	0.0163	5.7	4.3	0.189	0.142
52	Te	127.61	6.24	0.0295	0.0155	4.7	5	0.139	0.148
53	I	126.91	4.93	0.0234	0.0157	7.0	3.6	0.164	0.084
54	Xe	131.30	0.0059	2.7ᵃ	0.0152	35	4.3	95ᵃ	12ᵃ
55	Cs	132.91	1.873	0.0085	0.0150	28	20	0.238	0.170
56	Ba	137.36	3.5	0.0154	0.0145	1.2	8	0.018	0.123
57	La	138.92	6.19	0.0268	0.0143	8.9	15	0.239	0.403
58	Ce	140.13	6.78	0.0292	0.0142	0.73	9	0.021	0.263
59	Pr	140.92	6.78	0.0290	0.0141	11.3	4	0.328	0.116
60	Nd	144.27	6.95	0.0290	0.0138	46	16	1.33	0.464
61	Pm	145			0.0137	60			
62	Sm	150.35	7.7	0.0309	0.0133	5600	5	173	0.155
	Sm₂O₃	348.70	7.43	0.0128ᵇ	0.076	11,200	22.6	143	0.289
63	Eu	152	5.22	0.0207	0.0131	4300	8	89.0	0.166
	Eu₂O₃	352.00	7.42	0.0127ᵇ	0.063	8600	30.2	109	0.383

TABLE A.2. CROSS SECTIONS FOR NATURALLY OCCURRING ELEMENTS (continued)

Atomic No.	Element or Compound	Atomic or Mol. Wt.	Density, g/cm³	Nuclei per Unit Vol. ×10⁻²⁴	ξ	Microscopic Cross Section, barns		Macroscopic Cross Section, cm⁻¹	
						σ_a	σ_s	Σ_a	Σ_s
64	Gd	157.26	7.95	0.0305	0.0127	46,000		1403	
65	Tb	158.93	8.33	0.0316	0.0125	46		1.45	
66	Dy	162.51	8.56	0.0317	0.0122	950	100	30.1	3.17
	Dy₂O₃	372.92	7.81	0.0126ᵇ	0.019	2200	214	27.7	2.7
67	Ho	164.94	8.76	0.0320	0.0121	65	15	2.08	0.495
68	Er	167.27	9.16	0.0330	0.0119	173	7	5.71	0.233
69	Tm	168.94	9.35	0.0333	0.0118	127	12	4.23	0.293
70	Yb	173.04	7.01	0.0244	0.0115	37	8	0.903	0.359
71	Lu	174.99	9.74	0.0335	0.0114	112	5	3.75	0.277
72	Hf	178.5	13.3	0.0449	0.0112	105	5	4.71	0.316
73	Ta	180.95	16.6	0.0553	0.0110	21	14	1.16	0.930
74	W	183.86	19.3	0.0632	0.0108	19.2	11	1.21	0.783
75	Re	186.22	20.53	0.0664	0.0107	86	10	5.71	0.660
76	Os	190.2	22.48	0.0712	0.0105	15.3	9.3	1.09	0.550
77	Ir	192.2	22.42	0.0703	0.0104	440	20	30.9	0.814
78	Pt	195.09	21.37	0.0660	0.0102	8.8	14	0.581	0.489
79	Au	197	19.32	0.0591	0.0101	98.8	11	5.79	0.363
80	Hg	200.61	13.55	0.0407	0.0099	380	9	15.5	0.253
81	Tl	204.39	11.85	0.0349	0.0098	3.4		0.119	
82	Pb	207.21	11.35	0.0330	0.0096	0.170		0.006	
83	Bi	209	9.747	0.0281	0.0095	0.034		0.001	
84	Po	210	9.24	0.0265	0.0095				
85	At	211			0.0094				
86	Rn	222	0.0097	2.6ᵃ	0.0090	0.7			
87	Fr	223			0.0089				
88	Ra	226.05	5	0.0133	0.0088	20	12.6	0.266	0.369
89	Ac	227			0.0088	510			
90	Th	232.05	11.3	0.0293	0.0086	7.56		0.222	
91	Pa	231	15.4	0.0402	0.0086	200		8.04	
92	U	238.07	18.9	0.04783	0.0084	7.68	8.3	0.367	0.397
	UO₂	270.07	10	0.0223ᵇ	0.036	7.7	16.7	0.17	0.372
93	Np	237			0.0084	170			
94	Pu	239	19.7	0.0497	0.0083	1029	9.6	51.1	0.478

ᵃ Value has been multiplied by 10⁵.

ᵇ Molecules/cm³

TABLE A.3. CROSS SECTIONS FOR INDIVIDUAL ISOTOPES OF SPECIAL INTEREST
(2200-meters/sec neutrons)

Isotope	Cross section, barns	Reaction	Isotope	Cross section, barns	Reaction
Lithium-6	945	n, α	Uranium-235	$\begin{cases} 106 \\ 577 \end{cases}$	n, γ / Fission
Lithium-7	0.033	n, γ			
Boron-10	3813	n, α	Uranium-238	2.71	n, γ
Boron-11	<0.05	n, γ	Plutonium-239	$\begin{cases} 287 \\ 742 \end{cases}$	n, γ / Fission
Xenon-135	2.7×10^6	n, γ			
Uranium-233	$\begin{cases} 54 \\ 527 \end{cases}$	n, γ / Fission	Plutonium-240	300	n, γ
			Plutonium-241	$\begin{cases} 375 \\ 1025 \end{cases}$	n, γ / Fission

TABLE A.4.   TEN-GROUP (FAST REACTOR) CONSTANTS*

GROUP CHARACTERISTICS

u = lethargy
E_l = energy at lower group boundary
χ_i = fraction of the total neutrons originating with energy in the ith group

Group No.	1	2	3	4	5	6	7	8	9	10
u	0–1.5	1.5–2	2–3	3–4	4–5	5–6	6–7	7–11	8–16	Thermal
E_l (Mev)	2.231	1.353	0.498	0.183	0.0674	0.0248	9.12×10^{-3}	1.67×10^{-4}	1.125×10^{-6}	—
χ_i	0.3266	0.227	0.303	0.107	0.0283	0.0065	0.0022	0.0	0.0	0.0

MICROSCOPIC CROSS SECTIONS (BARNS)

σ_{tr} = transport cross section
$\sigma_{i \to i}$ = transfer cross section for scattering from any group i into the same group
$\sigma_{i \to i+n}$ = transfer cross section for scattering from group i into a higher lethargy group, $i + n$, where $n = 1, 2$, etc.

Group	$\nu\sigma_f$	σ_{tr}	$\sigma_{i \to i}$	$\sigma_{i \to i+1}$	$\sigma_{i \to i+2}$	$\sigma_{i \to i+3}$	$\sigma_{i \to i+4}$
			URANIUM-235				
1	3.646	4.50	1.096	0.773	0.880	0.260	0.100
2	3.290	4.52	1.474	1.173	0.350	0.100	0.050
3	3.041	5.10	3.129	0.467	0.120	0.030	0.010
4	3.338	7.50	5.699	0.199	0.035	0.015	—
5	4.079	9.50	7.391	0.0632	—	—	—
6	5.535	12.0	9.075	—	—	—	—
7	7.872	13.13	8.810	—	—	—	—
8	22.14	25.5	12.0	—	—	—	—
9	105.8	75.5	11.0	—	—	—	—
10	787.2	396.3	11.3	—	—	—	—
			PLUTONIUM-239				
1	6.566	4.50	0.929	0.581	0.66	0.20	0.07
2	6.009	4.80	1.62	0.805	0.23	0.065	0.025
3	5.488	5.50	2.86	0.494	0.13	0.030	0.010
4	4.863	8.00	5.87	0.199	0.035	0.015	—
5	4.752	9.50	7.46	0.0627	—	—	—
6	5.616	12.0	9.47	—	—	—	—
7	6.912	13.0	9.76	—	—	—	—
8	20.16	20.5	10.0	—	—	—	—
9	86.40	58.0	10.0	—	—	—	—
10	1584	790	10.0	—	—	—	—

* Data supplied by R. M. Kiehn

TABLE A.4. TEN-GROUP (FAST REACTOR) CONSTANTS (continued)

Group	$\nu\sigma_f$	σ_{tr}	$\sigma_{i\rightarrow i}$	$\sigma_{i\rightarrow i+1}$	$\sigma_{i\rightarrow i+2}$	$\sigma_{i\rightarrow i+3}$	$\sigma_{i\rightarrow i+4}$
			URANIUM-238				
1	1.540	4.450	1.1269	1.0130	1.20	0.35	0.15
2	1.125	4.403	1.5704	1.6246	0.50	0.15	0.05
3	0.073	5.010	4.0161	0.6339	0.15	0.05	—
4	—	6.970	6.5695	0.2055	0.05	—	—
5	—	9.000	8.6767	0.0733	—	—	—
6	—	12.05	11.633	0.0171	—	—	—
7	—	13.06	12.490	0.0106	—	—	—
8	—	13.00	11.500	—	—	—	—
9	—	20.00	14.000	—	—	—	—
10	—	10.00	8.000	—	—	—	—
			IRON				
1	—	2.0	0.950	0.428	0.44	0.13	0.05
2	—	2.0	1.353	0.494	0.12	0.03	—
3	—	2.4	2.186	0.170	0.03	0.01	—
4	—	3.0	2.889	0.106	—	—	—
5	—	3.0	2.886	0.106	—	—	—
6	—	3.0	2.888	0.100	—	—	—
7	—	4.0	3.850	0.133	—	—	—
8	—	7.5	7.408	0.0469	—	—	—
9	—	11.0	10.717	0.0134	—	—	—
10	—	13.0	11.300	—	—	—	—
			SODIUM				
1	—	2.0	1.68	0.258	0.044	0.013	0.005
2	—	2.0	1.65	0.339	0.010	0.005	—
3	—	3.0	2.75	0.254	—	—	—
4	—	3.8	3.48	0.321	—	—	—
5	—	3.2	2.93	0.271	—	—	—
6	—	5.0	4.58	0.422	—	—	—
7	—	5.0	4.58	0.422	—	—	—
8	—	10.0	9.79	0.209	—	—	—
9	—	3.3	2.93	0.374	—	—	—
10	—	3.5	3.22	—	—	—	—

TABLE A.5. PROPERTIES OF MODERATORS AT 78°F (20°C)

Material	Ordinary Water	Heavy Water (99.75% D_2O)	Beryllium	Beryllium Oxide	Graphite (Reactor grade)	Diphenyl (at 200°F)	Zirconium Hydride (ZrH_2)
Atomic (or molecular) weight	18	20	9	25	12	154	93.2
Density (g/cm³)	1.00	1.10	1.84	3.0	1.70	0.96	5.61
N (atoms or molecules per cm³) $\times 10^{-24}$	0.0334	0.0332	0.124	0.072	0.0855	0.0039	0.036
Thermal (2200 m/sec)							
σ_a (barns)	0.66	0.003	0.010	0.010	0.0037	2.5	0.84
σ_s (barns)	103	13.6	7.0	6.8	4.8	600	—
Σ_a (cm⁻¹)	0.022	8.5×10^{-5}	0.00123	0.00073	0.00032	0.01	0.030
Σ_s (cm⁻¹)	3.45	0.45	0.86	0.50	0.41	2.3	—
D (cm)	0.17	0.85	0.54	0.66	0.94	0.24	~0.2
L (cm)	2.76	100	21	30	54.2	4.9	~3.0
Epithermal							
ξ	0.93	0.51	0.206	0.17	0.158	0.81	0.84
σ_s (barns)	42	10.5	6.1	9.9	4.8	260	48.6
Σ_s (cm⁻¹)	1.40	0.35	0.75	0.72	0.41	1.0	1.75
$\xi\Sigma_s$ (cm⁻¹)	1.28	0.18	0.16	0.12	0.065	0.8	1.47
$\xi\Sigma_s/\Sigma_a$	58	21,000	130	163	200	80	49
Age to thermal							
τ (cm²)	31	120	85	100	350	58	27

TABLE A.6. PROPERTIES OF REACTOR COOLANTS

GASES

Temp. °F	Density lb$_m$/ft^3	Specific Heat Btu/(lb$_m$)(°F)	Viscosity $\times$ 10^5 lb$_m$/(ft)(sec)	Thermal Conductivity Btu/(hr)(ft)(°F)	Prandtl Number
Hydrogen (1 atm)					
0	0.0060	3.39	0.540	0.094	0.70
100	0.0049	3.42	0.620	0.110	0.69
200	0.0042	3.44	0.692	0.122	0.69
500	0.0028	3.47	0.884	0.160	0.69
1000	0.0019	3.51	1.160	0.208	0.70
1500	0.0014	3.62	1.415	0.260	0.71
2000	0.0011	3.76	1.64	0.307	0.72
3000	0.0008	4.02	1.72	0.380	0.66
Helium (1 atm)					
0	0.012	1.24	1.140	0.078	0.67
200	0.0083	1.24	1.480	0.097	0.69
400	0.0064	1.24	1.780	0.115	0.70
600	0.0052	1.24	2.02	0.129	0.72
800	0.0044	1.24	2.285	0.138	0.73
1000	0.0038	1.24	2.520	—	—
1500	0.0028	1.24	3.160	—	—
Dry air (1 atm)					
0	0.086	0.239	1.110	0.0133	0.73
100	0.071	0.240	1.285	0.0154	0.72
200	0.060	0.241	1.440	0.0174	0.72
300	0.052	0.243	1.610	0.0193	0.71
400	0.046	0.245	1.750	0.0212	0.69
500	0.0412	0.247	1.890	0.0231	0.68
Carbon dioxide (1 atm)					
0	0.132	0.184	0.88	0.0076	0.77
100	0.108	0.203	1.05	0.0100	0.77
200	0.092	0.216	1.22	0.0125	0.76
500	0.063	0.247	1.67	0.0198	0.75
1000	0.0414	0.280	2.30	0.0318	0.73
1500	0.0308	0.298	2.86	0.0420	0.73
Steam (1 atm)					
212	0.0372	0.451	0.870	0.0145	0.96
300	0.0328	0.456	1.000	0.0171	0.95
400	0.0288	0.462	1.130	0.0200	0.94
500	0.0258	0.470	1.265	0.0228	0.94
700	0.0213	0.485	1.555	0.0288	0.93
1000	0.0169	0.51	1.920	0.0388	0.91
1500	0.0126	0.56	2.47	0.057	0.87
2000	0.0100	0.60	3.03	0.076	0.86

TABLE A.6. PROPERTIES OF REACTOR COOLANTS (continued)

Liquids

Temp. °F	Vapor Pressure atm	Density lb_m/ft^3	Specific Heat $Btu/(lb_m)(°F)$	Viscosity $\times 10^3$ $lb_m/(ft)(sec)$	Thermal Conductivity $Btu/(hr)(ft)(°F)$	Prandtl Number
Water (sat. liq.)						
70	0.025	62.3	0.998	0.658	0.347	6.82
100	0.066	62.0	0.998	0.458	0.364	4.52
150	0.25	61.2	1.00	0.292	0.384	2.74
200	0.76	60.1	1.00	0.205	0.394	1.88
300	4.6	57.3	1.03	0.126	0.395	1.18
400	16.6	53.6	1.08	0.091	0.381	0.927
500	46.3	49.0	1.19	0.071	0.349	0.87
600	105.6	42.4	1.51	0.058	0.292	1.09
Sodium						
200	—	58.0	0.33	0.47	49.8	0.011
400	—	56.3	0.32	0.29	46.4	0.007
700	—	53.7	0.31	0.19	41.8	0.005
1000	0.01	51.2	0.30	0.14	37.8	0.004
1300	0.14	48.6	0.30	0.12	34.5	0.004
Sodium (56 wt %)- potassium alloy						
200	—	55	0.27	0.36	14.4	0.025
400	—	53	0.26	0.24	15.3	0.015
700	—	51	0.25	0.16	15.8	0.009
1000	0.03	48	0.25	0.12	16.3	0.007
1300	0.31	47	0.25	0.10	16.7	0.007
Sodium (22 wt %)- potassium alloy						
200	—	53	0.24	0.31	14.1	0.020
400	—	51	0.22	0.21	14.5	0.012
700	—	48	0.21	0.14	15.3	0.007
1000	0.05	46	0.21	0.10	15.8	0.005
1300	0.48	43	0.21	0.09	16.2	0.005
Mercury						
50	—	847	0.033	1.07	4.7	0.027
200	—	834	0.033	0.84	6.0	0.016
300	—	826	0.033	0.74	6.7	0.012
400	0.03	817	0.032	0.67	7.2	0.011
600	0.66	802	0.032	0.58	8.1	0.008
Lithium						
400	—	31.7	1.40	0.37	26.7	0.020
700	—	30.7	1.20	—	23.4	—
1000	—	29.8	1.00	—	17.6	—
1500	0.004	28.3	—	—	—	—
1800	0.04	27.5	—	—	—	—
Diphenyl						
200	—	60.8	0.42	0.672	0.078	3.6
400	0.29	55.2	0.56	0.25	0.073	1.9
600	3.2	49.0	0.65	0.14	0.051	1.8
800	15	40.9	0.69	0.095	0.060	1.1
900	270	34.6	0.69	0.081	0.057	0.98

TABLE A.7. PHYSICAL PROPERTIES OF SOME REACTOR MATERIALS
(Average values for preliminary calculations only; not to be used for design purposes)

Material	Temp. °F	Density lbm/ft³	Coeff. of Linear Thermal Expansion* per °F × 10⁶	Specific Heat Btu/(lbm)(°F)	Thermal Conductivity Btu/(hr)(ft)(°F)	Ultimate Tensile Strength psi	Yield Strength (Y), Compressive Strength (C), or Allowable Stress (A) psi	Young's Modulus 10⁶ psi	Poisson's Ratio
Aluminum (2S or 1100)	100	169	13.0	0.22	121	13,000	Y 5000	10	0.33
	300	—	13.3	0.23	126	7500	Y 3500	9.5	—
	500	—	14.0	0.24	128	3500	Y 2000	8.0	—
	700	—	14.5	0.25	131	1500	Y 1000	4.0	—
Beryllium (vacuum hot-pressed, hot extruded)	75	115	—	0.49	80	60,000	Y 39,000	44	0.024
	400	—	7.3	0.52	73	53,000	—	—	—
	800	114	8.3	0.56	61	46,000	—	—	—
Beryllium oxide (variable)	75	~175	—	0.30	120	15,000	C 100,000	40	0.34
	750	—	4.7	0.42	50	15,000	C 70,000	39	0.34
	1000	—	5.7	0.49	19	13,000	C 60,000	37	0.34
Graphite (average nuclear grade)	75	106	—	0.17	90†	2000	C 8400		—
	500	—	2.0†	0.30	68†	2300	—	1.3	—
	1000	—	2.8†	0.40	42†	2500	—	—	—
	1000	—	2.8†	0.40	42†	2500	—	—	—
	2000	—	3.5†	0.45	21†	2800	—	—	—
	4000	—	4.7†	—	18†	4000	—		—
Steel, carbon (A212-B)	75	490	—	0.12	30	70,000	Y 38,000	30	0.28
	700	—	5.7	0.14	25	66,000	A 16,000	26.5	—
	900	—	5.8	0.15	22	—	A 6500	25.0	—
	1000	—	5.8	0.16	20	—	A 2500	24.5	
Steel, carbon (A302-B)	75	490	—	0.12	30	80,000	Y 50,000	30	0.28
	700	—	5.7	0.14	25	80,000	A 27,000	26.5	
	900	—	5.8	0.15	22	—	A 13,200	25.0	
	1000	—	5.8	0.16	20	—	A 6200	24.5	

TABLE A.7. PHYSICAL PROPERTIES OF SOME REACTOR MATERIALS (continued)

Material	Temp. °F	Density lb_m/ft^3	Coeff. of Linear Thermal Expansion* per °F × 10^6	Specific Heat $Btu/(lb_m)(°F)$	Thermal Conductivity $Btu/(hr)(ft)(°F)$	Ultimate Tensile Strength psi	Yield Strength (Y), Compressive Strength (C), or Allowable Stress (A) psi	Young's Modulus 10^6 psi	Poisson's Ratio
Steel, stainless (Type 347)	75	500	—	0.12	9.0	75,000	Y 30,000	—	—
	300	—	9.2	—	9.4	66,000	Y 28,000	27.5	0.29
	500	—	9.4	—	—	61,000	—	26.1	0.30
	800	—	9.7	—	—	58,000	Y 22,000	24.1	0.31
	1000	—	10.3	—	12.9	56,000	—	22.8	0.32
	1400	—	11.0	—	—	36,000	—	20.0	0.34
Uranium (alpha phase)	75	1205	—	0.028	13.8	50,000	Y 20,000	28	0.24
	200	—	12	—	16.0	—	—	—	—
	500	—	15	0.033	16.7	35,000 (H)	Y 17,000	24	0.26
	1000	—	—	0.04	20.4	10,000 (H)	Y 5000	—	—
	1200	—	20	—	22.0	—	—	—	—
Uranium carbide (UC)	75	843	—	0.035	—	—	C 54,000	31	—
	1000	—	—	—	13.1	—	—	—	—
	1800	—	6.3	—	13.3	—	—	—	—
Uranium dioxide	75	684	—	0.06	5.3	—	C 140,000 (S)	26.5 (S)	—
	200	—	—	0.063	4.3	—	—	—	—
	500	—	5.0	0.07	2.9	—	—	—	—
	1000	—	5.6	—	2.2	—	—	24 (S)	—
	1500	—	—	—	1.8	—	—	—	—
	2000	—	7.1	0.08	—	—	—	—	—
Zirconium (crystal bar)	75	406	5.9†	0.070	12.2	29,000	Y 14,000	14.0	0.33
	200	—	—	0.072	11.8	26,000	—	13.2	—
	500	—	—	0.077	11.1	18,000	—	11.0	—
Zircaloy-2	75	409	—	0.071	6.7	71,000	Y 43,000	13.9	0.43
	200	—	—	—	6.9	59,000	Y 39,000	13.6	—
	400	—	—	—	7.1	41,000	Y 25,000	13.0	0.38
	600	—	3.6	—	7.2	31,000	Y 17,000	11.4	—

* Average values from 68°F to indicated temperatures   † Average of longitudinal and transverse properties   (H) Heat treated   (S) Sintered

INDEX

815

$$2.303 \times \text{Base } 10 = e^x$$